CW00953752

OOMYCETE GENETICS AND GENOMICS

OOMYCETE GENETICS AND GENOMICS

Diversity, Interactions, and Research Tools

Edited by

Kurt Lamour
The University of Tennessee
Knoxville, Tennessee

and

Sophien Kamoun
Sainsbury Laboratory
Norwich, United Kingdom

WILEY-BLACKWELL

A JOHN WILEY & SONS, INC., PUBLICATION

Copyright © 2009 by John Wiley & Sons, Inc. All rights reserved

Wiley-Blackwell is an imprint of John Wiley & Sons, formed by the merger of Wiley's global Scientific, Technical, and Medical business with Blackwell Publishing.

Published by John Wiley & Sons, Inc., Hoboken, New Jersey
Published simultaneously in Canada

No part of this publication may be reproduced, stored in a retrieval system, or transmitted in any form or by any means, electronic, mechanical, photocopying, recording, scanning, or otherwise, except as permitted under Section 107 or 108 of the 1976 United States Copyright Act, without either the prior written permission of the Publisher, or authorization through payment of the appropriate per-copy fee to the Copyright Clearance Center, Inc., 222 Rosewood Drive, Danvers, MA 01923, (978) 750-8400, fax (978) 750-4470, or on the web at www.copyright.com. Requests to the Publisher for permission should be addressed to the Permissions Department, John Wiley & Sons, Inc., 111 River Street, Hoboken, NJ 07030, (201) 748-6011, fax (201) 748-6008, or online at http://www.wiley.com/go/permission.

Limit of Liability/Disclaimer of Warranty: While the publisher and author have used their best efforts in preparing this book, they make no representations or warranties with respect to the accuracy or completeness of the contents of this book and specifically disclaim any implied warranties of merchantability or fitness for a particular purpose. No warranty may be created or extended by sales representatives or written sales materials. The advice and strategies contained herein may not be suitable for your situation. You should consult with a professional where appropriate. Neither the publisher nor author shall be liable for any loss of profit or any other commercial damages, including but not limited to special, incidental, consequential, or other damages.

For general information on our other products and services or for technical support, please contact our Customer Care Department within the United States at (800) 762-2974, outside the United outside the United States at (317) 572-3993 or fax (317) 572-4002.

Wiley also publishes its books in a variety of electronic formats. Some content that appears in print may not be available in electronic formats. For more information about Wiley products, visit our web site at www.wiley.com.

Library of Congress Cataloging-in-Publication Data:

Oomycete genetics and genomics: diversity, interactions, and research tools/edited by Kurt Lamour and Sophien Kamoun.
p. ; cm.
Includes bibliographical references and index.
ISBN 978-0-470-25567-4 (cloth)
1. Oomycetes Genetics. I. Lamour, Kurt. II. Kamoun, Sophien.
[DNLM: 1. Oomycetes–genetics. 2. Oomycetes–pathogenicity. 3. Animal Diseases–microbiology. 4. Host-Pathogen Interactions. 5. Infection–microbiology. 6. Plant Diseases–microbiology. QK 565.5 O59 2010]
QK565.5.O56 2010
579.5′4–dc22

2008055230

Printed in the United States of America

10 9 8 7 6 5 4 3 2 1

CONTENTS

FOREWORD

Francine Govers

Wageningen University, The Netherlands

The publication of this book is an important breakthrough; it is the first time that the existing knowledge on Oomycetes has been brought together in one volume. The Oomycetes, also known as water molds, comprise a diverse group of filamentous microorganisms that share many characteristics with Fungi (i.e., members of the taxonomic entity defined as Fungi). They have an absorptive mode of nutrition, grow by polarized hyphal extension, and their reproduction includes the formation of spores. Traditionally the Oomycetes have been presented in textbooks as a phylum in the kingdom Fungi and have been lumped together with other organisms of uncertain affinity as "lower fungi" or "zoosporic fungi." More recently, the Oomycetes have been placed in another kingdom, in some systems in the kingdom Stramenopila or the stramenopile lineage in the supergroup Chromalveolates; in others in the kingdom Chromista. Phylogenetics has clearly demonstrated that Oomycetes are not Fungi but instead, they are close relatives of heterokont algae. They have lost their plastids and have adopted a fungal-like lifestyle, absorbing their nutrients from the surrounding water or soil or invading the body of another organism to feed. In fact, Oomycetes are "algae in disguise" that qualify as fungi (i.e., organisms sharing the characteristics described above). It is my hope that this book will stimulate scientists, including mycologists, to adopt or rehabilitate Oomycetes as subjects of research.

Like Fungi, Oomycetes have a global distribution and prosper in quite diverse environments. Pathogenic species that live in association with plants, animals, or other microbes can be devastating and completely destroy their hosts. Their victims include natural forests, many crop plants, fish in fish farms, amphibians, and occasionally, humans. In contrast, saprophytic species that feed on decaying material are beneficial; they play important roles in the decomposition and recycling of biomass. Currently, at least 800 oomycete species are known, but depending on the definition of a species, this number might even reach 1500. Still the species richness is low when compared with the number of species of Fungi known to date (over 100,000), but very likely, there are many more Oomycetes out there to be discovered. In this respect, the genus *Phytophthora* is illustrative. In the last ten years, over 25 new species have been described, expanding the genus to at least 90 members. A few chapters in the first section in this book focus on the phylogeny of Oomycetes and the enormous diversity within the Oomycetes. The diversity at the species level is addressed in the second section.

Oomycete research has a long history. The type species of *Phytophthora, P. infestans,* was described 132 years ago by Anton de Bary, the founding father of plant pathology and the founder of modern mycology. This notorious plant pathogen was the cause of the severe late blight epidemic in Europe in the 1840s that resulted in the Irish potato famine and led to a turning point in history, the birth of Irish America. Today, late blight is still a major problem for potato growers worldwide. The same holds for downy mildew on grapes, which is another well-known oomycete disease that emerged in the nineteenth century. Less known is the serious outbreak of a disease in 1877 among the salmon in the rivers Conway and Tweed that spread into most of the rivers of the British Isles within two years. Again a water mold, *Saprolegnia ferax*, was to blame. In the last decade, the rise of industrial fish farming has gone hand-in-hand with the revival of Saprolegniosis as a major disease. The finding that Oomycetes can also cause a disease known as Pythiosis in humans is of a more recent date. The publication of this book is timely; with the (re-)emergence of oomycete diseases in hosts important for the world food economy, such as fish, soybean, potato, and other vegetable crops, or in hosts that shape the landscape (e.g., oak and alder) and inhabit unique ecological niches, interest in oomycete biology and pathology should be challenged, and the research should be intensified and strengthened. This book helps in identifying the challenges. The chapters dealing with sexual and asexual reproduction and interactions with plant hosts and animal hosts provide the necessary background but also point to gaps in our knowledge.

With the head title of this book, *Oomycete Genetics and Genomics*, the editors cross a frontier. Mentioning "Oomycete" and "Genetics" in one breath seems odd, and the search for an Oomycete in a genetics textbook is in vain. The genetics timeline begins with Gregor Mendel's discovery of the basic laws of genetics in 1865 and marks major milestones like the description of the double helix structure of DNA, the unraveling of the genetic code, and the first recombinant DNA experiments. Then, in the 1990s, genomics milestones start to appear: the first whole genome sequence of a prokaryote and a eukaryote, culminating in the human genome sequence in 2001, an event marked by the former U.S. President Bill Clinton as one that will change the history of mankind. Prominent organisms on the genetic timeline are the well-known models like *Escherichia coli*, yeast, *Caenorhabditis elegans*, or Arabidopsis but no Oomycete. David Shaw once called *Phytophthora* a "geneticist's nightmare." As exampled in this book we can now look beyond this nightmare. In recent years, many oomycete researchers have experienced that genomics gives rise to a bright morning with many new milestones at the horizon. To date, five oomycete genomes have been sequenced. The availability of genomics resources and technologies has changed the way we can address various longstanding biological questions and has certainly stimulated researchers to use genomics as an instrument to tackle Oomycetes. The chapters in the tools section of this book describe recent advances in technology aimed at either the functional analysis of individual genes or at overall genome-wide analyses.

This book will serve as an excellent introduction and a valuable resource for students and researchers at all levels. It echoes the enthusiasm of the oomycete research community. I advise the newcomers in this field to take part in this community and to join the oomycete molecular genetics network (OMGN; http://pmgn.vbi.vt.edu/). Last, but not least, I commend Kurt Lamour and Sophien Kamoun for taking the initiative to publish this book as well as the authors for their efforts in writing the chapters.

PREFACE

A bittersweet truth in our fast-paced genetic world is that an organism-specific book on genetics and genomics is outdated by the time it goes to print. This is especially true for the burgeoning field of Oomycete genetics and genomics where the foundations for genetic discovery have only recently been laid. Ten years ago who would have thought there would be genome sequences for multiple *Phytophthora* species, a *Hyaloperonospora*, and soon, a *Pythium*? Strangely enough, these reference genomes may themselves soon be viewed as archaic. Although only touched on in the chapter on genome sequencing, the ongoing quest to develop faster and less costly genetic sequencing has led to sequencing platforms that make it feasible to discover *all* of the changes between multiple whole genomes or transcriptomes — without the need for a reference genome. With so much forward momentum, there is *never* a good time to stop and take stock of where we've been. Our goal is to provide a useful overview of this fascinating group: a resource that can be handed to an incoming graduate student, a new colleague or a potential collaborator.

The book begins by presenting an overview of the evolutionary relationships within the Oomycetes. The diversity of life forms is astounding, and it is clear that additional taxa and sequences will continue to clarify this important area of research. The white blister rusts provide a good example of how genetic data can resolve relationships among morphologically similar yet evolutionarily distinct taxa — an important challenge in an age of costly quarantines and worldwide movement of pathogens. Interesting possibilities spring from the phylogenetic perspective, including the idea that terrestrial plant pathogens may have hitched an evolutionary ride from the open sea via nematodes and switched hosts to plant roots — on more than one occasion. For those interested in discovering whole new worlds of diversity, a plethora of lower oomycetes are waiting, particularly in the Tropics and Polar regions. A question lingering throughout is whether the parasitic lifestyle is derived from a saprophytic lifestyle or vice versa.

After the introductory chapters is an overview of asexual and sexual reproduction. This provides a useful framework for the next sections, which explore the population structure of representative species in natural populations, and the interactions with plant and animal hosts. These chapters range from overviews of the entrenched pathogen of potato, *Phytophthora infestans*, a staple crop and staple research area since the dawn of micro-organism research, to newly emerging invasive species such as the Sudden Oak Death pathogen, *Phytophthora ramorum*. For investigators familiar with the impact of Oomycetes on sessile organisms, the chapters on Oomycetes that attack fish, crustaceans, and humans should be particularly interesting. And finally, there are specific chapters describing the application and development of molecular tools to better understand these notoriously intractable organisms.

The response by the authors to contribute their work and perspective was overwhelmingly positive, and we are hopeful that this snapshot will stimulate new relationships and research.

CONTRIBUTORS

Ah-Fong, Audrey M. V., Department of Plant Pathology and Microbiology, University of California, Riverside, CA, USA.

Anderson, Victoria L., School of Biological Sciences, University of Aberdeen, Aberdeen, AB24 2TZ, Scotland, UK.

Andersson, M. Gunnar, Linneus Centre for Bioinformatics, Uppsala University, Uppsala, Sweden.

Avrova, Anna O., Plant Pathology Programme, Scottish Crop Research Institute, Errol Road, Invergowrie, Dundee DD2 5DA, UK.

Beakes, Gordon, School of Biology, Newcastle University, Newcastle upon Tyne, NE1 7RU, UK.

Belhaj, Khaoula, Department of Biology, University of Fribourg, CH-1700 Fribourg, Switzerland.

Birch, Paul, Division of Plant Sciences, College of Life Science, University of Dundee at SCRI, Errol Road, Invergowrie, Dundee DD2 5DA, UK.

Bottin, Arnaud, Université de Toulouse, UMR5546 CNRS, France.

Branciard, Lorelise, Department of Biology, University of Fribourg, CH-1700 Fribourg, Switzerland.

Cano, Liliana M., The Sainsbury Laboratory, John Innes Centre, Norwich, UK.

Cao, Keqiang, Agricultural University of Hebei, Baoding, China.

Cerenius, Lage, Department of Comparative Physiology, Uppsala University, Uppsala, Sweden.

Cooke, David E. L., Scottish Crop Research Institute, Dundee, Scotland.

Diéguez-Uribeondo, Javier, Departamento de Micología, Real Jardín Botánico CSIC, Plaza de Murillo 2, 28014 Madrid, Spain.

Dorrance, Anne, Ohio Agricultural Research and Development Center, The Ohio State University, Wooster, OH, USA.

Dumas, Bernard, Université de Toulouse, UMR5546 CNRS, France.

Forbes, Gregory A., International Potato Center, Lima, Peru.

Fry, William E., Cornell University, Ithaca NY, USA.

Gachon, Claire M.M., The Scottish Association for Marine Science, Dunstaffnage Marine Laboratory, Oban PA37 1QA, Scotland, UK.

Gaulin, Elodie, Université de Toulouse, UMR5546 CNRS, France.

Gijzen, Mark, Agriculture and Agri-Food Canada, London ON, Canada.

Göker, Markus, University of Tübingen, Institute of Systematic Biology and Mycology, Tübingen, Germany.

Goss, Erica M., Horticultural Crops Research Laboratory, USDA Agricultural Research Service, Corvallis, OR, USA.

Govers, Francine, Wageningen University, The Netherlands.

Grenville Briggs, Laura J., Aberdeen Oomycete Group, University of Aberdeen, College of Life Sciences and Medicine, Institute of Medical Sciences, Foresterhill, Aberdeen, AB25 2ZD, Scotland, UK.

Grünwald, Niklaus J. Horticultural Crops Research Laboratory, USDA Agricultural Research Service, Corvallis, OR, USA.

Hardham, Adrienne, Plant Cell Biology Group, School of Biology, Australian National University, Canberra ACT 2601, Australia.

Huitema, Edgar, The Sainsbury Laboratory, John Innes Centre, Norwich, UK.

Jacquet, Christophe, Université de Toulouse, UMR5546 CNRS, France.

Judelson, Howard, Department of Plant Pathology and Microbiology, University of California, Riverside, USA.

Kamoun, Sophien, Sainsbury Laboratory, Colney Lane, Norwich, NR4 7UH, UK.

Küpper, Frithjof C., The Scottish Association for Marine Science, Dunstaffnage Marine Laboratory, Scotland, UK.

Lamour, Kurt, Department of Entomology and Plant Pathology, University of Tennessee, Institute of Agriculture, Knoxville, TN, USA.

Martin, Frank, USDA-ARS Salinas, CA, USA.

Mauch, Felix, Department of Biology, University of Fribourg, CH-1700 Fribourg, Switzerland.

McLeod, Adele, Stellenbosch University, Stellenbosch, South Africa.

Mendoza, Leonel, Microbiology and Molecular Genetics, Biomedical Laboratory Diagnostics, Michigan State University, East Lansing, MI, USA.

Michelmore, Richard, The Genome Center and Department of Plant Sciences, University of California, Davis, USA.

Nusbaum, Chad, Broad Institute of MIT and Harvard, Cambridge, MA, USA.

Ochoa, Oswaldo, The Genome Center and Department of Plant Sciences, University of California, Davis, USA.

Parisy, Vincent, Department of Biology, University of Fribourg, CH-1700 Fribourg, Switzerland.

Phillips, Andrew J., School of Biological Sciences, University of Aberdeen, Scotland, UK.

Qutob, Dinah, Agriculture and Agri-Food Canada, London ON, Canada.

Rietman, Hendrik, Wageningen University, The Netherlands.

Robertson, Emma J., Albert Einstein College of Medicine, Department of Microbiology and Immunology, Bronx, NY, USA.

Savidor, Alon, Tel Aviv University, Israel.

Schlaich, Nikolaus, Department of Plant Physiology (BioIII), RWTH Aachen University, Aachen, Germany.

Schläppi, Klaus, Department of Biology, University of Fribourg, Fribourg, Switzerland.

Schornack, Sebastian, The Sainsbury Laboratory, John Innes Centre, Norwich, UK.

Secombes, Chris J., School of Biological Sciences, University of Aberdeen, Aberdeen, AB24 2TZ, Scotland, UK.

Sekimoto, Satoshi, Department of Botany, University of British Columbia, Vancouver, V6T 1Z4 Canada.

Si-Ammour, Azeddine, Department of Biology, University of Fribourg, Fribourg, Switzerland.

Slusarenko, Alan, Department of Plant Physiology (BioIII), RWTH Aachen University, Aachen, Germany.

Soderhall, Kenneth, Department of Comparative Physiology, Uppsala University, Norbyvägen 18A, 752 36 Uppsala, Sweden.

Strittmatter, Martina, The Scottish Association for Marine Science, Dunstaffnage Marine Laboratory, Oban PA37 1QA, Scotland, UK.

Thines, Marco, University of Hohenheim, Institute of Botany 210, Garbenstraße 30, D-70593 Stuttgart, Germany.

Torche, Samuel, Department of Biology, University of Fribourg, CH-1700 Fribourg, Switzerland.

Tyler, Brett M., Virginia Bioinformatics Institute, Virginia Polytechnic Institute and State University, Virginia 24061, USA.

van Damme, Mireille, The Sainsbury Laboratory, John Innes Centre, Norwich, UK.

van West, Pieter, Aberdeen Oomycete Group, University of Aberdeen, College of Life Sciences and Medicine, Institute of Medical Sciences, Foresterhill, Aberdeen, AB25 2ZD, Scotland, UK.

Vleeshouwers, Vivianne G. A. A., Wageningen University, The Netherlands.

Voglmayr, Hermann, Department of Systematic and Evolutionary Botany, University of Vienna, Rennweg 14, 1030 Wien, Austria.

Whisson, Stephen, Plant Pathology Programme, Scottish Crop Research Institute, Errol Road, Invergowrie, Dundee DD2 5DA, UK.

Wong, Joan, The Genome Center and Department of Plant Sciences, University of California, Davis, CA, USA.

Zody, Michael C., Broad Institute of MIT and Harvard, Cambridge, MA, USA.

1

THE EVOLUTIONARY PHYLOGENY OF OOMYCETES—INSIGHTS GAINED FROM STUDIES OF HOLOCARPIC PARASITES OF ALGAE AND INVERTEBRATES

Gordon W. Beakes

School of Biology, Newcastle University, Newcastle upon Tyne, United Kingdom

Satoshi Sekimoto

Department of Botany, University of British Columbia, Vancouver, Canada

> ... phylogenetic speculations, valueless though these are considered to be....may stimulate studies in the life-history, cytology, morphology etc.... and clear the way for laying the foundations of a more logical system of classification.
>
> —E. A. Bessey (1935), *A Textbook of Mycology*

1.1 INTRODUCTION

The unraveling of the evolutionary phylogeny of organisms has been given a tremendous impetus by the application of molecular techniques that have enabled biologists to, in effect, delve for phylogenetic clues in the DNA of organisms in a manner analogous to fossil hunters searching for physical evidence a century earlier. As pointed out by Bessey, a sound phylogenetic framework will hopefully inform and direct future exploration as well as provide a sound basis for classification. This is particularly pertinent in the era of bioinformatics, because this knowledge should help in choosing organisms that might be targeted for genome sequencing. The oomycetes are fungus-like heterotrophs that are saprophytes or parasites of diverse hosts in marine,

Oomycete Genetics and Genomics: Diversity, Interactions, and Research Tools
Edited by Kurt Lamour and Sophien Kamoun
Copyright © 2009 John Wiley & Sons, Inc.

freshwater, and terrestrial environments (Sparrow, 1960; Karling, 1981; Dick, 2001; Johnson et al., 2002). However, as a group, they are best known as devastating pathogens of plants.

Oomycetes are similar to the true fungi in that they produce complex branching, tip-growing, hyphal systems (forming mycelia) and have similar modes of nutrition and ecological roles (Richards et al., 2006). Summaries of the early speculations as to the likely evolutionary relationships of oomycetes to other organisms have been reviewed by Karling (1942), Dick (2001), and Johnson et al. (2002). Candidates cited as their likely ancestors have included amoebas, heterotrophic flagellates, diverse algal groups, and even chytrid fungi. However, most opinions tended to divide sharply between those, such as Scherffel, who considered oomycetes to have evolved from heterotrophic flagellates (Karling, 1942), and those like Bessey, who thought that photosynthetic algae were the more likely ancestors. In a seminal analysis, Bessey (1942) outlined two possible alternative evolutionary pathways within the oomycete lineage (Fig. 1.1a). In the first, it was suggested that oomycetes evolved from siphonaceous (coenocytic) algae and that they shared a common ancestor with the xanthophyte alga *Vaucheria*. The saprotrophic Saprolegniales were considered to be the most primitive order, which in turn gave rise to the Leptomitales, after which the lineage split and created the plant pathogenic Peronosporales along one branch and the holocarpic Lagenidiales along the other. The other scheme postulated that the most likely ancestor was an unknown "heterocont unicellular algae," which was ancestral to both the uniflagellate hyphochytrids and the biflagellate oomycetes. In this pathway, the holocarpic Olpidiopsidales were thought to be the most likely basal family and yielded the Lagenidiales. From these, the plant pathogenic Peronosporales diverged on one branch and the water moulds (Saprolegniales via the Leptomitales) on the other. In this review, we will summarize current views on the likely phylogeny and taxonomy of these organisms in the light of recent work that we have carried out on some of the less widely studied parasites of seaweeds, crustacea and nematodes.

1.2 ANIMAL OR VEGETABLE—WHERE DO OOMYCETES BELONG ON THE TREE OF LIFE?

The sequencing of conserved genes over the past two decades has led to a firm phylogenetic placement for most groups of living organisms. These studies have shown that the oomycetes are heterokonts (see Fig. 1.1b based on Cavalier-Smith and Chao, 2006; Tsui et al., 2008) within the chromalveolate "super kingdom" (Baldauf et al., 2000). The chromist section contains three, wholly or partially, photosynthetic lineages: the cryptomonads, haptophytes, and heterokonts, although the evidence for the inclusion of the former pair with the heterokonts is still not particularly strong (discussed by Harper et al., 2005). The alveolate section contains the parasitic apicocomplexa, phagotrophic ciliates, and mixotrophic dinoflagellates (Fig. 1.1b). The heterokonts/stramenopiles

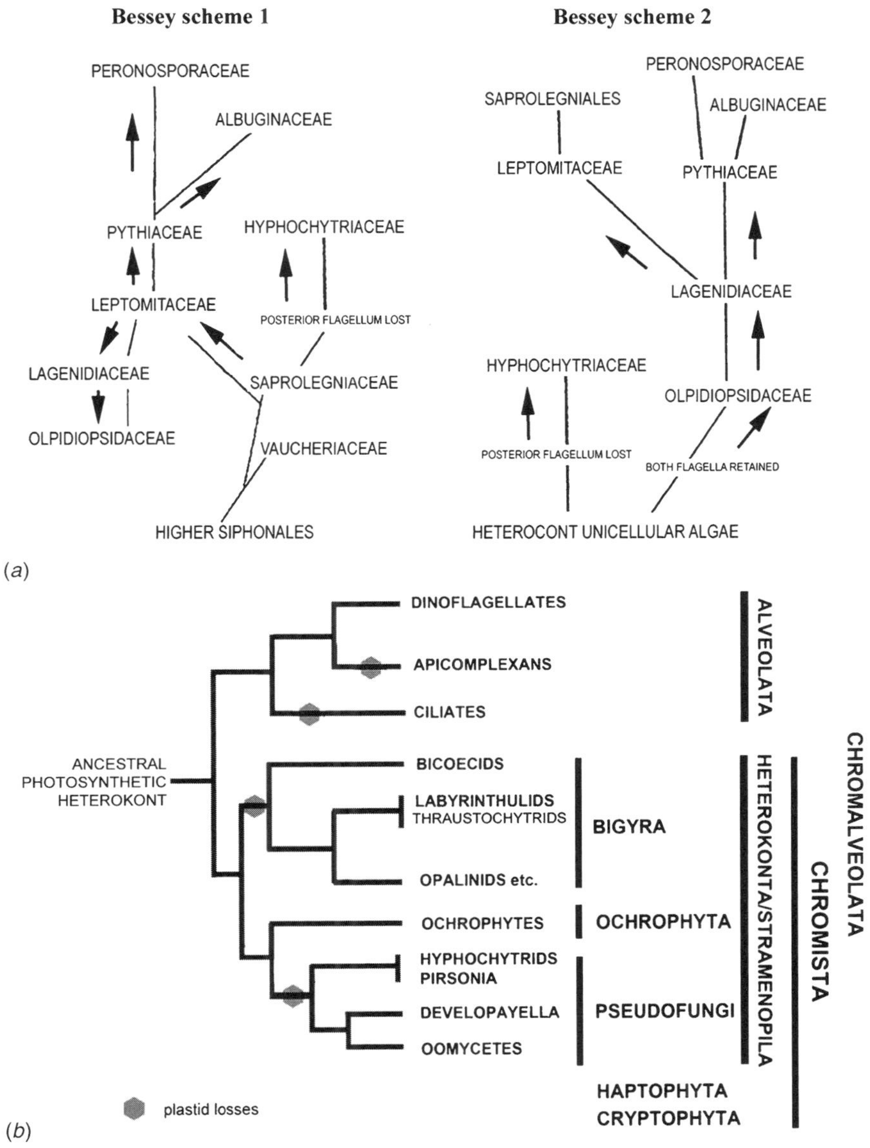

FIG. 1.1 Schematic summaries of the likely phylogenetic relationships of oomycetes and their relatives. (a) Schematic summary of two possible phylogenetic schemes showing the likely origins and family relationships within the oomycetes outlined by Bessey (1942). (b) Summary of the likely relationships between main classes and phyla within the Chromalveolata Superkingdom based on the terminology and information presented in Cavalier-Smith and Chao (2006) and Tsui et al. (2008).

(Fig. 1.1b) are an extraordinarily diverse assemblage (Cavalier-Smith and Choa, 2006) that encompasses both autotrophic and heterotrophic organisms, including the chlorophyll c-containing algae (diatoms, chrysophytes, xanthophytes, phaeophytes, etc.), free-living bacteriotrophic flagellates (bicoecids, etc.), a group of absorptive gut commensals/parasites (opalanids, proteromonads, and *Blastocystis*), as well as the fungal-like osmotrophic representatives (labyrinthulids, hyphochytrids, oomycetes, etc.). Recent multigene analyses have indicated that the Rhizaria (a very diverse group, including filose amoeboid organisms and flagellates) are the sister group to the "Stramenopiles," which has led to this lineage being renamed as the SAR (Stramenopile/Alveolate/Rhizaria) clade (Burki et al., 2007).

The first published phylogenetic trees, which are mostly based on nuclear-encoded ribosomal gene (SSU rDNA) sequences, showed that all the early branching heterokonts were nonphotosynthetic organisms, which suggested the late acquisition of plastids in the line (Leipe et al., 1996). Most recent evidence points to the whole chromalveolate lineage having developed from a common biflagellate (mastigonate) ancestor, which had acquired photosynthetic capabilities as a result of a single unique red algal enslavement (Patron et al., 2004; Harper et al., 2005; Cavalier-Smith and Chao, 2006). It is now thought that chloroplast loss has occurred many times within the lineage, including at least twice in the heterokont line (Fig. 1.1b; Cavalier-Smith and Chao, 2006; Tsui et al., 2008). Genomic data have also provided direct evidence for the photosynthetic ancestry of oomycetes with the discovery of vestigial plastid genes within the nuclear genome of *Phytophthora* (Lamour et al., 2007).

1.3 KINGDOM WARS AND FAMILY TIES—A CASE OF CONFLICTING NOMENCLATURE

There is still debate as to the correct (and taxonomically legal) kingdom/phylum/class names to be used for the lineage that contains the oomycetes. Dick (2001) formally proposed (and diagnosed) the kingdom Straminipila for the heterokont lineage, pointing out the incorrect etymological derivation of the by then widely used informal term "Stramenopile," which was first introduced by Patterson (1989) in reference to the "straw-like" flagellum hairs (mastigonemes) possessed by most members of this group. However, in their attempt to bring order and consistency to the naming of protists, algae, and fungi, Adl et al. (2005) forcefully argued for the continued use of the name Stramenopile for this lineage, although they side stepped the issue of assigning hierarchical taxonomic ranks. Cavalier-Smith and Chao (2006) in their review of the phylogeny of phagotrophic heterokonts considered Dick's kingdom Straminipila to be synonymous with the kingdom Chromista erected by Cavalier-Smith (1981); this is the name that is used in many current nomenclatural databases.

Which phylum the oomycetes should be placed in has been no less controversial. The name Heterokonta has been used, respectively, to define both a "phylum" (Dick, 2001) and an "infrakingdom" (Cavalier-Smith and Chao, 2006). The Heterokonta infrakingdom was split into three phyla (see Fig. 1.1b), the Ochrophyta (encompassing all photosynthetic heterokonts), Bygyra (thraustochytrids, labyrinthulids, opalinids, etc.) and Pseudofungi (Cavalier-Smith and Chao, 2006). This includes, in addition to the oomycetes, the anteriorly uniflagellate hyphochytrids and associated sister clade, the flagellate parasitoid *Pirsonia* (Kühn et al., 2004), and the free-living bacteriotrophic marine zooflagellate *Developayella*. The latter species usually forms the sister clade to the oomycetes in small ribosomal subunit phylogenetic trees (Figs. 1b, 1.2a; Leipe et al., 1996). Patterson (1999) introduced yet another name, Sloomycetes, for a clade that contains all the osmotrophic fungal-like heterokonts. Perhaps because of the plethora of conflicting higher level taxonomic schemes, it is not surprising that many review volumes and textbooks continue to afford the oomycetes/oomycota their own phylum status.

The separation of the photosynthetic ochrophyte and heterotrophic oomycete lineages into two parallel clades derived from a common ancestor (Fig. 1.1b) is supported in the most recent phylogenetic trees (e.g., Cavalier-Smith and Chao, 2006; Tsui et al., 2008). This makes evolutionary sense as it explains the often reciprocal host–pathogen relationships observed between members of these two groups. For instance, both the hyphochytrid *Anisopidium ectocarpi* and the oomycete *Eurychasma dicksonii* are parasites of ectocarpalean phaeophyte algae (Küpper and Müller, 1999) and *Pirsonia, Ectrogella*, and *Lagenisma* all infect centric marine diatoms (Kühn et al., 2004; Schnepf et al., 1977, 1978; Raghu Kumar, 1980), which suggests the coevolution of parasitism between these two heterokont lineages (Cavalier-Smith and Chao, 2006). Environmental SSU rDNA sequences derived from small nanoplanktonic organisms sampled from diverse marine locations and ecosystems have shown that many of these lineages not only cluster within existing stramenopile clades, such as the hyphochytrids and oomycetes, but also form many "novel stremenopile" clades whose identities largely remain a mystery (Massana et al., 2004, 2006). The inclusion of such environmental sequence data in phylogenetic analyses significantly alters the topography of the heterokont tree and suggests that the *Pirsonia*/hyphochytrid clade may not be related as closely to the oomycetes as shown in Fig. 1.1b, although they undoubtedly share a common ancestor (Massana et al., 2004, 2006). It is to be expected that a systematic multigene approach to determining phylogeny in this lineage, as well as a significantly increased taxon sampling, will result in a much better understanding of the precise branching relationships of these various groups.

1.4 THE NAME GAME—THE TAXONOMY OF "CROWN" OOMYCETES

The current taxonomic organization of the oomycetes has largely been forged by two eminent scholars of zoosporic fungi, Frederick Sparrow (Sparrow, 1960,

1976) and Michael Dick (Dick et al., 1984; Dick, 2001). In his encyclopedic treatise on aquatic fungi, Sparrow (1960) split the oomycetes into four orders, the Lagenidiales, Leptomitales, Peronosporales, and Saprolegniales. In his final synthesis, Sparrow (1976) suggested that all oomycetes could be assigned to one of two groups, which he informally termed "galaxies." Within the "saprolegnian galaxy," he placed the order Saprolegniales (in which he included the Leptomitaceae as a family) and introduced a new order the Eurychasmales, in which he placed many marine oomycete families. Within the "peronosporalean galaxy," he placed the Peronosporales (in which the Peronosporaceae, Pythiaceae, and Rhipidiaceae were included as families) and the holocarpic Lagenidiales.

Dick continued to refine oomycete classification culminating in his final synthesis, which he outlined in his *magnum opus* Straminipilous Fungi, in which he expanded the number of orders to around 12 (Dick, 2001). Sparrow (1976) had pointed out the inappropriateness of the name oomycete, which had been first introduced in 1879, and this was acted on by Dick (1998, 2001) who formally renamed the class the Peronosporomycetes. However, there has been a general reluctance to abandon the traditional name, and its retention does not apparently contravene the International Code of Nomenclature. Dick's major revision was substantially carried out before the advent of wide-ranging molecular studies and was based mostly on a scholarly reinterpretation of the available morphological and ecological data. The application of molecular methodologies has revolutionized understanding of the likely phylogenic relationships throughout biology, and it has become increasingly apparent that many of the more radical changes introduced by Dick (2001) are not supported by molecular data and will require revision.

For oomycetes, most molecular studies have used the sequences of either the nuclear-encoded SSU (Dick et al., 1999; Spencer et al., 2002), large ribosomal subunit (LSU) genes (Riethmüller et al., 1999, 2002; Petersen and Rosendahl, 2000; Leclerc et al., 2000) or associated internal spacer region (ITS) sequences (Cooke et al., 2000), or the mitochondrial-encoded cytochrome c oxidase subunit II (cox2) gene (Hudspeth et al., 2000; Cook et al., 2001; Thines et al., 2008). Phylogenetic sequence data for the oomycetes is still far from complete, and the current analyses should be viewed as work in progress. It is not possible, for instance, to assemble all species for which molecular data are available into a single all-encompassing tree. There are also significant gaps in data, particularly for many of the less economically important taxa and, particularly, for those holocarpic species that cannot be brought into laboratory culture.

The early molecular studies all supported both the monophyletic origins of the oomycetes (Riethmüller et al., 1999; Hudspeth et al., 2000; Petersen and Rosendahl, 2000) and the broad "galaxy split" proposed by Sparrow (1976), which were assigned formal subclass rank (Saprolegniomycetidae and Peronosporomycetidae) by Dick et al. (1999). However, it seems likely that these higher taxonomic ranks will also require major revision, particularly if the oomycetes are considered to be a phylum in their own right. The two main plant pathogenic orders, the Pythiales and Peronosporales, were also fairly well supported by sequence data (Cooke et al., 2000; Riethmüller et al., 2002; Hudspeth et al., 2003). Most analyses revealed the genus *Phytophthora* to be part of the Peronosporales rather than the Pythiales where it had traditionally been placed (Cooke et al., 2000; Riethmüller et al., 2002). Some larger genera of plant pathogenic oomycetes, such as *Phytophthora* (Cooke et al., 2000; Blair et al., 2008) and *Pythium* (Lévesque and de Cock, 2004), have been split into several clades, which ultimately may warrant at least genus-level separation. The K-clade of *Pythium* is phylogenetically interesting because it seems to form a clade that is intermediate between the Pythiales and Peronosporales orders as currently constituted (Lévesque and de Cock, 2004).

Another major surprise was the early divergence within this line of the white blister rusts (*Albugo*) and their clear separation from all other members of the Peronosporales (Fig. 1.2b; Petersen and Rosendahl, 2000; Hudspeth et al., 2003). They have now been placed in their own order, the Albuginales (Fig. 1.2b; Riethmüller et al., 2002; Voglmayr and Riethmüller, 2006). On the basis of their unusually long and unique COII amino acid sequence (derived from the cox2 gene analysis), Hudspeth et al. (2003) considered them to be the earliest diverging clade in the Peronosporomycetidae, and they have been assigned their own subclass rank, which is called Albugomycetidae in some analyses (Thines et al., 2008).

The Rhipidiales are a small group of saprotrophic species associated with submerged twigs and fruit, most of which show restricted thallus development, consisting of a basal cell, holdfasts, and constricted (jointed) hyphal branches (Sparrow, 1960). They are a phylogenetically significant group that sits at the cusp of the saprolegnian-peronosporalean clade divergence (Figs. 1.2 and 1.3). Dick (2001) proposed that they be given their own order and subclass status (Rhipidiales, Rhipidiomycetidae), although he acknowledged the limited data on which this was based. Unfortunately, *Sapromyces elongatus* is still the only representative of this clade to have been sequenced and is a species whose placement has proven problematic (compare Fig. 1.2a and b). It has been reported as the basal clade to the Peronospomycetidae in cox2 trees (Hudspeth et al., 2000) and the basal clade to the Saprolegniomycetidae in LSU rDNA trees (Riethmüller et al., 1999; Petersen and Rosendahl, 2000). In our SSU rDNA trees (Fig. 1.2a), it forms part of a clade together with the holocarpic nematode parasite *Chlamydomyzium*, which diverges before both the major subclasses. However, the derived COII amino acid sequence showed that *Sapromyces* has the same signature amino acid insertion-deletion (indel) sequence (LEF/T) as that found in members of the Pythiales in contrast to the YTD indel sequence found in members of the Leptomitaceae (Hudspeth et al., 2000, 2003; Cook et al., 2001). Other members of the genus, such as *C. oviparasiticum* (Glockling and Beakes, 2006a), are diplanetic and have K-bodies in their zoospores (saprolegnian characteristics) but release their zoospores into a transient vesicle (a peronosporalean characteristic). Nakagiri

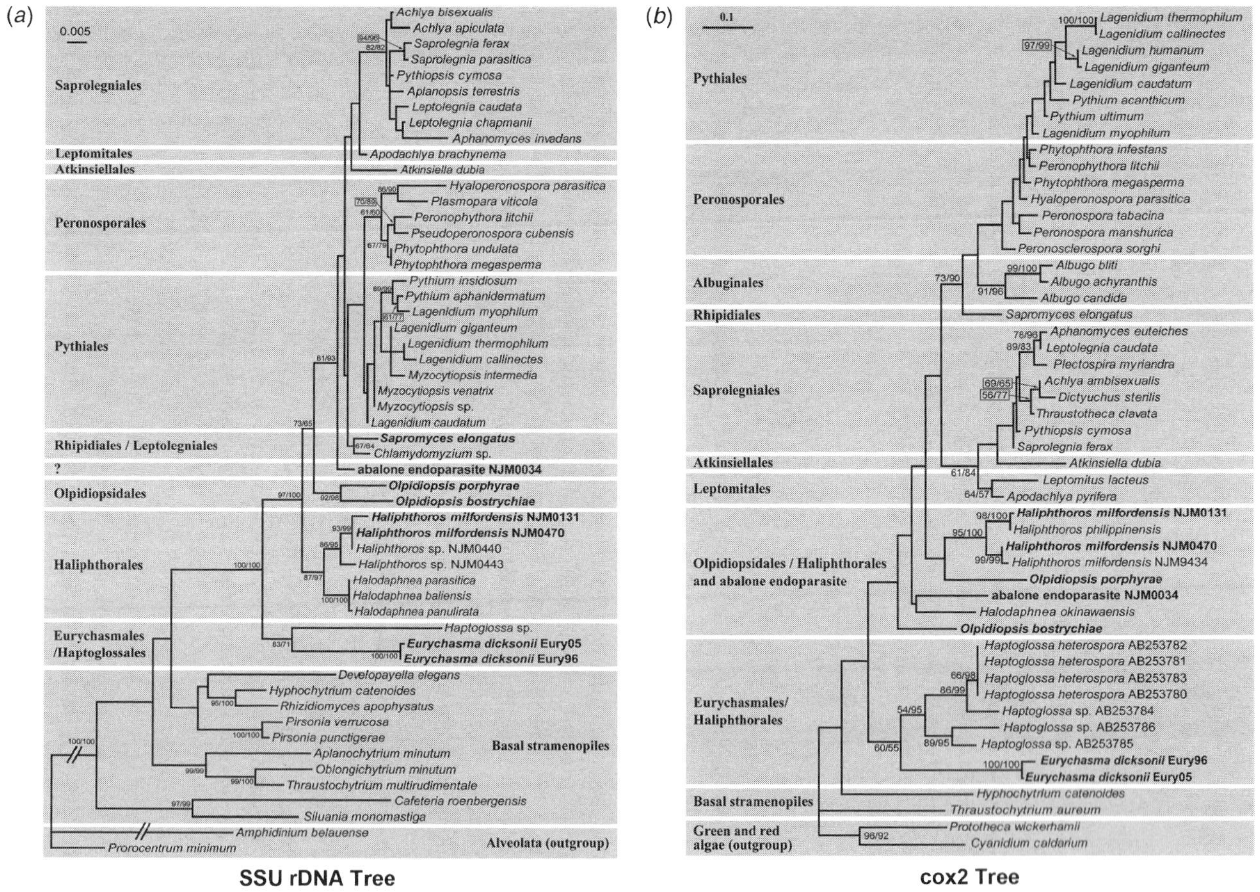

FIG. 1.2 Comparative nuclear (a) and mitochondria-encoded (b) phylogenetic trees of taxa in the Oomycete class and representative chromalveolates (and algae). 2(a) Maximum-likelihood (ML) tree (1,020 sites) based on small subunit (SSU) rDNA gene sequences. 2(b) Maximum-likelihood tree (167) sites based on 51 COII amino acid sequences. Organisms sequenced by Sekimoto (2008) are indicated in bold. ML and neighbor-joining (NJ) bootstrap values (100 and 2,000 replicates, respectively) above 50% are indicated above the internodes.

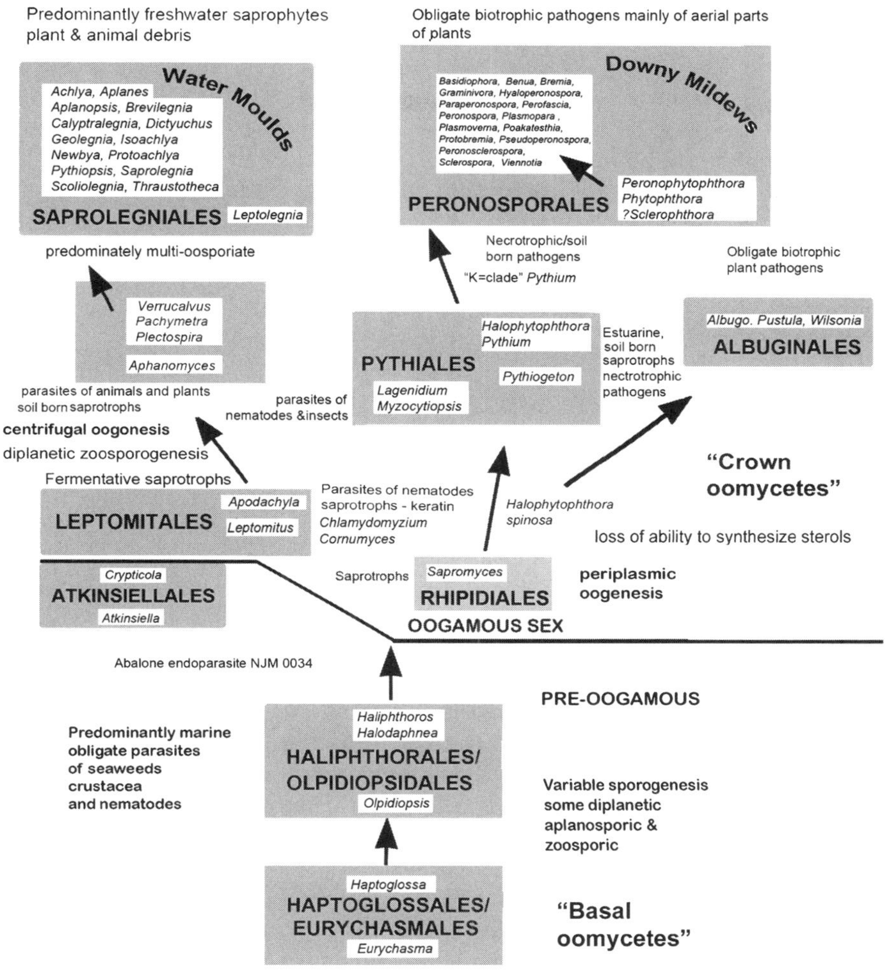

FIG. 1.3 Schematic summary of the likely phylogenetic relationships between the main orders within the oomycetes, based on molecular sequence data. The species listed are those for which sequence data are available. Some main ecological and morphological characteristics are also mapped onto this scheme. See the text for sources.

(2002) has also reported that *Halophytophthora spinosa* is not closely related to other members of the genus and apparently clusters close to *Sapromyces*.

The sequence data that support the early divergence of the Leptomitales clade in Saprolegniomycetidae comes from two taxa *Apodachlya* and *Leptomitus*, which are both members of the family Leptomitaceae (Riethmüller et al., 1999; Dick et al., 1999; Petersen and Rosendahl, 2000). This order, however, also includes the Leptolegnielliaceae, which contains many holocarpic genera, such as *Aphanomycopsis, Brevilegniella, Leptolegniella*, and the nematode parasite *Nematophthora*. *Cornumyces* was also tentatively included in this family by Dick (2001). *Cornumyces* isolates form a clade close to the Leptomitales at the base of the saprolegnian line (Inaba and Harayama, 2006) and also close to *Chlamydomyzium* when this species is included in the analyses (Inaba unpublished data). Unfortunately, no sequence data are available for any other of the genera in the Leptolegniellaceae. From the current, scant, molecular data, it seems that the clades located close to the point where the two main subclasses diverge (encompassing the Rhipidiales, Leptomitales, Atkinsiellales etc. Figs. 1.2 and 1.3) cannot be properly resolved until there has been far greater taxon and gene sampling.

1.5 ALL AT SEA—THE EARLIEST DIVERGING OOMYCETE CLADES

The first indication that some genera might fall outside the two main "crown" subclasses came from the study of Cook et al. (2001) who sequenced the cox2 gene for several parasites of marine crustaceans. Two genera, *Haliphthoros* (Fig. 1.4p) and *Halocrusticida* (Fig. 1.4n and o), which has been reclassified as *Halodaphnea* by Dick, 1998, 2001), formed a well-supported clade that diverged before the main crown subclasses (Cook et al., 2001). However, another enigmatic marine crustacean parasite, *Atkinsiella*, formed a deeply branched clade basal to the Saprolegniomycetidae. This study indicated that these obscure marine genera might hold the key to understanding the evolutionary origins of the oomycetes as a whole. This conclusion was reinforced when it was reported that *E. dicksonii*, which is a holocarpic parasite of brown seaweeds (Fig. 1.4a and b), was found to be the earliest diverging member of the oomycete lineage (Küpper et al., 2006).

A range of marine parasites of seaweeds and invertebrates was selected for an integrated study into their molecular phylogeny, morphological development, and ultrastructural characteristics (Sekimoto, 2008; Sekimoto et al., 2007, 2008a–c). Phylogenetic trees based on the SSU rDNA (Fig. 1.2a) and cox2 genes (Fig. 1.2b) revealed that most of these marine holocarpic species fell into one of two deeply branched early diverging clades, which we have termed "basal oomycetes" (Fig. 1.3). The first clade in both SSU rDNA (Fig. 1.2a) and cox2 gene (Fig. 1.2b) trees encompassed two genera, *Eurychasma* and *Haptoglossa* (Beakes et al., 2006; Hakariya et al., 2007; Sekimoto et al., 2008b). These two genera have few apparent morphological and structural features in common (cf. Fig. 1.4a,b, f–l) and would never have been linked without molecular data. These two genera may merit their own order status, the Eurychasmales and Haptoglossales, although they do seem to form a distinct clade, albeit showing long branch separation (Fig. 1.2a and b). *Eurychasma* is an obligate parasite of filamentous brown seaweeds, mostly in the Ectocarpales (Fig. 1.4a and b), but it has a broad host range (Küpper and Müller, 1999).

It will be interesting to determine whether the two as yet unsequenced enigmatic parasites of marine centric diatoms, *Ectrogella* (Raghu Kumar, 1980) and *Lagenisma* (Schnepf et al., 1977, 1978) also belong to this clade, as they also have a naked plasmodial infection stage.

Haptoglossa is an obligate parasite of rhabditid nematodes. Because of the apparent absence of mastigoneme hairs (Fig. 1.4f) and unique *Plasmodiophora*-like infection cells (Fig.1.4h–l; Beakes and Glockling, 1998), it was briefly considered to be related to the plasmodiophorids (Dick, 2001). *Haptoglossa* spp. show a remarkable and unsuspected diversity in their patterns of sporulation (Beakes and Glockling, 2002) and in the different types and micromorphology of the infection cells that are produced (Fig. 1.4h–l; Glockling and Beakes, 2000a and b, 2001, 2002). It seems to form an extremely diverse and deeply branching clade (Fig. 1.2b; Hakariya et al., 2007), which suggests that the Haptoglossaceae will undoubtedly require substantial taxonomic revision, employing both molecular sequencing and ultrastructural characterization.

The second basal clade (Fig. 1.2a and b) includes both parasites of red seaweeds (Fig.1.4c–e, m) and marine crustacea (Fig. 1.4n–p). The SSU rDNA tree suggests the two red seaweed parasites, *Olpidiopsis porphyrae* (Fig. 1.4c–e; Sekimoto et al., 2008a) and *Olpidiopsis bostrychiae* (Fig. 1.4m; Sekimoto, 2008; Sekimoto et al., 2009) form a separate clade from the crustacean parasites, *Haliphthoros* and *Halodaphnea* (syn. *Halocrusticida*) (Fig. 1.2a). However, in the cox2 tree, the two groups cannot be resolved from each other (Fig. 1.2b). In the SSU tree, *O. porphyrae* and *O. bostrychiae* are separated by a significant branch length from each other, which in other oomycete families would warrant genus-level distinction. *Haliphthoros* also requires splitting into more taxa, because the sequenced isolates fell into two well-separated clades (Fig. 1.2b), which were not coincidental with the two currently recognized taxa *Haliphthoros milfordensis* and *Haliphthoros phillipensis* (Sekimoto et al., 2007). Because of their very different host ranges and morphological differences, we suggest that the Olpidiopsidales and Haliphthorales probably merit being retained in separate orders, but more sequence data are required before these can be unequivocally defined. We also predict, from their overall morphological and ultrastructural similarities, that these two early diverging clades are likely to encompass other marine genera such as *Pontisma* and *Petersenia*. Although somewhat similar in its host preferences and morphology, *Atkinsiella dubia* does not seem to be within the Haliphthorales and has been assigned to its own order, the Atkinsiellales, by Sekimoto (2008). Dick (1998) transferred *Atkinsiella entomophaga,* a parasite of dipteran larvae (Martin, 1977), to the genus *Crypticola*, which had been created for *Crypticola clavulifera,* an entomopathogenic species isolated from mosquito larvae (Frances et al., 1989). Interestingly, the latter does seem to form a clade with *A. dubia* in cox2 analyses (D. Hudspeth, unpublished data).

Environmental sequences obtained from unidentified marine nanoflagellates have revealed four well-separated "stramenopile" sequences (RA010613.4,

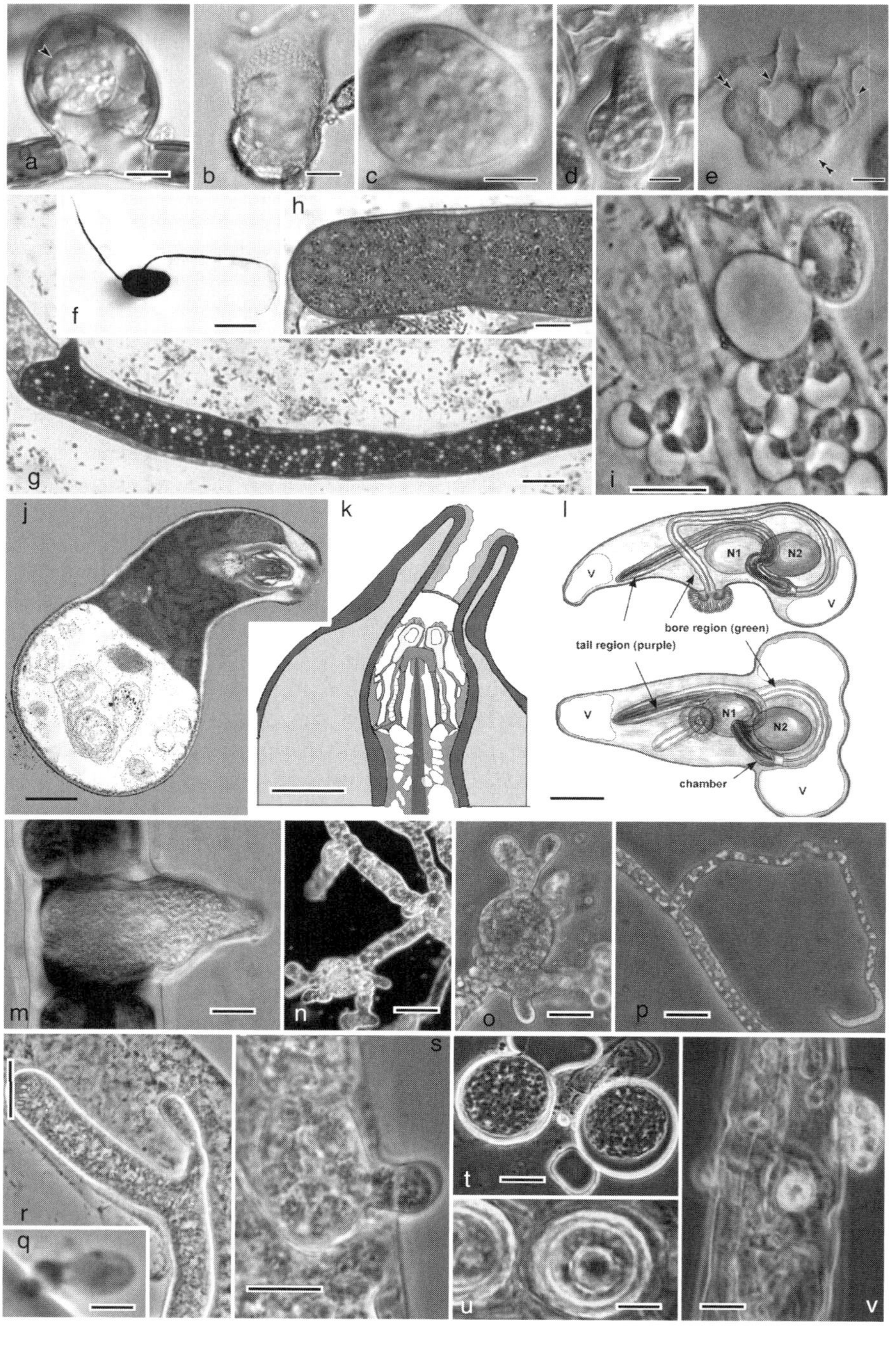
a
b
c
d
e
f
g
h
i
j
k
l
V
N1
N2
bore region (green)
tail region (purple)
chamber
m
n
o
p
q
r
s
t
u
v

FIG. 1.4 (See color insert) (a–v) Light micrographs (LMs), electron micrographs (EMs), and diagrams summarizing some morphological and structural characteristics of basal holocarpic oomycetes. (a) Differential interference contrast (DIC) LM of a nonwalled (plasmodial) stage of *Eurychasma* thallus development, showing hugely swollen host cell. Bar = 10 µm. (b) DIC LM of a fully differentiated *Eurychasma* sporangium, showing net-like layer of peripheral cysts. Bar = 10 µm. (c) DIC LM of spherical thallus of *Olpidiopsis porphyrae* after the sporangial wall has formed, showing well-scattered nuclei. Bar = 5 µm. (d) DIC LM of an immature thallus of *O. porphyrae*, showing zoospore initials and tapered apical discharge tube. Bar = 5 µm. (e) DIC LM showing multiple infection of a single host cell with *O. porphyrae* thalli. Empty sporangia (single arrowheads) and young plasmodial stage thalli (double arrowheads) can be observed. Bar = 5 µm. (f–l) Illustrations showing the morphological diversity shown by the nematode parasite *Haptoglossa*: (f) TEM wholemount of a zoospore of *Haptoglossa dickii*, showing smooth (nonmastigonate) anterior flagellum. Bar = 1 µm. (g) LM of a toluidine-blue-stained maturing thallus of *Haptoglossa polymorpha*, showing unbranched sausage-like thallus with a single terminal discharge tube, which has breached the nematode cuticle. Bar = 25 µm. (h) LM of a toluidine-blue-stained immature thallus of *H. polymorpha*, showing the dense nonvacuolated cytoplasm characteristic of this genus. Bar = 10 µm. (i) Phase contrast (PC) LM of mature thallus of *H. heteromorpha*, showing both large (upper) and small (lower) germinating aplanospores and showing typical tapering gun cell initials. Bar = 10 µm. (Courtesy of S. Glockling.) (j) Median TEM section of a gun cell of an unnamed *Haptoglossa* sp. Showing a basal vacuole and a recurved apex containing needle chamber. Bar = 1 µm. (k) Diagram showing a needle chamber at the apex of a mature gun cell of *H. dickii* and showing needle (dark purple) codes, investing cones (orange and green), and O-ring apparatus (yellow). Modified from Beakes and Glockling (1998). Bar = 0.5 µm. (l) Side and top LS views of small binucleate infection cell of *H. heteromorpha* [shown developing in (i) illustrating morphological diversity of such cells]. Color version of diagram available in Glockling and Beakes (2000b). Bar = 1 µm. (m) DIC LM of maturing thallus of *Olpidiopsis bostrychiae* infecting a single cell of filamentous red seaweed *Bostrychia*. Note the greatly expanded host cell and single apical discharge tube. Bar = 10 µm. (n–p) PC LM showing *in vitro* cultured thalli of the crustacean parasites *Halodaphnea panulirata* NJM9832 (n and o) and *Haliphthoros milfordensis* NJM 0470 (p) Note the irregularly branching bulbous growth form of the former the compared with more hypha-like thalli and discharge tubes, which contain differentiating zoospore initials in the latter. (n) Bar = 25 µm, (o) Bar = 10 µm, (p) Bar = 25 µm. (q, t) PC LM of a germinating cyst and branched tubular thallus of nematode parasite *Chlamydomyzium dictyuchoides*. Note the small appressorial-like pad produced by the germinating spore at the point of presentation (q) and rather frothy cytoplasm (r) (q) Bar = 5 µm (t) Bar = 20 µm. Courtesy of S. Glockling. (S) PC LM of mature sporangium of nematode parasite *Myzocytiopsis vermicola*, showing a small segmented thalli that contains fully differentiated zoospores and a discharge tube, with apical papillar plug, which forms an evanescent restraining vesicle around escaping zoospores. Bar = 10 µm. (t,u) PC LM of immature oospheres and adjacent antheridia (t) and fully mature oospores (u) of *M. vermicola*. Bar = 5 µm. Color versions are available from Glockling and Beakes (2006b). (v) PC LM of a mature thallus of the *Myzocytiopsis intermedia* showing hyphal-like thalli and external vesicles containing refractile clusters of differentiating zoospores. Bar = 10 µm. Courtesy of S. Glockling.

BL010320.2, BOLA320, CCW73) all located on the SSU rDNA tree between *E. dicksonii* and the crown oomycete clade (Massana et al., 2004, 2006). When included with the sequence data described here, these environmental sequences did not cluster within either of the two basal clades outlined above but formed two more novel clades between the *Haliphthoros*/*Halodaphnea* clade and the crown oomycetes (Sekimoto, 2008). This suggests that basal marine oomycetes are both more widespread and diverse than currently appreciated, and a concerted effort should be made to try and isolate, identify, and sequence as many of these taxa as possible.

1.6 ODD FELLOWS — WHERE DO THE LAGENIDIACEOUS NEMATODE PATHOGENS FIT IN?

The only holocarpic lagenidiaceous species that was included in early phylogenetic studies was the mosquito parasite *Lagenidium giganteum*, which was unambiguously shown to be within the *Pythium* clade (Dick et al., 1999; Petersen and Rosendahl, 2000; Hudspeth et al., 2000). Unfortunately, Dick (2001) argued that this species was the only valid representative of this long-established genus and proceeded to redistribute most of the other *Lagenidium* species among many newly created orders. He transferred the nematode parasites to the Myzocytiopsidales and the marine lagenidiaceous species to the Salilagenidiales, both of which he placed in the Saprolegniomycetidae. However, almost concurrently, Cook et al. (2001) reported that three marine lagenidiaceous parasites of crustacea were part of the same clade as *L. giganteum* and some *Pythium* species, which confirmed that both marine and terrestrial "lagenidiaceous" species were closely related as originally thought. Most nematode infecting *Myzocytiopsis* species (Fig. 1.4s–v) also seem to be closely related to *Lagenidium* spp. (Fig. 2a; Beakes et al., 2006), which suggests the Lagenidiaceae could form a discrete family within the Pythiales. The boundary among the genera *Lagenidium, Pythium*, and *Myzocytium* had always been ill-defined (discussed at length by Dick, 2001) and still requires additional gene and taxon sampling before this group can be properly resolved. It will be interesting to observe how these taxa relate to the various *Pythium* clades recently identified by Lévesque and De Cock (2004). There is no support for the orders Salilagenidiales and Myzocytiopsidales created by Dick (1998, 2001), and both should be rejected.

1.7 A PLACE FOR THE WATER MOLDS — A FISHY TALE

All genera in the water mold order the Saprolegniales form a well-defined clade (Dick et al., 1999; Riethmüller et al., 1999; Petersen and Rosendahl, 2000; Leclerc et al., 2000; Inaba and Tokumasu, 2002; Spencer et al., 2002). On the basis of their study, Dick et al. (1999) proposed creating a new family within the Saprolegniales called the Leptolegniaceae, in which he subsequently placed the genus *Aphanomyces* (Dick, 2001). This was almost certainly a premature decision, because in the LSU rDNA analysis of Petersen and Rosendahl (2000), *Aphanomyces* was found to form the first diverging clade in the Saprolegniales, whereas *Leptolegnia* continued to be associated with other genera of the saprotrophic water moulds.

Aphanomyces is a genus that includes many important pathogens of crustacea (e.g., *Aphanomyces astaci*; Dykstra et al., 1986), fish (e.g., *Aphanomyces invadans*; Lilley et al., 2003), and plant roots (e.g., *A. euteiches*; Johnson et al., 2002), which together with several genera of little studied soil/root inhabiting oomycetes, *Plectospira, Pachymetra* (Reithmüller et al., 1999), and *Verrucalvus* (Thines unpublished data) form a well-supported clade that is separate from other members of the Saprolegniaceae (Fig. 1.3). These genera, together with the unsequenced *Verrucalvus*, will probably merit their own as yet undescribed order (Fig. 1.3). Dick et al. (1984) had placed *Pachymetra* together with *Verrucalvus* in their own family called the Verrucalvaceae, which were then included with the graminocolous downy mildews in the order Sclerosporales. As a result, Dick et al. (1984) removed this group of well-known plant pathogens from the "peronosporalean line" to the Saprolegniales (Dick, 2001). However, recent molecular studies have shown that all the leaf-infecting genera of graminocolous downy mildews (e.g., *Peronosclerospora, Sclerospora*, etc.) are scattered among other downy mildew genera in the Peronosporales (Hudspeth et al., 2003; Göker et al., 2007; Thines et al., 2008). These graminaceous pathogens do belong to the Peronosporales, but there is no molecular support for retaining the Sclerosporales as a separate order or family.

Although the family Saprolegniaceae contains mostly saprotrophic species, some water molds are important pathogens of fish (e.g., *Saprolegnia parasitica*, Dieguez-Uribeondo et al., 2007). The first phylogenetic analysis that attempted to map traditional spore-release characters, which had been used to define genera in the water molds (Saprolegniaceae), was reported by Daugherty et al. (1998) using ITS sequence data. The familiar water mold genus *Saprolegnia*, which releases motile primary zoospores, seemed to form a separate clade from those genera, *Achlya, Thraustotheca*, and *Dictyuchus*, where the motile primary zoospore phase had been lost. However, this study was based on just a single sequence from each taxon, and it quickly became apparent that this was far too simplistic an overview of the Saprolegniaceae. When greater numbers of taxa were included in the phylogenetic analyses, it became apparent that the two largest and most familiar water mold genera *Achlya* and *Saprolegnia* did not form monophyletic taxa but had representatives scattered in several different "genus-level" clades (Leclerc et al., 2000; Inaba and Tokumasu, 2002, Spencer et al., 2002). It is now clear that the traditional generic classification of the Saprolegniaceae based on the pattern of zoospore discharge does not accurately reflect the underlying phylogenetic relationships in this family (Riethmüller et al., 1999; Inaba and Tokumasu, 2002; Spencer et al., 2002). Even in such a well-known genus as *Saprolegnia*, the application of molecular methods has proved problematic because many currently recognized taxa seem to be polyphyletic on ITS trees (Hulvey et al., 2007). A reclassification of the familiar water molds based on combined molecular and morphological characters is urgently required.

1.8 WHAT DOES IT ALL MEAN? EVOLUTIONARY PERSPECTIVES AND SPECULATIONS

The phylogenetic trees (Fig. 1.2a and b) clearly show that the earliest diverging oomycete genera are predominantly marine organisms. Even *Haptoglossa*, which is the only terrestrial genus in the "basal oomycete" assemblage, has been reported as a parasite of marine nematodes (Newell et al., 1977). This evidence is contrary to Dick's (2001) view that "all existing evidence points to a freshwater or terrestrial origin for the straminipilous fungi." Although "crown oomycetes" (see Fig. 1.3) are predominantly freshwater (the saprolegnian lineage) or terrestrial (the peronosporalean lineage), there are nevertheless a minority of marine representatives scattered throughout both lines. Some *Aphanomyces* sp. and many other genera of Saprolegniaceae have been isolated from estuarine ecosystems and can tolerate high or fluctuating salinities (Dykstra et al., 1986; Padgett, 1978). The Pythiales include many marine representatives, which include several *Lagenidium* (e.g., *Lagenidium callinectes*) and *Pythium* spp. (e.g., *Pythium porphyrae* and *Pythium grandisporangium*). Both *Myzocytiopsis vermicola* and *Gonimochaete latitubus* have been isolated from littoral marine nematodes (Newell et al., 1977). These observations suggest the intriguing possibility that the oomycetes may have migrated from the sea to the land (soil) along with their nematode hosts. Rhabditid nematodes are known from marine, estuarine, and terrestrial habitats (De Ley, 2006), which supports such a hypothesis. Host switching between soil-born nematodes and plants roots may have occurred at least twice, in *Aphanomyces* and in the Pythiales line. The exclusively marine genus *Halophytophthora*, which has papillate *Phytophthora*-like sporangia, forms a polyphyletic assemblage distributed among *Pythium* and *Phytophthora* species (Cooke et al., 2000; Nakagiri, 2002; Lévesque, unpublished data). It had been assumed that *Halophytophthora* had reacclimatized to the marine/estuarine environment (i.e., they were the oomycete equivalent of whales), but it is possible that they could represent vestiges of the original marine line (Nakagiri, 2002). If oomycetes had their origins in the open sea, it is in the estuarine benthic environments where they probably made the transition to becoming terrestrial saprotrophs and plant pathogens.

Recently, genomic studies have revealed that lateral gene transfer has occurred between the oomycetes and true fungi (Richards et al., 2006). It has even been suggested that their fungal-like growth form might have been acquired as a result of this. However, a complete morphological spectrum from simple spherical to ovoid thalli (Fig. 1.4a,c,d, and m), through unbranched sausage-like thalli (Fig. 1.4 g and h) to segmented branched thalli (Fig. 1.4r and s) and typical fine hyphal-like thalli (Fig. 1.4r), can all be found among these early diverging holocarpic parasites. It suggests that the fungus-like growth pattern may have evolved without the need to invoke gene transfer from true fungi. However, the body cavities of nematodes or invaded plant tissues may have provided suitable "closed environments," whereby true fungi and oomycetes could have come into close contact and exchanged genetic material.

All basal clade taxa (Fig. 1.2a and b) studied to date apparently lack sexual stages. Sparrow (1976) remarked that it seemed improbable that all marine oomycete genera could be genuinely asexual or "choose to live monastically" as he quaintly put it. He speculated that they probably had some form of nonoogamous sexual cycle. The best evidence in support of this comes from *Lagenisma coscinodisci*, which produces zoomeiospores that form sexual cysts that conjugate to form the zygote (Schnepf et al., 1977). Many *Haptoglossa* species produce both uninucleate and binucleate infection cells (Fig. 1.4l; Beakes and Glockling, 2000b, 2001, 2002), but we have no idea of how these fit into their overall life cycle. As the genes specifically associated with sexual reproduction in oomycetes are identified (e.g., Prakob and Judelson, 2007), it will be interesting to explore whether and where they may be expressed in these basal species. Oogamous sexual reproduction is clearly one of the major evolutionary developments that define crown oomycetes. Only the genus *Olpidiopsis* among those in the basal clades is reported to form oogonia and only then in freshwater species that parasitize water molds (Sparrow, 1960). The holocarpic differentiation of neighboring thallus compartments into antheridia and oogonia observed in the genus *Myzocytiopsis* (Fig.1.4t and u; Glockling and Beakes, 2006b) illustrates how oogamous reproduction probably evolved. The unioosporiate condition is clearly the most primitive form because it is prevalent in all orders except for the Saprolegniales (Fig. 1.3).

Another inescapable inference from the phylogenetic trees (Fig. 1.2) is that oomycetes have been "hard wired" for parasitism since their inception. Both basal-clade genera *Eurychasma* and *Haptoglossa* are obligate parasites, which cannot be cultured independently from their hosts. *E. dicksonii* is a wide-ranging parasite of phaeophyte seaweeds (Küpper et al., 1999). Related species are reported to infect both red and green seaweed hosts (Karling, 1981), which indicates that these may be fairly broad-spectrum parasites. At least one other major phylum, which is the apicocomplexa within the Chromalveolate lineage, is exclusively parasitic. Like many basal oomycetes, these are parasites of many invertebrate phyla, such as mollusks and arthropods, but they also infect all classes of vertebrates as well (Marquardt and Speer, 2001). Recent genomic analysis has revealed many significant similarities at the molecular level between parasitism in apicocomplexans and oomycetes (Robold and Hardham, 2005; Torto-Alalibo et al., 2005; Bhattacharjee et al., 2006; Talbot, 2007). These similarities are reinforced when one considers that the initial stages of thallus development in all basal oomycete parasites of marine algae are as unwalled plasmodia (Fig. 1.4a and c) located within a membrane-bound host vacuole (*Eurychasma, Olpidiopsis*; Sekimoto et al., 2008a–c; *Ectrogella*; Raghu Kumar, 1980; *Lagenisma,* Schnepf et al., 1978; *Petersenia*, Pueschel and van der Meer, 1985), which is equivalent to the parasitophorous vacuole of apicocomplexans (see Talbot, 2007). The flagellate parasitoid *Pirsonia* infects diatoms by means of an invasive pseudopodium that forms a "feeding" trophosome adjacent to the host protoplast, which it ingests by phagocytosis rather than by absorption (Schnepf and Schweikert, 1997). A *Pirsonia*-like parasitoid might have been the kind of organism that was ancestral to the oomycetes. It will certainly be interesting to find out more about the unknown novel stramenopile clades that have been shown to diverge just before the oomycetes (Massana et al., 2004, 2006). Many fundamental mechanisms associated with both infection (attachment to host) and host–parasite interaction (effector-protein delivery systems) are deeply embedded within the lineage and may have been present in the original flagellate root ancestor to all chromalveolates (Fig. 1.1b) perhaps even before the primary plastid acquisition event. Some present-day dinoflagellates are parasites of other chromalveolates and crustacea (Coats, 1999) and show that both parasitic and autotrophic lifestyles can coexist.

These molecular phylogenetic studies on holocarpic parasites of algae and invertebrates have provided a much clearer overview of the likely evolutionary and taxonomic relationships within the oomycetes, which we have summarized diagrammatically in Fig. 1.3. The scheme proposed by Bessey nearly 70 years ago (Fig. 1.1a), which suggested that the oomycetes evolved from a photosynthetic heterokont alga and that the holocarpic Olpidiopsidales and Lagenidiales were at the root of the lineage, has been shown using modern molecular methodologies to have been remarkably perceptive.

ACKNOWLEDGMENTS

We would like to acknowledge the sharing of phylogenetic information and/or helpful contributions made to our discussions by many of our colleagues who work on oomycete fungi, particularly Claire Gachon, Sally Glockling, Marcus Göker, Masateru Hakariya, Kishio Hatai, Shigeki Inaba, Frithjof Küpper, Andre Lévesque, Akira Nagakiri, Marco Thines, and Hermann Voglmayr. Our special thanks are extended to Daiske Honda, who mentored and guided the work on the marine parasites that formed part of the doctoral thesis of S. S.

REFERENCES

Adl SM, Simpson AG, Farmer MA, Anderson RA, Anderson OR, Barta SR, Bowser SS, Brogeroile G, Fensome RA, Fredericq S et al. 2005. The new higher level classification of the eukaryotes with emphasis on the taxonomy of protists. J Eukaryot Microbiol 52:399–451.

Baldauf SL, Roger AJ, Wenk-Siefert I, Doolittle WF 2000. A kingdom-level phylogeny of eukaryotes based on combined protein data. Science 290:972–977.

Beakes GW 1981. Ultrastructural aspects of oospore differentiation. In: Hohl H, Turian, G editors. The fungal spore: morphogenetic controls. Academic Press. New York. pp. 71–94.

Beakes GW 1987. Oomycete phylogeny: ultrastructural perspectives. In: Rayner ADM, Brasier CM, Moore D editors. Evolutionary biology of the fungi. Cambridge University Press, Cambridge, UK, pp. 405–421.

Beakes GW 1989. Oomycete fungi: their phylogeny and relationship to chromophyte algae. In: Green JP, Leadbeater BSC, Diver WL editors. The chromophyte algae: problems and perspectives. Clarendon Press, Oxford, UK, pp. 325–342.

Beakes GW, Glockling SL 1998. Injection tube differentiation in gun cells of a *Haptoglossa* species which infects nematodes. Fungal Genet Biol 24:45–68.

Beakes GW, Glockling SL 2000. An ultrastructural analysis of organelle arrangement during gun (infection) cell differentiation in the nematode parasite *Haptoglossa dickii*. Mycol Res 104:1258–1269.

Beakes GW, Glockling SL 2002. A comparative fine-structural study of dimorphic infection cells in the nematophagous parasite, *Haptoglossa erumpens*. Fungal Genet Biol 37:250–262.

Beakes GW, Glockling SL, James TY 2006. The diversity of oomycete pathogens of nematodes and its implications to our understanding of oomycete phylogeny. In: Meyer W, Pearce C editors. Proc. Eighth Interntl Mycol Cong. Medimond, Italy, pp. 7–12.

Bessey EA 1935. A textbook of mycology. P. Blakiston's Son & Co. Philadephia, PA.

Bessey EA 1942. Some problems in fungus phylogeny. Mycologia 34:355–379.

Bhattacharjee S, Hiller NL, Konstantinos L, Win J, Thirumala-Devi K, Young C, Kamoun S, Haldar K 2006. The malarial host-targeting signal is conserved in the Irish Potato famine pathogen. PLoS Path 2: e50.

Blair JE, Coffey MD, Park S-Y, Geiser DM, Kang S 2008. A multi-locus phylogeny for *Phytophthora* utilizing markers derived from complete genome sequences. Fungal Genet Biol 45:266–277.

Burki F, Shalchian-Tabrizi K, Minge M, Skjaeveland A, Nikolaev SI, Jakobsen KS, Pawlowski J 2007. Phylogenetics reshuffles the eukaryote supergroups. PLoS2: e790.

Cavalier-Smith T 1981. Eukaryote kingdoms: seven or nine? Biosystems 14:461–481.

Cavalier-Smith T, Chao EEY 2006. Phylogeny and megasystematics of phagotrophic heterokonts (Kingdom Chromista). J Mol Evol 62:388–420.

Coats DW 1999. Parasitic life styles of marine dinoflagellates. J Eukaryot Microbiol 46:402–409.

Cook KL, Hudspeth DSS, Hudspeth MES 2001. A cox2 phylogeny of representative marine peronosporomycetes (Oomycetes). Nova Hedwiga Beiheft 122:231–243.

Cooke DEL, Drenth A, Duncan JM, Wagels G, Brasier CM 2000. A molecular phylogeny of *Phytophthora* and related oomycetes. Fungal Genet Microbiol 30:17–32.

Daugherty J, Evans TM, Skillom T, Watson LE, Money NP 1998. Evolution of spore release mechanism in the Saprolegniaceae (Oomycetes): evidence form a phylogenetic analysis of internal transcribed spacer sequences. Fungal Genetic Microbiol 24:354–363.

De Ley P 2006. A quick tour of nematode diversity and the backbone of nematode phylogeny. WormBook http://wormbook.org.

Dick MW 1998. The species and systematic position of *Crypticola* in the Peronosporomycetes, and a new names for *Halocrusticida* and species therein. Mycol Res 102:1062–1066.

Dick MW 2001. Straminipilous fungi. Kluwer, Dordrecht, Germany.

Dick MW, Vick MC, Gibbings JG, Hedderson TA, Lopez Lastra CC 1999. 18S rDNA for species of *Leptolegnia* and other Peronosporomycetes: justification of the subclass taxa Saprolegniomycetidae and Peronosporomycetidae and division of the Saprolegniaceae *sensu lato* into the Leptolegniaceae and Saprolegniaceae. Mycol Res 103:1119–1125.

Dick MW, Wong PTW, Clark G 1984. The identity of the oomycete causing 'Kikuyu Yellow, with a reclassification of the downy mildews. Bot J Linn Soc 89:171–197.

Dieguez-Uribeondo J, Fregeneda-Grandes JM, Cerenius L, Perez-Iniesta M, Aller-Gancedo JM, Tellerıa MT, Soderhall K, Martına MP 2007. Re-evaluation of the enigmatic species complex *Saprolegnia diclina–Saprolegnia parasitica* based on morphological, physiological and molecular data. Funal Genet Biol 44: 585–601.

Dykstra DP, Noga EJ, Levine JF, Moye DE 1986. Characterization of the *Aphanomyces* species associated with ulcerative mycosis (UM) in menhaden. Mycologia 78:664–672.

Frances SP, Sweeney AW, Humber RA 1989. *Crypticola clavulifera* gen. et. sp. nov. and *Lagenidium giganteum*: oomycetes pathogenic for dipterans infesting leaf axils in an Australian rain forest. J Invertebr Pathol 54:103–111.

Glockling SL, Beakes GW 2000a. Two new *Haptoglossa* species (*H. erumpens* and *H. dickii*) infecting nematodes in cow manure. Mycol Res 104:100–106.

Glockling SL, Beakes GW 2000b. The ultrastructure of the dimorphic infection cells of *Haptoglossa heteromorpha* illustrates the developmental plasticity of infection apparatus structures in a nematode parasite. Can J Bot 78:1095–1107.

Glockling SL, Beakes GW 2001. Two new species of *Haptoglossa* from N.E. England, *H. northumbrica* and *H. polymorpha*. Bot J Linn Soc 136:329–338.

Glockling SL, Beakes GW 2002. Ultrastructural morphogenesis of dimorphic arcuate infection (gun) cells of *Haptoglossa erumpen*, an obligate parasite of *Bunonema* nematodes. Fungal Genet Biol 37:250–262.

Glockling SL, Beakes GW 2006a. Structural and developmental studies of *Chlamydomyzium oviparasiticum* from *Rhabditis* nematodes and in culture. Mycol Res 110:1119–1126.

Glockling SL, Beakes GW 2006b. An ultrastructural study of development and reproduction in the nematode parasite *Myzocytiopsis vermicola*. Mycologia 98:7–21.

Göker M, Voglmayr H, Riethmüller A, Oberwinkler F. 2007. How do obligate parasites evolve? A multi–gene phylogenetic analysis of downy mildews. Fungal Genet Biol 44:105–122.

Göker M, Voglmayr H, Riethmüller A, Weiss M, Oberwinkler, F 2003. Taxonomic aspects of Peronosporaceae inferred from Bayesian molecular phylogenetics. Can J Bot 81:672–683.

Hakariya M, Hirose D, Tokumasu S 2007. A molecular phylogeny of *Haptoglossa* species, terrestrial peronosporomycetes (oomycetes) endoparasitic on nematodes. Mycoscience 48:169–175.

Hakariya M, Masuyama N, Saikawa M 2002. Shooting of sporidium by "gun" cells in *Haptoglossa heterospora* and *H. zoospora* and secondary zoospore formation in *H. zoospora*. Mycoscience 43:119–125.

Harper JT, Waanders E, Keeling PJ 2005. On the monophyly of chromalveolates using a six-protein phylogeny of eukaryotes. Intl J Syst Evol Microbiol 55:487–496.

Hudspeth DSS, Nadler SA, Hudspeth MES 2000. A cox II molecular phylogeny of the Peronosporomycetes. Mycologia 92:674–684.

Hudspeth DSS, Stenger D, Hudspeth MES. 2003. A cox2 phylogenetic hyphothesis for the downy mildews and white rusts. Fungal Divers 13:47–57.

Hulvey JP, Padgett DE, Bailey JC 2007. Species boundaries within the Saprolegnia (Saprolegniales, Oomycota) based on morphological and DNA sequence data. Mycologia 99:421–429.

Inaba S, Hariyama S 2006. The phylogenetic studies on the genus *Cornumyces* (Oomycetes) based on the nucleotide sequences of the nuclear large subunit ribosomal RNA and the mitochondrially encoded cox2 genes. 8th Int. Mycol Cong. Handbook and Abstracts. p. 330.

Inaba S, Tokumasu S 2002. Phylogenetic relationships between the genus *Saprolegnia* and related genera inferred from ITS sequences. Abstracts 7th Int. Mycol Cong. Oslo. 687 p. 208.

Johnson TW, Seymour RL, Padgett DE 2002. Biology and systematics of the Saprolegniaceae. http://dl.uncw.edu/digilib/biology/fungi/taxonomy%20and%20 systematics/padgett%20book/ accessed April 2008.

Karling JS 1942. The simple holocarpic biflagellate phycomycetes. Columbia University Press, New York.

Karling JS 1981. Predominantly holocarpic and eucarpic simple biflagellate phycomycetes. J Cramer Vaduz.

Kühn SF, Medlin LK, Eller G 2004. Phylogenetic position of the parasitoid nanoflagellate *Pirsonia* inferred from nuclear-encoded small subunit ribosomal DNA and a description of *Pseudopirsonia* n. gen. and *Pseudopirsonia mucosa* (Drebes) comb. nov. Protist 155:143–156.

Küpper FC, Maier I, Müller DG, Loiseaux-de Goer S, Guillou L 2006. Phylogenetic affinities of two eukaryotic pathogens of marine macroalgae, *Eurychasma dicksonii* (Wright) Magnus and *Chytridium polysiphoniae* Cohn. Cryptogam Algol 27:165–184.

Küpper FC, Müller DG 1999. Massive occurrence of the heterokont and fungal parasites *Anisolpidium*, *Eurychasma* and *Chytridium* in *Pylaiella litoralis* (Ectocarpales, Phaeophyceae). Nova Hedwig 69:381–389.

Lamour KH, Win J, Kamoun S 2007. Oomycete genomics: new insights and future directions. FEMS Microbiol Lett 274:1–8.

Leclerc MC, Guillot J, Deville M 2000. Taxonomic and phylogenetic analysis of Saprolegniaceae (Oomycetes) inferred from LSU rDNA and ITS sequence comparisons. Anton Van Leeuk 77:369–377.

Leipe DD, Tong SM, Goggin CL, Slemenda SB, Pieniazek NJ, Sogin ML 1996. 16S-like rDNA sequences from *Developayella elegans*, *Labyrinthuloides haliotidis*, and *Proteromonas lacertae* confirm that the stramenopiles are a primarily heterotrophic group. Eur J Protist 33:369–377.

Lévesque CA, de Cook AW 2004. Molecular phylogeny and taxonomy of the genus *Pythium*. Mycol Res 108:1363–1383.

Lilley JH, Hart D, Panyawachira V, Kanachanakhan S, Chinabut S, Soderhall K, Cerenius K 2003. Molecular characterization of the fish-pathogenic fungus *Aphanomyces invadans*. J Fish Dis 26:263–275.

Marquardt WC, Speer CA 2001. Apicocomplexa. Encylopedia of Life Sciences (ELS). Wiley InterScience. http://mrw.interscience.wiley.com/emrw/9780470015902/els/article/a0001956/current/html, DOI: 10.1038/npg.els.0001956.

Martin WW 1977. The development and possible relationships of a new *Atkinsiella* parasitic in insect eggs. Am J Bot 64:760–769.

Massana R, Castresana J, Balagué V, Guillou L, Romari K, Groisillier A, Valentin K, Pedró-Alió C 2004. Phylogenetic and ecological analysis of novel marine stramenopiles. App Environ Microbiol 70:3528–3534.

Massana R, Terrado R, Forn I, Lovejoy C, Pedró-Alió C 2006. Distribution and abundance of uncultured heterotrophic flagellates in the world oceans. Envir Microbiol 8:1515–1522.

Müller DG, Küpper, FC, Küpper H 1999. Infection experiments reveal broad host ranges of *Eurychasma dicksonii* (Oomycota) and *Chytridium polysiphoniae* (Chytridiomycota), two eukaryotic parasites in marine brown algae (Phaeophyceae). Phycol Res 47:217–223.

Nakagiri A 2002. Diversity and phylogeny of *Halophytophthora* (Oomycetes). Abstracts 7th Int. Mycol. Cong. Oslo. 55 p. 19.

Newell SY, Cefalu R, Fell JW 1977. *Myzocytium*, *Haptoglossa* and *Gonimochaete* (fungi) in littoral marine nematodes. Bull Mar Sci 27:197–207.

Padgett DE 1978. Observations on the estuarine distribution of Saprolegniceae. Trans Br Mycol Soc 70:141–143.

Patron NJ, Rogers MB, Keeling PJ 2004. Gene replacement of fructose-1,6-bisphosphate aldolase supports the hypothesis of a single photosynthetic ancestor of chromalveolates. Eukaryot Cell 3:1169–1175.

Patterson DJ 1989. Chromophytes from a protistan perspective. In: Green JP, Leadbeater BSC, Diver WL editors. The chromophyte algae: problems and perspectives. Clarendon Press, Oxford, UK. pp. 357–379.

Patterson DJ 1999. The diversity of eukaryotes. Am Nat 65:S96–S124.

Petersen AB, Rosendahl S 2000. Phylogeny of the Peronosporomycetes (Oomycota) based on partial sequences of the large ribosomal subunit (LSU rDNA). Mycol Res 104:1295–1303.

Prakob W, Judelson HS 2007. Gene expression during oosporogenesis in heterothallic and homothallic *Phytophthora*. Fungal Genet Biol 44:726–739.

Pueschel CM, Van der Meer JP 1985. Ultrastructure of the fungus *Petersenia palmariae* (Oomycota) parasitic on the alga *Palmaria molis* (Rhodophyceae). Can J Bot 63: 409–418.

Raghu Kumar C 1980. An ultrastructural study of the marine diatom *Licmophora hyalina* and its parasite *Ectrogella perforans*. II. Development of the fungus in its host. Can J Bot 58:2557–2574.

Richards TA, Dacks JB, Jenkinson JM, Thornton CR, Talbot NJ 2006. Evolution of filamentous pathogens: gene exchange across eukaryote kingdoms. Curr Biol 16:1857–1864.

Riethmüller A, Voglmayr H, Göker M, Weiss M, Oberwinkler F 2002. Phylogenetic relationships of the downy mildews (Peronosporales) and related groups based on nuclear large subunit ribosomal DNA sequences. Mycologia 94:834–849.

Riethmüller A, Weiss M, Oberwinkler F 1999. Phylogenetic studies of Saprolegniomycetidae and related groups based on nuclear large subunit ribosomal DNA sequences. Can J Bot 77:1790–1800.

Robold A, Hardham AR 2005. During attachment *Phytophthora* spores secrete proteins containing thrombospondin type 1 repeats. Curr Genet 47:307–315.

Schnepf E, Deichgräber G, Drebes G 1977. Development and ultrastructure of the marine, parasitic oomcete, *Lagenisma coscinodisci* (Lagenidiales): sexual reproduction. Can J Bot 56:1315–1325.

Schnepf E, Deichgräber G, Drebes G 1978. Development and ultrastructure of the marine, parasitic oomycete, *Lagenisma coscinodisci* Drebes (Lagenidiales). The infection. Arch Microbiol 116:133–139.

Schnepf E, Schweikert M 1997. *Pirsonia*, phagotrophic nanoflagellate incertae sedis, feeding on marine diatoms: attachment, fine structure and taxonomy. Arch Protistenk 147:361–371.

Sekimoto S 2008. The taxonomy and phylogeny of the marine holocarpic oomycetes. Ph.D. Thesis, Graduate School of Natural Sciences, Konan University, Kobe.

Sekimoto S, Beakes GW, Gachon CMM, Müller DG, Küpper FC, Honda D 2008b. The development, ultrastructural cytology, and molecular phylogeny of the basal oomycete *Eurychasma dicksonii*, infecting the filamentous phaeophyte algae *Ectocarpus siliculosus* and *Pylaiella littoralis*. Protist 159:401–412.

Sekimoto S, Hatai K, Honda D 2007. Molecular phylogeny of an unidentified *Haliphthoros*-like marine oomycete and *Haliphthoros milfordensis* inferred from nuclear-encoded small and large subunit rDNA genes and mitochondrial-encoded cox2 gene. Mycoscience 48:212–221.

Sekimoto S, Kochkova TA, West JA, Beakes, GW, Honda D 2009. *Olpidiopsis bostrychiae*: a new species endoparasitic oomycete that infects *Bostrychia* and other red algae. Phycologia In Press.

Sekimoto S, Yokoo K, Kawamura Y, Honda D 2008a. Taxonomy, molecular phylogeny, and ultrastructural morphology of *Olpidiopsis porphyrae* sp. nov. (Oomycetes, stramenopiles), a unicellular obligate endoparasite of *Porphyra* spp. (Bangiales, Rhodophyta). Mycol Res 112:361–374.

Sparrow FK 1960. Aquatic phycomycetes, 2nd ed. University of Michigan Press, Ann Arbor.

Sparrow FK 1976. The present status of classification in biflagellate fungi. In: Gareth-Jones EB, editor. Recent advances in aquatic mycology. Elek Science, London, pp. 213–222.

Spencer MA, Vick MC, Dick MW 2002. Revision of Aplanopsis, Pythiopsis, and 'subcentric' *Achlya* species (Saprolegniaceae) using 18S rDNA and morphological data. Mycol Res 106:549–560.

Talbot NJ 2007. Deadly special deliveries. Nature 450:41–42.

Thines M, Göker M, Telle S, Ryley M, Mathur K, Narayana YD, Spring O, Thakur RP 2008. Phylogenetic relationships in graminicolous downy mildews based on cox2 sequence data. Mycol Res 112:345–351.

Torto-Alalibo T, Tian M, Gajendran K, Waugh ME, Van West P, Kamoun S 2005. Expressed sequence tags from the oomycete fish pathogen *Saprolegnia parasitica* reveal putative virulence factors. BMC Microbiol 5:46.

Tsui CKM, Marshall W, Yokoyama R, Honda D, Craven KL, Peterson PD, Lippmeier JC, Berbee ML 2008. Labyrinthulomycetes phylogeny and its implication for the evolutionary loss of chloroplasts and gain of ectoplasmic gliding. Molec. Phylogen Evol. In press.

Voglmayer H, Riethmüller A 2006. Phylogenetic relationships of *Albugo* species (white blister rusts) based on LSU rDNA sequence and oospore data. Mycol Res 110:75–85.

2

ECOLOGY OF LOWER OOMYCETES

Martina Strittmatter, Claire M.M. Gachon, and Frithjof C. Küpper

The Scottish Association for Marine Science, Dunstaffnage Marine Laboratory, Scotland, United Kingdom

2.1 INTRODUCTION

In contrast to oomycetes that infect terrestrial plants (and, to some extent, animals), their aquatic counterparts remain strongly understudied. This statement applies particularly to the group roughly outlined as "lower oomycetes," which mostly encompasses intracellular, holocarpic pathogens of marine algae and invertebrates. The shortcomings in their understanding can mostly be attributed to challenges in exploring their marine habitats, the relatively small communities of scientists studying marine taxa, and the often intracellular, biotrophic life history of these parasites.

However, it seems timely to devote more attention to these organisms—especially in the overall context of oomycete evolution and the impact of pathogen epidemics on natural communities. Recent studies have highlighted that oomycete and apicomplexan parasites share fundamental features of their infection mechanisms (Bhattacharjee et al., 2006)—in this regard, representatives closer to the basis of the oomycete lineage can be expected to offer more insight into the overall evolution of infection strategies than more highly evolved (but traditionally better studied) taxa. Also, there is increasing recognition of the impact of pathogens on the flow of energy and matter in ecosystems (Lafferty et al., 2006) and the overall structuring of natural communities (Packer and Clay, 2000). Although the epidemiology of many terrestrial plant pathogens is fairly well understood, there is an almost complete gap in our knowledge concerning the pathogens of marine algae and invertebrates—many of which are lower oomycetes. Increasing

Oomycete Genetics and Genomics: Diversity, Interactions, and Research Tools
Edited by Kurt Lamour and Sophien Kamoun
Copyright © 2009 John Wiley & Sons, Inc.

evidence suggests that climate change might be affecting epidemic patterns in the marine environment, with potentially devastating consequences on ecosystems (Harvell et al., 2002). However, the paucity of epidemiological data or of reliable baselines for wild populations may result in major environmental changes that remain undetected or unexplained (Harvell et al., 1999). Over the last decades, some well-documented events have revealed some dramatic consequences of epidemics in the sea, such as the iconic abrupt transition from a coral- to an algal-dominated ecosystem in Jamaica, which was driven by the disease-caused mass mortality of a predominant grazer, the urchin *Diadema antillarum* (Hughes, 1994). From such examples, clear parallels have developed with more comprehensive observations made on terrestrial ecosystems (Carey, 2000), which suggest that infectious agents may exert comparable—and so far underestimated—pressures at all levels of marine food webs as in their terrestrial counterparts.

In this context, the completion of genome sequences of marine organisms (e.g., the diatoms *Thalassiosira pseudonana* and *Phaeodactylum tricornutum*, the prasinophyte *Ostreococcus*, the sea urchin *Strongylocentrotus purpuratus*, and ongoing projects on the seaweeds *Porphyra sp.* and *Ectocarpus siliculosus*), combined with unprecedented molecular knowledge on many oomycete species, has opened technical opportunities that were unimaginable in the study of biotic interactions of marine organisms until a few years ago (Gachon et al., 2007).

2.2 DEFINITION OF LOWER OOMYCETES

Cook et al. (2001) presented the first molecular evidence for the existence of a basal sister group to the subclasses of Saprolegniomycetidae and Peronosporomycetidae (Dick et al., 1984; Sparrow, 1976) by analyzing *cox2* gene sequences of several marine, holocarpic oomycetes. Over the past few years, additional information was gained by sequencing molecular marker genes of *Eurychasma dicksonii* (Küpper et al., 2006; Sekimoto et al., 2008a), *Haptoglossa* (Hakariya et al., 2007), *Halodaphnea/Haliphthoros* (Sekimoto et al., 2007), as well as *Olpidiopsis* (Sekimoto, 2008; Sekimoto et al., 2008b), which grouped these species at the basis of the oomycete lineage. As the identification and recognition of this basal group of oomycetes is just in its beginning with only a handful of species analyzed using molecular markers, it is currently still difficult to define lower oomycete clades authoritatively (cf. Chapter 1 by Beakes and Sekimoto, in this volume). This problem becomes apparent when examining the species-rich genus *Olpidiopsis*: To date, only two marine species, *Olpidiopsis porphyrae* and *Olpidiopsis bostrychiae*, have been studied with molecular methods (Sekimoto, 2008; Sekimoto et al., 2008b), and it is questionable whether the genus is monophyletic as it contains both sexual and asexual species. This chapter covers the biology and ecology of organisms tentatively regrouped in the Haliphthorales, Olpidiopsidales, Haptoglossales and Eurychasmales by Sekimoto (2008; see also Chapter 1, Fig. 1.2): *Haptoglossa*, *Haliphthoros*, *Halodaphnea*, *O. porphyrae*, *O. bostrychiae*, and *E. dicksonii*. Based on morphological and ultrastructural data, it seems reasonable to assume that genera such as *Petersenia, Lagenisma, Pontisma, Sirolpidium*, and *Ectrogella* also belong to these two clades, although no molecular results are available yet.

It is also clear that the biodiversity of lower oomycetes is likely underestimated. Indeed, almost all recent sequencing projects carried out on ribosomal DNA libraries of marine microeukaryotes have revealed the existence of as yet unknown and uncultured novel marine stramenopiles, part of them branching as a sister clade to the oomycetes (Massana et al., 2002, 2004, 2006; Moon-van der Staay et al., 2001). The following review will focus on lower oomycetes as currently identified in molecular and ultrastructural studies, but it will also include species for which currently only morphological and ultrastructural data are available.

2.3 HABITAT, GEOGRAPHICAL EXTENSION, AND HOST SPECIFICITY

Lower oomycetes are predominantly, but not exclusively, parasites of marine organisms. In contrast to *Olpidiopsis, Petersenia*, and *Haptoglossa*, the species of *Eurychasma, Haliphthoros, Halodaphnea, Pontisma, Ectrogella, Sirolpidium*, and *Lagenisma* are exclusively found in marine environments; *Olpidiopsis*, *Petersenia*, and *Haptoglossa* species are also found in freshwater and terrestrial habitats.

E. dicksonii has been studied by numerous researchers since it was first described by Wright in 1877, who then termed it *Rhizophydium dicksonii*. Taxonomic revision by Wille (1899) and Löwenthal (1904) renamed the organism *Olpidium dicksonii* before the final revision by Magnus (1905). Based on the fairly extensive documentation of this marine oomycete (Table 2.1), it seems that *E. dicksonii* occurs predominantly in cold and temperate waters, which is supported by laboratory experiments on the temperature range of one *Eurychasma* isolate (Müller et al., 1999). The latter study showed that *Eurychasma* can infect its algal host in the range between 4°C and 23°C with an optimum temperature of 12°C. *E. dicksonii* has a broad host range that may even include two different algal lineages — Rhodophyta (red algae) and Phaeophyta (brown algae). Documentation of field-collected material revealed an infection of three different orders of red algae (Palmariales, Ceramiales, and Rhodymeniales; Jenneborg, 1977); however, it cannot be said with certainty whether the infections observed can be assigned to the same pathogen species as long as this has not been verified by culture studies with single, defined inocula identified by molecular methods. Among brown algae, it is well established that at least 13 different orders comprising 45 species are susceptible toward this basal oomycete, which has been shown in an extensive study under laboratory-controlled conditions by Müller et al. (1999). These algae vary dramatically in their morphology from filamentous species (Ectocarpales) to macrothallic species with compact/parenchymatic tissue (Desmarestiales and Laminariales). Hence, these observations make *E. dicksonii* one of the marine pathogens with the widest and most diverse host spectrum described so far.

TABLE 2.1 Geographic distribution of *E. dicksonii*.

Locality	Reference
Europe	
Great Britain (incl. Shetland Islands)	Aleem, 1950a, b; Gachon et al., 2009; Holmes, 1893; Küpper and Müller, 1999; Rattray, 1885
France (Brittany)	Cardinal, 1964; Dangeard, 1934; Feldmann, 1954; Sekimoto et al., 2008a
Germany (Helgoland)	Kuckuck, 1894
Ireland	Küpper and Müller, 1999; Wright, 1877
Norway (excluding Spitsbergen)	Foslie, 1891; Jaasund, 1965; Kjellman, 1883; Petersen, 1905
Spitsbergen (Svalbard)	Jenneborg, 1977; Kjellman, 1883; Svendsen, 1959
Sweden	Aleem, 1950b; Jenneborg, 1977; Rex, 1976
Finland	Skottsberg, 1911
Denmark (Mainland excl. Greenland and Faroe Islands)	Petersen, 1905; Rosenvinge and Lund, 1941; Sparrow, 1934
Greenland	Kjellman, 1883; Jónsson, 1904; Pedersén, 1976; Rosenvinge, 1893
Faroe Islands	Börgesen, 1902; Petersen, 1905
Italy	Giaccone and Bryce Derni, 1971; Hauck, 1878
Yugoslavia	Ercegovic, 1955
America	
USA (West Coast/California, Washington)	Saunders, 1898; Silva, 1957; Smith, 1942, 1944; Sparrow, 1969; Wynne 1972
USA (East Coast/Massachusetts)	Wilce et al., 1982
Mexico	Johnson and Sparrow, 1961
Argentina	Gachon et al., 2009
Falkland Islands	Gachon et al., 2009
Nova Scotia, Canada	Edelstein and McLachlan, 1967
Asia	
Northern Siberia	Kjellman, 1883
India	Misra, 1966
Japan	Konno and Tanaka, 1988
Africa	
Algeria	Feldmann and Feldmann, 1942
Australia	
Macquarie Island	Ricker, 1987
Kerguelen	Hariot, 1889

A second exclusively marine genus of lower oomycetes, *Halodaphnea*, is much less well documented than *Eurychasma*. Hitherto existing findings of this genus in marine animals are limited to Asia (Hideki, 2001; Nakamura and Hatai, 1995a). The typical hosts for various *Halodaphnea* species are listed in Table 2.2 and include egg and larval stages of crustaceans as well as abalone and a marine rotifer. Temperature studies on several *Halodaphnea* species

TABLE 2.2 Host range of the marine oomycete genera *Haliphthoros* and *Halodaphnea*.

Oomycete species	Host	Reference
Haliphthoros milfordensis	*Artemia salina* (brine shrimp)	Tharp and Bland, 1977
	Callinectes sapidus (blue crab)	Tharp and Bland, 1977
	Haliotis sieboldii (abalone)	Hatai, 1982
	Homarus:	
	H. americanus (American lobster)	Fisher et al., 1975
	H. gammarus (European lobster)	Abrahams and Brown, 1977
	Penaeus:	
	P. duorarum (pink shrimp)	Tharp and Bland, 1977
	P. japonicus (kuruma prawn)	Hatai et al., 1992
	P. monodon (black tiger prawn)	Chukanhom et al., 2003
	P. setiferus (white shrimp)	Tharp and Bland, 1977
	Portunus:	
	P. pelagicus (blue swimmer crab)	Nakamura and Hatai, 1995a
	P. trituberculatus (swimming crab)	Chukanhom et al., 2003
	Scylla serrata (mud crab)	Kaji et al., 1991
	Urosalpinx cinerea (oyster drill)	Vishniac, 1958
Haliphthoros sp.	*Jasus edwardsii* (spiny rock lobster)	Diggles, 2001
Haliphthoros philippinensis	*Penaeus monodon* (black tiger prawn)	Hatai et al., 1980
	Scylla serrata (mud crab)	Leaño, 2002
Halodaphnea hamanaensis	*Scylla serrata* (mud crab)	Bian and Egusa, 1980
Halodaphnea awabi	*Haliotis sieboldii* (abalone)	Kitancharoen et al., 1994
Halodaphnea okinawaensis	*Artemia salina* (brine shrimp)	Hideki, 2001
	Charybdis japonica (Asian paddle crab)	Hideki, 2001
	Chionoecetes opilio (snow crab)	Hideki, 2001
	Crangon cassiope	Hideki, 2001
	Portunus:	
	P. pelagicus (marine crab)	Nakamura and Hatai, 1995a
	P. trituberculatus (swimming crab)	Hideki, 2001
Halpodaphnea parasitica	*Brachionus plicatilis* (rotifer)	Nakamura and Hatai, 1994
Halodaphnea panulirata	*Panulirus japonicus* (spiny lobster)	Kitancharoen and Hatai, 1995

(*Halodaphnea panulirata, Halodaphnea dubia, Halodaphnea awabi*, and *Halodaphnea parasitica*) showed growth over a range of at least 15°C with an optimum of 20°C (*H. awabi* and *H. dubia*) and 25°C (*H. panulirata* and *H. parasitica*), respectively (Kitancharoen and Hatai, 1995; Kitancharoen et al., 1994; Nakamura and Hatai, 1994, 1995b).

A similar host range is attributed to the genus of *Haliphthoros* with its two described species *Haliphthoros milfordiensis* and *Haliphthoros philippinensis* (Table 2.2). These two species parasitize the egg, larval, and juvenile stages of various crustaceans as well as molluscs. Both species grow at temperatures between 20°C and 35°C and salinities between 10‰ and 40‰. Temperatures of 30°C and salinity of 30–35‰ are optimal for vegetative growth of *H. philippinensis*, whereas *H. milfordensis* grows best at 30‰ salinity and 30°C (Leaño, 2002). The zoospore production of *H. philippinensis* can be observed between 15°C and 35°C at a salinity of 30–35‰ and between 25°C and 30°C at 15–20‰ for *H. philippinensis* and *H. milfordensis*, respectively.

Haptoglossa, which is a genus endoparasitic in nematodes and rotifers, is found in soil, litter, decaying plant material (Davidson and Barron, 1973), and manure/dung of cows, horses, and rabbits (Glockling and Beakes, 2000a and b, 2001). One species, *Haptoglossa heterospora* has been reported to infect marine nematodes in the littoral zone (Newell et al., 1977). To date, 11 species of *Haptoglossa* are known (Dick, 2001; Glockling and Beakes, 2001). Host range studies on four *Haptoglossa* species isolated from rotifers and nematodes (Barron, 1989) revealed that *Haptoglossa intermedia, Haptoglossa mirabilis, Haptoglossa humicola*, and *Haptoglossa zoospora* can infect both nematodes and rotifers. Documentation of this genus is available from Canada (Estey and Olthof, 1965), the United States (Drechsler, 1940), the United Kingdom (Duddington 1954; Glockling and Beakes, 2000a and b, 2001), Denmark (Shepherd, 1956), and New Zealand (Hay, 1995).

The two species of *Olpidiopsis* characterized with molecular techniques, *O. porphyrae* and *O. bostrychiae*, are parasites of marine red algae. Laboratory-controlled studies (Sekimoto, 2008; Sekimoto et al., 2008b; West et al., 2006) revealed that *O. bostrychiae* can infect four species of the tropical mangrove red algal genus *Bostrychia*, four species of *Porphyra*, and *Stictosiphonia intricate*. *O. porphyrae* parasitizes exclusively red algae that belong to the order Bangiales (mainly *Porphyra* and two *Bangia* species). *O. porphyrae* infects the blade and conchocelis (sporophyte) stage of *Porphyra*; in contrast, *O. bostrychiae* has only been found to infect *Porphyra* conchocelis. However, the genus *Olpidiopsis* with its 53 known species (Dick, 2001) does not exclusively parasitize marine red algae but is also found in freshwater and terrestrial environments. *Olpidiopsis* hosts include a striking diversity of eukaryotes, which range from eucarpic oomycetes, angiosperm roots, freshwater green algae, diatoms, dinoflagellates, and angiosperms to terrestrial fungi and chytrids.

Petersenia palmariae (Pueschel and Vandermeer, 1985) infects the red alga *Palmaria mollis* on the Canadian Pacific Coast. Contrary to other pathogens discussed here, it seems to be host specific, neither infecting the closely related *Palmaria palmata* nor the filamentous *Ceramium rubrum*. However, a systematic screen of the host range does not seem to have been conducted so far. Other *Petersenia* species have been found hyperparasitic in *Achlya, Saprolegnia*, and *Pythium* (Dick, 2001).

The only two oomycete pathogens currently known to infect marine diatoms are *Ectrogella perforans* (Raghu Kumar, 1980a and b; Sparrow and Ellison, 1949; Zopf, 1884) and *Lagenisma coscinodisci* (Drebes, 1966). A related species, the hyperparasite *Ectrogella besseyi*, infects the pathogen of the freshwater green alga *Spirogyra* sp. and *Olpidiopsis schenkiana* (Sparrow and Ellison, 1949). The two endobiotic, marine algal parasites of the *Sirolpidium* and *Pontisma* (Petersen, 1905) species have been observed on Chlorophyta and on Rhodophyta, respectively (Dick, 2001).

Overall, many lower oomycetes seem to be generalist pathogens with a single species being able to infect various orders and families of hosts. Although many speculations have been formulated about the origins of the oomycete lineage, the evolutionary significance of such a broad host and habitat diversity among lower oomycetes remains to be fully assessed.

2.4 LIFE CYCLE AND BIOLOGY OF THE HOST–PATHOGEN INTERACTION

E. perforans and *P. palmariae* were among the first of the lower oomycetes to be studied in detail with regard to their life cycle and ultrastructure. *Ectrogella* zoospores encyst on the host cell wall, followed by the production of a germ tube that forms an appressorium and the penetration of the host cell wall by an infection peg. The pathogen cytoplasm is surrounded by the host cytoplasm (Raghu Kumar, 1980a), which initially constitutes a double-membrane envelope to which more membranes are added (Raghu Kumar, 1980b). Ultimately, all except the inner (= oomycete) membrane break down, and a cell wall is formed around the pathogen, which coincides with the disintegration of the host's cellular organization. The sporangium then becomes multinucleate with a peculiar "multitubular body." Subsequently, primary zoospores are formed that encyst, which leads to secondary zoospores that are released from the host. These either infect a new host or encyst (Raghu Kumar, 1980b).

In *Petersenia*, the infection starts with the penetration of a cortical cell of the host *P. mollis*. The pathogen then spreads to neighboring, inner tissues by promoting fusion of host cells or by dissolution of the pit plugs of the host cells, which creates a compound, confluent host cell lumen. Ultimately, a holocarpic sporangium with flagellated spores is formed. Flagella are later retracted (Pueschel and Vandermeer, 1985).

The life cycles of *E. dicksonii* (Fig. 2.1) and *O. porphyrae* have just recently been revisited in detailed ultrastructural studies by Sekimoto et al. (2008a and 2008b). The completion of *Eurychasma* development takes about 14 to 16 days under optimal growth conditions and is slightly dependent on the host organism (Müller et al., 1999; Sekimoto et al., 2008a). The developmental cycle can be divided into the following main steps: Attachment of the parasitic spore to the host surface, development of a parasitic thallus within the host cell, followed by sporangium differentiation, and spore release (Fig. 2.1). The first

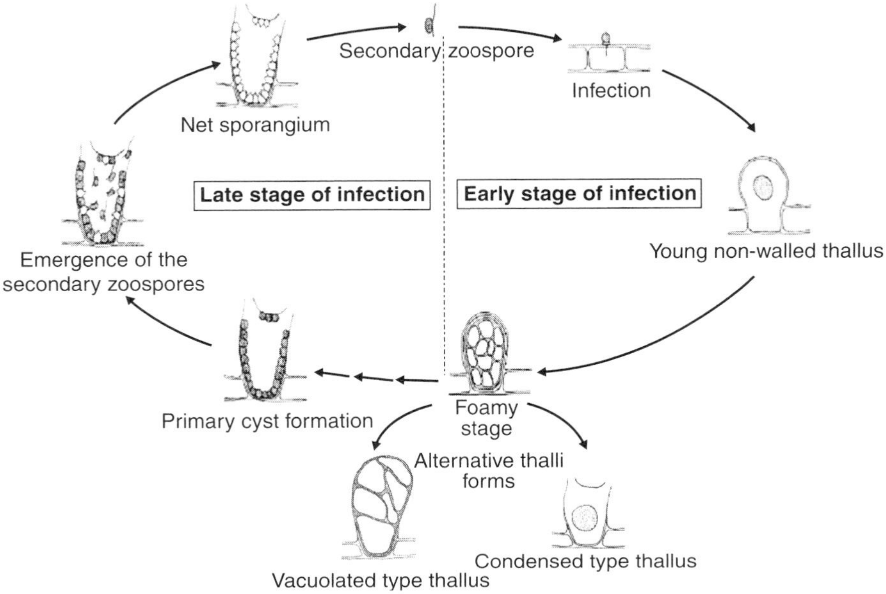

FIG. 2.1 Schematic life cycle of *Eurychasma dicksonii*, modified with permission from Sekimoto et al. (2008a).

contact between *E. dicksonii* and its host results in the encystment of a zoospore approximately 5 μm in diameter on the algal surface; it is mediated by a structure resembling an adhesorial pad (Sekimoto et al., 2008a). The subsequent penetration process is achieved by a germ tube that pierces the strong algal cell wall and transfers the parasitic cytoplasm into the host cell. The specific mechanism of penetration, however, is still unclear in contrast to appressoria-forming species of "higher" oomycetes. During the very early stage of *Eurychasma* infection, the parasitic thallus does not have a cell wall, is multinucleate, and is rich in vacuoles and mitochondria. Similar to *Eurychasma*, the early thallus of *O. porphyrae* also lacks a cell wall, as do the aforementioned species *P. palmariae* (Pueschel and Vandermeer, 1985), *L. coscinodisci* (Schnepf et al., 1978a), and *E. perforans* (Raghu Kumar, 1980b). The interface between *Eurychasma* and the algal cell is apparently defined by two membranes, one of algal and one of oomycete origin, and it shows a cell coat that consists of ribbon-like fibrils in the interspace. Algal host organelles, namely the nucleus and Golgi apparatus, are in close proximity to the *Eurychasma* thallus, which probably reflects an intimate host–pathogen interaction at the molecular level. The following stage is referred to as the "foamy stage" with an expanded parasite thallus that shows multiple uninucleate vacuolated units and a typical oomycete cell wall. By then, the algal cell is almost completely filled by the spherical *Eurychasma* thallus and undergoes a hypertrophic expansion. Similarly, a cell wall of the *O. porphyrae* thallus can also first be documented when the parasite almost completely fills the host cytoplasm. The uninucleate units supposedly correspond to preliminary primary aplanospores. The lack of any motile phase during primary spore differentiation of *Eurychasma* has been an interesting observation (Sekimoto et al., 2008a), as earlier studies had suggested this organism was diplanetic. The primary spores encyst to form a single layer at the bottom of the sporangium, whereas the upper part of the latter differentiates several discharge tubes. Secondary zoospores emerge from the primary cysts into the sporangium and complete the development of their flagella before being released from the sporangium. This stage of infection has been described by Petersen (1905) as the "net sporangium" stage. After the formation of flagella, the motile zoospores emerge from the sporangium via broad discharge openings. At this stage, it is frequently observed that not all zoospores are released, and these remaining spores encyst within the sporangium. These encysted spores can, in case of *O. porphyrae*, germinate within the sporangium, and infect neighboring host cells. In contrast to *E. dicksonii*, one narrow discharge tube develops in *O. porphyrae*, which pushes through the host cell wall. This significant feature incited Magnus (1905) to classify *Eurychasma* as a new genus.

The life cycle of the genus *Haptoglossa* is more complex compared with *Eurychasma* and *Olpidiopsis*. The species can be divided into aplanosporic or zoosporic types based on their sporogenesis (summarized by Hakariya et al., 2007). The infection structures in the genus *Haptoglossa* have been studied in detail because this genus uses highly specialized infection gun cells to invade the host (Barron, 1980; Beakes and Glockling, 1998, 2000; Glockling and Beakes, 2000c–e, 2002; Robb and Barron, 1982; Robb and Lee, 1986a and b). In general, encysted zoospores or aplanospores germinate and develop into gun cells. Although there are species-dependent variations in the apex of the gun cell, the overall organization of this infection apparatus is similar in all species investigated (Glockling and Beakes, 2000d). The apical part of the cell contains a needle chamber with a needle-like projectile, whereas the basal part contains a large vacuole that holds the cell under pressure (see Figs. 1.4j–l in Chapter 1). An inverted tubular system, in which the walled needle chamber is embedded, runs from the apical to the basal part of the cell. The cells are activated on a certain stimulus, which to date is not yet fully understood (see suggested model by Beakes and Glockling, 1998). When stimulated, the cell fires explosively by everting the tubular system, which causes the needle-like projectile to breach the cuticle of the host. The protoplastic content of the gun cell subsequently flows into the everted tube tail within the host body forming the sporidium (Glockling and Beakes, 2000e). The parasitic sporidium then differentiates into a single, nonseptate, cylindrical thallus. It could be observed that already the young thallus has a cell wall (Glockling and Beakes, 2000c) in contrast to *O. porphyrae* and *E. dicksonii*. The host can contain numerous parasitic thalli, which depends on the infection density. Mature sporangia form several dome-shaped papillae or evacuation tubes that eventually release zoospores or aplanospores (Glockling and Beakes, 2000d). However, *Haptoglossa erumpens* lacks these structures;

aplanospores are released from the sporangium by rupture of the sporangial wall and the host cuticle (Glockling and Beakes, 2000a). Aplanosporic species, with the exception of *H. erumpens*, are heterosporic and produce spores of two different sizes and karyologies, which develop into gun cells with different morphology (illustrated by Glockling and Beakes, 2002).

In contrast to the genera *Olpidiopsis, Eurychasma*, and *Haptoglossa* described above, *Haliphthoros* and *Halodaphnea* are not obligate-biotrophic organisms and can be grown on agar. Studies on their life cycle were therefore made *in vitro*. Hence, the exact mechanism of penetration in these genera remains unclear. However, it is unlikely that these oomycete pathogens of crustaceans and molluscs can directly penetrate the hard exoskeleton or shell, as it is apparent that mainly egg and larval stages are infested. The spores might enter the body of juvenile crustaceans through small wounds or by penetrating the membranes that connect the plates of the exoskeleton. In juvenile lobsters and crabs, an infection has been observed in gills and muscles of the walking legs that caused necrotic lesions (Fisher et al., 1978; Hatai et al., 1992). The vegetative growth of *Haliphthoros* is observed as branched, filamentous mycelium (Vishniac, 1958), followed by fragmentation of the thallus in small subthalli (Tharp and Bland, 1977). If detached from the thallus, independent growth is possible; otherwise, several fragments within one thallus are limited by septa or cytoplasmatic constrictions. These fragments then develop into a sporangium and form one discharge tube, which is usually straight or slightly curved. Uninucleate cytoplasmatic units within the immature sporangium are separated by small vacuoles and represent the prestage of the zoospores (Sekimoto, 2008). Released zoospores encyst after a motile phase and germinate (Overton et al., 1983). Compared with *Haliphthoros*, the hyphal thalli of *Halodaphnea* appear stouter with cross walls (Nakamura and Hatai, 1995a). In this genus, the highly vacuolated thalli differentiate into sporangia with one to three discharge tubes.

Two features shared among *E. perforans, P. palmariae, O. porphyrae, E. dicksonii, Haptoglossa, Haliphthoros*, and *Halodaphnea* become apparent when considering their life cycles. First, they are exclusively holocarpic, with the thallus differentiating completely into a sporangium as has also been documented in the diatom pathogen *L. coscinodisci* (Schnepf et al., 1978a). Second, no evidence of sexual reproduction has been observed in most of these species; this peculiarity has already been underlined in earlier studies (Sparrow, 1960). So far, a sexual stage has only been found in *L. coscinodisci* (Schnepf et al., 1978b), in which structurally indistinguishable spores, either vegetative or sexual, are formed in the zoosporangium after mitotic or meiotic division. The latter process seems to represent the main reproductive pattern in older cultures of *Lagenisma*. The genus *Olpidiopsis* has also been described as variable regarding sexuality with some species reproducing vegetatively and others sexually. Currently, all taxa of *Olpidiopsis* that infect marine red algae are considered to reproduce exclusively asexually (Dick, 2001). More investigation is clearly needed to assess the existence or absence of sexuality in lower oomycetes, as well as its significance in their life cycle.

Another characteristic of the species *E. dicksonii, O. porphyrae, O. bostrychiae, Halodaphnea, Haptoglossa, H. milfordensis, P. palmariae, E. perforans*, and *L. coscinodisci* is their biotrophic lifestyle. A closer distinction between biotrophy or hemibiotrophy, however, is difficult to make, especially when only one host cell is infected at a time. *Eurychasma-* or *Olpidiopsis*-infected algal cells will eventually die. Whether the complete (multicellular) host is killed greatly depends on the infection density considering that *Eurychasma* and *Olpidiopsis* do not propagate with hyphae. *O. bostrychiae* in culture has been observed to infect its host intensely and rapidly, which results in the death of the entire algal host culture. In *L. coscinodisci*, formation of the walled thallus starts after the death of the host cell (Schnepf et al., 1978a), which implies a hemibiotrophic lifestyle. Infected crustaceans usually do not survive *H. milfordensis* or *Halodaphnea* infection. Infected larvae first appear slow and then become unable to swim at later stages, showing extensive hyphal growth (Chukanhom et al., 2003; Roza and Hatai, 1999). Similarly, the survival rates of juvenile lobsters are low once the pathogen has entered the gills (Diggles, 2001). Whereas the species mentioned above include obligate-biotrophs (*Haptoglossa* sp., *Olpidiopsis* sp., *E. dicksonii*) and biotrophs (*H. milfordensis, Halodaphnea* sp.), it has been discussed whether *H. philippinensis* might be a saprophytic organism (Leaño, 2002).

If indeed the early oomycetes were holocarpic biotrophic pathogens, it follows that saprophytic, free-living species (e.g., some *Pythium and Phytophthora*) may have evolved from parasitic organisms. This theory challenges the widespread view that biotrophy is a highly specialized (and consequently derived) lifestyle, associated with a restriction of the ecological potentialities of the species (see for example Göker et al., 2007). It might also provide a long-sought explanation to the fact that the (mostly) obligate biotrophs downy mildews (*Peronospora, Plasmopara*) and the white rust *Albugo* apparently have independent geographic and temporal origins, as highlighted by Dick (2001). Within the next years, expanding genomic knowledge in several clades of the oomycete lineage and improved multigene phylogenies, along with sequencing projects in other pathogenic and nonpathogenic protists, should shed light on this paradox in terms of genome evolution.

2.5 ECONOMIC SIGNIFICANCE OF LOWER OOMYCETES

Porphyra sp., which is the red algal host of *O. porphyrae* and *O. bostrychiae*, is economically one of the most important seaweeds: Its thalli are used in the production of nori, which is a multibillion-dollar industry worldwide (Critchley et al., 2006). The leading producing countries are China, Japan, and Korea (Oohusa, 1993), whereas North American production has increased (Mumford, 1990). Large-scale cultivation of *Porphyra* is achieved by seeding nets with conchospores of the conchocelis (sporophyte) stage. After germination of the spores and initial growth in water tanks, the nets are then transferred into the sea, where the germlings differentiate into blade stage (gametophyte), which eventually is harvested and dried. An infection of *Porphyra* with *Olpidiopsis*, which is also known as chytrid blight disease, has been shown to cause decreases in nori production (Fujita, 1990). The infected cells of *Porphyra* blades collapse after release of *Olpidiopsis* zoospores and form spots, which most likely begin to rot and eventually form holes in the blade. *Porphyra* blades can also be parasitized by the oomycete *Pythium porphyrae* (red rot disease), simultaneously to *Olpidiopsis* infection (Ding and Ma, 2005). This double infection of the alga can result in rot of complete nets within a few days. As luster, color, and taste of dried nori determine the price of this product (Fujita, 1990), infections usually result in lower quality nori. *O. bostrychiae* and *O. porphyrae* are also known to infect the conchocelis reproductive stage (Sekimoto, 2008; Sekimoto et al., 2008b; West et al., 2006), hence it is likely that early stages of nori production are also affected even if no dedicated study has been conducted. Although the exact scale of economical loss in nori production caused by oomycetes like *P. porphyrae* is not known, it is clearly significant, and methods to detect early stages of red rot disease have been developed (Park, 2006; Park et al., 2001a and b; 2006). From the observations by Ding and Ma (2005), it can be inferred that *Porphyra*-infecting *Olpidiopsis* species have a comparable impact on nori production.

Likewise, various economically important crustaceans (lobster, blue crab, marine crab, kuruma prawn, pink shrimp, and mud crab) are plagued by *Haliphthoros* and *Halodaphnea*. Mud crab is renowned for its high proportion of edible, protein-rich meat and represents an important food source in Southeast Asia. Increasing concern about these diseases has paralleled the sharp expansion currently undergone by marine aquaculture to meet the global increase in food demand. Restrictions on wild fisheries have also contributed to an increased number of maricultures and sea farms. At these farms, reported problems are low larval survival rates of crustaceans in hatchery (Leaño, 2002; Roza and Hatai, 1999). Laboratory-controlled pathogenicity studies with *H. milfordensis* on mud crab eggs showed that about 10% of the eggs were infected 5 days past inoculation, which led to abortion of the egg mass (Leaño, 2002). Additionally, the larval stage of the mud crab *Scylla serrata* is known to be affected by *H. milfordensis*. The mortality of challenged larvae has reached up to 60% in two days after infection, whereas larvae at an earlier stage of development were even more susceptible (Roza and Hatai, 1999). A second problem in the survival of mud crab larvae is that it hosts at least two species of lower oomycetes (Table 2.2). Organisms isolated from infected larvae include *H. milfordensis* and *Halodaphnea sp.* (Roza and Hatai, 1999)—the latter shows a higher pathogenicity. Similar results were obtained when isolating oomycetes from infected larvae of the crab *Portunus pelagicus* (Nakamura and Hatai, 1995a). Mortality of larvae experimentally infected with *Halodaphnea okinawaensis* was reported to reach 70% at 25°C. Lobsters are affected by these pathogens during the juvenile stage with mortalities up to 40% (Diggles, 2001; Fisher et al., 1978; Kitancharoen and

Hatai, 1995), which has more severe effects on culture and breeding facilities regarding production costs. Diggles's study on *Haliphthoros*-infected spiny rock lobsters revealed however that lobsters seem to be resistant to this pathogen when having reached a certain carapace length, which is in agreement with observations made by other investigators (Fisher et al., 1978). The food sources of larval/juvenile lobsters and crabs could be potential carriers of disease as *Artemia salina* is parasitized by *H. milfordensis* and *H. okinawaensis*.

The impact of oomycete infections is currently far better understood for economically important freshwater organisms than marine species. Saprolegniales parasites of freshwater crayfish and fish (e.g., salmon, trout, and catfish), such as *Aphanomyces astaci* and *Saprolegnia parasitica*, are known to cause severe damage in aquacultures and wild populations (Edgerton et al., 2004; Phillips et al., 2008; van West, 2006). Apparently, current economic losses in crustacean farming caused by *Haliphthoros*/*Halodaphnea* infection have not reached the extent of their freshwater relatives. However, a possible reason for their poor documentation could be their more cryptic damages to egg, larval, and juvenile stages, whereas *S. parasitica* infects all life stages of fish development.

2.6 ECOLOGICAL IMPACT OF LOWER OOMYCETES

Although many aspects of the biology of host–pathogen interactions that involve lower oomycetes remain to be investigated, even less is known about their significance in ecosystem functioning. Some studies have highlighted the consequences of epidemics caused by eukaryotic pathogens on the abundance of primary producers in aquatic environments (Grahame, 1976; Holfeld, 1998; Tillmann et al., 1999). The inclusion of parasites in food web models leads to increased species richness, connectance (i.e., the percentage of possible links realized in a food web), and nestedness (the degree of structure and order in an ecosystem); it might have an important impact on ecosystem stability (Lafferty et al., 2006; Wood et al., 2007). This finding conversely implies that a good description of energy fluxes in food webs requires some understanding of host–parasite interactions at all levels of the trophic chain. Studies on freshwater chytrids, which infect primary producer phytoplankton, have revealed their contrasted influence on phytoplankton and zooplankton (Ibelings et al., 2004; Kagami et al., 2007). On the one hand, chytrids reduce algal blooms, which represent a food source for zooplankton. On the other hand, the fungal zoospores release a constitutive food source for crustaceans such as *Daphnia*, extracting (and making available to grazers) nutrients, which facilitates the flow of energy toward higher trophic levels.

Likewise, *E. dicksonii* probably plays a significant and similarly nuanced role in the biology and abundance of its hosts, as it might have more than an adverse effect on their fitness. Indeed, *E. dicksonii* has been found in many brown algal species in the field and was suggested to be responsible for the scarcity of *Striaria attenuata* on the west coast of Sweden (Aleem, 1955).

In contrast, a study by Wilce et al. (1982) that investigated the life history of the filamentous brown alga *Pylaiella littoralis* suggested that *Eurychasma* infection might have a positive influence on the vegetative propagation of the alga. They observed that parasitized host cells define a point of filament fragmentation after *Eurychasma* spores have been released from the sporangium. These fragments serve as vegetative propagules and represent the initial axis of new *Pylaiella* tufts. A massive fragmentation of the algal host correlated with a *Eurychasma* peak in winter was shown to define the seasonal vegetative reproduction peaks of *Pylaiella* and might cause so-called brown tides of free-floating masses of filaments of this normally sessile macroalga. *Eurychasma*-induced fragmentation of algal filaments might also affect other filamentous brown algae, especially if the parasite density on its host is high. In any case, the fragmentation of algal filaments causes drift and distribution of material, which creates new microcosms for endophytic and epiphytic species. Finally, although it is most unlikely that *E. dicksonii* has a significant impact on the life history of the sporophyte generation of kelps (Laminariales), its microscopic gametophytic generation is subject to high levels of *Eurychasma* infection in laboratory culture experiments. Therefore, *Eurychasma* epidemics in the field could influence the sexual reproduction and the recruitment of kelps, which are predominant primary producers of temperate and cold marine coastal ecosystems worldwide (Müller et al., 1999). The influence of parasitizing organisms on the life history of algae has also been demonstrated in a study of *Anisolpidium rosenvingei* (Küpper and Müller, 1999), which is a hyphochytrid that probably is one the closest nonoomycete relatives of lower oomycetes. The prevalence of this organism is linked to periods of host fertility and causes a decline of algal biomass by exclusively infecting reproductive cells of *Pylaiella*.

Documentation of a few emblematic outbreaks during the last decades has revealed the vulnerability of marine environments to the consequences of epidemics caused by pathogens (Hughes, 1994) at all—and sometimes even across—trophic levels. For example, the parasitic trematodes that infect a dominant herbivorous snail *Littorina littorea* are responsible for a 40% reduction of grazing in comparison with uninfected ones, which affected not only the biomass of edible algal species but also the overall composition of the intertidal macroalgal community. Hence, beyond simply affecting the fitness and abundance of their hosts, these parasites profoundly alter community structure (Wood et al., 2007), which is a well-known phenomenon for terrestrial pathogens (see for example Packer and Clay, 2000). The latter have also been shown extensively to influence the genetic structure of their host population on the long term, driving coevolution processes often described as an arms race (Allen et al., 2004). Moreover, epidemic patterns are affected by environmental changes and anthropogenic perturbations, which sometimes lead to breaking-news epidemics or mass mortalities (e.g., Harvell et al., 1999; Henricot and Prior, 2004; Hughes, 1994), but are also affected in more unexpected ways. For example, as a result of worldwide trade, introduced species are frequent in marine ecosystems. Because their native pathogens are likely not to be cointroduced, they are typically subject to a lower parasitic pressure compared with native species, which may account for their success and invasiveness (Borer et al., 2007; Torchin et al., 2003). Although our current knowledge about the ecology of lower oomycetes is little more than anecdotal or circumstantial evidence, all available data suggest that these parasites play an important — but so far mostly unrecognized — role in aquatic ecosystems. Hence, our group is developing tools to monitor lower oomycetes (mostly *E. dicksonii*) in the field, with the view of using them to assess the impact of epidemics in algal populations at various temporal and geographical scales (Gachon et al., 2009).

2.7 FINAL CONCLUSION

Lower oomycetes have attracted renewed interest during the last decade. Studies on their phylogeny, ultrastructure, life cycle, and ecology have significantly expanded our knowledge of these organisms, which represents an important resource for the understanding of oomycete biology and evolution. However, despite this increased focus, they still remain strongly understudied. In particular, the underlying molecular mechanisms of infections (and, conversely, host responses) are very poorly understood. The current accumulation of genomic and molecular data on an ever-expanding range of oomycete taxa will illuminate questions, such as the origins of pathogenicity or biotrophy and the overall evolution of infection strategies in the oomycete phylum. Also, wide parts of the Earth are strongly undersampled for these organisms, and a substantial range of biodiversity may remain to be discovered at this time — in particular, in the tropics and polar regions.

ACKNOWLEDGMENTS

M. S. gratefully acknowledges the ECOSUMMER Marie Curie Training Network for funding her PhD research. C. M. M. G. expresses her gratitude to the European Commission for a Marie Curie Fellowship (MEIF-CT-2006-022837). F. C. K. acknowledges funding from the U.K. Natural Environment Research Council (grant NE/D521522/1 and core strategic funding to the Scottish Association for Marine Science through the Oceans 2025 program, WP 4.5). We are grateful to Gordon W. Beakes and Satoshi Sekimoto for helpful discussions, sharing of unpublished results, and substantial work on phylogeny and ultrastructure of lower oomycetes. We also would like to thank Dieter G. Müller for pioneering work on *Eurychasma dicksonii* and Shady A. Amin for literature retrieval.

REFERENCES

Abrahams D, Brown WD 1977. Evaluation of fungicides for *Haliphthoros milfordensis* and their toxicity to juvenile European lobsters. Aquaculture 12:31–40.

Aleem AA 1950a. A fungus in *Ectocarpus granulosus* C. Agardh near Plymouth. Nature 165:119–120.

Aleem AA 1950b. The occurrence of *Eurychasma dicksonii* (Wright) Magnus in England and Sweden. Meddelanden från Göteborgs Botaniska Trädgård 18:239–245.

Aleem AA 1955. Marine fungi from the West coast of Sweden. Arkiv för Botanik 3:1–33.

Allen RL, Bittner-Eddy PD, Grenville-Briggs LS, Meitz SC, Rehmany AP, Rose CE, Beynon JL (2004). Host-parasite coevolutionary conflict between *Arabidopsis* and downy mildew. Science 306:1957–1960.

Barron GL 1980. A new *Haptoglossa* attacking rotifers by rapid injection of an infective sporidium. Mycologia 72:1186–1194.

Barron GL 1989. Host range studies for *Haptoglossa* and a new species, *Haptoglossa intermedia*. Can J Bot 67:1645–1648.

Beakes GW, Glockling SL 1998. Injection tube differentiation in gun cells of a *Haptoglossa* species which infects nematodes. Fungal Gen Biol 24:45–68.

Beakes GW, Glockling SL 2000. An ultrastructural analysis of organelle arrangement during gun (infection) cell differentiation in the nematode parasite *Haptoglossa dickii*. Mycol Res 104:1258–1269.

Bhattacharjee S, Hiller NL, Liolios K, Win J, Kanneganti TD, Young C, Kamoun S, Haldar K 2006. The malarial host-targeting signal is conserved in the Irish potato famine pathogen. PloS Path 2:453–465.

Bian BZ, Egusa S 1980. *Atkinsiella hamanaensis* sp. nov. isolated from cultivated ova of the mangrove crab, *Scylla serrata* (Forsskal). J Fish Dis 3:373–385.

Borer ET, Hosseini PR, Seabloom EW, Dobson AP 2007. Pathogen-induced reversal of native dominance in a grassland community. Proc Natl Acad Sci USA 104: 5473–5478.

Börgesen F 1902. The marine algae of the Faeröes. Bot Faeröes 2:339–532.

Cardinal A 1964. Etude sur les Ectocarpacées de la Manche. Nova Hedwigia Beiheft 15:86pp.

Carey C 2000. Infectious disease and worldwide declines of amphibian populations, with comments on emerging diseases in coral reef organisms and in humans. Environ Health Perspect 108:143–150.

Chukanhom K, Borisutpeth P, Van Khoa L, Hatai K 2003. *Haliphthoros milfordensis* isolated from black tiger prawn larvae (*Penaeus monodon*) in Vietnam. Mycoscience 44:123–127.

Cook KL, Hudspeth DSS, Hudspeth MES 2001. A cox2 phylogeny of representative marine peronosporomycetes (Oomycetes). Nova Hedwigia 231–243.

Critchley AT, Ohno M, Largo DB 2006. World seaweed ressources. An authoritative reference system (DVD). ISBN-10: 9075000804.

Dangeard P 1934. Sur la présence à Roscoff d'une Chytridiale parasite des Ectocarpacées, l'*Eurychasma dicksonii* (Wright) Magnus. Ann Protist 4:69–73.

Davidson JG, Barron GL 1973. Nematophagous fungi: *Haptoglossa*. Can J Bot 51: 1317–1323.

Dick MW 2001. Straminipilous fungi. Systematics of the Peronosporomycetes including accounts of the marine straminipilous protists, the Plamodiophorids and similar organisms. Kluwer, Dordrecht, Germany, 670pp.

Dick MW, Wong PTW, Clark G 1984. The identity of the oomycete causing 'Kikuyu Yellows,' with a reclassification of the downy mildews. Bot J Linn Soc 89: 171–197.

Diggles BK 2001. A mycosis of juvenile spiny rock lobster, *Jasus edwardsii* (Hutton, 1875) caused by *Haliphthoros* sp., and possible methods of chemical control. J Fish Dis 24:99–110.

Ding HY, Ma JH 2005. Simultaneous infection by red rot and chytrid diseases in *Porphyra yezoensis* Ueda. J Appl Phycol 17:51–56.

Drebes G 1966. Ein parasitischer Phycomycet (Lagenidiales) in *Coscinodiscus*. Helgoländer Wissenschaftliche Meeresuntersuchungen 13:426–435.

Drechsler C 1940. Three fungi destructive to free-living nematodes. J Wash Acad Sciences 30:240–254.

Duddington CL 1954. Nematode-destroying fungi in agricultrural soils. Nature 173: 500–501.

Edelstein T, McLachlan J 1967. Investigations of the marine algae of Nova Scotia. 3. Species new or rare to Nova Scotia. Can J 45:203–210.

Edgerton BF, Henttonen P, Jussila S, Mannonen A, Paasonen P, Taugbul T, Edsman L, Souty-Grosset C 2004. Understanding the causes of disease in European freshwater crayfish. Conserv Biol 18:1466–1474.

Ercegovic A 1955. Contribution à la connaissance des Ectocarpes (*Ectocarpus*) de l'Adriatique moyenne. Acta Adriatica 7:74pp.

Estey RH, Olthof THA 1965. The occurrence of nematophagous fungi in Quebec. Phytoprotection 46:14–17.

Feldmann J 1954. Inventaire de la Flore Marine de Roscoff: Algues, champignons, lichens et spermatophytes. Travaux de la Station Biologique de Roscoff Suppl. 6, Roscoff:152pp.

Feldmann J, Feldmann G 1942. Additions à la flore des algues marines d'Algérie. Bull Soc Hist Nat Afr Nord 33:230–245.

Fisher WS, Nilson EH, Shleser RA 1975. Effect of fungus *Haliphthoros milfordensis* on juvenile stages of American lobster *Homarus americanus*. J Inverteb Pathol 26:41–45.

Fisher WS, Nilson EH, Steenbergen JF, Lightner DV 1978. Microbial diseases of cultured lobsters — review. Aquaculture 14:115–140.

Foslie M 1891. Remarks on forms of *Ectocarpus* and *Pylaiella*. Tromsø Mus. Åsh. 14.

Fujita Y 1990. Diseases of cultivated *Porphyra* in Japan. In: Akatsuka I, editor. Introduction to Applied Phycology SPB Academic Publishing, Hague 177–190.

Gachon CMM, Day JG, Campbell CN, Pröschold T, Saxon RJ, Küpper FC 2007. The Culture Collection of Algae and Protozoa (CCAP): a biological resource for protistan genomics. Gene 406:51–57.

Gachon CMM, Strittmatter M, Kleinteich J, Müller DG, Küpper FC 2009. Detection of differential host susceptibility to the marine oomycete pathogen *Eurychasma dicksonii* by Real-Time PCR: not all algae are equal. Appl Environ Microbiol 75:322–328.

Giaccone G, Bryce Derni C 1971. Informazioni tassonomiche di elementi morfologici ed ecologici di stadi Ectocarpoidi presenti sulle coste Italiane. Atti Instit Veneto Sci Lett Art 130:39–81.

Glockling SL, Beakes GW 2000a. Two new species of *Haptoglossa, H. erumpens* and *H. dickii*, infecting nematodes in cow manure. Mycol Res 104:100–106.

Glockling SL, Beakes GW 2000b. Video microscopy of spore development in *Haptoglossa heteromorpha*, a new species from cow dung. Mycologia 92:747–753.

Glockling SL, Beakes GW 2000c. The ultrastructure of the dimorphic infection cells of *Haptoglossa heteromorpha* illustrates the developmental plasticity of infection apparatus structures in a nematode parasite. Can J Bot 78:1095–1107.

Glockling SL, Beakes GW 2000d. A review of the taxonomy, biology and infection strategies of "biflagellate holocarpic" parasites of nematodes. Fungal Divers 4:1–20.

Glockling SL, Beakes GW 2000e. An ultrastructural study of sporidium formation during infection of a rhabditid nematode by large gun cells of *Haptoglossa heteromorpha*. J Invertebr Pathol 76:208–215.

Glockling SL, Beakes GW 2001. Two new species of *Haptoglossa* from north-east England. Bot J Linn Soc 136:329–338.

Glockling SL, Beakes GW 2002. Ultrastructural morphogenesis of dimorphic arcuate infection (gun) cells of *Haptoglossa erumpens* an obligate parasite of *Bunonema* nematodes. Fungal Gen Biol 37:250–262.

Göker M, Voglmayr H, Riethmüller A, Oberwinkler F 2007. How do obligate parasites evolve? A multi-gene phylogenetic analysis of downy mildews. Fungal Genet Biol 44:105–122.

Grahame ES 1976. The occurrence of *Lagenisma coscinodisci* in *Palmeria hardmania* from Kingston harbour, Jamaica. Br Phycol J 11:57–61.

Hakariya M, Hirose D, Tokumasu S 2007. A molecular phylogeny of *Haptoglossa* species, terrestrial peronosporomycetes (oomycetes) endoparasitic on nematodes. Mycoscience 48:169–175.

Hariot P 1889. Champignons. Miss Sci Cap Horn Bot 5:173–200.

Harvell CD, Kim K, Burkholder JM, Colwell RR, Epstein PR, Grimes DJ, Hofmann EE, Lipp EK, Osterhaus A, Overstreet RM, Porter JW, Smith GW, Vasta GR 1999. Marine ecology, emerging marine diseases, climate links and anthropogenic factors. Science 285:1505–1510.

Harvell CD, Mitchell CE, Ward JR, Altizer S, Dobson AP, Ostfeld RS, Samuel MD 2002. Climate warming and disease risks for terrestrial and marine biota. Science 296:2158–2162.

Harvey WH 1849. Phycologia Britannica. II London: pp. 217–252.

Hatai K 1982. On the fungus *Haliphthoros milfordensis* isolated from temporarily held abalone (*Haliotis sieboldii*). Fish Pathol 17:199–204.

Hatai K, Bian BZ, Baticados CL, Egusa S 1980. Studies on the fungal diseases in crustaceans. 2. *Haliphthoros philippinensis* sp. nov. isolated from cultivated larvae of the jumbo tiger prawn (*Penaeus monodon*). Trans Mycol Soc of Jpn 21:47–55.

Hatai K, Rhoobunjongde W, Wada S 1992. *Haliphthoros milfordensis* isolated from gills of juvenile kuruma prawn *Penaeus japonicus* with black gill disease. Nippon Kingakukai Kaiho 33:185–192.

Hauck F 1878. Notiz über *Rhizophydium dicksonii* (Wright). Öst Bot Z 28.

Hay ES 1995. Endoparasites infecting nematodes in New Zealand. NZ J Bot 33: 401–407.

Henricot B, Prior C 2004. *Phytophthora ramorum*, the cause of sudden oak death or ramorum leaf blight and dieback. Mycologist 18:151–156.

Hideki Y 2001. Pathogenicity of *Halocrusticida okinawaensis* to larvae of five crustacean species. Suisan Zoshoku 49:115–116.

Holfeld H 1998. Fungal infections of the phytoplankton: seasonality, minimal host density, and specificity in a mesotrophic lake. New Phytol 138:507–517.

Holmes EM 1893. The occurrence of *Pylaiella varia* Kjellman in Scotland. Ann Scot Nat Hist 101–105.

Hughes TP 1994. Catastrophes, phase-shifts, and large-scale degradation of a Caribbean coral reef. Science 265:1547–1551.

Ibelings BW, De Bruin A, Kagami M, Rijkeboer M, Brehm M, van Donk E 2004. Host parasite interactions between freshwater phytoplankton and chytrid fungi (Chytridiomycota). J Phycol 40:437–453.

Jaasund E 1965. Aspects of the marine algal vegetation of North Norway. Botanica Gothoburg 4:174pp.

Jenneborg LH 1977. *Eurychasma* infection of marine-algae — Changes in algal morphology and taxonomical consequenses. Botanica Marina 20:499–507.

Johnson TW, Sparrow FK 1961. Fungi in oceans and estuaries. Weinheim: 668pp.

Jónsson H 1904. The marine algae of East Greenland. Arb Bot Have København 18:73pp.

Kagami M, de Bruin A, Ibelings BW, Van Donk E 2007. Parasitic chytrids: their effects on phytoplankton communities and food-web dynamics. Hydrobiologia 578:113–129.

Kaji S, Kanematsu M, Tezuka N, Fushimi H, Hatai K 1991. Effects of formalin bath for *Haliphthoros* infection on ova and larvae of the mangrove crab *Scylla serrata*. Nippon Suisan Gakkaishi 57:51–55.

Kitancharoen N, Hatai K 1995. A marine oomycete *Atkinsiella panulirata* sp. nov. from philozoma of spiny lobster, *Panulirus japonicus*. Mycoscience 36:97–104.

Kitancharoen N, Nakamura K, Wada S, Hatai K 1994. *Atkinsiella awabi* sp. nov. isolated from stocked abalone, *Haliotis sieboldii*. Mycoscience 35:265–270.

Kjellman FR 1883. The algae of the Arctic Sea. Kgl. svenska Vetensk. Akad Handl 20:351pp.

Konno K, Tanaka J 1988. *Eurychasma dicksonii*: new record a parasitic fungus on *Acinetospora crinita* brown alga in Japan. Bull Natl Sci Mus Ser B Bot 14:119–122.

Kuckuck P 1894. Bemerkungen zur marinen Algenvegetation von Helgoland. Wissenschaftliche Meeresuntersuchungen N.F. Abt. Helgoland 1:220–275.

Küpper FC, Maier I, Müller DG, Goer SL-D, Guillou L 2006. Phylogenetic affinities of two eukaryotic pathogens of marine macroalgae, *Eurychasma dicksonii* (Wright) Magnus and *Chytridium polysiphoniae* Cohn. Cryptogamie Algologie 27:165–184.

Küpper FC, Müller DG 1999. Massive occurrence of the heterokont and fungal parasites *Anisolpidium*, *Eurychasma* and *Chytridium* in *Pylaiella littoralis* (Ectocarpales, Phaeophyceae). Nova Hedwigia 69:381–389.

Lafferty KD, Dobson AP, Kuris AM 2006. Parasites dominate food web links. Proc Natl Acad Sci USA 103:11211–11216.

Leaño EM 2002. *Haliphthoros* spp. from spawned eggs of captive mud crab, *Scylla serrata*, broodstocks. Fungal Divers 9:93–103.

Löwenthal W 1904. Weitere Untersuchungen über Chytridineen. Archiv für Protistenkunde 5:225–228.

Magnus P 1905. Über die Gattung zu der *Rhizophydium dicksonii* gehört. Hedwigia 44:347–349.

Massana R, Guillou L, Diez B, Pedros-Alio C 2002. Unveiling the organisms behind novel eukaryotic ribosomal DNA sequences from the ocean. Appl Environ Microbiol 68:4554–4558.

Massana R, Castresana J, Balague V, Guillou L, Romari K, Groisillier A, Valentin K, Pedios-Alito C 2004. Phylogenetic and ecological analysis of novel marine stramenopiles. Appl Environ Microbiol 70:3528–3534.

Massana R, Terrado R, Forn I, Lovejoy C, Pedros-Alio C 2006. Distribution and abundance of uncultured heterotrophic flagellates in the world oceans. Environ Microbiol 8:1515–1522.

Misra JN 1966. Phaeophyceae in India. ICAR. New Delhi, India: 203pp.

Moon-van der Staay SY, De Wachter R, Vaulot D 2001. Oceanic 18S rDNA sequences from picoplankton reveal unsuspected eukaryotic diversity. Nature 409:607–610.

Müller DG, Küpper FC, Küpper H 1999. Infection experiments reveal broad host ranges of *Eurychasma dicksonii* (Oomycota) and *Chytridium polysiphoniae* (Chytridiomycota), two eukaryotic parasites of marine brown algae (Phaeophyceae). Phycol Res 47:217–223.

Mumford TF 1990. Nori cultivation in North America — Growth of the Industry. Hydrobiologia 204:89–98.

Nakamura K, Hatai K 1994. *Atkinsiella parasitica* sp. nov. isolated from a rotifer, *Brachionus plicatilis*. Mycoscience 35:383–389.

Nakamura K, Hatai K 1995a. Three species of Lagenidiales isolated from the eggs and zoeae of the marine crab *Portunus pelagicus*. Mycoscience 36:87–95.

Nakamura K, Hatai K 1995b. *Atkinsiella dubia* and its related species. Mycoscience 36:431–438.

Newell SY, Cefalu R, Fell JW 1977. *Myzocytium*, *Haptoglossa*, and *Gonimochaete* (fungi) in littoral marine nematodes. Bull Mar Sci 27:177–207.

Oohusa T 1993. Recent trends in nori products and markets in Asia. J Appl Phycol 5:155–159.

Overton SV, Tharp TP, Bland CE 1983. Fine structure of swimming, encysting, and germinating spores of *Haliphthoros milfordensis*. Can J Bot 61:1165–1177.

Packer A, Clay K 2000. Soil pathogens and spatial patterns of seedling mortality in a temperate tree. Nature 404:278–281.

Park CS 2006. Rapid detection of *Pythium porphyrae* in commercial samples of dried *Porphyra yezoensis* sheets by polymerase chain reaction. J App Phycol 18:203–207.

Park CS, Kakinuma M, Amano H 2001a. Detection of the red rot disease fungi *Pythium* spp. by polymerase chain reaction. Fish Sci 67:197–199.

Park CS, Kakinuma M, Amano H 2001b. Detection and quantitative analysis of zoospores of *Pythium porphyrae*, causative organism of red rot disease in *Porphyra*, by competitive PCR. J App Phycol 13:433–441.

Park CS, Kakinuma M, Amano H 2006. Forecasting infections of the red rot disease on *Porphyra yezoensis* Ueda (Rhodophyta) cultivation farms. J Appl Phycol 18:295–299.

Pedersén PM 1976. Marine, benthic algae from southernmost Greenland. Medd. Grønland 199:80.

Petersen HE 1905. Contributions à la connaisance des phycomycètes marins. Oversigt over det Kgl. Danske Videnskabernes Selskabs Forhandlinger 5:439–488.

Phillips AJ, Anderson VL, Robertson EJ, Secombes CJ, van West P 2008. New insights into animal pathogenic oomycetes. Trends Microbiol 16:13–19.

Pueschel CM, Vandermeer JP 1985. Ultrastructure of the fungus *Petersenia palmariae* (Oomycetes) parasitic on the alga *Palmaria mollis* (Rhodophyceae). Can J Bot 63:409–418.

Raghu Kumar C 1980a. An ultrastructural study of the marine diatom *Licmophora hyalina* and its parasite *Ectrogella perforans*. 1. Infection of host cells. Can J Bot 58:1280–1290.

Raghu Kumar C 1980b. An ultrastructural study of the marine diatom *Licmophora hyalina* and its parasite *Ectrogella perforans*. 2. Development of the fungus in its host. Can J Bot 58:2557–2574.

Rattray J 1885. Note on *Ectocarpus*. Trans R Soc Edinb 32:589–602.

Rex B 1976. Benthic vegetation in Byfjorden 1970–1973. The By Fjord: Marine botanical investigations SNV PM 684:153pp.

Ricker RW 1987. Taxonomy and biogeography of Macquarie Island seaweeds. British Museum Natural History, London, UK.

Robb EJ, Barron GL 1982. Nature's ballistic missile. Science 218:1221–1222.

Robb J, Lee B 1986a. Developmental sequence of the attack apparatus of *Haptoglossa mirabilis*. Protoplasma 135:102–111.

Robb J, Lee B 1986b. Ultrastructure of mature and fired gun cells of *Haptoglossa mirabilis*. Can J Bot 64:1935–1947.

Rosenvinge LK 1893. Grönlands havalger. Medd. Grönland 3:765–981.

Rosenvinge LK, Lund S 1941. The marine algae of Denmark. 2. Phaeophyceae. K. danske Videnskab. Selsk Biol Skr 1:79pp.

Roza D, Hatai K 1999. Pathogenicity of fungi isolated from the larvae of the mangrove crab, *Scylla serrata*, in Indonesia. Mycoscience 40:427–431.

Saunders DA 1898. Phycological memoirs. Proc Ca Acad Sci 3rd Ser Bot 1:147–168.

Schnepf E, Deichgräber G, Drebes G 1978a. Development and ulltrastructure of marine, parasitic oomycete, *Lagenisma coscinodisci* Drebes (Lagenidiales): thallus, zoosporangium, mitosis and meiosis. Arch Microbiol 116:141–150.

Schnepf E, Deichgräber G, Drebes G 1978b. Development and ultrastructure of marine, parasitic oomycete, *Lagenisma coscinodisci* (Lagenidiales): sexual reproduction. Can J Bot 56:1315–1325.

Sekimoto S 2008. The taxonomy and phylogeny of the marine holocarpic oomycetes. Dissertation of the Graduate School of Natural Science, Konan University, Kobe.

Sekimoto S, Beakes GW, Gachon CMM, Müller DG, Küpper FC, Honda D 2008a. The development, ultrastructural cytology, and molecular phylogeny of the basal oomycete *Eurychasma dicksonii*, infecting the filamentous phaeophyte algae *Ectocarpus siliculosus* and *Pylaiella littoralis*. Protist 159:299–318.

Sekimoto S, Hatai K, Honda D 2007. Molecular phylogeny of an unidentified *Haliphthoros*-like marine oomycete and *Haliphthoros milfordensis* inferred from nuclear-encoded small- and large-subunit rRNA genes and mitochondrial-encoded cox2 gene. Mycoscience 48:212–221.

Sekimoto S, Yokoo K, Kawamura Y, Honda D 2008b. Taxonomy, molecular phylogeny, and ultrastructural morphology of *Olpidiopsis porphyrae* sp. nov.

(Oomycetes, straminipiles), a unicellular obligate endoparasite of *Bangia* and *Porphyra* spp. (Bangiales, Rhodophyta). Mycol Res 112:361–374.

Shepherd AM 1956. A short survey of Danish nematophagous fungi. Friesia 5:396–408.

Silva PC 1957. Notes on Pacific marine algae. Madroño 14:41–51.

Skottsberg C 1911. Beobachtungen über einige Meeresalgen aus der Gegend von Tvärminne im südwestlichen Finnland. Acta Soc. Fauna Flora Fennica 34:18pp.

Smith GM 1942. Note on some marine brown algae of Monterey Peninsula, California. Am J Bot 29:645–653.

Smith GM 1944. Marine algae of the Monterey Peninsula, California. Stanford University, CA. 622pp.

Sparrow FK 1934. Observation on marine phycomycetes collected in Denmark. Dansk Botanisk Arkiv 8:1–24.

Sparrow FK 1960. Aquatic phycomycetes, 2nd ed. The University of Michigan Press.

Sparrow FK 1969. Zoosporic marine fungi from the Pacific Northwest (USA). Archiv für Mikrobiologie 66:129–146.

Sparrow FK 1976. The present status of classification in biflagellate fungi. Recent advances in aquatic mycology. Paul Elek Ltd., London, 213–222.

Sparrow FK, Ellison B 1949. *Olpidiopsis schenkiana* and its hyperparasite *Ectrogella besseyi* n. sp. Mycologia 41:28–35.

Svendsen P 1959. The algal vegetation of Spitsbergen. Norsk Polarinst. Skr 116:49pp.

Tharp TP, Bland CE 1977. Biology and host range of *Haliphthoros milfordensis*. Can J Bot 55:2936–2944.

Tillmann U, Hesse KJ, Tillmann A 1999. Large-scale parasitic infection of diatoms in the Northfrisian Wadden Sea. J Sea Res 42:255–261.

Torchin ME, Lafferty KD, Dobson AP, McKenzie VJ, Kuris AM 2003. Introduced species and their missing parasites. Nature 421:628–630.

van West P 2006. *Saprolegnia parasitica*, an oomycete pathogen with a fishy appetite: new challenges for an old problem. Mycologist 20:99–104.

Vishniac HS 1958. A new marine phycomycete. Mycologia 50:66–79.

West JA, Klochkova TA, Kim GH, Loiseaux-de Goer S 2006. *Olpidiopsis* sp., an oomycete from Madagascar that infects *Bostrychia* and other red algae: host species susceptibility. Phycol Res 54:72–85.

Wilce RT, Schneider CW, Quinlan AV, Bosch KV 1982. The life history and morphology of free-living *Pilayella littoralis* (L) Kjellm (Ectocarpaceae, Ectocarpales) in Nahant Bay, Massachusetts. Phycologia 21:336–354.

Wille N 1899. Om nogle Vandsoppe. Videnskabsselskabets Skrifter. I. Mathematisk-naturvidenskabelig Klasse 1899 3:1–14.

Wood CL, Byers JE, Cottingham KL, Altman I, Donahue MJ, Blakeslee AMH 2007. Parasites alter community structure. Proc Natl Acad Sci USA 104:9335–9339.

Wright EP 1877. On a species of *Rhizophydium* parasitic on species of *Ectocarpus* with notes on the fructification of the Ectocarpi. Trans R Irish Acad 26:369–379.

Wynne MJ 1972. Culture studies of Pacific coast Phaeophyceae. Mém Soc Bot Fr 129–144.

Zopf W 1884. Zur Kenntnis der Phycomyceten I. Zur Morphologie und Biologie der Ancylisteen und Chytridiaceen. Nova Acta Acad Leopold-Carol 47:143–236.

3

TAXONOMY AND PHYLOGENY OF THE DOWNY MILDEWS (PERONOSPORACEAE)

Marco Thines

University of Hohenheim, Institute of Botany 210, Stuttgart, Germany

Hermann Voglmayr

Department of Systematic and Evolutionary Botany, University of Vienna, Vienna, Austria

Markus Göker

University of Tübingen, Institute of Systematic Biology and Mycology, Tübingen, Germany

3.1 DOWNY MILDEW TAXONOMY

Taxonomy is something that any ecologist, geneticist, pathologist, and physiologist cannot go without. To understand and to communicate relationships of different species in any of these fields, a frame is needed that roughly expresses the phylogenetic relationships of the taxa examined. Although this can vary in different groups of organisms, members of one species are more closely related to each other than distinct species within a genus or distinct genera within a family. Such a hierarchical concept may have its flaws but can be used in scientific reasoning and is of a complexity a nonspecialist can still deal with. One major challenge of the modern global economy is to keep pathogens at bay and to avoid pandemic spreads. Of course, quarantine measures are hampered or made impossible if it is unknown which pathogen is responsible for causing a specific disease, or if the taxonomy for a species is inappropriate, for example, when several species are lumped into a composite species.

Oomycete Genetics and Genomics: Diversity, Interactions, and Research Tools
Edited by Kurt Lamour and Sophien Kamoun
Copyright © 2009 John Wiley & Sons, Inc.

An example for this is the epidemics of the *Ocimum basilicum* (basil) downy mildew. This pathogen was reported from Europe first as *Peronospora lamii* (Martini et al., 2003; Lefort et al., 2003). *P. lamii*, which is parasitic to *Lamium* (dead nettle), is a native species in almost any European country, and therefore it was assumed that the causal agent of basil downy mildew, on the basis of a broad species concept, was just some kind of specialized form. The pathogen subsequently spread from Switzerland to Italy, France, Germany, and other countries, and it was not until DNA sequences of the pathogen became available that it was realized that the species that occurred on basil was only distantly related to *P. lamii* and was therefore named *Peronospora* sp. (Belbahri et al., 2005). Because it was uncertain whether the pathogen could be conspecific with a known pathogen of sage (*Peronospora swinglei*), no name was attached to the new species. As a consequence, quarantine measures were not taken. Hence, *Peronospora* sp., which has only very recently been formally discribed as a new species, *P. belbahrii* (Thines et al., 2009), is now almost globally distributed and is, for instance, present in more than half of the seed lots tested by the diagnostics company PathoScan (personal communication).

Another example is *Plasmopara halstedii*, which clearly is a composite species (Novotel'nova, 1962). *P. halstedii* is included on the quarantine list for Australia, as Australia is the only sunflower-producing continent from which the downy mildew disease has not been reported. However, the downy mildew pathogen of *Arctotheca calendula* (capeweed) is not uncommon in Australia, and this species has been classified as *P. halstedii* (Eppo Datasheet on Quarantine Pests–*P. halstedii*). Therefore, it is difficult for the Australian government to negotiate with countries from where *P. halstedii* on sunflower has been reported, as these may claim that *P. halstedii* already occurs in Australia; regardless of the fact that these downy mildew pathogens are largely unrelated (Thines & Constantinescu, unpublished results).

The two examples outlined above demonstrate that for practical reasons, it is necessary to have a taxonomic system that includes the most current findings. Regarding scientific questions, it is also important to rely on a sound taxonomy. This becomes apparent when considering the genus *Hyaloperonospora*, which has attracted significant interest during the past decade, as it includes a well-studied pathogen of the model plant *Arabidopsis thaliana* (see 3.4.3).

3.2 THE PHYLOGENY AND TAXONOMY OF THE PERONOSPORACEAE

3.2.1 Before the Advent of Molecular Phylogenies (Pre-2000)

Traditionally, the downy mildews have been divided into two sections, the graminicolous downy mildews (GDM) and the Eudicot-infecting downy mildews (EDM). This divide has been postulated especially by Dick et al. (1984) and Dick (1995, 2001), who assumed an independent phylogenetic origin of these pathogens and placed them into different orders and even different subclasses. Recent molecular phylogenies (Riethmüller et al., 2002; Hudspeth et al., 2003; Göker et al., 2003, 2007; Thines et al., 2008) have shown that GDM and EDM are in fact closely related and that the placement into different families is not justified. Nonetheless, EDM and GDM are treated separately in the following passage, mainly for reasons of clarity. *Peronospora* (Corda, 1837), which is the first downy mildew genus of the EDM, was described more than 30 years after the description of the first genus of the Oomycota, *Albugo* (de Roussel, 1806). Only 6 years later, the genus *Bremia* followed (Regel, 1843), which was described on the basis of the regularly aggregated ultimate branchlets (Thines, 2006). In 1869, the third downy mildew genus *Basidiophora* was introduced (Roze and Cornu, 1869). *Basidiophora* differed from the preceding two genera in having club-shaped and mostly unbranched sporangiophores. From the description of *Plasmopara* in 1886 (Schröter, 1886) onward, the characteristics used for genus delimitation became more and more subtle and began to include biological features. *Plasmopara sensu* Schröter (1886) differs from *Peronospora* and *Bremia* in having monopodial sporangiophores that branch at right angles and in releasing zoospores during the germination of the asexual dispersal units. Consequently, *Basidiophora*, which also produces zoospores, was included in *Plasmopara*, and several species previously classified in *Peronospora* were transferred to this genus. Subsequently, *Pseudoperonospora* was segregated from *Peronospora* (Rostovzev, 1903). Although also having colored dissemination units and branching superficially dichotomous, this genus was differentiated from *Peronospora* by producing zoospores like *Plasmopara*; yet it was morphologically distinct from the latter genus. Four years later, Wilson (1907) described the sixth genus of the Peronosporaceae *Rhysotheca*. Wilson obviously assumed that Schröter (1886) had introduced *Plasmopara* as a modification of de Bary's *Peronospora* section *Plasmatoparae* (de Bary, 1863). Therefore, Wilson segregated the species germinating by releasing zoospores directly from the sporangia from those of *Plasmopara sensu* Wilson (Wilson, 1907), which germinated by the ejection of the whole cytoplasm, from which then zoospores develop. *Rhysotheca* never gained broad recognition, partly because of inconsistencies in the circumscription of this genus (Constantinescu et al., 2005). Seven years later, Wilson described the genus *Bremiella* (Wilson, 1914) on the basis of assumed differences in the morphology of the ultimate branchlets. Wilson (1914) stated that unlike in the other genera of the downy mildews, the tips of the ultimate branchlets of *Bremiella* were inflated. While producing zoospores on germination, the branching pattern of the sporangiophores was similar to that of *Bremia* and *Peronospora*. However, sporangiophore tips that widen at the apex and carry a distinct annulus are typical for a whole group of downy mildews, particularly the downy mildews with pyriform haustoria (DMPH) and some genera with plesiomorphic traits (Thines, 2006; Voglmayr and Thines, 2007; Thines et al., 2008). Nonetheless, the genus *Bremiella* was widely accepted until recently (Constantinescu, 1979, 1989, 1991a; Dick, 2001). The genus *Pseudoplasmopara* that was described in 1922 never gained recognition and was mostly ignored by subsequent authors. Although several new combinations were proposed for species of uncertain placement (Tao and Qin, 1982; Skalický, 1966), no additional genus was described for the EDM until 1989, when Constantinescu (1989) described the genus *Paraperonospora*. *Paraperonospora* differs from *Peronospora* in having uncolored dissemination units and globose to pyriform haustoria, and from *Plasmopara* by the production of conidia that germinate with a germ tube. *Paraperonospora* differs from both of these genera in having sporangiophores that are widening toward the ramifications. In 1998, Constantinescu (1998) segregated the monotypic genus *Benua* from *Basidiophora* based on the fact that sporangiophores in *Basidiophora*, emerge from stomata and are thick walled, whereas those of *Benua* are half immersed in the plant tissue and very thin walled. In contrast to *Basidiophora*, the ultimate branchlets in *Benua* are highly reduced. *Benua* was the last genus described in the EDM before the first comprehensive molecular phylogeny for the Peronosporaceae was published (Riethmüller et al., 2002).

3.2.2 A Global Phylogeny for the Downy Mildews and their Relatives: Taxonomic Consequences

The first molecular phylogeny based on a multigene approach was published by Göker et al. (2007). In this phylogenetic reconstruction, the monophyly of the downy mildews (DM) was highly supported. However, Göker and Stamatakis (2006) revealed that the high support received was most likely an artifact of a too-narrow search radius in popular maximum-likelihood software, such as PhyML and Treefinder. Göker and Stamatakis (2006), when applying RAxML (Stamatakis, 2006), did not receive any support for downy mildew monophyly but found *Phytophthora infestans* as a sister group to the downy mildews with pyriform haustoria (DMPH) without support. However, in the analyses of both Göker and Stamatakis (2006) and Göker et al. (2007), the monophyly of the clade that includes both the DM and *Phytophthora*, as well as the paraphyly of *Phytophthora*, received high bootstrap support.

The same result is shown in the present maximum-likelihood analysis of the concatenated cox2 and large ribosomal subunit (LSU) rDNA sequences (Fig. 3.2), which includes a more comprehensive sampling of outgroup species. The GenBank accession numbers of the sequences are provided in the supplementary material in the appendix to this chapter. Using Haliphthoraceae as the outgroup, Saprolegniomycetidae are revealed as the sister group of the remaining Oomycetes included in the analysis. The latter are subdivided into the Rhipidiales and a highly supported clade comprising Albuginales, Pythiales, and Peronosporales. In addition to *Phytophthora*, *Pythium* is revealed as paraphyletic with high support. In addition to the transfer of *Phytophthora* to Peronosporales suggested earlier (Göker et al., 2007), these results clearly indicate that not only *Phytophthora* but also *Pythium* taxonomy should be reconsidered. Although the phylogenetic relationships within Peronosporales are so far not fully resolved, it is obvious that the relationship of *Phytophthora* and the DM genera has interesting consequences regarding character evolution (Göker et al., 2007; Thines et al., 2007, Thines, 2009). In the multigene analyses of Göker et al. (2007), four groups of DM genera received significant bootstrap support. These groups were the DM with colored conidia (*Peronospora* and *Pseudoperonospora*); the brassicolous DM (*Perofascia* and *Hyaloperonospora*); which have been dealt with in detail above; the DMPH (*Basidiophora*, *Benua*, *Bremia*, *Paraperonospora*, *Plasmopara*, *Plasmoverna*, and *Protobremia*); and the GDMs with lasting sporangiophores (*Graminivora*, *Poakatesthia*, and *Viennotia*). The genus *Sclerophthora* could not be placed in any of these groups by Thines et al. (2008). Also, *Peronosclerospora* and *Sclerospora*, even though both genera were found to be embedded within the Peronosporales (Thines et al., 2008), could not be unambiguously placed in any of these groups.

3.3 SPECIES CONCEPTS IN DOWNY MILDEWS—OR THE PHILOSOPHY OF LUMPING VERSUS SPLITTING

Species concepts in downy mildews have been dominated by two opposing views. Gäumann (1918, 1923) advocated a very narrow species concept. He assumed that downy mildews are generally host–genus or even host–species specific. He substantiated his conclusion by taking many measurements, mainly of conidial dimensions. As Gäumann usually investigated only very few accessions of a specific species, it was later argued that his observations were based on too few samples and therefore underestimated the variability between different samples (Gustavsson, 1959b). In addition, his conclusions were not always tested with cross-infection experiments, even though observations in the field can be regarded as additional support for generally high host specificity. For instance, populations of *Capsella bursa-pastoris* (shepherd's purse) are often heavily infected with downy mildew on the rims of perfectly healthy cabbage fields. The opposing view has been formulated by Yerkes and Shaw (1959), who claimed that downy mildew species were, with a few exceptions, specific to a host family only. The latter view became common currency and so was followed by most non-taxonomists working with downy mildews. It continues to persist today among researchers who study downy mildews (e.g., the *Arabidopsis* downy mildew) (Knoth and Eulgem, 2008; Llorente et al., 2008; Rentel et al., 2008).

In the case of *Hyaloperonospora*, which can be considered an example for downy mildews in general, molecular phylogenetic reconstructions (Riethmüller et al., 2002; Voglmayr, 2003; Choi et al., 2003; Göker et al., 2004) have revealed that the tempting solution to accept only *Hyaloperonospora parasitica* or a very few species (Constantinescu and Fatehi, 2002) is not sensible. *Hyaloperonospora* is a genetically highly diverse genus, and Göker et al. (2004) have shown that the clusters that are most likely to be considered species are largely congruent with the species concept of Gäumann (1918, 1923). In many cases, the phylogenetic reconstructions revealed that even more species exist in *Hyaloperonospora* than Gäumann anticipated and that some host genera (e.g., *Cardamine* and *Sisymbrium*) are parasitized by at least four phylogenetically distinct species. Judging from the data available currently (Göker et al., 2004, 2009), it seems likely that several additional new species await their discovery in *Hyaloperonospora*.

3.4 THE BRASSICOLOUS DOWNY MILDEWS

3.4.1 Circumscription of *Hyaloperonospora* and *Perofascia* in Relation to Peronospora

The genera *Hyaloperonospora* and *Perofascia* were introduced by Constantinescu (Constantinescu and Fatehi, 2002) to accommodate the noncolored downy mildews that had previously been assigned to *Peronospora*. *Perofascia* is monotypic and so far only known from *Lepidium*; it differs from *Hyaloperonospora* in having hyphal haustoria and curved ultimate branchlets, in which the spiral curving is often less pronounced than in *Hyaloperonospora* (Constantinescu and Fatehi, 2002; Thines, 2006). Depending on the species concept applied, the genus *Hyaloperonospora* encompasses only one (Yerkes and Shaw, 1959), very few (Constantinescu and Fatehi, 2002), or based on a narrow species concept (Gäumann, 1923; Gustavsson, 1959a), roughly 100 species. Most of these species are parasitic to Brassicaceae, with only a few exceptions in Zygophyllaceae, Cistaceae, Resedaceae, Capperaceae, and Cleomaceae. *Hyaloperonospora* can be distinguished from *Peronospora* on the basis of three main morphological characteristics. First, *Hyaloperonospora* has uncolored conidia, whereas in *Peronospora*, the conidia are with very few exceptions grayish to brownish. Second, haustoria in *Peronospora* are usually hyphal, whereas in *Hyaloperonospora*, haustoria are globose to largely lobate when mature. Third, the ultimate branchlets are straight to curved in *Peronospora*, whereas they are often spiral in *Hyaloperonospora*.

Phylogenetically, *Hyaloperonospora* and *Peronospora* are very distinct, and it seems more likely that *Hyaloperonospora* is the sister genus to the graminicolous downy mildews (Göker et al., 2007) or the downy mildews with pyriform haustoria (Thines, 2007), than *Hyaloperonospora* being the sister genus to *Peronospora*. As the genus *Hyaloperonospora* can be easily distinguished from *Peronospora* both morphologically and phylogenetically, the generic name *Hyaloperonospora* should be applied for the species included in what Yerkes and Shaw (1959) classified as *Peronospora parasitica*.

3.4.2 Major Species Clusters in *Hyaloperonospora*

So far, molecular phylogenies (Göker et al., 2004, 2009) have revealed many distinct lineages in *Hyaloperonospora*, which mostly represent single species, although the backbone of the tree could not be well resolved. However, six clades could so far be identified, which are highly supported and harbor several distinct species from different hosts (Fig. 3.1). These are clade 1, mainly on *Cardamine* and its close relatives; clade 2, on *Draba* species; clade 3, on a variety of host species including *A. thaliana*; clade 4, on *Aurinia* and *Berteroa*; clade 5, on different *Sisymbrium* species; and clade 6, on variety of host genera but with a focus on *Brassica* and its relatives (tribe Brassiceae) in one subclade and on *Cardamine* and its relatives (subtribe Cardamininae) in the other subclade. *Hyaloperonospora brassicae*, which includes the economically important pathogens of *Brassica*, *Raphanus*, and *Sinapis*, is a polyphyletic construct, as specimens from these three host genera are phylogenetically distinct with maximum bootstrap support. The most noteworthy species in clade 6, which also includes *H. brassicae*, is *Hyaloperonospora tribulina*. As *Hyaloperonospora* is primarily parasitic to Brassicaceae (i.e., the most early diverging lineages are parasites of Brassicaceae), the occurrence of *Hyaloperonospora* on a Zygophyllaceae is the result of a host jump from Brassicaceae to Zygophyllaceae. From the molecular genetics perspective, *Hyaloperonospora* clade 3 is probably the most interesting because, although it does not include economically important species, this clade includes a parasite of the most intensely studied model plant *A. thaliana*. Also, *Arabidopsis arenosa* is among the hosts of clade 3. The investigation of the species closely related to *Hyaloperonospora arabidopsidis* offers the unique possibility to identify key pathogenicity genes involved in host jumps and pathogen establishment on a specific host. Thus, the general patterns of how plant pathogen evolution and adaptation to specific hosts takes place might be revealed. Similar to clade 6, a host jump across host families occurred within the clade 3, with *Hyaloperonospora crispula s.l.*, which is parasitic to *R. lutea* and *Reseda luteola* of the Resedaceae.

3.4.3 The Unstudied *H. parasitica*, the Hardly Known *H. brassicae*, and the Well Known *H. arabidopsidis*

Because of the widespread application of the broad species concept advocated by Yerkes and Shaw (1959), it was commonly believed that *Hyaloperonospora* on *Arabidopsis* was the same species as *Hyaloperonospora* on *Brassica* and *Hyaloperonospora* on *C. bursa-pastoris*, the type host of *H. parasitica sensu stricto*. This led to confusion and misconceptions over the taxonomy of the *Arabidopsis*-infecting downy mildew. Research on the genetics of the downy mildew pathogen of *Arabidopsis* has been justified by, among other arguments, the claims that this pathogen belongs to the same species as economically important lineages that infect *Brassica*, *Raphanus*, and *Sinapis* crops. In addition to the fact that *Arabidopsis* is the most extensively studied model plant, this claim, based on the approach of Yerkes and Shaw (1959), constituted a good argument for researchers to obtain funding for the investigation of what was thought to be "*Peronospora parasitica*" at that

0.05
H. lunariae on Lunaria rediviva 18-10
H. lunariae on Lunaria rediviva 46-03
H. lunariae on Lunaria rediviva 37-07
H. lunariae on Lunaria rediviva 20-11
H. lunariae on Erucastrum nasturtiifolium 36-07
H. lunariae on Erucastrum nasturtiifolium 36-09
H. brassicae on Sinapis alba 15-03
H. brassicae on Sinapis alba 36-08
H. brassicae on Armoracia rusticana 23-09
H. brassicae on Sinapis alba 14-03
H. sp. on Eruca sativa 52-06
H. brassicae on Sinapis arvensis 27-04
H. brassicae on Brassica napus 24-08
H. brassicae on Brassica napus 46-01
H. brassicae on Armoracia rusticana 21-11
H. brassicae on Armoracia rusticana 26-10
H. brassicae on Brassica napus 26-09
H. sp. on Diplotaxis tenuifolia 23-12
H. sp. on Lobularia maritima 22-10
H. sp. on Sisymbrium irio 52-05
H. sp. on Sisymbrium volgense 24-02
H. sp. on Diplotaxis tenuifolia 23-11
H. brassicae on Raphanus raphanistrum 12-09
H. teesdaliae on Teesdalia nudicaulis 23-02
H. sp. on Cardamine flexuosa 21-08
H. sp. on Cardamine amara 20-08
H. sp. on Cardamine flexuosa 20-04
H. sp. on Cardamine flexuosa 22-04
H. sp. on Cardamine amara 36-05
H. sp. on Cardamine hirsuta 21-07
H. sp. on Cardamine flexuosa 38-03
H. sp. on Cardamine pratensis 49-05
H. sp. on Cardamine pratensis 17-08
H. sp. on Cardamine hirsuta 05-01
H. sp. on Cardamine hirsuta 49-06
H. sp. on Cardamine hirsuta 36-11
H. sp. on Cardamine hirsuta 44-04
H. sp. on Cardamine pratensis 04-03
H. sp. on Nasturtium officinale 38-06
H. sp. on Nasturtium officinale 24-03
H. sp. on Cardamine sp. 21-01
6
H. galligena on Aurinia saxatilis 49-03
H. galligena on Aurinia saxatilis 20-12
H. galligena on Aurinia saxatilis 18-03
H. berteroae on Berteroa incana 24-06
H. berteroae on Berteroa incana 23-08
4
H. sp. on Iberis sempervirens 22-06
H. parasitica on Capsella bursa-pastoris 52-07
H. parasitica on Capsella bursa-pastoris 19-01
H. sp. on Descurainia sophia 44-07
H. sp. on Descurainia sophia 21-04
H. sp. on Descurainia sophia 49-02
H. arabidopsidis on Arabidopsis thaliana 49-10
H. arabidopsidis on Arabidopsis thaliana 17-01
H. arabidopsidis on Arabidopsis thaliana 41-01
H. arabidopsidis on Arabidopsis thaliana 24-04
H. arabidopsidis on Arabidopsis thaliana 36-03
H. cf. erophilae p.p. on Draba verna 49-12
H. cf. erophilae p.p. on Draba verna 17-05
H. cf. erophilae p.p. on Draba verna 41-06
H. cf. erophilae p.p. on Draba verna 43-01
H. cf. erophilae p.p. on Draba verna 19-04
H. cardaminopsidis on Arabidopsis arenosa 20-03
H. sp. on Arabis turrita 36-06
H. thlaspeos-arvensis on Noccaea caerulescens 20-05
H. thlaspeos-arvensis on Noccaea caerulescens 22-11
H. thlaspeos-perfoliati on Microthlaspi perfoliatum 17-04
H. thlaspeos-perfoliati on Microthlaspi perfoliatum 17-09
H. sp. on Lepidium ruderale 22-08
H. sp. on Reseda lutea 21-12
H. sp. on Reseda lutea 22-12
H. sp. on Reseda lutea 23-03
H. sp. on Aubrieta sp. 38-01
3
H. isatidis on Isatis tinctoria 22-05
H. thlaspeos-arvensis on Thlaspi arvense 15-01
H. thlaspeos-arvensis on Thlaspi arvense 37-06
H. thlaspeos-arvensis on Thlaspi arvense 07-10
H. thlaspeos-arvensis on Thlaspi arvense 49-11
H. sp. on Biscutella auriculata 51-02
H. sisymbrii-loeselii on Sisymbrium loeselii 23-10
H. sisymbrii-loeselii on Sisymbrium loeselii 23-05
H. sp. on Sisymbrium officinale 21-10
5
H. sp. on Lepidium draba 21-03
H. sp. on Lepidium draba 37-05
H. sp. on Lepidium draba 49-07
H. cheiranthi on Erysimum cheiri 22-03
H. cheiranthi on Erysimum crepidifolium 26-08
H. camelinae on Camelina sativa 21-06
H. camelinae on Camelina sativa 36-12
H. nesliae on Neslia paniculata 37-08
H. hesperidis on Hesperis matronalis 23-06
H. hesperidis on Bunias orientalis 21-09
H. hesperidis on Hesperis matronalis 20-07
H. hesperidis on Hesperis matronalis 23-01
H. niessleana on Alliaria petiolata 49-09
H. niessleana on Alliaria petiolata 03-01
H. niessleana on Alliaria petiolata 02-12
H. niessleana on Alliaria petiolata 37-01
H. niessleana on Alliaria petiolata 04-01
H. arabidis-alpinae on Arabis alpina 37-02
H. cf. erophilae p.p. on Draba verna 43-03
H. cf. erophilae p.p. on Draba verna 43-04
H. cf. erophilae p.p. on Draba verna 43-05
H. cf. erophilae p.p. on Draba verna 42-08
H. cf. erophilae p.p. on Draba verna 48-11
H. cf. erophilae p.p. on Draba verna 48-12
H. cf. erophilae p.p. on Draba verna 36-02
H. cf. erophilae p.p. on Draba verna 43-02
2
H. sp. on Cardamine bulbifera 46-08
H. sp. on Cardamine bulbifera 44-06
H. sp. on Cardamine bulbifera 21-05
H. sp. on Cardamine impatiens 38-04
H. sp. on Cardamine bulbifera 21-02
H. sp. on Cardamine impatiens 18-06
H. sp. on Cardamine impatiens 13-09
H. barbareae on Barbarea vulgaris 22-01
H. barbareae on Barbarea vulgaris 13-05
H. barbareae on Barbarea vulgaris 38-02
H. sp. on Rorippa palustris 20-01
H. sp. on Rorippa palustris 38-08
H. sp. on Rorippa palustris 38-07
H. sp. on Arabis soyeri 37-03
1
P. lepidii on Lepidium ruderale 22-08
P. lepidii on Lepidium ruderale 22-09

time. However, it was soon realized that the pathogens from different hosts (e.g., *Arabidopsis*, *Brassica*, and *Sinapis*) were not compatible regarding host range (Tör et al., 1994; Satou and Fukumoto, 1996). Therefore, different "races" of *Hyaloperonospora* on various hosts were named, assuming these would merely represent specialized forms of the same species. In the light of molecular phylogenetic reconstructions (Rehmany et al., 2000; Riethmüller et al., 2002; Voglmayr, 2003; Choi et al., 2003; Göker et al., 2004, 2009), this perception needs to change, as these studies revealed that the genus *Hyaloperonospora* contains a multitude of distinct species, which have partly gone through a long independent evolution. Considering this, the problems in relating results regarding pathogenicity genes and avirulence genes obtained from *H. arabidopsidis* to *H. brassicae* become understandable. Because of the high genetic divergence of *H. brassicae* and *H. arabidopsidis*, it is not surprising that pathogenicity genes, which are in common for *H. arabidopsidis* and *H. brassicae*, are hard to find.

H. arabidopsidis is the most intensely studied obligate biotrophic pathogen because the host genome was the first plant genome available whereas *H. brassicae* is still not well studied in many respects. For *H. parasitica*—a species so far known only from *C. bursa-pastoris*—only very few data are available. As this pathogen neither occurs on an economically important nor a genetically well-studied host, it could be assumed that this species will never attract considerable attention. However, it should be borne in mind that *C. bursa-pastoris* is probably the most widely distributed host for a *Hyaloperonospora* species and that *H. parasitica* is among the most often observed downy mildew diseases in natural populations in Europe and therefore is well suited for investigations on ecology and population structure of a downy mildew species.

3.5 THE DMPHs

Of the four clades mentioned before, the DMPHs are by far the most diverse group of DM. The morphology of the sporangiophores ranges from short, unbranched sporangiophores with thin walls in *Benua* to long, thick-walled sporangiophores with up to eight orders of branches in some *Plasmopara*

FIG. 3.1 Maximum-likelihood phylogenetic tree inferred with RAxML from concatenated internal spacer region (ITS) and large ribosomal subunit (LSU) rDNA sequences. The dataset represents a subset of the one analyzed by Göker et al. (2009). Technical details on the inference of this tree as well on the files used and on the origin of the sequences are provided in Göker et al. (2009). Note that the recognition of the clade numbers is based on the extended sampling used in Göker et al. (2009), whereas this figure only shows the specimens for which both ITS and LSU rDNA could be amplified. The numbers above the branches are bootstrap support values equal to or larger than 60% from 100 replicates. Abbreviations: *H.*, *Hyaloperonospora*; *P.*, *Perofascia*.

species. The characteristic feature of the group is the shape of haustoria, which are usually small and globose to pyriform, in contrast to all other genera of the downy mildews [see overviews in Voglmayr et al. (2004) and Göker et al. (2007)]. Although the shape of the haustoria probably represents a synapomorphy, the uncolored sporangia as well as the distinct annulus/widening at the tip of the ultimate branchlets likely represent synplesiomorphies of the group (Thines, 2006).

Within the DMPH, three lineages can be differentiated morphologically: *Plasmopara*, *Benua*, and a group of genera with sporangiophores broadening toward the ramifications to a different extent (Thines, 2006). The morphological characteristic of this group is most pronounced in *Paraperonospora* and less pronounced in *Bremia*. In molecular phylogenies, the affinities of *Basidiophora* and *Plasmoverna* are still equivocal, but the relationships of the genera *Paraperonospora*, *Protobremia*, *Novotelnova* (Voglmayr and Constantinescu, 2008), and *Bremia* are well resolved. From *Paraperonospora* to *Bremia*, a tendency of shortening the internodial parts between the outermost ramifications can be observed, which results in the usually regular aggregation of 4–6 ultimate branchlets in *Bremia*. Although the molecular phylogenetic results of Voglmayr et al. (2004) and Voglmayr and Constantinescu (2008) are in favor of a taxonomic solution with two monotypic genera (*Protobremia* and *Novotelnova*) between *Paraperonospora* and *Bremia*, it needs to be borne in mind that only a fraction of *Paraperonospora* species has so far been investigated phylogenetically, and the taxonomy of this group may need to be revised when more data become available.

Apart from the species cluster described above and the monotypic genus *Benua* (Constantinescu, 1998), a third distinct lineage is present in molecular phylogenies, which comprises all species of the genus *Plasmopara*. After the segregation of *Viennotia* (Göker et al., 2003), *Protobremia* (Voglmayr et al., 2004), *Plasmoverna* (Constantinescu et al., 2005), *Poakatesthia* (Thines et al., 2007), and *Novotelnova* (Constantinescu and Voglmayr, 2008), the genus *Plasmopara* is, to our current knowledge, a monophyletic assemblage. The genus *Plasmopara* is morphologically the most diverse genus of the Peronosporaceae, with a range from little-branched species with small sporangiophores (e.g., *Plasmopara pusilla*, and *Plasmopara skvortzovii*) to some species with sporangiophores as tall as 1 mm (e.g., *P. halstedii s.l.*). Two of the three economically relevant "species" of this genus (*P. halstedii s.l.*, *Plasmopara, nivea s.l.*, and *Plasmopara viticola*) are in fact composite species (*P. nivea s.l.*) or even polyphyletic (*P. halstedii s.l.*), as revealed by morphological and molecular phylogenetic investigations (Voglmayr et al., 2004; Komjáti et al., 2007). That *Plasmopara obducens*, which is an emergent pathogen on cultivated *Impatiens* species, is the sole *Plasmopara* species infective to various species of *Impatiens* is too simplified, considering the findings of Voglmayr and Thines (2007), who demonstrated that *Bremiella sphaerosperma* which is also parasitic to *Impatiens* is a member of the genus *Plasmopara* closely related to *P. obducens*.

3.6 THE DOWNY MILDEWS WITH COLORED CONIDIA

The downy mildews with colored conidia contain most of the economically important downy mildew species (e.g., *Peronospora tabacina*, *Peronospora farinosa*, *Peronospora destructor*, *Peronospora arborescens*, *Peronospora valerianellae*, *Peronospora cristata*, *Peronospora sparsa*, *Peronospora belbahrii*, *Pseudoperonospora cubensis*, *Pseudoperonospora humuli*), as well as most of the overall downy mildew species (roughly two thirds). Only two genera are so far known to possess colored conidia, which include *Peronospora* (more than 400 species; Constantinescu, (1991b)) and *Pseudoperonospora* (less than 10 species). *Pseudoperonospora* has retained some apparently plesiomorphic characters, such as the germination of the asexually formed dissemination units by zoospores. Although *Pseudoperonospora* and *Peronospora* also differ with respect to the surface ornamentation of the conidiosporangia (Thines, 2006), their germination is the most important characteristic to distinguish these two genera. But as germination by zoospores has also been reported for some *Peronospora* species like *P. sparsa* (Berkley, 1862), which was therefore considered a member of *Pseudoperonospora* by Skalický (1966), it may be that this differentiation between *Peronospora* and *Pseudoperonospora* can not be retained. The two economically important species in *Pseudoperonospora*, *P. cubensis*, and *P. humuli* are very closely related, and Choi et al. (2005) came to the conclusion that they should be considered a single species. Our own investigations, which include a variety of cross-infection experiments (Runge and Thines, unpublished results), confirm a very close relationship of these pathogens.

3.7 THE GDMs WITH LASTING SPORANGIOPHORES

The GDMs with lasting sporangia are probably the most curious group of downy mildews, which comprise the three monotypic genera *Graminivora*, *Poakatesthia*, and *Viennotia*. All three genera may be extremely rare; however, their geographic origins have only been poorly sampled for downy mildews of noncrop plants. *Poakatesthia* and *Viennotia* are so far only known from the type collections in Africa (Ethiopia and Guinea, respectively), and *Graminivora* is a pathogen of moderate abundance but widely distributed in Asia. The GDM with lasting sporangiophores are characterized by a high degree of morphological variation and by several apparently plesiomorphic characters not found in any other downy mildew (Thines, 2009). All three species of this group have previously been placed in either *Bremia* or *Plasmopara*, because morphology of the sporangiophores resembled these genera very closely. At superficial investigation, only *Viennotia*, with its conspicuous apophyses on the ultimate branchlets, differs significantly from the genus in which it was previously placed. A thorough investigation of the haustorium morphology, which is a character of high phylogenetic relevance (Fraymouth, 1956;

Constantinescu, 1989; Constantinescu and Fatehi, 2002; Göker et al., 2003, 2007; Voglmayr et al., 2004; Thines, 2006; Thines et al. 2006, 2007), revealed a *Phytophthora*-like feature present in this group: In *Poakatesthia*, haustoria may develop into intracellular mycelium, growing through several living host cells before entering the apoplast again (Thines et al., 2007). Similarly, an apparent plesiomorphy is that in *Viennotia*, sporangiophore growth is not determinate, but regular new outgrowth from places where sporangia have been disseminated is still possible in this genus (Thines, 2009).

3.8 GENERA OF UNRESOLVED PHYLOGENETIC POSITION

Thus far, only three genera cannot be placed in any of the monophyletic clusters of genera described above. These are the GDM of the genera *Peronosclerospora*, *Sclerospora*, and the still enigmatic genus *Sclerophthora*, which occupies an unresolved position with affinities to both downy mildews and *Phytophthora* (Thines et al., 2008). *Peronosclerospora* and *Sclerospora* are both pathogens of great economic importance in the semiarid tropics, where they pose a threat to millions of subsistence farmers. Among their hosts are *Pennisetum glaucum* (pearl millet), *Sorghum bicolor* (sorghum millet), *Zea mays* (mays), and *Saccharum officinarum* (sugarcane). Although *Sclerospora* seems to be restricted to *Setaria* and *Pennisetum* species, *Peronosclerospora* has a much wider host spectrum and may be found on a great variety of grasses with C4-metabolism (Kenneth, 1981; Jeger et al., 1998). Both *Peronosclerospora* and *Sclerospora* are characterized by evanescent sporangiophores. This means that sporangiophores are only thin walled and therefore dehydrate and collapse easily when exposed to dry conditions. Also, the sporangia are very fragile and collapse easily. Therefore, it is almost impossible to examine mature sporangiophores or sporangia from herbarized material, which thereby hampers systematic comparison of the different species described in these genera. Oospores are usually formed in great numbers in the leaves and stems of the host plants, and in *Peronosclerospora*, the leaves fray once the oospores are mature, which thereby releases millions of oospores that are disseminated by the wind (Bock et al., 1997; Jeger et al., 1998). *Peronosclerospora* is the only oomycete genus where the oospores have developed into aerial spores. Another difference between *Peronosclerospora* and *Sclerospora* is the germination of the sporangia. Germination takes place by releasing zoospores in *Sclerospora*, the conidia of *Peronosclerospora* germinate directly by producing a germ tube. A first molecular phylogeny that encompasses both genera (Thines et al., 2008) has shown that *Sclerospora* and *Peronosclerospora* are highly distinct. Whether these two genera form a monophyletic assemblage could not be inferred, although overall morphology suggests so. *Sclerophthora* has a similar distribution and host spectrum compared with *Peronosclerospora* and *Sclerospora*, but it extends much further in both respects. *Peronosclerospora* has a predominantly tropical to subtropical distribution and *Sclerospora* may also occur in temperate regions along with the host genus *Setaria*, but *Sclerophthora* reaches into cool temperate areas of southern Canada (Jones, 1955) and can also be found on a variety of grasses with C3-metabolism. *Sclerophthora* is characterized by hyphal sporangiophores, which is similar to *Phytophthora* (Thirumalachar et al., 1953; Waterhouse, 1963; Erwin and Ribeiro, 1996), and thick-walled oospores, which is similar to *Sclerospora* (Saccardo, 1890; Thirumalachar et al., 1953; Kenneth, 1981). Although it could be deduced from the molecular tree presented by Thines et al. (2008) that *Sclerophthora* is not to be placed within the Saprolegniomycetidae as assumed by Dick et al. (1989) or Dick (2001), but in the Peronosporales, its exact phylogenetic position remains unresolved.

3.9 FUTURE PERSPECTIVES

3.9.1 White Flecks in Downy Mildew Evolution

Although numerous molecular phylogenetic studies published during the past 7 years shed light on the generic relationships of the downy mildews, *Phytophthora*, and other oomycetes, many aspects of downy mildew evolution are still unknown. This is especially true for the relationships of the four clades and the three isolated genera discussed above. Also below the genus level, the resolution of the phylogenetic reconstructions is often not satisfactory (Voglmayr, 2003; Göker et al., 2004; Voglmayr et al., 2004), which makes it difficult to track the routes of evolution for the radiation of the downy mildew genera. In addition, only about one quarter of the downy mildew species described have been included in molecular phylogenies, and for only a bit more than 5% of them, phylogenies based on multiple genes have been computed (Göker et al., 2007). Therefore, although significant progress toward a natural classification (i.e., one that reflects common evolutionary origins) has been made, only a rough framework regarding the evolution of these pathogens is available. One of the most basic questions, that is, which host plants downy mildew evolution started from, is still not unequivocally resolved. Thines et al. (2007) assumed that downy mildew evolution might have started from hosts in Poaceae, but until convincing phylogenetic evidence from a multigene based phylogeny becomes available, other scenarios cannot be excluded.

Similarly, although a phylogenetic distinction can be made for most species of the downy mildews, the relationships of some closely related species are often not well resolved, which leaves a place for speculation of whether they should be considered separate species or not (García-Blázquez et al., 2008, Göker et al., 2009).

3.9.2 Suggestions for a Global Phylogenetic Approach for the Oomycota

To achieve a robust and well-supported phylogenetic hypothesis for the downy mildews and *Phytophthora*, a molecular phylogenetic approach that includes several genes and intergenic spacers easily sequenced is essential. Currently, the most common sequences deposited in GenBank for the Peronosporaceae are internal spacer region (ITS) sequences (more than 600), which are followed by partial sequences of the nuclear large and small ribosomal subunit and the mitochondrial *cox2* gene (more than 100 each). The two additional loci (*ß-tubulin* and *nadh1* genes) sequenced by Göker et al. (2007) are only available for less than 50 taxa. For *Phytophthora*, several additional loci were used in the study of Blair et al. (2008). In general, two criteria must be fulfilled by candidate loci. First, it must be possible to amplify them by universal primers, preferably primers that can be used throughout the Oomycota, while being specific enough to give no amplification from organisms that do not belong to the Straminipila. Second, the loci need to have different degrees of variability, so they could be used on a great variety of topics, which range from the supraordinal to the subspecies level. In addition, the loci should be easy to amplify from well-preserved herbarium specimens, as unlike in *Phytophthora*, no cultures on media are available for downy mildews. Of the loci (L10, *ß-tubulin*, *ef1a*, *enl*, *hsp90*, *tigA*, and SSU) used by Blair et al. (2008), especially the SSU, *hsp90*, L10, and *ß-tubulin* genes seem to fulfil these needs. Other loci have been brought into consideration, for example, the mitochondrial loci *cox1* (e.g., Sachay et al., 1993; Kroon et al., 2003; Martin and Tooley, 2003), *cox2* (e.g., Sachey et al., 1993; Hudspeth et al., 2000, 2003), and mtIGS (Wattier et al., 2003); it seems reasonable to add also these organelle loci to uncover possible reticulate evolution. In addition, both LSU and ITS sequences have been shown to be suited for phylogenetic analyses. As a conclusion, the following 10 genes are proposed for multigene-based phylogenetic reconstruction: ribosomal loci: SSU, ITS1-5.8S-ITS2, LSU D1-D3; other nuclear loci: *ß-tubulin*, L10, *hsp90*, *ef1a*; mitochondrial loci: *cox1*, *cox2*, mtIGS. For phylogenetic analyses with more species, the core genes SSU (400bp of the 3′ end), ITS1-5.8S-ITS2, *cox1*, *cox2*, and mtIGS are suggested. For very closely related species, it might be necessary to search for additional, highly variable genes, which possibly cannot be amplified by universal primers.

A word of caution, however, must be added regarding the use of highly variable noncoding loci such as the ITS rDNA for phylogenetic analysis. It is well known that the complicated indel structures present in such loci may lead to alignment artifacts (Mindell, 1991; Morrison and Ellis, 1997). In fact, unreliable statistical support caused by such an artifact has been presented in the important pioneering paper of Cooke et al. (2000) on the phylogeny of *Phytophthora* and its relatives. Our reexamination of their datasets (Table 3.1) indicates that the high support for the monophyly of a clade comprising *Hyalophytopthora batememanensis* and *Pythium vexans* is an artifact. In contrast to the original 5.8S rDNA and ITS2 alignment of Cooke et al. (2000), which was created by applying ClustalW (Thompson et al., 1994) and subsequent manual alterations, using other alignment software (Morgenstern, 1999; Lee et al., 2002; Edgar, 2004; Katoh et al., 2005) results in low support or

TABLE 3.1 Neighbor-joining bootstrap support for selected bipartitions (splits; indicating monophyla if the tree is rooted) in trees derived from the the two data matrices used by Cooke et al. (2000) after realigning the sequences.

Alignment software (and options used if deviating from the default values)	M749.NX: bootstrap values for the split separating *Halophytophthora batemanensis* and *Pythium vexans* from the remaining taxa	M751.NX: bootstrap values for the split separating *Peronospora sparsa*, *Phytophthora arecae*, *Phytophthora palmivora*, and *Phytophthora megakarya* from the remaining taxa
original Treebase alignment	100%	96%
Clustalw v1.81	95%	95%
Dialign v2.2.1 (-fa -n)	<50%	<50%
MAFFT v5.732 (–maxiterate 1000)	!70%	57%
MUSCLE v3.52 (-stable)	!69%	75%
POA v2 (-do_ progressive)	<50%	70%
POA v2	!70%	<50%
POA v2 (-do global)	!99%	<50%

"!" indicates bootstrap support for a conflicting grouping. Analysis was done with PAUP* (Swofford, 2002) applying 1,000 bootstrap replicates and HKY85 distances. The bootstrap values obtained with the original alignment downloaded from Treebase (http://www.treebase.org/) are very close to those presented by Cooke et al. (2000). Thus, the use of a potentially suboptimal phylogenetic inference technique has no impact on the results.

even high support for a conflicting clade. The latter is in agreement with the arrangement of the two species found in multigene analyses (Göker et al., 2007; Fig. 3.2). Cooke et al. (2000) also analyzed a complete ITS alignment of another taxon set. Reanalyses reveal that the support for a sister group relationship between three *Phytophthora* species and *P. sparsa* is also too high, as it completely breaks down if more sophisticated alignment software is applied (Table 3.1).

We conclude that at least ClustalW should not be applied to oomycete ITS sequences, unless very closely related taxa are examined; manual "improvements" also have not much value and usually make things worse rather than better. Using automated approaches to alignment avoids the problems of the investigator's bias and the lack of reproducibility that may be related to manual alignments (Gatesy et al., 1993). The most reliable approach seems to be "multiple analysis" (Lee et al., 2002), that is, running several alignment programs (or the same under different settings) and comparing the results (Kemler et al., 2006). Excluding alignment columns is also possible if performed in a reproducible manner, but it may lead to a loss of information (Gatesy et al., 1993).

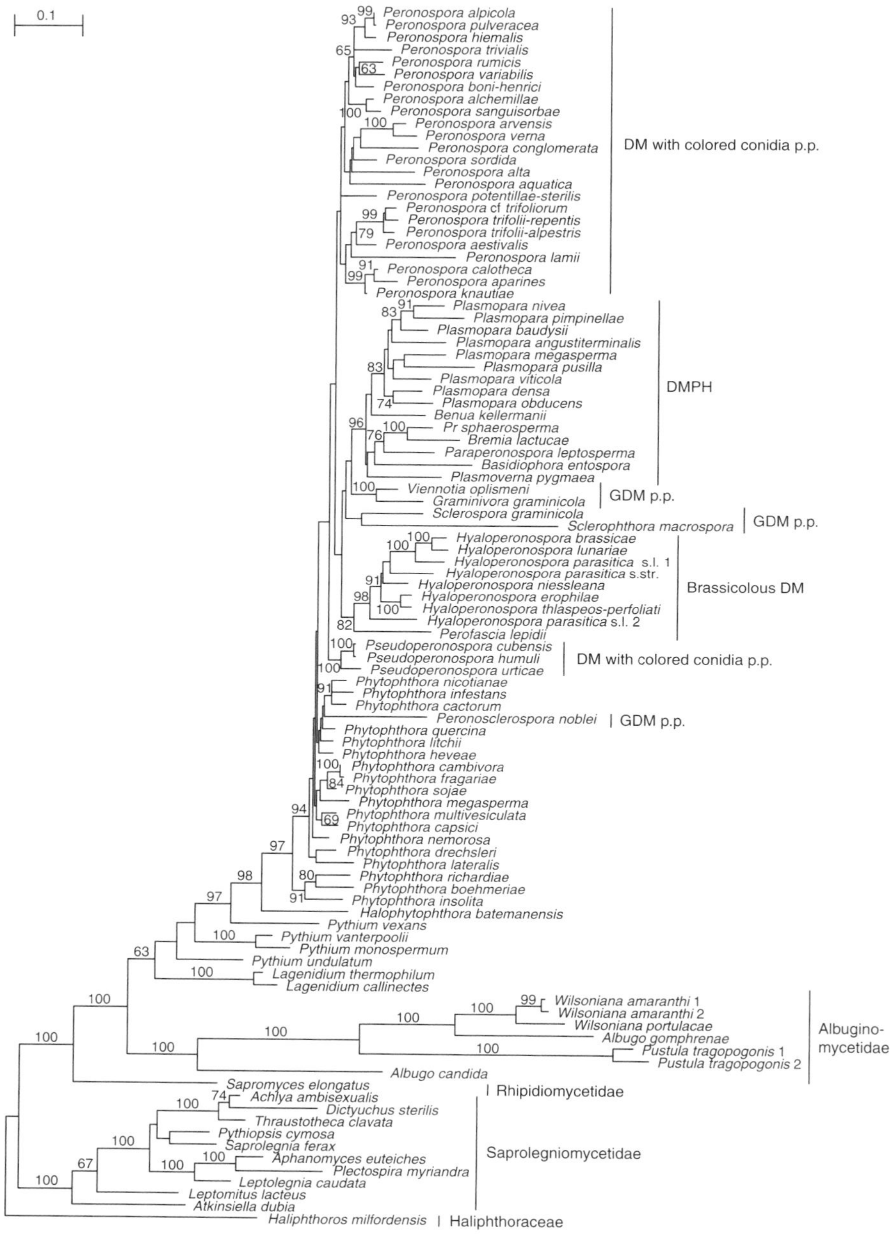

FIG. 3.2 Maximum-Likelihood phylogenetic tree inferred with RAxML under the GTRMIX model approximation from concatenated *cox2* and LSU rDNA sequences, which were aligned with POA v. 2 (Lee et al., 2002). The alignments are available on request from M.G. The numbers above the branches are fast bootstrap support values equal to or larger than 60% from 1,000 replicates. The GenBank accession numbers are provided in the supplementary material in the appendix to this chapter.

3.9.3 Prospects for Evolutionary Genomics

In the Peronosporaceae, the genomes of *P. infestans*, *Phytophthora ramorum*, and *Phytophthora sojae* are available, and annotated genome sequences of *H. arabidopsidis* are becoming available. With the advent of less expensive pyrosequencing techniques, the door to comparative evolutionary genomics has been pushed wide open. The downy pathogens and white blister rusts are promising candidates for research in this field. Downy mildews are mostly host–genus or even host–species specific, and these pathogens manipulate their hosts' metabolisms in a very sophisticated way (see later chapters in this book). Therefore, through the comparison of either closely related species on different hosts or several pathogen species that occur on the same host, fundamental insights can be gained both on the genetic changes involved in host adaptation and on speciation in general. By identifying the targets of pathogen effectors, the components and the function of the plant defence system can be elucidated to unparalleled depths, especially if data are considered not only from oomycetes but also from other biotrophic pathogens, such as rust fungi (Uredinales) or powdery mildews (Erysiphales).

For investigating the processes involved in speciation, the closely related species that parasitize the genus *Trifolium* (clover) and other Fabaceae are of particular interest (García-Blázquez et al., 2008). Numerous closely related species exist in this group, and cophylogenetic scenarios might be revealed. Another useful avenue to begin investigating speciation and the key processes involved in host specialization might be the *Hyaloperonospora* clade 3. It is ideal for comparative genomics, as in this clade two related species, *Hyaloperonospora cardaminopsidis* and *H. arabidopsidis*, exist on two closely related hosts, *A. arenosa* and *A. thaliana*, respectively. For two of these four species, a complete genome is available, so a tremendous amount of knowledge could be gained by affordable pyro-sequencing. Another member of clade 3 that is promising for evolutionary genomics is *H. crispula*, which has made a host jump across families and is parasitic to *Reseda* (Resedaceae). Also two additional species closely related to the well-studied *H. arabidopsidis* exist on annual hosts distantly related to *Arabidopsis*, *Hyaloperonospora erophilae* on *Draba verna* and *Hyaloperonospora thlaspeos-perfoliati* on *Microthlaspi perfoliatum*, and it may constitute equally useful targets for investigating host jumps, host adaptation and speciation.

For identifying the key mechanisms in plant defense and pathogen establishment, plants with a high amount of biotrophic pathogens of different phylogenetic position are especially suited. A possible candidate for this kind of investigation is *Tragopogon pratensis*. It is the host of a downy mildew, a white blister rust, two species of rust fungus, an anther smut, and a powdery mildew species, thus being host to almost any biotrophic group of pathogens parasitic to dicots. In the Brassicaceae, especially those species that are hosts of two different species of *Albugo* and a downy mildew are of interest in this respect [e.g., *C. bursa-pastoris*, which is affected by *H. parasitica s.str.*, *Albugo koreana* (Choi et al., 2007), and *Albugo candida*].

ACKNOWLEDGMENTS

M. T. wishes to thank the German Science Foundation (DFG) and the Landesstiftung Baden-Württemberg (Elite Program for Postdocs) for financial support. Financial support by the DFG for M. G. is gratefully acknowledged.

REFERENCES

de Bary A 1863. Recherches sur le développement de quelques champignons parasites. Annales des Sciences Naturelles, Botanique, Sér. 4 20:5–148.

Belbahri L, Calmin G, Pawlowski J, Lefort F 2005. Phylogenetic analysis and real time PCR detection of a presumbably undescribed *Peronospora* species on sweet basil and sage. Mycol Res 109:1276–1287.

Berkeley JM 1862. Fungi on rose leaves. Gardener's Chronicle 1862:307–308.

Blair JE, Coffey MD, Park S-Y, Geiser DM, Kang S 2008. A multi-locus phylogeny for *Phytophthora* utilizing markers derived from complete genome sequences. Fungal Genet Biol 45:266–277.

Bock CH, Jeger MJ, Fitt BDL, Sherington J 1997. Effect of wind on the dispersal of oospores of *Peronosclerospora sorghi* from sorghum. Plant Pathol 46:439–449.

Choi Y-J, Hong S-B, Shin H-D 2003. Diversity of the *Hyaloperonospora parasitica* complex from core brassicaceous hosts based on ITS rDNA sequences. Mycol Res 107:1314–1322.

Choi Y-J, Hong S-B, Shin H-D. 2005. A reconsideration of *Pseudoperonospora cubensis* and *P. humuli* based on molecular and morphological data. Mycol Res 109:841–848.

Choi Y-J, Shin H-D, Hong S-B, Thines M 2007. Morphological and molecular discrimination among *Albugo candida* materials infecting *Capsella bursa-pastoris* world-wide. Fungal Divers 27:11–34.

Constantinescu O 1979. Revision of *Bremiella* (Peronosporales). Trans Br Mycol Soc 72:510–515.

Constantinescu O 1989. *Peronospora* complex on Compositae. Sydowia 41:79–107.

Constantinescu O 1991a. *Bremiella sphaerosperma* sp. nov. and *Plasmopara borreriae* comb. nov. Mycologia 83:473–479.

Constantinescu O 1991b. An annotated list of *Peronospora* names. Thunbergia 15:1–110.

Constantinescu O 1998. A revision of *Basidiophora* (Chromista, Peronosporales). Nova Hedwig 66:251–265.

Constantinescu O, Fatehi J 2002. *Peronospora*-like fungi (Chromista, Peronosporales) parasitic on Brassicaceae and related hosts. Nova Hedwig 74:291–338.

Constantinescu O, Voglmayr H, Fatehi J, Thines M 2005. *Plasmoverna* gen. nov., and the taxonomy and nomenclature of *Plasmopara* (Chromista, Peronosporales). Taxon 54:813–821.

Cooke DEL, Drenth A, Duncan JM, Wagels G, Brasier CM 2000. A molecular phylogeny of *Phytophthora* and related Oomycetes. Fungal Genet Biol 30:17–32.

Corda ACJ 1837. Icones Fungorum Hucusque Cognitorum.

Dick MW. 1995. Sexual reproduction in the Peronosporomycetes (chromistan fungi). Can J Bot 73:S712–S724.

Dick MW 2001. The Peronosporomycetes. In: McLaughlin DJ, McLaughlin EG, Lemke PA, editors. The Mycota, Vol. VIIA. Berlin: Germany Springer. pp. 39–72.

Dick MW, Wong PTW, Clark G 1984. The identity of the oomycete causing 'Kikuyu Yellows', with a reclassification of the downy mildews. Bot J Linn Soc 89:171–197.

Dick MW, Croft BJ, Margarey RC, De Cock AWAM, Clark G 1989. A new genus of the Verrucalvaceae (Oomycetes). Bot J Linn Soc 99:97–113.

Edgar R 2004. MUSCLE: multiple sequence alignment with high accuracy and high throughput. Nuc Acids Res 32:1792–1797.

Erwin DC, Ribeiro OK 1996. *Phytophthora* diseases worldwide. St. Paul MN: APS Press.

Fraymouth J 1956. Haustoria of the *Peronosporales*. Trans Brit Mycol Soc 39:79–107.

García-Blázquez G, Göker M, Voglmayr H, Martín MP, Tellería MT, Oberwinkler F 2008. Phylogeny of *Peronospora*, parasitic on Fabaceae, based on ITS sequences. Mycol Res 112:502–512.

Gatesy J, DeSalle R, Wheeler W 1993. Alignment-ambiguous nucleotide sites and exclusion of systematic data. Mol Phylogenet Evol 2:152–157.

Gäumann E 1918. Über die Formen der *Peronospora parasitica* (Pers.) Fries. Beihefte zum Botanischen Centralblatt 35:395–533.

Gäumann E 1923. Beiträge zu einer Monographie der Gattung *Peronospora* Corda. Beiträge zur Kryptogamenflora der Schweiz 5:1–360.

Göker M, Voglmayr H, Gárcia-Blázquez G, Oberwinkler F 2009. Species delimitation in downy mildews: the case of *Hyaloperonospora* in the light of nuclear ribosomal ITS and LSU sequences. Mycol Res 113:308–325.

Göker M, Riethmüller A, Voglmayr H, Weiß M, Oberwinkler F 2004. Phylogeny of *Hyaloperonospora* based on nuclear ribosomal internal transcribed spacer sequences. Mycol Prog 3:83–94.

Göker M, Stamatakis A 2006. Maximum likelihood phylogenetic inference: An empirical comparison on a multi–locus dataset. Presented at the German Conference on Bioinformatics 2006, Tübingen, Germany. from http://diwww.epfl.ch/~stamatak/index–Dateien/publications/GCB2006_Poster.pdf, Accessed June 18, 2008.

Göker M, Voglmayr H, Riethmüller A, Oberwinkler F 2007. How do obligate parasites evolve? A multi-gene phylogenetic analysis of downy mildews. Fungal Gene Biol 44:105–122.

Göker M, Voglmayr H, Riethmüller A, Weiß M, Oberwinkler F 2003. Taxonomic aspects of Peronosporaceae inferred from Bayesian molecular phylogenetics. Can J Bot 81:672–683.

Gustavsson A 1959a. Studies on Nordic Peronosporas. I. Taxonomic revision. Opera Botanica 3:1–271.

Gustavsson A 1959b. Studies on Nordic Peronosporas. II. General account. Opera Botanica 3:1–61.

Hudspeth DSS, Nadler SA, Hudspeth MES 2000. A *cox2* molecular phylogeny of the Peronosporomycetes. Mycologia 92:674–684.

Hudspeth DSS, Stenger DC, Hudspeth MES 2003. A *cox2* phylogenetic hypothesis for the downy mildews and white rusts. Fungal Diver 13:47–57.

Jeger MJ, Gilijamse E, Bock CH, Frinking HD 1998. The epidemiology, variability and control of the downy mildews of pearl millet and sorghum, with particular reference to Africa. Plant Pathol 47:544–569.

Jones W 1955. Downy mildew of *Dactylis glomerata* caused by *Sclerophthora cryophila*. Can J Bot 33:350–354.

Katoh K, Kuma K, Toh H, Miyata T 2005. MAFFT version 5: improvement in accuracy of multiple sequence alignment. Nucle Acids Res 33:511–551.

Kemler M, Göker M, Begerow D 2006. Implications of molecular characters for the phylogeny of the Microbotryaceae (Basidiomycota: Urediniomycetes). BMC Evolutionary Biol 6:35.

Kenneth RG 1981. Downy mildews of graminaceous crops. In: Spencer DM, editor. The downy mildews. London, UK: Academic Press. pp. 367–394.

Knoth C, Eulgem T 2008. The oomycete response gene *LURP1* is required for defense against *Hyaloperonospora parasitica* in *Arabidopsis thaliana*. Plant J 55:53–64.

Komjáti H, Walcz I, Virányi F, Zipper R, Thines M, Spring O 2007. Characteristics of a *Plasmopara angustiterminalis* isolate from *Xanthium strumarium*. Eur J Plant Pathol 119:421–428.

Kroon LPNM, Bakker FT, Van den Bosch FT, Bonants PJM, Fliera WG 2004. Phylogenetic analysis of *Phytophthora* species based on mitochondrial and nuclear DNA sequences. Fungal Gen Biol 41:766–782.

Lee C, Grasso C, Sharlow M 2002. Multiple sequence alignment using partial order graphs. Bioinformatics 18:452–464.

Lefort F, Gigon V, Amos B 2003. Le mildiou s'étend. Déjà détecté dans des nombreux pays européens, *Peronospora lamii*, responsible du mildiou de basilic, a été observe en Suisse dans la region lémanique. Réussir Fruits et Légumes 223:66.

Llorente F, Muskett P, Sanchez-Vallet A, Lopez G, Ramos B, Sanchez-Rodriguez C, Jorda L, Parker J, Molina A 2008. Repression of the auxin response pathway increases *Arabidopsis* susceptibility to necrotrophic fungi. Mol Plant 1:496–509.

Martin FN, Tooley PW 2003. Phylogenetic relationships among *Phytophthora* species inferred from sequence analysis of mitochondrially encoded cytochrome oxidase I and II genes. Mycologia 95:269–284.

Martini P, Rapetti S, Bozzano G, Bassetti G 2003. Segnalazione in Italia di *Peronospora lamii* su basilico (*Ocimum basilicum* L). Proceedings of the meeting "Problemi fitopatologici emergenti e implicazioni per la difesa delle colture". San Remo. pp. 27–29.

Mindell DP 1991. Aligning DNA sequences: Homology and phylogenetic weighting. In: Miyamoto MM, Cracraft J, editors. Phylogenetic analysis of DNA sequences. Oxford University Press. New York. pp. 73–89.

Morgenstern B 1999. DIALIGN2: improvement of the segment-to-segment-approach to multiple sequence alignment. Bioinformatics 15:211–218.

Morrison DA, Ellis JT 1997. Effects of nucleotide sequence alignment on phylogeny estimation: a case study of 18S rDNAs of Apicomplexa. Molec Biol Evol 14:428–441.

Novotel'nova NS 1962. *Plasmopara halstedii* (Farl.) Berl. et De Toni kak sbornyj vid (Obosnovanie k taksonomicheskomu podrazdeleniyu roda *Plamopara* na slozhnotsvetnykh). Botanicheskij Zhurnal SSSR 47:970–981.

Regel E 1843. Beiträge zur Kenntnis einiger Blattpilze. Botanische Zeitung 1:665–667.

Rehmany AP, Lynn JR, Tör M, Holub EB, Beynon JL 2000. A comparison of *Peronospora parasitica* (downy mildew) isolates from *Arabidopsis thaliana* and *Brassica oleracea* using amplified fragment length polymorphism and internal transcribed spacer 1 sequence analyses. Fungal Gene Biol 30:95–103.

Rentel MC, Leonelli L, Dahlbeck D, Zhao B, Staskawicz BJ 2008. Recognition of the *Hyaloperonospora parasitica* effector ATR13 triggers resistance against oomycete, bacterial, and viral pathogens. Proc Nat Acad Sci USA 105:1091–1096.

Riethmüller A, Voglmayr H, Göker M, Weiß M, Oberwinkler F 2002. Phylogenetic relationships of the downy mildews (Peronosporales) and related groups based on nuclear large subunit ribosomal DNA sequences. Mycologia 94:834–849.

Rostovzev SJ 1903. Beiträge zur Kenntnis der Peronosporeen. Flora (Jena) 92:405–430.

Roze E, Cornu M 1869. Sur deux nouveaux types géneriques pour des familles Saprolegniées et des Péronosporées. Annales des Sciences Naturelles 5:72–91.

de Roussel HFA. 1806. Flore du Calvados et des terrains adjacents, composée suivant la méthode de M. Jussieu, comparée avec celle de Tournefort et de Linné. 2nd edn. Poisson, Caen, France.

Saccardo PA 1890. Fungi aliquot australiensis. Hedwig 29:154–156.

Sachay DJ, Hudspeth DSS, Nadler SA, Hudspeth MES 1993. Oomycete mtDNA: *Phytophthora* genes for cytochrome c oxidase use an unmodified genetic code and encode proteins most similar to plants. Exp Mycol 17:7–23.

Satou M, Fukumoto F 1996. The host range of downy mildew, *Peronospora parasitica*, from *Brassica campestris*, Chinese cabbage and turnip crops and *Raphanus sativus*, Japanese radish crop. Ann Phytopathol Soc Jpn 62:402–407.

Schröter J 1886. Fam. Peronosporacei. In: Cohn F, editor. Kryptogamen-Flora von Schlesien, Vol. 3. Breslau: Kern. pp. 228–252.

Skalický V 1966. Taxonomie der Gattungen der Familie Peronosporaceae. Preslia (Praha) 38:117–129.

Stamatakis A 2006. RAxML-VI-HPC: Maximum likelihood–based phylogenetic analyses with thousands of taxa and mixed models. Bioinformatics 22:2688–2690.

Swofford DL 2002. PAUP*: phylogenetic analysis using parsimony (*and other methods), Version 4.0 b10. Sinauer Associates, Sunderland, MA.

Tao J-F, Qin Y 1982. New species and new combinations of the genus *Bremiella* on *Compositae* of China. Acta Mycologica Sinica 1:61–67.

Thines M 2006. Evaluation of characters available from herbarium vouchers for the phylogeny of the downy mildew genera (Chromista, Peronosporales), with focus on scanning electron microscopy. Mycotaxon 97:195–218.

Thines M 2007. Characterisation and phylogeny of repeated elements giving rise to exceptional length of ITS2 in several downy mildew genera (Peronosporaceae). Fungal Genet Biol 44:199–207.

Thines M 2009. Bridging the Gulf: *Phytophthora* and Downy Mildews Are Connected by Rare Grass Parasites. PLoS ONE 4(3):e4790.

Thines M, Göker M, Oberwinkler F, Spring O 2007. A revision of *Plasmopara penniseti*, with implications for the host range of the downy mildews with pyriform haustoria. Mycol Res 111:1377–1385.

Thines M, Göker M, Spring O, Oberwinkler F 2006. A revision of *Bremia graminicola*. Mycol Res 11:646–656.

Thines M, Göker M, Telle S, Ryley M, Mathur K, Narayana YD, Spring O, Thakur RP 2008. Phylogenetic relationships of graminicolous downy mildews based on cox2 sequence data. Mycol Res 112:345–351.

Thines M, Telle S, Ploch S, Runge F 2009. Identity of the downy mildew pathogens of basil, coleus, and sage with implications for quarantine measures. Mycological Research, in press, http://dx.doi.org/10.1016/j.mycres.2008.12.005.

Thirumalachar MJ, Shaw CG, Narasimhan MJ 1953. The sporangial phase of the downy mildew of *Eleusine corcorana* with a discussion of the identity of *Sclerospora macrospora* Sacc. Bull Torrey Bot Club 80:299–307.

Thompson JD, Higgins DG, Gibson CWD 1994. Clustal W: improving the sensitivity of progressive multiple sequence alignment through sequence weighting, position-specific gap penalties and weight matrix choice. Nucleic Acids Res 22:4673–4680.

Tör M, Holub EB, Brose E, Musker R, Gunn N, Can C, Crute IR, Beynon JL 1994. Map positions of three loci in *Arabidopsis thaliana* associated with isolate-specific recognition of *Peronospora parasitica* (downy mildew). Mol Plant Microbe Interact 7:214–222.

Voglmayr H 2003. Phylogenetic study of *Peronospora* and related genera based on nuclear ribosomal ITS sequences. Mycol Res 107:1132–1142.

Voglmayr H, Constantinescu O 2008. Revision and reclassification of three *Plasmopara* species based on morphological and molecular phylogenetic data. Mycol Res 112:487–501.

Voglmayr H, Riethmüller A, Göker M, Weiß M, Oberwinkler F 2004. Phylogenetic relationships of *Plasmopara*, *Bremia* and other genera of downy mildews with pyriform haustoria based on Bayesian analysis of partial LSU rDNA sequence data. Mycol Res 108:1011–1024.

Voglmayr H, Thines M 2007. Phylogenetic relationships and nomenclature of *Bremiella sphaerosperma* (Chromista, Peronosporales). Mycotaxon 100:11–20.

Waterhouse GM 1963. Key to the species of *Phytophthora* de Bary. Mycological Paper 92. Commonwealth Mycological Institute Kew, Surrey.

Wattier RAM, Gathercole LL, Assinder SJ, Gliddon CJ, Deahl KL, Shaw DS, Mills DI 2003. Sequence variation of intergenic mitochondrial DNA spacers (mtDNA–IGS) of *Phytophthora infestans* (Oomycetes) and related species. Molec Ecol Notes 3: 136–138.

Wilson GW 1907. Studies in North American Peronosporales II. Phytophthorae and Rhysotheceae. Bull Torrey Bot Club 34:387–416.

Wilson GW 1914. Studies in north American Peronosporales — VI. Notes on miscellaneous species. Mycologia 6:192–210.

Yerkes WD, Shaw CG 1959. Taxonomy of *Peronospora* species on Cruciferae and Chenopodiaceae. Phytopathol 49:499–507.

APPENDIX

Specimens sequenced in the course of this study.

Organism	Host	DNA isolation no.	Collection (specimen-voucher)	Origin	cox2 GenBank accession number	LSU rDNA D1/D2/D3 GenBank accession number
Wilsoniana amaranthi (Schwein.) Y.-J. Choi, Thines & H.-D. Shin	*Amaranthus sp.*	MG 10-3	TUB	France, Alps, Giffre meadows nearby Samoens; 07/29/2000; leg. M. Göker	EU826090	EU826106
Wilsoniana amaranthi (Schwein.) Y.-J. Choi, Thines & H.-D. Shin	*Amaranthus sp.*	MG 9-11	COL (HNC Nr. 40/AR 335)	Colombia, Cundinamarca, Sopó; 09/06/1998; leg. A. Gil Correa, J. Gil Correa, J. Moehler and M. Piepenbring	EU826091	EU826107
Albugo candida (Pers.) Roussel	*Sisymbrium irio* L.	MG 7-3	LPB (HNB2590)	Bolivia, Dpto. La Paz, Prov. Manco, Kapak; 01/22/2000; leg. M. Piepenbring	EU826092	EU826108
Albugo gomphrenae (Speg.) Cif. & Biga	*Iresine diffusa* Humb. & Bonpl. ex Willd.	MG 7-6	TUB (AR 166)	Costa Rica, Prov. Limón, Bribri; 10/25/1992; leg. M. Piepenbring	EU826093	EU826109

(Continued)

APPENDIX (*Continued*)

Organism	Host	DNA isolation no.	Collection (specimen-voucher)	Origin	cox2 GenBank accession number	LSU rDNA D1/D2/D3 GenBank accession number
Peronosclerospora noblei (W. Weston) C.G. Shaw	*Sarga leiocladum* (Hack.) Spangler	MT891	HOH (HUH 891)	Australia, QLD, 26°50′51″S, 151°48′00″E; 01/11/2007; leg. M. Ryley	EU826094	EU826110
Peronospora alchemillae G.H. Otth	*Alchemilla vulgaris agg.*	MG 16-5	TUB (AR 222)	Austria, Tirol, Oberjoch, Ochsenalpe; 09/26/2000; leg. A. Riethmüller	EU826095	EU826111
Peronospora knautiae Fuckel	*Knautia sylvatica* (L.) Duby	MG 7-9	TUB (AR 163)	Austria, Tirol, Oberjoch; 10/09/1993; leg. M. Piepenbring	EU826096	EU826112
Plasmopara angustiterminalis Novot.	*Xanthium sp.*	MT X03A1	~	Laboratory Strain, cultivated on sunflower	EU826097	EU826113
Plasmopara baudysii Skalický	*Berula erecta* (Huds.) Coville	HV 571	WU	Austria, Niederösterreich, Gramatneusiedl; 08/02/2000; leg. H. Voglmayr	EU826098	~
Pseudoperonospora cubensis (Berk. and Curtis) Rostovzev	*Cucurbita pepo L.*	MT486	HOH (HUH 486)	Germany, Tübingen, Schwalldorf; 07/20/2002; Otmar Spring	EU826099	EU826114

(*Continued*)

Pustula tragopogonis (Pers.) Thines	*Cirsium oleraceum* Scop.	MG 9-3	TUB	Austria, nearby the border to Slovenia, Mariazell; 07/08/2000; leg. W. Maier	EU826100	EU826115
Pustula tragopogonis (Pers.) Thines	*Tragopogon pratensis L.*	MT 725	HOH (HUH 725)	Austria, Wien, 20 km east, street to Bratislava; 07/20/2005; Otmar Spring	EU826101	EU826116
Pythium vanterpoolii V. Kouyeas & H. *Kouyeas*	~	MG 53-3	Tübingen, F4252 = AR100	Germany, Baden-Württemberg, Kranzach; leg. A. Riethmüller	EU826102	EU826117
Saprolegnia ferax (Gruith.) Nees	~	MG 53-4	Tübingen, F4257 = AR6	Germany, Baden-Württemberg, Bad Buchau; leg. A. Riethmüller	EU826103	EU826118
Sclerophthora macrospora (Sacc.) Thirum., C.G. Shaw & Naras	*Zea mays L.*	MT892	HOH (HUH 892)	China, Yunnan, Kunming; summer 2004; leg. anonymous	EU826104	EU826119
Wilsoniana portulacae (DC.) Thines	*Portulaca oleracea L.*	MG 7-8	TUB (AR 164)	Costa Rica, San José, Tibas, patio Vanda; 08/19/1994; leg. M. Piepenbring	~	EU826120
Wilsoniana portulacae (DC.) Thines	*Portulaca oleracea L.*	MT640	HOH (HUH 640)	China, Yunnan, Jinghong; 09/02/2004; leg. M. Thines	EU826105	~

Sequence data used in this study.

Organism	cox2 GenBank accession number	LSU rDNA D1/D2/D3 GenBank accession number	Source
Achlya ambisexualis	AF086687	AF218202	GenBank
Albugo candida	EU826092	EU826108	This study
Aphanomyces euteiches	AF086692	AF235939	GenBank
Atkinsiella dubia	AF290312	AB285221	GenBank
Basidiophora entospora	DQ365699	AY035513	Göker et al. (2007) or one of our earlier studies
Benua kellermanii	DQ365700	DQ361226	Göker et al. (2007) or one of our earlier studies
Bremia lactuca	DQ365701	AY035507	Göker et al. (2007) or one of our earlier studies
Dictyuchus sterilis	AF086691	AF218193	GenBank
Graminivora graminicola	DQ365702	DQ195167	Göker et al. (2007) or one of our earlier studies
Haliphthoros milfordensis	AF290305	AB285218	GenBank
Halophytophthora batemanensis	DQ365703	DQ361227	Göker et al. (2007) or one of our earlier studies
Hyaloperonospora brassica	DQ365704	AY035503	Göker et al. (2007) or one of our earlier studies
Hyaloperonospora erophilae	DQ365705	AY271998	Göker et al. (2007) or one of our earlier studies
Hyaloperonospora lunariae	DQ365706	AY271997	Göker et al. (2007) or one of our earlier studies
Hyaloperonospora niessleana	DQ365707	AY035498	Göker et al. (2007) or one of our earlier studies
Hyaloperonospora parasitica s.l. 1	DQ365708	AY035505	Göker et al. (2007) or one of our earlier studies
Hyaloperonospora parasitica s.l. 2	DQ365709	AY272000	Göker et al. (2007) or one of our earlier studies
Hyaloperonospora parasitica s.str.	DQ365710	AY271996	Göker et al. (2007) or one of our earlier studies
Hyaloperonospora thlaspeos-perfoliati	DQ365711	AY271999	Göker et al. (2007) or one of our earlier studies
Lagenidium callinectes	AF290308	AB285217	GenBank
Lagenidium thermophilum	AF290304	AB285219	GenBank
Leptolegnia caudata	AF086693	AF218176	GenBank
Leptomitus lacteus	AF086696	AF119597	GenBank

(Continued)

Paraperonospora leptosperma	DQ365712	AY035515	Göker et al. (2007) or one of our earlier studies
Perofascia lepidii	DQ365713	DQ361228	Göker et al. (2007) or one of our earlier studies
Peronosclerospora noblei	EU826094	EU826110	This study
Peronospora aestivalis	DQ365714	AY035482	Göker et al. (2007) or one of our earlier studies
Peronospora alchemillae	EU826095	EU826111	This study
Peronospora alpicola	DQ365715	AY271990	Göker et al. (2007) or one of our earlier studies
Peronospora alta	DQ365716	AY035493	Göker et al. (2007) or one of our earlier studies
Peronospora aparines	DQ365717	AY035484	Göker et al. (2007) or one of our earlier studies
Peronospora aquatica	DQ365718	AY271991	Göker et al. (2007) or one of our earlier studies
Peronospora arvensis	DQ365719	AY035491	Göker et al. (2007) or one of our earlier studies
Peronospora boni-henrici	DQ365720	AY035475	Göker et al. (2007) or one of our earlier studies
Peronospora calotheca	DQ365721	AY035483	Göker et al. (2007) or one of our earlier studies
Peronospora cf. trifoliorum	DQ365722	AY035478	Göker et al. (2007) or one of our earlier studies
Peronospora conglomerata	DQ365723	AY271993	Göker et al. (2007) or one of our earlier studies
Peronospora hiemalis	DQ365724	AY271992	Göker et al. (2007) or one of our earlier studies
Peronospora knautiae	EU826096	EU826112	This study
Peronospora lamii	DQ365725	AY035494	Göker et al. (2007) or one of our earlier studies
Peronospora potentillae-sterilis	DQ365726	AY035486	Göker et al. (2007) or one of our earlier studies
Peronospora pulveracea	DQ365727	AY035470	Göker et al. (2007) or one of our earlier studies
Peronospora rumicis	DQ365728	AY035476	Göker et al. (2007) or one of our earlier studies
Peronospora sanguisorbae	DQ365729	AY035487	Göker et al. (2007) or one of our earlier studies
Peronospora sordida	DQ365730	AY271995	Göker et al. (2007) or one of our earlier studies
Peronospora trifolii-alpestris	DQ365731	AY271989	Göker et al. (2007) or one of our earlier studies
Peronospora trifolii-repentis	DQ365732	AY271988	Göker et al. (2007) or one of our earlier studies
Peronospora trivialis	DQ365733	AY035471	Göker et al. (2007) or one of our earlier studies
Peronospora variabilis	DQ365734	AY035477	Göker et al. (2007) or one of our earlier studies
Peronospora verna	DQ365735	AY271994	Göker et al. (2007) or one of our earlier studies
Phytophthora boehmeriae	DQ365736	DQ361229	Göker et al. (2007) or one of our earlier studies
Phytophthora cactorum	DQ365737	DQ361230	Göker et al. (2007) or one of our earlier studies

(Continued)

APPENDIX (*Continued*)

Organism	cox2 GenBank accession number	LSU rDNA D1/D2/D3 GenBank accession number	Source
Phytophthora cambivora	DQ365738	DQ361231	Göker et al. (2007) or one of our earlier studies
Phytophthora capsici	DQ365739	DQ361232	Göker et al. (2007) or one of our earlier studies
Phytophthora drechsleri	DQ365740	DQ361233	Göker et al. (2007) or one of our earlier studies
Phytophthora fragariae	DQ365741	DQ361234	Göker et al. (2007) or one of our earlier studies
Phytophthora heveae	DQ365742	DQ361235	Göker et al. (2007) or one of our earlier studies
Phytophthora infestans	DQ365743	AF119602	Göker et al. (2007) or one of our earlier studies
Phytophthora insolita	DQ365744	DQ361236	Göker et al. (2007) or one of our earlier studies
Phytophthora lateralis	DQ365745	DQ361237	Göker et al. (2007) or one of our earlier studies
Phytophthora litchii	DQ365746	AY035531	Göker et al. (2007) or one of our earlier studies
Phytophthora megasperma	DQ365747	DQ361238	Göker et al. (2007) or one of our earlier studies
Phytophthora multivesiculata	DQ365748	DQ361239	Göker et al. (2007) or one of our earlier studies
Phytophthora nemorosa	DQ365749	DQ361240	Göker et al. (2007) or one of our earlier studies
Phytophthora nicotianae	DQ365750	DQ361241	Göker et al. (2007) or one of our earlier studies
Phytophthora quercina	DQ365751	DQ361242	Göker et al. (2007) or one of our earlier studies
Phytophthora richardiae	DQ365752	DQ361243	Göker et al. (2007) or one of our earlier studies
Phytophthora sojae	DQ365753	DQ361244	Göker et al. (2007) or one of our earlier studies
Plasmopara angustiterminalis	EU826097	EU826113	This study
Plasmopara baudysii	EU826098	AY035517	LSU: Göker et al. (2007) or one of our earlier studies; COX: this study
Plasmopara densa	DQ365754	AY035525	Göker et al. (2007) or one of our earlier studies
Plasmopara megasperma	DQ365755	AY035516	Göker et al. (2007) or one of our earlier studies
Plasmopara nivea	DQ365756	AF119604	Göker et al. (2007) or one of our earlier studies
Plasmopara obducens	DQ365757	AY035522	Göker et al. (2007) or one of our earlier studies
Plasmopara pimpinellae	DQ365758	AY035519	Göker et al. (2007) or one of our earlier studies

(*Continued*)

Plasmopara pusilla	DQ365759	AY035521	Göker et al. (2007) or one of our earlier studies
Plasmopara viticola	DQ365760	AY035524	Göker et al. (2007) or one of our earlier studies
Plasmoverna pygmaea	DQ365761	AF119605	Göker et al. (2007) or one of our earlier studies
Plectospira myriandra	AF086694	AF218196	GenBank
Protobremia sphaerosperma	DQ365762	AY250150	Göker et al. (2007) or one of our earlier studies
Pseudoperonospora cubensis	EU826099	EU826114	This study
Pseudoperonospora humuli	DQ365763	AY035496	Göker et al. (2007) or one of our earlier studies
Pseudoperonospora urticae	DQ365764	AY035497	Göker et al. (2007) or one of our earlier studies
Pustula tragopogonis 1	EU826101	EU826116	This study
Pustula tragopogonis 2	EU826100	EU826115	This study
Pythiopsis cymosa	AF086689	DQ393490	GenBank
Pythium monospermum	DQ365765	AY035535	Göker et al. (2007) or one of our earlier studies
Pythium undulatum	DQ365766	AF119603	Göker et al. (2007) or one of our earlier studies
Pythium vanterpoolii	EU826102	EU826117	This study
Pythium vexans	DQ365767	DQ361245	Göker et al. (2007) or one of our earlier studies
Saprolegnia ferax	EU826103	EU826118	This study
Sapromyces elongatus	AF086700	AF235950	GenBank
Sclerophthora macrospora	EU826104	EU826119	This study
Sclerospora graminicola	DQ365768	AY035514	Göker et al. (2007) or one of our earlier studies
Thraustotheca clavata	AF086688	AF235951	GenBank
Viennotia oplismeni	DQ365769	AY035527	Göker et al. (2007) or one of our earlier studies
Albugo aff. gomphrenae	EU826093	EU826109	This study
Wilsoniana amaranthi 1	EU826091	EU826107	This study
Wilsoniana amaranthi 2	EU826090	EU826106	This study
Wilsoniana portulacae	EU826105	EU826120	This study

4

AN INTRODUCTION TO THE WHITE BLISTER RUSTS (ALBUGINALES)

Marco Thines
Institute of Botany 210, University of Hohenheim, Stuttgart, Germany

Hermann Voglmayr
Department of Systematic and Evolutionary Botany, University of Vienna, Vienna, Austria

4.1 THE ORDER ALBUGINALES—GENERAL ASPECTS

4.1.1 Biology of the White Blister Pathogens

The Albuginales are obligate biotrophic plant pathogens, and none of these organisms has been reported to be culturable without its respective host. Nath et al. (2001) noted that although the hyphae of *Albugo candida* grew out from infected mustard callus, they could obviously not take up enough nutrients to sustain their growth.

White blister disease caused by members of the Albuginaceae has been reported from a great variety of angiosperm hosts, which range from the basal order Piperales (*Albugo tropica*) to the highly complex order Asterales (e.g., *Pustula tragopogonis* and *Pustula spinulosa*). Recently, *Albugo macalpineana* has been described from Orchidaceae, which extends the host range of the Albuginaceae to the Liliidae. In general, the host ranges of downy mildews and white blister rusts are congruent (Dick, 2001, 2002a). The summary given by Dick (2002a) is not a complete compilation, as for example the Apiales are not given as harboring a white rust pathogen, although *Pustula hydrocotyles* is known from that order for a long time (Petrak, 1955). In fact, it may be assumed that the list of families previously reported to be parasitized by white blister rusts needs to be extended as our own observations from tropical

Oomycete Genetics and Genomics: Diversity, Interactions, and Research Tools
Edited by Kurt Lamour and Sophien Kamoun
Copyright © 2009 John Wiley & Sons, Inc.

specimens suggest. Although downy mildews seem to have their highest diversity in the temperate regions of the world (with the exception of the graminicolous downy mildews), the Albuginaceae seem to favor warm temperate to tropical climates (Dick, 2002b).

In contrast to the downy mildews, the white blister rusts produce their sporangia beneath the epidermis of their hosts, which make them independent from humid conditions for sporulation. Previously, it was thought that the blisters develop through a process that involves tearing off the epidermis from the mesophyll cells and later a rupturing of the epidermis because of the pressure exerted by the growing chains of sporangia beneath it (Gäumann, 1949; Webster and Weber, 2007). However, sporulation in the Albuginaceae is a highly sophisticated process (Heller and Thines, 2009). During sporulation, the epidermis of the host is enzymatically parted from the mesophyll without disrupting epidermal or mesophyll cells. Subsequently, sporogenous hyphae develop, which produce sporangia in a basipetal succession. The primary sporangia, which differ in various aspects from the secondary sporangia in all species of the Albuginaceae (Constantinescu and Thines, 2006), serve to dissolve the epidermal layer enzymatically once the secondary sporangia are ready for dispersal. As a consequence, the pustule develops fissures, and sporangia are disseminated as a white, flour-like powder. Remarkably, the sporangia are hydrophobic, which may be effected by the verrucose, striate, or reticulate surface ornamentation visible in scanning electron microscopy (SEM) (Thines and Spring, 2005; Voglmayr and Riethmüller, 2006).

The highly organized sporulation process and the fact that white blister rusts often grow systemically until flowering without causing visible symptoms on the host demonstrate that the Albuginaceae are well-adapted plant pathogens, with a much more complex and higher level of adaptation compared with the downy mildews.

4.1.2 Phylogeny of the Albuginaceae

The first molecular phylogenetic reconstruction that contains a member of the Albuginaceae revealed that the Albuginaceae are only distantly related to the downy mildews (Peronosporaceae), which were their assumed sister family (Cooke et al., 2000). Also, Riethmüller et al. (2002) and Hudspeth et al. (2003) came to similar conclusions, and Hudspeth et al. (2003) suggested the introduction of a separate order for the white blister rusts. The order Albuginales was formally introduced in Thines and Spring (2005), who considered morphological, cytological, and molecular phylogenetic data.

Obligate biotrophic dependence on angiosperms has evolved independently in the Peronosporaceae and the Albuginaceae, and the phylogenies of Hudspeth et al. (2003) and Thines et al. (2008) demonstrate that the Albuginales form a distinct clade at the basis of the "Peronosporacean Galaxy" (Sparrow, 1976). For the downy mildews, a close relative, which is hemibiotrophic and culturable, still exists with *Phytophthora* (Göker et al., 2007; Thines et al., 2007). *Phytophthora*, as currently circumscribed, is a paraphyletic assemblage with the downy mildews as a crown group. Even the monophyly of the downy mildews could so far not be ascertained, as the high support for downy mildew monophyly observed by Göker et al. (2007) was an artifact of a too narrow search radius [see discussion in Thines et al. (2007)]. This points out that obligate biotrophy in the downy mildews has evolved recently in comparison with the white blister rusts, which are an ancient phylogenetic lineage, without any closely related hemibiotrophic species. As no closely related hemibiotrophic species is known for the Albuginales, and the different lineages of the Albuginaceae have diverged early in oomycete evolution, it can be assumed that obligate biotrophy in this family has developed in ancient times and that the level of adaptation is much higher in this family, compared with the Peronosporaceae.

Already Riethmüller et al. (2002) noted that the white blister rusts contain several highly distinct lineages. Based on phylogeny, morphology, and ultrastructure, three of the four major lineages were assumed to deserve generic rank by Thines and Spring (2005). The monophyly of the newly described genera *Pustula* (white blister rusts of the Asteridae) and *Wilsoniana* (white blister rusts of a variety of Caryophyllidae) was confirmed by Voglmayr and Riethmüller (2006). A simplified tree of the oomycetes, drawing together the results of Hudspeth et al. (2003), Voglmayr and Riethmüller (2006), Göker et al. (2007), and Thines et al. (2008), is given in Fig. 4.1.

4.1.3 Taxonomy of a Character-Poor Group of Organisms

With *Aecidium candidum* (today *A. candida*), Persoon (in Gmelin, 1792) described the first species of the Albuginaceae. This species was first transferred to *Uredo* (Persoon, 1801), until it was accommodated in a genus of its own, *Albugo* (de Roussel, 1806). *Albugo* is the first genus that had been described in the Oomycota more than 30 years before the first downy mildew genus *Peronospora* was described by Corda (1837). During the following century, only a few new species were described, and Wilson (1907) listed 13 species of *Albugo*, which were mostly considered to be host–family specific. Biga (1955) recognized 40 species of *Albugo* and with the exception of the parasites of Convolvulaceae, he accepted only very few species on a single host family. According to Biga (1955), it is almost impossible to distinguish species considering only the morphology of the conidia. The similar morphology of sporangia and the uniformity of the club-shaped sporogenous hyphae in all species of the Albuginaceae is the main reason why, in contrast to the downy mildews, comparatively few species were introduced in the twentieth century. The monogeneric state of the Albuginaceae had not been questioned until the revision of Thines and Spring (2005). In the most recent monograph of the Albuginaceae (Choi and Priest, 1995), 44 species were distinguished in a tabular key, mostly following the species concept of Biga (1955). Choi and Priest (1995)

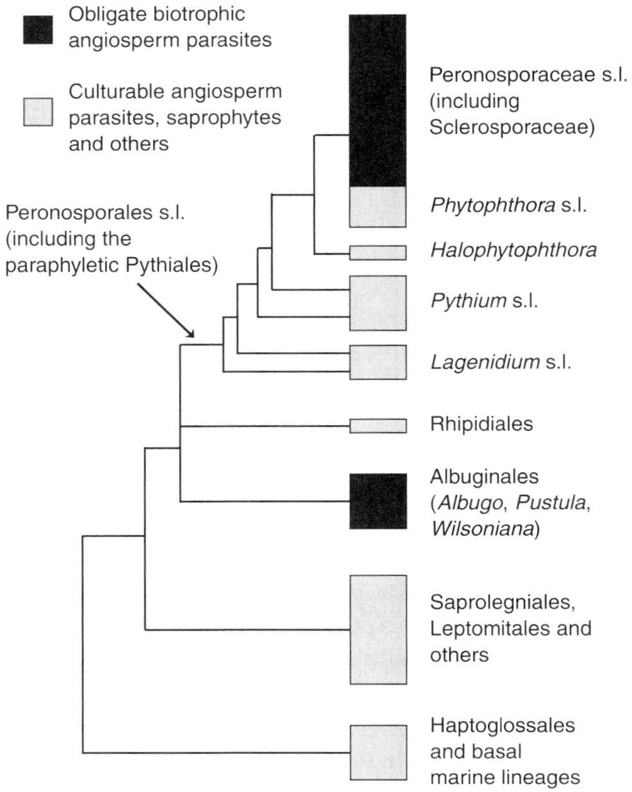

FIG. 4.1 Simplified phylogenetic tree for the Oomycota.

opted for a more detailed investigation of the species of the Albuginaceae and pointed out that the investigation of oospore ornamentation using SEM might give conclusive results regarding species delimitation. Thines and Spring (2005) demonstrated that the ultrastructure of the sporangia as observed in SEM reveals useful features that can be used to identify the major lineages of the Albuginaceae. Voglmayr and Riethmüller (2006) confirmed these results and showed that oospore ornamentation is valuable in distinguishing closely related species. By investigating the surface ornamentation of the resting organs in light microscopy (LM) and SEM, they could demonstrate that two distinct species *Wilsoniana amaranthi* and *Wilsoniana bliti* are present on hosts in the genus *Amaranthus*. Also for the white blister rusts on Brassicaceae, oospore surface ornamentation revealed characteristics useful for morphological distinction between phylogenetically distinct lineages, which led to the introduction of two new species of *Albugo* on Brassicaceae (Choi et al., 2007; Choi et al., 2008). In Fig. 4.2 an overview of the sporangial and oospore diversity in the Albuginaceae is given.

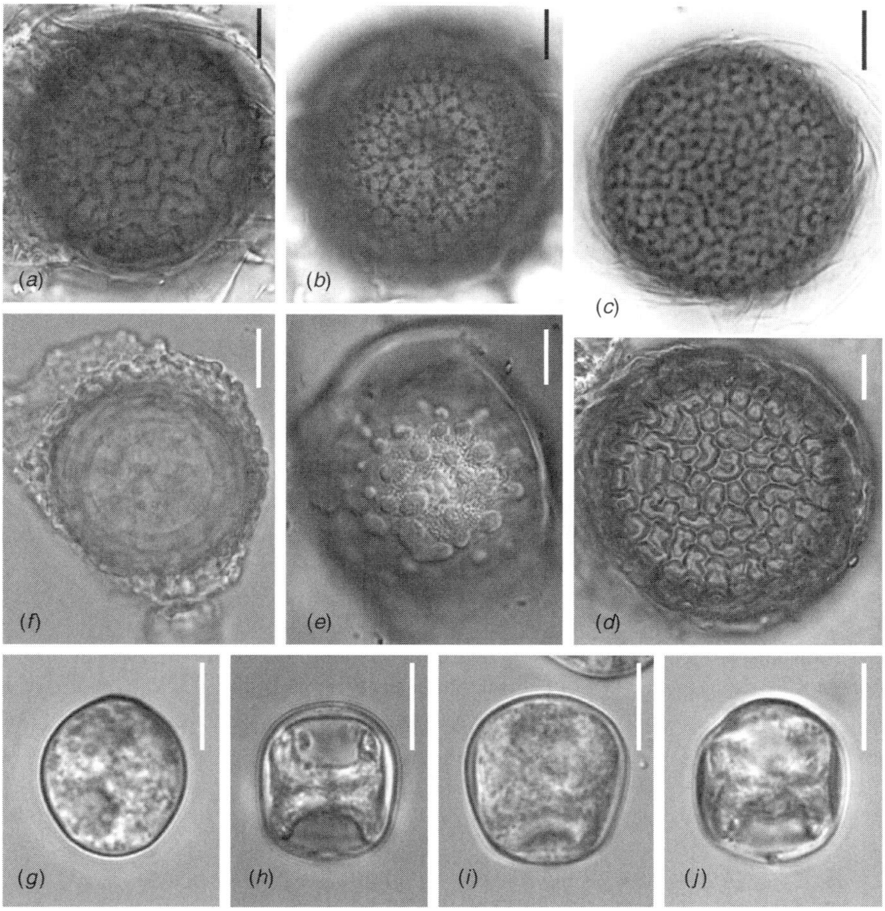

FIG. 4.2 Oogonia (a–f) and secondary sporangia (g–j) of selected Albuginaceae. Note the smooth (f), reticulate (d), coarsely (a) or finely reticulate verrucose (b, c) and tuberculate (e) oospore walls; the oogonium walls are smooth (a–e) or wrinkled verrucose (f). (a) *Albugo macalpineana*; (b, j) *Pustula tragopogonis*; (c) *Albugo lepigonii*; (d, i) *Wilsoniana portulacae*; (e, g) *Albugo candida*; (f, h) *Albugo ipomoeae-panduratae*. Bar = 10 μm.

4.2 *ALBUGO S.STR.* ON BRASSICACEAE

4.2.1 Biology and General Aspects

The most commonly known species of *Albugo* is probably *A. candida*, which is parasitic to a wide range of Brassicaceae. Especially on *Capsella bursa-pastoris* (shepherd's purse), the pathogen is common, and throughout the year, almost any larger population of the host will exhibit white blister symptoms on the leaves or stems of some plants. The pathogen can be found as early as March and as late as November and from Iceland to Australia. Besides *C. bursa-pastoris*, which is the type host (Choi et al., 2007), *A. candida* can be found on more than 200 species of several dozen genera of the Brassicaceae (Biga, 1955; Saharan and Verma, 1992) and also on Capparaceae. Its hosts include economically important crop species, such as *Brassica juncea, Brassica nigra, Brassica napus, Brassica oleracea*, and *Raphanus sativus*, and also ornamental plants like *Aurinia saxatilis, Lunaria annua*, and *Arabis alpina*. Notably, the model plant *Arabidopsis thaliana* is commonly infected by *Albugo* in wild populations.

In Brassicaceae, it has been shown that infections may be asymptomatic until flowering and that infection can be passed on to progeny without visible symptoms on the mother plants (Jacobson et al., 1998). In general, the widespread assumption that *Albugo* usually causes distortion and hypertrophy of stems in Brassicaceae needs thorough revision, as judging from our own observations, hypertrophy and distortion of stems may not be the most typical symptom of infection with *Albugo s.str*. Usually, it only reaches dramatic stages when a coinfection of *Albugo* and *Hyaloperonospora* is present. Nonetheless, *A. candida* may cause stunting and flower malformation when symptoms develop in the inflorescence of the plants, which causes a complete destruction of the infected organs. Usually, plants infected by *A. candida* exhibit only minor symptoms in parts less crucial for reproduction, and it is likely that infections remain cryptic for a long time and are often overlooked.

Recent studies (Voglmayr and Riethmüller, 2006; Choi et al., 2006, 2007, 2008) have shown that apart from *A. candida s.str*., several other species of *Albugo* on Brassicaceae exist, which had previously been overlooked. It is likely that more than a dozen distinct species are found on Brassicaceae and related families. Recent investigations have shown that at least three distinct species reside in *Cardamine* alone (Thines et al., unpublished results). The ecology of these species is still unknown, and their specific niche remains unknown. However, it is evident that the level of specialization in the white blister rusts parasitic to Brassicaceae is much higher than previously thought (Biga, 1955; Choi and Priest, 1995). We have to remain wary about the exact diversity, evolutionary history, and phylogenetic structure of the white blister rusts on Brassicaceae because isolates from less than 5% of the Brassicaceae species reported to be a host for *Albugo* have been investigated in molecular phylogenies. Clearly, more studies in this species complex are needed to shed light on these topics.

4.2.2 Species Concepts — Why Species Concepts for Downy Mildews Do Not Apply to White Blister Rusts

It can be ascertained that the concepts of species in *Albugo* on Brassicaceae cannot be extrapolated from species concepts for the downy mildews. This becomes apparent with the white blister rusts of *Capsella*. In Europe and most other parts of the world, *C. bursa-pastoris* is infected by *A. candida*. In Korea, however, the same plant species is parasitized by *A. koreana* (Choi et al., 2007). *Albugo koreana* has not been found on any other host plant so far, but *A. candida* is parasitic to a great variety of Brassicaceae species. This is markedly distinct from the situation in the downy mildews for example, in *Hyaloperonospora*, where the occurrence of two or even more specialised species with narrow host ranges on the same host genus is observed several times (Göker et al., 2004). Previously, it was thought that *Hyaloperonospora* has a wide host range (Yerkes and Shaw, 1959; Constantinescu and Fatehi, 2002) similar to *A. candida*. However, the investigations of Choi et al. (2003) and Göker et al. (2004) have demonstrated that *Hyaloperonospora* species are usually restricted to a very limited host range. The niche adaptation for *Albugo* on Brassicaceae should therefore not be limited to host specialization; thus, species concepts need to be reconsidered for these pathogens.

4.2.3 Major Species Clusters Within *Albugo s.str*

Albugo s.str., that is, *Albugo* parasitic to Brassicaceae and allied families, consists of two main clusters: *A. candida s.str.* and a cluster that contains a variety of distinct lineages. Besides *A. candida*, especially *Albugo lepidii* seems to be widespread, as well as one of the unnamed species of *Albugo* on *Cardamine*. The other distinct lineages, which include *A. koreana* and several species that await scientific description, have so far only been found in smaller geographic areas, although it might be expected that their distribution could correspond with the distribution of their respective hosts. A simplified phylogenetic tree that shows the relationships of the lineages of *Albugo s.str.* is given in Fig. 4.3.

4.3 OTHER LINEAGES OF THE ALBUGINACEAE

4.3.1 *Albugo s.l.* on Convolvulaceae

Apart from *Albugo s.str.* parasitic to Brassicaceae and related families, the white blister rusts of the Convolvulaceae comprise the second major lineage currently classified within the genus *Albugo*. White blister rusts commonly occur on numerous species of Convolvulaceae, which include the economically important *Ipomoea aquatica* (water spinach) and *Ipomoea batatas* (sweet potato). *Albugo* on members of the morning glory family exhibits exceptional wall thickenings in both sporangia and oospores (Mims and Richardson, 2003). These are probably a result of adaptation to dry subtropical climates, which reflects the distribution of the host family for this group of species.

Several species have been described as parasites of the Convolvulaceae, of which Biga (1955) accepted five in his monograph, and Choi and Priest (1995) included six in their tabular key. This is the highest number of individual species so far reported from any host family. Traditionally, *Albugo*

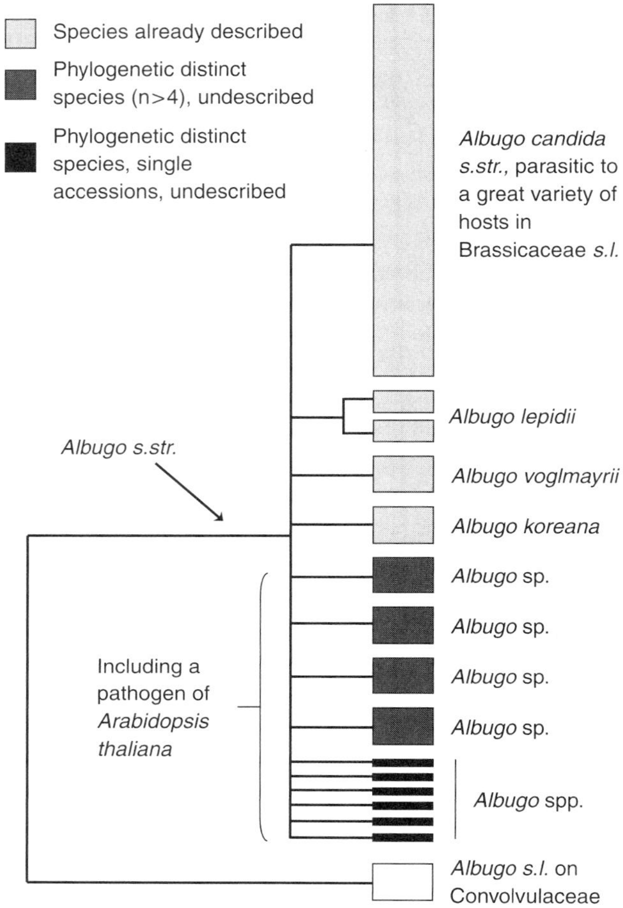

FIG. 4.3 Simplified phylogenetic tree for *Albugo s.str.*, based on both published and unpublished data.

ipomoeae-panduratae is considered to be the sole species infecting Convolvulaceae, which has a wide host range and distribution that ranges from North America to the inner tropics (Murkerji and Crichett, 1975; Anonymous, 1986). However, the investigations of Voglmayr and Riethmüller (2006) have shown that at least *Albugo evolvuli* and *A. ipomoeae-panduratae* are distinct species. For the other species of *Albugo* parasitic to Convolvulaceae, molecular phylogenetic data are still missing. Choi and Priest (1995) concluded that a revision of these pathogens is desperately needed, and Voglmayr and Riethmüller (2006) substantiated this demand by demonstrating that Wilson (1907) most likely mistook the persistent oogonial wall in *A. ipomoeae-panduratae* for the oospore wall, which is, in contrast to his claims, smooth in the specimens he investigated.

4.3.2 Pustula *on Asterids*

The genus *Pustula* has been erected by Thines (Thines and Spring, 2005) to accommodate the white blister rusts of Asterids. The type species is *Pustula chardiniae* parasitic to *Chardinia orientalis* (Asteraceae). *Pustula* differs from *Albugo s.str.* by the surface ornamentation of the sporangia, which consists of large, rounded verrucae in *A. candida* but is irregularly striate to reticulate in *Pustula* (Thines and Spring, 2005; Voglmayr and Riethmüller, 2006). In addition, oospores so far reported from *Pustula* are never bearing rounded tubercules, neither single nor fused, or smooth oospores, which are typical for *Albugo* on Brassicaceae and Convolvulaceae, respectively. Also, the sporangia of *Pustula* differ from those of *Albugo* in having conspicuous equatorial wall thickenings, which often appear lenticular in mature sporangia (Thines and Spring, 2005; Constantinescu and Thines, 2006, Heller and Thines, 2009). When dry, the equatorial ring remains rigid, whereas the other parts of the sporangia collapse, which results in an ash-tray-like appearance. In addition to these morphological characteristics, several aspects of fertilization clearly distinguish *Pustula* from *Albugo* and *Wilsoniana* (Stevens, 1899, 1901, 1904; Ruhland, 1904; Davis, 1904).

Of the species described in Asteraceae [for a synopsis, see Biga (1955)], only *P. tragopogonis* is widely accepted and thought to be parasitic to a great variety of asteraceous hosts (Biga, 1955; Choi and Priest, 1995), which range from *Scorzonera hispanica* (salsify) to *Helianthus annuus* (annual sunflower). In contrast to this assumption, the studies of Thines and Spring (2005) as well as Voglmayr and Riethmüller (2006) revealed a considerable level of genetic variation within *P. tragopogonis s.l.*, and it seems highly likely that, similar to *Albugo s.str.*, several hitherto overlooked species will be discovered. The most notable of these is the economically important white blister rust of *H. annuus*, which is phylogenetically distinct from both *P. tragopogonis s.str.* and *P. spinulosa*. *Pustula* on cultivated sunflower is a destructive pathogen in South Africa (van Wyk et al., 1995). It has also been reported from Australasia (Allen and Brown 1980, 1982), South America (Delhey and Kiehr-Delhey, 1985), and North America (Gulya, 2002), and it has been recently introduced into Germany (Thines et al., 2006). The spread of the pathogen is most likely linked to international seed trade with infected seeds, which is supported by the finding of Viljoen et al. (1999), who could detect *P. tragopogonis s.l.* inside sunflower seeds. Besides *P. chardiniae, P. tragopogonis*, and *P. spinulosa*, only one additional species *P. hydrocotyles* is currently classified within the genus *Pustula*.

Similar to the white blister pathogens of Convolvulaceae, the species described as parasites of Asteraceae are in need of thorough revision. It is likely that the other species of Albuginaceae parasitic to Asteraceae will also form distinct lineages within the genus *Pustula*. In addition, the genus *Pustula* can be found on a wide variety of Asterids and includes not only parasites of the closely related orders Apiales and Asterales, but also hosts in Gentianales and Solanales (Thines et al., unpublished results). Judging from the preliminary data of

Voglmayr and Riethmüller (2006) as well as our own unpublished results, it is likely that even more species exist in Asteraceae than in Brassicaceae.

4.3.3 *Wilsoniana* on Caryophyllales

The genus *Wilsoniana* contains several white blister pathogens parasitic to members of the Caryophyllales. *Wilsoniana* differs from *Albugo* by the ornamentation of the sporangia, which are irregularly striated and often have small rounded protuberances in the former genus. In contrast to *Albugo*, large, rounded protuberances have so far not been observed in any species parasitic to Caryophyllales. Sporangia in *Wilsoniana*, especially in the type species *Wilsoniana portulacae*, are often broadly pyriform and exhibit a slight crescent-moon-like wall thickening covering large proportions of the lateral walls and are most prominent above the equator of the sporangia. In addition, the fertilization in *Wilsoniana* differs largely from the fertilization in *Albugo* and also *Pustula* (Stevens, 1899, 1901, 1904; Ruhland, 1904; Davis, 1904).

Wilsoniana species are of little economic importance, although they can cause some harm in ornamental species like *Amaranthus cruentus* and *Boerhaavia* species, as well as in a limited range of crops, such as *Amaranthus bicolor*. The phylogenetic placement of *Albugo occidentalis* parasitic to *Spinacia oleracea* (spinach) is still unsettled. More investigations are needed to evaluate whether this species, along with its sister taxon *Albugo lepigonii*, is to be classified within *Wilsoniana*, as morphology suggests or constitutes a distinct lineage, as supported by cytological investigations of Stevens (1904). The white blister rust species parasitic to Caryophyllales have a worldwide distribution, although the highest diversity is observed in the tropics and subtropics.

Similar to the *Albugo* and *Pustula*, the existence of several distinct species on the same host genus or even the same host family has been doubted also for the species now classified within *Wilsoniana*. For example, neither Biga (1955) nor Choi and Priest (1995) accepted *Albugo amaranthi* and *Albugo bliti* as distinct species and in Thines and Spring (2005), only *A. bliti* was transferred to *Wilsoniana*. However, Voglmayr and Riethmüller (2006) have shown that these species can be distinguished both phylogenetically and on grounds of oospore morphology. As a consequence, *A. amaranthi* was later formally transferred to *Wilsoniana* by Choi et al. (2007). It can be expected that similar to the two lineages of the Albuginaceae discussed above, several species of *Wilsoniana* await discovery.

4.4 PRESENT STATE OF PHYLOGENETIC KNOWLEDGE AND NOTEWORTHY SPECIES OF UNKNOWN AFFINITY

So far, molecular phylogenetic investigations have focused on the white blister pathogens of the orders Asterales, Brassicales, Caryophyllales, and Solanales. However, numerous additional orders are infected by white blister pathogens: the Lamiales, Gentianales, and the unplaced family Boraginaceae within the Asteridae; the Rosales and Fabales within the Rosidae; and the unplaced Saxifragales and Ranunculales within the Eudicots *sensu* APG II (Bremer et al.,

2003). In addition, also the monocotyledonous order Asparagales and the basal angiosperm order Piperales include hosts parasitized by white blister pathogens. In a recent revision of the specimens deposited in the Herbarium G (Herbarium of the University of Geneva, Switzerland), hosts in families so far not recognized as hosts for white blister rusts have been found (Thines, unpublished data). It can be expected that in several, especially tropical, families, white blister disease will be newly reported in the near future. In general, the Albuginaceae have their highest diversity within the orders Asterales, Brassicales, Caryophyllales, and Solanales, on which more than three quarters of the hitherto described species parasitize.

Some species described are widespread, yet of uncertain taxonomic position. These include the economically important white blister pathogen of spinach. The weakly supported phylogenetic affinity to *Wilsoniana* found by Thines and Spring (2005) is contrasted by the moderately supported phylogenetic affinity to *Pustula* observed by Voglmayr and Riethmüller (2006). Also, Choi et al. (2007) could not infer a significantly supported phylogenetic placement of this pathogen.

The four species parasitic to hosts that do not belong to any of the orders of the Core-Eudicots *sensu* APG II (Bremer et al., 2003) are of particular interest as these might shed light on the evolutionary history of the Albuginaceae. Perhaps the most interesting species in this context is the recently described *A. macalpineana* (Walker and Priest, 2007) from *Pterostylis* (Greenhood Orchids). *A. macalpineana* is the only species so far reported from a monocotyledonous host, which has been carefully documented and deposited in an international herbarium. The phylogenetic position of the species is so far unresolved, and no recent collection is available for phylogenetic investigations, as all specimens of this species have been collected more than 80 years ago.

Other noteworthy species are *Albugo keeneri* and *Albugo eomeconis*, which are the sole species parasitic to the basal Eudicot family Papaveraceae. The distribution and taxonomic affinity of these species are so far unknown, and reports of these species are limited.

Only one species of *Albugo* has so far been reported from basal angiosperms, namely *A. tropica*, which is parasitic to a host in Piperales. Of *A. tropica*, only the type collection is known, which is deposited in FH (Farlow Herbarium, Harvard University, Cambridge, MA) and of which an isotype exists in G (Herbarium of the University of Geneva, Switzerland). Unusual for any other species of *Albugo* so far investigated is not only the isolated position of its host but also the bowl-like sporangium collapsing pattern in this species. Unfortunately, the phylogenetic position of this species could not yet be resolved, and recent collections are highly warranted for phylogenetic investigations.

4.5 FUTURE PERSPECTIVES

4.5.1 White Flecks in White Rust Evolution

So far, many aspects of white blister rust evolution are unresolved. It is known that the Albuginaceae occupy a basal position in all molecular phylogenetic investigations of the Peronosporomycetes published so far (Riethmüller et al., 2002; Hudspeth et al., 2003; Thines et al., 2008). However, the closest nonbiotrophic relatives of this group are still unclear, and in contrast to the downy mildews, affinities to other hemibiotrophic species are unknown. The white blister rusts might represent an ancient lineage, which has coevolved and radiated together with flowering plants. This is unlike the Peronosporaceae (including *Phytophthora*), which are the product of a more recent evolution from saprotrophic or necrotrophic to obligate biotrophic pathogens.

In Thines and Spring (2005), the Albuginaceae have been placed in an order and subclass of their own, because of their distinct morphology and molecular phylogenetic position compared with the Peronosporales and Pythiales. Whether the subclass Albuginomycetidae will be applied in the future is mainly a question of the level at which the taxonomic units of the Oomycota are separated. The Albuginales clearly belong to the "Peronosporacean Galaxy" (Sparrow, 1976) in a broader sense and can be opposed to the Peronosporales (including Peronosporaceae and Pythiales). The Rhipidiales (or Rhipidiomycetidae) seem to be basal to the Albuginales as revealed in the previous chapter (Chapter 3) of this book. However, as especially among Pythiales and Rhipidiales sampling is rather sparse, more sampling and sequencing of multiple genes are necessary to clarify the relationships of Albuginales, Peronosporales, and Rhipidiales.

Current knowledge of the phylogeny of individual Albuginaceae species is fragmentary. Only about a quarter of the species so far described within the Albuginaceae have been investigated in molecular phylogenetic analyses. In particular, the phylogenetic affinities of the species parasitic to Core-Eudicots should be investigated to evaluate the generic concept expressed in Thines and Spring (2005). To unravel the evolutionary history of the white blister rusts in general, the elucidation of the phylogenetic positions of the parasites of the more basal lineages of the angiosperms is of crucial importance.

Little is known about the driving forces in white blister rust diversification and radiation. It is also unclear which specific niche is occupied by *A. candida*, in relation to the more specialized species of this genus and whether these species are mutually exclusive for example, whether *A. koreana* and *A. candida* may co-occur in the same geographic region or even on the same host individual. Also, the host range of *A. candida s.str.* and the level of host specificity of specialized strains is so far unknown. From the results obtained so far, it can be concluded that *A. candida* may be expected to occur on almost any plant of the Brassicaceae and related families.

An additional unresolved aspect concerns the mutual influence of downy mildews and white blister rusts, which, especially in the case of *Albugo s.str.* and *Hyaloperonospora*, commonly co-occur on the same individual host plant. It is likely that these frequently occurring associations have left their footprints in the phylogeny and evolutionary strategies of these pathogens.

4.5.2 A Plethora of Unknown Species

Little is known about species diversity in the composite species *P. tragopogonis* and *A. candida*. Judging from the results published so far (Thines and Spring, 2005; Voglmayr and Riethmüller, 2006; Choi et al., 2006, 2007, 2008), there can be little doubt that the diversity of the Albuginaceae has been greatly underestimated in the past. If the findings of the studies listed above are roughly extrapolated to the overall biodiversity of the host plants, one has to come to the conclusion that both *P. tragopogonis* and *A. candida* may contain several dozen species that await discovery. It will be the privilege of future studies to unravel the true diversity in these composite species and to elucidate their evolutionary history, radiation, and niche adaptation.

4.5.3 Prospects for Evolutionary Genomics

The high level of adaptation and the long evolutionary history as obligate biotrophic pathogens render the Albuginaceae especially suited for investigations on the evolution of effectors involved in pathogenicity and adaptation to a specific environment. In this context, it will be interesting to understand how *A. candida* manages to parasitize a great variety of Brassicaceae, whereas other species are apparently restricted to a limited host range (Choi et al., 2007, 2008). The elucidation of this question by comparative genomics, of *A. candida* and *A. koreana* for example, will shed light on fundamental processes involved in host–pathogen interaction.

Another opportunity to understand the pathogen adaptation to their host and to identify the key targets in pathogen establishment and adaptation is the fact that the genetic model plant *A. thaliana* can by infected by both groups of obligate biotrophic oomycetes, white blister, and downy mildew pathogens (Eulgem et al., 2004; Borhan et al., 2008). By comparison of the targets in downy mildew and in white blister rust host–pathogen interaction (Borhan et al., 2001), basic insights into the defense mechanisms of plants and their evasion by the pathogens will be gained. Such comparative studies might also help to breed plants with a high level of resistance to various pathogens through selection of favorable key genes important for pathogen defense on a basic level.

ACKNOWLEDGMENTS

M. T. wishes to thank the German Science Foundation (DFG) and the Landesstiftung Baden-Württemberg (Elite Program for Postdocs) for financial support.

REFERENCES

Allen SJ, Brown JF 1980. White blister, petiole greying and defoliation of sunflowers caused by *Albugo tragopogonis*. Australas Plant Pathol 9:809.

Allen SJ, Brown JF 1982. White blister and petiole blight of sunflowers caused by *Albugo tragopogonis*. Proceedings of the 10th International Sunflower Conference. pp. 153–156.

Anonymous 1986. *Albugo ipomoeae-panduratae*. Distribution Maps of Plant Diseases 568. Egham, UK: CABI.

Biga MLB 1955. Riesaminazione delle specie del genre *Albugo* in base alla morfologia dei conidi. Sydowia 9:339–358.

Borhan MH, Brose E, Beynon JL, Holub EB 2001. White rust (*Albugo candida*) resistance loci on three *Arabidopsis* chromosomes are closely linked to downy mildew (*Peronospora parasitica*) resistance loci. Molec Plant Pathol 2:87–95.

Borhan MH, Gunn N, Cooper A, Gulden S, Tör M, Rimmer SR, Holub EB 2008. WRR4 encodes a TIR-NB-LRR protein that confers broad-spectrum white rust resistance in *Arabidopsis thaliana* to four physiological races of *Albugo candida*. Molec Plant Microbe Interact 21:757–768.

Bremer B, Bremer K, Chase MW, Reveal JL, Soltis DE, Soltis PS, Stevens PF, Anderberg AA, Fay MF, Goldblatt P, et al. 2003. An update of the Angiosperm Phylogeny Group classification for the orders and families of flowering plants: APG II. Bot J Linn Soc 141:399–436.

Choi D, Priest MJ 1995. A key to the genus *Albugo*. Mycotaxon 53:261–272.

Choi Y-J, Hong S-B, Shin H-D 2003. Diversity of the *Hyaloperonospora parasitica* complex from core brassicaceous hosts based on ITS rDNA sequences. Mycol Res 107:1314–1322.

Choi Y-J, Hong S-B, Shin H-D 2006. Genetic diversity within the *Albugo candida* complex (Peronosporales, Oomycota) inferred from phylogenetic analysis of ITS rDNA and COX2 mtDNA sequences. Molec Phylogene Evol 40:400–409.

Choi Y-J, Shin H-D, Hong S-B, Thines M 2007. Morphological and molecular discrimination among *Albugo candida* materials infecting *Capsella bursa-pastoris* worldwide. Fungal Divers 27:11–34.

Choi Y-J, Shin H-D, Thines M 2008. Evidence for uncharted biodiversity in the *Albugo candida* complex, with the description of a new species. Mycol Res 112:1327–1334.

Cooke DEL, Drenth A, Duncan JM, Wagels G, Brasier CM 2000. A molecular phylogeny of *Phytophthora* and related oomycetes. Fungal Genet Biol 30:17–32.

Constantinescu O, Fatehi J 2002. *Peronospora*-like fungi (Chromista, Peronosporales) parasitic on Brassicaceae and related hosts. Nova Hedwigia 74:291–338.

Constantinescu O, Thines M 2006. Dimorphism of sporangia in Albuginaceae (Chromista, Peronosporomycetes). Sydowia 58:178–190.

Corda ACJ 1837. Icones fungorum hucusque cogitorum 1:20.

Davis BM 1904. The relationships of sexual organs in plants. Bot Gazette 38: 241–264.

Delhey R, Kiehr-Delhey M 1985. Symptoms and epidemiological implications associated with oospore formation of *Albugo tragopogonis* on sunflower in Argentina. Proceedings of the 9th International Sunflower Conference. pp. 455–457.

Dick MW 2001. *Straminipilous Fungi: Systematics of the Peronosporomycetes Including Accounts of the Marine Straminipilous Protists, the Plasmodiophorids and Similar Organisms*. Kluwer Academic Publishers. Dordrecht, Germany.

Dick MW 2002a. Binomials in the Peronosporales, Sclerosporales and Pythiales. In: Spencer-Phillips PTN, Gisi U, Lebeda A, editors. Advances in downy mildew research Vol. 1. Kluwer Academic Publishers. Dordrecht, Germany. pp. 225–265.

Dick MW 2002b. Towards an understanding of the evolution of the downy mildews. In: Spencer-Phillips PTN, Gisi U, Lebeda A, editors. Advances in downy mildew research Vol. 1. Kluwer Academic Publishers, Dordrecht, Germany. pp. 1–57.

Eulgem T, Weigman VJ, Chang H-S, McDowell JM, Holub EB, Glazebrook J, Zhu T, Dangl JL 2004. Gene expression signatures from three genetically separable resistance gene signaling pathways for downy mildew resistance. Plant Physiol 135:1129–1144.

Gäumann E 1949. Die Pilze. Basel: Birkhäuser.

Gmelin JF 1792. Caroli a Linné, Systema Naturae per Regna Tria Naturae, Secundum Classes, Ordines, Genera, Species, cum Characteribus, Differentiis, Synonymis, Locis, 2.

Göker M, Riethmüller A, Voglmayr H, Weiß M, Oberwinkler F. 2004. Phylogeny of *Hyaloperonospora* based on nuclear ribosomal internal transcribed spacer sequences. Mycol Prog 3:83–176.

Göker M, Voglmayr H, Riethmüller A, Oberwinkler F 2007. How do obligate parasites evolve? A multi-gene phylogenetic analysis of downy mildews. Fungal Genet Biol 44:105–122.

Gulya TJ, Virany F, Appel J, Jardine D, Schwartz HF, Meyer R 2002. First report of *Albugo tragopogonis* on cultivated sunflower in North America. Plant Dis. 86:559.

Heller A, Thines M 2009. Evidence for the importance of enzymatic digestion of epidermal walls during subepidermal sporulation and pustule opening in white blister rusts (Albuginaceae). Mycol Res, In press.

Hudspeth DSS, Hudspeth MES, Stenger D 2003. A *cox2* phylogenetic hypothesis for the downy mildews and white rusts. Fungal Divers 13:47–57.

Jacobson DJ, LeFebvre SH, Ojerio RS, Berwald N, Heikkinen E 1998. Persistent, systemic, asymptomatic infections of *Albugo candida*, an oomycete parasite, detected in three wild crucifer species. Can J Bot 76:739–750.

Mims CW, Richardson EA 2003. Ultrastructure of the zoosporangia of *Albugo ipomoeae-panduratae* as revealed by conventional chemical fixation and high pressure freezing followed by freeze substitution. Mycologia 95:1–10.

Mukerji KG, Critchett C 1975. *Albugo ipomoeae-panduratae*. IMI Descriptions of Fungi and Bacteria No. 46. Egham: CABI.

Nath MD, Sharma SL, Kant U 2001. Growth of *Albugo candida* infected mustard callus in culture. Mycopathologia 152:147–153.

Persoon CH 1801. Synopsis methodica fungorum. Göttingen.

Petrak F 1955. Neue Mikromyzeten der Australischen Flora. Sydowia 9:559–570.

Riethmüller A, Göker M, Weiß M, Oberwinkler F, Voglmayr H 2002. Phylogenetic relationships of the downy mildews (Peronosporales) and related groups based on nuclear large subunit ribosomal DNA sequences. Mycologia 94:834–849.

de Roussel HFA 1806. Flore du Calvados. 2nd ed. Caen.

Ruhland W 1904. Studien über die Befruchtung der *Albugo lepigoni* etc. Jahrbücher der Wissenschaftlichen Botanik 39:135–166.

Saharan GS, Verma PR 1992. White rusts: a review of economically important species. Ottawa, CA: IDRC.

Sparrow FK 1976. The present status and classification of biflagellate fungi. In: Gareth-Jones EBG, editor. Recent advances in aquatic mycology. Elek Press, London UK. pp. 213–222.

Stevens FL 1899. The compound oosphere of *Albugo bliti*. Bot Gazette 28:149–176.

Stevens FL 1901. Gametogenesis and fertilization in *Albugo*. I-III. Bot Gazette 32:77–98.

Stevens FL 1904. Oogenesis and fertilization in *Albugo ipomoeae-panduranae*. Bot Gazette 38:300–302.

Thines M, Göker M, Oberwinkler F, Spring O 2007. A revision of *Plasmopara penniseti*, with implications for the host range of the downy mildews with pyriform haustoria. Mycol Res 111:1377–1385.

Thines M, Göker M, Telle S, Ryley M, Mathur K, Narayana YD, Spring O, Thakur RP 2008. Phylogenetic relationships of graminicolous downy mildews based on *cox2* sequence data. Mycol Res 112:345–351.

Thines M, Spring O 2005. A revision of *Albugo* (Chromista, Peronosporomycetes). Mycotaxon 92:443–458.

Thines M, Zipper R, Schäuffele D, Spring O 2006. Characteristics of *Pustula tragopogonis* (syn. *Albugo tragopogonis*) newly occurring on cultivated sunflower in Germany. J Phytopathol 154:88–92.

Viljoen A, van Wyk PS, Jooste WJ 1999. Occurrence of the white rust pathogen, *Albugo tragopogonis*, in seed of sunflower. Plant Dis 83:77.

Voglmayr H, Riethmüller A 2006. Phylogenetic relationships of *Albugo* species (white blister rusts) based on LSU rDNA sequence and oospore data. Mycol Res 110:75–85.

Walker J, Priest MJ 2007. A new species of *Albugo* on *Pterostylis* (Orchidaceae) from Australia: confirmation of the genus *Albugo* on a monocotyledonous host. Australas Plant Pathol 36:181–185.

Webster J, Weber RWS 2007. Introduction to fungi, 3rd Ed. Cambridge University Press, Cambridge, UK.

Wilson GW 1907. Studies in North American Peronosporales I, the genus *Albugo*. Bull Torrey Bot Club 34:61–84.

van Wyk PS, Jones BL, Viljoen A, Rong IH 1995. Early lodging, a novel manifestation of *Albugo tragopogonis* infection on sunflower in South Africa. Helia 18:83–90.

Yerkes WD, Shaw CG 1959. Taxonomy of the *Peronospora* species on Cruciferae and Chenopodiaceae. Phytopathol 49:499–507.

5

THE ASEXUAL LIFE CYCLE

ADRIENNE R. HARDHAM
Plant Cell Biology Group, Research School of Biological Sciences, The Australian National University, Canberra, ACT

5.1 INTRODUCTION

Asexual reproduction in the oomycetes consists of two main phases: sporangiogenesis, which is the formation of multinucleate sporangia, and zoosporogenesis, which is the formation of uninucleate, motile zoospores. Asexual sporangia and zoospores are the main dispersive agents for oomycete organisms, and in pathogenic species, are the most important means of initiating host infection. When asexual reproduction is induced, the mycelia produce sporangia at the apices of somatic hyphae or on the tips of branched sporangiophores. The mature sporangia may remain attached to the mycelium, or they may detach and function as dispersive propagules. Depending on the species and environmental conditions, the multinucleate sporangia germinate directly by forming hyphae or indirectly by forming biflagellate zoospores. Some obligate biotrophic species, such as *Hyaloperonospora arabidopsidis* or *Bremia lactuca*, have lost the ability to form zoospores and depend on sporangia for dissemination and establishment of infection.

Oomycete hyphae are generally coenocytic, although cross walls are formed at the bases of reproductive structures and in highly vacuolated hyphae in older regions of the mycelium. The hyphal diameter can vary widely, even within a single colony. Some oomycetes (e.g., Peronosporales species) also form asexual resting spores called chlamydospores that germinate through the production of germ tubes and sporangia.

In most species, sporangia produce and release zoospores whose directed motility plays an important role in enhancing the organism's chances of finding

Oomycete Genetics and Genomics: Diversity, Interactions, and Research Tools
Edited by Kurt Lamour and Sophien Kamoun
Copyright © 2009 John Wiley & Sons, Inc.

a suitable source of nutrients. These zoospores are ovoid in shape with two flagella that emerge from the center of a groove that runs along the ventral surface and gives the zoospores a kidney-shaped profile in cross section. On reaching a potential host or other nutrient source, the zoospores encyst. During encystment, the zoospores detach the two flagella, secrete extracellular matrix materials and change from ovoid wall-less cells to spherical walled cysts. Typically within half an hour of encystment, the cysts germinate, and if in the vicinity of a potential host, they attempt to penetrate the host surface. Cyst germination and germ tube growth return the organism to the vegetative state before reproduction is initiated once again. Some groups of oomycetes (e.g., the Saprolegniales) also produce pear-shaped, primary zoospores that have the two flagella attached at the anterior end of the cell. These primary zoospores are poor swimmers. Their encystment usually leads to production of the more typical, laterally flagellate secondary zoospores that are the sole zoospore type for most oomycetes.

5.2 SPORANGIOGENESIS

5.2.1 Induction of Sporulation and Upregulation of Gene Expression

Asexual sporulation in the oomycetes is induced by conditions that restrict growth, in particular low nutrient availability, but it is also affected by a variety of environmental factors, including temperature, light, aeration, humidity, carbon and nitrogen sources, and inorganic ions (Ali, 2005; Erwin and Ribeiro, 1996; Hohl, 1991; Kiryu et al., 2005). In some cases, the replacement of nutrient medium with water or a dilute salts solution will trigger sporulation *in vitro* (Chen and Zentmyer, 1970), whereas in other cases, sporangia form during continued incubation in staling nutrient medium (Cope and Hardham, 1994; Judelson and Blanco, 2005). During *in vitro* and *in planta* growth, mature sporangia are typically not observed until 1–3 days after the induction of sporulation or plant inoculation, however, they may form more rapidly (within 8–9 h) both in culture and in infected plants (Dearnaley et al., 1996; Maltese et al., 1995; Rumbolz et al., 2002) (Fig. 5.1a–l).

Early studies of *Achlya* and *Phytophthora* using pharmacological treatments with actinomycin D or cycloheximide showed that both RNA and protein synthesis are required for sporangium formation (Clark et al., 1978; Griffin and Breuker, 1969; Gwynne and Brandhorst, 1982; Timberlake et al., 1973). More recent characterization of individual or cohorts of genes, particularly in many species of *Phytophthora*, has confirmed the pharmacological results, identifying genes that are essential for sporangium formation and providing evidence of different patterns of gene expression during sporulation (Fig. 5.1e–g). *Pigpb1* and *piCdc14* are two genes whose expression has been shown to be required for sporangium development (Ah Fong and Judelson, 2003; Latijnhouwers and Govers, 2003). In both cases, transcription of these *Phytophthora infestans* genes is not detectable in vegetative mycelium but is highly upregulated during sporulation. *Pigpb1* codes for a G-protein β-subunit that is likely to play a role in signal transduction during sporulation (Latijnhouwers and Govers, 2003; Laxalt et al., 2002). *PiCdc14* codes for a phosphatase that regulates the cell cycle through interaction with a cyclin B protein complex (Ah Fong and Judelson, 2003). In both cases, the silencing of gene expression inhibits sporangium development.

A number of large-scale transcriptome studies have now identified genes that are expressed during sporangiogenesis (Judelson et al., 2008; Kim and Judelson, 2003; Randall et al., 2005). Expressed sequence tag (EST), macroarray, and microarray analyses have allowed these sporulation genes to be subcategorized according to their levels of expression in other stages of the asexual life cycle. Of 15,650 *P. infestans* genes, for example, 324 are highly expressed only in sporangia, 319 are highly expressed in sporangia and germinated cysts, and 629 are highly expressed in sporangia and cleaving sporangia (Judelson et al., 2008). It has also been demonstrated that the transcriptomes of sporulating and starving cultures overlap but are not identical (Cvitanich and Judelson, 2003; Judelson and Blanco, 2005). In *Phytophthora cinnamomi*, where it is possible to induce synchronous sporulation, changes in gene expression have been detected within half an hour of the induction of sporulation and cohorts of genes displaying specific patterns of transient expression have been identified (Marshall et al., 2001a and b; Narayan et al., unpublished observations). As details of the patterns of gene expression in developing sporangia continue to be unraveled, it will be important to remember the plasticity of subsequent development. Even after initiation of sporulation, the formation of sporangia can be inhibited as the mycelium reverts to a vegetative state (Dearnaley et al., 1996), and mature sporangia may germinate directly by forming a germ tube or indirectly by forming zoospores.

5.2.2 Morphological Development

Sporangiogenesis involves the differentiation of hyphal tips to form sporangia either directly through conversion of the hyphal apex into a sporangium (Fig. 5.1b–d) or indirectly through production of a sporangiophore on which sporangia subsequently develop (Figs. 5.1e–g and 5.2bb) (Alexopoulos et al., 1996). In the former case, mature sporangia tend to remain joined to the subtending mycelium, whereas in the latter, the sporangia may detach and function as dispersive propagules. In both cases, the mature sporangia become delimited from the adjoining hypha by a basal septum. Sporangiophore and sporangium formation begins with cessation of hyphal extension, loss of apical organelle stratification and movement of cytosol into the enlarging hyphal apex (Fig. 5.1a–d) (Allaway et al., 1998; Armbruster, 1982a; Heath and Kaminskyj, 1989). Formation of the basal septum coincides with attainment of the final size of the sporangium; this size can vary markedly even within a single isolate when grown under different conditions (Maltese et al., 1995). In the Saprolegniales,

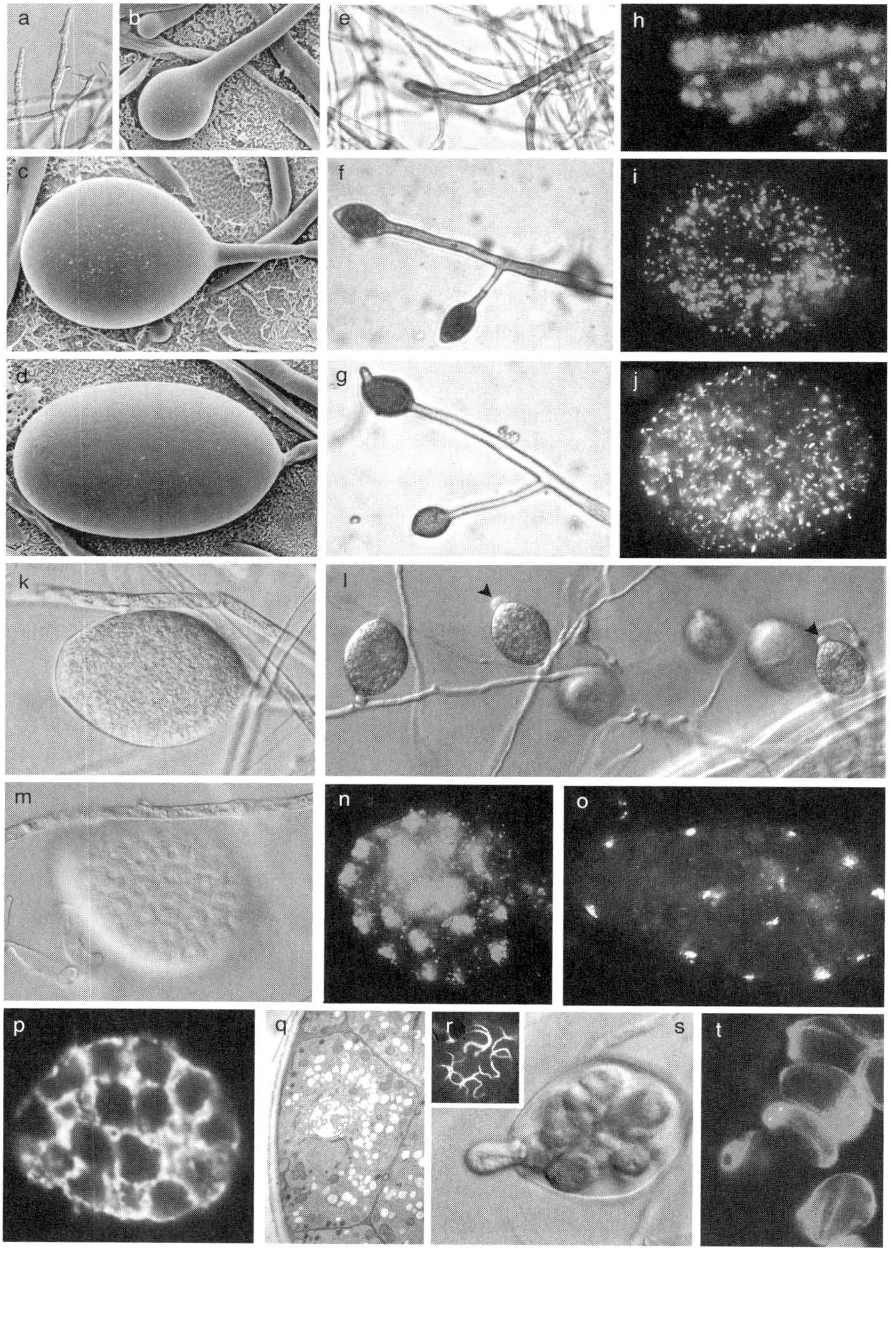
a
b
e
h
c
f
i
d
g
j
k
l
m
n
o
p
q
r
s
t

the sporangia are cylindrical expansions of the hyphal apex; in the Peronosporales, the sporangia are globose to lemon shaped (Fig. 5.1g and k–m). The wall of the sporangium is typically thicker than that of somatic hyphae and impermeable to molecules in the external solution (A. Hardham, unpublished observations). These features are likely to be important for mechanisms involved in subsequent release of the motile zoospores (see below). In Peronosporales species, the sporangial apex often differentiates as a specialized papilla that ruptures on zoospore discharge (Fig. 5.1l, r, and s).

FIG. 5.1 (See color insert) Development of oomycete sporangia. (a) Vegetative hyphae of *Phytophthora nicotianae*. Somatic hyphae are coenocytic and grow by tip growth. × 150. (b–d) CryoScanning electron microscopy (cryoSEM) images of sporangia developing at the apices of hyphae in *Phytophthora cinnamomi*. × 700, × 570, × 590, respectively. Reproduced with permission from Hardham et al. (1994). (e–g) Sporangia forming on sporangiophores in *Phytophthora infestans* expressing the β-glucuronidase (GUS) gene under the promoter of the *M90* gene that belongs to the Puf family of translational regulators. GUS expression indicative of M90 production occurs in hyphal tips prior to sporulation (e), in sporangiophores (f) and in mature sporangia (g). × 200. Reproduced with permission from Cvitanich and Judelson (2003). (h–j) Production of zoospore-specific components during sporulation. Immunofluorescence labelling shows the appearance of ventral vesicles in sporulating hyphae (h), and ventral vesicles (i) and packets of mastigonemes (j) in mature sporangia of *P. cinnamomi*. These components are randomly distributed within the sporangia before cleavage. × 1300, × 500, × 600, respectively. Micrographs courtesy of Dr M. Cope. (k, l) Mature, uncleaved *P. cinnamomi* sporangium grown *in vitro* (k) and *Phytophthora nicotianae* sporangia produced on the surface of a tobacco root (l). Arrowheads point to apical papillae. × 530, × 470, respectively. (k) Reproduced with permission from Hardham (2005). (m) Cleaved *P. cinnamomi* sporangium. Subdivision of the sporangial cytoplasm has defined uninucleate domains that will become the zoospores. The circular areas within each domain are nuclei or the bladder of the contractile vacuole. × 530. Reproduced with permission from Hardham (2005). (n, o) Immunolabelling of ventral vesicles (n) and mastigonemes (o) in cleaved sporangia of *P. cinnamomi*. During cleavage, the ventral vesicles become distributed along the ventral surface of the future zoospores, and the mastigonemes become clustered at the base of the flagella. × 500, × 610, respectively. Micrographs courtesy of Dr M. Cope. (p) Immunofluorescent labelling with monoclonal antibody Cpw-4 highlights the cleavage planes in a *P. cinnamomi* sporangium. × 530. Micrograph courtesy of Dr M. Cope. (q) Preparation of material by cryofixation and freeze substitution demonstrates the continuous nature of the cleavage membranes that delineate each uninucleate domain in cleaving sporangium in *P. cinnamomi*. The large central vacuole is the bladder of the developing contractile vacuole. Reproduced with permission from Hyde et al. (1991b). × 14,900. (r) Flagella formed during sporangial cleavage in *P. cinnamomi* immunofluorescently labelled with anti-β-tubulin antibodies. × 450. Micrograph courtesy of Dr M. Cope. (s) Zoospore release from a sporangium of *P. cinnamomi*. × 530. (t) Empty sporangia of *Plasmopara viticola* after release of zoospores. The sporangial wall is labelled with monoclonal antibody Cpw4. This antibody was raised against cell wall material in *P. cinnamomi*. × 1,000.

Although mitosis may occur within the sporangium during its formation (Maltese et al., 1995; Trigano and Spurr, 1987), most nuclei are transported from the hypha into the sporangium where they become regularly spaced throughout the sporangial cytoplasm (Hoch and Mitchell, 1975; Hohl and Hamamoto, 1967; Hyde et al., 1991a; Kevorkian, 1996; Oertel and Jelke, 1986). The nuclei are typically pear shaped with Golgi stacks and two basal bodies present near the narrow pole of each nucleus (Armbruster, 1982b; Chapman and Vujicic, 1965; Elsner et al., 1970; Gotelli, 1974; Hohl and Hamamoto, 1967; Hyde et al., 1991a; King et al., 1968; Lange et al., 1984, 1989; Schnepf et al., 1978; Williams and Webster, 1970). Nuclei near the sporangial wall have their narrow pole oriented toward the outer surface of the sporangium (Armbruster, 1982b; Elsner et al., 1970; Hyde et al., 1991a; Oertel and Jelke, 1986; Williams and Webster, 1970). Both the shape and distribution of nuclei in the sporangia are maintained by microtubules that emanate from the region of the basal bodies, fan across the nuclear surface, and form an aster-like array that radiates into the cytoplasm (Gotelli, 1974; Hoch and Mitchell, 1972b, 1975; Hyde et al., 1991a and b; Jelke et al., 1987; Lange et al., 1984, 1989; Schnepf et al., 1978; Williams and Webster, 1970). Pharmacological degradation of the microtubule arrays leads to an irregular distribution of spherical nuclei and formation of aberrantly shaped zoospores (Hyde and Hardham, 1993).

5.2.3 Synthesis of Zoospore-Specific Components

In addition to the usual range of eukaryotic organelles, such as nuclei, mitochondria, endoplasmic reticulum, Golgi bodies, microbodies, and lipid bodies, oomycete zoospores contain many components that function specifically in the motile spores. Most of these zoospore-specific components form in the hyphae during sporangiogenesis and move into the expanding sporangia (Fig. 5.1h–j). Other zoospore-specific components are assembled during sporangial cleavage (i.e., zoosporogenesis) (Fig. 5.1q and r) as described below.

Four zoospore-specific components are different categories of vesicles or membranous compartments that occur in the cortical cytoplasm of the zoospores (Beakes, 1989; Beakes et al., 1995; Hardham, 1995; Hardham and Hyde, 1997). Early studies of zoospores in different oomycete taxa led to a variety of appellations for these four categories of components, however, they are predominantly referred to as follows: (1) encystment, cyst coat, or dorsal vesicles; (2) K-bodies or ventral vesicles; (3) fibrillar or large peripheral vesicles; and (4) peripheral cisternae. Vesicles in the first three categories are spherical or ellipsoidal storage compartments; peripheral cisternae are flat, abutting disks that underlie most of the zoospore plasma membrane (Hardham, 1987). Structures similar to these four compartments are also found in the phylogenetically related, apicomplexan malarial zoites (de Souza, 2006; Morrissette and Sibley, 2002). Studies of *P. cinnamomi* have shown that the storage vesicles are synthesized at the Golgi apparatus during sporangiogenesis

(Dearnaley et al., 1996). The tubular hairs called mastigonemes that will adorn the anterior flagellum of the zoospores are also synthesized during sporangiogenesis but within special compartments of the endoplasmic reticulum (Figs. 5.1j and o, 5.2d) (Cope and Hardham, 1994). The peripheral cisternae, the flagella (including the addition of mastigonemes and other hairs on the flagellar surface), and the water expulsion vacuoles are assembled during zoosporogenesis (Figs. 5.1q and r) (Cope and Hardham, 1994; Hardham and Hyde, 1997; Hyde et al., 1991b). One of the likely advantages of having most zoospore components already present within the mature sporangia is that it allows zoosporogenesis to occur very rapidly (within 10–15 min in some species of *Phytophthora*).

5.3 ZOOSPOROGENESIS

5.3.1 Induction of Zoosporogenesis and Upregulation of Gene Expression

Cleavage of mature sporangia into uninucleate zoospores is favored under wet and cool conditions and is induced *in vitro* by a reduction in temperature. In *P. cinnamomi*, this cold shock has been shown to induce increases in both cytoplasmic calcium concentration and pH (Jackson and Hardham, 1996; Suzaki et al., 1996). Experimentally increasing cytoplasmic calcium concentration can induce cleavage in the absence of a cold shock, and inhibition of the increases in [Ca^{2+}] or pH prevent zoosporogenesis (Jackson and Hardham, 1996; Judelson and Roberts, 2002). Together, these results indicate that changes in [Ca^{2+}] and pH are part of the signal transduction cascade involved in triggering sporangial cleavage. In *P. infestans*, an experimentally induced increase in membrane rigidity, as would occur during cold shock, has been shown to stimulate zoospore release (Tani and Judelson, 2006). Analysis of the promoters of genes that are upregulated during zoosporogenesis has uncovered a seven-nucleotide motif, the cold box, that is required for the induction of expression of these genes (Tani and Judelson, 2006).

5.3.2 Sporangial Cleavage and Zoospore Formation

Early ultrastructural studies of sporangial cleavage in different oomycete genera suggested that a range of different mechanisms operated to subdivide the multinucleate sporangial cytoplasm (Hardham and Hyde, 1997). The mechanism most widely proposed entailed the alignment of large numbers of cleavage vesicles into flat planes between adjacent nuclei (Armbruster, 1982b; Hohl and Hamamoto, 1967; Lange et al., 1989; Lunney and Bland, 1976). However, an ultrastructural study of cleavage of *Phytophthora* sporangia that used rapid freezing and freeze substitution to prepare the material for microscopy revealed that sporangial cleavage is achieved through the development and expansion of paired membranous sheets that progressively partition the nuclei and associated cytoplasmic domains (Fig. 5.1m, p, and q) (Hyde et al., 1991a and b). The interpretation of these observations is that chemical fixation causes artifactual vesiculation of the cleavage membranes a phenomenon observed in other situations in which membranes are closely aligned (McCully and Canny, 1985). Observation of curved cleavage vacuoles in high-pressure frozen and freeze-substituted sporangia of *Albugo ipomoeae-panduratae* in a region occupied by clusters of vesicles near the basal bodies in chemically fixed material (Mims and Richardson, 2003) gives more evidence to suggest that expansion of cleavage planes is the mechanism operating during zoosporogenesis throughout the oomycetes.

During sporangial cytokinesis, subcellular components that were initially randomly distributed are transported to specific locations within the developing zoospores. The three categories of peripheral vesicles become associated with distinct domains of the zoospore plasma membrane, mitochondria take up a subcortical position, and most of the plasma membrane becomes underlain by the system of peripheral cisternae (Fig. 5.1m–q). In the absence of microtubules, the polarity of vesicle and mitochondrial distribution does not develop (Hyde and Hardham, 1993). During cytokinesis, actin microfilaments become associated with the zoospore surface (Jackson and Hardham, 1998). Their depolymerization disrupts sporangial subdivision but does not affect subcellular reorganization (Hyde and Hardham, 1993). As the cleavage membranes develop, flagellar axonemes assemble from the pair of basal bodies that reside close to the nuclear apex (Fig. 5.1r). The flagella extend within a plasma membrane-bound compartment that is continuous with the partitioning membranes (Cope and Hardham, 1994).

5.3.3 Zoospore Discharge

The sporangia of most oomycetes discharge their zoospores through rupture of the apical papilla of the sporangium (Fig. 5.1l, s, t). Depending on the species, the zoospores are released directly into the surrounding medium (as in *Saprolegnia* and *Aphanomyces:* Gay and Greenwood, 1966; Hoch and Mitchell, 1972a) or, at least initially, into an evanescent vesicle derived from the papillar cell wall material (as in *Phytophthora*, *Pythium*, and *Lagenidium*: Gisi et al., 1979; Gotelli, 1974; Webster and Dennis, 1967). Rupture of the papilla and movement of most zoospores out of the sporangium is driven by hydrostatic pressure that builds up within the sporangium (Gisi et al., 1979; Gisi, 1983; Hoch and Mitchell, 1973). The hydrostatic pressure could result from the accumulation of gel-like material or solutes in the extracellular space between the zoospores and between the zoospores and the sporangial wall (Gisi, 1983; Gisi and Zentmyer, 1980; MacDonald and Duniway, 1978; Money et al., 1988; Money and Webster, 1985). In either case, the sporangial cell wall must function as a semipermeable barrier that allows an influx of water but not an efflux of the osmoticum from the periplasm. Accumulation of osmotically active molecules within the extracellular space would tend to cause dehydration and shrinkage of the zoospores; indeed in *Phytophthora*, a reduction in zoospore volume is observed prior to their release. The shrinkage is, however, only slight, indicating that the high extracellular osmolarity must be counterbalanced by generation of a similar osmotic pressure within the zoospores. This could be achieved by the high concentrations of proline measured in the zoospore cytosol (Grenville-Briggs et al., 2005). On release into the hypoosmotic solution in the environment, the zoospores would need to quickly reduce the osmolarity of their cytosol, a process that may be achieved by proline expulsion. Subsequent reestablishment of normal levels of proline is consistent with observation of enhanced levels of expression of genes that encode proline biosynthetic enzymes in the zoospores (Ambikapathy et al., 2002).

Early pharmacological studies suggested that zoosporogenesis can occur without the need for RNA or protein synthesis (Clark et al., 1978; Penington et al., 1989). However, recent investigations of the transcriptome of *P. infestans* sporangia have revealed rapid and extensive changes in gene expression during zoosporogenesis in this species (Judelson et al., 2008; Tani et al., 2004). These changes that are likely to occur during zoospore formation in other oomycetes. It is possible that upregulation of expression of certain genes normally occurs during zoosporogenesis, but this transcription may not be an absolute requirement for cleavage. However, a recent study of a gene encoding an RNA-helicase that is highly expressed during zoosporogenesis has shown that silencing expression of this gene leads to aberrant zoospore formation (Walker et al., 2008), indicating that the increased expression of at least some genes is required for successful cleavage. The failure to prevent zoosporogenesis in the pharmacological studies may be due to the relative impermeability of the sporangial cell wall.

5.4 ZOOSPORE BIOLOGY AND FUNCTION

5.4.1 Osmoregulation and the Contractile Vacuole

Oomycete zoospores are not surrounded by a cell wall, and their outer surface is that of the plasma membrane covered by an extracellular carbohydrate layer (glycocalyx) that is typically rich in glucosyl and mannosyl residues (Hardham, 1989; Hardham et al., 1994). Because they lack a cell wall, zoospores cannot generate hydrostatic pressure, and water that diffuses into the cell down its chemiosmotic gradient must be actively pumped out to maintain cell size and integrity. This process is achieved through the operation of a contractile vacuole, often referred to as a water expulsion vacuole in the oomycete literature (Fig. 5.2a). In species of *Phytophthora* where it has been most studied, the contractile vacuole forms during the late stages of sporangial cleavage (Fig. 5.1q) and begins to function after cleavage is complete but before zoospore release (Mitchell and Hardham, 1999).

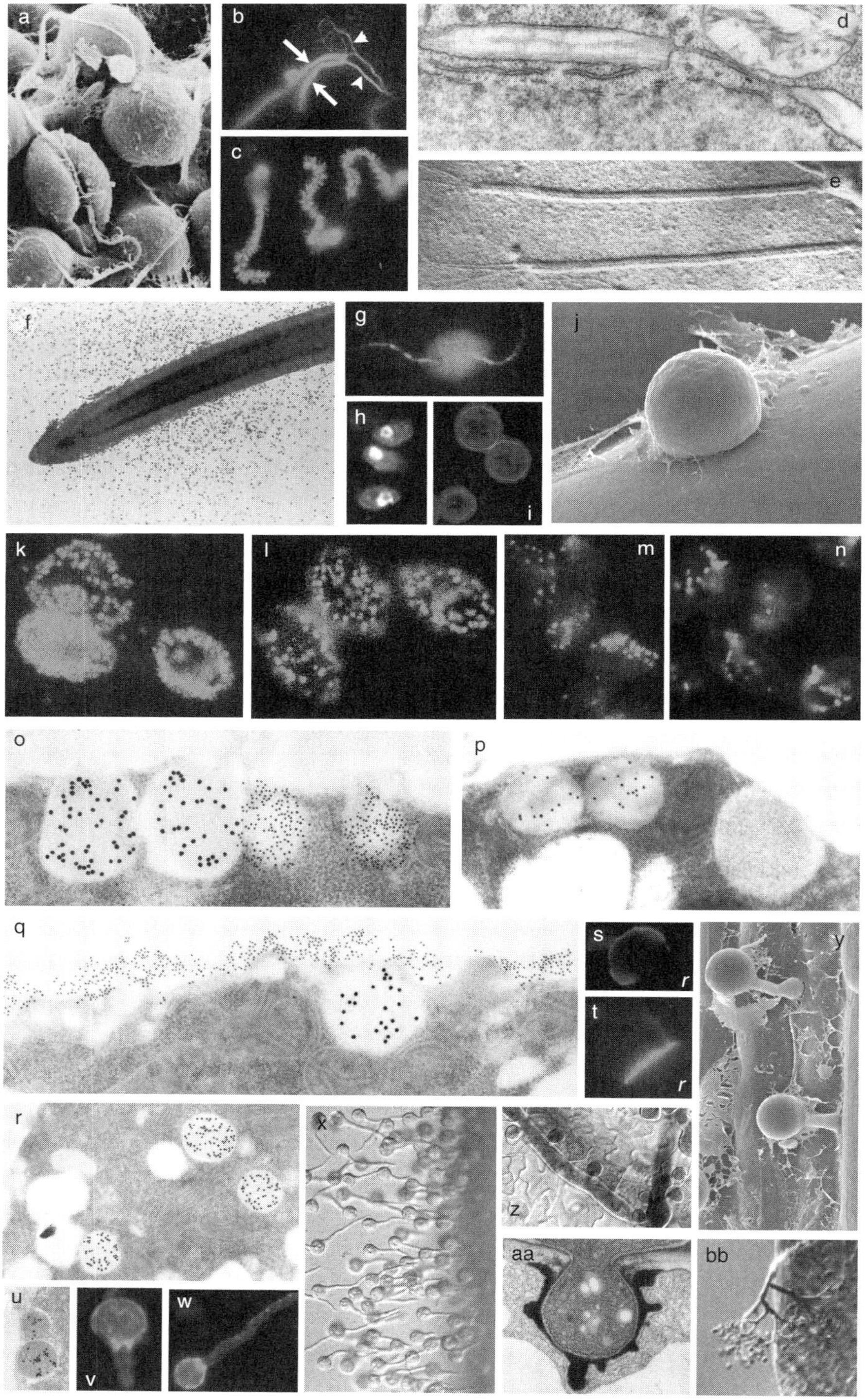
a
b
c
d
e
f
g
h
i
j
k
l
m
n
o
p
q
s
r
t
r
y
r
x
z
aa
bb
u
v
w

The contractile vacuole in oomycete zoospores has a similar structure to contractile vacuoles in a variety of other wall-less protists, including the apicomplexans *Plasmodium falciparum* and *Toxoplasma gondii* (Luo et al., 1999; Rodrigues et al., 2000). It consists of a central bladder surrounded by a reticulum of membranous tubules and vesicles known as the spongiome (Patterson, 1980). The membranes of the spongiome contain vacuolar H^+-ATPase molecules that transport protons into the lumen of the reticulum (Fig. 5.2h) (Fok et al., 1995; Mitchell and Hardham, 1999). The pH gradient and membrane potential thus generated is believed to power the accumulation of ions and solutes, which in turn leads to movement of water from the cytoplasm into the spongiome (Stevens and Forgac, 1997). Pharmacological inhibition of vacuolar H^+-ATPase interferes with contractile vacuole function (Mitchell and Hardham, 1999). By a mechanism that is still not clear, the water is passed from

FIG. 5.2 (See color insert) Structure and function of oomycete zoospores. (a) SEM of a biflagellate zoospore and walled cysts of *Phytophthora cinnamomi* on a root surface. × 1,200. Reproduced with permission from Hardham et al. (1994). (b) *P. cinnamomi* zoospore immunofluorescently labelled with anti-β-tubulin antibodies to show the two flagella (arrows) and microtubular flagellar roots (arrowheads) that run along the ventral groove. × 1,100. (c) Immunofluorescent labelling with monoclonal antibody Zg-4 shows the two rows of mastigonemes that adorn the anterior flagellum of three *P. cinnamomi* zoospores. × 600. (d, e) Transmission electron micrographs of tripartite, tubular mastigonemes in *P. cinnamomi.* Mastigonemes are synthesized in specialized regions of the endoplasmic reticulum (d). Shadowcast preparations show that each mastigoneme has a cluster of apical hairs, a tubular shaft with a beaded substructure (e) and a basal complex that attaches the hair to the flagellar axoneme. × 23,800, × 39,700, respectively. (f) *P. cinnamomi* zoospores are chemotactically and/or electrotactically attracted to the elongation zone of an onion root. × 15. (g) Immunofluorescence labelling with monoclonal antibody Zf-1, which was raised against *P. cinnamomi* antigens, reacts with the surface of flagella of *Plasmopara viticola*. When this antibody is added to motile *P. cinnamomi* zoospores, it induces zoospore encystment. × 850. (h) *P. nicotianae* zoospores labelled with polyclonal antibody S1B raised against *Nicotiana* plasma membrane H^+-ATPase (Morsomme et al., 1996) and showing reaction with the water-expulsion vacuole. × 370. (i) *P. cinnamomi* cysts stained with Calcofluor White to show the presence of β-1,3 glucans including cellulose in the cyst cell walls. × 570. (j) CryoSEM preparation of a young *P. cinnamomi* cyst on the surface of a tobacco root. × 1,500. Reproduced with permission from Hardham (2005). (k–m) *P. cinnamomi* zoospores labelled with monoclonal antibodies Lpv-1, Cpa-2 and Vsv-1 that react with high-molecular-weight glycoproteins (Lpv-1 and Cpa-2) or proteins (Vsv-1) stored in the large peripheral, dorsal and ventral vesicles, respectively. × 860, × 1,000, × 680, respectively. (n) Ventral vesicles in zoospores of the downy mildew pathogen, *P. viticola*, immunofluorescently labelled with Vsv-1. × 1,250. (o) Ultrathin section of a zoospore of *P. cinnamomi* after double immunogold labelling with Lpv-1-Au15 and Cpa-2-Au10 showing highly specific reactions with the contents of the large peripheral and dorsal vesicles, respectively. × 28,900. Micrograph courtesy of Dr. F. Gubler. (p) Ultrathin section of a zoospore of *P. cinnamomi* after immunogold labelling with Vsv-1-Au10. The large peripheral vesicle on the right is not labelled. × 23,000. Micrograph courtesy the spongiome into the bladder, which progressively expands before fusing with the plasma membrane and expelling the water within it. The bladder membrane, like the zoospore plasma membrane over the cell body (but not the flagella), contains P-type H^+-ATPase, vacuolar pyrophosphatase, and water-transporting aquaporins (Mitchell and Hardham, 1999; Paeper and Hardham, unpublished observations).

5.4.2 Motility and Chemotaxis

Oomycete zoospore flagella have a typical eukaryotic internal structure with nine microtubule doublets linked to each other and to a central pair of microtubules (Fig. 5.2a and b) (Hardham, 1987). The two flagella differ, however, in length and surface morphology, thus giving rise to the description of oomycete zoospores as being heterokont. The anterior (tinsel) flagellum is the shorter of the two and possesses two rows of tubular, tripartite hairs called

---

of **FIG. 5.2** (*Continued*) Dr. F. Gubler. (q, r) *P. cinnamomi* cysts fixed 1 min (q) or 5 min (r) after the induction of encystment and double immunogold labelled with Lpv-1-Au15 and Cpa-2-Au10. The clusters of small electron-lucent vesicles beneath the plasma membrane result from the vesiculation of the peripheral cisternae. While the contents of the dorsal vesicles are secreted onto the cell surface, the large peripheral vesicles do not undergo exocytosis but become randomly distributed within the cyst cytoplasm. × 24,000, × 12,400 respectively. Micrographs courtesy of Dr. F. Gubler. (s, t) Cyst of *P. cinnamomi* that has encysted on the surface of a root (*r*) double labelled with soybean agglutinin-rhodamine (s) and Vsv-1/sheep antimouse-FITC (t). The rhodamine-labelled lectin reacts with glycoproteins released from the dorsal vesicles to coat the outer dome of the cyst. The 220 kDa Vsv1 adhesin released from the ventral vesicles forms an adhesive pad between the cyst and the root. × 810. (u) *Phytophthora nicotianae* zoospore labelled with a polyclonal antibody that reacts with PnCcp, which is a 12-kDa protein that contains a complement control protein domain. The PnCcp protein resides in the large peripheral vesicles in the zoospore cortex. × 10,400. (v, w) *P. nicotianae* cysts fixed 45 min (v) and 90 min (w) after zoospore encystment and double labelled with monoclonal Lpv-1 and polyclonal PnCcp antibodies. While the large peripheral vesicles containing the Lpv-1 high molecular weight glycoproteins remain within the cyst cytoplasm, the PnCcp proteins have been secreted onto the cyst surface. Reproduced with permission from Škalamera and Hardham (2005). × 710, × 410, respectively. (x) *P. cinnamomi* germinated cysts growing chemotropically toward the surface of an onion root. × 190. (y) *P. cinnamomi* germinated cysts on the surface of an alfalfa root have formed appressorium-like swellings before penetration of the root surface along anticlinal walls between the epidermal cells. × 670. Reproduced with permission from Hardham (2007). (z) Haustoria produced by *Hyaloperonospora parasitica* in the epidermal cells of a cotyledon of *Arabidopsis thaliana* stained with lactophenol trypan blue. × 240. Reproduced with permission from Hardham (2007). (aa) Transmission electron micrograph of a haustorium of *Phytophthora sojae*. × 10,000. Micrographs courtesy of Dr. C. Mims. Reproduced with permission from Hardham (2007). (bb) Sporangiophore of *H. parasitica* produced on a susceptible ecotype of *A. thaliana*. × 120. Micrograph courtesy of Dr. D. Takemoto.

mastigonemes (Fig. 5.2c–e). As the anterior flagellum bends and propagates a sinusoidal wave from base to tip, the resultant motion of the mastigonemes creates a force that pulls the zoospore forward (Cahill et al., 1996; Jahn et al., 1964). The posterior (whiplash) flagellum projects behind the zoospore, occasionally bending abruptly to cause the spore to turn. Zoospores typically rotate as they swim in a tight helical path, with periods of straight swimming being interrupted by changes of direction.

Oomycete zoospores are attracted to the surface of potential hosts through detection of both chemical and electrical gradients (Fig. 5.2f) (Tyler, 2002; Van West et al., 2002). In plant pathogenic species, these tactic responses are generally nonspecific in that they guide zoospores to host and nonhost alike (Deacon and Donaldson, 1993), however, specific interactions also occur. Zoospores of *Phytophthora sojae* and *Aphanomyces cochlioides*, for example, are attracted to host roots through detection of flavones or isoflavones secreted by the host plants (Morris and Ward, 1992; Sakihama et al., 2004). In addition, zoospores can target particular regions of the plant, a feature that is likely to contribute to successful infection. The target zone is different for different pathogen species and also for different hosts and a single pathogen species (Deacon and Donaldson, 1993). For example, in rye grass, zoospores of *Pythium aphanidermatum* swim to the root hair zone, whereas zoospores of *Phytophthora palmivora* swim toward the root elongation zone (Gow, 2004; Morris and Gow, 1993). In contrast, zoospores of *Pythium dissotocum* are attracted to the root cap of cotton roots, the elongation zone of tomato roots, and the root hair zone of maize roots (Goldberg et al., 1989). Even more specific targeting is demonstrated by the ability of zoospores to swim to and settle preferentially in grooves between adjacent epidermal cells or next to stomata (Hardham, 2001, 2005, 2007; Kiefer et al., 2002; Soylu et al., 2003). This precise targeting achieved by zoospore motility and taxis clearly makes an important contribution to establishing infection in both root and foliar pathogens.

There is still little information on the nature of receptors responsible for detecting the chemical and electrical signals that guide chemotaxis and electrotaxis, although there is some evidence for their localization on the zoospore surface (Sakihama et al., 2004). Two genes involved in regulating zoospore motility have, however, now been identified. In *P. infestans*, the silencing of genes that encode the α-subunit of a trimeric G-protein and a bZIP transcription factor have been shown to interfere with zoospore motility (Blanco and Judelson, 2005; Latijnhouwers et al., 2004). In the silenced, transgenic isolates, instead of their normal swimming pattern, the zoospores turn more frequently or spin on the spot.

5.4.3 Encystment and Adhesion

Having reached a suitable location from which to initiate infection, the zoospores of at least some soil-borne phytopathogenic species adjust their swimming pattern so that their ventral surface, from which the flagella emerge, faces the potential host (Deacon and Donaldson, 1993; Hardham and Gubler, 1990). They maintain this orientation while swimming back and forth parallel to the root surface before becoming immotile and encysting. Encystment is a rapid process during which the flagella are detached, the spores round up, and adhesive, cell wall and other extracellular matrix materials are secreted from intracellular sources within the spore cortex. A range of physical and chemical stimuli, which include agitation, low temperatures, and millimolar concentrations of calcium or phosphatidic acid, can induce zoospore encystment. Evidence exists for a role of zoospore surface receptors, which are possibly located in the flagellar plasma membrane (Fig. 5.2g) (Bishop-Hurley et al., 2002; Hardham and Suzaki, 1986), and of the phospholipase D signal transduction pathway (Latijnhouwers et al., 2002).

Studies of *Phytophthora* zoospores have shown that regulated secretion events triggered during zoospore encystment involve exocytosis of the contents of many different types of vesicles in the zoospore cortical cytoplasm (Fig. 5.2j–t). One of the earliest responses that has been detected, occuring about 1 min after the onset of encystment, is vesiculation of the system of disk-like, peripheral cisternae that underlie the zoospore plasma membrane except beneath the groove along the ventral surface (Fig. 5.2q). Ultrastructural observations suggest that the vesicles derived from the peripheral cisternae fuse with the plasma membrane, thereby potentially achieving a rapid and extensive change in plasma membrane composition (Hardham, 1989). Immunogold labelling with a monoclonal antibody that reacts with the cyst cell wall has revealed the presence of the same epitope within the lumen of the peripheral cisternae, which has led to the suggestion that the peripheral cisternae membranes may contain proteins involved in the synthesis of cell wall components (Gubler and Hardham, 1991).

Accompanying vesiculation of the peripheral cisternae, the two categories of small peripheral vesicles also undergo exocytosis (Gubler and Hardham, 1988; Hardham and Gubler, 1990). Material secreted from one category of vesicle that occurs predominantly along the dorsal surface of the zoospores (the dorsal, cyst coat, or encystment vesicles) (Fig. 5.2l) forms a mucilage-like covering over the cysts and adjacent plant surface that may protect the cysts from desiccation (Fig. 5.2a and s) (Hardham, 2005, 2007). The second category of small peripheral vesicles is located predominantly underneath the ventral surface of the zoospore and is often concentrated along the ridges of the groove (Fig. 5.2m, n, p). In the Peronosporales, these vesicles are about 300 nm in diameter and are called ventral vesicles (Hardham and Gubler, 1990). In the Saprolegniales, they are larger and fewer in number and were named K-bodies because they lie close to the kinetosomes (basal bodies) (Hardham and Hyde, 1997; Lehnen and Powell, 1989). During encystment, ventral vesicle and K-body contents are secreted to form a pad of material between the cyst and the adjacent surface, and evidence shows that across the oomycetes, these vesicles are responsible for adhesion of the encysting spores (Fig. 5.2t) (Lehnen and Powell, 1989; Robold and Hardham, 2005). Cloning of the gene *PcVsv1* that encodes a 200-kDa protein contained in the ventral vesicles in *P. cinnamomi* has revealed that it contains 47 copies of the thrombospondin type I repeat (TSR1), which is a conserved 50-amino-acid motif found in a range of adhesive molecules secreted by mammalian cells and malarial parasites (Robold and Hardham, 2005). In malarial parasites, the TSR1-containing adhesins are stored in small vesicles called micronemes and, just as in oomycete zoospores, they are rapidly secreted during the initial stages of host infection (Tomley and Soldati, 2001). Immunolabelling and EST studies have shown that homologes of the PcVsv1 ventral vesicle adhesin occur in other *Phytophthora* species, as well as in species of *Pythium*, *Plasmopara*, and *Albugo* (Fig. 5.2n), which indicates that the Vsv1 adhesin is a spore adhesive used throughout the plant pathogenic oomycetes.

The third category of peripheral vesicles in oomycete zoospores is called large peripheral or fibrillar vesicles (Fig. 5.2k, o–r) (Hardham and Hyde, 1997). The large peripheral vesicles are about 800 nm in diameter and occur throughout the cortex except along the groove. Immunolabelling studies, which employed a monoclonal antibody directed toward a set of three high-molecular-weight glycoproteins stored within the large peripheral vesicles of *P. cinnamomi*, showed that these vesicles are not exocytosed during zoospore encystment but become internalized and randomly distributed throughout the cyst cytoplasm (Fig. 5.2r) (Gubler and Hardham, 1990). Cloning of the 3′ half of the *PcLpv* genes revealed that, like PcVsv1, these proteins also contain multiple copies of a conserved domain (Marshall et al., 2001b). In the case of PcLpv, the repeated domain consist of 534 amino acids, and variation in the number of repeats leads to the three polypeptides observed in immunoblots. Immunocytochemical studies indicate that the Lpv proteins are degraded during cyst germination and are likely to serve as a source of protein during early germling growth (Gubler and Hardham, 1990). Given the clear evidence that the large peripheral vesicles are not secreted during encystment of *Phytophthora* zoospores, it recently came as a surprise to discover that at least one protein constituent of these vesicles, PnCcp, is secreted during encystment (Fig. 5.2u–w) (Škalamera and Hardham, 2006). Studies of the selective secretion of this potential adhesive protein are underway.

Loss of motility as well as the rounding up and secretion of cortical vesicles during zoospore encystment is quickly followed by the formation of a cellulosic cell wall (Fig. 5.2i) that can withstand cell turgor within 5–10 min after the onset of encystment. As the cell wall develops, the pulsing of the contractile water expulsion vacuole slows down. As turgor pressure increases, the vacuole disappears.

5.5 CYST GERMINATION AND INITIATION OF INFECTION

Encystment triggers a new program of mRNA and protein synthesis in cysts of *Saprolegnia* and *Phytophthora* (Andersson and Cerenius, 2002; Avrova et al., 2003;

Judelson and Tani, 2007; Krämer et al., 1997; Shan et al., 2004). Pharmacological studies have shown that both transcription and translation are required for cyst germination (Penington et al., 1989). More recently, the silencing of the *NIFC* genes in *P. infestans* has illustrated that these transcriptional regulators are required for cyst germination (Judelson and Tani, 2007). Of the hundreds of genes that are upregulated during cyst germination, many are involved in protein and energy synthesis, and others function in plant cell wall degradation or defense against the host's oxidative burst (Avrova et al., 2003; Ebstrup et al., 2005; Görnhardt et al., 2000; Shan et al., 2004). In at least *Phytophthora* and *Pythium*, the site of cyst germination is predetermined, and the germ tube is produced from the center of the ventral surface (Deacon and Donaldson, 1993; Hardham and Gubler, 1990). The germ tube thus emerges from the side of the spore that faces the host and grows chemotropically toward a suitable penetration site (Fig. 5.2x). Preferential targeting and settling of the zoospores in the grooves between the epidermal cells is often followed by penetration of the host surface preferentially at the anticlinal walls (Fig. 5.2y) (Enkerli et al., 1997; Hardham, 2001; Slusarenko and Schlaich, 2003). In the biotrophic foliar pathogens *Plasmopara viticola* and *Albugo candida*, whose zoospores target stomatal complexes, the hyphae enter the leaf via stomatal pores (Gindro et al., 2003; Kiefer et al., 2002; Soylu, 2004). In *P. viticola,* actin and microtubules are required for cyst germination (Riemann et al., 2002).

5.5.1 Development of Appressoria

Oomycete hyphae may develop appressorium-like swellings as they penetrate the plant surface (Fig. 5.2y) (Bircher and Hohl, 1997; Grenville-Briggs et al., 2005; Slusarenko and Schlaich, 2003; Soylu and Soylu, 2003). In *P. infestans*, the appressorium-like swelling is separated from the subtending hypha by a cross wall, however, the degree of differentiation of oomycete appressoria does not seem to be as great as shown by the appressoria of some fungi such as the *Magnaporthe* or *Colletotrichum* species. As in fungi, the formation of oomycete appressoria-like swellings is induced by factors such as surface topography and hydrophobicity (Bircher and Hohl, 1997) and is accompanied by changes in patterns of gene expression and protein synthesis (Ebstrup et al., 2005; Grenville-Briggs et al., 2005). Evidence that, like fungal appressoria, oomycete appressoria-like cells are specialized to exert pressure that facilitates penetration of the plant surface has come from recent studies of the CesA family of cellulose synthase genes in *P. infestans* (Grenville-Briggs et al., 2008). In this pathogen, the expression of *CesA1-3* and cellulose synthesis is required for the formation of functional appressoria-like cells and for pathogenicity.

5.5.2 Secretion of Cell-Wall Degrading Enzymes

Although appressoria are specialized for the production of pressure to force their way physically through the plant cell wall, even the most highly differentiated fungal appressoria also employ chemical means to aid invasion (Howard et al., 1991). In the oomycetes, as in fungal phytopathogens, the invading hyphae synthesize and secrete enzymes that degrade plant cell wall components such as pectin, cellulose, and xyloglucans (Boudjeko et al., 2006). Recent EST projects have begun to catalog the genes that encode cell wall degrading enzymes, although to date only a small number have been characterized in any detail (Costanzo et al., 2006; Götesson et al., 2002; McLeod et al., 2003; Torto et al., 2002; Yan and Liou, 2005). In *Phytophthora*, these enzymes are typically encoded by large multigene families.

Like fungal hyphae, oomycete hyphae grow by tip growth. Tip growth involves the secretion of new membrane, wall material and other extracellular materials within the dome of the hyphal apex. Materials to be secreted are packaged into small apical vesicles at the Golgi apparatus and are transported to the hyphal apex along cytoskeletal actin and microtubule networks (Heath, 1987; Heath and Kaminskyj, 1989; Walker et al., 2006). Before fusing with the plasma membrane, the apical vesicles form a cluster, the so-called vesicle supply center within the apical dome. Mathematical modeling of fungal and oomycete growth predicts that the shape of the hyphal apex as well as the speed and direction of expansion will be determined by regulating the movement of the cluster of vesicles and the rate of their fusion with the plasma membrane at the hyphal tip (Dieguez-Uribeondo et al., 2004). Vesicle fusion contributes to plasma membrane expansion, providing membrane proteins such as channels, receptors and enzymes needed for uptake of nutrients and synthesis of cell wall microfibrils (Loprete and Hill, 2002). In addition to the secretion of extracellular matrix components, including adhesives and wall materials (Gaulin et al., 2002; Shapiro and Mullins, 2002) as well as the enzymes that degrade the plant cell wall, oomycete hyphae also secrete proteins that protect the pathogen against plant defense molecules. Such counter-defense proteins include glucanase and protease inhibitors (Damasceno et al., 2008; Rose et al., 2002; Tian et al., 2005).

5.6 GROWTH AND REPRODUCTION IN THE HOST

After penetrating the epidermis, hyphae ramify through the underlying host tissues, growing both intercellularly and intracellularly. In the case of necrotrophic species, host colonization is accompanied by host cell death to allow the pathogen access to sources of nutrients from within the dying or dead cells. In the case of biotropic species, such as the downy mildews and white rusts in the genera *Plasmopara*, *Hyaloperonospora*, *Bremia* and *Albugo*, disruption of the host is minimized because it is important that the integrity of the host cells is maintained so the pathogen can establish a stable relationship with living host cells. This is also the case during the initial biotrophic phase of hemibiotrophic *Phytophthora* species. During biotrophic growth, hyphal contact with a suitable host parenchyma cell triggers the formation of haustoria, specialized feeding cells, within the living host. For foliar pathogens, the haustoria form in mesophyll cells (Fig. 5.2z); for root pathogens, they form in cortical cells (Fig. 5.2aa). During haustorium development, the host cell wall is locally degraded, and a thin penetration hypha grows though the wall to invaginate the plant plasma membrane as it enters and continues to expand within the cell (Enkerli et al., 1997). At maturity, oomycete haustoria may be elongated, spherical or lobed. They contain the usual complement of organelles, although in a few cases, nuclei do not migrate into the haustorium from the subtending hypha (Soylu et al., 2003; Soylu, 2004).

In both compatible and incompatible interactions, the initial host cell invasion by oomycete pathogens is accompanied by rapid changes in the organization of the host cell (Takemoto et al., 2003). Most of these initial responses are likely to be associated with the basal host defense response. In susceptible hosts, where these defenses fail to halt pathogen ingress, haustoria continue to develop. For haustoria to function in the uptake of nutrients from the living host cells in which they have formed, they must be able to orchestrate extensive changes in host cell organization and metabolism. Recent research indicates that they do this by transporting specialized effector proteins into the host cell (Whisson et al., 2007), see also Chapter 18. The resulting changes to host cell metabolism favor export of nutrients from the plant cell and their uptake by the haustoria. Successful colonization of the host by necrotrophic and biotrophic oomycetes culminates in sporulation. Within 1–3 days, asexual chlamydospores or sporangia form on the plant surface or sexual spores may form within the host tissues (Chambers et al., 1995) and the pathogen's life cycle begins again.

ACKNOWLEDGMENTS

I would like to thank Drs. M. Cope, F. Gubler, H. Judelson, C. Mims, and D. Takemoto for allowing me to use their published or unpublished micrographs, and Dr. M. Boutry for providing antibodies to the plasma membrane H^+-ATPase.

REFERENCES

Ah Fong AMV, Judelson HS 2003. Cell cycle regulator Cdc14 is expressed during sporulation but not hyphal growth in the fungus-like oomycete *Phytophthora infestans*. Molec Microbiol 50:487–494.

Alexopoulos CJ, Mims CW, Blackwell M 1996. Introductory mycology. John Wiley & Sons: New York. p. 867.

Ali EH 2005. Morphological and biochemical alterations of oomycete fish pathogen *Saprolegnia parasitica* as affected by salinity, ascorbic acid and their synergistic action. Mycopathologia 159:231–243.

Allaway WG, Ashford AE, Heath IB, Hardham AR 1998. Vacuolar reticulum in oomycete hyphal tips: an additional component of the Ca^{2+} regulatory system? Fungal Genet Biol 22:209–220.

Ambikapathy J, Marshall JS, Hocart CH, Hardham AR 2002. The role of proline in osmoregulation in *Phytophthora nicotianae*. Fungal Genet Biol 35:287–299.

Andersson MG, Cerenius L 2002. Pumilio homologue from *Saprolegnia parasitica* specifically expressed in undifferentiated spore cysts. Eukaryotic Cell 1:105–111.

Armbruster BL 1982a. Sporangiogenesis in three genera of the Saprolegniaceae. I. Pre-sporangium hyphae to early primary spore initial stage. Mycologia 74:433–459.

Armbruster BL 1982b. Sporangiogenesis in three genera of the Saprolegniaceae. II Primary spore initial to secondary spore inital stage. Mycologia 74:975–999.

Avrova AO, Venter E, Birch PRJ, Whisson SC 2003. Profiling and quantifying differential gene transcription in *Phytophthora infestans* prior to and during the early stages of potato infection. Fungal Genet Biol 40:4–14.

Beakes GW 1989. Oomycete fungi: their phylogeny and relationship to chromophyte algae. In: Green JC, Leadbeater BSC, Diver WL, editors. The chromophyte algae: problems and perspectives. Clarendon Press, Oxford, UK. pp. 325–342.

Beakes GW, Burr AW, Wood SE, Hardham AR 1995. The application of spore surface features in defining taxonomic versus ecological groupings in Oomycete fungi. Can J Bot 73:S701–S711.

Bircher U, Hohl HR 1997. Environmental signalling during induction of appressorium formation in *Phytophthora*. Mycol Res 101:395–402.

Bishop-Hurley SL, Mounter SA, Laskey J, Morris RO, Elder J, Roop P, Rouse C, Schmidt FJ, English JT 2002. Phage-displayed peptides as developmental agonists for *Phytophthora capsici* zoospores. Appl Environ Microbiol 68:3315–3320.

Blanco FA, Judelson HS 2005. A bZIP transcription factor from *Phytophthora* interacts with a protein kinase and is required for zoospore motility and plant infection. Molec Microbiol 56:638–648.

Boudjeko T, Andème-Onzighi C, Vicré M, Balangé A-P, Ndoumou DO, Driouich A 2006. Loss of pectin is an early event during infection of cocoyam roots by *Pythium myriotylum*. Planta 223:271–282.

Cahill DM, Cope M, Hardham AR 1996. Thrust reversal by tubular mastigonemes: immunological evidence for a role of mastigonemes in forward motion of zoospores of *Phytophthora cinnamomi*. Protoplasma 194:18–28.

Chambers SM, Hardham AR, Scott ES 1995. *In planta* immunolabelling of three types of peripheral vesicles in cells of *Phytophthora cinnamomi* infecting chestnut roots. Mycol Res 99:1281–1288.

Chapman JA, Vujicic R 1965. The fine structure of sporangia of *Phytophthora erythroseptica* Pethyb. J Gen Microbiol 41:275–282.

Chen D-W, Zentmyer GA 1970. Production of sporangia by *Phytophthora cinnamomi* in axenic culture. Mycologia 62:397–402.

Clark MC, Melanson DL, Page OT 1978. Purine metabolism and differential inhibition of spore germination in *Phytophthora infestans*. Can J Microbiol 24:1032–1038.

Cope M, Hardham AR 1994. Synthesis and assembly of flagellar surface antigens during zoosporogenesis in *Phytophthora cinnamomi*. Protoplasma 180:158–168.

Costanzo S, Ospina-Giraldo MD, Deahl KL, Baker CJ, Jones RW 2006. Gene duplication event in family 12 glucosyl hydrolase from *Phytophthora* spp. Fungal Gen Biol 43:707–714.

Cvitanich C, Judelson HS 2003. A gene expressed during sexual and asexual sporulation in *Phytophthora infestans* is a member of the Puf family of translational regulators. Eukaryot Cell 2:465–473.

Damasceno CMB, Bishop JG, Ripoll DR, Win J, Kamoun S, Rose JKC 2008. Structure of the glucanase inhibitor protein (GIP) family from Phytophthora species suggests coevolution with plant endo-beta-1,3-glucanases. Molec Plant-Microbe Interact 21:820–830.

Deacon JW, Donaldson SP 1993. Molecular recognition in the homing response of zoosporic fungi, with special reference to *Pythium* and *Phytophthora*. Mycol Res 97: 1153–1171.

Dearnaley JDW, Maleszka J, Hardham AR 1996. Synthesis of zoospore peripheral vesicles during sporulation of *Phytophthora cinnamomi*. Mycol Res 100:39–48.

Dieguez-Uribeondo J, Gierz G, Bartnicki-Garcia S 2004. Image analysis of hyphal morphogenesis in Saprolegniaceae (Oomycetes). Fungal Gene Biol 41:293–307.

Ebstrup T, Saalbach G, Egsgaard H 2005. A proteomics study of in vitro cyst germination and appressoria formation in *Phytophthora infestans*. Proteomics 5:2839–2848.

Elsner PR, VanderMolen GE, Horton JC, Bowen CC 1970. Fine structure of *Phytophthora infestans* during sporangial differentiation and germination. Phytopathol 60:1765–1772.

Enkerli K, Hahn MG, Mims CW 1997. Ultrastructure of compatible and incompatible interactions of soybean roots infected with the plant pathogenic oomycete *Phytophthora sojae*. Can J Bot 75:1493–1508.

Erwin DC, Ribeiro OK 1996. *Phytophthora* diseases worldwide. APS Press. St. Paul, Minnesota. p. 562.

Fok AK, Aihara MS, Ishida M, Nolta KV, Steck TL, Allen RD 1995. The pegs on the decorated tubules of the contractile vacuole complex of *Paramecium* are proton pumps. J Cell Sci 108:3163–3170.

Gaulin E, Jauneau A, Villalba F, Rickauer M, Esquerré-Tugayé MT, Bottin A 2002. The CBEL glycoprotein of *Phytophthora parasitica* var. *nicotianae* is involved in cell wall deposition and adhesion to cellulosic substrates. J Cell Sci 115:4565–4575.

Gay JL, Greenwood AD 1966. Structural aspects of zoospore production in *Saprolegnia ferax* with particular reference to the cell and vacuolar membranes. In: Madelin MF, editor. The fungal spore. Butterworths, London, UK. pp. 95–108.

Gindro K, Pezet R, Viret O 2003. Histological study of the responses of two *Vitis vinifera* cultivars (resistant and susceptible) to *Plasmopara viticola* infections. Plant Physiol Biochemi 41:846–853.

Gisi U 1983. Biophysical aspects of the development of *Phytophthora*. In: Erwin DC, Bartnicki-Garcia S, Tsao PH, editors. Phytophthora. Its biology, taxonomy, ecology, and pathology. American Phytopathological Society. St Paul, MN. pp. 109–119.

Gisi U, Hemmes DE, Zentmyer GA 1979. Origin and significance of the discharge vesicle in *Phytophthora*. Exp Mycol 3:321–339.

Gisi U, Zentmyer GA 1980. Mechanism of zoospore release in *Phytophthora* and *Pythium*. Exp Mycol 4:362–377.

Goldberg NP, Hawes MC, Stanghellini ME 1989. Specific attraction to and infection of cotton root cap cells by zoospores of *Pythium dissotocum*. Can J Bot 67:1760–1767.

Görnhardt B, Rouhara I, Schmelzer E 2000. Cyst germination proteins of the potato pathogen *Phytophthora infestans* share homology with human mucins. Molecu Plant-Microbe Interact 13:32–42.

Gotelli D 1974. The morphology of *Lagenidium callinectes* II. zoosporogenesis. Mycologia 66:846–858.

Götesson A, Marshall JS, Jones DA, Hardham AR 2002. Characterization and evolutionary analysis of a large polygalacturonase gene family in the oomycete plant pathogen *Phytophthora cinnamomi*. Molecu Plant-Microbe Interact 15:907–921.

Gow NAR 2004. New angles in mycology: studies in directional growth and directional motility. Mycol Res 108:5–13.

Grenville-Briggs LJ, Anderson VL, Fugelstad J, Avrova AO, Bouzenzana J, Williams A, Wawra S, Whisson SC, Birch PRJ, Bulone V, Van West P 2008. Cellulose synthesis in *Phytophthora infestans* is required for normal appressorium formation and successful infection of potato. Plant Cell 20:1–19.

Grenville-Briggs LJ, Avrova AO, Bruce CR, Williams A, Whisson SC, Birch PRJ, Van West P 2005. Elevated amino acid biosynthesis in *Phytophthora infestans* during appressorium formation and potato infection. Fungal Genet Biol 42:244–256.

Griffin DH, Breuker C 1969. Ribonucleic acid synthesis during the differentiation of sporangia in the water mold *Achlya*. J Bacteriol 98:689–696.

Gubler F, Hardham AR 1988. Secretion of adhesive material during encystment of *Phytophthora cinnamomi* zoospores, characterized by immunogold labelling with monoclonal antibodies to components of peripheral vesicles. J Cell Sci 90:225–235.

Gubler F, Hardham AR 1990. Protein storage in large peripheral vesicles in *Phytophthora* zoospores and its breakdown after cyst germination. Exp Mycol 14:393–404.

Gubler F, Hardham AR 1991. The fate of peripheral vesicles in zoospores of *Phytophthora cinnamomi* during infection of plants. In: Mendgen K, Lesemann D-E, editors. Electron microscopy of plant pathogens. Springer-Verlag. Berlin, Germany. pp. 197–210.

Gwynne DI, Brandhorst BP 1982. Alterations in gene expression during heat shock of *Achlya ambisexualis*. J Bacteriol 149:488–493.

Hardham AR 1987. Ultrastructure and serial section reconstruction of zoospores of the fungus *Phytophthora cinnamomi*. Exp Mycol 11:297–306.

Hardham AR 1989. Lectin and antibody labelling of surface components of spores of *Phytophthora cinnamomi*. Aust J Plant Physiol 16:19–32.

Hardham AR 1995. Polarity of vesicle distribution in oomycete zoospores: development of polarity and importance for infection. Can J Bot 73:S400–S407.

Hardham AR 2001. The cell biology behind *Phytophthora* pathogenicity. Australas Plant Pathol 30:91–98.

Hardham AR 2005. *Phytophthora cinnamomi*. Molec Plant Pathol 6:589–604.

Hardham AR 2007. Cell biology of plant-oomycete interactions. Cell Microbiol 9:31–39.

Hardham AR, Cahill DM, Cope M, Gabor BK, Gubler F, Hyde GJ 1994. Cell surface antigens of *Phytophthora* spores: biological and taxonomic characterization. Protoplasma 181:213–232.

Hardham AR, Gubler F 1990. Polarity of attachment of zoospores of a root pathogen and pre-alignment of the emerging germ tube. Cell Biol Int Rep 14:947–956.

Hardham AR, Hyde GJ 1997. Asexual sporulation in the Oomycetes. Adv Bot Res 24:353–398.

Hardham AR, Suzaki E 1986. Encystment of zoospores of the fungus, *Phytophthora cinnamomi*, is induced by specific lectin and monoclonal antibody binding to the cell surface. Protoplasma 133:165–173.

Heath IB 1987. Preservation of a labile cortical array of actin filaments in growing hyphal tips of the fungus *Saprolegnia ferax*. Eur J Cell Biol 44:10–16.

Heath IB, Kaminskyj SGW 1989. The organization of tip-growth-related organelles and microtubules revealed by quantitative analysis of freeze-substituted oomycete hyphae. J Cell Sci 93:41–52.

Hoch HC, Mitchell JE 1972a. The ultrastructure of *Aphanomyces euteiches* during asexual spore formation. Phytopathol 62:149–160.

Hoch HC, Mitchell JE 1972b. The ultrastructure of zoospores of *Aphanomyces euteiches* and of their encystment and subsequent germination. Protoplasma 75:113–138.

Hoch HC, Mitchell JE 1973. The effects of osmotic water potentials on *Aphanomyces euteiches* during zoosporogenesis. Can J Bot 51:413–420.

Hoch HC, Mitchell JE 1975. Further observations on the mechanisms involved in primary spore cleavage in *Aphanomyces euteiches*. Can J Bot 53:1085–1091.

Hohl HR 1991. Nutrition. In: Ingram DS, Williams PH, editors. Advances in plant pathology volume 7 *Phytophthora infestans*, the cause of late blight of potato. 7. Academic Press. London, UK. pp. 52–83.

Hohl HR, Hamamoto ST 1967. Ultrastructural changes during zoospore formation in *Phytophthora parasitica*. Am J Bot 54:1131–1139.

Howard RJ, Ferrari MA, Roach DH, Money NP 1991. Penetration of hard substrates by a fungus employing enormous turgor pressures. Proc Nat Acad Sci USA. 88:11281–11284.

Hyde GJ, Gubler F, Hardham AR 1991a. Ultrastructure of zoosporogenesis in *Phytophthora cinnamomi*. Mycol Res 95:577–591.

Hyde GJ, Hardham AR 1993. Microtubules regulate the generation of polarity in zoospores of *Phytophthora cinnamomi*. Eur J Cell Biol 62:75–85.

Hyde GJ, Lancelle S, Hepler PK, Hardham AR 1991b. Freeze substitution reveals a new model for sporangial cleavage in *Phytophthora*, a result with implications for cytokinesis in other eukaryotes. J Cell Sci 100:735–746.

Jackson SL, Hardham AR 1996. A transient rise in cytoplasmic free calcium is required for the induction of cytokinesis in zoosporangia of *Phytophthora cinnamomi*. Eur J Cell Biol 69:180–188.

Jackson SL, Hardham AR 1998. Dynamic rearrangement of the filamentous actin network occurs during zoosporogenesis and encystment in the oomycete *Phytophthora cinnamomi*. Fungal Gene Biol 24: 24–33.

Jahn TL, Landman MD, Fonseca JR 1964. The mechanism of locomotion of flagellates. II. function of the mastigonemes of *Ochromonas*. J Protozool 11:291–296.

Jelke E, Oertel B, Böhm KJ, Unger E 1987. Tubular cytoskeletal elements in sporangia and zoospores of *Phytophthora infestans* (Mont.) de Bary (Oomycetes, Pythiaceae). J Basic Microbiol 27:11–21.

Judelson HS, Ah-Fong AMV, Aux G, Avrova AO, Bruce C, Calkir C, da Cunha L, Grenville-Briggs L, Latijnhouwers M, Ligterink W, et al. 2008. Gene expression profiling during asexual development of the late blight pathogen *Phytophthora infestans* reveals a highly dynamic transcriptome. Molec Plant-Microbe Interact 21: 433–447.

Judelson HS, Blanco FA 2005. The spores of *Phytophthora*: weapons of the plant destroyer. Nat Rev Microbiol 3:47–58.

Judelson HS, Roberts S 2002. Novel protein kinase induced during sporangial cleavage in the oomycete *Phytophthora infestans*. Eukaryot Cell 1:687–695.

Judelson HS, Tani S 2007. Transgene-induced silencing of the zoosporogenesis-specific NIFC gene cluster of *Phytophthora infestans* involves chromatin alterations. Eukaryot Cell 6:1200–1209.

Kevorkian AG 1996. Studies in the Leptomitaceae. II Cytology of *Apodachlya brachynema* and *Sapromyces reinschii*. Mycologia 27:274–285.

Kiefer B, Riemann M, Buche C, Kassemeyer H-H, Nick P 2002. The host guides morphogenesis and stomatal targeting in the grapevine pathogen *Plasmopara viticola*. Planta 215:387–393.

Kim KS, Judelson HS 2003. Sporangia-specific gene expression in the oomycete phytopathogen *Phytophthora infestans*. Eukaryot Cell 2:1376–1385.

King JE, Colhoun J, Butler RD 1968. Changes in the ultrastructure of sporangia of *Phytophthora infestans* associated with indirect germination and ageing. Trans Br Mycol Soc 51:269–281.

Kiryu Y, Blazer VS, Vogelbein WK, Kator H, Shields JD 2005. Factors influencing the sporulation and cyst formation of *Aphanomyces invadans*, etiological agent of ulcerative mycosis in Atlantic menhaden, *Brevoortia tyrannus*. Mycologia 97:569–575.

Krämer R, Freytag S, Schmelzer E 1997. *In vitro* formation of infection structures of *Phytophthora infestans* is associated with synthesis of stage specific polypeptides. Eur J Plant Pathol 103:43–53.

Lange L, Edén U, Olson LW 1989. Zoosporogenesis in *Pseudoperonospora cubensis*, the causal agent of cucurbit downy mildew. Nord J Bot 8:497–504.

Lange L, Olson LW, Safeeulla KM 1984. Pearl millet downy mildew (*Sclerospora graminicola*): zoosporogensis. Protoplasma 119:178–187.

Latijnhouwers M, Govers F 2003. A *Phytophthora infestans* G-protein B-subunit is involved in sporangium formation. Eukaryot Cell 2:971–977.

Latijnhouwers M, Ligterink W, Vleeshouwers VGAA, Van West P, Govers F 2004. A G alpha subunit controls zoospore motility and virulence in the potato late blight pathogen *Phytophthora infestans*. Molec Microbiol 51:925–936.

Latijnhouwers M, Munnik T, Govers F 2002. Phospholipase D in *Phytophthora infestans* and its role in zoospore differentiation. Molecular Plant-Microbe Interactions 15:939–946.

Laxalt AM, Latijnhouwers M, van Hulten M, Govers F 2002. Differential expression of G protein a and b subunit genes during development fo *Phytophthora infestans*. Fungal Gene Biol 36:137–146.

Lehnen LP, Powell MJ 1989. The role of kinetosome-associated organelles in the attachment of encysting secondary zoospores of *Saprolegnia ferax* to substrates. Protoplasma 149:163–174.

Loprete DM, Hill TW 2002. Isolation and characterization of an endo-(1,4)-beta-glucanase secreted by *Achlya ambisexualis*. Mycologia 94:903–911.

Lunney CZ, Bland CE 1976. An ultrastructural study of zoosporogenesis in *Pythium proliferum* de Bary. Protoplasma 88:85–100.

Luo SH, Marchesini N, Moreno SNJ, Docampo R 1999. A plant-like vacuolar H+-pyrophosphatase in *Plasmodium falciparum*. FEBS Lett 460:217–220.

MacDonald JD, Duniway JM 1978. Influence of the matric and osmotic components of water potential on zoospore discharge in *Phytophthora*. Phytopathol 68:751–757.

Maltese CE, Conigliaro G, Shaw DS 1995. The development of sporangia of *Phytophthora infestans*. Mycol Res 99:1175–1181.

Marshall JS, Ashton AR, Govers F, Hardham AR 2001a. Isolation and characterization of four genes encoding pyruvate, phosphate dikinase in the oomycete plant pathogen *Phytophthora cinnamomi*. Curr Genet 40:73–81.

Marshall JS, Wilkinson JM, Moore T, Hardham AR 2001b. Structure and expression of the genes encoding proteins resident in large peripheral vesicles of *Phytophthora cinnamomi* zoospores. Protoplasma 215:226–239.

McCully ME, Canny MJ 1985. The stabilization of labile configurations of plant cytoplasm by freeze substitution. J Microsc 139:27–33.

McLeod A, Smart CD, Fry WE 2003. Characterization of 1,3-b-glucanase and 1,3;1,4-b-glucanase genes from *Phytophthora infestans*. Fungal Gene Biol 38:250–263.

Mims CW, Richardson EA 2003. Ultrastructure of the zoosporangia of *Albugo ipomoeae-panduratae* as revealed by conventional chemical fixation and high pressure freezing followed by freeze substitution. Mycologia 95:1–10.

Mitchell HJ, Hardham AR 1999. Characterisation of the water expulsion vacuole in *Phytophthora nicotianae* zoospores. Protoplasma 206:118–130.

Money NP, Webster J 1985. Water stress and sporangial emptying in *Achlya* (Saprolegniaceae). Bot J Linn Soc 91:319–327.

Money NP, Webster J, Ennos R 1988. Dynamics of sporangial empyting in *Achlya intricata*. Exp Mycol 12:13–27.

Morris BM, Gow NAR 1993. Mechanism of electrotaxis of zoospores of phytopathogenic fungi. Phytopathol 83:877–882.

Morris PF, Ward EWB 1992. Chemoattraction of zoospores of the soybean pathogen, *Phytophthora sojae*, by isoflavones. Physiological and Molecular Plant Pathology 40:17–22.

Morrissette NS, Sibley LD 2002. Cytoskeleton of apicomplexan parasites. Microbiol Molec Biol Revi 66:21–38.

Morsomme P, de Kerchove d'Exaerde A, Goffeau A, Boutry M 1996. Single point mutations in various domains of a plant plasma membrane H^+-ATPase expressed in *Saccharomyces cerevisiae* increase H^+-pumping and permit yeast growth at low pH. EMBO J 15:55113–5526.

Oertel B, Jelke E 1986. Formation of multinucleate zoospores in *Phytophthora infestans* (Mont.) de Bary (*Oomycetes, Pythiaceae*). Protoplasma 135:173–179.

Patterson DJ 1980. Contractile vacuoles and associated structures: their organization and function. Biol Rev 55:1–46.

Penington CJ, Iser JR, Grant BR, Gayler KR 1989. Role of RNA and protein synthesis in stimulated germination of zoospores of the pathogenic fungus *Phytophthora palmivora*. Exp Mycol 13:158–168.

Randall TA, Dwyer RA, Huitema E, Beyer K, Cvitanich C, Kelkar H, Fong AMVA, Gates K, Roberts S, Yatzkan E, et al. 2005. Large-scale gene discovery in the oomycete *Phytophthora infestans* reveals likely components of phytopathogenicity shared with true fungi. Molec Plant-Microbe Interact 18:229–243.

Riemann M, Büche C, Kassemaeyer H-H, Nick P 2002. Cytoskeletal responses during early development of the downy mildew of grapevine (*Plasmopara viticola*). Protoplasma 219:13–22.

Robold AV, Hardham AR 2005. During attachment *Phytophthora* spores secrete proteins containing thrombospondin type 1 repeats. Curr Genet 47:307–315.

Rodrigues CO, Scott DA, Bailey BN, de Souza W, Benchimol M, Moreno B, Urbina JA, Oldfield E, Moreno SNJ 2000. Vacuolar proton pyrophosphatase activity and pyrophosphate (PP_i) in *Toxoplasma gondii* as possible chemotherapeutic targets. Biochem J 349:737–745.

Rose JKC, Ham K-S, Darvill AG, Albersheim P 2002. Molecular cloning and characterization of glucanase inhibitor proteins: coevolution of a counterdefence mechanism by plant pathogens. Plant Cell 14:1329–1345.

Rumbolz J, Wirtz S, Kassemeyer HH, Guggenheim R, Schafer E, Buche C 2002. Sporulation of *Plasmopara viticola*: Differentiation and light regulation. Plant Biol 4:413–422.

Sakihama Y, Shimai T, Sakasai M, Ito T, Fukushi Y, Hashidoko Y, Tahara S 2004. A photoaffinity probe designed for host-specific signal flavonoid receptors in phytopathogenic Peronosporomycete zoospores of of *Aphanomyces cochlioides*. Arch Biochem Biophys 432:145–151.

Schnepf E, Deichgräber G, Drebes G 1978. Development and ultrastructure of the marine, parasitic Oomycete, *Lagenisma coscinodisci* Drebes (Lagenidiales). Thallus, zoosporangium, mitosis, and meiosis. Arch Microbiol 116:141–150.

Shan W, Marshall JS, Hardham AR 2004. Stage-specific expression of genes in germinated cysts of *Phytophthora nicotianae*. Molec Plant Pathol 5:317–330.

Shapiro A, Mullins JT 2002. Hyphal tip growth in *Achlya bisexualis*. I. Distribution of 1,3-b-glucans in elongating and non-elongating regions of the wall. Mycologia 94:267–272.

Škalamera D, Hardham AR 2006. PnCcp, a *Phytophthora nicotianae* protein containing a single complement control protein module, is sorted into large peripheral vesicles in zoospores. Australas Plant Pathol 35:593–603.

Slusarenko AJ, Schlaich NL 2003. Downy mildew of *Arabidopsis thaliana* caused by *Hyaloperonospora parasitica* (formerly *Peronospora parasitica*). Molec Plant Pathol 4:159–170.

de Souza W 2006. Secretory organelles of pathogenic protozoa. Anais da Academia Brasileira de Ciencias 78:271–291.

Soylu EM, Soylu S 2003. Light and electron microscopy of the compatible interaction between *Arabidopsis* and the downy mildew pathogen *Peronospora parasitica*. J Phytopath 151:300–306.

Soylu S 2004. Ultrastructural characterisation of the host-pathogen interface in white blister-infected *Arabidopsis* leaves. Mycopathologia 158:457–464.

Soylu S, Keshavarzi M, Brown I, Mansfield JW 2003. Ultrastructural characterisation of interactions between *Arabidopsis thaliana* and *Albugo candida*. Physiol Molec Plant Pathol 63:201–211.

Stevens TH, Forgac M 1997. Structure, function and regulation of the vacuolar (H^+)-ATPase. Annu Rev Cell Dev Biol 13:779–808.

Suzaki E, Suzaki T, Jackson SL, Hardham AR 1996. Changes in intracellular pH during zoosporogenesis in *Phytophthora cinnamomi*. Protoplasma 191:79–83.

Takemoto D, Jones DA, Hardham AR 2003. GFP-tagging of cell components reveals the dynamics of subcellular re-organization in response to infection of *Arabidopsis* by oomycete pathogens. Plant J 33:775–792.

Tani S, Judelson HS 2006. Activation of zoosporogenesis-specific genes in *Phytophthora infestans* involves a 7-nucleotide promoter motif and cold-induced membrane rigidity. Eukaryo Cell 5:745–752.

Tani S, Yatzkan E, Judelson HS 2004. Multiple pathways regulate the induction of genes during zoosporogenesis in *Phytophthora infestans*. Molec Plant-Microbe Interact 17:330–337.

Tian M, Benedetti B, Kamoun S 2005. A second kazal-like protease inhibitor from *Phytophthora infestans* inhibits and interacts with the apoplastic pathogenesis-related protease P69B of tomato. Plant Physiol 138:1785–1793.

Timberlake WE, McDowell L, Cheney J, Griffin DH 1973. Protein synthesis during the differentiation of sporangia in the water mold *Achlya*. J Bacteriol 116:67–73.

Tomley FM, Soldati DS 2001. Mix and match modules: structure and function of microneme proteins in apicomplexan parasites. Trends Parasitol 17:81–88.

Torto TA, Rauser L, Kamoun S 2002. The *pipg1* gene of the oomycete *Phytophthora infestans* encodes a fungal-like endopolygalacturonase. Curr Genet 40:385–390.

Trigano RN, Spurr HW Jr 1987. The development of the multinucleate condition of *Peronospora tabacina* sporangia. Mycologia 79:353–357.

Tyler BM 2002. Molecular basis of recognition between Phytophthora pathogens and their hosts. Ann Rev Phytopathol 40:137–167.

Van West P, Morris BM, Reid B, Appiah AA, Osborne MC, Campbell TA, Shepherd SJ, Gow NAR 2002. Oomycete plant pathogens use electric fields to target roots. Molecu Plant-Microbe Interact 15:790–798.

Walker CA, Köppe M, Grenville-Briggs LJ, Avrova AO, Horner NR, McKinnon AD, Whisson SC, Birch PRJ, Van West P 2008. A putative DEAD-box RNA-helicase is required for normal zoospore development in the late blight pathogen *Phytophthora infestans*. Fungal Genet Biol 45:954–962.

Walker SK, Chitcholtan K, Yu YP, Christenhusz GM, Garrill A 2006. Invasive hyphal growth: An F-actin depleted zone is associated with invasive hyphae of the oomycetes *Achlya bisexualis* and *Phytophthora cinnamomi*. Fungal Gene Biol 43:357–365.

Webster J, Dennis C 1967. The mechanism of sporangial discharge in *Pythium middletonii*. New Phytol 66:307–313.

Whisson SC, Boevink P, Moleleki L, Avrova AO, Morales JG, Gilroy AM, Armstrong MR, Grouffaud S, Van West P, Chapman S, et al. 2007. A translocation signal for delivery of oomycete effector proteins into host plant cells. Nature 450:115–119.

Williams WT, Webster RK 1970. Electron microscopy of the sporangium of *Phytophthora capsici*. Can J Bot 48:221–227.

Yan H-Z, Liou R-F 2005. Cloning and analysis of *pppg1*, an inducible endopolygalacturonase gene from the oomycete plant pathogen *Phytophthora parasitica*. Fungal Gene Biol 42:339–350.

6

SEXUAL REPRODUCTION IN OOMYCETES: BIOLOGY, DIVERSITY, AND CONTRIBUTIONS TO FITNESS

Howard S. Judelson
University of California, Riverside, California

6.1 INTRODUCTION

The defining attribute of the class Oomycotina is the thick-walled oospore, which is the culmination of sexual reproduction. These spores are important to the biology of most oomycetes because they can survive inhospitable environments such as freezing or dry conditions and can resist microbial degradation. The sexual cycle also enhances fitness by providing a mechanism for genetic variation. The features of sexual reproduction have also been useful to systematists for distinguishing oomycetes from true fungi. Several characteristics unique to oomycetes include their distinct female and male gametangia (oogonia and antheridia, respectively), a diploid vegetative stage with gametangial meiosis, and the use of secondary metabolites to regulate sexual compatibility and differentiation. In contrast, most true fungi are haploid and employ proteinaceous mating hormones.

Oomycetes exhibit substantial variation in the details of sexual reproduction, with such differences underlying current species classification schemes. This chapter will illustrate much of that diversity. For more extensive information, the reader is referred to the literature, especially the enormous contributions from Dick et al. (2001a and b). The importance of the sexual cycle in the life and disease cycles of selected oomycetes will also be discussed.

Oomycete Genetics and Genomics: Diversity, Interactions, and Research Tools
Edited by Kurt Lamour and Sophien Kamoun
Copyright © 2009 John Wiley & Sons, Inc.

6.2 SEXUAL DEVELOPMENT AND OOSPORE FORMATION

Both homothallic and heterothallic mating systems exist within the oomycetes, with the former predominating across most orders. Homothallics are capable of sexual reproduction in single culture, whereas matings between heterothallics require distinct sexual compatibility types (also called mating or induction types). These are called A1 and A2 in *Phytophthora*, B1 and B2 in *Bremia*, and P1 and P2 in *Plasmopara* (Gallegly, 1968; Michelmore and Sansome, 1982; Wong et al., 2001).

The representative structures made during sexual development are shown in Fig. 6.1. This begins with the formation of male and female gametangia, which in most species appear as swollen hyphal tips (Fabritius et al., 2002; Hemmes, 1983). The oogonial initials usually form without septation from the rest of the thallus and are considered female because they provide most cytoplasm for the oospores, which develop within oogonia. In contrast, antheridia are normally delimited by a septum. Pairing is apparently facilitated by an adhesive secreted by the antheridium; after contact, both organs swell although this is more extreme for the oogonium. Synchronous meioses occur within each gametangium, and a fertilization tube or pore develops between antheridium and oogonium. A single haploid nucleus is then transmitted from a male to a female. Oogonia typically contain about 10 gametic nuclei, although most degrade as only one diploid zygotic nucleus is established. After fertilization, a plug partitions the rest of the thallus from the oogonial cytoplasm, which becomes rich in lipid bodies, proteins, and β-linked glucans. Also formed is a phosphate-rich birefringent vacuolar body called the ooplast, which represents about half of the oospore volume and largely defines its geometry (Howard, 1971).

Maturation of the oospore involves establishing a thick multilayered wall, which in many species pulls away from the exterior of the oogonium (the oosphere) to give an endospore-like appearance. The ribosomes and cytochromes disappear during maturation, which indicates that the metabolism is very reduced (Hemmes, 1983; Leary et al., 1974). The combination of a low metabolism, a thick wall, and a lipid-rich cytoplasmic matrix makes oospores extremely effective resting structures.

6.3 INTERSPECIES VARIATION IN OOSPOROGENESIS

Sexual development proceeds as outlined above in most species. However, enormous diversity exists in the orientation and origin of antheridia and oogonia (Fig. 6.1). These can be diclinous (from separate thalli) or monoclinous (same thallus), with variations of the latter including the formation of gametangia from discrete branches of the thallus or the antheridium emerging from the oogonial stalk. Gametangia may also develop terminally, subterminally, or in an intercalary position of hyphae, and they typically emerge from hyphae that are indistinguishable from vegetative hyphae. However, exceptions exist in the Myzocytiopsidales and some Pythiales, in which sexual hyphae become septate

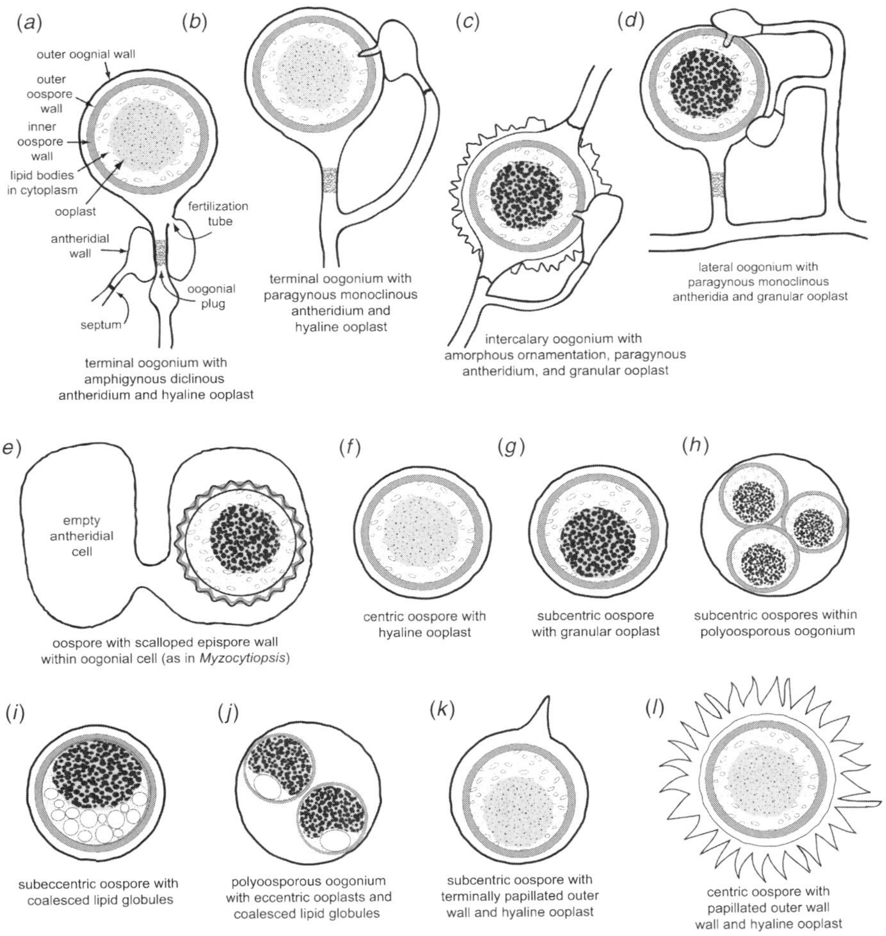

FIG. 6.1 Diversity of sexual structures. Illustrated are examples of variation in gametangial geometry, oospore structure, and ornamentation. Gametangia and oospores from oomycetes with hyphal growth patterns are shown in panels (a–d), and a species with globular growth is shown in panel (e). Panels (f–j) present centric, subcentric, subeccentric, and eccentric oospore geometries. Panels (c), (e), (k), and (l) show several ornamentation patterns. These represent only a subset of possible configurations. For a more comprehensive description see Dick (2001b), who describes 12 forms of the outer oogonial wall, 8 types of oospore walls, and 7 patterns of protoplasmic organization.

with adjacent cells becoming gametangial donors. Gametangial couplings can also be amphigynous, in which the antheridium encircles the oogonial initial, or paragynous, in which the antheridium enters away from the oogonial stalk.

The pattern of oospore development is also variable. For example, whereas most Peronosporales, Pythiales, Rhipidales, and Sclerosporales, as well as some

Leptomitales and Saprolegniales, produce single oospores within each oosphere, some in the latter two orders make multiple oospores. In most species, the fertilization tube is simple and small, although these are branched in polyoosporous species and are often large in species that have globular vegetative stages. Oospores form either from the perimeter inward (centripetal) or outward from the center (centrifugal). The lipid globules within an oospore are either single or numerous, or uniformly distributed or noncentric, and ooplasts can be solid and hyaline or fluid and granular. Although the ooplast is usually spherical, in species that make large lipid globules, it often distorts into a kidney-like shape. The resulting orientations of oosporic cytoplasm are termed centric, eccentric, subcentric, and subeccentric, and these arrangements are of significant taxonomic value (Dick, 2001a and b). Most oospores have diameters between 15 and 50 μm.

Both oogonial and oospore (epispore) walls also show much diversity. In some species, these are smooth and spherical, but in others, they are ornamented with outgrowths such as papillae and spines, coated with mucilage, or pitted. The oospore wall may also retract from the oogonial wall, or it may abut the latter. Walls are usually hyaline but can contain a brownish pigment in some species. Divergence is also observed in wall thickness. For example, in most Pythiales, this ranges between 0.2 and 1 μm, and it ranges between 3 and 6 μm in most Peronosporales and Saprolegniales (Hemmes, 1983; Ruben and Stanghellini, 1978; Vercesi et al., 1999). However, there are exceptions and thicknesses may vary with oospore age (Ho et al., 2002).

The distinct oospore structures likely reflect species-specific functional differences. For example, the mucilage that extends from oospores in some species may serve an ecological function by allowing oogonia to attach to organic matter (Dick, 2001). The varying sizes of oospore components may reflect differential use of nutrients at germination. For example, species that germinate through zoospore-releasing germ sporangia may rely more heavily on stored lipids. Conversely, species in which germination occurs through rapidly elongating germ tubes may be more dependent on glucans stored in the wall.

6.4 REGULATION OF SEXUAL DEVELOPMENT

Multiple factors determine when and how sexual development occurs. These factors include whether isolates are homothallic or heterothallic, their disposition to generate male or female gametangia, and the environment. Physiological, biochemical, and genetic studies have demonstrated complexity in the sexual cycle, with much remaining to be learned. Unfortunately, only a few species have been subjected to detailed analysis.

6.4.1 Sexual Dimorphism, Bisexuality, and Relative Sexuality

A complication in the sexual cycle is imposed by the nature of oomycetes to form distinct male and female gametangia. With heterothallics, parents must not only be of opposite mating type but also competent to form antheridia and oogonia. Strains of many heterothallic species are bisexual, but others are male or female, which limits the number of productive interactions. Although bisexuality is implicit to homothallism, these can also outcross and during such interactions may exhibit sexual preferences.

Whereas some strains can serve only as antheridial or oogonial donors, in others sexuality is variable and is described by a system of relative sexuality. This is best studied in the *Achlya* water molds, in which a strain can be classified on a gradient of maleness or femaleness such that it acts male with a stronger female and acts female with a stronger male (Barksdale and Lasure, 1973). Such behavior is related to the production of the hormones oogonial and antheridiol, which are described later. Relative sexuality is also described in *Phytophthora infestans*, where it is manifested as a quantitative character with strains that produce varying fractions of male and female gametangia in the same culture (Gotoh et al., 2005; Judelson, 1997). Sexual preference is unrelated to mating type and is a polygenic trait (Judelson, 1997; Lasure and Griffin, 1975).

6.4.2 Male and Female Pheromones

Most data on this topic come from *Achyla*, in which early studies observed attraction between strong male and female strains (Raper, 1951). Additional analyses revealed that the production of gametangia and subsequent stages of oosporogenesis are regulated by the diffusible pheromones antheridiol (produced by the female) and oogoniol (from the male; Barksdale, 1969; Barksdale and Lasure, 1973; McMorris, 1978). These C_{29} steroids are active against both heterothallic and homothallic *Achlya* (Fig. 6.2a and b). However, homothallism does not seem to result from producing both hormones, as many homothallics do not secrete antheridiol.

Little is known about functionally equivalent pheromones from other oomycetes, although there is evidence for them in several species, including *Dictyus*, *Pythium*, and *Saprolegnia* (Bishop, 1940; Couch, 1926; Gall and Elliott, 1985). Antheridiol and oogoniol lack activity outside of *Achlya*, but it was suggested that other oomycetes also use steroid-based sex pheromones. This is because species that cannot synthesize sterols, such as *Lagenidium*, *Phytophthora*, and *Pythium*, fail to produce oospores without being provided with sterols, which may be the precursors (Domnas and Warner, 1991; Kerwin and Duddles, 1989). However, this theory is controversial because sterols have other effects, which include promoting fatty acid uptake, membrane fluidity, and vegetative growth (Kerwin and Washino, 1986b). For example, membrane fluidity is presumably a critical factor in gametic fusion.

6.4.3 Regulation by Mating Type

Most information about mating type comes from heterothallic species of *Phytophthora*, in which A1 and A2 types are distinguished by the differential

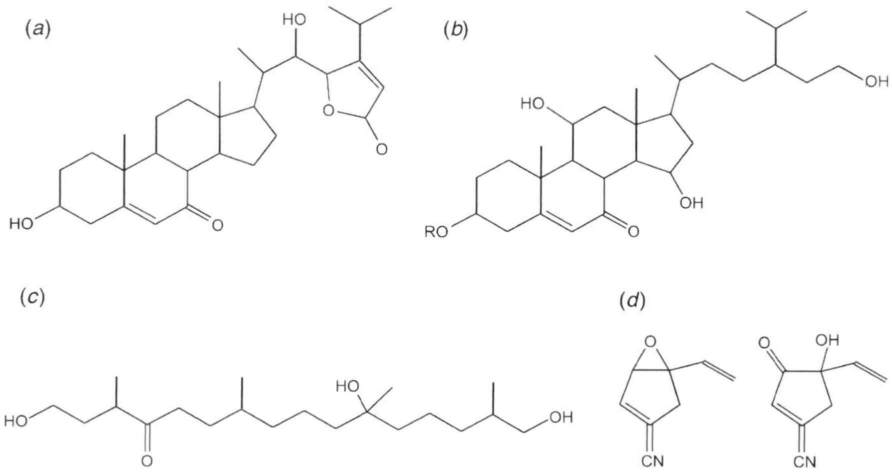

FIG. 6.2 Compounds affecting sexual development. (a) antheridiol. (b) oogoniol, with R indicating the variable group. (c) α1 hormone from *Phytophthora nicotianae*. (d) homothallins produced by *Trichoderma koningii* that mimic the α1 hormone.

production and detection of hormones (Ko, 1988). This establishes a system called hormonal heterothallism, where perception of the opposite hormone establishes compatibility. Once compatibility is realized, gametangia develop that can form both selfed and outcrossed couplings. This is illustrated in Fig. 6.3b–e, which shows interactions in a culture that contains an A1 strain of *P. infestans* that expresses the *GUS* transgene and a nontransgenic A2 strain; the four patterns indicate both the bisexuality of the strains and their abilities to self and hybridize.

This form of chemical heterothallism is unique in biology. Heterothallism elsewhere usually involves sexual dimorphism or self-incompatibility, as in true fungi. In contrast, there is no barrier to selfing in heterothallic oomycetes if both hormones are present. It should be stressed that mating type hormones function distinctly from compounds such as antheridiol, which act only if compatibility is established.

In *P. infestans* and *Phytophthora parasitica* a single nuclear locus that determines mating type was delineated by classic genetics, with A1 and A2 behaving as heterozygote and homozygote, respectively (Fabritius and Judelson, 1997; Judelson et al., 1995). The genes within the region were first identified by chromosome walking from linked DNA markers and later by mining sequenced genomes (Randall et al., 2003). Although the precise determinant of mating type is not yet known, the region has good synteny among heterothallics such as *P. infestans, P. parasitica,* and *Phytophthora ramorum*. The locus is rearranged in the homothallic species *Phytophthora sojae*, but whether this is connected to mating behavior is unknown.

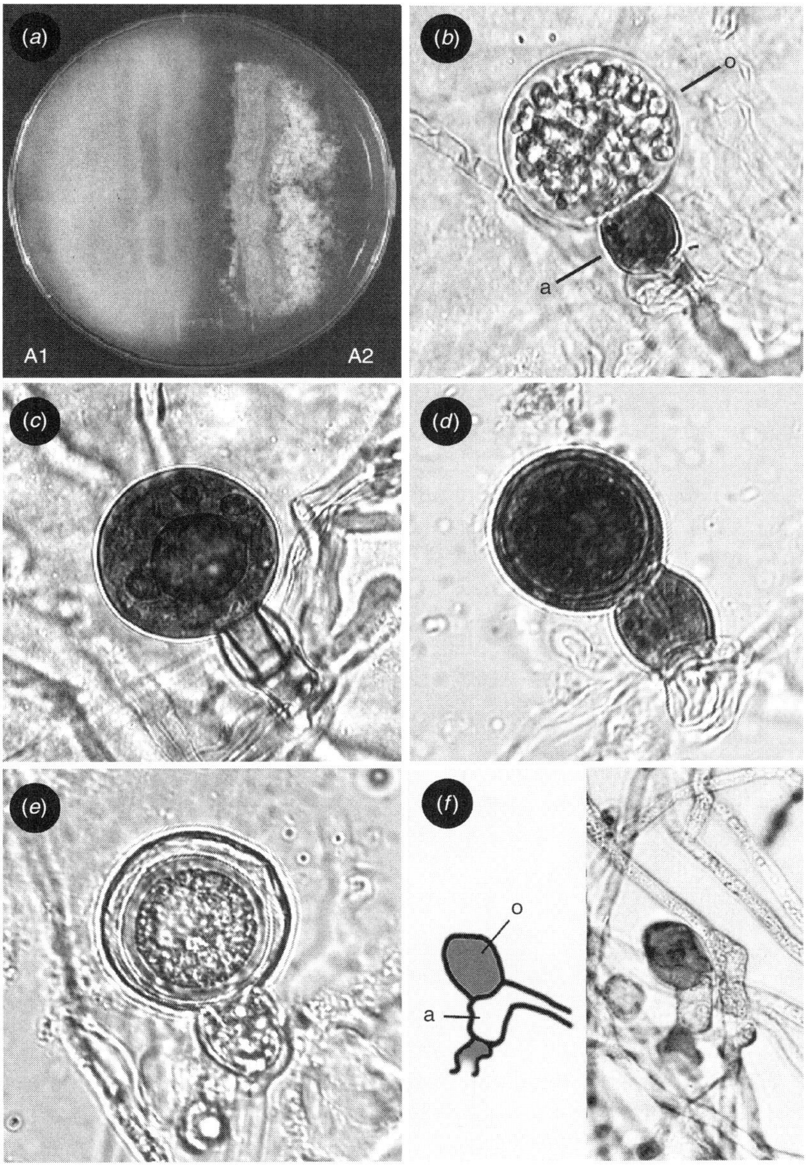

FIG. 6.3 Matings in *Phytophthora infestans* involving parents with GUS reporter genes. (a) coculture of A2 strain (right) with A1 strain constitutively expressing GUS from the *ham34* promoter. (b–e), Four types of gametangial interactions from culture in panel (a), stained for GUS (dark color). Panels (b) and (c) represent hybrid oospores formed from antheridia (a) and oogonia (o) of the two parents, and panels (d) and (e) are selfs. (f) young gametangial interaction involving a strain expressing GUS under control of mating-specific *M90* promoter. Expression is noted in a very young oogonium. Portions reprinted by permission from Cvitanich and Judelson (2003) and Judelson (1997).

Several chromosomal aberrations have been described in connection with the mating type locus, although not all are exhibited by each strain. In *P. infestans*, these include balanced lethals, repressed recombination, translocations, heteromorphic regions, and segregation patterns that are skewed toward one mating type or result in self-fertile progeny (Judelson, 1996; Judelson et al., 1995; Randall et al., 2003; van der Lee et al., 2004). In *Bremia* and *Phytophthora*, cytogenetic analyses associated the locus with reciprocal translocations (Michelmore and Sansome, 1982; Sansome, 1980). The roles of such anomalies will remain obscure until the locus is molecularly defined.

An alternative model for mating type was recently proposed based on fusion experiments between *Phytophthora* strains that carry nuclear and mitochondrial markers for drug resistance (Gu and Ko, 2005). In that study, the mating type and a mitochondrial marker seemed to be linked. How this relates to the data on nuclear inheritance is unclear, because three independent groups associated nuclear loci with mating type (Judelson et al., 1995; van der Lee et al., 2004; Zhang et al., 2006). Unfortunately, the fusion study did not employ molecular markers to determine unambiguously the origins of the fusions and exclude spontaneous mutations or contaminants. Nevertheless, it is conceivable that certain nuclear and mitochondrial genotypes are coinherited as in some other taxa (Berlin and Ellegren, 2001).

The hormone specific to A1 strains of *Phytophthora nicotianae* (α1) was recently identified as an acyclic oxygenated C_{20} compound resembling a diterpene, which acts at nanomolar concentrations (Fig. 6.2c; Qi et al., 2005). The α1 hormone has been synthesized, which should facilitate studies of the mating pathway including its receptor through binding studies (Yajima et al., 2008). The A2 hormone (α2) is not purified, but is also believed to be a slightly polar small compound. The hormones seemed to be fairly conserved throughout *Phytophthora*, because an A1 of one species can usually induce oospores in A2 strains of another and vice versa (Ko, 1988). This has major ecological consequences because multiple oomycetes frequently share the same environmental niche.

6.4.4 Environmental Features Regulating Sexual Development

Generalizations concerning external factors that stimulate oosporogenesis are challenging, reflecting the diversity of oomycetes. Nonetheless, the process can be strongly affected by the composition of media, temperature, light, and pH (Elliott, 1983; Johnson et al., 2002). Such factors often relate to the ecological niche occupied by the species. For example, oosporogenesis by many water molds is favored by salinity levels consistent with their normal estuarine habitats (Harrison and Jones, 1975; Johnson et al., 2002). In many plant pathogens, low carbon/nitrogen ratios typical of infected hosts can be stimulatory (Frinking et al., 1984; Leal et al., 1967). Interestingly, conditions that favour sexual and asexual sporulation are often negatively correlated, and in some species asexual sporulation is repressed within mating zones (Fabritius et al., 2002; Inaba and Morinaka, 1983; Pegg and Mence, 1970).

Certain environmental factors stimulate selfing within otherwise heterothallic oomycetes, in other words they trigger secondary homothallism. These include aging, physical damage, fungicides, host-plant exudates, and volatile "homothallins" from *Trichoderma* (Fig. 6.2d; Brasier, 1971; Groves and Ristaino, 2000; Jayasekera et al., 2007; Reeves and Jackson, 1974; Savage et al., 1968). The number of oospores generated is much less than in normal matings, however. In a study of selfed oospores from aged *P. infestans* cultures, at least some segregated nuclear markers indicated that they resulted from meiosis and not ameiotic apomixis (Smart et al., 1998). Secondary homothallism is also reported in other oomycetes, such as downy mildews (Michelmore and Sansome, 1982; Pushpavathi et al., 2006). Such phenomena are distinct from secondary homothallism in true fungi, which usually results from heterokaryosis. Instead, in oomycetes this seems to be caused by breakdown in the mating type regulation system. This may confer an evolutionary advantage by enabling heterothallics to sometimes form oospores for resisting harsh environments, in addition to the genetic benefits of recombination.

6.5 GERMINATION OF OOSPORES

A "ripening" period is typically required before germination can occur. For example, in *Phytophthora*, germination rates increase with age, up to a few months (Ribeiro, 1983). This can vary, however, even in the same genus. For example, although it was found that at least 8 days of ripening were needed in *Achlya hypogyna*, the oospores of *Achlya recurva* can germinate immediately after formation (Johnson et al., 2002). Most oospores also go through a long period of constitutive dormancy that may last several years. This reversible hypometabolic state may be caused by self-inhibitors or barriers to nutrient entry (Sussman and Douthit, 1973). Consequently, many oospores fail to germinate even in favorable environments. This may enhance fitness by preventing the simultaneous germination of an oospore population at inopportune times. The apparent record for oospore survival, 25 years, belongs to the onion downy mildew agent *Peronospora destructor* (McKay, 1957).

Various factors reportedly stimulate germination, such as plant or soil extracts, light, and alternating humidity and temperature regimes. However, these do not affect all oomycetes (Johnson et al., 2002; Nelson, 1990; Shang et al., 2000; Van Der Gaag and Frinking, 1997). Also suggested to enhance germination in some species is degradation of the oospore wall, which would normally occur because of microbial action. This can be mimicked in the laboratory by enzymes, which is convenient when performing controlled crosses in some species (Shattock et al., 1986). Crosses have been used for experimental studies of traits such as mating behavior, fungicide resistance, and virulence, although low or asynchronous rates of germination of oospores can be a challenge (Judelson and Roberts, 1999; Judelson et al., 1995; MacGregor et al., 2002; van der Lee et al., 2004; Whisson et al., 1995).

When germination does occur, the cytoplasm of the oospore returns to an active configuration. This involves dispersing lipid bodies into the cytoplasm and ooplast contents into the "fingerprint" vacuoles characteristic of oomycetes (Hemmes, 1983). In many species, the glucan reserves are also mobilized, thinning the oospore wall (Ruben and Stanghellini, 1978). At least four modes of germination are noted. These entail producing a single germ hypha that bears an apical sporangium, which in most species can release zoospores like normal asexual propagules, branched hyphae with terminal sporangia, and branched or unbranched hyphae that lack terminal sporangia. Each pattern may be observed within a species, with germ sporangia favored when nutrients are present (Henry and Stelfox, 1968; Johnson et al., 2002; Ribeiro, 1983).

6.6 MOLECULAR ANALYSES OF THE SEXUAL CYCLE

Only a few experiments have been performed that shed light on the molecular events in sexual reproduction. Early work in *Lagenidium* and *Phytophthora* demonstrated that calcium plays an important role (Domnas and Warner, 1991; Elliott, 1988; Kerwin and Washino, 1986a), and that a decrease in cyclic adenosine monophosphate (cAMP) levels is necessary for antheridium development (Kerwin and Washino, 1984). In *Achlya*, antheridiol and oogoniol were shown to induce many mRNAs and proteins, which include certain heat shock proteins that may be part of steroid receptor complexes (Brunt et al., 1998; Horton and Horgen, 1985; Timberlake, 1976). Biochemical evidence for an antheridiol receptor in male strains was also obtained (Riehl and Toft, 1984). It is unfortunate that such studies in *Achlya* have not been repeated with the more advanced technical and genomic tools of today.

Most data on gene expression during the sexual cycle now come from *P. infestans*, where a microarray study showed that 108 of 15,644 genes tested are induced more than 10-fold during mating (Prakob and Judelson, 2007). Of these, about 30% are expressed exclusively during mating, and most are highly transcribed during oosporogenesis in the homothallic species *Phytophthora phaseoli*. The genes encode proteins with predicted roles in regulation, transport, metabolism, meiosis, structure, and other activities. Many encode RNA-regulating factors including *Puf* and KH-domain RNA binding proteins, RNAses and regulators of RNAses, and decapping enzymes, which implies a major role for post-transcriptional regulation. Also induced are several adenosine triphosphate-binding cassette (ABC) transporters and sterol carrier-like proteins, which might participate in hormone signaling, and enzymes involved in lipid metabolism or transport, such as lipoxygenases and ketoacyl reductases, which may participate in forming lipid bodies during oosporogenesis or using lipids during germination. The latter proteins are consistent with data from *Lagenidium* indicating that lipid metabolism is required for antheridial induction, gametangial fusion, and oospore maturation (Kerwin and Washino, 1986b). Interestingly, 18 genes induced during oosporogenesis are also upregulated during asexual sporulation. These might function in shared activities such as establishing dormancy and nutrient reserves. Three RNA-binding proteins were in this class, which suggests that RNA sequestration may be a shared feature of spore dormancy.

Although microarray data provide a global picture of gene expression during oosporogenesis, they do not indicate where the genes are expressed or if they are specific to antheridia or oogonia. Such information can be obtained using transgenic *P. infestans* that express fusions between the mating-induced sequences and a reporter (Cvitanich and Judelson, 2003). This was done with several genes predicted to encode RNA- and sterol-binding proteins, which were found to be induced in young gametangia (Fig. 6.3f).

6.7 EVOLUTION OF MATING SYSTEMS

Several data suggest that homothallism, not heterothallism, is the ancestral state. Not only are most oomycetes homothallic, but also homothallic genera typically occupy basal clades in orders that contain both mating systems (Riethmueller et al., 2002). In genera like *Phytophthora* and *Pythium* that include both systems, heterothallism may have evolved multiple times or occasionally reverted to self-fertility. How this occurred should become evident once mating type loci are identified.

Another trait that has evolved in several species is a loss of the ability to produce oospores. For example, *Pythium* includes several asexual species, such as the hyphal swelling (HS) group. HS isolates were suggested to be male forms evolved from the otherwise closely related homothallic species *Pythium ultimum* (Francis and St. Clair, 1997). *Phytophthora gonapodyides* also seems to be asexual in nature. It is self-sterile, yet forms oospores with A2 isolates of other *Phytophthora* species (Brasier et al., 1993). It is unclear whether *P. gonapodyides* lost its A2 members from mutation or a demographic sweep. Another asexual species is *Aplanopsis terrestris*, which does not produce antheridia, but generates oospores that bear ameiotic diploid nuclei (Dick, 1972). Because most *Aplanopsis* relatives are homothallic, the loss of a sexual lifestyle was suggested to reflect its adaptation to stable environments (Brasier et al., 2003).

The evolving components of mating systems are also likely factors in speciation. Proteins expressed specifically during mating in *P. infestans* have evolved faster than average, as observed with reproductive proteins in plants, true fungi, and animals (Swanson and Vacquier, 2002). For example, the mating-specific M96 protein has less than 50% amino acid similarity between different *Phytophthora* species, compared with about 80% for other proteins (Cvitanich et al., 2006). If complementary sequences between parents are required for fertility, then divergence would lead to reproductive isolation.

6.8 IMPORTANCE OF THE SEXUAL CYCLE IN NATURE

Oospores function in natural and agricultural ecosystems as structures for surviving harsh environments and infectious propagules, as well as part of a process that maintains fitness. Oosporogenesis by saprophytes is widely described and assumed important, although there is sparse evidence on the relative contributions of oospores compared with asexual spores to proliferation (Johnson et al., 2002). Better data exist for several plant pathogens, which show that the importance of sexual spores depends on the biology of the host and pathogen, agronomic practices, and the climate. There is also evidence for a role of overwintering oospores in *Lagenidium* infections of insects (Kerwin and Washino, 1988).

With the plant pathogenic species, oospores typically develop within the host and then incorporate into the soil or surface litter along with the dying plant. New plants at that site can then become infected, and oospores can also be transported elsewhere by fungus gnats, wind, or infected seeds (Jarvis et al., 1993; Jeger et al., 1998). Because they are resistant to control efforts such as fumigation, the presence of oospores in soil is a major challenge to agriculture.

Some plant diseases mainly use oospores as inoculum, whereas others also employ the relatively short-lived asexual spores. Diseases that rely on oospores alone are typically monocyclic, with one round of infection per season. Those that also use asexual spores are commonly polycyclic, with oospores and asexual spores causing initial and subsequent infections, respectively. One example of a disease that relies primarily on oospores is sorghum downy mildew, which is caused by the homothallic species *Peronosclerospora sorghi*. The species can theoretically produce asexual spores, but this only occurs in a narrow climactic range. Consequently, oospores are the principal inoculum in most regions (Jeger et al., 1998). Another case in which oospores play a main role is root rot of soybean caused by *P. sojae*, in which most primary lesions develop from overwintered oospores. Asexual spores are produced later in the season, but their infections are limited because plants become resistant with age (Schmitthenner, 1999). Oosporogenesis is also important in *P. sojae* because recombination has generated new pathotypes that overcome *Rps* resistance genes in soybean (Leitz et al., 2000).

Examples of diseases that usually involve both oospores and asexual spores are grape downy mildew caused by *Plasmopara viticola*, cucurbit blights caused by *Phytophthora capsici*, and potato late blight caused by *P. infestans*. All three oomycetes are heterothallic. In the United States, the two mating types of *P. capsici* occur at similar frequencies, and oospores are the normal primary inoculum, with asexual spores spreading disease later in the season (Hausbeck and Lamour, 2004). The situation is slightly more complex with *P. viticola*. Most primary infections of the grape in Europe and northeastern United States develop from oospores. Depending on the climate, some secondary infections from asexual spores may also occur (Gobbin et al., 2005; Kennelly et al., 2004). However, in regions such as Western Australia, oosporogenesis is rare because of an uneven distribution of the two mating types (Killigrew et al., 2005). This illustrates how the population dynamics of a pathogen, as well as the inherent biology of the species, must be considered when assessing the potential of sexual reproduction. A similar situation exists with *P. infestans*. In many regions of the world, both mating types are present, and oospores are an important inoculum along with asexual spores and infected tubers (Grunwald and Flier, 2005). However, until recently, only the A1 mating type was found in most of the world where the sexual cycle did not occur; this is still the case in some regions (Fry, 2008; Knapova and Gisi, 2002).

Many heterothallic species have experienced unequal global distributions of their two mating types in addition to *P. infestans*. Besides influencing the inoculum available for disease, this may adversely affect fitness. For example, many isolates of *P. infestans* from regions that contain the A1 mating type display poor fertility and are less aggressive than those from regions in which the sexual cycle has been active (Fry, 2008). This supports the premise that sexual reproduction maintains genomic integrity (Mooney, 1995).

6.9 KNOWLEDGE GAPS AND FUTURE DIRECTIONS

Sexual reproduction has proved to be an important component of the life cycles of many oomycetes, which addresses several fundamental issues in biology, such as self-/non-self-recognition, intercellular communication, cell-type differentiation, and evolution. Nevertheless, the sexual cycle remains poorly understood, especially at the molecular level. Research has also been fragmentary, with studies of different stages of reproduction scattered among species. Limitations in the knowledge base are not surprising considering that it is difficult to isolate large amounts of tissue highly enriched in gametangia or oospores with high rates of synchronous germination. The recent establishment of genomics resources, the development of gene-silencing methods, and the availability of purified hormones has the potential to accelerate studies of sexual development, although challenges will remain.

Important goals of future research include obtaining a molecular understanding of how the extracellular signals determine compatibility and induce the dimorphic gametangia, the chemical nature of those signals as only a few have been studied to date, the role of physical contact between gametangia in signaling later stages of development, how the standard eukaryotic cell-signaling and cell-cycle pathways participate in oosporogenesis, and the mechanism that establishes oospore dormancy. Learning how homothallic and heterothallic mating systems evolved is also an important goal. In addition, learning how germination occurs is an imperative as this may lead to new strategies for managing the pathogens. For example, the viability of soil populations of oospores might be reduced prior to planting by compounds that stimulate germination. In addition, interference with signaling processes *in planta* might block oospore formation and applying mating hormone analogs might divert isolates away from vegetative growth and thus block lesion expansion.

REFERENCES

Barksdale AW 1969. Sexual hormones of *Achlya* and other fungi. Science 166:831–837.

Barksdale AW, Lasure LL 1973. Induction of gametangial phenotypes in *Achlya*. Bull Torrey Bot Club 100:199–202.

Berlin S, Ellegren H 2001. Evolutionary genetics. Clonal inheritance of avian mitochondrial DNA. Nature 413:37–38.

Bishop H 1940. A study of sexuality in *Sapromyces reinschii*. Mycologia 32:505–529.

Brasier CM 1971. Induction of sexual reproduction in single A2 isolates of *Phytophthora* species by *Trichoderma viride*. Nat New Biol 231:283.

Brasier CM, Cooke DE, Duncan JM, Hansen EM 2003. Multiple new phenotypic taxa from trees and riparian ecosystems in *Phytophthora gonapodyides–P. megasperma* ITS Clade 6, which tend to be high-temperature tolerant and either inbreeding or sterile. Mycol Res 107:277–290.

Brasier CM, Hamm PB, Hansen EM 1993. Cultural characters, protein patterns and unusual mating behaviour of *Phytophthora gonapodyides* isolates from Britain and North America. Mycol Res 97:1287–1298.

Brunt SA, Borkar M, Silver JC 1998. Regulation of hsp90 and hsp70 genes during antheridiol-induced hyphal branching in the oomycete *Achlya ambisexualis*. Fungal Genet Biol 24:310–324.

Couch JN 1926. Heterothallism in *Dictyus*, a genus of the water molds. Ann Bot 40:849–881.

Cvitanich C, Judelson HS 2003. A gene expressed during sexual and asexual sporulation in *Phytophthora infestans* is a member of the Puf family of translational regulators. Eukaryot Cell 2:465–473.

Cvitanich C, Salcido M, Judelson HS 2006. Concerted evolution of a tandemly arrayed family of mating-specific genes in *Phytophthora* analyzed through inter- and intraspecific comparisions. Molec Genet Genom 275: 169–184.

Dick MW 1972. Morphology and taxonomy of the oomycetes with special reference to Saprolegniaceae Leptomitaceae and Pythiaceae, part 2 cytogenetic systems. New Phytol 71:1151–1159.

Dick MW 2001a. The Peronosporomycetes. In: McLaughlin DJ, McLaughlin EG, Lemke PA, editors. The mycota. Springer Heidelberg, Germany. pp. 39–72.

Dick MW 2001b. Straminipilous Fungi. Kluwer Academic Publishers, Dordrecht.

Domnas AJ, Warner SA 1991. Biochemical activities of entomophagous fungi. Crit Rev Microbiol 18:1–13.

Elliott CG 1983. Physiology of sexual reproduction in *Phytophthora*. In: Erwin DC, Bartnicki-Garcia S, Tsao PH, editors. Phytophthora, its biology, taxonomy, ecology, and pathology. APS Press. St. Paul, MN. pp. 71–80.

Elliott CG 1988. Stages in oosporogenesis of phytophthora sensitive to inhibitors of calmodulin and phosphodiesterase. Trans Brit Mycol Soc 90:187–192.

Fabritius A-L, Cvitanich C, Judelson HS 2002. Stage-specific gene expression during sexual development in *Phytophthora infestans*. Molec Microbiol 45:1057–1066.

Fabritius A-L, Judelson HS 1997. Mating-type loci segregate aberrantly in *Phytophthora infestans* but normally in *Phytophthora parasitica*: implications for models of mating-type determination. Curr Genet 32:60–65.

Francis DM, St. Clair DA 1997. Population genetics of *Pythium ultimum*. Phytopathol 87:454–461.

Frinking HD, Harrewijn JL, Geerds CF 1984. Factors governing oospore production by *Peronospora farinosa* f. sp. *spinaciae* in cotyledons of spinach [*Spinacia oleracea*]. Neth J Plant Pathol 91:215–224.

Fry WE 2008. *Phytophthora infestans:* the plant (and R gene) destroyer. Molec Plant Pathol 9:385–402.

Gall AM, Elliott CG 1985. Control of sexual reproduction in *Pythium sylvaticum*. Trans Brit Mycol Soc 84:629–636.

Gallegly ME 1968. Genetics of pathogenicity of *Phytophthora infestans*. Ann Rev Plant Pathol 6:375–396.

Gobbin D, Jermini M, Loskill B, Pertot I, Raynal M, Gessler C 2005. Importance of secondary inoculum of *Plasmopara viticola* to epidemics of grapevine downy mildew. Plant Pathol 54:522–534.

Gotoh K, Akino S, Kiyoshi T, Naito S 2005. Sexual mating preferences in vitro and in planta among Japandes isolates of *Phytophthora infestans*. J Gen Plant Pathol 71:29–32.

Groves CT, Ristaino JB 2000. Commercial fungicide formulations induce *in vitro* oospore formation and phenotypic change in mating type in *Phytophthora infestans*. Phytopathol 90:1201–1208.

Grunwald NJ, Flier WG 2005. The biology of *Phytophthora infestans* at its center of origin. Ann Rev Phytopathol 43:171–190.

Gu YH, Ko WH 2005. Evidence for mitochondrial gene control of mating types in *Phytophthora*. Can J Microbiol 51:934–940.

Harrison JL, Jones EBG 1975. The effect of salinity on sexual and asexual sporulation of members of the Saprolegniaceae. Trans Brit Mycol Soc 65:389–394.

Hausbeck MK, Lamour KH 2004. *Phytophthora capsici* on vegetable crops: research progress and management challenges. Plant Dis 88:1292–1303.

Hemmes DE 1983. Cytology of *Phytophthora*. In: Erwin DC, Bartnicki–Garcia S, Tsao PH, editors. Phytophthora, its biology, taxonomy, ecology, and pathology. APS Press. St. Paul, MN. pp. 9–40.

Henry AW, Stelfox D 1968. Comparative behavior of the oospores and oogonia of *Phytophthora citricola* during germination on an artificial medium. Can J Bot 46:1419–1421.

Ho HH, Zeng HC, Zheng FC 2002. *Phytophthora insolita* on Hainan Island. Bot Bull Acad Sin 43:227–230.

Horton JS, Horgen PA 1985. Synthesis of an antheridiol-inducible polypeptide during sexual morphogenesis of *Achlya ambisexualis*. Can J Biochem Cell Biol 63:355–365.

Howard KL 1971. Oospore types in the Saprolegniaceae. Mycologia 63:679–686.

Inaba T, Morinaka T 1983. The relationship between conidium and oospore production in soybean leaves infected with *Peronospora manshurica*. Ann Phytopathl Soc Jpn 49:554–557.

Jarvis WR, Shipp JL, Gardiner RB 1993. Transmission of *Pythium aphanidermatum* to greenhouse cucumber by the fungus gnat *Bradysia impatiens* (Diptera: Sciaridae). Ann Appl Biol 122:23–29.

Jayasekera AU, McComb JA, Shearer BL, Hardy GE 2007. In planta selfing and oospore production of *Phytophthora cinnamomi* in the presence of *Acacia pulchella*. Mycol Res 111:355–362.

Jeger MJ, Gilijamse E, Bock CH, Frinking HD 1998. The epidemiology, variability and control of the downy mildews of pearl millet and sorghum, with particular reference to Africa. Plant Pathol 47:544–569.

Johnson TW, Seymour RL, Padgett DE 2002. Biology and Systematics of the Saprolegniaceae. Wilmington: University of North Carolina.

Judelson HS 1996. Genetic and physical variability at the mating type locus of the oomycete, *Phytophthora infestans*. Genetics 144:1005–1013.

Judelson HS 1997. Expression and inheritance of sexual preference and selfing potential in Phytophthora infestans. Fungal Genet Biol 21:188–197.

Judelson HS, Roberts S 1999. Multiple loci determining insensitivity to phenylamide fungicides in *Phytophthora infestans*. Phytopathol 89:754–760.

Judelson HS, Spielman LJ, Shattock RC 1995. Genetic mapping and non-Mendelian segregation of mating type loci in the oomycete, *Phytophthora infestans*. Genet 141:503–512.

Kennelly MM, Eugster C, Gadoury DM, Smart CD, Seem RC, Gobbin D, Gessler C 2004. Contributions of oosporic inoculum to epidemics of grapevine downy mildew (*Plasmopara viticola*). Phytopathol 94:S50.

Kerwin JL, Duddles ND 1989. Reassessment of the role of phospholipids in sexual reproduction by sterol-auxotrophic fungi. J Bacteriol 171:3831–3839.

Kerwin JL, Washino RK 1984. Cyclic nucleotide regulation of oosporogenesis by *Lagenidium giganteum* and related fungi. Exp Mycol 8:215–224.

Kerwin JL, Washino RK 1986a. Oosporogenesis by *Lagenidium giganteum:* induction and maturation are regulated by calcium and calmodulin. Can J Microbiol 32:663–672.

Kerwin JL, Washino RK 1986b. Regulation of oosporogenesis by *Lagenidium giganteum* promotion of sexual reproduction by unsaturated fatty-acids and sterol availability. Can J Microbiol 32:294–300.

Kerwin JL, Washino RK 1988. Field evaluation of Lagenidium giganteum (Oomycetes: Lagenidiales) and description of a natural epizootic involving a new isolate of the fungus. J Med Entomol 25:452–460.

Killigrew BX, Sivasithamparam D, Scott ES 2005. Absence of oospores of downy mildew of grape caused by *Plasmopara viticola* as the source of primary inoculum in most western Australian vineyards. Plant Dis 89:777.

Knapova G, Gisi U 2002. Phenotypic and genotypic structure of *Phytophthora infestans* populations on potato and tomato in France and Switzerland. Plant Pathol 51:641–653.

Ko WH 1988. Hormonal heterothallism and homothallism in *Phytophthora*. Ann Rev Phytopathol 26:57–73.

Lasure LL, Griffin DH 1975. Inheritance of sex in *Achlya bisexualis*. Am J Bot 62: 216–220.

Leal JA, Gallegly ME, Lilly VG 1967. The relation of the carbon-nitrogen ratio in the basal medium to sexual reproduction in species of *Phytophthora*. Mycologia 59:953–964.

Leary JV, Roheim JR, Zentmyer GA 1974. Ribosome content of various spore forms of *Phytophthora* spp. Phytopathol 64:404–408.

Leitz RA, Hartman GL, Pedersen WL, Nickell CD 2000. Races of *Phytophthora sojae* on soybean in Illinois. Plant Dis 84:487.

MacGregor T, Bhattacharyya M, Tyler B, Bhat R, Schmitthenner AF, Gijzen M 2002. Genetic and physical mapping of *Avr1a* in *Phytophthora sojae*. Genetics 160:949–959.

McKay R 1957. The longevity of the oospores of onion downy mildew *Peronospora destructor* (Berk.) Casp. Sci Proc R Dublin Soc, New Series 27:295–307.

McMorris TC 1978. Antheridiol and the oogoniols steroid hormones which control sexual reproduction in *Achlya*. Phil Trans R Soc London B Biol Sci 284:459–470.

Michelmore RW, Sansome ER 1982. Cytological studies of heterothallism and secondary homothallism in *Bremia lactucae*. Trans Brit Mycol Soc 79:291–298.

Mooney SM. 1995. H. J. Muller and R. A. Fisher on the evolutionary significance of sex. J Hist Biol 28:133–149.

Nelson EB 1990. Exudate molecules initiating fungal responses to seeds and roots. Plant Soil 129:61–74.

Pegg GF, Mence MJ 1970. The biology of peronospora-viciae on pea: laboratory experiments on the effects of temperature relative humidity and light on the production germination and infectivity of sporangia. Ann Appl Biol 66:417–428.

Prakob W, Judelson HS 2007. Gene expression during oosporogenesis in heterothallic and homothallic *Phytophthora*. Fungal Genet Biol 44:726–739.

Pushpavathi B, Thakur RP, Rao KC 2006. Fertility and mating type frequency in Indian isolates of *Sclerospora graminicola*, the downy mildew pathogen of pearl millet. Plant Dis 90:211–214.

Qi J, Asano T, Jinno M, Matsui K, Atsumi K, Sulcagami Y, Ojika M 2005. Characterization of a *Phytophthora* mating hormone. Science 309:1828.

Randall TA, Ah Fong A, Judelson H 2003. Chromosomal heteromorphism and an apparent translocation detected using a BAC contig spanning the mating type locus of *Phytophthora infestans*. Fungal Genet Biol 38:75–84.

Raper JR 1951. Sexual hormones in *Achlya*. Am Sci 39:110–120.

Reeves RJ, Jackson RM 1974. Stimulation of sexual reproduction in *Phytophthora cinnamomi* by damage. J Gen Microbiol 84:303–310.

Ribeiro OK 1983. Physiology of asexual sporulation and spore germination in *Phytophthora*. In: Erwin DC, Bartnicki-Garcia S, Tsao PH, editors. Phytophthora, its biology, taxonomy, ecology, and pathology. APS Press. St. Paul, MN. pp. 55–70.

Riehl RM, Toft DO 1984. Analysis of the steroid receptor of *Achlya ambisexualis*. J Biol Chem 259:15324–15330.

Riethmueller A, Voglmayr H, Goeker M, Weiss M, Oberwinkler F 2002. Phylogenetic relationships of the downy mildews (Peronosporales) and related groups based on nuclear large subunit ribosomal DNA sequences. Mycologia 94:834–849.

Ruben DM, Stanghellini ME 1978. Ultrastructure of oospore germination in *Pythium aphanidermatum*. Amer J Bot 65:491–501.

Sansome E 1980. Reciprocal translocation heterozygosity in heterothallic species of *Phytophthora* and its significance. Nature 241:344–345.

Savage EJ, Clayton CW, Hunter JH, Brenneman JA, Laviola C, Gallegly ME 1968. Homothallism heterothallism and interspecific hybridization in the genus *Phytophthora*. Phytopathol 58:1004–1021.

Schmitthenner AF 1999. Phytophthora rot of soybean. In: Hartman GF, Sinclair JB, Rupe JC, editors. Compendium of soybean diseases, 4th ed. APS Press. St Paul, MN. pp. 39–42.

Shang H, Grau CR, Peters RD 2000. Oospore germination of *Aphanomyces euteiches* in root exudates and on the rhizoplanes of crop plants. Plant Dis 84:994–998.

Shattock RC, Tooley PW, Fry WE 1986. Genetics of *Phytophthora infestans:* determination of recombination, segregation, and selling by isozyme analysis. Phytopathol 76:410–413.

Smart CD, Willmann MR, Mayton H, Mizubuti ESG, Sandrock RW, Muldoon AE, Fry WE 1998. Self-fertility in two clonal lineages of *Phytophthora infestans*. Fungal Genet Biol 25:134–142.

Sussman AS, Douthit HA 1973. Dormancy in microbial spores. Ann Rev Plant Physiol 24: 311–352.

Swanson WJ, Vacquier VD 2002. The rapid evolution of reproductive proteins. Nat Rev Genet 3:137–144.

Timberlake WE 1976. Alterations in RNA and protein synthesis associated with steroid hormone induced sexual morphogenesis in the water mold *Achlya*. Deve Biol 51:202–214.

Van Der Gaag D, Frinking HD 1997. Factors affecting germination of oospores of *Peronospora viciae* f. sp. *pisi* in vitro. Eur J Plant Pathol 103:573–580.

van der Lee T, Testa A, Robold A, van ‘t Klooster JW, Govers F 2004. High density genetic linkage maps of *Phytophthora infestans* reveal trisomic progeny and chromosomal rearrangements. Genetics 157:949–956.

Vercesi A, Tornaghi R, Sant S, Burruano S, Faoro F 1999. A cytological and ultrastructural study on the maturation and germination of oospores of *Plasmopara viticola* from overwintering vine leaves. Mycol Res 103:193–202.

Whisson SC, Drenth A, MacLean DJ, Irwin JAG 1995. *Phytophthora sojae* avirulence genes, RAPD, and RFLP markers used to construct a detailed genetic linkage map. Mol Plant Microbe Interact 8:988–995.

Wong FP, Burr HN, Wilcox WF 2001. Heterothallism in *Plasmopara viticola*. Plant Pathol 50:427–432.

Yajima A, Qin U, Zhou X, Kawanishi N, Xiao X, Wang J, Zhang D, Wu Y, Nukadu T, Yabuta G 2008. Synthesis and absolute configuration of hormone alpha1. Nat Chem Biol 4:235–237.

Zhang X-Z, Seo H-W, Ahn W-G, Kim B-S 2006. SCAR marker linked with A1 mating type locus in *Phytophthora infestans*. J Microbiol Biotech 165:724–730.

7

POPULATION GENETICS AND POPULATION DIVERSITY OF *PHYTOPHTHORA INFESTANS*

William E. Fry
Cornell University, Ithaca, New York
Niklaus J. Grünwald
USDA, Corvallis, Oregon
David E.L. Cooke
Scottish Crop Research Institute, Dundee, Scotland
Adele McLeod
Stellenbosch University, Stellenbosch, South Africa
Gregory A. Forbes
International Potato Center, Lima, Peru
Keqiang Cao
Agricultural University of Hebei, Baoding, China

7.1 INTRODUCTION

During most of the twentieth century, our knowledge of the population genetics of *Phytophthora infestans* was dominated by two issues: the lack of sexual reproduction (except in central Mexico) and the diversity of races that overcame *R* genes. However, during the final two decades of that century, it also became clear that major migrations had and were occurring. These migrations have continued into the twenty-first century and have changed dramatically the population structure of this pathogen in many locations.

Oomycete Genetics and Genomics: Diversity, Interactions, and Research Tools
Edited by Kurt Lamour and Sophien Kamoun
Copyright © 2009 John Wiley & Sons, Inc.

In many of these locations, the population that was dominant before the migrations of the late twentieth century can no longer be detected. However, a few regions of the world seem to have been protected from these migrations, and populations there retain the genotypes and characteristics of the populations that presumably have been in place for many years. Although the pathogen remains asexual in many locations, in a few regions, strong evidence now suggests that sexual recombination is contributing to the population structure and to the dynamics of plant disease epidemics.

This chapter will investigate the structure and dynamics of populations of *P. infestans* in many regions worldwide. There seem to be a few locations in which the populations may be very similar at the beginning of the twenty-first century to what they were earlier in the twentieth century. For example, it seems that the populations in Mexico (the source of the recent migrants) remain consistent with earlier understanding. However, there are significant changes in Europe, South America, North America north of Mexico, and recently also from Asia.

7.2 MARKERS

To analyze populations, one must have genetic markers. Initially the only markers were phenotypic, but with development of molecular technologies, genotypic markers are now available [see (Cooke and Lees, 2004) for an excellent overview of markers]. The earliest markers were mating type (A1 or A2) and specific virulence. If an isolate makes antheridia and oogonia in the presence of a known A2 isolate, but not with an A1, it is an A1 mating type. Although mating type determination is typically robust, there can be some "leakiness" in the process. This is because other factors can sometimes induce the formation of gametangia. These factors include aged cultures, various substrates (oats, lima beans, V-8 juice, etc.), physical wounding, and *Trichoderma viride* (Brasier, 1971; Reeves and Jackson, 1974; Smart et al., 2000). As might be expected, some strains are more responsive to these stimuli than are other strains (Smart et al., 2000). There have been reports of mating type "change" (Groves and Ristaino, 2000; Ko, 1994), but instead of a mating type change in a particular isolate, it seems most likely that there was the stimulation of gametangia production by factors other than a hormone (Goodwin and Drenth, 1997). These other factors are not nearly so efficient at inducing the formation of gametangia, and mating type remains a relatively robust marker. Specific virulence can be used as a marker to characterize populations, but this trait seems to be evolving rapidly (Goodwin et al., 1995b), so its use in analysis of diverse populations for relatedness is limited. The virulence characteristics of the pathogen population were important in terms of predicting whether a specific *R* gene would be effective. [Disappointingly, resistance-breaking isolates are often detected very soon after release of a potato cultivar with *R* gene resistance (Malcolmson, 1969)]. However, it can be informative to determine whether the virulence diversity is different or not different when comparing two populations of *P. infestans* (Drenth et al., 1994; Sujkowski et al., 1996).

In the final two decades of the twentieth century, resistance to metalaxyl/mefenoxam became available as a phenotypic marker. During the 1970s and into the 1980s, this chemical was wonderfully effective in suppressing diseases caused by oomycetes, which include the late blight of potatoes and tomatoes caused by *P. infestans*. Unfortunately, resistance in some populations began to occur at the end of the 1970s and in the 1980s (Dowley and O'Sullivan, 1981). Resistance is conditioned by a single dominant gene, but it is modified by other genes (Fabritius et al., 1997; Judelson and Roberts, 1999; Lee et al., 1999).

The most durable and readily used markers to assess populations of *P. infestans* have been molecular (Cooke and Lees, 2004). In approximate chronological order of their use, these are allozymes, restriction fragment length polymorphisms (RFLPs), mitochondrial haplotypes, amplified fragment length polymorphisms (AFLPs), and simple sequence repeats (SSRs or microsatellites). These markers have been used to distinguish among populations and to document the origin of exotic strains and major migration events.

7.3 TWENTIETH-CENTURY POPULATIONS AND MIGRATIONS

Before the 1980s, populations around the world (except for those in central Mexico) contained only A1 mating types. Furthermore, these populations were dominated by a single clonal lineage of *P. infestans*, which was designated the US-1 clonal lineage (Goodwin et al., 1994b). This lineage was defined by mating type (A1), allozyme genotype (86/100 and 92/100 at two loci, *Glucose-6-phosphate isomerase* and *Peptidase*, respectively), by a particular RFLP fingerprint pattern as detected by RG-57, and by the Ib mtDNA haplotype (sensu Griffith and Shaw, 1998). This worldwide situation was disrupted by at least two different migrations from Mexico in the latter quarter of the twentieth century—events that changed (and are changing) the population structure of this organism worldwide.

The first migration was to Europe and was first signaled by the discovery of isolates of the A2 mating type in Switzerland (Hohl and Iselin, 1984). These reports were followed by analyses using molecular markers (mtDNA, allozymes, and RLFP fingerprinting) to demonstrate that indeed by the 1980s and 1990s there were isolates with characteristics different from those previously reported in Europe. Surprisingly, the "new" strains displaced the previous ones rapidly—in some places within just a few years (Drenth et al., 1994; Fry and Goodwin, 1997b; Spielman et al., 1991; Sujkowski et al., 1994). It is now believed that the pathway of this migration was via a shipment of potato tubers from Mexico to Europe in the summer of 1976 (Niederhauser, 1991). The motivation for this highly unusual shipment was to satisfy a shortage of potatoes due to under production in Europe during the 1976 season. (Normally, shipments from Mexico to Europe were not allowed because of fear of the A2 mating type.)

This shipment must have contained a population of *P. infestans* that was diverse for mating type and contained novel alleleles. Because European producers send seed tubers to many locations throughout the world as well as throughout Europe, this population was widely distributed in Europe and subsequently to South America, North Africa, and Asia. However, Australia and Sub-Saharan Africa may have been largely protected from this migration (see below).

The second migration event(s) was from Mexico to the United States and Canada, and it involved a very different mechanism with different genotypes of *P. infestans*. By analysis of specific isolates from Northwest Mexico, it was possible to determine that these same clonal lineages from northwest Mexico were detected in the United States and Canada (Goodwin et al., 1994a; Goodwin et al., 1998, 1992, 1996). Although both A1 and A2 mating types were introduced into the United States, the population (especially on potatoes) is very simple, which consists of only a few clonal lineages that remain distinct (Fry and Goodwin, 1997a; Fry, unpublished results). In contrast to the European situation, the migration into the United States resulted in only a small number of diverse lineages—perhaps only a few lineages were introduced. However, in concert with the situation in Europe, the introduced strains have displaced the previous population, which now can no longer be detected in the United States (Fry and Goodwin, 1997; Fry, unpublished results).

7.4 CURRENT POPULATIONS OF *P. INFESTANS*

7.4.1 Mexico—(Niklaus J. Grünwald)

The populations of *P. infestans* in Mexico are distinctly different from those in other areas of the world regarding several aspects of the biology, natural history, genetics, and evolution of this pathogen. The populations in the Central highlands of Mexico provide all the hallmarks of a sexual population that coevolved with native *Solanum* as would be expected for a plant pathogen population at its center of origin (Grünwald and Flier, 2005). The two mating types occur in a 1:1 ratio (Goodwin et al., 1992; Grünwald et al., 2001). Populations are sexual given that oospores are formed ubiquitously on leaves, stems, and tubers of both native and cultivated *Solanum* (Fernandez-Pavia et al., 2002; Flier et al., 2001; Gallegly and Galindo, 1958). As expected for sexually reproducing populations, neutral genetic markers including allozymes, RFLP, AFLP, and microsatellite markers show that populations are in Hardy-Weinberg equilibrium and highly diverse (Flier et al., 2003; Goodwin et al., 1992; Grünwald et al., 2001; Tooley et al. 1985). Other features of the Mexican *P. infestans* population are described below and highlight distinct differences to populations elsewhere in world.

The biology of *P. infestans* is best understood for the Toluca Valley located at the high altitude of 2,600 m in the Central highlands southwest of Mexico City. Potatoes are grown in the valley using intensive production methods.

The production systems are characterized by use of high inputs of fungicides, fertilizer, and a few cultivars, such as Alpha that have no known *R* genes (Flier et al., 2003; Grünwald et al., 2001). Potatoes are also grown on the mountains surrounding the valley up to an altitude of approximately 3,500 m. These production systems are less intensive and often consist of growing landrace varieties that harbor some *R* genes (Flier et al., 2003; Grünwald et al., 2001). Wild *Solanum* species are restricted mostly to the edges and clearings in pine and fir forests of the mountains up to an altitude of about 3,800 m (Flier et al., 2003; Grünwald et al., 2001; Rivera-Pena, 1990a–d). Many of these *Solanum* species occur in small populations. Late blight epidemics in Toluca are driven by the rainy season, which typically starts in mid-to-late June. Once commercial potatoes emerge and rains have commenced, epidemics progress steadily on susceptible cultivars such as Alpha (Grünwald et al., 2002, 2000). Epidemics on native *Solanum* species, however, rarely occur before the beginning of September because of the altitude, patchiness, and spatial isolation of these plants (Grünwald and Flier, 2005; Rivera-Pena, 1990a and c).

The distinct differences in environments, namely the cultivated high-input potato crops in the valley, the low-input land-race potato cultivation in the mountains surrounding the valley, and the native tuber-bearing *Solanum* species, might result in distinct and well-differentiated populations of *P. infestans*. As would be expected from knowledge of the natural history of *Solanum* in Toluca, *P. infestans* populations on the wild *Solanum* species are the most differentiated and show more private alleles and evidence for reduced gene flow relative to populations among cultivated potatoes (Flier et al., 2003; Grünwald et al., 2001). Populations of *P. infestans* in the Toluca Valley best fit the model of a metapopulation with distinct genetic differentiation into populations based on the host plant and potato production system (Grünwald and Flier, 2005).

Interestingly, populations in northern Mexico are clonal, whereas those in central Mexico are sexual (Goodwin et al., 1992). These Northern populations are similar to the U.S. populations being dominated by one to a few clones (Fry et al., 1993), and it is likely that the current U.S. populations migrated from northern Mexico.

Resistance to the fungicide metalaxyl has served as a selectable marker for characterizing the population structure of *P. infestans* in Mexico. Potato production in Mexico relies heavily on fungicides given that severe late blight epidemics occur yearly (Grünwald et al., 2002; Grünwald et al., 2000). Metalaxyl had been used in the past and continues to be used today. Remarkably, Toluca harbors a population of *P. infestans* with a baseline level of resistance to metalaxyl. Resistance to metalaxyl evolved rapidly after introduction of metalaxyl (Matuszak et al., 1994) but then reverted to resistance levels found prior to introduction (Grünwald et al., 2001). Toluca also offers a unique laboratory for studying fungicide resistance evolution in the field. Selection for metalaxyl resistance evolved under experimental conditions within a single field season (Grünwald et al., 2006). Repeated metalaxyl application imposed purifying selection and reduced genotypic diversity in experimental plots. Similar patterns of directional selection were not observed for other fungicides, which include azoxystrobin, dimethomorph, fluazinam, and propamocarb (Grünwald et al., 2006).

Mexico is also a center of diversity for tuber-bearing *Solanum* species, which has implications for the coevolution and population biology of *P. infestans* (Hawkes, 1990; Hijmans and Spooner, 2001; Rivera-Pena and Molina-Galan, 1989; Spooner et al., 1991, 2000; Spooner and Hijmans, 2001). All of the native tuber-bearing *Solanum* species can be infected by *P. infestans* (Grünwald and Flier, 2005; Lozoya-Saldana et al., 1997; Rivera-Pena, 1990b; Rivera-Pena and Molina-Galan, 1989). The first 11 *R* genes described in the pregenomic era all were derived from two endemic Mexican species, namely *Solanum demissum* and *Solanum stoloniferum* (Mills and Niederhauser, 1953; Niederhauser, 1991; Niederhauser et al., 1954). It is apparent that *P. infestans* coevolved with endemic *Solanum* species. Isolates with different combinations of the 11 pathogen effector genes compatible with the corresponding *Solanum R* genes are generally present in every growing season in Toluca (Rivera-Pena, 1990). The genetic diversity for virulence is high, and all known *R* genes are overcome (Grünwald and Flier, 2005; Mills and Niederhauser, 1953; Niederhauser et al., 1954; Rivera-Pena, 1990b; Tooley et al., 1986). Because of this coevolutionary history, central Mexico provides a rich source of germplasm for breeding potatoes for durable resistance using both race-specific and race-nonspecific resistance (Grünwald et al., 2002).

One intriguing aspect of the evolution of *P. infestans* in Mexico is the existence of the following two endemic sister taxa: *Phytophthora mirabilis*, which is a close relative that can hybridize with *P. infestans* (Goodwin and Fry, 1994) and that infects the Mexican weedy host *Mirabilis jalapa*, and *Phytophthora ipomoeae*, which occurs on two endemic morning glory species *Ipomoea longipedunculata* and *Ipomoea purpurea* (Badillo-Ponce et al., 2004; Flier et al., 2002). Both *P. ipomoeae* and *P. mirabilis* show polymorphism for allozymes as well as RFLP or AFLP loci, and it is likely that these two populations are sexual as well (Grünwald and Flier, 2005). Other close relatives not endemic to Mexico include *Phytophthora phaseoli* and *Phytophthora andina*. *P. phaseoli* is known to affect lima beans in the United States, but to date it has not been found in Mexico. *P. andina* is a newly described species that seems to be admixed with current *P. infestans* populations in South America (Gomez-Alpizar et al., 2007) and has not been documented in Mexico. Clearly, *P. andina* has shared ancestry with *P. infestans* (Gomez-Alpizar et al., 2007), but it is not clear whether this occurred through sexual recombination, speciation, or hybridization. More work is needed to establish that *P. andina* and *P. infestans* have South America as a center of origin. The absence of sexual reproduction, presence of clonality, and host adaptation in South America are in conflict with the idea of a center of origin. Currently, the most parsimonious explanation for the evolutionary history, which is congruent with the known natural history of *P. infestans* in

TABLE 7.1 Biology of *P. infestans* in South America and Central Mexico.

Trait	Central Mexico	South America
Reproductive mode	sexual	clonal
Mating type ratio 1:1	yes	no
Oospores formed	yes	no
Race structure	diverse	?
Demonstrated coevolution with endemic *Solanum*	yes	no
Reproductive mode of endemic sister taxa	*Phytophthora ipomoeae:* sexual *Phytophthora mirabilis*: sexual	*Phytophthora andina*: clonal

Mexico, is that Mexico remains a center of origin of *P. infestans* and is a center of evolution of some of clade 1c taxa (Table 7.1).

7.4.2 Europe — (David E. L. Cooke)

As a significant component of the global potato industry, Europe has no doubt played a role as both recipient and donor of *P. infestans* genotypes with far-reaching influences on global population diversity. Records of the devastating spread of potato blight over the summer of 1845 leave little doubt about the arrival date of *P. infestans* (e.g., Bourke, 1964). All European isolates of the pathogen tested prior to the 1980s were of the A1 mating type, but the theory that a single A1 lineage, which is now known as US-1, dominated the population for the whole period has been challenged. A distinguishing trait of the US-1 lineage is its Ib mtDNA haplotype, and yet all nineteenth- and early twentieth-century herbarium samples analyzed from Europe to date were of haplotype Ia (May and Ristaino, 2004). This implies a previously unconsidered second wave of *P. infestans* population displacement by the US-1 lineage at some stage after the mid-nineteenth century. Such displacements are, as we shall see, now a familiar theme in Europe. The first records of A2 in Europe were in 1981 in Switzerland, the United Kingdom, and the Netherlands (Frinking et al., 1987; Hohl and Iselin, 1984; Tantius et al., 1986) with reports from many other countries over subsequent years. It is now clear that new A1 mating type strains arrived at this time, and the complete displacement of US-1 by the "new" population has been well documented across Europe (Drenth et al., 1994; Fry and Goodwin, 1997).

Clearly one of the main impacts of the arrival of the A2 mating type on the population genetics of *P. infestans* was the possible transition to a sexual breeding population. Research groups have continued to monitor the status of European populations with a range of increasingly powerful molecular tools (Cooke and Lees, 2004). Although the overall picture is of an increasing prevalence of mixed mating type populations and increasing diversity, there are significant regional differences in population structure across the region from the 1990s to the present day. This complex pattern highlights the need to standardize genotyping methods and data collection into a single format to allow a meaningful comparison of objective data at a range of spatial and temporal scales. A summary of the population history for the past 30 years follows, along with information on recent dramatic transition to the A2 mating type and the mechanism that has been developed in Europe to standardize, collate, and display *P. infestans* population data.

Although all regions have recorded the arrival of the A2 mating type, its frequency has varied markedly. In Northern Ireland, for example, despite the appearance of A2 strains in the 1980s and 1990s, it was not found in surveys of over 200 isolates from 1998 to 2002 (Cooke et al., 2006). This same result is mirrored in the Republic of Ireland (Griffin et al., 2002). Although isolation from the European mainland might account for the low A2 frequency in Ireland, it does not explain the low frequency from 1992–1996 in France. Of 485 isolates, only 2.5% were of the A2 mating type, and these were mostly collected in private gardens on potato and tomato from 1995 to 1996 (Lebreton et al., 1998). Similarly, in the United Kingdom, among 1,864 samples collected between 1985 and 1988, the A2 frequency ranged from 2% to 11% (Shattock et al., 1990). No clear trends were noted, and a larger survey 10 years later revealed only 3% A2 (Day et al., 2004). Interestingly, the RG57 and mtDNA haplotypes of A2 isolates collected in 1981 were identical to those from the 1995–1998 collection, which indicates the maintenance of a clonal A2 lineage at low frequency in the population for almost 20 years (Day et al., 2004). In Scotland over the same period, 19% of the population was A2 (Cooke et al., 2003), which fell to less than 1% in 2003–2004 (data not shown). In the Netherlands however, a detailed survey of a small region revealed an A2 frequency of 4%, 19%, and 56% over the three seasons 1994, 1995, and 1996, respectively (Zwankhuizen et al., 2000). In Poland, the A2 frequency has fluctuated from 44% in 1989 down to 12% in 1991, and it increased to 40% again in 1998 (Kapsa, 2001; Sujkowski et al., 1994). In Hungary, limited sampling has also revealed a high frequency of A2 in 1998 (Bakonyi et al., 2002). In Norway and Finland in 1993–1996, the A2 frequency was 25% and 15%, respectively (Hermansen et al., 2000), with a marked increase in Finland in the year 2000 (Lehtinen et al., 2007). A range of 36% to 49% A2 was observed across all Nordic countries in 2003 (Lehtinen et al., 2008).

The tantalizing and important question is as follows: Are these populations reproducing sexually? Several lines of indirect evidence have been followed. First, high levels of genetic diversity have provided strong support for sexual recombination in some regions of Europe over specific time periods, e.g., in the Netherlands in the 1980s and 1990s (Drenth et al., 1994; Zwankhuizen et al., 2000), Poland in the late 1980s and early 1990s (Sujkowski et al., 1994) and Norway and Finland in the mid-1990s (Bruberg et al., 1999). Second, a balanced mating type ratio and the presence of both mating types in a high proportion of sampled outbreaks as shown in the Nordic countries in 2003 (Lehtinen et al., 2008) is suggestive of a sexual population, especially when coupled with field observations consistent with a soil-borne source of inoculum, such as those in Finland (Lehtinen and Hannukkala, 2004), Sweden (Andersson et al., 1998), and the Netherlands (Evenhuis et al., 2007). Conclusive proof of the role of oospores has proved surprisingly elusive with the most convincing case to date being of primary disease foci in a Swedish potato field (Widmark et al., 2007) and some observations backed up by AFLP analysis in the Netherlands (Evenhuis et al., 2007). Conversely, analyses of other populations indicate a dominance of asexual clones and numerous rare genotypes (Cooke et al., 2003, 2006; Day et al., 2004; Knapova and Gisi, 2002). Careful scrutiny of AFLP diversity revealed a clear substructure of the U.K. population that corresponded to lineages defined by mating type and mtDNA haplotype (Purvis et al., 2001). This and other observations, such as the historical absence of metalaxyl resistance in European A2 lineages (Cooke et al., 2003), is consistent with an absence of sexual processes that would reassort such traits. SSR markers (Lees et al., 2006) have been used to analyze over 2,600 isolates from more than 800 outbreaks in Great Britain (GB) from 2003 to 2007 and demonstrated that eight or fewer clones dominated the population. There was little evidence that the sexual phase made a significant contribution to the population structure (Cooke et al., 2007b, 2008).

Since 2004, a sweep in the Western European *P. infestans* population has been noted. An A2 lineage, which was initially defined by RG57 analysis (Shaw et al., 2007) and subsequently by SSR markers was termed genotype_13 (13_A2) and has increased in frequency from 12% to 71% of the GB population over three seasons since 2005 (Cooke et al., 2008). This clone was first recorded in a starch-growing region of the Netherlands in 2004 (Cooke et al., 2007b), and studies into its origin are ongoing. Preliminary evidence suggests it is fit and aggressive (unpublished results), and this is certainly borne out by its rapid spread in the Netherlands (Van Raaij et al., 2007), France (Detourne et al., 2007), and Germany (www.eucablight.org). There are also indications that it is resistant to metalaxyl (Shaw et al. 2007; Detourne et al., 2007). This seems to be yet another dramatic shift in the European population, and although it is reminiscent of the spread of US-8 in the United States (Fry and Goodwin, 1997), it is clear that 13_A2 is unrelated to US-8 (unpublished results). Its spread, sexual compatibility with A1 lineages, and impact on blight management and the population structure are currently under investigation.

From the above, it is clear that scaling up the analysis of *P. infestans* populations from a national to an international scale is important if we are to understand the mechanisms and implications of change over time and space. Attempts to do this on the basis of mtDNA, isozymes, and RG57 fingerprint have proved challenging in diverse European populations. Advances in *P. infestans* genomics, however, have underpinned the development of powerful codominant SSR markers (Knapova and Gisi, 2002; Lees et al., 2006) that offer new opportunities. It was with this in mind that Eucablight, which is a

European-Union-funded Concerted Action project, was established (www.eucablight.org) to coordinate the efforts of many research teams across Europe to standardize and collate the wealth of *P. infestans* data available into a single comprehensive database (Cooke et al., 2007a). The database currently comprises over 18,000 isolates from 22 European countries with the data displayed via a powerful web interface. The addition of SSR data is ongoing and provides a powerful means of comparing populations on a European scale. There is now a need to bridge between datasets based on RG57, isozymes, and mtDNA haplotypes, and contemporary data based on SSRs to build a comprehensive picture of the European population structure. An expansion of the database tools into South and Central America has already been released, and additional funding is being sought to expand the geographic range and the analysis tools, for example, to enable searches for specific genotypes (or their closest relatives), to examine phylogenetic relationships between and within populations, and to integrate the database with software that will allow a full analysis of the population genetics at a range of scales. Currently, sequence data can be associated with database entries — an important feature as we develop systems to understand the diversity and selection pressure on key *P. infestans* effectors within the population.

7.4.3 Africa — (Adele McLeod)

Information on the population genetics of *P. infestans* in Africa is just beginning to emerge. Late blight seems to have been introduced only relatively recently in some countries. For example, in Kenya and Uganda, potatoes had been grown for more than a half century before blight was first observed and reported in 1941 (Nattrass and Ryan, 1951). The earliest information on characteristics of *P. infestans* isolates in Africa prior to the 1980s is that the A1 mating type was present in South Africa in 1958 (Smoot et al., 1958).

Although both potatoes and tomatoes are important hosts of *P. infestans* in Africa, potatoes are more important because many countries have significant areas devoted to potato production. Egypt, Morocco, Cameroon, Ethiopia, Kenya, Uganda, and South Africa have production in the range of 50,000–100,000 ha (http://faostat.fao.org/site/567/default.aspx). In most of these countries, potatoes are grown year round, which along with favorable climatic conditions make late blight control difficult (McLeod, 1998; Olanya et al., 2001; Sedegui et al., 2000). In each country, a few cultivars dominate the production, and these are typically susceptible to late blight. Except for Egypt (with 195,000 ha of tomatoes), tomatoes generally occupy much less production area than do potatoes. Ethiopia, Uganda, and South Africa have less than 7,000 ha under tomato production, whereas Cameroon, Kenya, Morocco, and Tanzania have more than 16,000 ha under tomato production (http://faostat.fao.org/site/567/default.aspx).

The movement of seed tubers likely influences the *P. infestans* population structure of each country. For Morocco, Europe contributes up to 36% (up to

40,000 tons) of the total seed potatoes (Hammi et al., 2001, 2002; Sedegui et al., 2000), so it is expected that populations in Morocco might resemble those in Europe. In Kenya, most farmers (~90%) use their own seeds and only ~10% of farmers obtain their seeds from formal sources (Nyankanga et al., 2004). In South Africa, seed potatoes were imported from Europe into the late 1980s or perhaps early 1990s. Subsequently, the Plant Improvement law in South Africa has only allowed the importation of *in vitro* material or G0 seed (greenhouse grown) (Mike Holtzhausen, National Plant and Plant Product Inspection services, personal communication).

P. infestans populations in northern Africa (Morocco and Egypt) are distinct from those in the rest of Africa. In these countries, both A1 and A2 mating types have been reported (Baka, 1997; El-Korany, 1994; Shaat, 2002; Hammi et al., 2001, 2002; Sedegui et al., 2000). Some diversity based on allozyme and mating type has been identified in Morocco (Sedegui et al., 2000). There may also be change within these populations over different years (Hammi et al., 2001, 2002; Sedegui et al., 2000) although data are limited. The presence of both A1 and A2 mating type isolates within the same field has been reported in four different regions in Morocco (Hammi et al., 2001; Sedegui et al., 2000), which creates at least a theoretical potential for sexual reproduction.

Further south, in Rwanda, Uganda, Kenya, Tanzania, Burundi, and Ethiopia, most data suggest that only the A1 mating type has been detected, and two mitochondrial haplotypes, Ia and Ib, are present in these populations. The one exception is from Mukalazi et al. (2001) who report possible diversity in mating type, and these possibilities need more investigation. In Rwanda, the US-1 clonal lineage formed an important component of the population structure in samples from the mid-1980s. Three A1 mating type genotypes distinct from US-1 were also detected (Goodwin et al., 1994). Two of these genotypes were later designated as genotypes RW-1 and RW-2 by Forbes et al. (1998). For the third genotype, which is here named RW-3, the specific RG-57 fingerprint pattern is unknown, but allozyme and mating type data are available (Goodwin et al., 1994). In Uganda and Kenya, from 1995 to 2001, only the US-1 clonal lineage (including variants) was detected (Ochwo et al., 2002; Vega-Sanchez et al., 2000). A current survey (2007–2008) involves collections from Uganda, Kenya, Rwanda, Burundi, and Tanzania (Pule et al., 2008). To date, only A1 mating types have been detected. Initial genotype data suggest that the US-1 clonal lineage may still dominate populations in these countries (Pule et al., 2008). However, in Kenya a different genotype KE-1 has also been detected. Based on RG-57 fingerprinting, *Gpi* genotype, mitochondrial haplotype (Ia), and mating type data, the KE-1 genotype (Pule et al., 2008) seems most closely related to genotypes RW-1 to RW-3 (Forbes et al., 1998: Gavino and Fry, 2002). The mitochondrial haplotype Ia has also been detected in A1 mating type isolates in Ethiopia, which borders Kenya (Schiessendopper and Molnar, 2002).

In South Africa, extensive *P. infestans* characterization studies have been conducted (McLeod et al., 2001; Pule et al., 2008). The first population study was conducted in the late 1990s and clearly established that the US-1 clonal lineage was the only genotype detected on tomatoes and potatoes in South Africa. The second study was initiated in 2007, and it is still in progress. Initial results suggest that the US-1 clonal lineage is still predominant (Pule et al., 2008).

In addition to potato and tomato, there are other hosts in Africa. Garden huckleberry (*Solanum scabrum* Mill.) is a host in Cameroon (Fontem et al., 2004), and petunia is a host in South Africa (McLeod and Coertze, 2006). Both petunia and garden huckleberry may play important roles in the population biology of *P. infestans* because infections have been observed consistently on each host. Additionally, *Solanum incanum* L. and a *Solanum* species closely related to *Solanum panduraeforme* Drege, were reported in 1951 as hosts of *P. infestans* by Nattrass and Ryan (1951); these reports should be confirmed. And, although they are not observed to be infected in the field, *Solanum macrocarpon, Solanecio biafrae, Ageratum conyzoides*, and *Dichrocephala integrifolia* were demonstrated to be infected in detached leaflet inoculation studies conducted in Cameroon (Fontem et al., 2004).

There seems to be host specialization within some populations. Tomato specialization was documented within the US-1 clonal lineage in Uganda and Kenya (Vega-Sanchez et al., 2000). Tomato specialized isolates show a highly biotrophic growth with little or no necrosis while sporulating abundantly. In contrast, the potato isolates caused dark pigmentation and less abundant sporulation in tomato leaflets (Vega-Sanchez et al., 2000). Mukalazi et al. (2001) also found that Ugandan isolates from potatoes caused larger lesions on potato leaflets than isolates from tomatoes. Additionally, half of the potato isolates were unable to infect tomato leaflets. In Morocco, there also seems to be tomato specialization, because tomatoes isolates seemed to be much more virulent on tomatoes than on potatoes (Hammi et al., 2002).

Metalaxyl-resistant individuals are often present and sometimes dominant in African populations. In South Africa, resistant individuals were reported in the US-1 clonal lineage, but metalaxyl resistance was mainly restricted to populations from potatoes grown within the southern coastal regions, with no resistant isolates being detected on tomatoes (McLeod et al., 2001). High frequencies (50–80%) of metalaxyl resistance have been reported in Uganda and Kenya (Kankwatsa et al., 2003; Mukalazi et al., 2001; Olanya et al., 2001). The first report of metalaxyl-resistant isolates in Morocco was in 1996 (Sedegui et al., 1997), and subsequently various levels of resistance have been reported (Hammi et al., 2002; Sedegui et al., 2000). In Cameroon, metalaxyl-resistant isolates were found on all the known hosts of *P. infestans* in this region, which includes the potato, tomato, and garden huckleberry (Fontem et al., 2005).

7.4.4 South America — (Gregory A. Forbes)

Until 1998, the known population structure of *P. infestans* in South America was simple, comprising only a few clonal lineages (Forbes et al., 1998). The one exception to this was Argentina, where there was more diversity in the population with five multilocus genotypes that occurred among 15 isolates coming from potatoes (Forbes et al., 1998). All the genotypes from Argentina were A2 but were different from the A2 lineage BR-1, which was predominant on potatoes in Brazil, Bolivia, Uruguay, and Paraguay. This situation of A2 lineages dominating on potatoes in the southern part of South America stopped at the Bolivian/Peruvian border in the area of Lake Titicaca. From that point north through Peru, Ecuador, Colombia, and Venezuela, the A1 lineage EC-1 was dominant on potatoes (Forbes et al., 1998). This lineage had been first identified and characterized in Ecuador (Forbes et al., 1997). Other lineages were also found on the potato in Peru (PE-3 and PE-7) but in lesser frequency and primarily in the southern part of the country. The other lineages were often associated with the production of native potato varieties (Garry et al., 2005; Perez et al., 2001).

Throughout South America, the US-1 lineage dominated tomato production. This lineage was assumed to have been common or dominant throughout South America on both potatoes and tomatoes until it was displaced by the lineages mentioned above, which presumably came from Europe (Forbes et al., 1997, 1998). In the 1998 review, Chile was unique in South America in that it had only the US-1 lineage on both the potato and tomato. Presumably, strict quarantine regulations, mountains to the east and desert to the north have acted together to stop migration of new populations effectively into Chile.

The situation described above has remained stable for most parts of South America for the last decade. We are not aware of population studies per se in Argentina; however, in a recent study on the stability of host resistance, A2 isolates were identified (Adriana Andreu, personal communication). BR-1 seems to still be dominant in Brazil (Reis et al., 2005) and Uruguay (Deahl et al., 2003). Similarly, there is no evidence of change in the pathogen population in Chile, which is still dominated by the US-1 clonal lineage on the potato (Rivera et al., 2002). We do not have published evidence that US-1 attacks tomatoes in Chile, but that is the most probable case. Despite the displacement of US-1 on potatoes in other South American countries, this lineage has remained the dominant population on tomatoes (Adler et al., 2004; Lima et al., 2008). There is no clear evidence for change in the pathogen population on potatoes in the Northern Andes, which is still predominated by the EC-1 lineage (Oliva et al., 2008); however, there is reason to believe that a new population is now present in Venezuela that is characterized by the Ia mtDNA haplotype, which is not typical of EC-1 (G. Fermin, personal communication). Forbes et al. (1997) earlier hypothesized that the EC-1 lineage had been introduced in Venezuela, which imports seed potatoes, and then it moved south through the roughly contiguous potato-growing regions of the Andes. It will be interesting to evaluate whether the new Venezuelan population also migrates south.

Although there is no evidence for change in the pathogen populations that attack potatoes and tomatoes in the last decade in the northern Andes

(Colombia, Ecuador, and Peru), significant research in the area has demonstrated that much greater diversity exists than had been previously thought, and this diversity is associated with other hosts of the pathogen. The appraisal of a simple pathogen population genetic structure in the Andes began to change in 1995 when four *Phytophthora* isolates were collected from blighted leaves of a plant then identified as *S. brevifolium*. This caused significant interest because all four were A2 mating type when exposed *in vitro* to *P. infestans* A1 tester strains, and previously no A2s had been found north of Bolivia. From 1996 to 1999, a total of 53 isolates were collected from hosts then identified as *S. brevifolium* and *S. tetrapetalum*. Based on a peculiar RFLP fingerprint with limited polymorphism, a new mtDNA haplotype, and an A2 mating type, these isolates were classified in a new clonal lineage named EC-2 (Ordonez et al., 2000).

In a more recent study, which included a systematic sampling in Ecuador of foliar *Phytophthora* pathogens from various wild and cultivated hosts of the genus *Solanum*, Adler et al. (2004) reported the presence of yet another clonal lineage (EC-3) on tree tomato (*S. betaceum*), which is a native Andean fruit crop. These authors also found numerous isolates that attack plants in the *Anarrichomenum* complex that do not correspond to the EC-2 lineage, because they were A1 mating type and Ia mtDNA haplotype. In the study by Adler et al. (2004), AFLP fingerprinting confirmed the presence of host-specific groups of isolates, but the taxonomic status of these *Phytophthora* strains attacking several nontuber-bearing *Solanum* spp. remained unresolved (Adler et al., 2004).

Kroon et al. (2004) used parts of the β-tubulin, translation elongation factor 1-alpha, NADH-4, and Cox-1 genes to construct a high-resolution molecular phylogeny for the genus *Phytophthora*. This demonstrated that the EC-2 clonal lineage was closely related to *P. infestans, P. ipomoeae*, and *P. mirabilis* yet not identical to them, which stimulated discussion on the origin of these *Phytophthora* isolates in Ecuador. Currently, a genetically diverse population of pathogens that attacks several *Solanum* hosts in the Northern Andes is being proposed as a new species *Phytophthora andina* (R. Oliva, personal communication). At first view, this may have limited relevance to a review of *P. infestans*; however, studies to date indicate that the two species *P. infestans* and *P. andina* are very closely related (Gomez-Alpizar et al., 2007; Kroon et al., 2004). This leads to an evolutionary conundrum, because according to the most widely accepted hypotheses, *P. infestans* originated in Mexico (along with three more species of its clade) (Fry, 2008), and *P. andina* has only been found in South America.

7.4.5 Asia (Keqiang Cao)

Populations of *P. infestans* in Asia are diverse, and some are clearly in transition. The US-1 lineage has been reported in several countries (Guo et al., 2008b; Kato and Naito, 2001; Koh et al., 1994), and in some cases, this is the only lineage reported. However, recent analyses of herbarium specimens suggest that the US-1 lineage only arrived in the mid-twentieth century (Ristaino and Hu, 2008). In some locations, the US-1 lineage is still dominant. For example, a collection in 2002–2003 in Vietnam detected only the US-1 lineage, but more recent collections suggest that changes are now occurring (Le et al., 2008). In contrast, in other countries, this lineage is or apparently has been replaced by other lineages. A2 mating types (as well as A1 mating types) have now been reported from many areas, which include Bangladesh (Forbes et al., 2004), China (Zhang et al., 1996), India (Singh et al., 1994), Indonesia (Nishimura et al., 1999), Israel (Grinberger et al., 1989), Nepal (Shrestha, 1998), Pakistan (Ahmad and Mirza, 1995), Siberia (Elansky et al., 2001), and Thailand (Nishimura et al., 1999), with all four major mtDNA haplotypes being represented (Ristaino and Hu, 2008). In several locations, the US-1 lineage has recently become much less dominant (Akino et al., 2008; Guo et al., 2008a). There were reports of A2 mating type strains in Japan as early as 1989 (Mosa et al., 1989) and by 1991 in Korea (cited by Koh et al., 1994). Recently, there has been a decline of the A2 mating type in Korea (Park et al., 2008). The A2 strain in Japan (JP-1) was dominant into the 1990s, but recently other A1 clonal lineages (JP-2, JP-3, and JP-4) have appeared. It is postulated that JP-3 and JP-4 might have developed via sexual recombination (Akino et al., 2008). In those cases where there is sufficient information, it seems that populations remain clonal.

7.4.6 United States and Canada — (William E. Fry)

Before the 1980s, the populations in the United States and Canada were dominated by the US-1 clonal lineage, but subsequent migrations have completely changed that situation. The first hints of changes in the United States were reported by Deahl in reports of metalaxyl resistance (Deahl et al., 1993) and A2 mating types (Deahl et al., 1991). Subsequent analysis revealed that these and other isolates were distinct from the US-1 clonal lineage and came from Mexico (Goodwin et al., 1994). These isolates contained characteristics that were different from those present in the previous population: Some were resistant to metalaxyl; some were specialized to either potatoes or tomatoes; some were more aggressive to their hosts than the previous population, and some were A2 mating type (new to the United States and Canada) (Goodwin and Fry, 1994; Goodwin et al., 1995, 1996, 1998; Kato et al., 1997; Lambert and Currier, 1997).

One of the most striking features of the current population in the United States and Canada (with a possible exception of the Fraser Valley, British Columbia, Canada) is its very simple structure. After the initial detection of migrations and very serious epidemics of the mid-1990s (Fry and Goodwin, 1997), the US-8 clonal lineage has quickly become the dominant lineage on potatoes — to the extent that subsequent to the late 1990s it is only this lineage that has been detected on potatoes (unpublished results). During the early and mid-1990s, the US-6 and US-11 lineages were also detected on potatoes. On tomatoes, the situation is rather different. Initially, there were several diverse clonal lineages detected on tomatoes, which include the US-6, US-7, US-11, and US-17 lineages. However, populations on tomatoes seem to be dynamic with strains detected in one year not necessarily being the same as those from the previous year. However, in each epidemic population sampled, there has been only one genotype (Fry, unpublished). For example, isolates from an epidemic on tomatoes on Long Island in 2007 were all the same (unnamed) A1 clonal lineage (unpublished results). The origin of the diversity in populations on tomatoes is unknown. This diversity may have come from migration, but it might also have come from sporadic sexual reproduction.

Both mating types now occur in the United States and Canada, but the occurrence of clonal populations on both potatoes and tomatoes results in spatial separation of mating types and would seem to effectively preclude sexual reproduction. However, this situation could change, and the possibility for sexual recombination and production of oospores exists — even if the probability is very low. Indeed, observations in the 1990s suggested that some populations might have contained individuals from a sexual population (Gavino et al., 2000; Goodwin et al., 1998). It is clear that strains in the United States and Canada during the 1990s were (and presumably remain) sexually compatible (Mayton et al., 2000). However, the current situation (up to 2007) is that a persisting sexual population in the United States or Canada has not been documented.

7.5 CONCLUDING COMMENTS

It is very clear that there have been major changes in the structures of *P. infestans* populations during the past several decades. Host specialization has been recorded in many populations. Sexual reproduction is now recorded outside of central Mexico, but most populations remain dominated by asexual reproduction. Migration has been a major factor in these changes as well as sexual recombination and mutation/selection. For example, the developing predominance of an A2 lineage in Europe seems likely to have resulted from sexual reproduction followed by selection. The practical importance of these changes in populations has important implications for management of the late blight disease and epidemics. Scientists now are paying close attention to changes in population structure and are sharing information about the dominant forms in their locations. Neutral markers are being used to determine whether sexual recombination is likely to be occurring and if so, the role of oospores in the epidemiology is being investigated so that management efforts can be appropriately directed.

REFERENCES

Adler N, Erselius LJ, Chacon MG, Flier WG, Ordonez ME, Kroon LPNM, Forbes GA 2004. Genetic diversity of *phytophthora infestans* sensu lato in Ecuador provides new insight into the origin of this important plant pathogen. Phytopathol 94:154–162.

Ahmad I, Mirza JI 1995. Occurrence of A2 mating type of *Phytophthora infestans* in Pakistan. Proc. National Seminar on Research and Development of Potatoes in Pakistan. Islamabad, Pakistan.

Akino S, Kato M, Baba Y, Hirotomi D, Mizunuma K, Hagiwara H, Kondo N 2008. Spatial and temporal genotypic changes in Japanese isolates of *Phytophthora infestans* 1997–2007. Proc. 3rd International Late Blight Conference. Beijing, International Potato Center, Lima, Peru.

Andersson B, Sandstrom M, Stromberg A 1998. Indications of soil borne inoculum of *Phytophthora infestans*. Potato Res 41:305–310.

Badillo-Ponce G, Fernandez-Pavia SP, Grünwald NJ, Garay-Serrano E, Rodriguez-Alvarado G, Lozoya-Saldana H 2004. First report of blight on *Ipomoeoa purpurea* caused by *Phytophthora ipomoeae*. Plant Dis 88:1283.

Baka ZAM 1997. Mating type, nuclear DNA content and isozyme analysis of Egyptian isolates of *Phytophthora infestans*. Folia Microbiol 42:613–620.

Bakonyi J, Laday M, Dula T, Ersek T 2002. Characterization of *Phytophthora infestans* isolates from Hungary. Eur J Plant Pathol 108:139–146.

Bourke PMA 1964. Emergence of potato blight, 1843–46. Nature 203:805–808.

Brasier CM 1971. Induction of sexual reproduction in single A2 isolates of *Phytophthora* species by *Trichoderma viride*. Nature New Biol 231:283.

Bruberg MB, Hannukkala A, Hermansen A 1999. Genetic variability of *Phytophthora infestans* in Norway and Finland as revealed by mating type and fingerprint probe RG57. Mycol Res 103:1609–1615.

Cooke DEL, Lees AK 2004. Markers, old and new, for examining *Phytophthora infestans* diversity. Plant Pathol 53:692–704.

Cooke DEL, Lees AK, Hansen EM, Lassen P, Andersson B, Bakonyi J 2007a. EUCABLIGHT one year on: an update on the European blight population database. Workshop of an European network for the development of an integrated control strategy for late blight. PPO special report No. 12.

Cooke DEL, Lees AK, Shaw DS, Taylor M, Prentice MWC, Bradshaw NJ, Bain RA 2007b. Survey of GB blight populations. 10th Workshop of an European network for the development of an integrated strategy for late blight.

Cooke DEL, Lees AK, Shaw DS, Taylor M, Prentice MWC, Bradshaw NJ, Bain RA 2008. The status of GB blight populations and the threat of oospores. Crop Protection in Northern Britain 2008.

Cooke DEL, Young V, Birch PRJ, Toth R, Gourlay R, Day JP, Carnegie SF, Duncan JM 2003. Phenotypic and genotypic diversity of Phytophthora *infestans* populations in Scotland 1995–97. Plant Pathol 52:181–192.

Cooke LR, Carlisle DJ, Donaghy C, Quinn M, Perez FM, Deahl KL 2006. The Northern Ireland Phytophthora infestans population 1998–2002 characterized by genotypic and phenotypic markers. Plant Pathol 553:320–330.

Day JP, Wattier RAM, Shaw DS, Shattock RC 2004. Phenotypic and genotypic diversity in *Phytophthora infestans* on potato in Great Britain, 1995–98. Plant Pathol 53:303–315.

Deahl KL, DeMuth SP, Pelter G, Ormrod DJ 1993. First report of resistance of *Phytophthora infestans* to metalaxyl in Eastern Washington and Southwestern British Columbia. Plant Dis 77:429.

Deahl KL, Goth RW, Young R, Sinden SL, Gallegly ME 1991. Occurrence of the A2 mating type of *Phytophthora infestans* in potato fields in the United States and Canada. Am Potato J 68:717–726.

Deahl KL, Pagani MC, Vilaro FL, Perez FM, Moravec B, Cooke LR 2003. Characteristics of *Phytophthora infestans* from Uruguay. Eur J Plant Pathol 109:277–281.

Detourne D, Dubois L, Duvauchelle S 2007. The evolution of *Phytophthora infestans* in France mating type, metalaxyl resistance. Agrifood Research Working Papers 142. MTT Agrifood Research, Finland: 17.

Dowley LJ, O'Sullivan E 1981. Metalaxyl-resistant strains of *Phytophthora infestans* Mont. de Bary in Ireland. Potato Res 24:417–421.

Drenth A, Tas ICQ, Govers F 1994. DNA fingerprinting uncovers a new sexually reproducing population of *Phytophthora infestans* in the Netherlands. Euro J Plant Pathol 100:97–107.

El-Korany AE 1994. Pathological studies on late blight of potato caused by *Phytophthora infestans*. Ph.D. thesis. Department of Agricultural Botany, Faculty of Agriculture. Ismalia, Egypt, Suez Canal University.

Elansky S, Smirnov A, Dyakov Y, Dolgova A, Filippov A, Kozlovsky B, Kozlovskaya I, Russo P, Smart C, Fry W 2001. Genotypic analysis of Russian isolates of *Phytophthora infestans* from the Moscow region, Siberia and Far East. J Phytopathol 149:605–611.

Evenhuis B, Turkensteen LJ, Flier WG, 2007. Monitoring primary sources of inoculum of *Phytophthora infestans* in the Netherlands 1999–2005. 10th Workshop of an European network for the development of an integrated control strategy for late blight., PPO special report no. 12.

Fabritius AL, Shattock RC, Judelson HS 1997. Genetic analysis of metalaxyl insensitivity loci in *Phytophthora infestans* using linked DNA markers. Phytopathol 8710:1034–1040.

Fernandez-Pavia SP, Grünwald NJ, Fry WE 2002. Formation of *Phytophthora infestans* oospores in nature on tubers in central Mexico. Plant Dis 86:73.

Flier WG, Grünwald NJ, Fry WE, Turkensteen LJ 2001. Formation, production and viability of oospores of *Phytophthora infestans* from potato and *Solanum demissum* in the Toluca Valley, central Mexico. Mycol Res 105:998–1006.

Flier WG, Grünwald NJ, Kroon LPNM, Sturbaum AK, van den Bosch TBM, Garay-Serrano E, Lozoya-Saldana H, Fry WE, Turkensteen LJ 2003. The population structure of *Phytophthora infestans* from the Toluca Valley of central Mexico suggests genetic differentiation between populations from cultivated potato and wild *Solanum* spp. Phytopathol 93:382–390.

Flier WG, Grünwald NJ, Kroon LPNM, van den Bosch TBM, Garay-Serrano E, Lozoya-Saldana H, Bonants PJM, Turkensteen LJ 2002. *Phytophthora ipomoeae*, a new homothallic species causing late blight on *Ipomoeae longipedunculata* in the Toluca Valley of central Mexico. Mycol Res 106:848–856.

Fontem DA, Olanya OM, FNB 2004. Reaction of certain Solanaceous and Asteraceous plant species to inoculation with *Phytophthora infestans* Cameroon. J Phytopathol 152:331–336.

Fontem DA, Olanya OM, Tsopmbeng GR, Owona MAP 2005. Pathogenicity and metalaxyl sensitivity of *Phytophthora infestans* isolates obtained from garden huckleberry, potato and tomato in Cameroon. Crop Protection 24:449–456.

Forbes GA 2004. Global overview of late blight. Proc Regional Workshop on Potato Late Blight for East and Southest Asia and the Pacific. Yezin Agricultural University, Myanmar.

Forbes GA, Escobar XC, Ayala CC, Revelo J, Ordonez ME, Fry BA, Doucett K, Fry WE 1997. Population genetic structure of *Phytophthora infestans* in Ecuador. Phytopathol 87:375–380.

Forbes GA, Goodwin SB, Drenth A, Oyarzun P, Ordonez ME, Fry WE 1998. A global marker database for *Phytophthora infestans*. Plant Dis 82:811–818.

Frinking HD, Davidse LC, Limburg H 1987. Oospore formation by *Phytophthora infestans* in host tissue after inoculation with isolates of opposite mating type found in the Netherlands. Netherlands J Plant Pathol 93:147–149.

Fry WE 2008. *Phytophthora infestans*, the crop and *R* gene destroyer. Molec Plant Pathol 9:385–402.

Fry WE, Goodwin SB 1997a. Re-emergence of potato and tomato late blight in the United States. Plant Dis 8112:1349–1357.

Fry WE, Goodwin SB 1997b. Resurgence of the Irish Potato Famine Fungus. Bioscience 47:363–371.

Fry WE, Goodwin SB, Dyer AT, Matuszak JM, Drenth A, Tooley PW, Sujkowski LS, Koh YJ, Cohen BA, Spielman LJ et al. 1993. Historical and recent migrations of *Phytophthora infestans*: chronology, pathways, and implications. Plant Dis 77:653–661.

Gallegly ME, Galindo J 1958. Mating types oand oospores of *Phytophthora infestans* in nature in Mexico. Phytopathol 48:274–277.

Garry G, Forbes GA, Salas A, Santa Cruz M, Perez WG, Nelson RJ 2005. Genetic diversity and host differentiation among isolates of Phytophthora infestans from cultivated potato and wild solanaceous hosts in Peru. Plant Pathol 546:740–748.

Gavino PD, Fry WE 2002. Diversity in and evidence for selection on the mitochondrial genome of *Phytophthora infestans*. Mycologia 94:781–793.

Gavino PDCD, Smart CD, Sandrock RW, Miller JS, Hamm PB, Lee TY, Davis RM, Fry WE 2000. Implications of sexual reproduction for *Phytophthora infestans* in the United States: generation of an aggressive lineage. Plant Dis 84:731–735.

Gomez-Alpizar L, Carbone I, Ristaino JB 2007. An Andean origin of *Phytophthora infestans* inferred from mitochondrial and nuclear gene genealogies. PNAS 104:3306–3311.

Goodwin SB, Cohen BA, Deahl KL, Fry WE 1994a. Migration from northern Mexico was the probable cause of recent genetic changes in populations of *Phytophthora infestans* in the United States and Canada. Phytopathol 84:553–558.

Goodwin SB, Cohen BA, Fry WE 1994b. Panglobal distribution of a single clonal lineage of the Irish potato famine fungus. Proc Nat Acad Sci USA 91:11591–11595.

Goodwin SB, Drenth A 1997. Origin of the A2 mating type of *Phytophthora infestans* outside Mexico. Phytopathol 87:992–999.

Goodwin SB, Fry WE 1994a. Continued migration of A2 mating type, metalaxyl-resistant genotypes of *Phytophthora infestans* in the eastern United States and Canada. Phytopathol 8411:1371.

Goodwin SB, Fry WE 1994b. Genetic analysis of interspecific hybrids between *Phytophthora infestans* and *P. mirabilis*. Experimental. Mycology 18:20–32.

Goodwin SB, Smart CD, Sandrock RW, Deahl KL, Punja ZK, Fry WE 1998. Genetic change within populations of *Phytophthora infestans* in the United States and Canada during 1994 to 1996: Role of migration and recombination. Phytopathol 88:939–949.

Goodwin SB, Spielman LJ, Matuszak JM, Bergeron SN, Fry WE 1992. Clonal diversity and genetic differentiation of *Phytophthora infestans* populations in northern and central Mexico. Phytopathol 82:955–961.

Goodwin SB, Sujkowski LS, Dyer AT, Fry BA, Fry WE 1995a. Direct detection of gene flow and probable sexual reproduction of *Phytophthora infestans* in northern North America. Phytopathol 85:473–479.

Goodwin SB, Sujkowski LS, Fry WE 1995b. Rapid evolution of pathogenicity within clonal lineages of the potato late blight disease fungus. Phytopathol 85:669–676.

Goodwin SB, Sujkowski LS, Fry WE 1996. Widespread distribution and probable origin of resistance to metalaxyl in clonal genotypes of *Phytophthora infestans* in the United States and western Canada. Phytopathol 86:793–800.

Griffin D, O'Sullivan E, Harmey MA, Dowley LJ 2002. DNA fingerprinting, metalaxyl resistance and mating type deermination of the *Phytophthora infestans* in the Republic of Ireland. Potato Res 45:25–36.

Griffith GW, Shaw DS 1998. Polymorphisms in *Phytophthora infestans*: four mitochondrial haplotypes are detected after PCR amplification of DNA from pure cultures or from host tissue. App Environ Microbiol 64:4007–4014.

Grinberger M, Kadish D, Cohen Y 1989. Occurrence of the A2 mating type and oospores of *Phytophthora infestans* in potato crops in Israel. Phytoparasitica 17: 197–204.

Groves CT, Ristaino JB 2000. Commercial fungicide formulations induce in vitro oospore formation and phenotypic change in mating type in *Phytophthora infestans*. Phytopathol 90:1201–1208.

Grünwald NJ, Cadena-Hinojosa MA, Rubio-Covarrubias O, Rivera-Pena A, Niederhauser JS, Fry WE 2002. Potato cultivars from the Mexican national potato program: sources and durability of resistance against late blight. Phytopathol 92:688–693.

Grünwald NJ, Flier WG 2005. The biology of *Phytophthora infestans* at its center of origin. Ann Rev Phytopathol 431:171–190.

Grünwald NJ, Flier WG, Sturbaum AK, Garay-Serrano E, van den Bosch TBM, Smart CD, Matuszak JM, Lozoya-Saldana H, Turkensteen LJ, Fry WE 2001. Population structure of *Phytophthora infestans* in the Toluca Valley region of central Mexico. Phytopathol 91:882–890.

Grünwald NJ, Romero-Montes G, Lozoya-Saldana H, Rubio-Covarrubias OA, Fry WE 2002. Potato late blight management in the Toluca Valley: field validation of SimCast modified for cultivars with high field resistance. Plant Dis 86:1163–1168.

Grünwald NJ, Rubio-Covarrubias OA, Fry WE 2000. Potato late-blight management in the Toluca Valley: forecasts and resistant cultivars. Plant Dis 844:410–416.

Grünwald NJ, Sturbaum AK, Montes GR, Serrano EG, Lozoya-Saldana H, Fry WE 2006. Selection for fungicide resistance within a growing season in field populations of *Phytophthora infestans* at the center of origin. Phytopathol 96:1397–1403.

Guo J, van der Lee T, Qu D, Yao Y, Gong X, Liang D, Xie K, Wang X, Govers F 2008a. *Phytophthora infestans* isolates from northern China show high virulence diversity but low genotypic diversity. Plant Biology 11:57–67

Guo LY, Zhu XQ, Liu G, Hu J, Ristaino JB 2008b. Genetic diversity of *Phytophthora infestans* from China. Global Initiative on Late Blight. 3rd International Meeting. Beijing, International Potato Center, Lima, Peru.

Hammi A, Bennani A, Ismaili AE, Msatef Y, Serrhini MN 2001. Production and germination of oospores of *Phytophthora infestans* Mont. de Bary in Morocco. Eur J Plant Pathol 107:553–556.

Hammi A, Msatef Y, Bennani A, Ismaili AE, Serrhini MN 2002. Mating type, metalaxyl resistance and aggressiveness of *Phytophthora infestans* Mont. de Bary in Morocco. J Phytopathol 150:289–291.

Hawkes JG 1990. The potato: evolution, biodiversity and genetic resources. Washington, DC, Smithsonian Institution Press.

Hermansen A, Hannukkala A, Hafskjold Naerstad R, Brurberg MB 2000. Variation in populations of *Phytophthora infestans* in Finland and Norway: mating type, metalaxyl resistance and virulence phenotype. Plant Pathol 49:11–22.

Hijmans RJ, Spooner DM 2001. Georgraphic distribution of wild potato species. Am J Bot 88:2101–2112.

Hohl HR, Iselin K 1984. Strains of *Phytophthora infestans* from Switzerland with A2 mating type behavior. Trans Br Mycol Soc 83:529–530.

Iram S, Ahmad I 2008. Review of metalaxyl resistance in *Phytophthora infestans* from Pakistan. 3rd International Late Blight Conference. Beijing, International Potato Center, Lima, Peru.

Judelson HS, Roberts S 1999. Multiple loci determining insensitivity to phenylamide fungicides in *Phytophthora infestans*. Phytopathol 899:754–760.

Kankwatsa P, Hakiza JJ, Olanya M, Kidanemariam HM, Adipala E 2003. Efficacy of different fungicide spray schedules for control of potato late blight in southwestern Uganda. Crop Protection 22:545–552.

Kapsa J 2001. Incidence of late blight *Phytophthora infestans* in potato crops and its control in Poland in 1995–1999. Workshop of an European network for the development of an integrated control strategy for late blight. Applied research for arable farming and field production of vegetables. Lelystad, the Netherlands.

Kato M, Mizubuti ESG, Goodwin SB, Fry WE 1997. Sensitivity to protectant fungicides and pathogenic fitness of clonal lineages of *Phytophthora infestans* in the United States. Phytopathol 87:973–978.

Kato M, Naito S 2001. Change of predominate genotypes of *Phytophthora infestans* in Tokachi district, Hokkaido, Japan and differences of lesion productivity to the field resistant cultivar 'Matilda' among genotypes. Journal of the Agricultural University of Hebei 24 2:11–15.

Knapova G, Gisi U 2002. Phenotypic and genotypic structure of *Phytophthora infestans* populaitons on potato and tomato in France and Switzerland. Plant Pathol 51:641–653.

Ko WH 1994. An alternative possible origin of the A2 mating type of *Phytophthora infestans* outside Mexico. Phytopathol 84:1224–1227.

Koh YJ, Goodwin SB, Dyer AT, Cohen BA, Ogoshi A, Sato N, Fry WE 1994. Migrations and displacements of *Phytophthora infestans* populations in east Asian countries. Phytopathol 84:922–927.

Kroon LPNM, Bakker FT, van den Bosch GBM, Bonants PJM, Flier WG 2004. Phylogenetic analysis of *Phytophthora* species basesd on mitochondrial and nuclear DNA sequences. Fungal Genet Biol 41:766–782.

Lambert DH, Currier AI 1997. Differences in tuber rot development for North American clones of *Phytophthora infestans*. Am Potato J 74:39–43.

Le VH, Ngo XT, Pham TX, Hermansen A 2008. Late blight in Vietnam; pathogen population, host specficity and control. 3rd International Late Blight Conference. Beijing, International Potato Center, Lima, Peru.

Lebreton L, Laurent C, Andrivon D 1998. Evolution of *Phytophthora infestans* populations in the two most important potato production areas of France during 1992–96. Plant Pathol 47:427–439.

Lee TY, Mizubuti E, Fry WE 1999. Genetics of metalaxyl resistance in *Phytophthora infestans*. Fungal Genet Biol 26:118–130.

Lees AK, Wattier R, Shaw DS, Sullivan L, Williams NA, Cooke DEL 2006. Novel microsatellite markers for the analysis of *Phytophthora infestans* populations. Plant Pathol 553:311–319.

Lehtinen A, Hannukkala A 2004. Oospores of *Phytophthora infestans* in soil provide an important new source of primary inoculum in Finland. Agricul Food Sci 13:399–410.

Lehtinen A, Hannukkala A, Andersson B, Hermansen A, Le VH, Naerstad R, Brurberg MB, Nielsen BJ, Hansen JG, Yuen J 2008. Phenotypic variation in Nordic populations of *Phytophthora infestans* in 2003. Plant Pathol 57:227–234.

Lehtinen A, Hannukkala A, Rantanen T, Jauhiainen L 2007. Phenotypic and genetic variation in Finnish potato-late blight populations, 1997–2000. Plant Pathol 56: 480–491.

Lima MA, Maffia LA, Barreto RW, Mizubuti ESG 2008. *Phytophthora infestans* in a subtropical region: survival on tomato debris, temporal dynamics of airborne sporangia and alternative hosts. Plant Pathol in press.

Lozoya-Saldana H, Hernandez A, Flores R, Bamberg J 1997. Late blight on wild *Solanum* species in the Toluca Valley in 1996. Am Potato J 74:445.

Malcolmson JF 1969. Races of *Phytophthora infestans* occurring in Great Britain. Trans Br Mycol Soc 53:417–423.

Matuszak JM, Fernandez-Elquezabal J, Gu W-K, Villarreal-Gonzalez M, Fry WE 1994. Sensitivity of *Phytophthora infestans* populations to metalaxyl in Mexico: distribution and dynamics. Plant Dis 78:911–916.

May KJ, Ristaino JB 2004. Identity of the mtDNA haplotypes of *Phytophthora infestans* in historical specimens from the Irish Potato Famine. Mycol Res 108:471–479.

Mayton H, Smart CD, Moravec BC, Mizubuti ESG, Muldoon AE, Fry WE 2000. Oospore survival and pathogenicity of single oospore recombinant progeny from a cross involving the US-8 and US-17 lineages of *Phytophthora infestans*. Plant Dis 84:1190–1196.

McLeod A 1998. Characterization of *Phytophthora infestans* populations in South Africa. Department of Plant Pathology. Stellenbosch, South Africa, University of Stellenobosch: 107.

McLeod A, Coertze S 2006. First report of *Phytophthora infestans* on Petunia x hybrida in South Africa. Plant Dis 90:1550.

McLeod A, Denman S, Sadie A, Denner FDN 2001. Characterization of South African isolates of *Phytophthora infestans*. Plant Dis 85:287–291.

Mills WR, Niederhauser JS 1953. Observations on races of *Phytophthora infestans* in Mexico. Phytopathol 43:454–455.

Mosa AA, Kato M, Sato N, Kobayashi K, Ogoshi A 1989. Occurrence of the A2 mating type of *Phytophthora infestans* on potato in Japan. Ann Phytopathol Soc Jpn 55:615–620.

Mukalazi J, Adipala E, Sengooba T, Hakiza JJ, Olanya M, Kidanemariam HM 2001. Metalaxyl resistance, mating type and pathogenicity of *Phytophthora infestans* in Uganda. Crop Protection 20:379–388.

Nattrass RM, Ryan M 1951. New hosts of *Phytophthora infestans* in Kenya. Nature 168: 85–86.

Niederhauser JS 1991. *Phytophthora infestans*: The Mexican connection. In Phytophthora. Lucas JA, Shattock RC, Shaw DS, Cooke LR editors. Cambridge University Press, Cambridge UK. pp. 25–45.

Niederhauser JS, Cervantes J, Servin L 1954. Late blight in Mexico and its implications. Phytopathol 44:406–408.

Nishimura R, Sato K, Lee WH, Singh UP, Chang TT 1999. Distribution of *Phytophthora infestans* in seven Asian countries. Ann Phytopathol Soc Jpn 65: 66–75.

Nyankanga RO, Wien HC, Olanya M, Ojiambo PS 2004. Farmers' cultural practices and management of potato late blight in Kenya highlands: implications for development of integrated disease management. Int J Pest Manage 50:135–144.

Ochwo MKN, Kamoun S, Adipala E, Rubaihayo PR, Lamour K, Olanya M 2002. Genetic diversity of *Phytophthora infestans* Mont. de Bary in the eastern and western highlands of Uganda. J Phytopathol 150:541–542.

Olanya M, Adipala E, Hakiza JJ, Kedera JC, Ojiambo PS, Mukalazi J, Forbes GA, Nelson R 2001. Epidemiology and population dynamics of *Phytophthora infestans* in sub-Saharen Africa: progress and constraints. African Crop Sci J 9:185–193.

Oliva RF, Flier WG, Kroon LPNM, Ristaino JB, Forbes GA 2008. *Phytophthora andina* a newly identified heterothallic pathogen of Solanaceous hosts in the Andean highlands. Phytopathol. In press.

Ordonez ME, Hohl HR, Velasco A, Ramon MP, Oyazun PJ, Smart CD, Fry WE, Forbes GA, Erselius LJ 2000. A novel A2 population of *Phytophthora* similar to *P. infestans* attacks wild *Solanum* species in Ecuador. Phytopathol 90:197–202.

Park KH, Cheon JU, Kim JS, Ham YI, Ryu KY, Cha BJ 2008. Occurrence characteristic, mating type, and chemical resistance of *Phytophthora infestans* in Korea. 3rd International Late Blight Conference. Beijing, International Potato Center, Lima, Peru.

Perez WG, Gamboa JS, Falcon YV, Coca M, Raymundo RM, Nelson RJ 2001. Genetic structure of Peruvian populations of *Phytophthora infestans*. Phytopathol 9110: 956–965.

Pule BB, Meitz J, Thompson A, Fry WE, Meyer KL, Wakahiu M, Senkesha N, McLeod A 2008. Characterization of *Phytophthora infestans* populations from selected central, eastern, and southern Africa. Third International Late Blight Conference Beijing. International Potato Center, Lima, Peru.

Purvis AI, Pipe ND, Day JP, Shattock RC, Shaw DS, Assinder SJ 2001. AFLP and RFLP RG57 fingerprints can give conflicting evidence about the relatedness of isolates of *Phytophthora infestans*. Mycol Res 105:1321–1330.

Reeves RJ, Jackson RM 1974. Stimulation of sexual reproduction in *Phytophthora* by damage. J Gene Microbiol 84:303–310.

Reis A, Ribeiro FHS, Maffia LA, Mizubuti ESG 2005. Sensitivity of brazilian isolates of *Phytophthora infestans* to commonly used fungicides in tomato and potato crops. Plant Dis 89:1279–1284.

Ristaino JB, Hu C-H 2008. Historical and current population biology of late blight from the Irish potato famine to SE Asia, Russia and China. 3rd International Late Blight Conference. Beijing, International Potato Center, Lima, Peru.

Rivera-Pena A 1990a. Wild tuber-bearing species of *Solanum* and incidence of *Phytophthora infestans* Mont. de Bary on the western slopes of the volcano Nevado de Toluca. 2. Distribution of *Phytophthora infestans*. Potato Res 33:331–347.

Rivera-Pena A 1990b. Wild tuber-bearing species of *Solanum* and incidence of *Phytophthora infestans* Mont. de Bary on the western slopes of the volcano Nevado de Toluca. 3. Physiological races of *Phytophthora infestans*. Potato Res 33:349–355.

Rivera-Pena A 1990c. Wild tuber-bearing species of *Solanum* and incidence of *Phytophthora infestans* Mont. de Bary on the western slopes of the volcano Nevado de Toluca. 4. Plant development in relation to late blight infection. Potato Res 33:469–478.

Rivera-Pena A 1990d. Wild tuber-bearing species of *Solanum* and incidence of *Phytophthora infestans* Mont. de Bary on the Western slopes of the volcano Nevado de Toluca. 5. Type of resistance to *P. infestans*. Potato Res 33:479–486.

Rivera-Pena A, Molina-Galan J 1989. Wild tuber-bearing species of *Solanum* and incidence of *Phytophthora infestans* Mont. de Bary on the western slopes of the volcano Nevado de Toluca. 1. *Solanum* species. Potato Res 32:181–195.

Rivera V, Riveros F, Secor G 2002. Characterization of a *Phytophthora infestans* population in Chile. Late blight: managing the global threat, Hamburg, Germany. International Potato Center, Lima, Peru.

Schiessendopper E, Molnar O 2002. Characterization of *Phytophthora infestans* populations in sub-Saharen Africa as a basis for simulation modeling and integrated management. Late Blight: managing the global threat, Hamburg, Germany. International Potato Center, Lima, Peru.

Sedegui M, Carroll RB, Morehart AL 1997. First report from Morocco of *Phytophthora infestans* isolates with metalaxyl resistance. Plant Dis 81:831.

Sedegui M, Carroll RB, Morehart AL, Evans TA, Kim SH, Lakhdar R, Arifi A 2000. Genetic structure of the *Phytophthora infestans* population in Morocco. Plant Dis 842:173–176.

Shaat MMN 2002. Detection of mating types of potato late blight pathogen, *Phytophthora infestans* Mont. de Bary, in El-Minia Governorate, Egypt. Ass J Agricul Sci 33:161–175.

Shattock RC, Shaw DS, Fyfe AM, Dunn JR, Loney KH, Shattock JA 1990. Phenotypes of *Phytophthora infestans* collected in England and Wales from 1985 to 1988: mating type, response to metalaxyl and esoenzyme analysis. Plant Pathol 39:242–248.

Shaw DS, Nagy ZA, Evans D, Deahl K 2007. The 2005 population of Phytophthora infestans in Great Britain: the frequency of A2 mating type has increased and new molecular genotypes have been detected. Proc 10th workshop of an European network for the development of an integrated control strategy for late blight PPO special report No. 12. 137–143.

Shrestha SK, Shrestha K, Kobayashi K, Kondo N, Nishimura R, Sato K, Ogoshi A 1998. First report of A1 and A2 mating types of *Phytophthora infestans* on potato and tomato in Nepal. Plant Dis 829:1064.

Singh BP, Bhat MN 2008. *Phytophthora infestans* population structure in India. 3rd International Late Blight Conference. Beijing. International Potato Center, Lima, Peru.

Singh BP, Roy S, Bhattacharyya SK 1994. Occurrence of the A2 mating type of *Phytophthora infestans* in India. Potato Res 37:227–231.

Smart CD, Mayton H, Mizubuti ESG, Willmann MR, Fry WE 2000. Environmental and genetic factors influencing self-fertility in *Phytophthora infestans*. Phytopathol 90:987–994.

Smoot JJ, Gough FJ, Lamey HA, Eichenmuller JJ, Gallegly ME 1958. Production and germination of oospores of *Phytophthora infestans*. Phytopathol 48:165–171.

Spielman LJ, Drenth A, Davidse LC, Sujkowski LJ, Gu WK, Tooley PW Fry WE 1991. A second world-wide migration and population displacement of *Phytophthora infestans?* Plant Pathol 40:422–430.

Spooner DM, Bamberg J, Hjerting JP, Gomez J 1991. Mexico, 1988 potato germplasm collecting expedition and uility of the Mexican potato species. Am Potato J 68:29–43.

Spooner DM, Hijmans RJ 2001. Potato systematics and germplasm collecting, 1989–2000. Am J Potato Res 78:237–268.

Spooner DM, Rivera-Pena A, van den Berg RG, Schuler K 2000. Potato germplasm collecting expedition to Mexico in 1997: taxonomy and new germplasm resources. Am J Potato Res 77:261–270.

Sujkowski LS, Goodwin SB, Dyer AT, Fry WE 1994. Increased genotypic diversity via migration and possible occurrence of sexual reproduction of *Phytophthora infestans* in Poland. Phytopathol 84:201–207.

Sujkowski LS, Goodwin SB, Fry WE 1996. Changes in specific virulence in Polish populations of *Phytophthora infestans*: 1985–1991. Eur J Plant Pathol 102:555–561.

Tantius PH, Fyfe AM, Shaw DS, Shattock RC 1986. Occurrence of the A2 mating type and self-fertile isolates of *Phytophthora infestans* in England and Wales. Plant Pathol 35:578–581.

Tooley PW, Fry WE, Villarreal Gonzalez MJ 1985. Isozyme characterization of sexual and asexual *Phytophthora infestans* populations. J Heredity 76:431–435.

Tooley PW, Sweigard JA, Fry WE 1986. Fitness and virulence of *Phytophthora infestans* isolates from sexual and asexual populations. Phytopathol 76:1209–1212.

Van Raaij HMG, Evenhuis A, Van den Bosch GBM, Forch MG, Spits HG, Kessel GJT, Flier WG 2007. Monitoring virulence and mating type of *Phytophthora infestans* in the Netherlands in 2004 and 2005. Proc. 10th Workshop of an European network for the development of an integrated control strategy for late blight., PPO special report No. 12. 281–284.

Vega-Sanchez ME, Erselius LJ, Rodriguez AM, Bastidas O, Hohl HR, Ojiambo PS, Mukalazi J, Vermeulen T, Fry WE, Forbes GA 2000. Host adaptation to potato and tomato within the US-1 clonal lineage of *Phytophthora infestans* in Uganda and Kenya. Plant Pathol 49:531–539.

Widmark AK, Andersson B, Cassel-Lundhagen A, Sandstrom M, Yuen JE 2007. *Phytophthora infestans* in a single field in southwest Sweden early in spring: symptoms, spatial distribution and genotypic variation. Plant Pathol 564:573–579.

Zhang Z, Li Y, Tian S, Zhu J, Wang J, Song B 1996. The occurrence of potato late blight pathogen *Phytophthora infestans* A2 mating type in China. J Agricul Uni Hebei 194:62–66.

Zhu JH, Yang ZH, Shen JW, Yao GS, Liu R, Zhang ZM 2008. Microsatellite genotypic analysis of *Phytophthora infestans* in investigated area of China. 3rd International Late Blight Conference. Beijing. International Potato Center, Lima, Peru.

Zwankhuizen M, Govers F, Zadoks JC 2000. Inoculum sources and genotypic diversity of *Phytophthora infestans* in Southern Flevoland, The Netherlands. Eur J Plant Pathol 106:667–680.

8

PHYTOPHTHORA CAPSICI: SEX, SELECTION, AND THE WEALTH OF VARIATION

KURT LAMOUR
Department of Entomology and Plant Pathology, Institute of Agriculture, The University of Tennessee, Knoxville, Tennessee

8.1 INTRODUCTION

Phytophthora capsici is found worldwide attacking solanaceous and cucurbit hosts, which include pepper, tomato, eggplant, cucumber, pumpkin, melon, and squash, and is often reported after sudden and severe epiphytotics (Critopoulos, 1955; Crossan et al., 1954; Kreutzer, 1937; Kreutzer et al., 1940; Leonian, 1922; Tompkins, 1937, 1941; Weber, 1932; Wiant, 1940). More recently, *P. capsici* has caused significant damage to Lima and snap beans; these crops were previously shown to be nonhosts (Davidson et al., 2002; Polach and Webster, 1972). *P. capsici* is considered an invasive species, and it is uncertain how it is spread long distances or where it originated. The two earliest reports of *P. capsici* in New Mexico and Florida suggest *P. capsici* was introduced to these locations on contaminated pepper seed (Leonian, 1922; Weber, 1932). To date, no studies have been published on the ability of *P. capsici* to survive in seed.

In many areas, the most serious epidemics occur during the warm rainy months often after a significant portion of the production costs have been incurred. For many cucurbit varieties, the fruit are the most susceptible portion of the plant. For example, the roots, crown, and vines of common commercial varieties of pickling cucumbers are relatively tolerant of *P. capsici*, whereas the fruit are extremely susceptible. In Michigan, it is not uncommon to discover a

Oomycete Genetics and Genomics: Diversity, Interactions, and Research Tools
Edited by Kurt Lamour and Sophien Kamoun
Copyright © 2009 John Wiley & Sons, Inc.

large proportion of fruit covered in a fine white powder (which is actually massive numbers of deciduous sporangia) under a healthy leaf canopy (Hausbeck and Lamour, 2004). An additional factor that adds to the overall losses at the end of the growing season is the hemibiotrophic nature of the infection process. The initial stages of infection are biotrophic often showing no symptoms (Fig. 8.1). Depending on the temperature, there is a lag of at least 24 h before the initial necrotic lesions are visible. Significant additional losses occur when growers harvest what seems to be a healthy crop only to have lesions appear on the fruit in transit to a distribution or processing facility (Wiant, 1940). In the United States, many growers have experienced the added insult of paying freight costs for an unsalable product.

Although *P. capsici* is referred to as a soil-borne pathogen, there can be copious above-ground spore production (Fig. 8.1). To help growers in Michigan better understand the explosive potential of this pathogen, the amount of viable sporangia on the surface of a naturally infected spaghetti squash fruit was estimated. Zoospores were released from the sporangia collected from ten 1-cm^2 sections across a single 15-cm diameter lesion and counted using a hemocytometer. A conservative estimate indicated that this infected squash could produce more than 300 million motile zoospores (K. Lamour, unpublished). Clearly, infected fruit are "time bombs," which only require rain or irrigation water to allow massive spore dispersal.

In many areas of the world, *P. capsici* is the most important factor that limits vegetable production (Erwin and Ribeiro, 1996; Hausbeck and Lamour, 2004; Hwang and Kim, 1995; Islam et al., 2004). Although there are ongoing efforts to devise novel management strategies, currently no control measures can fully protect a susceptible crop from serious damage when the environmental conditions are favorable (wet and warm). For most susceptible crops, few or no resistant varieties are available, and once *P. capsici* has been introduced to a field, the best way to limit disease is to control moisture. In dry areas (e.g., Southwest United States) where irrigation is essentially the sole source of moisture, disease can be managed through judicious irrigation practices (Cafe Filho et al., 1995). At less arid locations, rainfall provides an ideal vehicle for spore dispersal, and a late summer thunderstorm can quickly spread sufficient inoculum to destroy an entire crop. Growers are encouraged to rotate to nonsusceptible crops, to plant on well-drained sites, and in some cases to employ chemical controls—but, in many cases, even when these preventative measures are adopted, there is significant loss.

8.2 EXPANSION AND CONTRACTION OF THE *P. CAPSICI* CONCEPT

In 1922, Leon Leonian described *P. capsici* after observing a blight of peppers at the New Mexico Experiment Station (Leonian, 1922). The isolates had distinctive mycelium with tuberous outgrowths, deciduous oval sporangia with

FIG. 8.1 (See color insert) *Phytophthora capsici* on peppers and squash in the United States and Peru. (a) *Capsicum pubescens* (Rocoto) production in the cleared forest of the high jungle in Peru. (b) Typical loose cuticle on Rocoto fruit infected with *Phytopthora capsici*. (c) Intact tissue at center of a *P. capsici* lesion on winter squash colonized during the biotrophic phase of the infection. (d) Healthy squash plant harboring fruit infected with *P. capsici* that are producing a plethora of deciduous sporangia.

prominent apical thickening (papilla) and long pedicels, and produced oospores in a single culture. Infection was observed exclusively on the fruit, and aerial plant parts and the isolates did not produce significant root rot under field or greenhouse conditions. This is notably different from subsequent reports where oospores are rarely observed in single culture and root rot of pepper is common, particularly in the southwestern United States where the disease was first described (Erwin and Ribeiro, 1996). Nevertheless, the shape of the sporangia and the ability to infect pepper proved sufficiently distinctive to allow accurate identifications during the 50 years after the 1922 species description.

In the 1980s and 1990s, *Phytophthora* isolates recovered from a variety of perennial woody tropical hosts (e.g., cacao, black pepper, and macadamia) were added to *P. capsici* based on production of similar sporangia (deciduous with long pedicels), self-infertility, and the production of amphigynous oospores when mated (Mchau and Coffey, 1995; Tsao and Alizadeh, 1987). Most of these isolates are not pathogenic to pepper, have sporangia with tapered versus rounded bases, and often produce asexual chlamydospores. Molecular analyses of a worldwide collection using isozymes clearly separated the woody tropical isolates from the isolates recovered from vegetable hosts, but continuous variation of all available morphological characters made it difficult to define clearly the species boundaries (Mchau and Coffey, 1995). In 2001, Aragaki and Uchida revised *P. capsici* putting the tropical isolates into a separate species — *Phytophthora tropicalis*. The separation was based on the tapered shape of the sporangia and the inability to infect pepper seedlings. Recent genetic analyses strongly support the separation of *P. capsici* and *P. tropicalis* as distinct evolutionary lineages (Donahoo and Lamour, 2008). Although most *P. capsici* isolates are from annual (or annualized) vegetable hosts and most *P. tropicalis* are from woody perennial hosts, there is no absolute separation between *P. tropicalis* and *P. capsici* based on morphological criteria (e.g., chlamydospores and growth temperatures), host preference, or geographical origin. As discussed briefly in the section on molecular markers, a sequence analysis of 40 single-copy genes for *P. capsici* and *P. tropicalis* genomes reveals a high proportion of fixed nucleotide substitutions, and the development of species specific genetic markers will greatly facilitate proper identifications for these closely related, yet evolutionarily distinct, sibling species. To date, there have been few detailed genetic studies of *P. tropicalis* (or what were previously considered *P. capsici*) isolates recovered from woody perennial hosts.

8.3 POPULATION STRUCTURE

As described, *P. capsici* has the potential for explosive polycyclic disease development. In the United States, this results in field populations with a highly dynamic population structure that often shifts from high genotypic diversity at the beginning of the growing season to a few clonal lineages by the end (Lamour and Hausbeck, 2001a). Although fine-scale population studies are limited, reports for populations at locations outside of the United States indicate that the population dynamics may be different depending on the geographical location.

8.3.1 North America

In 1990, a survey of pepper and squash fields in North Carolina provided a first glimpse of the diversity present in U.S. populations (Ristaino, 1990). Both mating types were recovered from a single pepper field and from a single squash plant, and isolates varied continuously for morphological characters and virulence on pepper (Ristaino, 1990). This study prompted more investigations of the genetic diversity of *P. capsici* with an aim to better understand survival, spread, and response to fungicidal selection pressure. In Michigan, fields and above-ground water sources (e.g., rivers, streams, and ponds) have been sampled spatially throughout the course of multiple growing seasons, and a detailed picture of the epidemiology of *P. capsici* has emerged (Gevens et al., 2007; Lamour and Hausbeck, 2002). Isolates recovered at the beginning of the growing season are genotypically diverse with clonal lineages increasing in frequency as the growing season progresses (Lamour and Hausbeck, 2001a). In general, it appears that clonal lineages are confined to discrete locations (e.g., contiguous production fields) (Lamour and Hausbeck, 2003). From 2002 to 2005, more than 250 isolates of *P. capsici* were baited from surface irrigation sources (river systems, ponds, and ditches) in Michigan and characterized for mating type, mefenoxam sensitivity, and amplified fragment length polymorphism (AFLP) genotype (Gevens et al., 2007). Mefenoxam-insensitive isolates were common, and both mating types were recovered from many different sites. Overall, the isolates had diverse genotypes, and there was no indication that *P. capsici* could survive in surface water over the winter months. Clearly, contaminated surface water can contribute to the spread of *P. capsici* on a local or even regional scale.

The simplest characters that have been used to measure isolate diversity are mating type and sensitivity to the fungicide mefenoxam (Lamour and Hausbeck, 2000). Mefenoxam, which is the the older formulation of metalaxyl, is translocated systemically in the plant and is extremely effective at halting the growth of *P. capsici* (Davidse et al., 1991). The mode of action is thought to be site specific, and insensitivity is common in populations where there is a history of mefenoxam or metalaxyl use. Crosses between sensitive, intermediately sensitive, and fully insensitive isolates indicate that mefenoxam insensitivity is conditioned by a single codominant gene of major effect (Lamour and Hausbeck, 2000). Not surprisingly, analyses of populations where mefenoxam (or metalaxyl) has been applied often reveal both intermediate and fully insensitive isolates (French-Monar et al., 2006; Lamour and Hausbeck, 2000;

Parra and Ristaino, 2001). In Michigan, an investigation of the dynamics of a field population during and 2 years after mefenoxam selection pressure provided insight into the importance of sex to adaptation in a field population (Lamour and Hausbeck, 2001a). At the site investigated, mefenoxam was applied through a trickle irrigation system under black plastic mulch to summer and zucchini squash, which provided a particularly effective application of the fungicidal selection pressure. In 1998, when the fungicide was still being applied, 63 intermediate or fully insensitive isolates were recovered. Both mating types were recovered in roughly equal proportions, and the genotypic diversity of the population was high (>90%). Based on the assumption that there was a single codominant allele for mefenoxam insensitivity, the frequency for the insensitivity allele was estimated to be 84%. During 1998, because of intense selection pressure, the wild-type-sensitive allele was present only in the heterozygous, intermediately sensitive isolates. During 1999 and 2000, squash were planted at this same site, the use of mefenoxam was halted, and a total of 236 isolates were recovered. The A1 and A2 mating types were recovered in roughly equal proportions; of the 175 isolates with unique genotypes, only 2 were fully sensitive to mefenoxam. The proportion of isolates in the six possible mefenoxam sensitivity/mating type categories did not deviate significantly for the expectations of a population in Hardy-Weinberg equilibrium—indicating the insensitivity allele acted as expected for a neutral genetic polymorphism. The study provided strong evidence that a significant cost is not associated with mefenoxam insensitivity. More importantly, it showed how a novel allele can be integrated into a huge array of genotypic backgrounds via sexual recombination (Lamour and Hausbeck, 2001a).

The genetic diversity in U.S. populations is most likely the result of a strong selection for the thick-walled sexual oospore that serves as dormant inoculum. In the northern United States, the selection for oospores is likely driven by the winter and the lack of available host material. In the more southern areas where plants are growing year round, this may not be the case. *P. capsici* has been recovered from multiple weed hosts (Carolina geranium, American black nightshade, and common purslane) in Florida, and the life history may differ from what has been recorded for locations further north (French-Monar et al., 2006).

Analyses of U.S. *P. capsici* populations reveal a composite picture where (1) genotypic diversity is high, although clonal lineages increase significantly within a growing season; (2) long-distance dispersal of clonal lineages is limited, although there is high probability that contaminated surface water contributes to dispersal; (3) dormant inoculum can persist for extended periods (at least 4 years) outside of susceptible host material; and (4) outcrossing and oospore survival is sufficient to stave off severe population bottleneck, as pools of neutral variation are maintained between growing seasons and extended crop rotations. Most importantly, it seems once *P. capsici* has been introduced to a new location, it is difficult or impossible to eradicate.

8.3.2 Peru

8.3.2.1 Coastal Populations. The population structure of *P. capsici* in Peru differs dramatically from the United States. In Peru, the production of peppers has increased in the last few decades to produce peppers for paprika. In 2005, a set of 23 *P. capsici* isolates recovered from widely separated coastal growing areas was assembled by Liliana Aragon-Caballero and Walter Apaza-Tapia at the diagnostic clinic at the National Agricultural University, La Molina (Lima, Peru). Surprisingly, all the isolates had the same mating type (A2), and all but six isolates had the same multilocus AFLP profile, which indicated that a single clonal lineage (dubbed PcPE-1) was widely dispersed. Of the six remaining isolates, five were of one genotype (PcPE-2), and one had a unique profile (PcPE-3). In 2006 and 2007, an additional 204 isolates were recovered from 10 different pepper fields along the Supe River in the Barranca Valley. Again, all the isolates were of the A2 mating type with the same AFLP profile as the PcPE-1 clonal lineage recovered at multiple coastal locations in 2005 (Hurtado-Gonzáles et al., 2008).

The recovery of the PcPE-1 clonal lineage across 3 years and dispersed across a wide geographical area is starkly different from the situation in the United States. Clearly, some key factor varies from the areas sampled thus far in the United States. The climatic conditions allow continuous cropping at some sites in Peru, and some of the clonal survival and dispersal is thought to be caused by this. For example, continuous cropping occurs at higher elevations in the Barranca Valley, and inoculum may be dispersed to sites in the lower valley via the Supe River. However, the dispersal of PcPE-1 beyond the Barranca Valley on such a large scale is difficult to explain. Tracking current and past movements of the PcPE-1 clonal lineage is potentially difficult and will require polymorphisms that have occurred within the PcPE-1 genetic background. Analyses of six single-nucleotide polymorphism (SNP) markers in the PcPE-1 lineage revealed a few isolates that had switched to homozygosity at two of the SNP loci, similar to the fingerprint changes recorded for *Phytophthora infestans* and *Phytophthora ramorum* clonal lineages (see Chapters 7 and 9). The application of additional SNP and fast-evolving microsatellite markers may be helpful to characterize current and past patterns of movement and eventually to identify the mechanism(s) for dispersal.

An option that is not open to U.S. producers because of the huge amount of variation in natural populations of *P. capsici* is the possibility of identifying varieties of pepper that are more tolerant or even resistant to the PcPE-1 clonal lineage. Currently, a research project has been initiated to begin screening a wide variety of commercially viable pepper accessions to assess their relative susceptibility to representative isolates of the PcPE-1 lineage from widely different locations.

8.3.2.2 Amazonian Populations. Peppers grown east of the Andes Mountains in Peru are often cultivated on a smaller scale because of the rugged terrain.

In 2008, a sampling foray was organized between researchers at the University of Tennessee and the National Agricultural University to collect *P. capsici* from pepper fields in the high jungle near the town of Oxapampa in the Department of Pasco. The main pepper grown in this area is *Capsicum pubescens* referred to locally as Rocoto. Rocoto fruit are shaped like bell peppers with thick fleshy walls and are generally consumed fresh. Many fields are at sites where the forest has been cut and burned and the ground is not cultivated (Fig. 8.1). Infection by *P. capsici* is referred to as "pela pela" (peeling, peeling) disease because the fruit are often the site of infection, and a common symptom is a peeling cuticle (Fig. 8.1). Over 200 isolates of *Phytophthora* were recovered from infected fruit at 11 different sites, and preliminary analyses confirm that more than 95% of the isolates are *P. capsici* with a few isolates of *P. tropicalis* (K. Lamour, unpublished). Both mating types were isolated from multiple locations, and although the genotypic diversity is much higher than the coastal regions, it seems that a limited number of clonal lineages (both A1 and A2) are widely dispersed and dominate the population structure. Local farmers report that disease by *P. capsici* is steadily increasing, and until recently, the Rocoto fields were productive for as long as 5 years. Currently, it is difficult to produce more than a single crop before the sites are abandoned because of disease.

8.3.2.3 Korea. Phytophthora blight of pepper is one of the most devastating diseases of pepper in Korea, and the epidemiology is well understood. Fertile land suitable for pepper production is limited, and growers often repeatedly crop peppers at the same location. Similar to the United States, the disease is most severe during the rainy summer months. Hwang and Kim (1995) provide an excellent overview of the epidemiology and control efforts in Korea. They list seven factors that predispose peppers to infection by *P. capsici*, which include; long monocropping, poorly drained clay-loam soils, lack of organic fertilization, plantings on acidic soil and during long rainy seasons with low temperatures, low efficacy of fungicide application, and lack of any highly resistant pepper cultivars (Hwang and Kim, 1995).

8.4 GENETICS AND GENOMICS

8.4.1 Genome Sequencing

Thus far, detailed genetic and genomic investigations of *P. capsici* have been limited. Until very recently, only a few nuclear gene sequences were available in GenBank. A *P. capsici* genome project was launched in 2005 and is expected to be complete in 2009. A hybrid strategy is being employed using both traditional and next-generation sequencing (see Chapter 27 for an overview), and the objectives are to generate a draft genome sequence and to develop a molecular marker resource. Although the genome sequence is not yet complete, approximately 55,000 expressed sequence tag (EST) sequences have been deposited in GenBank from a library of mixed life stages, which include sporangia, swimming zoospores, germinating cysts, and mycelium grown in rich media and under starvation conditions.

Because of the high genetic diversity of *P. capsici*, the isolate (LT1534) used for sequencing was developed through a series of relatively mild inbreeding crosses. The goal was to produce an isolate with a reduced number of heterozygous sites to aid the assembly process. LT1534 is an oospore progeny from the second of two consecutive backcrosses to an isolate of *P. capsici* recovered from the crown of a pumpkin plant during 2004 in Tennessee (CBS 12157, referred to as LT263). The initial F1 population was produced by crossing LT263 to an isolate recovered from a cucumber fruit in Michigan in 1997 (LT 51, CBS 121656). The focus of inbreeding was on LT263 because it has the following attributes: (1) rapid growth at room temperature, (2) rapid production of large numbers of sporangia, and (3) the production of viable oospores when crossed to a variety of tester isolates. The inheritance of genetic markers (AFLP and SNP) confirmed the recombinant nature of the oospore progeny as well as the incremental decrease in the number of segregating markers consistent with the expected loss of heterozygosity caused by inbreeding (Hurtado and Lamour, 2009).

In addition to the backcrosses, a series of successive sibling crosses were initiated from the second backcross progeny to explore the limits of inbreeding (Hurtado and Lamour, 2009). Four successive sibling crosses were analyzed before the production of viable oospores ceased. All the inbreeding crosses produced some oospores that germinated to produce apomictic progeny, which are isolates with multilocus genotypes identical to one or the other parent isolate. Apomixis played an increasingly significant role with additional inbreeding with *P. capsici* (Hurtado and Lamour, 2009). Apomixis was also documented in multiple interspecific crosses between *P. capsici* and *P. tropicalis* and may be a factor in maintaining reproductive isolation between these species (Donahoo and Lamour, 2008).

8.4.2 Molecular Markers

Thus far, the genetic diversity of *P. capsici* has primarily been described using anonymous markers (Islam et al., 2004; Lamour and Hausbeck, 2001b, 2003). A goal of the current *P. capsici* genome project is to develop a codominant genetic marker resource. The resource will consist of a database of verified polymorphic sites. Over the past year, a panel of five isolates has been selected for additional characterization including four *P. capsici* isolates recovered from pepper and cucurbit hosts in Michigan, Tennessee, Pennsylvania, and Peru, as well as an isolate of *P. tropicalis* recovered from a (nursery-grown) rhododendron in Tennessee. Portions of single-copy genes are amplified from each isolate, and the polymerase chain reaction (PCR) products sequenced directly. Thus far, over 40 genes have been amplified and resequenced. As expected, there is significant variation in *P. capsici* and an even greater level of variation maintained discretely between *P. capsici* and *P. tropicalis*. Among the four *P. capsici* isolates, there is a polymorphic site approximately every 200 bases, whereas comparison to *P. tropicalis* adds an additional fixed difference roughly every 100 bases. This difference reflects the evolutionary distance between these two species. Although this analysis is preliminary, it appears there is more than enough variation within this small cohort of *P. capsici* isolates to mark any unique gene or genetic region in the genome (K. Lamour, unpublished).

8.5 INTERACTIONS WITH HOSTS

In 1971, Polach and Webster described the interaction of 24 field isolates of *P. capsici* with tomato, eggplant, squash, watermelon, and six lines of pepper (Polach and Webster, 1972). Fourteen pathogenic strains were described, which includes isolates that could not infect any of the hosts and an isolate from New Mexico that could infect them all (Polach and Webster, 1972). They also presented data on 391 single-oospore progeny from crosses between different pathogenic strains that show recombination between factors conditioning pathogenicity concluding that "pathogenicity to the various hosts is controlled by separate genes or gene systems for each host" (Polach and Webster, 1972). Subsequent investigations corroborate these findings, and it appears that the high level of genetic variation measured by neutral markers corresponds to an equally high level of phenotypic variation (Kim and Hwang, 1992; Oelke and Bosland, 2003). Considerable research has been conducted to develop resistance to *P. capsici* primarily in pepper but also in other hosts (Barksdale et al., 1984; Hartman and Huang, 1993; Hwang and Kim, 1995; Hwang and Hwang, 1993; Johnston et al., 2002; Ogundiwin et al., 2005). Many factors interfere with detailed studies in the *P. capsici*–host interaction, which include the previously discussed high level of variability among isolates; variation in the host material, which is often hybrid in origin (Gevens et al., 2006); age- and position-related effects in the host plants (Biles et al., 1993; Kim et al., 1989); and dosage effects with inoculum (Palloix et al., 1988). Investigators have overcome some of these hurdles by developing recombinant inbred lines for studying the *P. capsici*–pepper interaction, many of which are developed using crosses that involve Criollo de Morelos-334 from Mexico, which is fully resistant to *P. capsici* (Bonnet et al., 2007; Sy et al., 2008).

8.6 FUTURE DIRECTIONS

The near completion of genome resources for *P. capsici* means the future prospects for genetic research are bright. Valuable areas of future research and key questions include the following:

- How are nonsynonymous-base substitutions distributed throughout the genome(s)? Sequencing the global transcriptome for multiple isolates

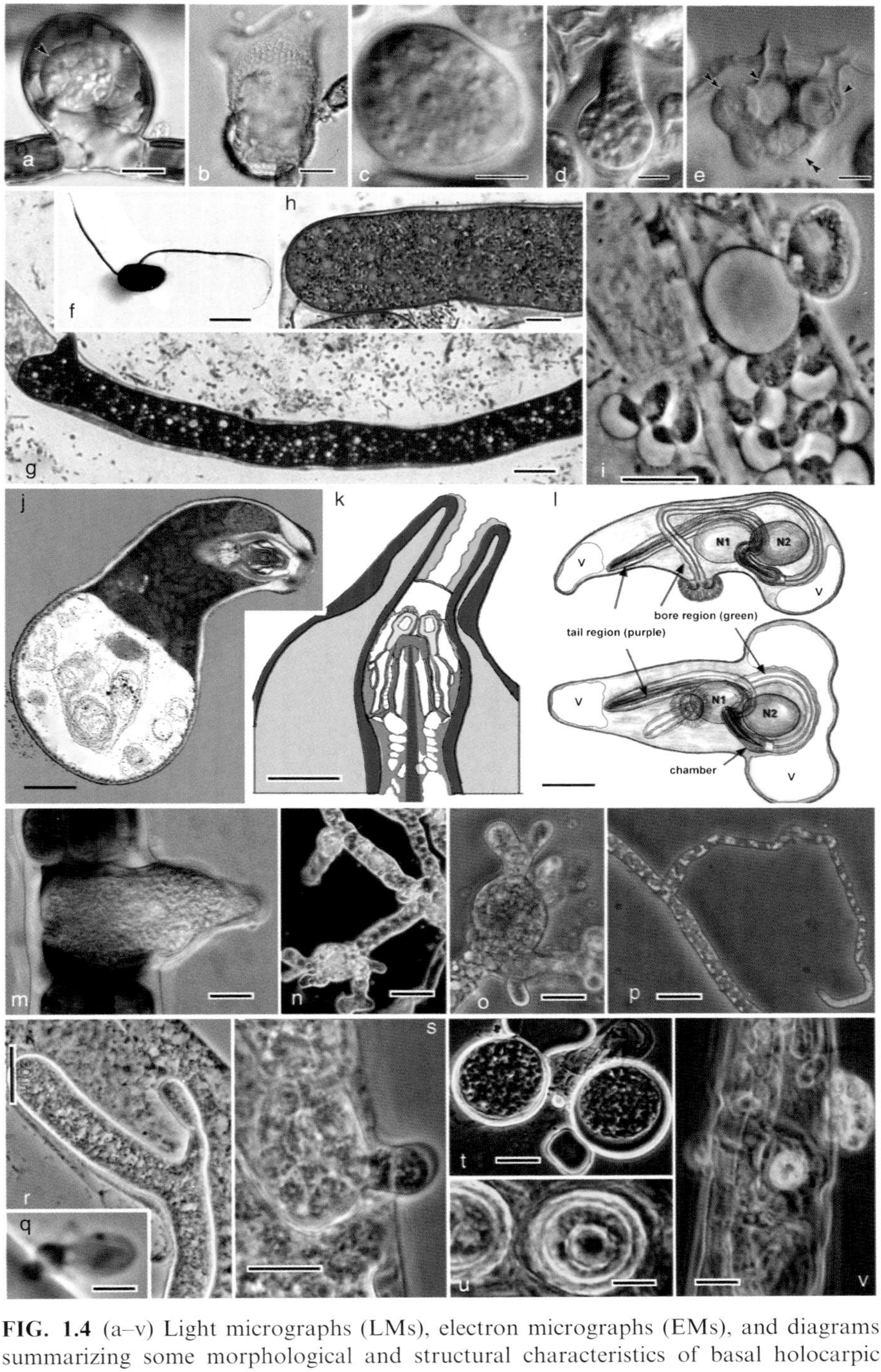

FIG. 1.4 (a–v) Light micrographs (LMs), electron micrographs (EMs), and diagrams summarizing some morphological and structural characteristics of basal holocarpic oomycetes.

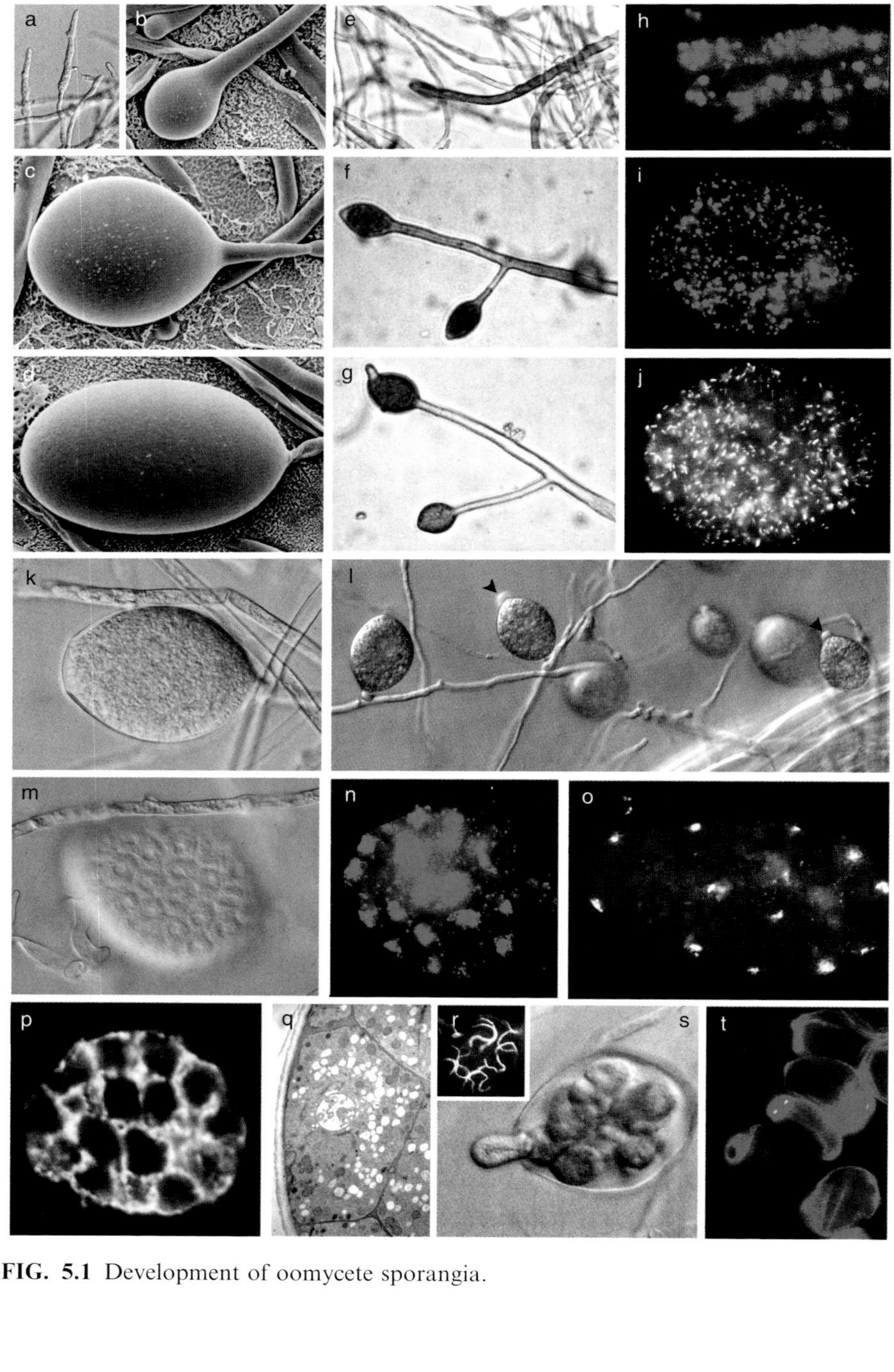

FIG. 5.1 Development of oomycete sporangia.

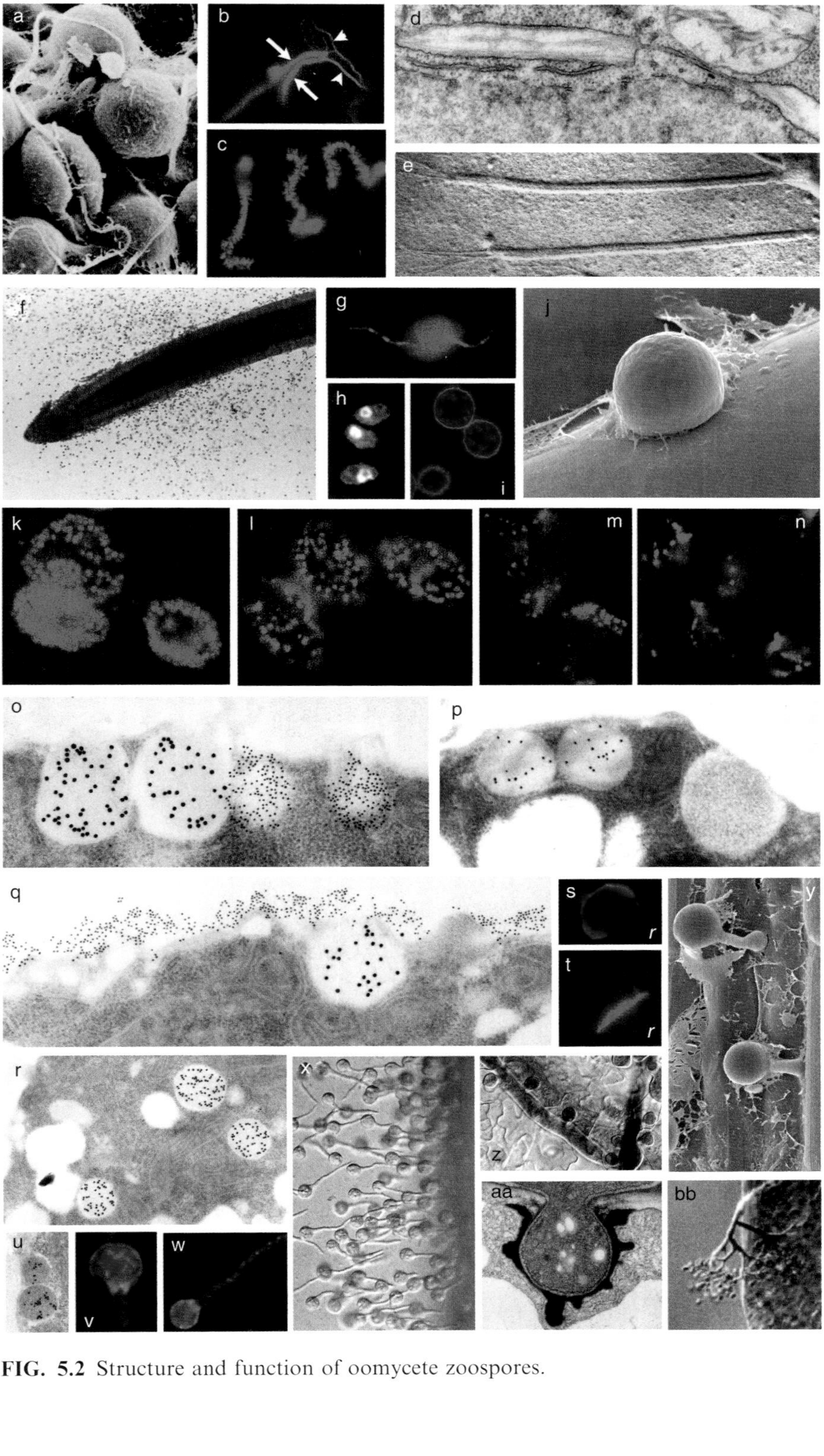

FIG. 5.2 Structure and function of oomycete zoospores.

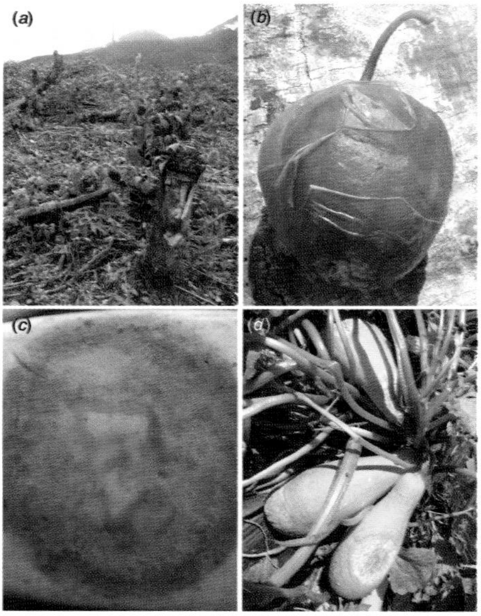

FIG. 8.1 *Phytophthora capsici* on peppers and squash in the United States and Peru. (a) *Capsicum pubescens* (Rocoto) production in the cleared forest of the high jungle in Peru. (b) Typical loose cuticle on Rocoto fruit infected with *Phytopthora capsici*. (c) Intact tissue at center of a *P. capsici* lesion on winter squash colonized during the biotrophic phase of the infection. (d) Healthy squash plant harboring fruit infected with *P. capsici* that are producing a plethora of deciduous sporangia.

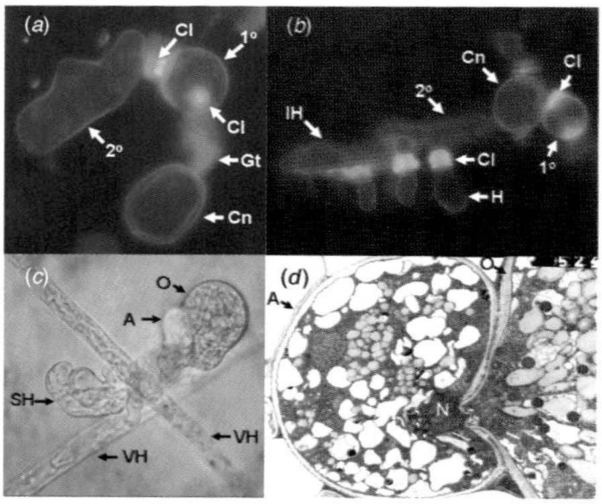

FIG. 12.1 (a, b) Infection structures of *Bremia lactucae*. Lettuce cotyledons were stained with 4′,6-diamidino-2-phenylindole (DAPI) 20 h (a) or with aniline blue 30 h (b) after infection with a compatible isolate of *B. lactucae*. Cn: Conidium, Gt: Germ tube on epidermal surface, Cl: Host callose, 1°: Primary infection vesicle within epidermal cell, 2°: Secondary infection vesicle within epidermal cell, IH: intracellular hypha, H: haustorium. Note the gradient of callose deposition and haustorial size in (b). (c, d) Sexual reproduction of *Bremia lactucae*. (c) Development of sexual hyphae (SH) at the point of contact between vegetative hyphae (VH) and the elaboration of the oogonium (O) and antheridium (A). (d) Dissolution of the antheridial (A) and oogonial (O) walls just prior to migration of an antheridial nucleus (N) into the oogonium.

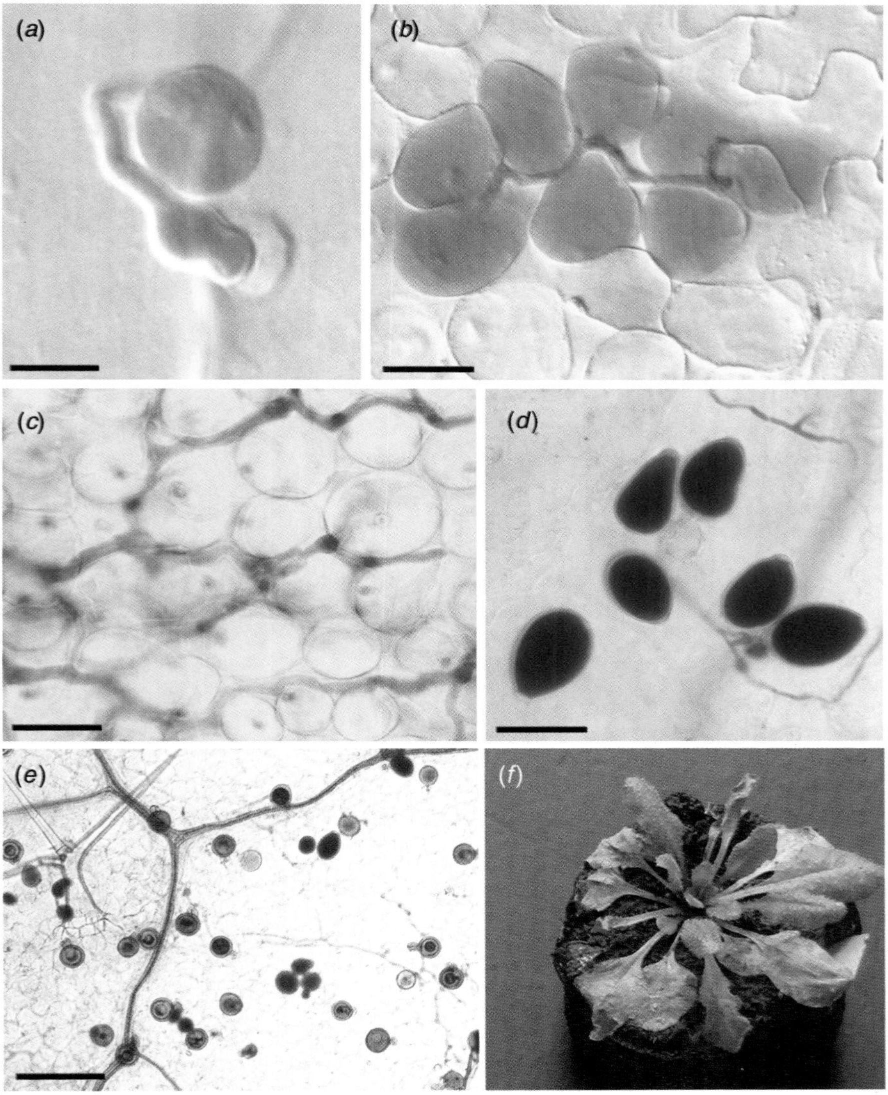

FIG. 16.1 Compatible and incompatible interactions of Arabidopsis with *P. brassicae*. Arabidopsis plants were spray inoculated with a zoospore suspension of *P. brassicae* isolate D. (a, b) Incompatible interaction. (c–f) Compatible interaction. (a–e) Micrographs of lactophenol-trypan blue-stained leaves. (a) Germinated cyst which has formed an appresorium above the junction of two epidermal cells to penetrate in between these cells into the leave of the resistant accession Ws. The cell on the right has initiated the formation of a papillae. Bar = 5 μm. (b) Hypersensitive reaction of Arabidopsis accession Ws (12 hpi). The cells in contact with the hyphae have undergone an HR and are stained blue because of retention of trypan blue. Bar = 45 μm. (c) Hyphae spreading in the intercellular space of the susceptible mutant *pad2-1* during the biotrophic phase of the interaction. The mycelium is visible as a dark blue network. Bar = 55 μm. (d) Zoosporangia on the surface of a leaf of a susceptible Arabidopsis plant 4 dpi. Bar = 60 μm. (e) Infected leaf of *pad 2-1* with oospores and a few zoosporangia 7 dpi. Bar = 250 μm. (f) The leaves of the mutant *pad2-1* are disintegrated 7 dpi.

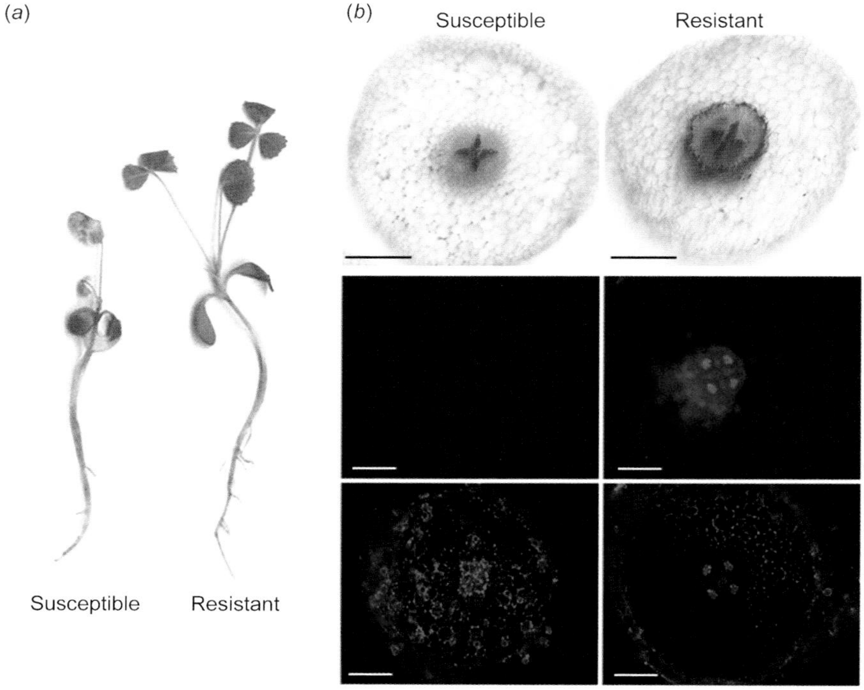

FIG. 17.2 Interaction between *A. euteiches* and the model legume *Medicago truncatula*. (a) Differential symptoms between the A17 resistant line and F83005.5 susceptible lines, observed 15 days after *A. euteiches* inoculation. *In vitro* infection assays were performed on 2-week-old *M. truncatula* plants by drop inoculation of the roots with a zoospore suspension. Root browning detected in the resistant line was linked with an accumulation of defense compounds. Plant growth was generally not altered by pathogen infection. By contrast, the susceptible line showed typical root rot symptoms (soft brown macerated tissues). Fast pathogen development led to the onset of secondary symptoms on the aerial parts (such as leaf wilting) and finally to plant death. (b) Transversal root sections of inoculated susceptible and resistant *M. truncatula* lines. Top: staining with phloroglucinol to visualize lignin accumulation (6 days postinoculation). A ring of lignin was observed on the outer cell walls of pericycle cells only in A17 roots. In F83005.5 only xylem cells were stained. Middle and bottom: Root sections (15 dpi) were observed either in autofluorescence in blue range under ultraviolet light (middle), or in epifluorescence to visualize *A. euteiches* mycelium, stained by a fluorescein isothiocyanate—wheat germ agglutinin (bottom). Defense-associated fluorescent soluble phenolic compounds (in blue) were detected only in the cortex cells of the A17 resistant line. (Note the autofluorescence of vascular cells.) Bars, 400 μm.

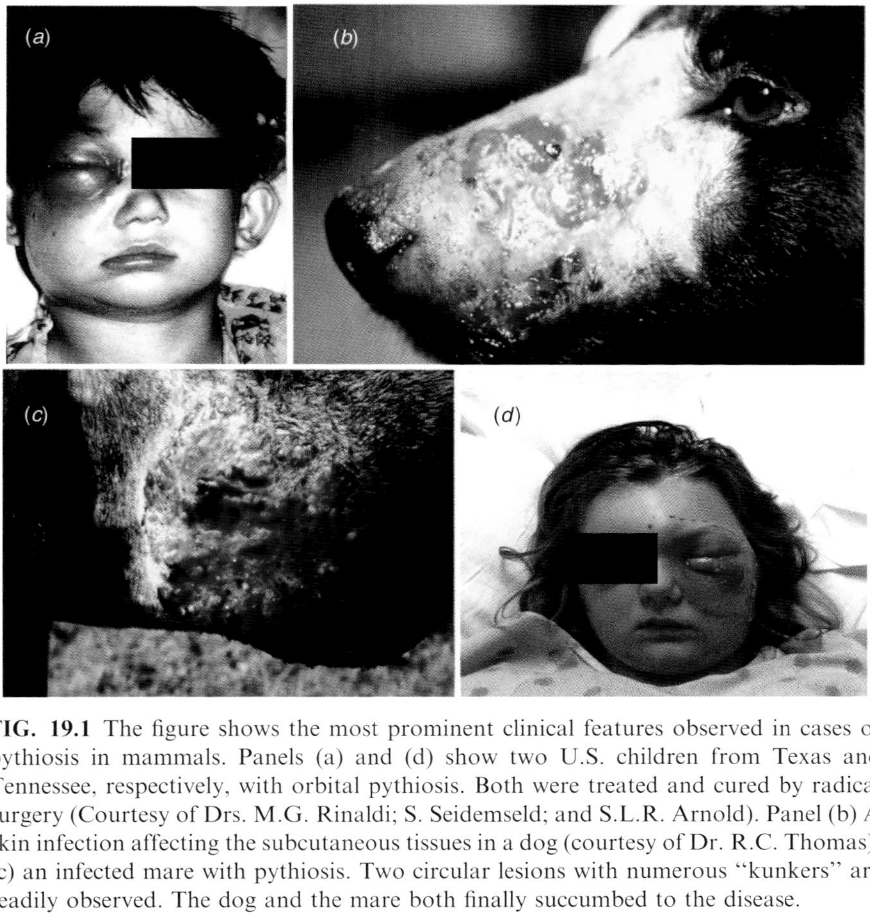

FIG. 19.1 The figure shows the most prominent clinical features observed in cases of pythiosis in mammals. Panels (a) and (d) show two U.S. children from Texas and Tennessee, respectively, with orbital pythiosis. Both were treated and cured by radical surgery (Courtesy of Drs. M.G. Rinaldi; S. Seidemseld; and S.L.R. Arnold). Panel (b) A skin infection affecting the subcutaneous tissues in a dog (courtesy of Dr. R.C. Thomas). (c) an infected mare with pythiosis. Two circular lesions with numerous "kunkers" are readily observed. The dog and the mare both finally succumbed to the disease.

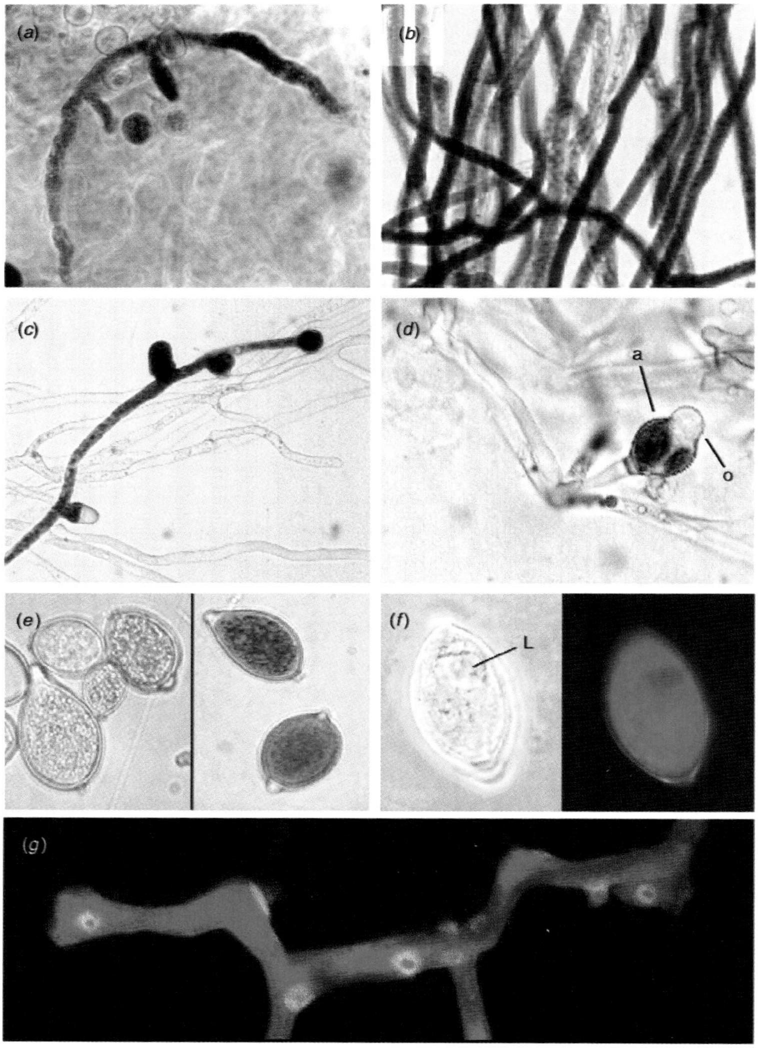

FIG. 22.1 Expression of reporter genes. (a) Staining of GUS in conidiophore of *Bremia lactucae* emerging from lettuce cotyledon, driven by the constitutive *ham34* promoter (Judelson and Michelmore, unpublished). (b) GUS in stable transformant of *Phytophthora infestans* under control of *ham34* promoter. (c) Stage-specific expression of GUS using sporulation-specific *Cdc14* promoter in *P. infestans*, showing expression in sporangiophore and developing sporangia. (d) GUS under control of mating-specific *M25* promoter in *P. infestans*, with staining evident in an antheridium (a) through which an oogonial initial (o) is emerging (Niu and Judelson, unpublished). (e) GUS expressed from zoosporogenesis-specific *NIFC* promoter in *P. infestans*, which is stained for activity in freshly harvested sporangia (left) or chilled sporangia initiating zoospore formation (right). (f) Sporangium from transgenic *P. infestans* expressing green fluorescent protein (GFP) from the *ham34* promoter, which is shown under brightfield (left) and fluorescence (right) conditions; the unstained central vacuole is labeled (l). (g) *P. infestans* expressing GFP from constitutive promoter and mRFP fused to the Avr3A effector protein, showing delivery of the latter to haustoria (courtesy of Whisson and Boevink).

during key life stages will allow assessment of both transcript abundance and base substitutions. The base substitutions can be mapped to validate (or suggest revision of) the genome assembly and can also reveal genes under selection. Genes with a higher proportion of amino acid changing substitutions are likely under some form of natural selection and may be important for differential pathogenicity and virulence.

- How are functional gene variations partitioned among populations, and how does this relate to host, chemical, or other adaptations? This will be particularly interesting for populations in the United States that are attacking snap and lima beans as well as for comparisons between U.S. and Peruvian populations where the life histories differ significantly.
- What are the functional differences between *P. capsici* and the sibling species *P. tropicalis*? The recent finding that interspecific crosses are possible opens the door for identification of the genetic components driving the preference for woody perennial versus annual vegetable hosts.
- Can *P. capsici* be spread on seed? Evidence increasingly points to this possibility. Considering the long-term implications after the introduction of *P. capsici* to new locations—this may be the most important question awaiting an answer.

ACKNOWLEDGMENTS

I would like to thank all the undergraduate and graduate students who have worked in my laboratory thus far. Also, I would like to thank my mentor and friend Dr. Mary Hausbeck for encouragement and the chance to embark on this wonderful journey of discovery.

REFERENCES

Barksdale TH, Papavizas GC, Johnston SA 1984. Resistance to foliar blight and crown rot of pepper caused by *Phytophthora capsici*. Plant Dis 68:506–509.

Biles CL, Wall MM, Waugh M, Palmer H 1993. Relationship of Phytophthora fruit rot to fruit maturation and cuticle thickness of New Mexican-type peppers. Phytopathol 83:607–611.

Bonnet J, Danan S, Boudet C, Barchi L, Sage-Palloix A, Caromel B, Palloix A, Lefebvre V 2007. Are the polygenic architectures of resistance to *Phytophthora capsici* and *P. parasitica* independent in pepper? Theoret Appl Genet 115:253–264.

Cafe Filho AC, Duniway JM, Davis RM 1995. Effects of the frequency of furrow irrigation on root and fruit rots of squash caused by *Phytophthora capsici*. Plant Dis 79:44–48.

Critopoulos PD 1955. Foot rot of tomato incited by *Phytophthora capsici*. Torrey Bot Club Bull 82:168–182.

Crossan DF, Haasis FA, Ellis DE 1954. *Phytophthora* blight of summer squash. Plant Dis Rep 38:557–559.

Davidse LC, van den Berg-Velthuis GCM, Mantel BC, Jespers ABK 1991. Phenylamides and *Phytophthora*. In: Lucas JA, Shattock RC, Shaw DS, Cooke LR, editors. *Phytophthora*. British Mycological Society, Cambridge, UK. pp. 349–360.

Davidson CR, Carroll RB, Evans TA, Mulrooney RP 2002. First report of *Phytophthora capsici* infecting lima bean (*Phaseolus lunatus*) in the Mid-Atlantic Region. Plant Dis 86:1049.

Donahoo RS, Lamour KH 2008. Interspecific hybridization and apomixis between *Phytophthora capsici* and *P. tropicalis*. Mycologia: 100:911–920.

Erwin DC, Ribeiro OK 1996. Phytophthora diseases worldwide. The American Phytopathological Society. St. Paul, MN. pp. 562.

French-Monar RD, Jones JB, Roberts PD 2006. Characterization of *Phytophthora capsici* associated with roots of weeds on Florida vegetable farms. Plant Dis 90:345–350.

Gevens A, Donahoo RS, Lamour KH, Hausbeck MK 2007. Characterization of *Phytophthora capsici* from Michigan surface irrigation water. Phytopathol 97:421–428.

Gevens AJ, Ando K, Lamour KH, Grumet R, Hausbeck MK 2006. Development of a detached cucumber fruit assay to screen for resistance and effect of fruit age on susceptibility to infection by *Phytophthora capsici*. Plant Dis 90:1276–1282.

Hartman GL, Huang YH 1993. Pathogenicity and virulence of *Phytophthora capsici* isolates from Taiwan on tomatoes and other selected hosts. Plant Dis 77:588–591.

Hausbeck MK, Lamour KH 2004. *Phytophthora capsici* on vegetable crops: research progress and management challenges. Plant Dis 88:1292–1303.

Hurtado OP, Lamour KH 2009. Evidence for inbreeding and apomixis in close crosses of *Phytophthora capsici*. Plant Path (in press).

Hwang BK, Kim CH 1995. *Phytophthora* blight of pepper and its control in Korea. Plant Dis 79:221–227.

Hwang JS, Hwang BK 1993. Quantitative evaluation of resistance of Korean tomato cultivars to isolates of *Phytophthora capsici* from different georgraphic areas. Plant Disease 77:1256–1259.

Islam SZ, Babadoost M, Lambert KN, Ndeme A 2004. Characterization of *Phytophthora capsici* isolates from processing pumpkins in Illinois. Plant Dis 89:191–197.

Johnston SA, Kline WL, Fogg ML, Zimmerman MD 2002. Varietal resistance evaluation for control of Phytophthora blight of pepper. Phytopathol 92:S40.

Kim ES, Hwang BK 1992. Virulence to Korean pepper cultivars of isolates of *Phytophthora capsici* from different geographic areas. Plant Dis 76:486–489.

Kim YJ, Hwang BK, Park KW 1989. Expression of age-related resistance in pepper plants infected with *Phytophthora capsici*. Plant Dis 73:745–747.

Kreutzer WA 1937. A *Phytophthora* rot of cucumber fruit. Phytopathol 27:955.

Kreutzer WA, Bodine EW, Durrell LW 1940. Cucurbit diseases and rot of tomato fruit caused by *Phytophthora capsici*. Phytopathol 30:972–976.

Lamour KH, Hausbeck MK 2000. Mefenoxam insensitivity and the sexual stage of *Phytophthora capsici* in Michigan cucurbit fields. Phytopathol 90:396–400.

Lamour KH, Hausbeck MK 2001a. The dynamics of mefenoxam insensitivity in a recombining population of *Phytophthora capsici* characterized with amplified fragment length polymorphism markers. Phytopathol 91:553–557.

Lamour KH, Hausbeck MK 2001b. Investigating the spatiotemporal genetic structure of *Phytophthora capsici* in Michigan. Phytopathol 91:973–980.

Lamour KH, Hausbeck MK 2002. The spatiotemporal genetic structure of *Phytophthora capsici* in Michigan and implications for disease management. Phytopathol 92:681–684.

Lamour KH, Hausbeck MK 2003. Effect of crop rotation on the survival of *Phytophthora capsici* and sensitivity to mefenoxam. Plant Dis 87:841–845.

Leonian LH 1922. Stem and fruit blight of peppers caused by *Phytophthora capsici sp. nov*. Phytopathol 12:401–408.

Mchau GRA, Coffey MD 1995. Evidence for the existence of two subpopulations in *Phytophthora capsici* and a redescription of the species. Mycol Res 99:89–102.

O. Hurtado-Gonzáles O, Aragon-Caballero L, Apaza-Tapia W, Donahoo R, Lamour K 2008. Survival and spread of *Phytophthora capsici* in coastal Peru. Phytopathol 98:688–694.

Oelke LM, Bosland PW 2003. Differentiation of race specific resistance to *Phytophthora* root rot and foliar blight in *Capsicum annuum*. J Am Soc Hort Sci 128:213–218.

Ogundiwin EA, Berke TF, Massoudi M, Black LL, Huestis G, Choi D, Lee S, Price JP 2005. Construction of 2 intraspecific linkage maps and identification of resistance QTLs for *Phytophthora capsici* root-rot and foliar-blight diseases of pepper (Capsicum annuum L.). Genome 48:698–711.

Palloix A, Daubeze AM, Pochard E 1988. Phytophthora root rot of pepper: influence of host genotype and pathogen strain on the inoculum density-disease severity relationships. J Phytopathol 123:25–33.

Parra G, Ristaino JB 2001. Resistance to mefenoxam and metalaxyl among field isolates of *Phytophthora capsici* causing Phytophthora blight of bell pepper. Plant Dis 85:1069–1075.

Polach FJ, Webster RK 1972. Identification of strains and inheritance of pathogenicity in *Phytophthora capsici*. Phytopathol 62:20–26.

Ristaino JB 1990. Intraspecific variation among isolates of *Phytophthora capsici* from pepper and cucurbit fields in North Carolina. Phytopathol 80:1253–1259.

Sy O, Steiner R, Bosland PW 2008. Recombinant inbred line differentials identifies race-specific resistance to Phytophthora root rot in *Capsicum annuum*. Phytopathol 98:867–870.

Tompkins CM 1937. Phytophthora rot of honeydew melon. J Agr Res 54:933–944.

Tompkins CM 1941. Root rot of pepper and pumpkin caused by *Phytophthora capsici*. J Agr Res 63:417–426.

Tsao PH, Alizadeh A 1988. Recent advances in the taxonomy and nomenclature of the so-called "*Phytophthora palmivora*" MF4 occurring on cocoa and other tropical crops. Proc 16th International Cocoa Research Conference Proceedings. Santo Domingo. pp. 441–445.

Weber GF 1932. Blight of peppers in Florida caused by *Phytophthora capsici*. Phytopathol 22:775–780.

Wiant JS 1940. A rot of winter queen watermelons caused by *Phytophthora capsici*. J Agr Res 60:73–88.

9

EVOLUTION AND GENETICS OF THE INVASIVE SUDDEN OAK DEATH PATHOGEN *PHYTOPHTHORA RAMORUM*

Niklaus J. Grünwald and Erica M. Goss

Horticultural Crops Research Laboratory, U.S. Department of Agriculture, Agricultural Research Service, Corvallis, Oregon

9.1 INTRODUCTION

Phytophthora ramorum (Werres, De Cock, and Man in't Veld) is an emerging pathogen of oaks and tanoak, among other trees, as well as of an increasing number of woody and herbaceous perennials (Frankel, 2008; Grünwald et al., 2008b; Rizzo et al., 2002, 2005; Werres et al., 2001). It causes rapid mortality in tanoak (*Lithocarpus densiflorus*) and coast live oak (*Quercus agrifolia*) and has been responsible for the precipitous decline of forest populations of these species in coastal California (Rizzo et al., 2002, 2005). Its infection of common nursery crops has served as a means of dispersal across the United States (Ivors et al., 2006; Prospero et al., 2007). It is feared that *P. ramorum* will spread and colonize dominant oaks in the eastern and southern United States (Tooley and Kyde, 2007) and heathland plants in the United Kingdom (Brasier et al., 2004a and b).

The rapid onset and death of infected tanoaks in Marin County, California, in the early 1990s raised alarm and led to the term Sudden Oak Death. The cause of the outbreak was not known until 2000 when an unknown *Phytophthora* species was isolated from cankers on dying trees (Rizzo et al., 2002). It was soon recognized to be the same pathogen that causes tip dieback, leaf blight, and stem cankers on *Rhododendron* and *Viburnum* in European

Oomycete Genetics and Genomics: Diversity, Interactions, and Research Tools
Edited by Kurt Lamour and Sophien Kamoun
Copyright © 2009 John Wiley & Sons, Inc.

greenhouse settings (Werres et al., 2001). This prompted an examination of the nursery stock in the infested areas in California, and infected *Rhododendron* were found (Frankel, 2008). The economic impact of losses from *P. ramorum* in the United States is estimated to be in the tens of millions of dollars because of the direct loss of nursery and ornamental crops, the decrease of property values because of the dead/dying trees, as well as the cost of monitoring, tracking, and eradicating the disease (Cave et al., 2005; Dart and Chastagner, 2007; Frankel, 2008).

9.2 BIOLOGY OF *P. RAMORUM*

P. ramorum has an extraordinarily wide host range, which currently includes over 109 plant species and continues to expand (Denman et al., 2005; Hansen et al., 2005; Tooley and Kyde, 2007; Tooley et al., 2004). An up-to-date list of confirmed hosts is maintained by the U.S. Department of Agriculture Animal and Plant Health Inspection Service (USDA APHIS) (http://www.aphis.usda gov/ppg/ispm/pramorum/). Symptoms of infection by *P. ramorum* include stem cankers, twig dieback, and/or foliar lesions, depending on the host (Davidson et al., 2003).

P. ramorum is heterothallic and thus requires two mating types to form oospores; yet controlled mating in culture produces few oospores that are unusually slow to develop (Brasier and Kirk, 2004; Werres et al., 2001). Thus far, there is no evidence of oospore formation in nursery settings where both mating types of the pathogen have been found (Grünwald et al., 2008a). *P. ramorum* does produce large and abundant chlamydospores (Werres and Kaminsky, 2005; Werres et al., 2001), which are thought to serve as resting structures that allow the pathogen to survive adverse conditions and may be particularly important for survival in soil if oospores are indeed missing from its life cycle (Shishkoff, 2007; Tooley et al., 2008).

9.3 POPULATION GENETICS

The known populations of *P. ramorum* are currently limited to Europe and North America, where they are structured into three distinct clonal lineages (Table 9.1) (Grünwald et al., 2008b). Population genetic analyses using microsatellites and amplified fragment length polymorphisms (AFLPs) have shown them to be genetically distinct and asexually reproducing yet clearly conspecific (Ivors et al., 2004, 2006; Mascheretti et al., 2008; Prospero et al., 2007). In 2006, an informal agreement within the *P. ramorum* research community assigned names to each clonal lineage: EU1, NA1, and NA2 (Grünwald et al., 2008b). The EU1 clonal lineage was first identified in European nurseries, which reveal some genotypic diversity within the lineage (Werres et al., 2001) (Fig. 9.1a). It is the only lineage found in Europe to date,

TABLE 9.1 Descriptors differentiating the currently known clonal lineages of *Phytophthora ramorum* (Brasier, 2003; Ivors et al., 2004, 2006; Prospero et al., 2007).

Lineage	Distribution	Mating type	Growth rate[a]	Colony type[b]
NA1	U.S. nurseries and forests	A2	slow	appressed
NA2	U.S. nurseries	A2	fast	aerial
EU1	EU and U.S. nurseries	A1 (A2 rare)[c]	fast	aerial

[a]Colony growth rate on V8 agar as described in Brasier (2003).

[b]Mycelial growth habit on V8 agar at room temperature.

[c]Isolates of the A2 mating type have been found in Belgium at very low frequencies (Werres and De Merlier, 2003).

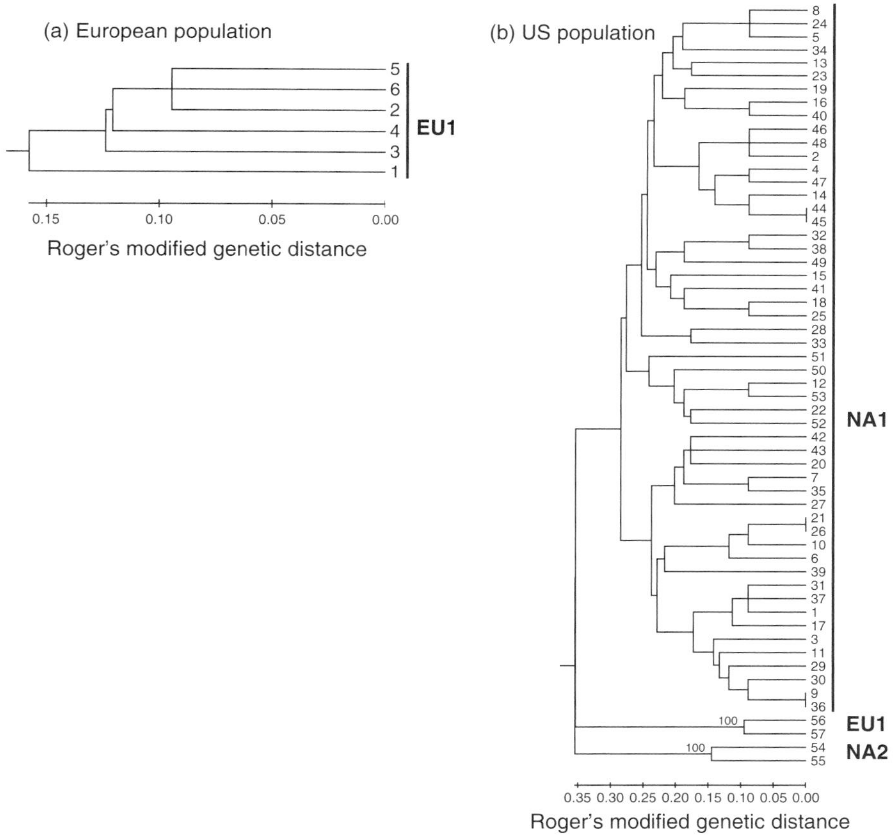

FIG. 9.1 Unweighted pair group method with arithmetic mean (UPGMA) dendrograms illustrating the genetic distance observed between multilocus microsatellite genotypes found in (a) European and (b) U.S. nurseries. Support values greater than 70% using 1,000 bootstrap samples are shown. Multilocus genotype 2 in the European nursery population is identical to multilocus genotype 56 in the U.S. population.

but it is now regularly found in nurseries on the West Coast of the United States where only two EU1 genotypes have been identified to date (Grünwald et al., 2008a; Hansen et al., 2003; Ivors et al., 2006; Mascheretti et al., 2008) (Fig. 9.1b). Clonal lineage NA1 is responsible for the wildland infestations in California and Oregon, and many nursery infections in North America. NA1 also shows considerable genotypic diversity (Ivors et al., 2006; Mascheretti et al., 2008; Prospero et al., 2007) (Fig. 9.1b). The third clonal lineage NA2 has a much more limited distribution, and to date, it only reveals two microsatellite, multilocus genotypes in the United States (Fig. 9.1b). The NA2 lineage has only been isolated from nurseries in North America (Ivors et al., 2006). All NA1 and NA2 isolates that have been tested are of mating type A2, although because mating is slow and laborious only a small fraction of isolates have actually been tested (Table 9.1). Interestingly, EU1 is predominately mating type A1; yet the A2 mating type has been identified at a very low frequency in EU1 isolates in Belgium (Werres and De Merlier, 2003).

There is currently no evidence for sexual reproduction in the nurseries in which both mating types are present. The genotypic diversity observed within lineages NA1, EU1, and NA2 follows the stepwise mutation model for microsatellite loci for asexually reproducing, genetically isolated lineages (Ivors et al., 2006; Prospero et al., 2007) (Fig. 9.1); yet the levels of heterozygosity observed in *P. ramorum* are consistent with an outcrossing species (Tyler et al., 2006). In addition, when each haplotype of several nuclear genes in each clonal lineage was cloned and sequenced, we found that the two haplotypes of each lineage did not necessarily cluster together, and the genealogical relationships among the lineages changed with each gene (Fig. 9.4; Goss et al., 2009). This suggests that the clonal lineages are descended from a sexually recombining population (see below).

The clonal lineages show phenotypic variation in terms of colony morphology, growth rate (Table 9.1), and aggressiveness on certain hosts. EU1 isolates have faster average growth rates in culture and larger chlamydospore size, although the growth rate of NA1 isolates can equal that of EU1 isolates (Brasier, 2003; Brasier et al., 2006; Werres and Kaminsky, 2005). NA1 subcultures have been noted to exhibit phenotypic variation and intrinsic instability in colony morphology and vegetative growth rates, which has not been observed in the EU1 lineage (Brasier et al., 2006). Abnormally shaped sporangia have also been observed in single-colony segments of NA1 isolates (Werres and Kaminsky, 2005). EU1 isolates produced larger lesions than NA1 isolates on cut stems of *Quercus rubra* (Brasier et al., 2006) and detached *Rhododendron* leaves (McDonald and Grünwald, 2006), but no differences in aggressiveness were observed on a variety of other host species (Denman et al., 2005; Tooley et al., 2004). Variation in aggressiveness among isolates within both NA1 and EU1 lineages has been documented on coast live oak seedlings, detached bay laurel leaves, and *Q. rubra* cut stems (Brasier et al., 2006; Hüberli et al., 2005). None of these studies have observed genotype-specific (gene-for-gene-like) variation in virulence. Much less work has been done on

NA2, but it seems to exhibit a relatively rapid growth rate in culture similar to EU1 isolates (Ivors et al., 2006).

Population structure and variation within clonal lineages has been examined using AFLP and fast-evolving microsatellites. There are also numerous simple sequence repeats in the *P. ramorum* genome sequence that have not yet been screened for variation (see below; Garnica et al., 2006; Tyler et al., 2006). AFLP and microsatellite studies initially indicated that the EU1 lineage was more variable than the NA1 lineage (Ivors et al., 2004, 2006). However, Prospero et al. (2007) found several tetranucleotide repeats that exhibit long allele sizes and are hypervariable within NA1 but not EU1. These loci also show nearly uniform homozygosity in NA1, which is surprising given the levels of heterozygosity observed at other microsatellite loci and at the DNA sequence level. We have observed loss of heterozygosity at two other microsatellite loci where the differences between allele sizes are large and therefore more likely caused by mitotic recombination than mutation. Mitotic recombination by crossing over has been documented at microsatellite loci in *Phytophthora cinnamomi* (Dobrowolski et al., 2003), and mitotic gene conversion has been observed in *Phytophthora sojae* (Chamnanpunt et al., 2001).

The hypervariable microsatellites in NA1 significantly differentiate the nursery and forest populations in Oregon and have shown that the genotypic composition of nursery populations changed from 2003 to 2004 (Prospero et al., 2007). These microsatellite loci were also used to reconstruct the probable origin of the sudden oak death epidemic in California (Mascheretti et al., 2008). Microsatellite variation has also proven valuable for rapid and accurate diagnosis of clonal lineage and is being used to monitor nursery finds in the United States, particularly now that both mating types are being found together (Grünwald et al., 2008a).

9.4 EVOLUTION

Within the genus *Phytophthora*, *P. ramorum* falls into clade 8c, which is a subclade of the second most basal *Phytophthora* group (Blair et al., 2008), that it shares with *Phytophthora lateralis*, *Phytophthora hibernalis*, and *Phytophthora foliorum* (Blair et al., 2008; Cooke et al., 2000; Martin and Tooley, 2003) (Fig. 9.2). *P. lateralis* is well supported by molecular phylogenetics as the closest known relative to *P. ramorum*; yet it remains unclear whether *P. hibernalis* or *P. foliorum* is the sister taxon of the *P. ramorum*/*P. lateralis* lineage (Blair et al., 2008; Ivors et al., 2004; Kroon et al., 2004; Martin and Tooley, 2003). *P. ramorum*, *P. lateralis*, and *P. foliorum* are all thought to be exotic in their known ranges (Donahoo et al., 2006; Erwin and Ribeiro, 1996; Ivors et al., 2004). The origin of *P. lateralis* is unknown, and it is clearly not native to the natural range of Port Orford-cedar, which is the host it kills in the Pacific Northwest (Hansen et al., 2000). It seems to have been introduced through shipments and plantings of ornamental *Chamaecyparis* species or

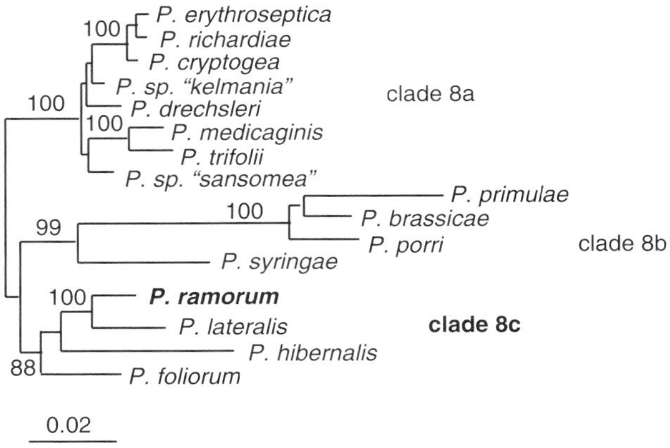

FIG. 9.2 Phylogeny of *Phytophthora* clade 8 using seven nuclear loci, modified from Blair et al. (2008). Maximum likelihood was used to infer the phylogeny and bootstrap support values. The scale is given in substitutions per site.

perhaps rhododendrons (on which it is not a severe pathogen) (Hansen et al., 2000). *P. foliorum* was recently discovered as a new pathogen on azalea with limited genotypic diversity based on AFLP analysis, which indicates an exotic origin (Donahoo et al., 2006). *P. hibernalis* is known to be an important pathogen of citrus fruit and occasionally foliage in Australia (Erwin and Ribeiro, 1996); yet relatively few isolates exist in culture collections. The 8c species are fairly genetically distant from each other; their centers of origin are unknown, and new *Phytophthora* species are continually being discovered. Therefore, it is likely that the closest relative of *P. ramorum* is unknown to science.

Another tree pathogen, *Phytophthora kernoviae* was recently found in the United Kingdom during routine surveys for *P. ramorum*; it causes similar disease symptoms that include dieback of rhododendron and necrotic, occasional bleeding, and cankers on beech trees. However, *P. kernoviae* is not closely related to *P. ramorum* and belongs to clade 10. Thus, it is basal to the *Phytophthora* phylogeny (Blair et al., 2008).

Because *P. ramorum* is structured into three clonal lineages in its introduced range, there are obvious questions regarding the relationships among the lineages. Are they ancient or recent? Could they be different genotypes introduced from a single population, or are they more likely to have originated from three different populations? An initial sequencing of the *cox1* mitochondrial gene in *P. ramorum* isolates from the three clonal lineages indicated that NA2 has the oldest mitochondrial haplotype (Ivors et al., 2006). Sequencing of seven mitochondrial loci in the three lineages has confirmed this hypothesis (Martin, 2008) (Fig. 9.3). We have recently sequenced five nuclear loci to examine the relationships among the lineages in the nuclear genome (Goss et al., 2009). We observed heterozygosity at several of the loci, this was

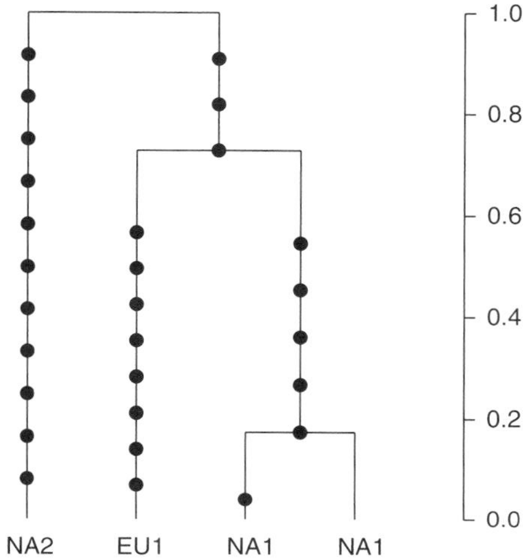

FIG. 9.3 Coalescent genealogy for seven concatenated mitochondrial gene loci based on data by Martin (2008). Trees were generated in genetree, which assumes no recombination, and were scaled to a time to the most common recent ancestor (TMRCA) of 1.0. The dots on the branches represent mutations in the genealogy (segregating sites among the sequences).

the extent of the variation observed among isolates within the lineages. In other words, there were one or two alleles per gene in an isolate, depending on whether the gene was homozygous or heterozygous, and all isolates within a lineage were identical to one another. This suggests that a single representative of each lineage was introduced to North America and Europe and that most or all the diversity observed within lineages at microsatellite loci has emerged since introduction. We found that relationships among lineages changed across loci and even within loci, which suggests that recombination had played a role in structuring the observed variation (Fig. 9.4). However, the last 11% to 25% of the history of the *P. ramorum* lineages was estimated to have been spent in isolation from each other based on coalescent analyses. A rough calculation puts 11% of their history at 165,000–500,000 years, depending on the synonymous substitution rate used. Therefore, it seems unlikely that the three lineages originated from a single location or population; in fact, they may have been introduced from three geographically isolated populations.

9.5 GENOME

A 7X draft whole genome sequence for a NA1 isolate from California was published in 2006 along with that of the soybean pathogen *P. sojae* (Tyler et al., 2006). The high-profile nature of *P. ramorum* allowed the funding to be put

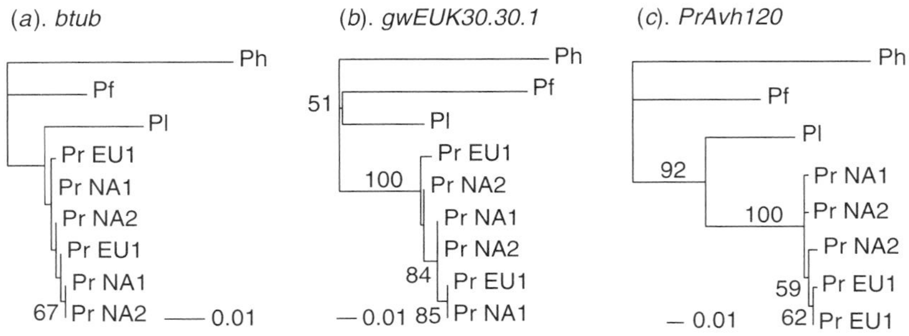

FIG. 9.4 Maximum likelihood phylogenetic trees inferred from the DNA sequences of three nuclear genes: (a) beta-tubulin, (b) hypothetical protein gwEUK30.30.1 in the JGI genome annotation containing a glycosyl transferase group 1 domain, and (c) an avirulence gene homolog. Bootstrap values are shown for branches that had greater than 50% support. Species names are abbreviated: Ph, *Phytophthora hibernalis*; Pf, *Phytophthora foliorum*; Pl, *Phytophthora lateralis*; Pr, *Phytophthora ramorum*. Each haplotype observed for *P. ramorum* (labeled by lineage) as well as a single representative sequence for the other three species are shown. Modified from Goss et al., 2009.

forward for a genome sequence only a few years after being formally described. However, relatively little is known about the biology of *P. ramorum* compared with model systems such as *P. sojae* and *P. infestans*. Most functional studies that follow up the genome sequences will be conducted in established model species. *P. sojae* and *P. infestans* are fairly host specific, whereas *P. ramorum* infects many different genera of woody plants and may rely on different strategies for infection. For example, *P. ramorum* can infect leaves, roots, shoots, and xylem via penetration of the tree bark (Davidson et al., 2003; Parke et al., 2007). Thus, the *P. ramorum* genome may differ from host-specific *Phytophthora* genomes in genes involved in host colonization and pathogenicity.

The *P. ramorum* genome size is 65 Mb, which is smaller than *P. sojae* (95 Mb) and *P. infestans* (240 Mb), but it is approximately the same size as *P. capsici* (Lamour et al., 2007). Comparison of the *P. ramorum* and *P. sojae* genomes showed secreted proteins to be evolving faster than the rest of the proteome and family specific expansion of pathogenicity-related genes (Tyler et al., 2006). Among proteins that may be important to the necrotrophic stage of infection, *P. ramorum* has more protease genes but fewer of the other groups of hydrolytic enzymes compared with *P. sojae*. Among effectors (not including RXLR-class effectors, see below), the necrosis and ethylene-inducing protein (NPP) family shows an expansion in *P. ramorum*, whereas *P. sojae* contains relatively more of the necrosis-inducing PcF and Crn families, which are also large and diverse families in *P. infestans* (Kamoun, 2006).

9.5.1 Avirulence Gene Homologs

All avirulence genes cloned from Oomycetes to date have contained the conserved amino acid motif RXLR (Arg, any amino acid, Leu, Arg). Secreted proteins with this motif comprise a large and diverse family of potential effectors in the *P. ramorum* genome (Jiang et al., 2008; Tyler et al., 2006; Win et al., 2007). Jiang et al. (2008) placed the number of RXLR-class effectors in *P. ramorum* at 374; Whisson et al. (2007) estimated 314, and Win et al. (2007) identified a minimum of 181 and a maximum of 531. To date, none of these putative effectors have been validated in *P. ramorum*.

An analysis of paralogous RXLR-class effectors in *P. ramorum* showed many to be under detectable positive selection, often localized to the C-terminal region of the protein (Win et al., 2007). Of 59 groups of closely related paralogs examined, 41 showed evidence of positive selection, and half of these had two to four members clustered in the genome (within 100 kb). The *P. sojae* genome contained a comparable number of putative RXLR-class effectors, but only 28 groups of closely related paralogs were identified, of which 18 had experienced positive selection and only 5 of these were clustered. These data indicate that *P. ramorum* may have experienced a more recent expansion of RXLR-type effectors than *P. sojae*.

We examined a group of four paralogs in *P. ramorum* (PrAvh60, PrAvh68, PrAvh108, and PrAvh205), which were identified as avirulence gene homologs (Avhs) by Jiang et al. (2008) and were chosen for examination because microarray data indicated that two of the four genes and two other related paralogs (PrAvh121 and PrAvh120) were expressed in sporangia. We were able to clone only a single homolog to the four *P. ramorum* paralogs from *P. lateralis* and *P. hibernalis*, which suggests that this group is a result of gene duplication subsequent to *P. ramorum*–*P. lateralis* speciation (Fig. 9.5). Win et al. (2007) found that for those *P. ramorum* effectors containing an RXLR motif and with close paralogs, three quarters of them had between 70% and 99% amino acid identity to their closest paralog. The most diverged pair out of the four genes that we examined have 72% amino acid identity (in NA1 isolate Pr102), and thus, they fall in the lower end of the above range. Therefore, many RXLR expansions may indeed have occurred since speciation with *P. lateralis*, although one would expect rates of evolution to vary widely among effector genes.

The above four paralogs seem to have diverged independently by point mutation. In contrast, a second group of Avh paralogs shows evidence of recombination or gene conversion among loci (Fig. 9.6). Three of these genes are clustered in the genome with a fourth on the same scaffold approximately 150 kb from the cluster. This group also shows high levels of divergence localized to the C-terminal region (see branch lengths in Fig. 9.6) and positively selected amino acids (Win et al., 2007). Therefore, the shuffling of variation among closely related paralogs may also be an important mechanism in the evolution of *P. ramorum* effector genes.

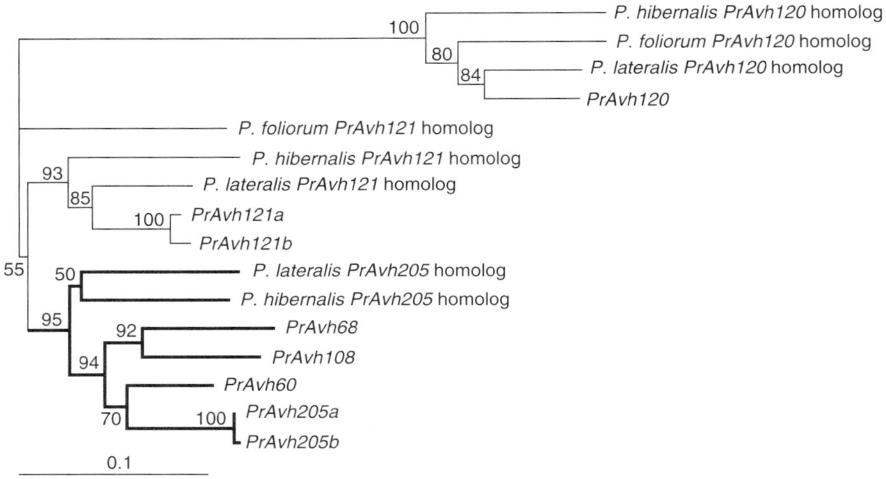

FIG. 9.5 Maximum likelihood genealogy of a subfamily of avirulence gene homologs in *Phytophthora* clade 8c. *PrAvh* gene sequences are from *Phytophthora ramorum* NA1 isolate Pr102. Genes are named according to Jiang et al. (2008). For heterozygous genes, the two alleles are distinguished by the suffixes "a" and "b". The *Phytophthora lateralis* and *Phytophthora hibernalis* homologs to *P. ramorum* Avhs 60, 68, 108, and 205 (bolded clade) seem to be syntenic to 205 based on flanking sequence similarity. Bootstrap support is shown for all branches.

9.5.2 Simple Sequence Repeats in the Genome

Simple sequence repeats (SSRs), which are also known as microsatellites or short tandem repeats, are by definition tandemly repeated DNA sequences of one to six base pairs in length. With mutation rates from about 10^{-3} to 10^{-6} per locus per gamete per generation, SSRs are thought to have higher mutation rates than most other areas of a genome (Primmer et al., 1996; Schug et al., 1997; Weber and Wong, 1993), which makes them very polymorphic markers that have found many useful applications in fields as varied as physical mapping, population genetics, and evolution (Avise, 2004; Goldstein and Schlötterer, 1999; Hennequin et al., 2001). To create a resource for future efforts with *P. ramorum*, the abundance and distribution of SSRs in *P. ramorum* was characterized and contrasted to *P. sojae* (Grünwald, Tripathy, Ivors, and Lamour, unpublished data). These SSR loci and their locations are available at http://web.science.oregonstate.edu/bpp/labs/grunwald/resources.htm.

The criteria for including an SSR in the analysis were that it have at least two consecutive repeat units (or one unit of 12 bp for mononucleotides) and be greater than 12 bp long. A total of 2,125 perfect and compound SSR loci were observed in the genome of *P. ramorum* and 3,349 in *P. sojae*. Mononucleotide repeats were the most abundant microsatellite repeats, followed by dinucleotide repeats (Table 9.2). Pentanucleotide repeats were the least common motif,

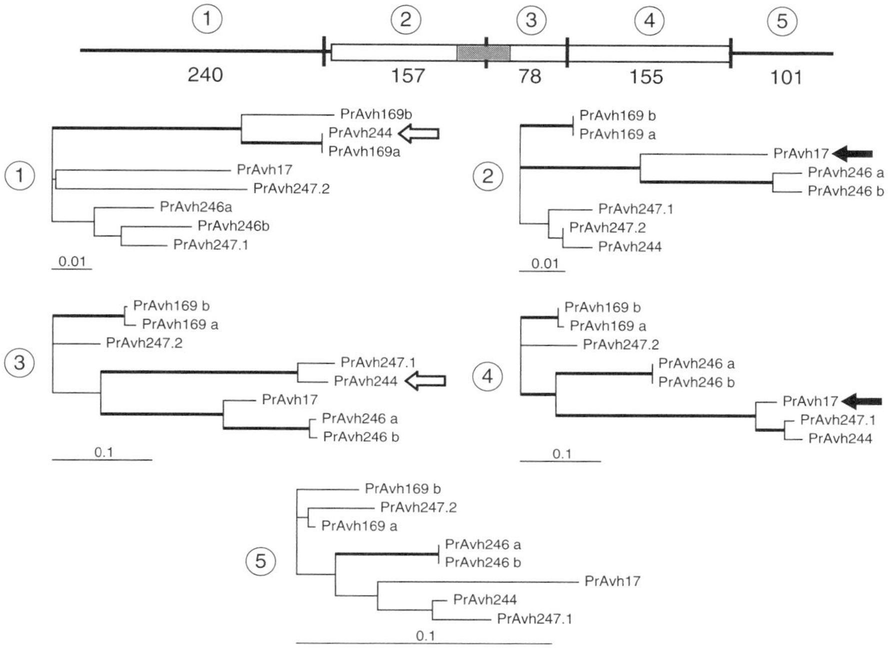

FIG. 9.6 Changing relationships across an alignment of a paralogous group of avirulence gene homologs in a *Phytophthora ramorum* NA1 isolate. A schematic of the alignment is shown at the top, with the boxed region indicating the Avh gene from start codon to stop codon. The shaded region is bounded by the RXLR and dEER motifs. Possible recombination breakpoints (vertical lines) were inferred using Chimera, MaxChi, and GENECONV in the software package RDP. The neighbor-joining trees were constructed using the alignments around four hypothetical breakpoints. Segment 1: 5′ flanking sequence, 2: targeting domain plus RXLR; 3: dEER plus first part of functional domain; 4: remaining C-terminal functional domain; and 5: 3′ flanking sequence. Branches with 95% bootstrap support or greater are shown in bold. Two genes with changed relationships among different NJ trees are highlighted with arrows. Genes are named according to Jiang et al. (2008). For heterozygous loci the two alleles are labeled with "a" or "b", whereas PrAvh247.1 and PrAvh247.2 are likely two different loci that were merged into a single locus in the *P. ramorum* genome sequence. Note the difference in scale for trees of segment 2 versus 3 and 4, which reflect higher levels of divergence in the C-terminal region.

whereas heptanucleotide repeats were more abundant than pentanucleotide repeats (Table 9.2). The overall density of SSRs was similar in the two species but was biased by the relatively high density of mononucleotide repeats found in *P. ramorum* (Table 9.2). These densities are considerably lower than those reported for fungal genomes, which ranged from 6,676 bp/Mb for *Neurospora crassa* to 767 bp/Mb for *Encephalitozoon cuniculi* (Karaoglu et al., 2005), and even more so compared with other eukaryotic genomes (Toth et al., 2000).

TABLE 9.2 Frequency and density of 1–6-bp microsatellite repeats in the genomes of *Phytophthora ramorum* and *Phytophthora sojae*.

Motif length	*P. ramorum* Frequency	%	Density	*P. sojae* Frequency	%	Density	Density ratio (*Pr*/*Ps*)
1	1,125	52.9	258.7	1,221	36.5	265.4	1.0
2	558	26.3	72.2	989	29.5	137.1	0.5
3	290	13.6	30.0	742	22.2	56.8	0.5
4	80	3.8	10.9	171	5.1	16.0	0.7
5	14	0.7	1.5	59	1.8	4.8	0.3
6	58	2.7	4.8	167	5.0	9.3	0.5
Total	2,125	100.0	378.9	3,349	100.0	360.4	1.1

Density is expressed as sequence length in bp for motifs of length of 1–6-bp divided by genome size (Mb).

SSRs were classified into canonical classes by repeat as described by Jurka and Pethiyagoda (1995). Briefly, repeats with unit patterns that were circular permutations and/or reverse complements of each other were grouped together as one kind. *P. ramorum* has considerably higher densities of both the A and C mononucleotide repeat classes than *P. sojae*. For the dinucleotide motifs, the density of AC was higher in *P. ramorum*, whereas the density of AG and AT was higher in *P. sojae*, and CG was the rarest dinucleotide repeat. AC is the most abundant dinucleotide repeat in animals (Bachtrog et al., 1999; Katti et al., 2001), whereas AG and AT dominate in several fungal genomes (Karaoglu et al., 2005). The trinucleotide repeats ACG and AGC occurred at highest density in both *Phytophthora* species. Tetranucleotide motif density was greater than 1 bp/Mbp for motifs AAAC, ACAG, and AGGC in *P. ramorum* and ACAG, ACTC, and AGAT in *P. sojae*.

Across the two species, the average repeat number varied between 13 and 15 for mononucleotide repeats (using a cutoff of 12 bp), 7 and 11 for dinucleotide repeats, 6 and 9 for trinucleotide repeats, and 5 and 45 for tetranucleotide repeats. Although most motifs were repeated 5–7 times, repeats ACCC, ACCG, and AGCG were repeated on average up to 45 times and appeared more variable.

Most SSRs were found in intergenic regions (Table 9.3). The percentage of SSR loci found in coding regions was higher in *P. sojae* than *P. ramorum* (Table 9.3). Trinucleotides and hexanucleotides were almost as abundant in exons as in intergenic regions (Table 9.3). The frequency distribution of trinucleotide repeats in exons and intergenic areas was not significantly different in *P. ramorum* ($\chi^2 = 0.77$; $P = 0.38$); yet it differed significantly in *P. sojae* ($\chi^2 = 6.4$; $P < 0.011$).

TABLE 9.3 Location of microsatellite loci in the genomes of *Phytophthora ramorum* and *Phytophthora sojae* with respect to exons, introns, and intergenic regions.

		Phytophthora ramorum		*Phytophthora sojae*	
Motif length	Region	Frequency	%	Frequency	%
1	Exons	4	0.4	11	0.9
	Introns	73	6.5	99	8.1
	Intergenic	1,047	93.1	1,110	91.0
2	Exons	19	3.4	45	4.6
	Introns	76	13.6	154	15.6
	Intergenic	462	82.9	789	79.9
3	Exons	108	37.4	252	34.0
	Introns	69	23.9	221	29.8
	Intergenic	112	38.8	269	36.3
4	Exons	—	—	9	5.3
	Introns	2	2.5	27	15.8
	Intergenic	78	97.5	135	78.9
5	Exons	—	—	2	3.4
	Introns	—	—	5	8.5
	Intergenic	14	100.0	52	88.1
6	Exons	21	36.2	62	37.3
	Introns	13	22.4	36	21.7
	Intergenic	24	41.4	68	41.0
Total	Exons	381	11.4	381	11.4
	Introns	542	16.2	542	16.2
	Intergenic	2,423	72.3	1,737	81.7

9.5.3 Mitochondrial Genome

The mitochondrial genome of *P. ramorum* has been sequenced in both an NA1 isolate and an EU1 isolate (Martin, 2008; Martin et al., 2007). The NA1 mitochondrial genome is 39,314 bp in size and smaller than the 42,977 bp *P. sojae* mtDNA genome (Martin et al., 2007). The *P. ramorum* genome is more similar in size to that of *P. infestans*, which is thought to be about 37,922 to 37,957 bp for the mitochondrial lineages characterized to date (Avila-Adame et al., 2006). *P. ramorum* and *P. sojae* both have the same gene order except for a small 1,150 bp inverted repeat, with two additional inversions relative to *P. infestans* (Martin et al., 2007). Inversions in other Stramenopile mitochondrial genomes typically represent much larger proportions of the genome than the inverted repeat observed in *P. ramorum*.

The mitochondrial genome contains 37 protein- and rRNA-encoding genes, which include 18 respiratory chain proteins, 16 ribosomal proteins, rRNAs for the large and small ribosomal subunits, and an import protein (Avila-Adame et al., 2006; Martin et al., 2007). A total of 26 tRNA genes specify 20 amino acids. About 11.5% of the genome is noncoding. The mtDNA GC content is 22%, and all genes lack introns (Martin et al., 2007).

9.6 CONCLUSIONS

P. ramorum, which affects plants as divergent as coast live oaks, redwood, bay laurel, rhododendron, and lilac, clearly differs from other model Oomycetes in having a broad host range that consists of many woody ornamentals and trees. Other model Oomycetes such as *P. sojae*, which affect only legumes and mostly soybean (Tyler, 2007), or *P. infestans*, which infect mostly Solanaceous hosts including the cultivated and wild tuber bearing *Solanum* (Grünwald and Flier, 2005), have family-specific host ranges. The *P. ramorum* effector and enzyme secretome necessary to invade woody ornamentals and trees might thus be different from that of *P. sojae* and *P. infestans*. We expect to find novel functional, host–pathogen interaction traits specific to *P. ramorum*. Despite the fact that *P. ramorum* was only formally described to science in 2001 (Werres et al., 2001), we have come a long way in understanding its genome. It seems as though *P. ramorum* will continue to be a particularly interesting candidate for future genetic analyses.

ACKNOWLEDGMENTS

This work was supported in part by funds from USDA ARS CRIS Project 5358-22000-034-00D, the USDA Forest Service Pacific Southwest Research Station Sudden Oak Death research program, and a training fellowship from the Oomycete Molecular Genetics Research Collaboration Network. We thank Meg Larsen and Caroline Press for expert technical assistance and the Oregon State University Center for Genome Research and Biocomputing (CGRB) for use of the bioinformatics computer cluster. Mention of trade names or commercial products in this manuscript are solely for the purpose of providing specific information and do not imply recommendation or endorsement by the U.S. Department of Agriculture.

REFERENCES

Avila-Adame C, Gomez-Alpizar L, Zismann V, Jones KM, Buell CR, Ristaino JB 2006. Mitochondrial genome sequences and molecular evolution of the Irish potato famine pathogen, *Phytophthora infestans*. Curr Gene 49:39–46.

Avise JC 2004. Molecular markers, natural history, and evolution. Sinauer Associates. Sunderland, MA. p. 684

Bachtrog D, Weiss S, Zangerl B, Brem G, Schlotterer C 1999. Distribution of dinucleotide microsatellites in the *Drosophila melanogaster* genome. Mole Biol Evol 16:602–610.

Blair JE, Coffey MD, Park S-Y, Geiser DM, Kang S 2008. A multi-locus phylogeny for *Phytophthora* utilizing markers derived from complete genome sequences. Fungal Gene Biol 45:266–277.

Brasier C 2003. Sudden oak death: *Phytophthora ramorum* exhibits transatlantic differences. Mycol Res 107:258–259.

Brasier C, Denman S, Brown A, Webber J 2004a. Sudden Oak Death (*Phytophthora ramorum*) discovered on trees in Europe. Mycol Res 108:1108–1110.

Brasier CM, Denman S, Rose J, Kirk SA, Hughes KJD, Griffin RL, Lane CR, Inman AJ, Webber JF 2004b. First report of ramorum bleeding canker on *Quercus falcata*, caused by *Phytophthora ramorum*. Plant Pathol 53:804.

Brasier C, Kirk S 2004. Production of gametangia by *Phytophthora ramorum* in vitro. Mycol Res 108:823–827.

Brasier C, Kirk S, Rose J 2006. Differences in phenotypic stability and adaptive variation between the main European and American lineages of *Phytophthora ramorum*. Progress in Research on *Phytophthora* Diseases of Forest Trees. Forest Research, Farnham. pp. 166–173.

Cave GL, Randall-Schadel B, Redlin SC 2005. Risk analysis for *Phytophthora ramorum* Werres, de Cock & In't Veld, causal agent of Phytophthora Canker (Sudden Oak Death), Ramorum Leaf blight, and Ramorum Dieback. USDA APHIS report. pp. 77.

Chamnanpunt J, Shan WX, Tyler BM 2001. High frequency mitotic gene conversion in genetic hybrids of the oomycete *Phytophthora sojae*. Proc Nat Acad Sci USA 98:14530–14535.

Cooke DEL, Drenth A, Duncan JM, Wagels G, Brasier CM 2000. A molecular phylogeny of *Phytophthora* and related oomycetes. Fungal Genet Biol 30:17–32.

Dart NL, Chastagner GA 2007. Estimated economic losses associated with the destruction of plants due to *Phytophthora ramorum* quarantine efforts in Washington State. Plant Health Progress.

Davidson JM, Werres S, Garbelotto M, Hansen EM, Rizzo DM 2003. Sudden oak death and associated diseases caused by *Phytophthora ramorum*. Plant Health Progress.

Denman S, Kirk SA, Brasier CM, Webber JF 2005. In vitro leaf inoculation studies as an indication of tree foliage susceptibility to *Phytophthora ramorum* in the UK. Plant Pathol 54:512–521.

Dobrowolski MP, Tommerup IC, Shearer BL, O'Brien PA 2003. Three clonal lineages of *Phytophthora cinnamomi* in Australia revealed by microsatellites. Phytopathol 93:695–704.

Donahoo R, Blomquist CL, Thomas SL, Moulton JK, Cooke DEL, Lamour KH 2006. *Phytophthora foliorum* sp nov., a new species causing leaf blight of azalea. Mycol Res 110:1309–1322.

Erwin DC, Ribeiro OK 1996. Phytophthora diseases worldwide. APS Press St. Paul, MN. pp. 562.

Frankel SJ 2008. Sudden oak death and *Phytophthora ramorum* in the USA: a management challenge. Australas Plant Pathol 37:19–25.

Garnica DP, Pinzon AM, Quesada-Ocampo LM, Bernal AJ, Barreto E, Grünwald NJ, Restrepo S 2006. Survey and analysis of microsatellites from transcript sequences in

Phytophthora species: frequency, distribution, and potential as markers for the phylum Oomycota. BMC Genom 7:245.

Goldstein DB, Schlötterer C 1999. Microsatellites, evolution and applications. Goldstein DB, Schlötterer C, editors. Oxford University Press, New York. p. 352.

Goss EM, Carbone I, Grünwald NJ 2009. Ancient isolation and independent evolution of the three clonal lineages of the exotic sudden oak death pathogen *Phytophthora ramorum*. Mol Ecol 18:1161–1174.

Grünwald NJ, Flier WG 2005. The biology of *Phytophthora infestans* at its center of origin. Ann Rev Phytopathol 43:171–190.

Grünwald NJ, Goss EM, Larsen MM, Press CM, McDonald VT, Blomguist C, Thomas SL 2008a. First report of the European lineage of *Phytophthora ramorum* on *Viburnum* and *Osmanthus* spp. in a California nursery. Plant Dis 92:314–314.

Grünwald NJ, Goss EM, Press CM 2008b. *Phytophthora ramorum*: a pathogen with a remarkably wide host-range causing sudden oak death on oaks and ramorum blight on woody ornamentals. Mole Plant Pathol. 9:729–740.

Hansen EM, Goheen DJ, Jules ES, Ullian B 2000. Managing Port-Orford-Cedar and the introduced pathogen *Phytophthora lateralis*. Plant Dis 84:4–10.

Hansen EM, Parke JL, Sutton W 2005. Susceptibility of Oregon forest trees and shrubs to *Phytophthora ramorum*: a comparison of artificial inoculation and natural infection. Plant Dis 89:63–70.

Hansen EM, Reeser PW, Sutton W, Winton LM 2003. First report of A1 mating type of *Phytophthora ramorum* in North America. Plant Dis 87:1267–1267.

Hennequin C, Thierry A, Richard GF, Lecointre G, Nguyen HV, Gaillardin C, Dujon B 2001. Microsatellite typing as a new tool for identification of *Saccharomyces cerevisiae* strains. J Clin Microbiol 39:551–559.

Hüberli D, Harnik T, Meshriy M, Miles L, Garbelotto M 2005. Phenotypic variation among *Phytophthora ramorum* isolates from California and Oregon. In: Frankel SJ, Shea PJ, Haverty MI, editors. Proceedings of the sudden oak death second science symposium: the state of our knowledge. Monterey, CA. pp. 131–134.

Ivors K, Garbelotto M, Vries IDE, Ruyter-Spira C, Hekkert BT, Rosenzweig N, Bonants P 2006. Microsatellite markers identify three lineages of *Phytophthora ramorum* in US nurseries, yet single lineages in US forest and European nursery populations. Mole Ecol 15:1493–1505.

Ivors KL, Hayden KJ, Bonants PJM, Rizzo DM, Garbelotto M 2004. AFLP and phylogenetic analyses of North American and European populations of *Phytophthora ramorum*. Mycol Res 108:378–392.

Jiang RHY, Tripathy S, Govers F, Tyler BM 2008. RXLR effector reservoir in two *Phytophthora* species is dominated by a single rapidly evolving superfamily with more than 700 members. Proc Nat Acad Sci, USA 105:4874–4879.

Jurka J, Pethiyagoda C 1995. Simple repetitive DNA sequences from primates: compilation and analysis. J Molec Evol 40:120–126.

Kamoun S 2006. A catalogue of the effector secretome of plant pathogenic Oomycetes. Ann Rev Phytopathol 44:41–60.

Karaoglu H, Lee CMY, Meyer W 2005. Survey of simple sequence repeats in completed fungal genomes. Molec Biol Evol 22:639–649.

Katti MV, Ranjekar PK, Gupta VS 2001. Differential distribution of simple sequence repeats in eukaryotic genome sequences. Molec Biol Evol 18:1161–1167.

Kroon LPNM, Bakker FT, van den Bosch GBM, Bonants PJM, Flier WG 2004. Phylogenetic analysis of *Phytophthora* species based on mitochondrial and nuclear DNA sequences. Fungal Genet Biol 41:766–82.

Lamour KH, Win J, Kamoun S 2007. Oomycete genomics: new insights and future directions. FEMS Microbiol Lett 274:1–8.

Martin FN 2008. Mitochondrial haplotype determination in *Phytophthora ramorum*. Curr Genet 54:23–34.

Martin FN, Bensasson D, Tyler BM, Boore JL 2007. Mitochondrial genome sequences and comparative genomics of *Phytophthora ramorum* and *P. sojae*. Curr Genet 51:285–296.

Martin FN, Tooley PW 2003. Phylogenetic relationships of *Phytophthora ramorum*, *P. nemorosa*, and *P. pseudosyringae*, three species recovered from areas in California with sudden oak death. Mycol Res 107:1379–1391.

Mascheretti S, Croucher PJP, Vettraino A, Prospero S, Garbelotto M 2008. Reconstruction of the Sudden Oak Death epidemic in California through microsatellite analysis of the pathogen *Phytophthora ramorum*. Molec Ecol 17:2755–2768.

McDonald V, Grünwald NJ 2006. Evaluation of fitness of *Phytophthora ramorum* on six *Rhododendron* cultivars. Phytopathol 96:S75.

Parke JL, Oh E, Voelker S, Hansen EM, Buckles G, Lachenbruch B 2007. *Phytophthora ramorum* colonizes tanoak xylem and is associated with reduced stem water transport. Phytopathol 97:1558–1567.

Primmer CR, Saino N, Moller AP, Ellegren H 1996. Directional evolution in germline microsatellite mutations. Nature Gene 13:391–393.

Prospero S, Hansen EM, Grünwald NJ, Winton LM 2007. Population dynamics of the sudden oak death pathogen *Phytophthora ramorum* in Oregon from 2001 to 2004. Molec Ecol 16:2958–2973.

Rizzo DM, Garbelotto M, Davidson JM, Slaughter GW, Koike ST 2002. *Phytophthora ramorum* as the cause of extensive mortality of *Quercus* spp. and *Lithocarpus densiflorus* in California. Plant Dis 86:205–214.

Rizzo DM, Garbelotto M, Hansen EM 2005. *Phytophthora ramorum*: Integrative research and management of an emerging pathogen in California and Oregon forests. Ann Rev Phytopathol 43:309–335.

Schug MD, Mackay TF, Aquadro CF 1997. Low mutation rates of microsatellite loci in *Drosophila melanogaster*. Nature Genet 15:99–102.

Shishkoff N 2007. Persistence of *Phytophthora ramorum* in soil mix and roots of nursery ornamentals. Plant Dis 91:1245–1249.

Tooley PW, Browning M, Berner D 2008. Recovery of *Phytophthora ramorum* following exposure to temperature extremes. Plant Dis 92:431–437.

Tooley PW, Kyde KL 2007. Susceptibility of some Eastern forest species to *Phytophthora ramorum*. Plant Dis 91:435–438.

Tooley PW, Kyde KL, Englander L 2004. Susceptibility of selected Ericaceous ornamental host species to *Phytophthora ramorum*. Plant Dis 88:993–999.

Toth G, Gaspari Z, Jurka J 2000. Microsatellites in different eukaryotic genomes: survey and analysis. Gen Res 10:967–981.

Tyler BM 2007. *Phytophthora sojae*: root rot pathogen of soybean and model oomycete. Molec Plant Pathol 8:1–8.

Tyler BM, Tripathy S, Zhang X, Dehal P, Jiang RHY, Aerts A, Arredondo FD, Baxter L, Bensasson D, Beynon JL, et al. 2006. *Phytophthora* genome sequences uncover evolutionary origins and mechanisms of pathogenesis. Science 313:1261–1266.

Weber JL, Wong C 1993. Mutation of human short tandem repeats. Human Molec Genet 2:1123–1128.

Werres S, De Merlier D 2003. First detection of *Phytophthora ramorum* mating type A2 in Europe. Plant Dis 87:1266–1266.

Werres S, Kaminsky K 2005. Characterization of European and North American *Phytophthora ramorum* isolates due to their morphology and mating behaviour in vitro with heterothallic *Phytophthora* species. Mycol Res 109:860–871.

Werres S, Marwitz R, Veld W, De Cock A, Bonants PJM, De Weerdt M, Themann K, Ilieva E, Baayen RP 2001. *Phytophthora ramorum* sp. nov., a new pathogen on *Rhododendron* and *Viburnum*. Mycol Res 105:1155–1165.

Whisson SC, Boevink PC, Moleleki L, Avrova AO, Morales JG, Gilroy EM, Armstrong MR, Grouffaud S, van West P, Chapman S and others 2007. A translocation signal for delivery of oomycete effector proteins into host plant cells. Nature 450:115–118.

Win J, Morgan W, Bos J, Krasileva KV, Cano LM, Chaparro-Garcia A, Ammar R, Staskawicz BJ, Kamoun S 2007. Adaptive evolution has targeted the C-terminal domain of the RXLR effectors of plant pathogenic Oomycetes. Plant Cell 19: 2349–2369.

10

PHYTOPHTHORA SOJAE: DIVERSITY AMONG AND WITHIN POPULATIONS

ANNE DORRANCE
Department of Plant Pathology, The Ohio State University

NIKLAUS J. GRÜNWALD
Horticultural Crops Research Laboratory, U.S. Department of Agriculture—Agricultural Research Service, Corvallis, Oregon

10.1 INTRODUCTION

Soybean production is increasing around the world and not surprisingly so are the reports of soybean diseases caused by *Phytophthora sojae* (Kaufmann and Gerdemann), including Phytophthora seed, root, and stem rot. *P. sojae* is a diploid oomycete, which is homothallic and is limited to primarily one host: the soybean. It is soilborne, and when soils are saturated, this oomycete produces nondehiscent sporangia, which are nonpapillate. The development and spread of zoospores and subsequent infection of soybean roots is favored by saturated soil conditions (Dorrance et al., 2007). Oospores are readily produced in susceptible soybean roots and can survive in crop residues and soil for many years (Schmitthenner, 1985).

Throughout the season, soybeans are susceptible to *P. sojae,* but seed rot, damping off, and seedling blight result in the greatest economic loss caused by individual plant loss combined with the additional costs associated with replanting a field. For late-season infections during the reproductive growth stages, disease development of root and stem rot results in an overall yield decline caused by both plant loss and reduced plant vigor. Foliar leaf spots do not develop under field conditions. In the past 10 years, there were numerous cases of significant levels of Phytophthora root and stem rot in almost all of the

Oomycete Genetics and Genomics: Diversity, Interactions, and Research Tools
Edited by Kurt Lamour and Sophien Kamoun
Copyright © 2009 John Wiley & Sons, Inc.

world's soybean production regions. In Japan, the incidence of root and stem rot ranged from 2% to 46% in specific fields from 2002 to 2004 along with reports of increasing yield losses in the Hyogo soybean production region (Sugimoto et al., 2006). In South America, yield losses as high as 60% to 70% were reported in Argentina (Vallone et al., 1999). Estimated losses in the United States have exceeded 1 million tons annually since 2000 (Wrather et al., 2003; Wrather and Koenning, 2006).

10.2 MANAGEMENT OF *P. SOJAE*

Phytophthora root and stem rot is managed predominately through the deployment of cultivars with single dominant resistance genes (*Rps*). In the United States, it is generally believed that the *P. sojae* populations have adapted to many of these *Rps* genes and that these genes are no longer effective in some regions (Grau et al., 2004). Pathogens and hosts that interact in a gene-for-gene system require a clear means to characterize virulence pathotypes (sensu races) with well-defined and well-characterized differentials. Fortunately, several well-defined sets of differentials are available to characterize *P. sojae* isolates, with each differential having only one *Rps* gene. These were developed in several different soybean genetic backgrounds, including Bedford, Clark, Williams, and Harosoy (Dorrance et al., 2004). Most *Rps* genes are well characterized and in many cases were mapped in the soybean genome to six different loci on molecular linkage groups F, G, J, and N (Demirbas et al., 2001; Diers et al., 1992; Gordon et al., 2004; Lohnes and Schmitthenner, 1997; Weng et al., 2001). As a result of this well-characterized resource, extensive assessments of pathotype diversity have been completed across most soybean-producing regions. Prior to 2000, 55 races were described based on the response of eight soybean differentials for *Rps*1a, *Rps*1b, *Rps*1c, *Rps*1d, *Rps*1k, *Rps*2, *Rps*3a, *Rps*6, and *Rps*7 after inoculation (Grau et al., 2004). More recent surveys that describe the pathotypes included additional differentials that represent regional uncharacterized sources of resistance or additional *Rps* genes (*Rps*3b, *Rps*3c, *Rps*4, *Rps*5, or *Rps*8) were completed in Argentina (Barreto et al., 1995), China (Yang et al., 2003), Iran (Garmaroodhi et al., 2007; Mohammadi et al., 2007), Japan (Sugimoto et al., 2006), and the United States (Abney et al., 1997; Dorrance et al., 2003; Grau et al., 2004, Jackson et al., 2004; Malvik et al., 2004; Nelson et al., 2008).

10.3 *P. SOJAE* PATHOTYPES WORLDWIDE

Twelve different virulence pathotypes were identified among 42 isolates of *P. sojae* collected from Heilongjiang and Jilin Provinces in Northeast China (Yang et al., 2003). *P. sojae* was also characterized from the Huangehe-Huaihe basin and the Yangtze basin. There were reportedly "abundant" virulence pathotypes from both regions but with the greatest number from the Yangtze basin (Zhu et al., 2004). In Iran, two studies evaluated virulence pathotypes from *P. sojae* isolates recovered from symptomatic plants collected from 2001 to 2004 (Garmaroodi et al., 2007) and from plants and soil from 1998 to 2005 (Mohammadi et al., 2007). In both studies, the pathotype of most isolates collected consisted primarily of virulence to *Rps*7 (race 1), followed by those with virulence to *Rps*1a and *Rps*7 (race 3) (Garmaroodi et al., 2007; Mohammadi et al., 2007). Two isolates were also characterized with virulence *Rps*1a, *Rps*1c, and *Rps*7 (race 4) and one *Rps*6 and *Rps*7 (race 13) (Garmaroodi et al., 2007). Sugimoto et al. (2006) evaluated 51 *P. sojae* isolates collected from Japan from 2002 to 2004 on six soybean cultivars that originated from Japan and on four U.S. differentials that contained *Rps*1a + *Rps*7 (Harosoy63), *Rps*1d (PI103091), and *Rps*6 (Altona). Interestingly, 90% of the *P. sojae* isolates collected from three regions in Japan had virulence to the *Rps*6-containing differential; 33% had virulence to *Rps*1a, whereas *Rps*1d conferred resistance to all isolates. In contrast, differentials with *Rps*1a, *Rps*1c, *Rps*1d, *Rps*1k, *Rps*3a, and *Rps*6 were all resistant to 46 *P. sojae* isolates collected in Argentina (Barreto et al., 1995).

During the past 10 years, the distribution of pathotypes in the United States was assessed. Seventy two different pathotypes were identified in a collection of *P. sojae* isolates from Ohio with 54 pathotypes identified from isolates collected from one field based on eight single *Rps* gene differentials (Dorrance et al., 2003). This was an increase in diversity reported from assessments taken 10 to 20 years earlier (Schmitthenner et al., 1994) and was thought to be the result of recent *Rps*-gene deployment. Similar trends were also reported from Illinois, Indiana, Missouri, North Dakota, and South Dakota as more *P. sojae* isolates were identified that had virulence to the most recently deployed *Rps* genes in those states (Table 10.1) (Abney et al., 1997; Dorrance et al., 2003; Malvik and Grunden, 2004; Nelson et al., personal communication; English and Sweets, personal communication; Draper et al., personal communication). Albeit in some instances widespread virulence to *Rps* genes that were never thought to be

TABLE 10.1 Summary of *P. sojae* virulence pathotypes indicating the percentage of isolates found in the United States from isolates collected from soil or symptomatic plants during 1998 to 2007 that have virulence on tested *Rps* gene differential.

State	Collection	*Rps*3a	No. of fields	*Rps*6	No. of isolates	*Rps*1a	*Rps*1c	*Rps*1k
OH	1997–2004	118	656	94	72	75	49	46
IL	2002–2003	24	121	58	41	37	7	9
IA	2002–2003	41	63	74	52	55	11	24
IN	2000–2003	190	200	80	40	50	13	47
MO	2000–2003	58	60	67	53	55	47	36
ND	2002–2004	252	153	31	18	5.5	0.3	2
SD	2002–2004	46	124	98	77	75	81	41

TABLE 10.2 Summary of *Phytophthora sojae* virulence pathotypes indicating the percentage of isolates found in Canada collected from soil or symptomatic plants from 2002 to 2004 that have virulence on the tested *Rps* gene differential.

Province	Collection	No. of fields	No. of isolates	*Rps*1a	*Rps*1c	*Rps*1k	*Rps*3a	*Rps*6
Ontario	2001–2002	18	72	100	27	75	Nt	86
Quebec	2001–2002	78	208	100	28	64	Nt	73

deployed in northern germplasm was also identified (Abney et al., 1997; Dorrance et al., 2003). In the southern United States, Jackson et al. (2004) identified seven different pathotypes from 23 fields from soil and plant samples in Arkansas. They also concluded that the soybeans currently marketed in this region have effective *Rps* genes for the predominant pathotypes, which is a key function of pathotype surveys. *P. sojae* pathotype diversity was also evaluated in Ontario and Quebec Canada (Anderson and Tenuta, unpublished data) (Table 10.2). The Ontario population is very similar to reports from the Midwest, but the Quebec population has a very high level of diversity in that none of the *Rps* genes were effective against this population. This finding is believed to be very unusual in that soybeans have only been produced in this region for a few years (Anderson, personal communication).

10.4 MECHANISMS FOR CHANGE IN PATHOTYPE

Although assessments of pathotype diversity are essential to determine which resistance gene may be the most effective in a given region, it does not necessarily relate to population genetic diversity. Case in point, *P. sojae* in Argentina, where only one pathotype was found but there was a high diversity among isolates both within and between soybean production regions based on seven random applications of polymorphic DNA (RAPDs) that identified 35 polymorphic bands (Galley et al., 2007). How a pathogen is changing or adapting is key to developing effective and durable disease-management strategies. McDonald and Linde (2002) outlined schemes that predict which pathogens have the greatest potential to adapt to resistance genes based on their population genetics. For *Phytophthora* spp., pathotype diversity may develop through many different mechanisms either alone or in combination, for example, mutation, mitotic recombination, parasexual recombination, inter-specific hybridization, or migration (Goodwin, 1997). Based on these different mechanisms of genetic change, McDonald and Linde (2002) proposed that pathogen populations that are high risk have mixed reproductive systems, a high potential for gene flow, large effective populations and high mutation rates. In addition, they are believed to adapt quickly to any selection pressure such as a resistance gene, whereas populations that reproduce asexually, have low potential for gene flow, small effective population sizes, and low mutation rates are considered low-risk pathogens. Based on the current data, *P. sojae* falls into the medium-risk pathogen category with low gene/genotype flow as it is a soilborne pathogen, is homothallic with low outcrossing, has a medium/moderate population size, and has a low mutation rate. For some regions, the pathotype diversity is much higher than expected for a medium-risk pathogen. The influence of outcrossing within these regional populations is unknown with only a few exceptions, and how related each of these populations are to each other is largely unknown.

10.5 GENETIC DIVERSITY OF *P. SOJAE*

A few studies have evaluated the genetic diversity of *P. sojae*, but these were limited by either the total number of isolates or by the scope of the regions examined. The first genetic analysis of *P. sojae* was related to its taxonomy and classification with *Phytophthora medicaginis* (E.M. Hansen and D.P. Maxwell, 1991), *Phytophthora trifolii* (E.M. Hansen and D.P. Maxwell, 1991), and *Phytophthora megasperma* (Drechsler) as all were classified under *P. megasperma* for a period of time. Based on this first classic study to assess genetic variation among isolates of *P. sojae*, Nygaard et al. (1989) found no variation in isozymes among isolates within the *P. megasperma* group but did identify variation between groups (sensu *P. medicaginis*, *P. megasperma*, and *P. sojae*). This was confirmed recently by an investigation where 157 *P. sojae* isolates were evaluated for diversity among 10 isozymes with cellulose acetate electrophoresis; no diversity was found (deSilva and Dorrance, unpublished data). Förster et al. (1989) reported that there was variation in mitochondrial DNA based on restriction fragment length polymorphisms (RFLPs) among the *P. sojae* (sensu *P. megasperma* f.sp. *glycinea*), *P. medicaginis* (sensu *P. megasperma* f.sp. *medicaginis*), and *P. megasperma* groups. In addition, within the *P. sojae* isolates evaluated in this study were four distinct RFLP patterns, and these were not linked to pathotype or geographic region.

Of the population genetic studies of *P. sojae* that were completed (Table 10.3), numerous mechanisms responsible for diversity within *P. sojae* populations have been proposed, which include mutation (Förster et al., 1994; Drenth et al., 1996), heterokaryon formation (Layton and Kuhn, 1990), as well as outcrossing (Bhat and Schmitthenner, 1993; Whisson et al., 1994; Förster et al., 1994). Mutation is the most likely force that drives pathotype diversity in Australia based on the studies of Drenth et al. (1996), in which 95% of the *P. sojae* isolates evaluated from Australia all had the same multilocus RFLP chromosomal genotype. The very low levels of genetic diversity in this region at the time continue to point to one probable introduction with mutation as the primary means of continued change.

In the United States, there are several reports that support either mutation or outcrossing or both contributing to genetic diversity. Förster et al. (1994)

TABLE 10.3 List of population genetic analysis of *Phytophthora sojae* populations from soybean production regions worldwide, marker techniques, and general conclusions.

Molecular marker technique	No. of isolates/geographic region	Conclusion	Citation
RFLP nuclear and mitochondrial	33 United States,[a] 4 Australia, 9 Canada, 2 Japan	Four Goups 1/4 RFLPs very similar and most likely differences caused by mutation, 3/4; groups more variation among isolates and could be caused by rare outcrosses	Förster et al., 1994
RAPD	55 United States[b]	Four Groups	Meng et al., 1999
RAPD	32 Argentina	High level of variability both between and within regions	Galley et al., 2007
RFLP	84 Australia, 15 United States	Australian isolates were 95% identical, mutation driving force in this region	Drenth et al., 1996
Sequence of ITS1 and ITS2	8 China, 9 North America	Variation among ITS1 and ITS2 for isolates within China	Xu et al., 2007
ISSR	65 2 provinces, 28 from soil collected from ships, 18 North America	6 groups, Genetic diversity was lower within 2 provinces than U.S. population and a low identity with U.S. population	Wang et al., 2007
EST-SSRs, Genomic–SSRs	34 Colombia	Low level of variability, predominantly monomorphic markers	Garnica, 2007

[a]indicates 33 isolates originated from seven different production regions in Arkansas, Illinois, Indiana, Ohio, Mississippi, Virginia, and Wisconsin.
[b]indicates 55 isolates originated from Indiana (7), Illinois (4), Iowa (26), Ohio (2), and Minnesota (16), which collectively represented 9 different pathotypes.

evaluated 48 isolates (33 isolates from the United States and additional isolates from Australia, Canada, and Japan) using nuclear and mitochondrial RFLP markers; in all, they identified 4 groups. Isolates in the first group were all very similar to *P. sojae* race 1 (vir 7), and the pathotypes were all very similar. The remaining three groups had higher levels of molecular variation. These data led them to hypothesize that new races arose by mutation (group I) and rare outcrosses may contribute to the variation among existing races in addition to clonal evolution (groups II to IV).

Four distinct groups were also identified in an analysis of 55 isolates from Illinois, Indiana, Iowa, and Minnesota by Meng et al. (1999) using RAPD markers. Again, there was diversity within pathotype (race) classifications and geographic regions. In South America, RAPD markers were also used to evaluate 32 isolates from different production regions in Argentina (Galley et al., 2007). Seven of 20 decanucleotide primers amplified 49 fragments, of which 35 were polymorphic. Based on this analysis, they detected a high level of variability even among isolates from the same geographic region.

Recently, inter-simple sequence repeats (ISSRs) were used to compare *P. sojae* isolates from China and the United States (Wang et al., 2007). ISSRs use a single primer composed of microsatellite sequence anchored at the 3′ or 5′ end with 2-4 arbitrary; often the degenerate nucleotides and these sequences are abundant throughout most genomes (Fang and Roose, 1997). For this study, *P. sojae* isolates were collected from two provinces, Heilongjiang and Fujian, as well as a collection of 18 isolates from the United States. These isolates were selected based on their diversity as well as from isolates presumed to have originated from the United States from soils collected from the ships that deliver soybeans to China. Wang et al. (2007) concluded that there was more diversity among the U.S. isolates than either of the two Chinese provinces; this obvious conclusion is based on the sampling strategies that were used. Nevertheless, the 13 ISSRs proved to be repeatable with high levels of variation among the different isolates reported 71 of 90 bands were polymorphic from Heilongiang; 60 of 83 from Fujian, and 84 of 102 from the United States. More interesting is a study that evaluated the sequence of the internal spacer (ITS) region (both ITS-1 and ITS-2) of the internal transcribed spacer region among *P. sojae* isolates collected in China and from the United States. Previous studies have not identified variation within *P. sojae* in the United States (Cooke et al., 2000); however, there were distinct differences among the isolates collected in China that may not be caused by sequencing error (Xu et al., 2007).

10.6 A NEW STRATEGY TO ASSESS DIVERSITY

In all of these previous studies, it is difficult to address the mechanisms that drive the genetic change in *P. sojae* for several reasons. First, the number of isolates was few (most studies had less than 50 isolates), several geographic regions were represented, and the markers used in most of the analyses were dominant. In many cases, including some recent studies from China, comparisons were made to U.S. isolates that were already known to be genotypically distinct. With these contrasting conclusions and continuing rapid evolution of pathotypes, it is necessary to document the genetic variation on a larger number of isolates within smaller geographic units. The purpose is to determine whether these original conclusions that indicate *P. sojae* is a medium-risk pathogen because of limited outcrossing and low mutation rate will pertain to an individual region or to all regions where Phytophthora root rot is a limiting soybean production factor. To identify the true mechanisms of genetic change in this diploid organism, codominant markers that exploit polymerase chain reaction (PCR) based strategies would be especially useful.

Simple sequence repeats (SSRs) were identified from transcript sequences from *Phytophthora infestans*, *P. sojae*, and *Phytophthora ramorum* (Garnica et al., 2006), as well as from genome sequences of *P. sojae* and *P. ramorum* (Tyler et al., 2006). SSRs were identified in the genome, and those with greater than 15 repeats were selected. Primers that were 18 to 20 base pairs in length were designed flanking the SSRs using Primer3 (http://frodo.wi.mit.edu/cgi-bin/primer3/primer3_www.cgi) as a main program and also FastPCR (http://www.biocenter.helsinki.fi/bi/Programs/fastpcr.htm). One-hundred and fifty-five unique primer pairs were evaluated on six standard isolates of *P. sojae* representing many pathotypes: race 1 (vir 7), 2 (1b, 7), 3 (1a, 7), 4 (1a, 1c, 7), 7(1a, 3a, 6, 7), and 25 (1a, 1b, 1c, 1k, 7). Of these, only 38 primer pairs were polymorphic, and these were evaluated on 33 *P. sojae* isolates collected from 19 locations in Ohio, which includes 9 isolates from one field. The 21 microsatellite markers (SSRs) that were polymorphic among *P. sojae* isolates and developed from *P. sojae* race 2 sequences obtained from http://phytophthora.vbi.vt.edu/ are described in Table 10.4. There were 53 alleles identified among these 33 *P. sojae* isolates, and the observed number of alleles ranged from 2 to 6 with an average of 2.5 alleles per locus. The sequences for these SSRs were assigned to 14 different super contigs, whereas three were not assigned. Thus, these molecular markers sample many different areas of the *P. sojae* genome and will be useful for population genetic analysis to address the questions posed earlier.

Population genetic parameters for these 21 SSRs on the 33 isolates from Ohio were calculated with POPGENE version 1.31 (Yeh and Boyle, 1997). Another interesting finding of this small regional population was that all the loci significantly deviated from Hardy-Weinberg equilibrium ($P < 0.001$). The observed heterozygosity was 0.0 for 15 of the 21 loci, and the overall mean was 0.015. The observed heterozygosity for the remaining six loci ranged from 0.0294 to 0.0882. Although heterozygosity was low for all loci, this was expected for a soilborne species that is homothallic and selfing. However, these results do indicate that outcrossing is occurring in these fields with subsequent selfing. This result was unexpected because of the limited size of the study (only nine isolates from the same 100-acre field), but this is the second time this laboratory has identified putative heterozygotes from the same field. Layton

TABLE 10.4 Primer sequences and characteristics for polymorphic microsatellite loci developed from the *Phytophthora sojae* sequence. Locus name, GenBank Accession no., primer sequence, repeat motif, size range, no. of alleles, observed heterozygosity (H_o), expected heterozygosity (H_E), determined by genotyping 37 isolates from Ohio and sequenced isolate race 2; scaffold that the sequence was derived from and the super contig where sequence was found.

Locus name	GenBank Accession no.	Primer sequences (5′–3′)[a]	Repeat motif	Size (bp) (Race 2)	No. of alleles	H_o	H_E	Scaffold	Super Contig[b]
PS01	EF667485	F: TGATGGGAGATGGCTACAGG R: TCGCAACGACAGATTGATG	$(GACACT)_{49}$	200–450 (419)	3	0.000	0.516	62:98848–99141	9
PS04	EF654114	F: CTTCCCATCACTCCGACAAG R: TTGACACTGCCTCCTACACG	$(TTC)_{30}$	250–320 (307)	3	0.000	0.574	21:26031–26150	31
PS05	EF667486	F: GAAACAATCAACCGAACAACG R: ATAGGAGGGCAAACTGGATG	$(TCAG)_{34}$	200–300 (263)	3	0.000	0.641	1:758827–758962	3
PS06	EF667487	F: AACTACCTTCGCTCGTGGTG R: CCTATCGCCTGGAAAATGTG	$(TC)_{31}$	160-200 (214)	2	0.000	0.058	11:803579–803640	6
PS07	EF667488	F: TCCTTAGCTTCCGGTTAAGC R: TCTCATTTTGGCCTGGAAAC	$(CT)_{29}$ $(CT)_{12}$	220–240 (234) 200 (200)	2 1	0.000	0.365	1079:5987–6044	NI
PS10	EF667489	F: CGACGAAGAACAACATTACTTG R: ATGAAACCGAACCAAACCTG	$(CAAAC)_{27}$	150–250 (228)	6	0.000	0.639	18:19679–19813	NI
PS12	EF667490	F: GCTGCTTGTTGCTGTTGTTG R: GCGGGTGTTTGGAGAGTATC	$(GCTGTT)_{23}$	220–306 (306)	3 2	0.000	0.585	48:119393–119530	32
PS16	EF667491	F: AATCTGACTTGGACGCTGTG R: GCTTAGTGTTTTGGGTTACGC	$(ATTAT)_{20}$	400–500 (469)	2	0.000	0.468	17:198275–198374	20
PS17	EF667497	F: GAAGCCGAAGACGAAAAGAG R: TGAAAAAGTGACCAAACAGTGG	$(GGA)_{19}$	180–230 (213)	2	0.000	0.444	46:153503–153559	32
PS18	EF667492	F: TCCAGGTTGAGGTGACTTG R: GAAGAGCGTGAGCAGGAAC	$(TC)_{18}$	170–200 (185)	2	0.000	0.482	65:225298–225333	13
PS19	EF667493	F: CGTGATGGAGCACTCAGAAG R: CGCAATCTTCCTGCTAATGG	$(AG)_{18}$	225–275 (254)	2	0.000	0.332	69:43959–43994	6

(*Continued*)

TABLE 10.4. Continued

Locus name	GenBank Accession no.	Primer sequences (5′–3′)[a]	Repeat motif	Size (bp) (Race 2)	No. of alleles	H_o	H_E	Scaffold	Super Contig[b]
PS20	EF667494	F: AAATCCAACCAGCCTTACCC R: CGTGCTTCATGTCTGCACTAC	$(CT)_{18}(ATCT)_{12}$	150–225 (184)	3	0.063	0.437	87:139912–139947	19
PS24	EF667495	F: GTCATTTCCCTCGCTCACAC R: ACACTGGCAACAAGCAACAG	$(CT)_{16}TCAT(CT)_3$	220–260 (252)	2	0.000	0.403	40:38319–38350	36
PS25	EF667496	F: AGCGTTGGTGTTGACGAC R: TAGCGAAACCTGGCAAATG	$(CT)_{16}TT(CTTT)_8$	350–400 (366)	2	0.059	0.163	11:580306–580337	8
PS27	EF667498	F: TTCAGATTCCGTGAGCATTG R: CGGCTTGGTCTCTTAGCTTC	$(AG)_{16}$	275–325 (287)	3	0.033	0.541	45:333463–333494	8
PS29	EF667499	F: CCACTGAAGCGAGGTAGAGG R: GTAGCACAAAATCCGTCTGC	$(TAC)_{15}$	250–280 (273)	2	0.088	0.507	15:588315–588359	NI
PS30	EF667500	F: ACGAAGTCCAACCATAAATCG R: TGAAAGATGACCCCAGGATG	$(AGTC)_{15}$	270–300 (300)	3	0.000	0.520	31:498678–498737	29
PS33	EF667501	F: CTGCTAGTGCCGTTCGTTG R: TAAAAGGGCTGCTCAAATCG	$(AT)_{15}$	240–267 (267)	2	0.042	0.489	199:28445–28474	2
PS36	EF667502	F: CAAAAATCATCAGCACCTTCG R: TAGCCAAAAGAGCGACAACC	$(CT)_{28}$	150–220 (209)	3	0.000	0.458	71:143091–143144	46
PS37	EF667503	F: ACGAGCCCACGGAAGAGTTC R: CTGGATGATGTGCGGGTTTG	$(GAG)_{17}$	200–370 (204)	2	0.000	0.183	21:370854–370901.2	31
PS38	EF667504	F: AGTGGGCGTTTGCTTGTG R: TCGCTGTGGTTCCTCTTCTC	$(GTAGAA)_{18}$	245–270 (245)	2	0.029	0.504	60:332948–333043	7

[a] F, forward primer; R, reverse primer.

[b] NI, not identified.

and Kuhn (1990) demonstrated that heterokaryons/heterozygotes can develop *in planta* using two isolates, which were each resistant to a different antibiotic. They inoculated both susceptible and resistant soybean genotypes as well as reisolated *P. sojae*, which was then resistant to both antibiotics. This technique was repeated several times, as several laboratories subsequently crossed different strains of *P. sojae* to characterize the expression of avirulence genes. This strategy has been used toward developing maps of the genome as well as for map-based cloning of avirulence loci (MacGregor et al., 2002; May et al., 2002; Tyler et al., 1995; Whisson et al., 1995, 2004). How often outcrossing is occurring in fields within a region and what impact it is having on the generation of new virulence pathotypes remains to be determined.

In addition to the evidence for outcrossing, significant linkage disequilibria ($P<0.05$) were observed for 47.1% of the comparisons between the SSR loci among these *P. sojae* isolates. This may reflect that *P. sojae* was introduced into the United States more than 50 years ago, or it may be related to the clonal nature of this soilborne pathogen. Within species where selfing is the primary means of reproduction, certain "good" combinations of loci across the genome or favorable alleles that have been selected for over time can account for high levels of linkage disequilibrium within populations (McDonald and Linde, 2002). It will be interesting to observe as more populations around the world are evaluated with this same set of SSRs whether there is significant linkage disequilibrium across all of these loci.

Mutation (Förster et al., 1994) and mitotic gene conversion (Chamnanpunt et al., 2001) have also been proposed as potential mechanisms for genetic change in the avirulence loci. It has been well documented through many studies, and most laboratories will agree, that *P. sojae* virulence pathotypes are very unstable in culture (Long et al., 1975; Rutherford et al., 1985; Schmitthenner et al., 1994). The same isolate can have a different response (resistant or susceptible) in successive generations from single zoospore or oospore cultures on one to nine and one or two differentials, respectively (Jia, 2003). In one generation, five resistance genes may confer resistance, whereas in the next generation they may not, and the plants are susceptible. These changes in pathotype can develop both through mutations or transcriptional control of the avirulence gene (Shan et al., 2004). Chamnanpunt et al. (2001) proposed that with a high frequency of gene conversions, this would result in a large reservoir of diversity within each individual strain. Thus, it could respond more easily to environmental stress or selection pressures.

This study in Ohio was completed on a small sample set (33 isolates). To assess accurately the clonal nature of *P. sojae* within fields and the role played by outcrossing or mutation, a larger dataset with more isolates is needed. This work is currently in progress. These types of studies will help to determine whether alternative management strategies should be incorporated into the current management scheme. As soybean regions around the world continue to focus on high-yielding types instead of locally adapted soybean genotypes, and if individual *P. sojae* populations continue to mature and diversify with regard to pathotype, management strategies will need to be modified. Several key questions should be addressed. Is more effort required for deploying stacked *Rps* genes in soybean cultivars? Should there be regional deployment of *Rps* genes to allow for rotation of *Rps* genes as is currently done in wheat for wheat leaf rust? Should more effort be made to supply data continually on new pathotypes and response of putative new sources of resistance to regional *P. sojae* populations? Should efforts to incorporate *Rps* genes be abandoned and companies urged to focus on quantitatively inherited resistance?

10.7 CONCLUSION

It also must be recognized that the forces that drive genetic diversity with each of these unique populations across these numerous soybean production regions around the world could in fact be different. For example, in Australia where several studies on the population have occurred and *P. sojae* can only be found in a limited region, there is strong evidence that this was introduced as a few clones, and that mutation is the force that continues to drive genetic change. In contrast, the Quebec population is interesting in that soybeans have only recently been produced in this region, but the pathotype diversity reported is much greater than in any other region. Also, many *Rps* genes that are effective in other areas are not effective against this population. In addition, a new study from China reports variability within the ITS region among isolates from China, as well as a high level of polymorphism among isolates from two regions with ISSRs. This study raises several interesting issues. Now with an efficient molecular marker system in place to assess diversity, more intensive regional comparisons can be made without redeveloping a new set of markers for each study. Larger studies with more isolates sampled more intensively within regions should and can now take place.

ACKNOWLEDGMENTS

A. D. would like to express deep appreciation to the many colleagues that have provided access to unpublished manuscripts and to the great collegiality in the North Central Region *Phytophthora* Working Group. We would also like to recognize Sue Ann Berry, who has single-handedly killed acres of soybeans through assessing pathotypes of Ohio's *P. sojae* populations, as well as A. deSilva, Karl Bluemel, Peerapat Roongsattham, Diana Garnica, Silvia Restrepo, Brett Tyler, Sucheta Tripathy, and Maria Andrea Ortega, who have contributed to these studies. Data for this study were funded through the soybean check-off from the Ohio Soybean Council and the North Central Soybean Research Program.

REFERENCES

Abney TS, Melgar JC, Richards TL, Scott DH, Grogan J, Young J 1997. New races of *Phytophthora sojae* with *Rps*1-d virulence. Plant Dis 81:653–655.

Barreto D, Stegman de Gurfinkel B, Fortugno C 1995. Races of *Phytophthora sojae* in Argentina and reaction of soybean cultivars. Plant Dis 79:599–600.

Bhat RG, Schmitthenner AF 1993. Genetic crosses between physiological races of *Phytophthora sojae*. Exp Mycol 17:122–129.

Chamnanpunt J, Shan WX, Tyler BM 2001. High frequency mitotic gene conversion in genetic hybrids of the oomycete *Phytophthora sojae*. Proc Nat Acad Sci, USA 98:14530–14535.

Cooke DEL, Drenth A, Duncan JM, Wagels G, Brasier CM 2000. A molecular phylogeny of Phytophthora and related Oomycetes. Fungal Genet Biol 30:17–32.

Demirbas A, Rector BG, Lohnes DG, Fioritto RJ, Graef GL, Cregan PB, Shoemaker RC, Specht JE 2001. Simple sequence repeat markers linked to the soybean *Rps* genes for Phytophthora resistance. Crop Sci 41:1220–1227.

Diers BW, Mansur L, Imsande J, Shoemaker RC 1992. Mapping *Phytophthora* resistance loci in soybean with restriction fragment polymorphism markers. Crop Sci 32:377–383.

Dorrance AE, Jia H, Abney TS 2004. Evaluation of soybean differentials for their interaction with *Phytophthora sojae*. Plant Health Progress.

Dorrance AE, McClure SA, de Silva A 2003. Pathogenic diversity of *Phytophthora sojae* in Ohio soybean fields. Plant Dis 87:139–146.

Dorrance AE, Mills D, Robertson AE, Draper MA, Giesler L, Tenuta A 2007. Phytophthora root and stem rot of soybean. The Plant Health Instructor.

Dorrance AE, Schmitthenner AF 2000. New sources of resistance to *Phytophthora sojae* in the soybean plant introductions. Plant Dis 84:1303–1308.

Drechsler C 1931. A crown rot of hollyhocks caused by *Phytophthora megasperma* n. sp. J Wash Acad Sci 21:513–526.

Drenth A, Whisson SC, Maclean DJ, Irwin JAG, Obst NR, Ryley MJ 1996. The evolution of races of *Phytophthora sojae* in Australia. Phytopathol 86:163–169.

Fang DQ, Roose ML 1997. Identification of closely related citrus cultivars with inter-simple sequence repeat markers. Theoret Appl Genet 95:408–417.

Förster H, Kinscherf TG, Leong SA, Maxwell DP 1989. Restriction fragment length polymorphisms of the mitochondrial DNA of *Phytophthora megasperma* isolated from soybean, alfalfa, and fruit trees. Can J Bot 67:529–537.

Förster H, Tyler BM, Coffey MD 1994. *Phytophthora sojae* races have arisen by clonal evolution and by rare outcrosses. Molec Plant Microte Interat 7:780–791.

Galley M, Ramos AM, Dokmetzian D, Lopez SE 2007. Genetic variability of *Phytophthora sojae* isolates from Argentina. Mycologia 99:877–883.

Garmaroodi HS, Mirabolfathy M, Babai H, Zeinali H 2007. Physiological races of *Phytophthora sojae* in Iran and race-specific reaction of some soybean cultivars. J Agricultural Sci Technol 9:243–249.

Garnica DP 2007. *Microsatélites en Phytophthora spp.: exploración in silico — evaluación in vivo*. Laureated Master thesis. Universidad de los Andes, Bogotá, Colombia.

Garnica DP, Pinzón AM, Quesada-Ocampo LM, Bernal AJ, Barreto E., Grünwald NJ, Restrepo S 2006. Survey and analysis of microsatellites from transcript sequences in Phytophthora species: frequency, distribution, and potential as markers for the genus. BMC Genom 7:245.

Goodwin SB 1997. The population genetics of Phytophthora. Phytopathol 87:462–473.

Gordon SG, St. Martin SK, Dorrance AE 2004. Mapping *Rps*8 a gene for resistance to Phytophthora root and stem rot in soybean. In: crop science society of America annual meeting, ASA-CSA-SS, Madison, WI. pp. 320.

Grau CR, Dorrance AE, Bond J, Russin JS 2004. Fungal diseases. In: Boerma HR, Specht JE, editors. Soybeans: improvement, production, and uses, 3rd ed. pp. 679–763. Agronomy Monograph no. 16.

Hansen EM, Maxwell DP 1991. Species of the *Phytophthora megasperma* complex. Mycologia 83:376–381.

Jackson TA, Kirkpatrick TL, Rupe JC 2004. Races of *Phytophthora sojae* in Arkansas soybean fields and their effects on commonly grown soybean cultivars. Plant Dis 88:345–351.

Jia H 2003. An evaluation of protocols for stabilizing pathotypes of *Phytophthora sojae* isolates and characterizing race-specific resistant genes in soybean plant introductions. M.S. Thesis. Ohio State University Columbus, Ohio. pp. 78.

Kaufmann MJ, Gerdemann JW 1958. Root and stem rot of soybean caused by *Phytophthora sojae* n. sp. Phytopathology 48:201–208.

Layton AC, Kuhn DN 1990. *In planta* formation of heterokaryons of *Phytophthora megasperma* f. sp. *glycinea*. Phytopathol 80:602–606.

Lohnes DG, Schmitthenner AF 1997. Position of the Phytophthora resistance gene *Rps*7 on the soybean molecular map. Crop Sci 37:555–556.

Long M, Keen NT, Ribeiro OK 1975. *Phytophthora megasperma* var. *sojae*: Development of wild-type strains for genetic research. Phytopathol 65:592–597.

MacGregor T, Bhattacharyya M, Tyler B, Bhat R, Schmitthenner AF, Gijzen M 2002. Genetic and physical mapping of *Avr1a* in *Phytophthora sojae*. Genet 160:949–959.

Malvick DK, Grunden E 2004. Traits of soybean-infecting *Phytophthora* populations from Illinois agricultural fields. Plant Dis 88:1139–1145.

May KJ, Whisson SC, Zwart RS, Searle IR, Irwin JAG, Maclean DJ, Carroll BJ, Drenth A 2002. Inheritance and mapping of eleven avirulence genes in *Phytophthora sojae*. Fungal Genet Biol 37:1–12.

McDonald BA, Linde C 2002. Pathogen population genetics, evolutionary potential, and durable resistance. Ann Rev Phytopathol 40:349–379.

Meng XQ, Shoemaker RC, Yang XB 1999. Analysis of pathogenicity and genetic variation among *Phytophthora sojae* isolates using RAPD. Mycol Res. 103:173–178.

Mohammadi A, Alizadeh A, Mirabolfathi M, Safaie N 2007. Changes in racial composition of *Phytophthora sojae* in Iran between 1998 and 2005. J Plant Protection Res 47:29–33.

Nelson BC, Mallik I, McEwen D, Christianson T 2008. Pathotypes, distribution and metalaxyl sensitivity of *Phytophthora sojae* in North Dakota. Plant Dis 92:1062–1066.

Nygaard SL, Elliott CK, Cannon SJ, Maxwell DP 1989. Isozyme variability among isolates of *Phytophthora megasperma*. Phytopathol 79:773–780.

Rutherford FS, Ward EWB, Buzzell RI 1985. Variation in virulence in successive single-zoospore propagations of *Phytophthora megasperma* f.sp. *glycinea*. Phytopathol 75:371–374.

Ryley MJ, Obst NR, Irwin JAG, Drenth A 1988. Changes in racial composition of *Phytophthora sojae* in Australia between 1979 and 1996. Plant Dis 82:1048–1054.

Schmitthenner AF 1985. Problems and progress in control of Phytophthora root rot of soybean. Plant Dis 69:362–368.

Schmitthenner AF, Hobe M, Bhat RG 1994. *Phytophthora sojae* races in Ohio over a 10-year interval. Plant Dis 78:269–276.

Shan W, Cao M, Leung D, Tyler B 2004. The *Avr1b* locus of *Phytophthora sojae* encodes an elicitor and a regulator required for avirulence on soybean plants carrying resistance gene *Rps*1b. Molecular Plant Microbe Interact 17:394–403.

Sugimoto T, Yoshida S, Aino M, Watanabe K, Shiwaku K, Sugimoto M 2006. Race distribution of *Phytophthora sojae* on soybean in Hyogo, Japan. J Gen Plant Pathol 72:92–97.

Tyler BM 2007. *Phytophthora sojae*: Root rot pathogen of soybean and model oomycete. Molec Plant Pathol 8:1–8.

Tyler BM, Förster H, Coffey MD 1995. Inheritance of avirulence factors and restriction fragment length polymorphism markers in outcrosses of the oomycete *Phytophthora sojae*. Molec Plant Microbe Interact 8:515–523.

Tyler BM, Tripathy S, Zhang X, Dehal P, Jiang RHY, Aerts A, Arredondo FD, Baxter L, Bensasson D, Beynon JL, et al. 2006. Phytophthora genome sequences uncover evolutionary origins and mechanisms of pathogenesis. Science 313:1261–1266.

Vallone S, Botta G, Ploper D, Grijalba P, Gally M, Barreto D, Perez B 1999. Incidencia de *Phytophthora sojae* en cultivos de soja en las regiones Pampeana Norte y Noroccidental de Argentina. In: Actas Mercosja 99. Rosario, Argentina. June 1:21–23.

Wang, Z, Wang Y, Zhang Z, Zheng X 2007. Genetic relationships among Chinese and American isolates of *Phytophthora sojae* by ISSR markers. Biodivers Sci 15: 215–223.

Weng C, Yu K, Anderson TR, Poysa V 2001. Mapping genes conferring resistance to Phytophthora root rot of soybean, *Rps*1a and *Rps*7. J Hered 92:442–446.

Whisson SC, Basnayake S, Maclean DJ, Irwin JAG, Drenth A 2004. *Phytophthora sojae* avirulence genes Avr4 and Avr6 are located in a 24kb, recombination rich region of genomic DNA. Fungal Genet Biol 41:62–74.

Whisson SC, Drenth A, Maclean DJ, Irwin JAG 1994. Evidence for outcrossing in *Phytophthora sojae* and linkage of a DNA marker to two avirulence genes. Curr Genet 27:77–82.

Whisson SC, Drenth A, Maclean DJ, Irwin JAG 1995. *Phytophthora sojae* avirulence genes, RAPD, and RFLP markers used to construct a detailed genetic linkage map. Molec Plant Microbe Interact 6:998–995.

Wrather JA, Koenning SR 2006. Estimates of disease effects on soybean yields in the United States 2003 to 2005. J Nematol 38:173–179.

Wrather JA, Koenning SR, Anderson TR 2003. Effect of diseases on soybean yields in the United States and Ontario (1999–2002). Plant Health Progress.

Xu P, Han Y, Wu J, Lv H, Qiu L, Chang R, Jin L, Wang J, Yu A, Chen C, et al. 2007. Phylogenetic analysis of the sequences of rDNA internal transcribed spacer (ITS) of *Phytophthora sojae*. J Genetics Genomics 34:180–188.

Xu X, Lü H, Qu J, Yang Q 2003. *Phytophthora sojae* races in northeast of China and virulence evaluation of the isolates. J Northeast Agricult University 10:97–100.

Yang XB, Ruff RL, Meng XQ, Workneh F 1996. Races of *Phytophthora sojae* in Iowa soybean fields. Plant Dis 80:1418–1420.

Yeh FC, Boyle TB 1997. Population genetic analysis of co-dominant and dominant markers and quantitative traits. Belgian J Bot 129:157.

Zhang X, Schuering C, Tripathy S, Xu Z, Wu C, Ko A, Ken Tian S, Arredondo F, Lee M, Santos FA, et al. 2006. An integrated BAC and genome sequence physical map of *Phytophthora sojae*. Molec Plant Microbe Interact 12:1302–1310.

Zhu Z, Wang H, Wang X, Chang R, Wu X 2004. Distribution and virulence diversity *Phytophthora sojae* in China. Agricult Sci China 3:116–123.

11

PYTHIUM GENETICS

FRANK MARTIN
USDA-ARS, Salinas, California

11.1 INTRODUCTION

When viewed in comparison with its sister taxa *Phytophthora*, a limited amount of information is available on the genetics of *Pythium* spp. Perhaps the primary reason for this is the different types of plant diseases these two genera cause, which in turn has driven the research effort. *Phytophthora* spp. have been more associated with lethal diseases responsible for significant epidemics, such as late blight caused by *Phytophthora infestans*, jarrah die back in Australia caused by *Phytophthora cinnamomi*, or the more recent impact of *Phytophthora ramorum* on coastal California forest ecosystems. In contrast, whereas *Pythium* spp. can cause pre-emergence damping off, they are also associated with root pruning that causes reduced vigor of more mature plants. Although this leads to reduction in plant growth and subsequent yield and requires implementation of control measures, these pathogens generally are not lethal to mature plants. Hence, the observer may not be aware of the presence of the pathogen even though significant reductions in yield may occur. Given the number of species in the genus, their broad host range and distribution, and the lack of defined visible symptoms of disease, it is likely that the impact of *Pythium* spp. on causing yield reductions is underreported and not fully appreciated.

Estimates of the number of species in the genus *Pythium* have been as high as 120, although based on the similarity of morphological features and internal spacer (ITS) data, some of these may be conspecific (Lévesque and de Cock, 2004). Members of the genus occupy ecological niches that range from facultative plant pathogens to soil saprophytes in terrestrial and fresh water or marine aquatic environments. Based on *in vitro* studies, several species have

Oomycete Genetics and Genomics: Diversity, Interactions, and Research Tools
Edited by Kurt Lamour and Sophien Kamoun
Copyright © 2009 John Wiley & Sons, Inc.

been reported as mycoparasites and been evaluated as biological control agents for managing plant pathogenic *Pythium* spp. (such as *Pythium oligandrum* and *Pythium nunn*). One species (*Pythium insidiosum*) is also a mammalian pathogen that causes a disease referred to as pythiosis. Species such as *Pythium ultimum* have been evaluated for the production of polyunsaturated fatty acids for use as a human dietary supplement (Gandhi and Weete, 1991). Although some species can have a somewhat narrow host range, others such as *Pythium aphanidermatum* or *P. ultimum* have been reported to have a broad host range and cause significant losses on a wide range of economic crop plants. For a broader overview on the ecology and biology of the genus, see Martin and Loper (1999); for an overview of host–pathogen interactions, see Martin (1995d); for taxonomic descriptions and keys, see van der Plaats-Niterink (1981) or Dick (1990); and for phylogenetic relationships, see Lévesque and de Cock (2004).

11.2 TAXONOMY AND PHYLOGENY

The morphological classification of species is based on features such as sporangial shape and size, oogonia size and ornamentation, the presence of space between the oogonia and oospore wall, and antheridial numbers and mode of attachment (Van der Plaats-Niterink, 1981). With the large number of species and the range in values for morphological features that can be found among isolates of the same species, the taxonomic classification of some species based on morphology alone can be a daunting endeavor. Molecular criteria, such as restriction fragment length polymorphism (RFLP) analysis of the ITS region of the rDNA (Chen, 1992; Chen et al., 1992; Wang and White, 1997; Kageyama et al., 2005; Rafin et al., 1995), the mitochondrially encoded *cox2* gene (Kageyama et al., 2005; F. Martin, unpublished), total mitochondrial DNA (Martin and Kistler, 1990; Martin, 1990a), and SSCP analysis of the ITS region (Kong et al., 2004), as well as sequence analysis of the ITS region (Matsumoto et al., 1999; Lévesque and Cock, 2004) and the *cox2* gene (Martin, 2000), have helped to simplify identification of cultures to a species level. The use of these molecular criteria are especially important for those isolates that do not form oospores, and hence they cannot be morphologically classified to a species level (Martin, 1990a; Kageyama et al., 1998, 2003; Rafin et al., 1995). Currently, there is an effort to sequence the *cox1* gene to include the genus *Pythium* in the Barcode of Life initiative, which should provide another molecular tool for isolate identification (Lévesque, personal communication).

Although taxonomic criteria such as sporangial shape correlate with phylogenetic grouping in a broader sense, several other features do not. Using a limited number of species, Briard et al. (1995) reported that grouping of species based on 28S rDNA sequence data correlated with sporangial morphology with two main clades differentiating species with filamentous/inflated and spherical/globose sporangia. Matsumoto et al. (1999) confirmed this finding with sequence analysis of the ITS region for a larger number of species; he also observed that features such as oogonial ornamentation and heterothallism were polyphyletic. Similar results were also observed for analysis performed with the mitochondrially encoded *cox2* gene with *P. oligandrum* placed intermediate between the two main clades representing sporangial morphology (Martin, 2000). This analysis also concluded that mycoparasitism and the presence of predominantly linear mitochondrial chromosomes were polyphyletic. The most comprehensive phylogenetic analysis of the genus published to date included 116 described species and was based on the ITS region with portions of the adjacent large ribosomal subunit sequenced for some species (Lévesque and de Cock, 2004). This analysis confirmed prior observations on phylogenetic relationships using a much broader number of species [many of which were morphologically characterized for the monograph by Van der Plaats-Niterink (1981)]. One notable observation of this work was the more comprehensive analysis that placed species with contiguous sporangia (*Pythium acanthicum*, *Pythium periplocum*, and *P. oligandrum*) intermediate between the filamentous/lobulate and globose/spherical sporangia clades. More recent work with *Pythium* and *Phytophthora* by Villa et al. (2006) using combined datasets for the ITS, β-tubulin and *cox2* gene identified several *Pythium* spp. that clustered independent of the 2 major sporangial morphology clades for this genus. In the ITS phylogeny they were also on the same clade as *Phytophthora* spp. (this grouping was also observed in the ITS phylogeny of Lévesque and de Cock, 2004). This observation led the authors to suggest these species may be the intermediates in the evolution from *Pythium* to *Phytophthora*. A new genus for these species has been suggested by Lévesque, Robideau, Abad, and de Cock (Lévesque, personal communication).

11.3 SEXUAL REPRODUCTION

Most species in the genus are homothallic and can form oospores in single culture; however, there are seven species that are heterothallic (*Pythium catenulatum, Pythium flevoense, Pythium heterothallicum, Pythium intermedium, Pythium macrosporum, Pythium splendens*, and *Pythium sylvaticum*). Generally, these species need opposite mating types for oospore formation; however, exceptions to this have been reported for some of these species. Homothallic isolates have been observed for *P. catenulatum* (Hendrix and Campbell, 1969) and *P. sylvaticum* (Campbell and Hendrix, 1967). Van der Plaats-Niterink (1968) reported a low level of selfing among European field isolates of *P. sylvaticum*, and these were predominantly oogonial isolates. Homothallism in this species was explored in greater detail by Pratt and Green (1971, 1973), who reported that homothallism was commonly encountered with isolates from the United States but more so among antheridial than oogonial isolates. Pairings of opposite mating types in various combinations revealed variation in the strength of sexual reactions depending on the parental isolates.

Some pairings resulted in a weak line of oogonia formation at the point of contact between the two cultures, whereas other pairings resulted in heavy oogonia production. Differential levels of oospore maturation were also observed, but this did not always correlate with the strength of sexual response. Some of the progeny from crosses between opposite mating types were homothallic; this characteristic exhibited variation when hyphal tip cultures were selected from these homothallic isolates and evaluated for their ability to self (Pratt and Green, 1973). From experimentation using a polycarbonate membrane to separate opposite mating types, Gall and Elliott (1985) reported that *P. sylvaticum* responded to hormonal regulation of sexual reproduction in a similar manner as *Achlya* spp. When the hyphae of opposite mating types start to approach each other, the oogonial hyphae produced a hormone that stimulated the antheridial hyphae to start to form antheridia, which in turn produce a hormone that stimulated the formation of oogonia on the maternal parent hyphae. When isolates are paired with a polycarbonate membrane in between and the oogonial isolate was removed after stimulation of antheridial formation, it was concluded that the antheridial isolate was capable of producing oogonia (unfortunately the experimentation was not continued long enough to determine whether these matured into viable oospores).

Similar results were observed when opposite mating types of *P. splendens* were paired on either side of a polycarbonate membrane; the oogonial isolate was stimulated to self, and there was a range in the number of oospores produced depending on the isolates that were paired (Guo and Ko, 1991). Following long-term storage and recovery of cultures from single hyphal swellings, mating type changes were observed; one oogonial isolate was found to produce homothallic sectors on culture medium. Although most oospores from this sector gave rise to progeny with an oogonial mating type, a few isolates were antheridial or capable of stimulating outcrossing with either oogonial or antheridial isolates. Interestingly, single-hyphal-swelling isolates of this later category segregated into strictly oogonial or antheridial heterothallic isolates.

Although homothallic species can self in single culture, data from *in vitro* crosses with *P. ultimum* (Francis and St-Clair, 1993) and *Pythium irregulare* (Harvey et al., 2001) indicates that outcrossing between isolates of the same species can also occur. The results from examining field populations supports that outcrossing is occurring in nature. Francis et al. (1994) observed field isolates of *P. ultimum* that were heterozygous for specific molecular loci, and these traits segregated in expected Mendelian ratios when isolates were selfed. One isolate examined had three alleles at one locus, which suggests that polyploidy, aneuploidy, or heterokaryosis had occurred. Harvey et al. (2000, 2001) observed similar results with field isolates of *P. irregulare* from Australia; there were heterozygous loci that segregated when an isolate was selfed. These observations can have important implications for the genus as a whole, as the ability of homothallic isolates to outcross provides a mechanism for the generation of additional genetic variation within a population. As will be discussed below, it is believed that intraspecific outcrossing of homothallic species is responsible for the generation of intraspecific polymorphisms in electrophoretic karyotype (EK).

As noted, there are isolates of heterothallic species that can form oospores in single culture, but there are also isolates of homothallic species that are no longer able to self. Van der Plaats-Niterink (1981) placed isolates that were incapable of sexual reproduction into specific groups based on the morphology of the sporangia/hyphal swelling, but whereas these groups may share similarities in the morphology of the asexual reproductive structures, they do no necessarily represent phylogenetic groupings. For example, hyphal swelling isolates of *P. ultimum* incapable of selfing have been identified (Saunders and Hancock, 1994; Martin, 1990a; Kageyama et al., 1998) and Francis and St. Clair (1993) demonstrated they are capable of outcrossing with another *P. ultimum* isolate. Based on morphology alone, these isolates cannot be separated from other isolates that are phylogenetically distinct based on mitochondrial gene sequence data (Martin, unpublished). Likewise, Vasseur et al. (2005) identified some group F isolates (filamentous sporangia, no oospores) that, based on identical ITS sequence data, could be classified as the homothallic species *Pythium dissotocum* and Davison et al. (2003) identified group F isolates that were *Pythium sulcatum*.

Interspecific hybrids similar to what has been found with *Phytophthora alni* have not been reported in *Pythium*, but a recent report on *Pythium mercuriale* sp. nov. may be indicative of this happening (Belbahri et al., 2008). From sequencing cloned ITS amplicons, it was found that each individual isolate had as many as three to five sequences that differed not only in single nucleotide polymorphisms (SNPs) but also in insertion/deletion events. Phylogenetic analysis with these sequences revealed a placement in clade K of Lévesque and de Cock (2004) intermediate between other *Pythium* species and *Phytophthora*, with no significant identity with other species in the genus, which makes it difficult to identify the "parental" isolates if it represents an interspecific hybrid. Heterozygosity was also observed for the nuclear encoded genes elongation factor 1α and β-tubulin. Surprisingly, heterozygosity was also observed for the mitochondrially encoded *nadh*1 gene with five forms in one isolate that differed by a total of seven SNPs. From working with *cox2* gene sequences for several *Pythium* spp., polymorphic forms in a single isolate were not observed, although this may be because of direct sequencing of polymerase chain reaction (PCR) amplified products masking SNPs that were present in low amounts rather than sequencing cloned amplicons (F. Martin, unpublished). The fact that oogonia and antheridia are rarely formed by *P. mercuriale* and that mature oospore are not observed may be indicative of some level of incompatibility in the sexual reproductive cycle.

Unlike *Phytophthora*, the individual mating types in *Pythium* are primarily sexually dimorphic; they function as either the oogonial (maternal) or antheridial (paternal) parent (Papa et al., 1967; Pratt and Green, 1971). However, some isolates of *P. sylvaticum* have been reported that could function as either oogonial or antheridial depending on the isolates they were paired with (Pratt and Green, 1971; Gall and Elliott, 1985). From analysis of outcrossing with *P. sylvaticum*, the maternal parent contributes the mitochondrial DNA to progeny (Martin, 1989).

11.4 GENETIC ANALYSIS

Genetic analysis of oospore progeny for inheritance of specific traits has received limited attention in the genus. In a study to clarify whether the life cycle of *Pythium* was diploid, Dennett and Stanghellini (1977) induced a mutation with ethyl methansulfonate in an isolate of *P. aphanidermatum* for chloramphenicol resistance and observed a 1:5.83 segregation of resistance in single oospore progeny. Guo and Ko (1994, 1995) also examined segregation of growth rate and drug tolerance in progeny of a homothallic isolate of *P. splendens* and found they were invariant in isolations of single hyphal swellings. However, the single-oospore progeny did exhibit variation in these phenotypes. Some single-oospore cultures did not exhibit wild-type colony morphology, and when single hyphal swellings recovered from these isolates were examined, they also exhibited variation in growth rate and drug tolerance. This mitotic variation was observed following three successive generations of cultures from single hyphal swellings, which led the authors to suggest this process could be an evolutionary advantage by providing a mechanism for the generation of variation in a species that generally requires opposite mating types for sexual reproduction.

Insight on the processes that may be involved in the generation of this variation in *P. splendens* as well as addressing the intraspecific variation in EK for the genus as a whole (Martin, 1995a) can be gleaned from some of the work that has been done with crosses of *P. sylvaticum*. Strictly heterothallic parental isolates were crossed, and single-oospore hybrid progeny was examined for marker inheritance and EK (Martin, 1995c). Marker inheritance confirmed the hybrid nature of the progeny. Whereas most inheritance ratios fulfilled Mendelian expectations, several did not. When these isolates were examined by pulsed field gel electrophoresis (PFGE), all the progeny EKs were different from each other and from the parental isolates. Coupled with Southern analysis using individual genes, 80% of the progeny chromosome-sized bands were found to be nonparental in size or grouping of specific coding regions. It was believed that length mutations, translocations, and variation in ploidy were responsible for the observed karyotypic variation. Although summation of chromosomal-sized bands revealed totals that ranged from 29.1 to 39 Mb, this method is not accurate for determining genome size, as it is not known how many chromosomal-sized bands are represented by brighter staining bands, and errors will be introduced by estimation of individual band sizes from size standards. However, flow cytometry measurements of isolated nuclei confirmed a 10% variation in DNA content among different progeny isolates

(F. Martin, unpublished). One interesting observation from the flow cytometry analysis was that one isolate had a second peak with a 43% higher relative DNA content. This was consistently observed even after multiple hyphal tips were subcultured to make sure the culture originated from a single isolate. This double peak is believed to represent mitotic instability and variation in ploidy that started in one nucleus, which led to the generation of a stable heterokaryon. When crossing various isolates of *P. sylvaticum*, it is common to observe different levels of spontaneous abortion of oogonia as well as abortive germinants (the oospore germinates but stops growing shortly thereafter); perhaps differences in ploidy or EK between parental isolates preventing proper chromosomal pairing in the fertilized oogonium is responsible for this (Martin, 1995c).

One interesting feature of the progeny of the above described outcross of strictly heterothallic isolates of *P. sylvaticum* was that, depending on the cross analyzed, between 5% and 10% were homothallic. One of these isolates was selected and selfed with single-oospore progeny analyzed by the same approaches that were used for the heterothallic crosses (Gavino, 1994). Although there was a low rate of oospore germination, enough progeny were examined to demonstrate that similar events occurred with these selfed progeny as were observed with the outcrossed progeny. There was non-Mendelian inheritance for some markers and, based on EKs and Southern analysis, 65% of the chromosomal-sized bands were nonparental for size or grouping of coding regions with differences attributed to length mutations, translocations, and variation in ploidy. Similar behavior of selfed progeny was observed for hybrid isolates of *P. ultimum* obtained by Francis and St. Clair (1993) from *in vitro* crosses (Francis and Martin, unpublished). The EK of the selfed progeny were nonparental for the numbers and sizes of chromosomal-sized bands, and non-Mendelian inheritance of some markers was observed. This variation in progeny EK was observed following successive selfings for several generations with all selfed progeny exhibiting EK polymorphisms compared with its parental isolate. So the occurrence of one outcrossing was responsible for the generation of EK polymorphisms in all selfed progeny in the successive generations that were examined.

Selfing of field isolates of some homothallic species also was found to generate variations in EK. For example, with *P. aphanidermatum* (a species with mono- and diclinous antheridia), all progeny had the same EK as the parental isolate, whereas for one isolate of *P. oligandrum* (a species in which antheridia are rarely observed), only 1 of 28 progeny exhibited any polymorphisms in EK, which was the loss of a putative 0.95 Mb supernumerary chromosome encoding rDNA (Martin, 1995a). But selfings of two additional isolates revealed size variation for a chromosomal-sized band less than 1 Mb in size in all progeny isolates. A few progeny exhibited polymorphisms in bands between 1.4 and 1.6 Mb in size, or in the largest chromosomal-sized band that also encoded rDNA (ca. 3.6 Mb; Martin, unpublished). In contrast, for *Pythium spinosum* (a species with monoclinous and diclinous antheridia),

61% of the single-oospore progeny had a polymorphic EK for at least one chromosomal-sized band. The isolate used in this analysis was the same that also exhibited heterozygosity in the rDNA (Martin, 1990b), which was interpreted to mean that outcrossing had occurred at some point in the parental lineage.

To understand fully the contribution of meiotic instability on the generation of intraspecific variation, it is important to know the contribution of mitotic instability as well. Although infrequently published, from personal experience and conversations with colleagues it is not unheard of for species such a *Pythium graminicola* or *Pythium myriotylum* to lose the ability to discharge zoospores following repeated culture transfers over time. Some species can also lose the ability to sporulate when kept in cornmeal agar/mineral oil slants for extended periods of time. The mitotic stability of *P. sylvaticum* and *P. oligandrum* was evaluated by transferring cultures on a nutrient-poor (water agar) and -rich medium (PDA) on a twice-weekly basis for 22 weeks (Martin, unpublished). Although changes in EK were not observed for *P. sylvaticum*, for one subcultured isolate of *P. oligandrum* size polymorphisms were observed in the 0.97-Mb chromosomal-sized band, and another isolate had an increase in the size of the largest chromosomal sized band (ca. 3.6 Mb), both of which encoded the rDNA. Similar polymorphisms were observed in single-zoospore isolates as well. The results of restriction digests with rare cutting enzymes that did not cut within the rDNA repeat and Southern analysis of PFGE gels with rDNA suggested this size polymorphism could have been caused by a variation in the number of rDNA repeat units. The mitotic instability in the number of rDNA repeats had been previously been reported in eumycotan fungi such as *Neurospora crassa* (Butler, 1992; Butler and Metzenberg, 1992) and *Coprinus cinerea* (Pukkila and Skzynia, 1993). A hyphal swelling isolate of *P. ultimum* was also found to be mitotically unstable during subculturing, as it lost specific chromosomal-sized bands and RFLP makers (cited in Francis and St. Clair, 1997). Based on the EK and the presence of specific RFLP markers, it was believed that at the beginning of the experimentation this isolate was heterozygous and had a higher ploidy than other isolates.

11.5 POPULATION BIOLOGY

An examination of several isolates of an individual species can often identify variation in phenotypic traits, such as growth rate, the ability to form sporangia, discharge zoospores, the frequency of oogonial abortion or formation of mature oospores, and the frequency of oospore germination. Depending on the species, intraspecific variation in specific physiological or molecular criteria can be found as well. For example, using nine isozyme loci, Barr et al. (1996) separated 97 isolates of *P. ultimum* into 10 multilocus genotypes. An analysis of the *cox2* gene for this species complex clearly separated *P. ultimum* var *ultimum* from *P. ultimum* var. *sporangiiferum* and placed many hyphal swelling (HS) isolates in with *P. ultimum* var. *ultimum* (Martin, 2000). However, some HS isolates formed a distinct, closely affiliated clade, and an isolate of *P. ultimum* var. *sporangiiferum* clustered separately from other isolates of this species. Although Francis et al. (1994) observed a differential grouping of isolates representing these three taxonomic groupings based on RAPD analysis, the differences were minor enough that they concluded they are not genetically distinct. It is possible that more than one species may fit within the morphological taxonomic classification of the species. Barr et al. (1996) reported the isolate that grouped separately from the other *P. ultimum* isolates based on isozymes analysis also had a lower temperature optima. Tojo et al. (1998) reported on similar lower temperature isolates that differed from *P. ultimum* by the size of their oogonia and antheridia as well as grouping following analysis of isozymes and RAPD data. With this in mind, an additional analysis with a larger representative sample of isolates from geographically diverse areas that will allow for the evaluation to be performed at a population level is needed to clarify the taxonomy of this species complex. The presence of heterozygous isozyme loci (Barr et al., 1996) and RFLP markers (Francis et al., 1994) in isolates of *P. ultimum* indicates that intraspecific outcrossing had occurred in field isolates of this homothallic species. The 37% mean genetic distance among a collection of 12 isolates reported by Garzon et al. (2005b) for amplified fragment length polymorphism (AFLP) data supports this finding as well.

Similar results have been observed for *P. irregulare*. Barr et al. (1997) separated isolates into two distinct groups based on 11 isozyme loci. Using ITS sequence data, Matsumoto et al. (2000) identified four groupings with two groups each clustered on one of two clades. A similar ITS grouping of isolates was also observed by Lévesque and de Cock (2004) and by Garzon et al. (2005a), with one of the clusters more closely affiliated with *P. sylvaticum* than the other *P. irregulare* clade. Lévesque and de Cock (2004) also observed a grouping of *Pythium regulare* and *Pythium cylindrosporum* with some isolates of *P. irregulare* in their ITS phylogeny. The *cox2* gene phylogeny also indicated two groupings for isolates morphologically classified as this species (Martin, 2000). A subsequent analysis by Garzon et al. (2007) led to the classification of isolates on the ITS clade most proximal to *P. irregulare sensu stricto* as a new species (*Pythium cryptoirregulare*). It is interesting to note that based on an analysis of AFLP data, Garzon et al. (2005a) concluded that speciation of *P. irregulare* is in progress and that *P. irregulare* and *P. cryptoirregulare* were cryptic species with the potential for limited gene flow between them. Garzon et al. (2005a) also observed a high level of intraspecific variation among isolates of *P. irregulare* with 52% of the AFLP markers polymorphic. Harvey et al. (2000, 2001) had a series of interesting studies on variation in populations of *P. irregulare* in South Australia. A RFLP analysis of 92 isolates collected from monocot hosts in seven locations revealed a high level of genetic variation within the collection ($D_{\Gamma} = 0.502$), and the presence of heterozygous loci and their segregation in selfed progeny led to the conclusion that outcrossing may have been responsible for this. Interestingly, there was significant genetic differentiation among populations, which suggests there was limited gene flow between geographically separated populations with the differences observed believed to be caused by genetic drift (Harvey et al., 2000). Furthermore, although there was some indication that specific genotypes exhibited differential pathogenicity on barley and wheat, there was no firm evidence of host specialization. Using 34 isolates collected from cereals, medic, and subclover, Harvey et al. (2001) observed that isolates grouped more according to host species rather than geographic origin, especially for isolates recovered from medic. Although differences in pathogenicity on the three hosts were observed, this did not always correlate with the host from which the isolate was recovered.

Intraspecific variation in genetic markers has been observed for some other *Pythium* spp. as well. Although isolates of *P. spinosum* from Oman exhibited high levels of genetic similarity (99%) among themselves in AFLP analysis, when compared with isolates from the Netherlands, South Africa, and Japan, the average level of similarity was reduced to 54%. This finding led Al Sa'di et al. (2008b) to conclude that the presence of *P. spinosum* in Oman was the result of a single introduction. High levels of intraspecific variation were also observed for *Pythium porphyrae*, which is the pathogen responsible for disease of the marine plant *Porphyra yezoensis*. Using RAPD markers, Park et al. (2003) observed isolates grouped into three clades with a maximum dissimilarity coefficient of 0.70 (0.24 among isolates from Japan but 0.13 among isolates from Korea). Likewise using RAPD markers, Davison et al. (2003) reported that isolates of *P. sulcatum* recovered from carrots in Australia clustered into two distinct clades with differences in growth rate and mean oospore diameter observed between them. All Australian isolates also differed from the type culture as the oospores tended to be more plerotic. There is evidence that isolates of *P. myriotylum* recovered from cocoyam are distinct from isolates recovered from other hosts, as most cocoyam isolates grouped in a separate AFLP clade and had a single nucleotide polymorphism in the ITS region at a specific base that was correlated with pathogenicity on cocoyam (Perneel et al., 2006). In contrast to these levels of variation, limited intraspecific polymorphisms have been observed for *P. aphanidermatum* using RAPD analysis (Herrero and Klemsdal, 1998). Likewise, AFLP analysis exhibited a limited variation, with Garzon et al. (2005b) reporting a 15% distance among a range of isolates, whereas Al-Sa'di et al. (2008a) observed 96.6% genetic similarity among 89 isolates of *P. aphanidermatum* from Oman and 94.1% similarity in comparisons of these cultures with 20 other isolates collected from other parts of the world. Interestingly, most variation among populations in Oman was attributed to geographic recovery of the isolate, which indicates a limited spread of the pathogen.

A detailed analysis of the population genetics of a species within a single field has not been performed, but there have been several studies that demonstrate the potential for this type of research to be conducted. In addition to studies using RFLP (Francis et al., 1994; Harvey et al., 2000, 2001) and AFLP analysis

(Garzon et al., 2005b) discussed above, several PCR techniques have been reported that should be useful for studies on population genetics. Vasseur et al. (2005) reported on several primers for amplification of intersimple sequence repeats (ISSR) for use with *Pythium* group F isolates that, based on ITS sequence data, seemed to be *P. dissotocum*. These primers generated banding profiles that separated 23 isolates into 11 distinct molecular clusters that were not reflective of geographic location of recovery or host plant from which they were recovered. The ability of these primers to amplify sequences from other species was not reported. Primers for microsatellite analysis of *P. aphanidermatum, P. irregulare*, and *P. cryptoirregulare* have been recently reported and found to be useful for differentiating isolates into specific molecular clusters (Lee and Moorman, 2008). Given that some of them also amplify sequences in other *Pythium* spp. (albeit in some cases this was restricted to phylogenetically related species), they should facilitate additional studies on the population structure of a *Pythium* spp. in the field. With the current effort to sequence the genome of *P. ultimum*, additional markers for investigating populations should be forthcoming.

11.6 TRANSFORMATION

Limited research on molecular transformation has been conducted with *Pythium* spp. Weiland (2003) reported on the transformation of *P. aphanidermatum* using vectors developed for *P. infestans* (Judelson et al., 1991). Electroporation generated stable transformants where the vector DNA was integrated into chromosomal DNA and inherited through zoospore and oospore progeny, although the transformation frequency was relatively low (0.1 to 0.4 transformants from 10^5 protoplasts per microgram DNA). Using vectors developed for *P. infestans*, Horner and van West were able to transform *P. oligandrum* (van West, personal communication). They also were able to express a green fluorescent protein construct in *P. oligandrum* as well as functionally characterize a gene family by creating stably silenced lines using the protocols described by Judelson et al. (1991). Repeated attempts were made at transformation of *P. sylvaticum* by polyethylene glycol (PEG) treatment of protoplasts using a range of vectors that were effective for *P. infestans* (Judelson et al., 1991) without success (Martin, unpublished). During the course of this work, it was found that resistance to the aminoglycoside class of antibiotics commonly used as selective markers for transformants (geneticin and kanamycin) can be generated by sublethal enrichment (Martin and Semer, 1997), so this needs to be taken into account when deciding on the antibiotic concentration to use during protoplast regeneration. Given the greater success of the transformation system with the related genus *Phytophthora*, it is likely that additional work with *Pythium* spp. would yield an improved transformation system. The recent report by McLeod et al. (2008) evaluating transient expression of a reporter gene to identify the optimum vector and transformation protocol for several oomycetes (including *P. aphanidermatum*) should provide valuable information in this regard.

11.7 GENOMICS

Based on the results of PFGE for many *Pythium* species, it seems that the genome can be fairly plastic with regard to the number and size of chromosomes as well as the total genome size. Some aspects of this were previously discussed for the heterothallic species *P. sylvaticum* in the section on genetic analysis. Additional examination of 64 field isolates for nine homothallic species revealed that whereas the general ranges in sizes for chromosomal-sized bands may be consistent among isolates of a species, extensive polymorphisms in the number of bands and their sizes were observed (Martin, 1995a). Flow cytometry measurements of isolated nuclei supported intraspecific differences in total genome sizes as well. For example, there was a 25% variation in DNA content among 11 field isolates of *P. oligandrum* (Martin, unpublished). Although the amount of repetitive DNA has not been determined for a *Pythium* spp., the results from flow cytometry, along with the results from the progeny analysis of the heterothallic species *P. sylvaticum* (Martin, 1995c) and the outcrossed isolates of *P. ultimum* (Francis and Martin, unpublished), indicate that variation in ploidy contributes to differences in genome sizes. From the results of PFGE and Southern analysis of progeny of *P. sylvaticum*, deletions and translocations also may be other mechanisms that contribute to EK polymorphisms.

The sequencing of the nuclear genome of *P. ultimum* is currently in progress, but unfortunately at the time of this writing information is not available to comment on the progress. Completion of this project will provide an important resource for expanding the research that can be performed with the genus. The ability to target specific regions for additional analysis in other isolates should help to clarify the processes that contribute to genome plasticity in this species, especially because outcrossing between isolates of this species has been performed, and hybrid progeny are available (Francis and St. Clair, 1993; Francis et al., 1994). It would be particularly interesting to examine expressed sequence tag (EST) libraries or genomes of other species with different ecologies so comparative genomics could be used to identify genes associated with pathogenicity or stimulating host–defense reactions. For example, an analysis of an isolate of *P. oligandrum* that is capable of biological control could identify not only the pathogenicity genes that this species lacks (or does not express) but also the genes it possesses that are associated with inducing the host–defense reactions that are responsible for disease control. Likewise, an analysis of *P. insidiosum* would help to identify genes associated with mammalian pathogenicity and perhaps lead to more effective disease-management strategies. As of this writing, 14,560 EST sequences are deposited in GenBank, with 9,727 from vegetative hyphae of *P. ultimum* and 4,661 from vegetative hyphae of

P. oligandrum. The numbers for *P. oligandrum* are low in comparison with *P. ultimum*, but they may provide some insight into the differences between these species. Given the genomic sequencing that has been done in the closely related genus *Phytophthora* (*Phytophthora capsici, P. infestans, P. ramorum*, and *Phytophthora sojae*) and a downy mildew (*Hyaloperonospora parasitica*), analysis of the genomic data from *P. ultimum* should provide a broader framework for understanding the importance of specific genes in moderating the host–pathogen interaction as well as genome organization and evolution for the Oomycetes in general.

11.8 VIRULENCE AND HOST RANGE

Many differences in virulence and host range are observed among *Pythium* spp., but an understanding of the mechanisms responsible for these observations and the genetics that control them are unknown. Elucidation of this could lead to the development of more effective pathogen-control approaches. For example, although some *Pythium* spp. like *P. ultimum* and *P. aphanidermatum* have a broad host range and can attack a wide variety of host species, other species such as *Pythium arrhenomanes* and *Pythium graminicola* are more restricted in host range and tend to infect only monocot grasses. The reasons for this more restricted host range are not known, but several suggestions have been made, which include differences in root exudates responsible for zoospore attraction (reviewed in Martin and Loper, 1999), electrotaxis of zoospores (Morris and Gow, 1993; van West et al., 2002), or the ability of the zoospores to encyst and infect the host root (reviewed in Martin, 1995d). Mitchell and Deacon (1986) investigated the ability of zoospores of four species to be attracted to, encyst on, and infect roots of different plant species. Species that tended to infect monocot grasses (*P. arrhenomanes* and *P. graminicola*) were found to be attracted to and encyst on roots of wild grasses more so than dicotyledonous plants. Species with a broad host range (*P. aphanidermatum* and *P. ultimum*) did not exhibit a preferential response to the host. More recently, Raftoyannis and Dick (2006) examined zoospore attraction of 10 *Pythium* spp. to the roots of seven host plants and concluded that whereas there is a differential response among species and host for zoospore accumulation and encystment, there was a limited correlation for only a few species between the density of encysted zoospores and subsequent disease severity. There have been several reports on the potential involvement of pathogen-produced toxins or indole acetic acid (IAAs) on the expression of disease (reviewed in Martin, 1995d), but this work has focused on the use of culture filtrates and has not adequately addressed toxin involvement in determining host range or contributing to virulence. Perhaps an improvement in DNA transformation techniques with *Pythium* spp. or examination of gene expression with microarrays once the *P. ultimum* genome has been sequenced would lead to a closer examination of this possibility.

Another mechanism involved with host range determination may be a differential host response to colonization by a particular *Pythium* species. An elicitor from *P. aphanidermatum* has been found to stimulate the expression of host-defense reactions in cell cultures of *Daucus carota* (Glassgen et al., 1998; Koch et al., 1998) and *Coleus blumeri* (Szabo et al., 1999). A single elicitor protein, PaNie, was subsequently identified and found to stimulate host–defense reactions not only in *D. carota* cell lines but also in intact leaves of *Nicotiana tabacum, Lycopersicon esculentum*, and *Arabidopsis thaliana* (Veit et al., 2001). Interestingly, host–defense responses were not observed with the monocot plants *Zea mays*, *Avena sativa*, or *Tradescantia zebrina*, even though the dicotyledonous plants tested and the first two monocots listed are reported to be susceptible to the pathogen. Both ethylene- and jasmonic-acid-regulated host–defense responses have been found to be involved with susceptibility to infection by some *Pythium* spp. Tobacco lines that had a defective ethylene signal perception component (*Etr1-1*), and hence were not sensitive to ethylene, were susceptible to *P. sylvaticum*, whereas wild-type plants were not (Knoester et al., 1998). This enhanced susceptibility to *Pythium* spp. was not modified by prior plant treatment with chemicals known to induce resistance (Geraats et al., 2007). A similarly modified *Etr1-1 A. thaliana* line was more susceptible to infection by three *Pythium* species, as was another modified *A. thaliana* line (*Ein2*) that is defective in ethylene signal transduction (Geraats et al., 2003). Similarly, a mutant *A. thaliana* line (*fad*) deficient in the production of a jasmonic precursor was more susceptible to infection by *Pythium mastophorum*, whereas this pathogen had limited effect on wild-type plants (Vijayan et al., 1998). Another mutant *A. thaliana* line that was resistant to jasmonic acid (*jar1*) was also more susceptible to infection by *P. irregulare* compared with the wild-type plant (Staswick et al., 1998). In response to wounding or treatment with methyl jasmonate or ethylene, *A. thaliana* produces a 23-aa peptide called *At*PEP1 that activates the expression of several genes associated with host–defense reactions (Huffaker et al., 2006). Generation of *A. thaliana* lines that constitutively express a *At*Pep1 precursor gene PROPEP1 have enhanced resistance to *P. irregulare* (Huffaker et al., 2006), as did mutant plants that constitutively express PROPEP2 (Huffaker and Ryan, 2007). The expression of several host–defense genes by *At*PEP1 was blocked in mutant plants that had nonfunctional jasmonate/ethylene and salicylate pathways (Huffaker and Ryan, 2007).

As noted previously, several *Pythium* spp. can protect plants from pathogenic *Pythium* spp. by colonizing plant roots; perhaps the most examined is *P. oligandrum*. A variety of papers has reported on the efficacy of specific isolates in disease control (Martin and Hancock, 1987; Al Rawahi and Hancock, 1998; McQuilken et al., 1990; Vesely, 1979); some isolates were identified that protected tomato roots from damping off in field evaluations as effectively as metalaxyl treatments (Martin, 1999). Benhamou et al. (1997) observed that prior tomato root colonization by *P. oligandrum* protected from subsequent infection by *Fusarium oxysporum* f. sp. *radicis-lycopersici*. This protection was associated with colonization of the cortical cells of the root without causing extensive damage, and it induced host–defense reactions. Rey et al. (1998) observed that *P. oligandrum* colonized the surface of tomato roots and then subsequently penetrated the cortical regions and grew on through to the vascular stele without causing necrosis of host cells. However, host cells exhibited changes associated with colonization. Le Floch et al. (2005) subsequently conducted a time-course study and reported a rapid colonization throughout the cortex of the tomato roots by *P. oligandrum* with changes in the appearance of the hyphae occurring 9 hours after infection. However, a less extensive level of root colonization was observed for a different isolate of *P. oligandrum* that was as effective in the field as fungicide treatments for protecting tomato from damping off (Martin, unpublished). An *in situ* enzyme-linked immunosorbent assay (ELISA) staining procedure using alkaline phosphatase endogenously produced by *P. oligandrum* was used to visualize the hyphae during colonization *in vitro*. Growth was limited to the first several layers of cortical cells with the hyphal of *P. oligandrum* having a stunted, irregular growth habit in the cells.

Root colonization by *P. oligandrum* can also induce a physiological response by the plant. LeFloch et al. (2005) observed that colonization induced the accumulation of phenolic compounds 3 h after inoculation of tomato roots with phytoalexin production, which started 14 h after colonization. There are several potential *P. oligandrum*-produced products that may be responsible for stimulating the host–defense reaction. Picard et al. (2000) identified an elicitin, which is called oligandrin, that can induce resistance in tomato plants to infection by *Phytophthora parasitica*. Subsequently, it has been found to induce resistance in tomato to *F. oxysporum* f.sp. *radicis-lycopersici* (Benhamou et al., 2001), tobacco to *P. parasitica*, and a phytoplasma (Lherminier et al., 2003) and grape to *Botrytis cinerea* (Mohamed et al., 2007). The effect on the host has been an alteration of physical attributes of the host cells as well as changes in regulation of specific genes (Picard et al., 2000; Benhamou et al., 2001; Lherminier et al., 2003; Mohamed et al., 2007). Elicitins are commonly encountered in *Phytophthora* spp. but are uncommon in *Pythium* spp.; they also have been recovered from *Pythium vexans, Pythium oedochilium, Pythium marsipium, Pythium ostracodes, P. sylvaticum*, and *P. aphanidermatum* (Panabieres et al., 1997; Gayler et al., 1997; Lascombe et al., 2007; Veit et al., 2001).

In addition to oligandrin, there are other inducers of host–defense reactions produced by *P. oligandrum*. Takenaka et al. (2003) identified several cell wall protein (CWP) fractions that can reduce damping off severity of sugar beet caused by *Rhizoctonia solani* and infection of wheat by *Fusarium graminearum* when applied to the roots. Sugar beet root treatments induced an increased production of phenylalanine ammonia lyase, chitinase, and cell-wall-bound phenolic compounds. The treatment of sugar beets roots also induced resistance to *Aphanomyces cochliodes* and increased expression of genes associated with host–defense reactions (Takenaka et al., 2006). The treatment of tomato roots with CWP reduced the severity of disease caused by *Ralstonia solanacearum* that was believed to be associated with activation of an ethylene-dependent signaling pathway, as there was an increase in ethylene production followed by enhanced expression of three ethylene-inducible host–defense-related genes and two genes associated with an ethylene-dependent signaling pathway (Hase et al., 2006). Takahashi et al. (2006) examined tomato gene expression following CWP treatments by microarray analysis and identified increased expression of genes associated with ethylene and jasmonic acid signaling pathways, some of which were also induced following treatment with the host–defense activator 3-allyloxy-1,2-benzisothiazole-1,1-dioxide (PBZ). Subsequent analysis by Takenaka et al. (2006) identified two cell wall glycoproteins (POD-1 and POD-2), which represent a novel class of elicitins that were responsible for inducing resistance.

More clarification of the processes associated with *P. oligandrum*-induced resistance in plants and the genes that control it may lead to the development of specific molecular markers that can be used to select more effective isolates for use as a biocontrol agent. This information also could be used for the transformation of highly rhizosphere-competent isolates to enhance their efficacy in inducing the host–defense response by upregulating the appropriate genes. Knowledge of how the elicitors of *P. oligandrum* and plant pathogenic *Pythium* species interact with the host could also lead to the development of novel compounds that can be applied to field-grown plants in an effort to prime the host–defense reaction to the point where the incidence of disease can be managed. Last, clarification of the interaction between plant pathogenic *Pythium* spp. and their hosts at a molecular level could lead to improved disease-management strategies by plant transformation or by identifying specific plant characteristics to be selected for in conventional breeding programs.

11.9 CHARACTERISTICS OF SPECIFIC MOLECULAR REGIONS COMMONLY EXAMINED IN *PYTHIUM* SPP.

11.9.1 Ribosomal RNA

Whereas the small and large subunit of the rRNA gene is present in the rDNA repeat unit along with the 5.8 S gene, the 5S rRNA gene can be in different locations depending on the phylogenetic placement of the species. Belkhiri et al. (1992) initially reported that in species with filamentous sporangia, the gene is present in the rDNA repeat unit approximately 1 kb downstream from the large-subunit rRNA gene and is encoded on the opposite strand as the other rRNA genes. But for most species with globose or unknown sporangia, it is disbursed in tandem arrays outside of the rDNA repeat. A more detailed analysis with more species was recently reported by Bedard et al. (2006) with similar conclusions, but for some species (such as *Pythium monospermum* and *Pythium apleroticum*), the 5S gene was not in an inverted orientation. Based on the observation that most *Phytophthora* spp. had the 5S rRNA gene in the noninverted orientation in the rDNA repeat unit, it was concluded that having the gene in the rDNA repeat represented the ancestral state (Bedard et al., 2006). Because of the levels of sequence divergence among the 5S coding region and the intervening spacer regions in species where the 5S gene is not part of the rDNA repeat unit, these tandem arrays have been useful for construction of species-specific probes for some species with globose sporangia (Klassen et al., 1996).

The number of copies of the rDNA repeat unit present in the genome of a *Pythium* spp. has not been reported, but a PFGE analysis of several species revealed that within a species it can be encoded on a variable number of chromosomal-sized bands (Martin, 1995a). For example, in *P. ultimum* or *P. sylvaticum*, the rDNA can be encoded on two to five chromosomal-sized bands that range in size from 2.2 to approximately 5 Mb. Differences in hybridization intensity of rDNA probes relative to other genes following Southern analysis (Martin, 1995c) indicate there might also be variation in copy number among isolates of *P. sylvaticum*, although this has not been experimentally verified.

One feature of the rDNA of *Pythium* that facilitates research on this region using purified DNA that is it often can be recovered as a distinct band intermediate between the mitochondrial and nuclear DNA following cesium chloride–bisbenzimide ultracentrifugation (Klassen et al., 1987). Using this purified rDNA for analysis, multiple forms of the repeat unit were found to be present in the same homothallic *Pythium* isolate (Martin, 1990b). With *Pythium paroecandrum*, two major forms were differentiated by a 0.27-kb indel in the intergenic spacer (IGS) region approximately 1 kb from the 3′ end of the large subunit rRNA gene, which were recovered when the region was cloned. With this species and with many others, there also were several minor forms present that differed by the addition/loss of a restriction site or additional indels in this same region, the latter of which provided a stepladder appearance of bands following restriction analysis. Most of the size difference between bands was approximately 60 bp, but for one isolate of *P. irregulare*, the unit size was approximately 90 bp. Multiple forms of the rDNA repeat unit with length heterogeneity were also found for *P. ultimum* (with intervals of approximately 0.4 kb; Klassen and Buchko, 1990; Buchko and Klassen, 1990) and *P. pachycaule* (approximately 200 bp; Belkhiri and Klassen, 1996). Interestingly, the primary location of the length variation for all examples was approximately 1 kb from the 3′ end of the large ribosomal subunit; for *P. pachycaule*, the size variation was caused by a tandem repeat of the 5S gene with a pseudogene located in the same region. A similar heterogeneity in the ITS region of the rDNA repeat unit based on addition/loss of restriction sites has been recently reported for *Pythium* group F isolates (Vasseur et al., 2005) *P. mercuriale* (Belbahri et al. 2008) and *P. helicoides* (Kageyama et al., 2007), with the exception that this latter species variability was also found in a single-zoospore progeny of a single isolate. The heterothallic species *P. sylvaticum* also exhibited heterogeneity in the rDNA repeat, but this is not unexpected for an outcrossing species (Martin, 1995c).

11.9.2 Mitochondrial DNA

Mitochondrial DNAs in *Pythium* map to a circular orientation and have the unusual feature of a large inverted repeat (IR) that represents between 75% and 80% of the genome size (McNabb et al., 1987; McNabb and Klassen, 1988; Martin, 1991b). These large IRs help to stabilize the genome from rearrangements and are likely the reason RFLP analysis of the mtDNA is useful for species identification in the genus (Martin and Kistler, 1990; Martin, 1990a). In between the repeated sequences are single-copy unique regions referred to as the small and large unique region based on their relative size differences. Rather than having a circular mapping genome, isolates of some species (such *P. oligandrum*) have a linear genome with the termini mapping in the small unique region. When run on an agarose gel, a distinct band for the linear genomes was observed (Martin, 1991b; Martin, 1995b). Cutting the terminal regions with a restriction enzyme and running them on a denaturing gel confirmed there were hairpin loops at the termini. Although most species examined have circular mapping genomes, PFGE analysis revealed that essentially all isolates examined had at least low levels of linear chromosomes present as unit-length monomers or concatenated multimers. It was suggested that these linear forms could perhaps be generated during mitochondrial replication if there was an origin of replication close to the small unique region (Martin, 1995b).

IRs are also observed in oomycetes such as *Achlya* spp. (Hudspeth et al., 1983; Boyd et al., 1984; Shumard et al., 1986), *Aplanopsis terrestris, Leptolegnia caudata, Sapromyces elongatus* (McNabb and Klassen, 1988), and *Saprolegnia ferax* (Grayburn et al., 2004). For these species, the IR represents greater than 37% of the genome size and encodes the large and small ribosomal RNA as well as several mitochondrial genes. Although it is uncommon to find an IR in *Phytophthora*, a small one has been reported in *Phytophthora megasperma* (0.5 to 0.9 kb; Shumard-Hudspeth and Hudspeth, 1990) and *P. ramorum* (1,150 bp, Martin et al., 2007; Martin, 2008). Based on sequence analysis of the *P. ramorum* mitochondrial genome, this IR is unlike that observed in other oomycetes because of its small size (5.8% of the genome), and it encodes only a putative *orf37*.

In an effort to learn more about genome stability and the evolutionary processes that contribute to mitochondrial genome divergence in *Pythium*, a project has been undertaken to sequence the entire mitochondrial genome from a range of species (Martin and Richardson, unpublished). Although the work is still in progress, the data from 18 genomes sequenced thus far indicate several major differences in gene order that correspond to the phylogenetic division between species with spherical or inflated/lobulate sporangia. For species with spherical hyphal swellings (*P. ultimum, P. sylvaticum, P. nunn*) the *cox 1* and *2* gene cluster is present in the large unique region compared with being in the small unique or adjacent IR sequences for species with inflated/lobulate sporangia (*P. graminicola, P. oligandrum*). Likewise, the *nad6* gene is in the IR for all species, but it is closer to the small unique region for the species with spherical hyphal swellings and is in an opposite orientation closer to the large unique region for those species with inflated/lobulate zoosporangia. Although it is present in approximately the same location in the genome, the *nad5* gene also has an opposite orientation depending on the above noted phylogenetic groupings. The IR in the species examined thus far represent between 71.4% and 84.0% of the total genome size.

Although the IR is thought to function as a stabilizing influence on the genome (recombination would occur between opposite arms of the repeat), interspecific comparisons of gene order revealed inversions and translocations in the unique and IR regions contributed to genome divergence. Also, the rate of evolutionary divergence of an individual gene seems to be more of a function of the gene itself rather than its placement in the IR compared with the unique regions (Martin, unpublished). Consistent gene order differences are observed in comparison with the mitochondrial genomes of 11 *Phytophthora* spp., which are proving to be useful for the development of genus- and species-specific diagnostic markers (Martin, unpublished).

11.9.3 Extrachromosomal Genetic Elements

P. irregulare is the only species in the genus in which double-stranded RNA (dsRNA) has been reported. Klassen et al. (1991) observed that two isolates from Canada had 28-nm virus-like particles in their sporangia when viewed under transmission electron microscopy (TEM). On agarose gels, there were between 5 and 6 distinct bands migrating at a rate corresponding to between 1.8 and 6.0 kb from DNA size standards with different sizes of dsRNAs for each isolate. In contrast, Gillings et al. (1993) observed isometric virus-like particles approximately 45 nm in dsRNA-infected isolates of *P. irregulare* from Australia. When the dsRNAs were run on a gel, the bands ranged in size from 1.0 to 6.0 kb. Although the dsRNA patterns for an individual isolate remained stable following subculturing, single zoospore culture, and storage for 2 years, there was variation in banding profiles among isolates, even from a group of isolates collected from the same field location. Whereas Gillings et al. (1993) reported that dsRNAs were not recovered from 14 additional *Pythium* spp., isolates of the phylogenetically related species *P. sylvaticum* from North America have been found to have one to five dsRNAs below 6 kb in size (Martin, unpublished).

Circular mitochondrial plasmids also have been found in the genus *Pythium*, however, only in species with inflated, filamentous sporangia such as *P. aphanidermatum*, *P graminicola*, *P. torulosum*, and a species with an ornamented oogonia (Martin, 1991a). Southern analysis revealed no homology among these plasmids or with circular plasmids from *Neurospora* spp. For most isolates, the plasmids were retained following subculturing and in asexual and sexual single-spored progeny; however, for some isolates they were lost following prolonged storage and subculturing (an isolate of *P. vanterpoolii* and a few isolates of *P. aphanidermatum*). With 42% of the 24 cultures of *P. aphanidermatum* from geographically diverse areas that contain plasmids (three size categories were observed), this was the species in which plasmids were most commonly encountered (Martin, unpublished). A wide range of isolates and species spanning all phylogenetic groupings were examined as part of this project, but none of the species with spherical/globose sporangia were found to have circular plasmids. A portion of one plasmid from *P. aphanidermatum* was sequenced, and there was no homology with mitochondrial genomic sequences from several of the *Pythium* spp. examined. Surprisingly, the only significant homology with sequences in GenBank was with several unassigned open reading frames (ORFs) in the mitochondrial genome of *P. infestans* (haplotype IIa and IIb) and *P. sojae* (78–89% on a sequence level and 68–87% identity on a translated amino acid level; F. Martin, unpublished). Given that circular mitochondrial plasmids have not been identified in *Phytophthora* spp., this result presents some intriguing questions on their potential involvement in evolution of the mitochondrial genome.

REFERENCES

Al Rawahi AK, Hancock JG 1998. Parasitism and biological control of *Verticillium dahliae* by *Pythium oligandrum*. Plant Dis 82:1100–1106.

Al Sa'di AM, Drenth A, Deadman ML, Aitken E-AB 2008a. Genetic diversity, aggressiveness and metalaxyl sensitivity of *Pythium aphanidermatum* populations infecting cucumber in Oman. Plant Pathol 57:45–56.

Al Sa'di AM, Drenth A, Deadman ML, Cock A-WAM, Al Said FA, Aitken E-AB 2008b. Genetic diversity, aggressiveness and metalaxyl sensitivity of *Pythium spinosum* infecting cucumber in Oman. J Phytopathol 156:29–35.

Barr DJS, Warwick SI, Desaulners NL 1996. Isozyme variation, morphology, and growth response to temperature in *Pythium ultimum*. Can J Bot 74:753–761.

Barr DJS, Warwick SI, Desaulniers NL 1997. Isozyme variation, morphology, and growth response to temperature in *Pythium irregulare*. Can J Bot 75:2073–2081.

Bedard J-EJ, Schurko AM, Cock A-WAM, Klassen GR 2006. Diversity and evolution of 5S rRNA gene family organization in *Pythium*. Mycol Res 110:86–95.

Belbahri L, McLeod A, Paul B, Calmin G, Moralejo E, Spies CFJ, Botha WJ, Clemente A, Descals E, Sanchez-Hernandez E, Lefort F 2008. Intraspecific and within-isolates sequence variation in the ITS rRNA gene region of *Pythium mercuriale* sp. nov. (*Pythiaceae*). FEMS Microbiol Lett 284:17–27.

Belkhiri A, Buchko J, Klassen GR 1992. The 5S ribosomal RNA gene in *Pythium* species: Two different genomic locations. Mol Biol Evol 9:1089–1102.

Belkhiri A, Intengan H, Klassen GR 1997. A tandom array of 5S ribosomal RNA genes in *Pythium irregulare*. Gene 186:155–159.

Belkhiri A, Klassen GR 1996. Diverged 5 s rRNA sequences adjacent to 5 s rRNA genes in the rDNA of *Pythium pachycaule*. Curr Genet 29:287–292.

Benhamou N, Belanger RR, Rey P, Tirilly Y 2001. Oligandrin, the elicitin-like protein produced by the mycoparasite *Pythium oligandrum*, induces systemic resistance to Fusarium crown and root rot in tomato plants. Plant Physiol Biochem 39:681–696.

Benhamou N, Rey P, Cherif M, Hockenhull J, Tirilly Y 1997. Treatment with the mycoparasite *Pythium oligandrum* triggers induction of defense-related reactions in tomato roots when challenged with *Fusarium oxysporum* f.sp. *radicis-lycopersici*. Phytopathol 87:108–122.

Boyd DA, Hobman TC, Gruenke SA, Klassen GR 1984. Evolutionary stability of mitochondrial DNA organization in *Achlya*. Can J Biochem Cell Biol 62:571–576.

Bradshaw-Smith RP, Craig GD, Biddle AJ 1991. Glasshouse and field studies using *Pythium oligandrum* to control fungal food rot pathogens of peas. Aspects Appl Biol 27:347–350.

Briard M, Dutertre M, Rouxel F, Brygoo Y 1995. Ribosomal RNA sequence divergence within the Pythiaceae. Mycol Res 99:1119–1127.

Buchko J, Klassen GR 1990. Detection of length heterogeneity in the ribosomal DNA of *Pythium ultimum* by PCR amplification of the intergenic region. Curr Genet 18:203–205.

Butler DK 1992. Ribosomal DNA is a site of chromosome breakage in aneuploid strains of *Neurospra*. Genetics 131:581–592.

Butler DK, Metzenberg RL 1992. Expansion and contraction of the nucleolus organizer region of *Neurospora*: changes originate in both proximal and distal segments. Genetics 126:325–333.

Campbell WA, Hendrix FF 1967. A new heterothallic *Pythium* from southern United States. Mycologia 59:274–278.

Chen W 1992. Restriction fragment length polymorphisms in enzymatically amplified ribosomal DNAs of three heterothallic *Pythium* species. Phytopathol 82:1467–1472.

Chen W, Hoy JW, Schneider RW 1992. Species-specific polymorphism in transcribed ribosomal DNA of five *Pythium* species. Exp Mycol 16:22–34.

Davison EM, MacNish GC, Murphy PA, McKay AG 2003. *Pythium* spp. from cavity spot and other root diseases of Australian carrots. Australa Plant Pathol 32:455–464.

Dennett CW, Stanghellini ME 1977. Genetic and cytological evidence for a diploid life cycle in *Pythium aphanidermatum*. Phytopathol 67:1134–1141.

Dick MW 1990. Key to *Pythium*. University of Reading Press. Reading, UK. pp. 64.

Floch Gl, Benhamou N, Mamaca E, Salerno MI, Tirilly Y, Rey P 2005. Characterisation of the early events in atypical tomato root colonisation by a biocontrol agent, *Pythium oligandrum*. Plant Physiol Biochem 43:1–11.

Francis DM, Gehlen MF, St Clair DA 1994. Genetic variation in homothallic and hyphal swelling isolates of *Pythium ultimum* var. *ultimum* and *P. ultimum* var. *sporangiferum*. Mol Plant Microb Interact 7:766–775.

Francis DM, St-Clair DA 1993. Outcrossing in the homothallic oomycete, *Pythium ultimum*, detected with molecular markers. Curr Genet 24:100–106.

Francis DM, St.Clair DA 1997. Population genetics of *Pythium ultimum*. Phytopathol 87:454–461.

Gall AM, Elliott CG 1985. Control of sexual reproduction in *Pythium sylvaticum*. Trans Brit Mycol Soc 84:629–636.

Gandhi SR, Weete JD 1991. Production of the polyunsaturated fatty acids arachidonic acid and eicosapentaenoic acid by the fungus *Pythium ultimum*. J Gen Microbiol 137:1825–1830.

Garzon CD, Geiser DM, Moorman GW 2005a. Amplified fragment length polymorphism analysis and internal transcribed spacer and coxII sequences reveal a species boundary within *Pythium irregulare*. Phytopathol 95:1489–1498.

Garzon CD, Geiser DM, Moorman GW 2005b. Diagnosis and population analysis of *Pythium* species using AFLP fingerprinting. Plant Dis 89:81–89.

Garzon CD, Yanez JM, Moorman GW 2007. *Pythium cryptoirregulare*, a new species within the *P. irregulare* complex. Mycologia 99:291–301.

Gavino PD 1994. Self fertility and its contribution to genetic variation in the heterothallic Oomycete, *Pythium sylvaticum*. M.S. Thesis, University of Florida. Gainesville, FL.

Gayler KR, Popa KM, Maksel DM, Ebert DL, Grant BR 1997. The distribution of elicitin-like gene sequences in relation to elicitin protein secretion within the class Oomycetes. Molecular Plant Pathology On-Line [http://www.bspp.org.uk/mppol/] 1997/0623gayler.

Geraats B-PJ, Bakker P-AHM, Lawrence CB, Achuo EA, Hofte M, Loon L-CV 2003. Ethylene-insensitive tobacco shows differentially altered susceptibility to different pathogens. Phytopathol 93:813–821.

Geraats B-PJ, Bakker P-AHM, Linthorst H-JM, Hoekstra J, Loon L-CV 2007. The enhanced disease susceptibility phenotype of ethylene-insensitive tobacco cannot be counteracted by inducing resistance or application of bacterial antagonists. Physiol Mol Plant Pathol 70:77–87.

Gillings MR, Tesoriero LA, Gunn LV 1993. Detection of double-stranded RNA and virus-like particles in Australian isolates of *Pythium irregulare*. Plant Pathol 42:6–15.

Glassgen WE, Rose A, Madlung J, Koch W, Gleitz J, Seitz HU 1998. Regulation of enzymes involved in anthocyanin biosynthesis in carrot cell cultures in response to treatment with ultraviolet light and fungal elicitors. Planta 204:490–498.

Grayburn WS, Hudspeth DSS, Gane MK, Hudspeth MES 2004. The mitochondrial genome of *Saprolegnia ferax*: organization, gene content, and nucleotide sequence. Mycologia 96:980–987.

Guo LY, Ko WH 1991. Hormonal regulation of sexual reproduction and mating type change in heterothallic *Pythium splendens*. Mycol Res 95:452–456.

Guo LY, Ko WH 1994. Growth rate and antibiotic sensitivities of conidium and selfed-oospore progenies of heterothallic *Pythium splendens*. Can J Bot 72:1709–1712.

Guo LY, Ko WH 1995. Continuing variation in successive asexual generations of *Pythium splendens* following sexual reproduction. Mycol Res 99:1339–1344.

Harvey PR, Butterworth PJ, Hawke BG, Pankhurst CE 2000. Genetic variation among populations of *Pythium irregulare* in southern Australia. Plant Pathol 49:619–627.

Harvey PR, Butterworth PJ, Hawke BG, Pankhurst CE 2001. Genetic and pathogenic variation among cereal, medic and sub-clover isolates of *Pythium irregulare*. Mycol Res 105:85–93.

Hase S, Shimizu A, Nakaho K, Takenaka S, Takahashi H 2006. Induction of transient ethylene and reduction in severity of tomato bacterial wilt by *Pythium oligandrum*. Plant Pathol 55:537–543.

Hendrix FF, Campbell WA 1969. Heterothallism in *Pythium catenulatum*. Mycologia 61:639–641.

Herrero ML, Klemsdal SS 1998. Identification of *Pythium aphanidermatum* using the RAPD technique. Mycol Res 102:136–140.

Hudspeth MES, Shumard DS, Bradford JR, Grossman LI 1983. Organization of *Achlya* mtDNA: a population with two orientation and a large inverted repeat containing the rRNA genes. Proc Nat Acad Sci USA 80:142–146.

Huffaker A, Pearce G, Ryan CA 2006. An endogenous peptide signal in Arabidopsis activates components of the innate immune response. Proc Nat Acad Sci USA 103:10098–10103.

Huffaker A, Ryan CA 2007. Endogenous peptide defense signals in Arabidopsis differentially amplify signaling for the innate immune response. Proc Nat Acad Sci USA 104:10732–10736.

Judelson HS, Tyler BM, Michelmore RW 1991. Transformation of the Oomycete pathogen, *Phytophthora infestans*. Molec Plant-Microbe Inter 4:602–607.

Kageyama K, Nakashima A, Kajihara Y, Suga H, Nelson EB 2005. Phylogenetic and morphological analyses of *Pythium graminicola* and related species. J Gen Plant Pathol 71:174–182.

Kageyama K, Senda M, Asano T, Suga H, Ishiguro K 2007. Intra-isolate heterogeneity of the ITS region of rDNA in *Pythium helicoides*. Mycol Res 111:416–423.

Kageyama K, Suzuki M, Priyatmojo A, Oto Y, Ishiguro K, Suga H, Aoyagi T, Fukui H 2003. Characterization and identification of asexual strains of *Pythium* associated with root rot of rose in Japan. J Phytopathol 151:485–491.

Kageyama K, Uchino H, Hyakumachi M 1998. Characterization of the hyphal swelling group of *Pythium*: DNA polymorphisms and cultural and morphological characteristics. Plant Dis 82:218–222.

Klassen GR, Balcerzak M, deCock AWAM 1996. 5S ribosomal RNA gene spacers as species specific probes for eight species of *Pythium*. Phytopathol 86:581–587.

Klassen GR, Buchko J 1990. Subrepeat structure of the intergenic region in the ribosomal DNA of the oomycetous fungus *Pythium ultimum*. Curr Genet 17:125–127.

Klassen GR, Kim WK, Barr DJS, Desaulners NL 1991. Presence of double-stranded RNA in isolates of *Pythium irregulare*. Mycologia 83:657–661.

Klassen GR, McNabb SA, Dick MW 1987. Comparison of physical maps of ribosomal DNA repeating units in *Pythium*, *Phytophthora* and *Apodachlya*. J Gen Microbiol 133:2953–2959.

Knoester M, van Loon LC, van den Heuvel J, Henning J, Bol JF, Linthorst, HJM 1998. Ethylene-insensitive tobacco lacks nonhost resistance against soil-borne fungi. Proc Nat Acad Sci USA 95:1933–1937.

Koch W, Wagner C, Seitz HU 1998. Elicitor-induced cell death and phytoalexin synthesis in *Daucus carota* L. Planta 206:523–532.

Kong P, Richardson PA, Moorman GW, Hong CX 2004. Single-strand conformational polymorphism analysis of the ribosomal internal transcribed spacer 1 for rapid species identification within the genus *Pythium*. FEMS Microbiol Lett 240:229–236.

Lascombe MB, Retailleau P, Ponchet M, Industri B, Blein JP, Prange T 2007. Structure of sylvaticin, a new alpha-elicitin-like protein from *Pythium sylvaticum*. Acta Crystallographica Sec D, BiolCrystal 63:1102–1108.

Lee S, Moorman GW 2008. Identification and characterization of simple sequence repeat markers for *Pythium aphanidermatum*, *P. cryptoirregulare*, and *P. irregulare* and the potential use in *Pythium* population genetics. Curr Genet 53:81–93.

Lévesque CA, Cock A-WAM 2004. Molecular phylogeny and taxonomy of the genus *Pythium*. Mycol Res 108:1363–1383.

Lherminier J, Benhamou N, Larrue J, Milat ML, Boudon-Padieu E, Nicole M, Blein JP 2003. Cytological characterization of elicitin-induced protection in tobacco plants infected by *Phytophthora parasitica* or phytoplasma. Phytopathol 93:1308–1319.

Martin FN 1989. Maternal inheritance of mitochondrial DNA in sexual crosses of *Pythium sylvaticum*. Curr Genet 16:375–376.

Martin FN 1990a. Taxonomic classification of asexual isolates of *Pythium ultimum* based on cultural characteristics and mitochondrial DNA restriction patterns. Exp Mycol 14:47–56.

Martin FN 1990b. Variation in the ribosomal DNA repeat unit within single-oospore isolates of the genus *Pythium*. Genome 33:585–591.

Martin FN 1991a. Characterization of circular mitochondrial plasmids in three *Pythium* species. Curr Genet 20:91–97.

Martin FN 1991b. Linear mitochondrial molecules and intraspecific mitochondrial genome stability in a species of *Pythium*. Genome 34:156–162.

Martin FN 1995a. Electrophoretic karyotype polymorphisms in the genus *Pythium*. Mycologia 87:333–353.

Martin FN 1995b. Linear mitochondrial genome organization *in vivo* in the genus *Pythium*. Curr Genet 28:225–234.

Martin FN 1995c. Meiotic instability of *Pythium sylvaticum* as demonstrated by inheritance of nuclear markers and karyotype analysis. Genetics 139:1233–1246.

Martin FN 1995d. *Pythium*. In: Singh US, Kohmoto K, Singh RP, editors. Pathogenisis and host specificity in plant diseases. Histopathological, biochemical, genetic, and molecular basis. Elsevier. Tarrytown, NY. pp. 17–36.

Martin FN 1999. Biocontrol of fungal soilborne pathogens by *Pythium oligandrum*. US Patent 5,961,971. 10-5-1999.

Martin FN 2000. Phylogenetic relationships among some *Pythium* species inferred from sequence analysis of the mitochondrially encoded cytochrome oxidase II gene. Mycologia 92:711–727.

Martin FN 2008. Mitochondrial haplotype determination in the oomycete plant pathogen *Phytophthora ramorum*. Curr Genet 54:23–34.

Martin FN, Bensasson D, Tyler BM, Boore JL 2007. Mitochondrial genome sequences and comparative genomics of *Phytophthora ramorum* and *P. sojae*. Curr Genet 51:285–296.

Martin FN, Hancock JG 1987. The use of *Pythium oligandrum* for biological control of preemergence damping-off caused by *P. ultimum*. Phytopathol 77:1013–1020.

Martin FN, Kistler HC 1990. Species-specific banding patterns of restriction endonuclease- digested mitochondrial DNA from the genus *Pythium*. Exp Mycol 14:32–46.

Martin FN, Loper JE 1999. Soilborne plant diseases caused by *Pythium* spp.: ecology, epidemiology, and prospects for biological control. Crit Rev Plant Sci 18:111–181.

Martin FN, Semer CR 1997. Selection of drug-tolerant strains of *Pythium sylvaticum* using sublethal enrichment. Phytopathol 87:685–692.

Matsumoto C, Kageyama K, Suga H, Hyakumachi M 1999. Phylogenetic relationships of *Pythium* species based on ITS and 5.8S sequences of the ribosomal DNA. Mycoscience 40:321–331.

Matsumoto C, Kageyama K, Suga H, Hyakumachi M 2000. Intraspecific DNA polymorphisms of *Pythium irregulare*. Mycol Res 104:1333–1341.

McLeod A, Fry BA, Zuluaga AP, Myers KL, Fry WE 2008. Toward improvements of oomycete transformation protocols. J Eukaryot Microbiol 55:103–109.

McNabb SA, Boyd DA, Belkhiri A, Dick MW, Klassen GR 1987. An inverted repeat comprises more than three-quarters of the mitochondrial genome in two species of *Pythium*. Curr Genet 12:205–208.

McNabb SA, Klassen GR 1988. Uniformity of mitochondrial DNA complexity in oomycetes and the evolution of the inverted repeat. Exp Mycol 12:233–242.

McQuilken MP, Whipps JM Cooke RC 1990. Control of damping-off in cress and sugar-beet by commercial seed-coating with *Pythium oligandrum*. Plant Pathol 39: 452–462.

Mitchell RT, Deacon JW 1986. Differential (host-specific) accumulation of zoospores of *Pythium* on roots of graminacious and non-graminacious plants. New Phytol 102:113–122.

Mohamed N, Lherminier J, Farmer MJ, Fromentin J, Beno N, Houot V, Milat ML, Blein JP 2007. Defense responses in grapevine leaves against *Botrytis cinerea* induced by application of a *Pythium oligandrum* strain or its elicitin, oligandrin, to roots. Phytopathol 97:611–620.

Morris BM, Gow NAR 1993. Mechanism of electrotaxis of zoospores of phytopathogenic fungi. Phytopathol 83:877–882.

Panabieres F, Ponchet M, Allasia V, Cardin L, Ricci P 1997. Characterization of border species among Pythiaceae: several *Pythium* isolates produce elicitins, typical proteins from *Phytophthora* spp. Mycol Res 101:1459–1468.

Papa KE, Campbell WA, Hendrix FF 1967 Sexuality in *Pythium sylvaticum*: heterothallism. Mycologia 59:589–595.

Park C, Kakinuma M, Sakaguchi K, Amano H 2003. Genetic variation detected with random amplified polymorphic DNA markers among isolates of the red rot disease fungus *Pythium porphyrae* isolated from *Porphyra yezoensis* from Korea and Japan. Fisheries Sci 69:361–368.

Perneel M, Tambong JT, Adiobo A, Floren C, Saborio F, Lévesque A, Hofte M 2006. Intraspecific variability of *Pythium myriotylum* isolated from cocoyam and other host crops. Mycol Res 110:583–593.

Picard K, Ponchet M, Blein JP, Rey P, Tirilly Y, Benhamou N 2000. Oligandrin. A proteinaceous molecule produced by the mycoparasite *Pythium oligandrum* induces resistance to *Phytophthora parasitica* infection in tomato plants. Plant Physiol 124:379–395.

Pratt RG, Green RJ 1971. The taxonomy and heterothallism of *Pythium sylvaticum*. Can J Bot 49:273–279.

Pratt RG, Green RJ 1973. The sexuality and population structure of *Pythium sylvaticum*. Can J Bot 51:429–436.

Pukkila PJ, Skzynia C 1993. Frequent changes in the number of reiterated ribisomal RNA genes throughout the life cycle of the basidiomycete *Coprinus cinereus*. Genetics 133:203–211.

Rafin C, Brygoo Y, Tirilly Y 1995. Restriction analysis of amplified ribosomal DNA of *Pythium* spp. isolated from soilless culture systems. Mycol Res 99:277–281.

Raftoyannis Y, Dick MW 2006. Zoospore encystment and pathogenicity of *Phytophthora* and *Pythium* species on plant roots. Microbiol Res 161:1–8.

Rey P, Benhamou N, Wulff E, Tirilly Y 1998. Interactions between tomato (*Lycopersicon esculentum*) root tissues and the mycoparasite *Pythium oligandrum*. Physiol Mol Plant Pathol 53:105–122.

Saunders GA, Hancock JG 1994. Self-sterile isolates of *Pythium* mate with self-fertile isolates of *Pythium ultimum*. Mycologia 86:660–666.

Shumard DS, Grossman LI, Hudspeth MES 1986. *Achlya* mitochondrial DNA: gene localization and analysis of inverted repeats. Mol Gen Genet 202:16–23.

Shumard-Hudspeth DS, Hudspeth MES 1990. Genetic rearrangements in *Phytophthora* mitochondrial DNA. Curr Genet 17:413–415.

Staswick PE, Yuen GY, Lehman CC 1998. Jasmonate signaling mutants of Arabidopsis are susceptible to the soil fungus *Pythium irregulare*. Plant J 15:747–754.

Szabo E, Thelen A, Petersen M 1999. Fungal elicitor preparations and methyl jasmonate enhance rosmarinic acid accumulation in suspension cultures of *Coleus blumei*. Plant Cell Reports 18:485–489.

Takahashi H, Ishihara T, Hase S, Chiba A, Nakaho K, Arie T, Teraoka T, Iwata M, Tugane T, Shibata D, et al. 2006. Beta-cyanoalanine synthase as a molecular marker for induced resistance by fungal glycoprotein elicitor and commercial plant activators. Phytopathol 96:908–916.

Takenaka S, Nakamura Y, Kono T, Sekiguchi H, Masunaka A, Takahashi H 2006. Novel elicitin-like proteins isolated from the cell wall of the biocontrol agent *Pythium oligandrum* induce defense-related genes in sugar beet. Mol Plant Pathol 7:325–339.

Takenaka S, Nishio Z, Nakamura Y 2003. Induction of defense reactions in sugar beet and wheat by treatment with cell wall protein fractions from the mycoparasite *Pythium oligandrum*. Phytopathol 93:1228–1232.

Tojo M, Nakazono E, Tsushima S, Morikawa T, Matsumoto N 1998. Characterization of two morphological groups of isolates of *Pythium ultimum* var. *ultimum* in a vegetable field. Mycoscience 39:135–144.

Van der Plaats-Niterink AJ 1968. The occurrence of *Pythium* in the Netherlands I. heterothallic species. Acta Bot Neer 17:320–329.

Van der Plaats-Niterink AJ 1981. *Monograph of the genus Pythium*. Studies in mycology No. 21. Baarn, The Netherlands: Centraalbureau Voor Schimmelcultures pp. 242.

van West P, Morris BM, Reid B, Appiah AA, Osborne MC, Campbell TA, Shepherd SJ, Gow N-AR 2002. Oomycete plant pathogens use electric fields to target roots. Mol Plant Microbe Interact 15:790–798.

Vasseur V, Rey P, Bellanger E, Brygoo Y, Tirilly Y 2005. Molecular characterization of *Pythium* group F isolates by ribosomal-and intermicrosatellite-DNA regions analysis. Eur J Plant Pathol 112:301–310.

Veit S, Worle JM, Nurnberger T, Koch W, Seitz HU 2001. A novel protein elicitor (PaNie) from *Pythium aphanidermatum* induces multiple defense responses in carrot, Arabidopsis, and tobacco. Plant Physiol 127:832–841.

Vesely D 1979. Use of *Pythium oligandrum* to protect emerging sugar beet. In: Schippers B, Gams W, editors. Soil-borne plant pathogens. Academic Press. London, UK. pp. 593–595.

Vijayan P, Shockey J, Lévesque CA, Cook RJ, Browse J 1998. A role for jasmonate in pathogen defense of Arabidopsis. Proc Nat Acad Sci USA 95:7209–7214.

Villa NO, Kageyama K, Asano T, Suga H 2006. Phylogenetic relationships of *Pythium* and *Phytophthora* species based on ITS rDNA, cytochrome oxidase II and beta -tubulin gene sequences. Mycologia 98:410–422.

Wang PH, White JG 1997. Molecular characterization of *Pythium* species based on RFLP analysis of the internal transcribed spacer region of ribosomal DNA. Physiol Mol Plant Pathol 51:129–143.

Weiland JJ 2003. Transformation of *Pythium aphanidermatum* to geneticin resistance. Curr Genet 42:344–352.

12

BREMIA LACTUCAE AND LETTUCE DOWNY MILDEW

RICHARD MICHELMORE, OSWALDO OCHOA, AND JOAN WONG
The Genome Center and Department of Plant Sciences, University of California, Davis, California

12.1 INTRODUCTION

Lettuce downy mildew, which is caused by *Bremia lactucae* Regel, is the most important disease of one of the most valuable vegetable crops. As such, it has been the subject of basic and applied research for many decades. This has generated a wealth of genetic and physiological data on the interaction between this oomycete and its host. *B. lactucae* was also the subject of early applications of molecular approaches to the oomycetes; however, molecular characterization of *B. lactucae* stalled in the mid-1990s as efforts focused on the host side of the interaction as well as on more tractable bacterial–plant interactions. Genomic approaches, particularly the advent of efficient DNA sequencing, now offer opportunities for rapid advancement of our understanding of *B. lactucae* once again.

This review provides a synopsis of the classic and molecular data to describe lettuce downy mildew as well as to access to the techniques available to manipulate *B. lactucae*. In addition, it considers the future directions afforded by molecular and genomic approaches.

12.2 TAXONOMY, HOST RANGE, AND IMPORTANCE

Bremia is considered a distinct genus within the Peronosporaceae, although the current status of discrete species within the *B. lactucae* complex is unclear

Oomycete Genetics and Genomics: Diversity, Interactions, and Research Tools
Edited by Kurt Lamour and Sophien Kamoun
Copyright © 2009 John Wiley & Sons, Inc.

(Skidmore and Ingram, 1985; Voglmayr et al., 2004; Voglmayr and Constantinescu, 2008). *B. lactucae* has been reported to infect more than 200 species from 40 genera within the Compositae (Asteraceae) (Crute, 1981; Lebeda et al., 2002, 2008b). Most of its hosts belong to two closely related tribes, the *Lactuceae* and *Cardueae*. Analysis of partial sequences of the large (28S) subunit rDNA region from 32 isolates of *B. lactucae* from 28 plant species indicated the existence of three distinct clades (Voglmayr et al., 2004). These data suggested that the adaptive radiation started on *Lactuceae* hosts and subsequently transitioned to *Cardueae* species. Isolates of *B. lactucae* are highly host specific and there is little evidence of gene flow between isolates from distinct hosts (Voglmayr et al., 2004; Lebeda et al., 2008a). A recent analysis of the ribosomal internal transcribed spacer region (ITS) of eight isolates of *B. lactucae* from five host species revealed one of the largest ITS regions reported for any species (Choi et al., 2007). The large size of the ITS region was because of the presence of nine 179- to 194-bp repeats within the ITS2 region. All the isolates of *B. lactucae* studied had nine repeats; however, there was considerable sequence variation in the repeats within and between isolates, which indicates that they could be useful markers for population genetic and epidemiological studies.

The only host of economic importance infected by *B. lactucae* is cultivated lettuce, *Lactuca sativa* L. Lettuce is one of the 10 most valuable crops in the United States with an annual value of over $2.7 billion, and it has a similar value in Europe (USDA-NASS, 2008). *B. lactucae* can cause severe disease on lettuce and is one of the major pathogens of lettuce worldwide. Control strategies rely on a combination of chemical protectants and genetic resistance. However, there are few effective systemic chemicals, notably metalaxyl (Ridomil) and fosetyl aluminium (Aliette), which can be rendered ineffective by changes in *B. lactucae* (Crute et al., 1987; Schettini et al., 1991; Brown et al., 2004). Downy mildew resistance (*Dm*) genes can provide high levels of resistance, and introgression of new *Dm* genes from wild species into cultivated lettuce has been a major activity of many lettuce-breeding programs since the 1920s. However, *Dm* genes have often been rapidly overcome by changes in the pathogen. The control of lettuce downy mildew therefore requires a continual supply of new *Dm* genes. The advent of molecular markers for *Dm* genes provides opportunities for more rapid introgression and new strategies for gene deployment to provide more durable resistance.

12.3 CULTURE

Like all members of the Peronosporaeae, *B. lactucae* is an obligate biotroph and cannot currently be cultured axenically. The reasons for this are not clear; presumably, it is because of the lack of the capability to synthesize one or more compounds that it obtains from its host. However, attempts at supplementing media with a wide variety of candidate nutrients have not resulted in successful axenic culture. The inability to culture *B. lactucae* axenically is unlikely to be caused by a missing physical requirement because culture of surface-sterilized, infected cotyledons on various media can result in a narrow fringe of mycelium growing out over the medium from the host. However, if this mycelium is severed from the plant tissue, then growth stops. The availability of the genomic sequences of downy mildews will provide insights into their biosynthetic capabilities; this should lead to clues as to which pathways are missing and thereby hopefully will result in their axenic culture.

Even though axenic culture is not possible, *B. lactucae* can be readily cultured on lettuce seedlings and manipulated in the laboratory. Both the asexual lifecycle of 1 to 3 weeks and the sexual cycle of several weeks' to many years' duration can be performed in the laboratory. The asexual cycle allows the facile clonal propagation of individual genotypes in closed containers. Conidia are washed from infected seedlings, pelleted, and resuspended to remove a water-soluble inhibitor of germination (Sargent, 1976), and then they are sprayed onto fresh week-old lettuce seedlings. Incubation at ~15°C with diurnal light and dark periods results in profuse sporulation after 1 week. Sporulation is induced by the onset of the dark period and can be inhibited by light (Raffray and Sequeira, 1971). Coinoculation with isolates of opposite mating type results in a switch from asexual to sexual reproduction and the production of large numbers of oospores (Michelmore and Ingram, 1980; Michelmore, 1981). Sowing of lettuce seeds in a suspension of ripened oospores or infested soil results in low frequencies of infected seedlings, each representing individual sexual progeny; the conidia from individual infected seedlings can then be propagated asexually and phenotyped (Michelmore and Ingram, 1981b). It is also possible to derive large numbers of lines from single conidia (Michelmore and Ingram, 1982). *B. lactucae* can be stored for many years at −80°C either as pellets of conidia or as sporulations on lettuce seedlings. Oospores can survive dry for over 10 years.

12.4 HOST-PATHOGEN INTERACTIONS

The classic genetics of the interaction between *B. lactucae* and lettuce is one of the most extensively characterized gene-for-gene plant–pathogen relationships (Crute and Johnson, 1976; Norwood et al., 1983; Michelmore et al., 1984; Norwood and Crute, 1984; Hulbert and Michelmore, 1985; Farrara et al., 1987, Ilott et al., 1987, 1989; Lebeda et al., 2008a). The characterization of resistance genes in lettuce has been facilitated by parallel studies of avirulence genes in *B. lactucae* (see below). At least 27 major *Dm* genes or resistance (R) factors are now known (Farrara et al., 1987; Bonnier et al., 1994; Maisonneuve et al., 1994; Jeuken and Lindhout, 2002). Many additional resistant accessions have been identified, but their resistance has not been extensively characterized genetically (e.g., Farrara and Michelmore, 1987; Gustafsson, 1989; Bonnier et al., 1994; Lebeda and Zinkernagel, 2003; Beharav et al., 2006). Many hundred *Dm* genes with specificity to *B. lactucae* will likely be identified as more *Dm* genes are characterized from these and other sources.

There can be a wide range of interaction phenotypes. These range from little or no reaction when some accessions of *L. saligna* are challenged, to varying extents of host necrosis and intensities of sporulation on other species of *Lactuca* (Lebeda et al., 2008a and b). A fully compatible interaction results in profuse sporulation from chlorotic lesions that are usually delimited by the leaf vasculature. Systemic infection occurs only extremely rarely on colonization of the apical meristem of young seedlings. Asexual sporulation occurs when sporangiophores emerge through the stomata and therefore tends to be more profuse on the abaxial leaf surface. Interaction phenotypes are influenced by both environmental and genetic factors as well as by plant physiology and development (e.g., Judelson and Michelmore, 1992; Nordskog et al., 2007; Lebeda et al., 2008b). Some interaction phenotypes are influenced by temperature; several *Dm* genes are less effective at lower temperatures. Temperature-shift experiments indicated that the determinants of specificity are present in most host cells and expressed throughout pathogen development (Judelson and Michelmore, 1992). Most characterized *Dm* genes confer high levels of resistance that involve a rapid hypersensitive response (HR). This may be a consequence of characterized *Dm* genes encoding extreme phenotypes that have been identified and used by breeders. Some *Dm* genes, such as *Dm6*, confer incomplete resistance (Crute and Norwood, 1978). Similarly, there may be gene dosage effects in which heterozygotes of some *Dm* genes (e.g., *Dm18*) also confer incomplete resistance (Maisonneuve et al., 1994). On the pathogen side, different isolates of *B. lactucae* can exhibit different levels of incompatibility to the same *Dm* gene (Ilott et al., 1989). There are also resistance genes of minor effect that confer resistance in adult plants but not seedlings (field resistance: Eenink et al., 1983; Jeuken and Lindhout, 2002). Many genes of minor effect will probably be identified in the future through quantitative trait locus (QTL) analysis using molecular markers. Cosegregation with candidate genes and RNAi analysis will reveal whether such quantitative resistance is fundamentally different from that already characterized or whether it is determined by sequences similar to *Dm* genes but with lower penetrance and/or delayed expression.

B. lactucae forms well-defined infection structures. The conidium forms a single germ tube of variable length and then an appressorium. Typically, it penetrates the cuticle directly and establishes primary and secondary infection vesicles in epidermal cells (Sargent et al., 1973) (Fig. 12.1a and b) rather than entering via the stomata as many other downy mildews do. In a compatible interaction, the secondary vesicle forms an intracellular hypha within the epidermal cell, and then a coenocytic intercellular mycelium becomes established. Simple pyriform haustoria are elaborated from the mycelium that invaginate but do not penetrate the plasmalemma of host epidermal and mesophyll cells. Observation of freshly prepared, vacuum-infiltrated, infected leaves reveals major cytoplasmic fluxes in and out of the haustoria, which

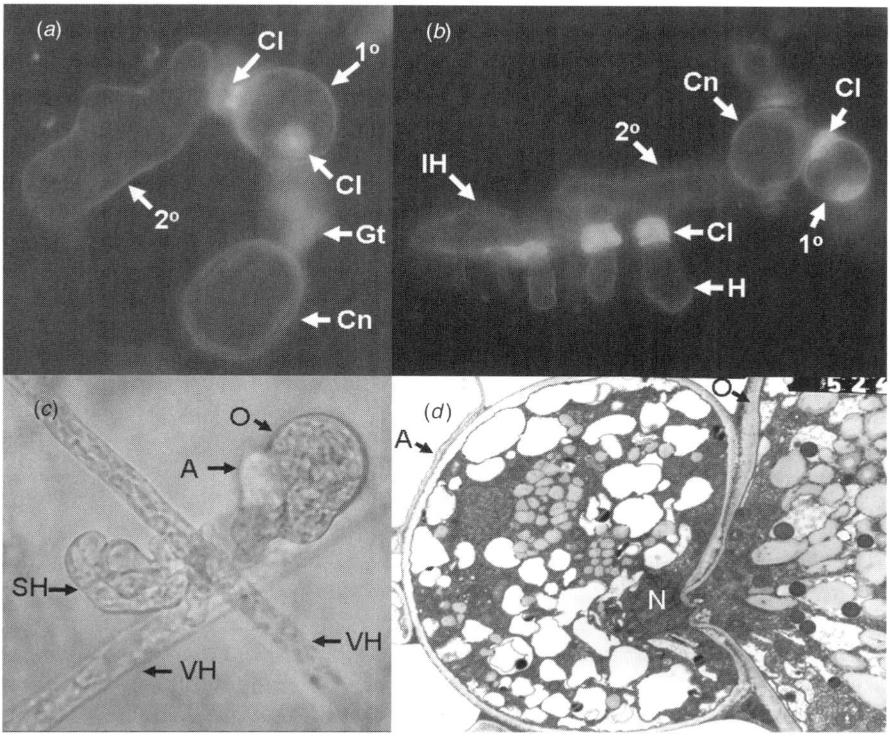

FIG. 12.1 (See color insert) (a, b) Infection structures of *Bremia lactucae*. Lettuce cotyledons were stained with 4′,6-diamidino-2-phenylindole (DAPI) 20 h (a) or with aniline blue 30 h (b) after infection with a compatible isolate of *B. lactucae*. Cn: Conidium, Gt: Germ tube on epidermal surface, Cl: Host callose, 1°: Primary infection vesicle within epidermal cell, 2°: Secondary infection vesicle within epidermal cell, IH: intracellular hypha, H: haustorium. Note the gradient of callose deposition and haustorial size in (b). (c, d) Sexual reproduction of *Bremia lactucae*. (c) Development of sexual hyphae (SH) at the point of contact between vegetative hyphae (VH) and the elaboration of the oogonium (O) and antheridium (A). (d) Dissolution of the antheridial (A) and oogonial (O) walls just prior to migration of an antheridial nucleus (N) into the oogonium.

indicates that the interiors of the haustorium and mycelium are a shared cellular compartment.

In incompatible reactions, the rapid cell death associated with the HR is initiated often as soon as the secondary infection vesicle develops (Maclean et al., 1974). The timing and extent of HR varies depending on the genes involved and the environmental conditions (Lebeda et al., 2008a). Several ultrastructural and biochemical studies have indicated that the HR involved in the resistance of lettuce to *B. lactucae* is similar to the HR described in other plant–pathogen interactions, but it includes the induction of phytoalexins characteristic of the Compositae (Maclean and Tommerup, 1979; Woods et al., 1988; Bennett et al., 1996; Bestwick et al., 1998; Lebeda et al., 2008b). Few data are available on the genes involved in determining resistance other than the *Dm* genes. Homologs of genes encoding signaling and resistance response proteins are present in expressed sequence tag (EST) libraries from *Lactuca* spp. (http://compgenomics.ucdavis.edu). Therefore, signal transduction and resistance responses are likely to be similar to those of other plants. Transcriptional profiling of either compatible or incompatible interactions has not yet been performed; however, an Affymetrix oligonucleotide array for lettuce has recently become available that enables such studies (van Leeuwen et al., unpublished).

The resistance exhibited by *Lactuca saligna* is interesting and may be different from that exhibited by related *Lactuca* species. Most accessions of *L. saligna* seem to be resistant to *B. lactucae*. However, there is minimal reaction at the cytological and biochemical level in contrast to incompatible interactions in *L. sativa*, *Lactuca serriola*, and *Lactuca virosa*, which suggests that it may have a different molecular basis (Lebeda et al., 2002, 2008b). Genetic analysis of the resistance in *L. saligna* using an interspecific cross with *L. sativa* indicated both major qualitative as well as multiple quantitative loci, but the underlying genes have yet to be characterized (Jeuken and Lindhout, 2002). In addition, several major genes for resistance have been introgressed into *L. sativa* from *L. saligna* (Michelmore, Ochoa, and Truco, unpublished); however, there is no evidence that they will prove any more durable than those that originate from *L. serriola*. One such gene (*R32*) that was introgressed from *L. saligna* turned out to have the same resistance specificity as *Dm18* that had been independently introgressed from *L. serriola* (Maisonneuve et al., 1994) and is a nucleotide binding site-leucine rich repeat (NBS-LRR)-encoding gene (Wroblewski et al., 2007). Both resistances have been rendered ineffective in parallel by changes in the pathogen.

12.5 CLASSIC AND MOLECULAR GENETICS OF HOST RESISTANCE

The process of germplasm screens for new sources of resistance, and the subsequent introgression of resistance into cultivated lettuce over the last 80 years has identified more than 27 *Dm* genes or resistance factors as well as many resistant accessions whose resistance has yet to be genetically characterized (see above for references). Thus far, 18 phenotypic *Dm* genes have been mapped. Most of these are located in one of five major clusters (Hulbert and Michelmore, 1985; Farrara et al., 1987; Bonnier et al., 1994; Kesseli et al., 1994; Witsenboer et al., 1995). The largest of these clusters contains at least nine resistance specificities to downy mildew as well as resistance to root aphid (Kesseli et al., 1994). The second largest cluster determines at least five downy mildew specificities as well as resistance to the root-infecting downy mildew

Plasmopara lactuca-radicis, and the hypersensitive reaction to Turnip Mosaic Virus (Witsenboer et al., 1995). Mapping of additional *Dm* genes is a major ongoing activity to determine the genomic architecture of disease resistance in lettuce.

Dm3 was cloned by a combination of map-based and candidate gene approaches. It encodes a large coiled-coil (CC)-NBS-LRR protein (Meyers et al., 1998a; Shen et al., 1998, 2002). Dm3 is large, and it contains nearly double the number of LRRs compared with NBS-LRR proteins characterized from other species (McHale et al., 2006). *Dm3* maps to the largest cluster of resistance genes and is a member of the *Resistance Gene Candidate 2* (*RGC2*) multigene family. The *RGC2* locus is dynamic, and the *RGC2* copy number can vary within *L. sativa* and *L. serriola* from 12 to over 30 depending on the genotype (Meyers et al., 1998a and b; Kuang et al., 2004). A sequence analysis of paralogs from *L. sativa*, *L. serriola*, and *L. saligna* indicated that the *RGC2* cluster evolves by a birth-and-death mechanism (Michelmore and Meyers, 1998; Kuang et al., 2004). *RCG2* genes exhibit two distinct patterns of evolution. Type I genes are evolving rapidly because of frequent sequence exchange between paralogs; they are consequently extensive chimeras, and specific sequences rarely occur (Kuang et al., 2006). Type II genes only rarely undergo sequence exchanges with other lineages and occur more frequently in nature (Kuang et al., 2004). Some Type II lineages exhibit trans-specific polymorphisms indicating that balancing selection has maintained diversity at this locus.

Numerous haplotypes of the *RGC2* cluster have been identified. In one study, 51 different haplotypes were identified in 74 accessions from diverse geographical origins (Sicard et al., 1999). In a second study, 366 haplotypes were observed in 709 accessions from the Eastern Mediterranean center of diversity for *L. serriola* (Kuang et al., 2008). All accessions carried *RGC2* genes, even though some do not carry detectable *Dm* specificities. The large number of different haplotypes indicates that there are at least several hundred distinct *Dm* genes in *Lactuca* spp. and that wild germplasm will continue to be a rich source of new *Dm* genes that can be introgressed and pyramided using molecular markers. Although there are many haplotypes at the *RGC2* locus, it is not highly recombinogenic. The *RGC2* locus has a recombination rate 18 times lower than the genome-wide average (Chin et al., 2001). This is probably the consequence of reduced pairing during meiosis between haplotypes because of structural heterozygosity.

The structural diversity at the *RGC2* locus reflects spontaneous rearrangements at resistance loci. Different rates of spontaneous losses of resistance have been observed for several *Dm* genes (Chin et al., 2001). Rates of 10^{-3} to 10^{-4} spontaneous mutations per generation were observed for *Dm1*, *Dm3*, and *Dm7*. However, no spontaneous mutations were detected for *Dm5/8*. Spontaneous mutations at the *Dm3* locus were frequently associated with large deletions that resulted from unequal crossing over. Not all spontaneous mutations involved large deletions; one spontaneous loss of *Dm3* resistance was the result of a gene conversion event between the LRR-encoding regions of similar paralogs (Chin et al., 2001). A spontaneous loss of *Dm7* resistance involved a small deletion within the LRR-encoding region (McHale, unpublished). A lettuce plant is capable of producing several thousand seeds per generation; such spontaneous mutation rates indicate that an average of one novel haplotype at a resistance locus is generated per plant every generation.

A combination of searches of EST sequences and polymerase chain reaction (PCR) with degenerate oligonucleotides primers has been used to identify *RGC*s. These approaches have resulted in the identification of 90 CC-NBS-LRR-encoding genes and 106 TIR-NBS-LRR-encoding genes from *L. sativa* and *L. serriola* (Shen et al., 1998; McHale and Michelmore, unpublished). These fall into 20 distinct families of NBS-LRR-encoding genes, which are being mapped relative to phenotypic resistances to diverse diseases including downy mildew. Each *Dm* specificity is associated with at least one *RGC*. Often distinct *RGC*s are clustered; sometimes both CC and TIR types are closely linked, which makes it difficult to infer the causal gene.

RNA interference (RNAi) is being used to determine the causal relationships between *RGC* genes and *Dm* phenotypes using a library of transgenic lettuce lines silenced for each *RGC* family. Each interfering hairpin RNA (ihpRNA) construct contains a fragment of the targeted *RGC* gene fused to a fragment of the β-glucuronidase (*GUS*) gene; *Agrobacterium*-mediated transient assays for GUS activity allow the identification of silenced lines. Initial experiments demonstrated sequence similarity for several *Dm* specificities that mapped to the major cluster of *Dm* genes (Wroblewski et al., 2007). An ihpRNA construct designed to the LRR-encoding region of *Dm3* was used to induce posttranscriptional gene silencing of the *RGC2* family. This indicated that the genetically defined *Dm18* locus had at least two specificities, only one of which was silenced by the ihpRNA derived from the LRR-encoding region of *Dm3*. Silenced transgenic tester lines were crossed to lettuce accessions that carry other resistance genes previously mapped to the *RGC2* locus; an analysis of progeny indicated that two additional resistance specificities to *B. lactucae*, *Dm14* and *Dm16*, as well as resistance to lettuce root aphid (*Pemphigus bursarius* L.), *Ra*, are encoded by *RGC2* family members.

12.6 MATING SYSTEM

B. lactucae is diploid for most of its life cycle. It is predominantly heterothallic with two distinct mating types, which are designated B_1 and B_2 (Michelmore and Ingram, 1980; Michelmore and Sansome, 1982). Coinoculation with isolates of both mating types results in the suppression of asexual sporulation and profuse oospore production. This allows controlled crosses between isolates with defined phenotypes. When the hyphae of opposite mating type come into physical contact, the morphology of the hyphae changes; there is a proliferation of short hyphae, and clusters of gametangia are formed at points of contact (Michelmore and Ingram, 1981a) (Fig. 12.1c). Multiple synchronous meioses occur in both the oogonium and periclinal antheridium; a single nucleus is transferred from the antheridium to the oogonium to effect fertilization (Michelmore and Sansome, 1982) (Fig. 12.1d). Oogonia and antheridia are probably produced by both mating types as in *Phytophthora* spp.; however, there has not been any investigation of maleness and femaleness in *B. lactucae*. Such studies would be greatly aided by the ability to label an individual mating type with a cytological marker such as green fluorescent protein (GFP) or GUS.

The mating type seems to be determined by two haplotypes at a single genetic locus. The mating type segregates in approximately 1:1 ratios in sexual progeny (Michelmore and Ingram, 1981b; Norwood et al., 1983; Michelmore et al., 1984; Sicard et al., 2003). The genetic data are consistent with B_1 compatibility type being conferred by a homozygous recessive condition and the B_2 mating type by a heterozygous condition. However, the molecular basis of the mating type in *B. lactucae* awaits characterization as it does for all oomycetes, and the current data do not preclude a more complicated situation such as double heterozygotes and balanced lethals, as has been proposed for *Phytophthora infestans* (Fabritius and Judelson, 1997). Nothing is known of the chemical communication involved in sexual reproduction of *B. lactucae*. There is no evidence of highly diffusible mating hormones because close physical contact seems to be required for the production of gametangia.

Some isolates of *B. lactucae* can produce oospores even when cultured alone. This is caused by secondary homothallism, which possibly results from trisomy of the mating type determinants (Michelmore and Ingram, 1982; Michelmore and Sansome, 1982). These isolates function predominantly as B_2 types in that they tend to reproduce asexually except when they come into contact with B_1 types, upon which abundant oospores are produced. When secondarily homothallic isolates are cultured at high inoculum densities, sparse oospore production occurs. Single conidial analysis demonstrated that homothallic isolates are unstable, and self-sterile B_1 and B_2 components can somatically segregate from a self-fertile progenitor. This somatic segregation involves transitory heterokaryosis as evidenced by abnormal hyphal growth similar to but less compact than the sexual hyphae (Michelmore and Ingram, 1982).

The prevalence of each mating type and the importance of the sexual cycle seem to vary in nature. Isolates of both mating types occur frequently in Europe and in New York State, although the B_2 type is sometimes more common (Michelmore and Ingram, 1980; Gustafsson et al., 1985; Yuen and Lorbeer, 1987; Lebeda and Blok, 1990; Petrželová and Lebeda, 2003; Lebeda et al., 2008a). The occurrence of both mating types is consistent with the sexual cycle, which generates the wide variety of virulence phenotypes that have been detected in these regions. In California, the B_2 mating type predominates, and B_1 isolates have been identified extremely rarely. The one B_1 isolate analyzed genetically from California had reduced fertility (Ilott et al., 1987).

12.7 SOMATIC VARIATION

There are only limited data that assess the importance of the sexual cycle in other parts of the world; however, phenotypic and molecular data suggest that *B. lactucae* can exhibit extensive somatic variation. A restriction fragment length polymorphism (RFLP) analysis of 25 isolates from diverse geographical origins worldwide revealed multiple ploidy levels and somatic variants (Hulbert and Michelmore, 1988). European isolates were diploid and heterozygous at approximately 44% of their loci; these isolates had highly variable genotypes consistent with the occurrence of the sexual cycle in Europe. In contrast, most isolates from Japan, Wisconsin, and Australia had more than two alleles at multiple RFLP loci, which indicates that they were either polyploids or stable heterokaryons (hyperploids). Sympatric non-European isolates with similar virulence phenotypes had similar genotypes, which indicates that they had been developed by the somatic loss of alleles. One hyperploid California isolate was the result of the fusion of two diploid California isolates of the same mating type; this was the first evidence for natural somatic fusion in the Oomycetes.

Virulence phenotype data for California isolates over the past 25 years indicate that the California population that infects cultivated lettuce is predominantly asexual; a limited spectrum of pathotypes predominates and persists from year to year (Ochoa and Michelmore, unpublished). However, even in the absence of the sexual cycle, the California population has still been able to change in fungicide (in)sensitivity and virulence phenotype in response to the deployment of new *Dm* genes. Fungicide insensitivity in both Europe and California occurred through somatic changes in isolates of the most common virulence phenotype (Crute et al., 1987; Schettini et al., 1991; Brown et al., 2004). The molecular events underlying these somatic changes in fungicide sensitivity and virulence phenotype are unknown but should be tractable as soon as the molecular basis of these phenotypes is determined.

12.8 GENETICS OF AVIRULENCE

Consistent with a gene-for-gene relationship, avirulence to specific *Dm* genes is usually inherited as single dominant unlinked loci (Michelmore and Ingram, 1981b; Norwood et al., 1983; Michelmore et al., 1984; Norwood and Crute, 1984; Ilott et al., 1987, 1989). A critical analysis of the gene-for-gene interaction between *B. lactucae* and lettuce was made that involved numerous crosses between 20 isolates of diverse worldwide geographical origins in parallel with studies of host resistance (Farrara et al., 1987; Ilott et al., 1989). Most data were consistent with a gene-for-gene interaction. Avirulence was predominantly determined by dominant alleles at unlinked loci, although phenotypes sometimes varied depending on the genetic background of the host and pathogen. Hyperploidy and gene-dosage effects could account for some segregation anomalies. One hundred and twenty-five tests that involved 19 crosses were made to test for complementation between avirulence (*Avr*) loci (Ilott et al., 1989). There was no case in which all progeny were avirulent to a specific *Dm* gene when both parental isolates had been virulent; therefore, avirulence to individual *Dm* genes was determined at the same locus in all isolates tested. Crosses were also made between avirulent and virulent isolates to search for dominant inhibitors of avirulence. Good but not unequivocal data were obtained for an inhibitor locus epistatic to *Avr5/8*; there was no evidence for inhibitors of other *Avr* loci despite earlier indications of more inhibitor loci (Ilott et al., 1989). Therefore, there currently is little evidence for inhibitor loci in *B. lactucae*, which is unlike the situation in phytopathogenic bacteria (Abramovitch et al., 2003; Espinosa et al., 2003; Jamir et al., 2004; Fu et al., 2007).

12.9 GENETIC MAPPING

Because of the outcrossing nature of *B. lactucae*, isolates are highly heterozygous. Segregation is analyzed in F_1 progeny; maps have to be constructed for each parent, and linkage groups can only be integrated when loci within them segregate in both parents. The first genetic linkage map of *B. lactucae* was constructed using the segregation of 53 RFLP loci, 8 *Avr* loci, and the mating type locus (Hulbert et al., 1988). A total of 70 F_1 individuals were analyzed from two crosses that involved three isolates of distinct geographic origins with one parent in common. Thirteen small linkage groups included 35 RFLP loci and one *Avr* gene; however, many markers were unlinked. The construction of a more comprehensive genetic map was hindered by the ambiguous phase of the alleles in the parents and a low number of markers because of the labor-intensive marker technology available at the time. The more diverse of the two crosses was expanded to 97 F_1 progeny to facilitate the identification of the phase of the parental alleles and to improve the power to detect linkage. This allowed the construction of more extensive genetic maps using 347 PCR-based amplified fragment length polymorphism (AFLP) markers and 83 RFLP loci as well as six *Avr* genes and the mating type locus (Sicard et al., 2003). All six *Avr* genes were mapped to different linkage groups, which is consistent with the lack of linkage observed in classic segregation analysis of *Avr* loci (Ilott et al., 1989). Four *Avr* loci were located at the ends of linkage groups; telomeric locations of *Avr* loci may contribute to the instability of avirulence phenotypes in *B. lactucae*.

12.10 KARYOTYPE AND CHROMOSOMAL ASSIGNMENT OF MARKERS

At least seven or eight chromosome pairs could be resolved at meiosis using light microscopy (Michelmore and Sansome, 1982); however, individual chromosomes of *B. lactucae* are too small to be resolved clearly. Analysis by pulse-field gel electrophoresis (PFGE) revealed a minimum of seven chromosomes that ranged in size from 3 to at least 8 Mb and a variable set of linear polymorphic molecules (Francis and Michelmore, 1993). Two distinct classes of molecules were confirmed by genetic and hybridization analyses. The larger, greater than 2 Mb class of molecules is constant in size and number and presumably represents the true chromosomes. A total of 25 RFLP probes could be successfully hybridized to these chromosomal molecules (Sicard et al., 2003). Twenty-three of these had been mapped and represented 16 linkage groups. Therefore, two consensus and seven parent-specific linkage groups could be assigned to chromosomes. Linkage to RFLP markers resulted in three *Avr* loci being assigned to chromosomes. Together, the physical and genetic data indicate that there are at least 10 chromosomes in *B. lactucae*. The smaller, 0.3 to 1.6 Mb class of molecules is inherited in a non-Mendelian manner, highly polymorphic, and variable in number; hybridization with probes derived from these molecules indicated that they are related in sequence (Francis and Michelmore, 1993). These small polymorphic molecules seem to be large linear plasmids or B chromosomes. None of the RFLP markers hybridized to these small molecules; therefore, there was no evidence that variability in these small molecules is involved in changes in specificity of *B. lactucae*.

12.11 EPIDEMIOLOGY

The availability of lettuce lines with well-defined resistance genes led to the development of a standard differential set of 24 resistant lines (Michelmore and Crute, 1982). These have been used widely to determine the virulence phenotypes of isolates of *B. lactucae* throughout the world. In 1998, the International *Bremia* Evaluation Board (IBEB) was established to standardize the nomenclature of isolates of *B. lactucae* (van Ettekoven and van der Arend, 1999). This has been a very successful collaboration between breeders in seed companies, government institutes, and academic researchers. The seeds of the differential set of lettuce lines and reference isolates of *B. lactucae* are maintained and distributed to researchers and commercial breeders. The isolates are assigned a sextet code on the basis of their reaction to the differential set (Limpert et al., 1994). The IBEB meets annually to review virulence phenotype data (mainly from Europe) and to assign new race designations, if warranted. The virulence phenotypes of over 2,000 isolates have been determined since 1999 (http://www.plantum.nl/ibeb.html). Important European isolates are assigned to BL races by the IBEB (currently up to BL26). California isolates are assigned to Pathotypes by University of California, Davis (currently up to Pathotype VIII). Assignment to these groupings is, however, for pragmatic utility rather than comprehensive categorization of all isolates. A race or pathotype designation is awarded to a group of isolates when isolates with the same virulence phenotype are observed from multiple locations in multiple years, and they can overcome most or all currently deployed *Dm* genes. Rare isolates with unique virulence phenotypes are frequently observed, but because they do not become established widely, they are not awarded race or pathotype status. Lettuce cultivars are described on the basis of the races and/or pathotypes to which they are resistant. The composition of the differential set has been periodically expanded as new *Dm* genes became available. This has allowed changes in pathogen virulence to be monitored in response to *Dm* gene deployment (http://www.plantum.nl/ibeb.html). However, the recent increase in the number of new *Dm* genes being used is going to require a much larger set of resistant lines to phenotype isolates.

The deployment of increased numbers of *Dm* genes, especially the introgression of different *Dm* genes into different lettuce types, is fragmenting the selection pressure on the pathogen population. Consequently, an increasing diversity of virulence phenotypes is being detected from commercial fields; also, anecdotal evidence suggests that the breakdown of resistance may be slowed. Determining the *Dm* gene complement of cultivars will become increasingly challenging as more *Dm* genes are deployed in the lettuce crop. The diagnosis and selection of *Dm* genes will be helped by molecular markers derived from cloned resistance genes. In addition, the cloning of the cognate avirulence genes and the generation of a diagnostic set of isogenic strains of *Agrobacterium tumefaciens* will allow transient expression of individual avirulence proteins and the functional identification of specific *Dm* genes.

It is not clear how much gene flow occurs between populations of *B. lactucae* growing on cultivated *L. sativa* and those growing on wild *L. serriola*. *B. lactucae* is highly host specific, and cross-inoculation experiments have demonstrated that isolates from *L. serriola* could only rarely infect *L. sativa* (Lebeda et al., 2008a). However, an extensive analysis of (a)virulence factors present in 313 isolates from *L. serriola* sampled over 7 years in the Czech Republic demonstrated that virulence to overcome most of the 28 resistance specificities tested was present in the isolates from *L. serriola* (Lebeda et al., 2008a). Virulence to overcome resistances of *L. serriola* origin was more common than virulence to overcome resistances originating from *L. sativa*. Also, coinfections with isolates from both species can result in oospore formation, which indicates the potential for novel hybrids and gene flow. Isolates from *L. serriola* exhibited high diversity in virulence phenotype. Populations of *L. serriola* exhibited parallel high levels of heterogeneity for resistance. Nearly a third of the 16 populations analyzed were completely susceptible to all 10 BL races tested, whereas only 10% were resistant to all 10; the remainder had heterogeneous reactions to the test isolates (Lebeda et al., 2008a). Parallel studies of *B. lactucae* from wild and cultivated hosts using molecular markers for (a)virulence genes are now needed; these should include populations collected from eastern Turkey and Armenia, which is the potential center of diversity for *L. serriola* (Kuang et al., 2008).

12.12 GENOME SIZE AND COMPOSITION

Several methods have been used to estimate the physical size of the *B. lactucae* genome. These have included dot-blot reconstructions that compared hybridizations to low-copy cloned DNA fragments and genomic DNA, DNA–DNA reassociation kinetics assayed by hydroxyapatite chromatography, and totaling chromosomal sizes determined by contour-clamped homogeneous electric field (CHEF) gel electrophoresis (Francis et al., 1990; Francis and Michelmore, 1993). All of these three methods gave similar estimates of 50 Mb; however, this estimate may be too low. The genome size estimates for *Aspergillus nidulans* and *Arabidopsis thaliana* that were used as controls in the genomic reconstruction experiments were 17 and 52 Mb, respectively, at the time of the analysis; their genome sizes have now been shown by genome sequencing to be closer to 30 and 125 Mb, respectively (Arabidopsis Genome Initiative, 2000; Galagan et al., 2005). Therefore, the genome size estimate for *B. lactucae* should probably be increased to approximately 100 Mb. This is consistent with estimates made using Feulgen absorbance cytophotometry of 70 to 144 Mb, depending on the isolate analyzed (Voglmayr and Greilhuber, 1998), and it is comparable with sizes estimated for other members of the Peronosporales (Govers and Gijzen, 2006). A more accurate estimation of the genome size of *B. lactucae* awaits sequencing and assembly of the complete genome.

The genome of *B. lactucae* is therefore approximately three to five times that of *Saccharomyces cerevisiae*. Consistent with this, DNA reassociation kinetics indicated that the nuclear DNA of *B. lactucae* is approximately 21% high-copy, 38% intermediate-copy, and 35% low-copy sequences (Francis et al., 1990). The low-copy sequences are interspersed with high-copy repeats as shown by hybridization analysis of random genomic λ clones to genomic DNA.

12.13 TRANSFORMATION

Efforts to transform *B. lactucae* began in the late 1980s. These techniques required the characterization of several genes from *B. lactucae* because no oomycete regulatory sequences were available at the time (Judelson and Michelmore, 1989, 1990). Although attempts to transform *B. lactucae* were not successful, the promoters and terminators of *Hsp70* and a constitutively, highly expressed single-copy gene *HAM34* from *B. lactucae* were critical to the subsequent successful transformation of several other oomycetes (Judelson and Michelmore, 1991; Judelson et al., 1991, 1992, 1993a and b). These regulatory sequences resulted in better gene expression than that achieved using several promoters from *Phytophthora* spp. and still provide the basis for expressing selectable markers and genes of interest in a wide range of oomycetes (Judelson and Ah-Fong, 2008). Interestingly, although *HAM34* is constitutively and highly expressed in *B. lactucae*, its sequence provides no clues as to its function.

It is present in *P. infestans* (Win et al., 2005), but not in the currently available sequence of *Hyaloperonospora arabidopsidis*.

The transformation of *B. lactucae* has yet to be achieved. Several methods to introduce DNA into *B. lactucae* were tried. Protoplasting that proved successful for *P. infestans* (Judelson et al., 1991) is not an option because of the obligate biotrophic nature of *B. lactucae*. Attempts to use *A. tumefaciens* were not successful but are worth retrying with better selectable markers. The most promising method was the microprojectile bombardment of conidia. Subsequent infection of lettuce seedlings yielded conidiophores and conidia that stained positive for GUS activity (Judelson and Ah-Fong, 2008) (see Fig. 12.1). However, this may have been transient expression without chromosomal integration, and no stable transformants of *B. lactucae* were obtained.

Efforts to transform *B. lactucae* were halted in 1990 in favor of working on the more tractable *P. infestans* (Judelson et al., 1991). It is time to resume experiments to transform *B. lactucae* using better selectable markers. Earlier experiments attempted to use selection for G418 resistance conferred by the *NPTII* gene. Elevated rates of sporulation were observed on transgenic lettuce cotyledons expressing *NPTII* and floating on a solution of G418 (Judelson and Michelmore, unpublished). However, this may have been caused by transient expression, like the GUS activity described above, or the detoxification of G418 by the transgenic lettuce. Selectable markers are needed that allow the growth of transformants without detrimental effects on wild-type lettuce. Insensitivity to fungicides such as metalaxyl offers this possibility, but no gene conferring fungicide insensitivity in oomycetes has yet been cloned.

12.14 MOLECULAR GENETICS AND GENOMICS

Genomic approaches are only just beginning for *B. lactucae*. Several EST and genomic libraries have been made from a variety of sources, which include conidia, germlings, and infected lettuce leaves. ESTs from a cDNA library made after subtraction of mock-inoculated leaf material from heavily infected leaves had a bimodal distribution of guanine-cytosine (GC) content indicative of sequences from lettuce and *B. lactucae* (unpublished results). At least a third of these sequences were judged to be of *B. lactucae* origin based on their GC content as well as their dissimilarity to plant and similarity to oomycete sequences. These putative *B. lactucae* unigenes had an average GC content of 50%. A normalized cDNA library enriched for *B. lactucae* sequences has been generated by hybridizing cDNA prepared from heavily infected leaves to *B. lactucae* genomic DNA using a protocol developed by J. Jones (Sainsbury Lab., Norwich, UK). The construction of fosmid and bacterial artificial chromosome (BAC) libraries is underway. Extensive genomic sequence is also being generated using the Illumina Genome Analyzer.

Sequences from all available libraries are being searched for candidate effectors using multiple bioinformatics approaches. Sequences with similarity to genes encoding known (a)virulence genes in other oomycetes have yet to be identified. However, this is not overly surprising as such genes may be evolving rapidly. Searches for sequences containing a canonical secretion signal peptide and RXLR amino acid motif have revealed several candidate effectors. These candidates are currently being tested for their ability to elicit an HR on a differential set of resistant cultivars using *Agrobacterium*-mediated transient assays (Wroblewski et al., 2005).

12.15 FUTURE DIRECTIONS

B. lactucae has been ranked as one of the high-priority plant pathogens targeted for genome sequencing since 2002 (American Phytopathological Society, 2006); this has yet to occur but should happen soon. A combination of the new generation of sequencing technologies and conventional Sanger sequencing will provide large amounts of genomic sequence information for *B. lactucae*. The challenge will be genome assembly and annotation. This will be facilitated by the increasing number of sequenced oomycete genomes. Similarly, *B. lactucae* will provide another reference genome for annotation and the development of gene models for oomycetes. We will particularly focus on sequencing the gene space of multiple isolates of *B. lactucae*. This will provide an expedient and cost-efficient approach to the identification of effector proteins and other types of molecules involved in determining specificity, fungicide insensitivity, and mating type.

The sequencing of the gene space of multiple isolates of *B. lactucae* will result in the identification of numerous single nucleotide polymorphisms (SNPs). The new generation of marker technologies, such as the Illumina Goldengate SNP assay, will enable analysis of large populations for variation at effector loci and in other genes important to virulence. Such studies will help to elucidate the mechanisms of variation underlying changes in virulence and fungicide insensitivity as well as to reveal the patterns of gene flow within and between populations of *B. lactucae* in different parts of the world. Sequencing the genome of *B. lactucae* and comparing it with the genomes of nonbiotrophic oomycetes will also indicate the extent to which its genome has become reduced as a consequence of its biotrophic mode of nutrition. The identification of missing biosynthetic capabilities may allow the development of media for axenic culture.

ACKNOWLEDGMENTS

The work described here has resulted from the efforts of many people in the RWM and other laboratories over the past 25 years. We thank them all for their contributions. Financial support has come from multiple sources and includes sustained support from the California Lettuce Research Board and the United States Department of Agriculture (USDA) Cooperative State Research, Education, and Extension Service (CSREES) National Research Initiative.

REFERENCES

Abramovitch RB, Kim YJ, Chen S, Dickman MB, Martin GB 2003. *Pseudomonas* type III effector AvrPtoB induces plant disease susceptibility by inhibition of host programmed cell death. EMBO J 22:60–69.

American Phytopathological Society 2006. Microbial genomic sequencing: perspectives of the American Phytopathological Society (Revised 2006). http://199.86.26.56/members/ppb/PDFs/MicrobialGenomicsSeq06.pdf. Accessed 28 Aug 2008.

Arabidopsis Genome Initiative 2000. Analysis of the genome sequence of the flowering plant *Arabidopsis thaliana*. Nature 408:796–815.

Beharav A, Lewinsohn D, Lebeda A, Nevo E 2006. New wild *Lactuca* genetic resources with resistance against *Bremia lactucae*. Genet Reso Crop Evolut 53:467–474.

Bennett M, Gallagher M, Fagg J, Bestwick C, Paul T, Beale M, Mansfield J 1996. The hypersensitive reaction, membrane damage and accumulation of autofluorescent phenolics in lettuce cells challenged by *Bremia lactucae*. Plant J 9:851–865.

Bestwick CS, Brown IR, Mansfield JW 1998. Localized changes in peroxidase activity accompany hydrogen peroxide generation during the development of a nonhost hypersensitive reaction in lettuce. Plant Physiol 118:1067–1078.

Bonnier FJK, Reinink K, Groenwold R 1994. Genetic analysis of *Lactuca* accessions with new major gene resistance to lettuce downy mildew. Phytopathol 84:462–468.

Brown S, Koike ST, Ochoa OE, Laemmlen F, Michelmore RW 2004. Insensitivity to the fungicide fosetyl-aluminum in California isolates of the lettuce downy mildew pathogen, *Bremia lactucae*. Plant Dis 88:502–508.

Chin DB, Arroyo-Garcia R, Ochoa OE, Kesseli RV, Lavelle DO, Michelmore RW 2001. Recombination and spontaneous mutation at the major cluster of resistance genes in lettuce (*Lactuca sativa*). Genetics 157:831–849.

Choi YJ, Hong SB, Shin HD 2007. Extreme size and sequence variation in the ITS rDNA of *Bremia lactucae*. Mycopathologia 163:91–95.

Crute IR 1981. The host specificity of *Peronosporaceous* fungi and the genetics of the relationship between host and parasite. In: Spencer DM, editor. The Downy Mildews. Academic Press. London, UK. pp. 237–250.

Crute IR, Johnson AG 1976. The genetic relationship between races of *Bremia lactucae* and cultivars of *Lactuca sativa*. Ann Appl Biol 83:125–137.

Crute IR, Norwood JM 1978. Incomplete specific resistance to *Bremia lactucae* in lettuce. Ann Appl Biol 89:467–474.

Crute IR, Norwood JM, Gordon PL 1987. The occurrence, characteristics and distribution in the United Kingdom of resistance to phenylamide fungicides in *Bremia lactucae* (lettuce downy mildew). Plant Pathol 36:297–315.

Eenink AH, Groenwold R, Bijker W 1983. Partial resistance of lettuce to downy mildew (*Bremia lactucae*). 4. Resistance after natural, semi-artificial and artificial infestation and examples of mutual interference of resistance levels. Euphytica 32:139–149.

Espinosa A, Guo M, Tam VC, Fu ZQ, Alfano JR 2003. The *Pseudomonas syringae* type III-secreted protein HopPtoD2 possesses protein tyrosine phosphatase activity and suppresses programmed cell death in plants. Molec Microbiol 49:377–387.

Fabritius AL, Judelson HS 1997. Mating type loci segregate aberrantly in *Phytophthora infestans* but normally in *Phytophthora parasitica*: implications for models of mating-type determination. Curr Genet 32:60–65.

Farrara BF, Ilott TW, Michelmore RW 1987. Genetic analysis of factors for resistance to downy mildew (*Bremia lactucae*) in species of lettuce (*Lactuca sativa* and *L. serriola*). Plant Pathol 36:499–514.

Farrara BF, Michelmore RW 1987. Identification of new sources of resistance to downy mildew in *Lactuca* spp. Hortscience 22:647–649.

Francis DM, Hulbert SH, Michelmore RW 1990. Genome size and complexity of the obligate fungal pathogen, *Bremia lactucae*. Exp Mycol 14:299–309.

Francis DM, Michelmore RW 1993. Two classes of chromosome-sized molecules are present in *Bremia lactucae*. Exp Mycol 17:284–300.

Fu ZQ, Guo M, Jeong BR, Tian F, Elthon TE, Cerny RL, Staiger D, Alfano JR 2007. A type III effector ADP-ribosylates RNA-binding proteins and quells plant immunity. Nature 447:284–288.

Galagan JE, Calvo SE, Cuomo C, Ma LJ, Wortman JR, Batzoglou S, Lee SI, Basturkmen M, Spevak CC, Clutterbuck J, et al. 2005. Sequencing of *Aspergillus nidulans* and comparative analysis with *A. fumigatus* and *A. oryzae*. Nature 438: 1105–1115.

Govers F, Gijzen M 2006. *Phytophthora* genomics: The plant destroyers' genome decoded. Molec Plant-Microbe Interact 19:1295–1301.

Gustafsson I 1989. Potential sources of resistance to lettuce downy mildew (*Bremia lactucae*) in different *Lactuca* species. Euphytica 40:227–232.

Gustafsson M, Liljeroth E, Gustafsson I 1985. Pathogenic variation and sexual reproduction in Swedish populations of *Bremia lactucae*. Theoret Appl Genet 70:643–649.

Hulbert SH, Hott TW, Legg EJ, Lincoln SE, Lander ES, Michelmore RW 1988. Genetic analysis of the fungus, *Bremia lactucae*, using restriction fragment length polymorphisms. Genetics 120:947–958.

Hulbert SH, Michelmore RW 1985. Linkage analysis of genes for resistance to downy mildew (*Bremia lactucae*) in lettuce (*Lactuca sativa*). Theoret Appl Genet 70:520–528.

Hulbert SH, Michelmore RW 1988. DNA restriction fragment length polymorphism and somatic variation in the lettuce downy mildew fungus, *Bremia lactucae*. Molec Plant-Microbe Interact 1:17–24.

Ilott TW, Durgan ME, Michelmore RW 1987. Genetics of virulence in California populations of *Bremia lactucae* (lettuce downy mildew). Phytopathol 77:1381–1386.

Ilott TW, Hulbert SH, Michelmore RW 1989. Genetic analysis of the gene-for-gene interaction between lettuce (*Lactuca sativa*) and *Bremia lactucae*. Phytopathol 79: 888–897.

Jamir Y, Guo M, Oh HS, Petnicki-Ocwieja T, Chen SR, Tang XY, Dickman MB, Collmer A, Alfano JR 2004. Identification of *Pseudomonas syringae* type III effectors that can suppress programmed cell death in plants and yeast. Plant J 37:554–565.

Jeuken M, Lindhout P 2002. *Lactuca saligna*, a non-host for lettuce downy mildew (*Bremia lactucae*), harbors a new race-specific *Dm* gene and three QTLs for resistance. Theoret Appl Genet 105:384–391.

Judelson H, Ah-Fong AMV 2008. Progress and challenges in oomycete transformation. This volume.

Judelson HS, Coffey MD, Arredondo FR, Tyler BM 1993a. Transformation of the oomycete pathogen *Phytophthora megasperma* f. sp. *glycinea* occurs by DNA integration into single or multiple chromosomes. Curr Genet 23:211–218.

Judelson HS, Dudler R, Pieterse CMJ, Unkles SE, Michelmore RW 1993b. Expression and antisense inhibition of transgenes in *Phytophthora infestans* is modulated by choice of promoter and position effects. Gene 133:63–69.

Judelson HS, Michelmore RW 1989. Structure and expression of a gene encoding heat-shock protein Hsp70 from the Oomycete fungus *Bremia lactucae*. Gene 79: 207–217.

Judelson HS, Michelmore RW 1990. Highly abundant and stage-specific mRNAs in the obligate pathogen *Bremia lactucae*. Molec Plant-Microbe Interact 3:225–232.

Judelson HS, Michelmore RW 1991. Transient expression of genes in the oomycete *Phytophthora infestans* using *Bremia lactucae* regulatory sequences Curr Genet 19: 453–459.

Judelson HS, Michelmore RW 1992. Temperature and genotype interactions in the expression of host resistance in lettuce downy mildew. Physiol Molec Plant Pathol 40:233–245.

Judelson HS, Tyler BM, Michelmore RW 1991. Transformation of the Oomycete pathogen, *Phytophthora infestans*. Molec Plant-Microbe Interact 4:602–607.

Judelson HS, Tyler BM, Michelmore RW 1992. Regulatory sequences for expressing genes in oomycete fungi. Molec Gen Genet 234:138–146.

Kesseli RV, Paran I, Michelmore RW 1994. Analysis of a detailed genetic linkage map of *Lactuca sativa* (Lettuce) constructed from RFLP and RAPD markers. Genetics 136:1435–1446.

Kuang H, Ochoa OE, Nevo E, Michelmore RW 2006. The disease resistance gene *Dm3* is infrequent in natural populations of *Lactuca serriola* due to deletions and frequent gene conversions at the *RGC2* locus. Plant J 47:38–48.

Kuang HH, van Eck HJ, Sicard D, Michelmore R, Nevo E 2008. Evolution and genetic population structure of prickly lettuce (*Lactuca serriola*) and its RGC2 resistance gene cluster. Genetics 178:1547–1558.

Kuang H, Woo SS, Meyers BC, Nevo E, Michelmore RW 2004. Multiple genetic processes result in heterogeneous rates of evolution within the major cluster disease resistance genes in lettuce. Plant Cell 16:2870–2894.

Lebeda A, Blok I 1990. Sexual compatibility types of *Bremia lactucae* isolates originating from *Lactuca serriola*. Netherlands J Plant Pathol 96:51–54.

Lebeda A, Petrzelova I, Maryska Z 2008a. Structure and variation in the wild-plant pathosystem: *Lactuca serriola–Bremia lactucae*. Eur J Plant Pathol 122:127–146.

Lebeda A, Pink DAC, Astley D 2002. Aspects of the interactions between wild *Lactuca* spp. and related genera and lettuce downy mildew (*Bremia lactucae*). In: Spencer-Phillips PTN, Gisi U, Lebeda A, editors. Advances in Downy Mildew Research. Kluwer Academic. Dordrecht, Germany. pp. 85–117.

Lebeda A, Sedlárová M, Petrivalský M, Prokopová J 2008b. Diversity of defence mechanisms in plant-oomycete interactions: A case study of *Lactuca* spp. and *Bremia lactucae*. Eur J Plant Pathol 122:71–89.

Lebeda A, Zinkernagel V 2003. Characterization of new highly virulent German isolates of *Bremia lactucae* and efficiency of resistance in wild *Lactuca* spp. germplasm. J Phytopathol 151:274–282.

Limpert E, Clifford B, Dreiseitl A, Johnson R, Muller K, Roelfs A, Wellings C 1994. Systems of designation of pathotypes of plant pathogens. J Phytopathol 140:359–362.

Maclean DJ, Sargent JA, Tommerup IC, Ingram DS 1974. Hypersensitivity as a primary event in resistance to fungal parasites. Nature 249:186–187.

Maclean DJ, Tommerup IC 1979. Histology and physiology of compatibility and incompatibility between lettuce and the downy mildew fungus, *Bremia lactucae* Regel. Physiol Plant Pathol 14:291–312.

Maisonneuve B, Bellec Y, Anderson P, Michelmore RW 1994. Rapid mapping of two genes for resistance to downy mildew from *Lactuca serriola* to existing clusters of resistance genes. Theoret Appl Genet 89:96–104.

McHale L, Tan X, Koehl P, Michelmore RW 2006. Plant NBS-LRR proteins: adaptable guards. Genome Biol 7:212.

Meyers BC, Chin DB, Shen KA, Sivaramakrishnan S, Lavelle DO, Zhang Z, Michelmore RW 1998a. The major resistance gene cluster in lettuce is highly duplicated and spans several megabases. Plant Cell 10:1817–1832.

Meyers BC, Shen KA, Rohani P, Gaut BS, Michelmore RW 1998b. Receptor-like genes in the major resistance locus of lettuce are subject to divergent selection. Plant Cell 10:1833–1846.

Michelmore RW 1981. Sexual and asexual sporulation in the downy mildews. In: Spencer DM, editor. The Downy Mildews. Academic Press. London, UK. pp. 165–181.

Michelmore RW, Crute IR 1982. A method for determining the virulence phenotype of isolates of *Bremia lactucae*. Trans B Mycol Soc 79:542–546.

Michelmore RW, Ingram DS 1980. Heterothallism in *Bremia lactucae*. Trans B Mycol Soc 75:47–56.

Michelmore RW, Ingram DS 1981a. Origin of gametangia in heterothallic isolates of *Bremia lactucae*. Trans B Mycol Soc 76:425–432.

Michelmore RW, Ingram DS 1981b. Recovery of progeny following sexual reproduction of *Bremia lactucae*. Trans B Mycol Soc 77:131–137.

Michelmore RW, Ingram DS 1982. Secondary homothallism in *Bremia lactucae*. Trans B Mycol Soc 78:1–9.

Michelmore RW, Meyers BC 1998. Clusters of resistance genes in plants evolve by divergent selection and a birth-and-death process. Gen Res 8:1113–1130.

Michelmore RW, Norwood JM, Ingram DS, Crute IR, Nicholson P 1984. The inheritance of virulence in *Bremia lactucae* to match resistant factors 3, 4, 5, 6, 8, 9, 10 and 11 in lettuce (*Lactuca sativa*). Plant Pathol 33:301–315.

Michelmore RW, Sansome ER 1982. Cytological studies of heterothallism and secondary homothallism in *Bremia lactucae*. Trans B Mycol Soc 79:291–297.

Nordskog B, Gadoury DM, Seem RC, Hermansen A 2007. Impact of diurnal periodicity, temperature, and light on sporulation of *Bremia lactucae*. Phytopathol 97:979–986.

Norwood JM, Crute IR 1984. The genetic control and expression of specificity in *Bremia lactucae* (lettuce downy mildew). Plant Pathol 33:385–399.

Norwood JM, Michelmore RW, Crute IR, Ingram DS 1983. The inheritance of specific virulence of *Bremia lactucae* (downy mildew) to match resistance factors 1, 2, 4, 6 and 11 in *Lactuca sativa* (lettuce). Plant Pathol 32:177–186.

Petrželová I, Lebeda A 2003. Distribution of compatibility types and occurrence of sexual reproduction in natural populations of *Bremia lactucae* on wild *Lactuca serriola* plants. Acta Phytopathol Entomol Hungarica 38:43–52.

Raffray JB, Sequeira L 1971. Dark induction of sporulation in *Bremia lactucae*. Can J Bot 49:237–239.

Sargent JA 1976. Germination of spores of *Bremia lactucae*. Ann Appl Biol 84:290–294.

Sargent JA, Tommerup IC, Ingram DS 1973. The penetration of a susceptible lettuce variety by the downy mildew fungus *Bremia lactucae* Regel. Physiol Plant Pathol 3:231–239.

Schettini TM, Legg EJ, Michelmore RW 1991. Insensitivity to metalaxyl in California populations of *Bremia lactucae* and resistance of California lettuce cultivars to downy mildew. Phytopathol 81:64–70.

Shen KA, Chin DB, Arroyo-Garcia R, Ochoa OE, Lavelle DO, Wroblewski T, Meyers BC, Michelmore RW 2002. *Dm3* is one member of a large constitutively expressed family of nucleotide binding site-leucine-rich repeat encoding genes. Molec Plant-Microbe Interact 15:251–261.

Shen KA, Meyers BC, Islam-Faridi MN, Chin DB, Stelly DM, Michelmore RW 1998. Resistance gene candidates identified by PCR with degenerate oligonucleotide primers map to clusters of resistance genes in lettuce. Molec Plant-Microbe Interact 11:815–823.

Sicard D, Legg E, Brown S, Babu NK, Ochoa O, Sudarshana P, Michelmore RW 2003. A genetic map of the lettuce downy mildew pathogen, *Bremia lactucae*, constructed from molecular markers and avirulence genes. Fungal Genet Biol 39:16–30.

Sicard D, Woo SS, Arroyo-Garcia R, Ochoa O, Nguyen D, Korol A, Nevo E, Michelmore R 1999. Molecular diversity at the major cluster of disease resistance genes in cultivated and wild *Lactuca* spp. Theoret Appl Genet 99:405–418.

Skidmore DI, Ingram DS 1985. Conidial morphology and specialization of *Bremia lactucae* Regel (Peronosporaceae) on hosts in the family Compositae. Bot J Linn Soc 91:503–522.

USDA-NASS 2008. Crop values 2007 summary. Washington, D.C.: US Department of Agriculture, National Agricultural Statistics Service. http://usda.mannlib.cornell.edu/usda/current/CropValuSu/CropValuSu-02-14-2008.pdf. Accessed 28 Aug 2008.

van Ettekoven K, van der Arend AJM 1999. Identification and denomination of "new" races of *Bremia lactucae*. In: Lebeda A, Krístková E, editors. Eucarpia leafy vegetables '99 Olomouc. Palacký University. Czech Republic. pp. 171–175.

Voglmayr H, Constantinescu O 2008. Revision and reclassification of three *Plasmopara* species based on morphological and molecular phylogenetic data. Mycol Res 112:487–501.

Voglmayr H, Greilhuber J 1998. Genome size determination in Peronosporales (Oomycota) by Feulgen image analysis. Fungal Genet Biol 25:181–195.

Voglmayr H, Rietmuller A, Goker M, Weiss M, Oberwinkler F 2004. Phylogenetic relationships of *Plasmopara*, *Bremia* and other genera of downy mildew pathogens with pyriform haustoria based on Bayesian analysis of partial LSU rDNA sequence data. Mycol Res 108:1011–1024.

Win J, Kanneganti TD, Torto-Alalibo T, Kamoun S 2005. Computational and comparative analyses of 150 full-length cDNA sequences from the oomycete plant pathogen *Phytophthora infestans*. Fungal Genet Biol 43:20–33.

Witsenboer H, Kesseli RV, Fortin MG, Stanghellini M, Michelmore RW 1995. Sources and genetic structure of a cluster of genes for resistance to three pathogens in lettuce. Theoret Appl Genet 91:178–188.

Woods AM, Didehvar F, Gay JL, Mansfield JW 1988. Modification of the host plasmalemma in haustorial infections of *Lactuca sativa* by *Bremia lactucae*. Physiol Molec Plant Pathol 33:299–310.

Wroblewski T, Piskurewicz U, Tomczak A, Ochoa O, Michelmore RW 2007. Silencing of the major family of NBS-LRR-encoding genes in lettuce results in the loss of multiple resistance specificities. Plant J 51:803–818.

Wroblewski T, Tomczak A, Michelmore R 2005. Optimization of *Agrobacterium*-mediated transient assays of gene expression in lettuce, tomato and *Arabidopsis*. Plant Biotechn J 3:259–273.

Yuen JE, Lorbeer JW 1987. Natural and experimental production of oospores of *Bremia lactucae* in lettuce in New York. Plant Dis 71:63–64.

13

DOWNY MILDEW OF *ARABIDOPSIS* CAUSED BY *HYALOPERONOSPORA ARABIDOPSIDIS* (FORMERLY *HYALOPERONOSPORA PARASITICA*)

NIKOLAUS L. SCHLAICH AND ALAN SLUSARENKO
Department of Plant Physiology (BioIII), RWTH Aachen University, Aachen, Germany

13.1 INTRODUCTION

Downy mildew of *Arabidopsis thaliana* is caused by the Oomycete plant pathogen *Hyaloperonospora parasitica* (formerly *Peronospora parasitica*). Currently, a new analysis of the phylogenetic relationships of Oomycetes is underway (see Chapters 1 and 12 in this volume by Göker et al.), and after a short interim as *Hyaloperonospora arabidopsis*, it was proposed that *H. parasitica* be reassigned to *Hyaloperonospora arabidopsidis* (Gäumann), Göker, Riethmüller, Voglmayr, Weiß & Oberw., which again matches the-*sidis* ending proposed by Gäumann in 1918 (Göker et al., 2004) (Göker, personal communication).

It is interesting to consider the historical timeline concerning the human view of this disease, which has been eating *Arabidopsis* for an evolutionarily long period, oblivious to our humble deliberations to define it accurately. Thus, Lindau (1901) reported *P. parasitica* as a pathogen of *Stenophragma thalianum* (L.) Cel., which is a synonym for *Arabidopsis thaliana* (L.) Heynh. (Lindau, 1901). Until 1918, most downy mildews that infected brassicaceous hosts were assigned to *P. parasitica*. Gäumann used host range and conidial morphology to define 49 new species of *Peronospora* infecting various brassicas, which include *P. arabidopsidis*; and thus brought the total of brassica-infecting species to 52 (Gäumann, 1918). In this publication, Gäumann refers to *P. arabidopsidis* as synonymous with *P. parasitica*

Oomycete Genetics and Genomics: Diversity, Interactions, and Research Tools
Edited by Kurt Lamour and Sophien Kamoun
Copyright © 2009 John Wiley & Sons, Inc.

Pers. f. *sisymbrii-thaliani* Schneider 1865 (*nom. nud. in sched.*). *Sisymbrium thalianum* (L.) Gay and *Arabidopsis thaliana* (L.) Heynh. are synonyms; thus, this is probably the earliest known report of this disease. However, the narrow species concept introduced by Gäumann was not accepted by Yerkes and Shaw, who cited the many morphological similarities among the downy mildew isolates that attack brassicas and Gäumann's species were "lumped" together again as *P. parasitica*, while recognizing that there were some isolate-specific differences (Yerkes and Shaw, 1959). Many brassica-infecting downy mildews were placed in the novel genus *Hyaloperonospora* by Constantinescu and Fatehi based on morphology and ITS1, ITS2, and 5.8S rDNA sequence analyses (Constantinescu and Fatehi, 2002). The genus also contains other species parasitizing specific brassicas (e.g., *Hyaloperonospora lepidii-perfoliati* on *Lepidium*) and members of other families (e.g., *Hyaloperonospora floerkea*). The latest revisions are based on molecular phylogenetic methods, and the reader is referred to Chapters 1 and 12 in this volume for more details. The Index Fungorum (http://www.indexfungorum.org/names/Names.asp) lists the current name of downy mildew isolates that infect *Arabidopsis* as *H. arabidopsidis*. Thus, the entry reads as follows:

Current Name: ***Hyaloperonospora arabidopsidis* (Gäum.) Göker, Riethm., Voglmayr, Weiss & Oberw.** [as '*arabidopsis*'], *Mycol. Prog.* **3**(2): 89 (2004)

Synonymy: ***Peronospora arabidopsidis* Gäum.** (1918)

Because of the nature of the data used in this latest taxonomic revision, it is likely that this latest name change will be stable. Therefore, we have decided to use *Hyaloperonospora arabidopsidis* throughout this review, and we hope that this will be taken up by others working with this pathosystem.

One unfortunate consequence of these recent name changes is that the resistance gene designations no longer bear any relation to the name of the organism itself. Thus, the recognition of *Peronospora parasitica* (*RPP*) gene designation, through the changes from *P. parasitica* to *H. parasitica* to *H. arabidopsidis*, is no longer intuitively connected with the downy mildew pathogen from its current name. This perhaps illustrates the desirability for better communication among all researchers working with an organism so that changes in one area do not result in unforeseen, irritating consequences for researchers in another area.

Downy mildew of *Arabidopsis* is an obligate biotrophic infection that, in common with other downy mildew diseases, typically requires high humidity and low temperatures for infection and sporulation (Channon, 1981; Slusarenko and Schlaich, 2003). Details of the disease cycle and photographs of infection structures have been published previously, and the reader is referred to these earlier works for details (Koch and Slusarenko, 1990; Mauch-Mani and Slusarenko, 1993; Slusarenko and Schlaich, 2003). The aerial parts of the plant are typically those infected, but it should be noted that in the field, the pathogen overwinters as oospores in leaf debris from the previous season and a new infection cycle begins via oospores germinating and infecting roots. How organ-specificity in this pathosystem may be regulated is discussed in a later section.

The naming convention for field isolates of the pathogen is explained in Slusarenko and Schlaich (2003) and was originally published by Dangl et al. (1992). Briefly, a four-letter system is used in which the first two letters designate the location where the isolate was collected (e.g., EM for East Malling, UK) and the last two letters designate a susceptible *Arabidopsis* ecotype (e.g., WA for Wassilewskija), which results in the EMWA isolate. The pathogen and host show a classic gene-for-gene relation with one another (Dangl et al., 1992; Mauch-Mani and Slusarenko, 1993; Holub et al., 1994), and both resistance genes and pathogen avirulence genes have been cloned (see later). Systemic acquired resistance against the pathogen was reported early on (Uknes et al., 1992); thus, with all the advantages the host has to offer, such as collections of mutants, ease of genetic transformation, known genome sequence, and so on, this pathosystem has become a choice model for addressing problems in molecular plant pathology (Slusarenko and Schlaich, 2003). However, the obligate biotrophic nature of the pathogen has meant that progress with *Hyaloperonospora* has not moved at the same pace as that with the host or other more genetically tractable pathogens. Nevertheless, several pathogen effectors have been cloned recently (Allen et al., 2004; Rehmany et al., 2005), and the *H. arabidopsidis* genome has been sequenced (http://genome.wustl.edu/pub/organism/Fungi/Hyaloperonospora_parasitica/assembly/Hyaloperonospora_parasitica-2.0/).

This review will concentrate on the most recent research developments with this pathosystem, and the reader is encouraged to refer to earlier reviews that contain much information, which in the interests of brevity is not repeated here (Slusarenko and Schlaich, 2003; Holub, 2008).

13.2 ISSUES PERTAINING TO THE BIOTROPHIC LIFE STYLE

Because *H. arabidopsidis* cannot be grown independently of its host, several standard research approaches are very difficult. Apart from the obvious one that it is simply tedious and labor intensive to keep a continual supply of living plants and growth room space available for routine propagation of the pathogen, many other actions become difficult or impossible. Thus, it is difficult to get the pathogen sterile to have sterile inoculum to work in tissue culture (Hermanns et al., 2002, 2003), it is difficult to carry out genetic transformation and selection to yield stable transformants, and it is time consuming to obtain sufficient pathogen biomass as a starting point for nucleic acid extractions or biochemical analyses.

A characteristic of biotrophic hyphal pathogens is the differentiation of specialized organs called haustoria. It is generally assumed that haustoria, which penetrate through the cell wall and invaginate the host plasma membrane, serve as the main interface for nutrient uptake by the pathogen. Indeed, this has been convincingly demonstrated for a rust fungus (*Uromyces fabae*)

(Voegele and Mendgen, 2003). Nevertheless, these authors also report that, although the haustoria are the main site of carbohydrate uptake, amino acids are also taken up along the intercellular hyphae. Haustoria are present in representatives of basidiomycete and ascomycete fungi (e.g., rusts and powdery mildews) but certainly evolved independently in the Oomycetes, which are not true fungi. Thus, although morphologically similar (see Fig. 13.1), there is also potential functional divergence between these organs in the different groups (Spencer-Phillips, 1997). For example, in contrast to the rust *U. fabae*, the uptake of sugars by intercellular hyphae of pea downy mildew (*Peronospora viciae*) has been demonstrated (Clark and Spencer-Phillips, 1993, 2004; El Gariani and Spencer-Phillips, 2004). It should also be considered that although the rate of nutrient uptake via haustoria might be high compared with that by intercellular hyphae, the latter have a much greater relative surface area and may thus still contribute significantly to nutrition. Interestingly, uptake mechanisms may differ between the two interfaces because it was shown for *P. viciae* that *p*-chloromercuribenzene sulfonate (PCMBS), which is a thiol reagent that inhibits the plant plasmalemma sucrose transporter, blocked sucrose uptake by haustoria but not by intercellular hyphae (Spencer-Phillips, 1997).

In addition to a role in nutrient uptake, it is suggested that fungal haustoria can be involved in the suppression of host defenses and in redirecting host metabolism as well as being major sites of biosynthesis for the pathogen (Voegele and Mendgen, 2003). It is possible that oomycete haustoria have analogous functions. Another possible function is that haustoria might be a major site for the transfer of fungal effector molecules into plant cells. The RXLR motif was recently identified as a motif downstream of a secretory signal present in oomycete effector proteins transferred into plant cells (see later) (Birch et al., 2006, 2008). Although haustoria may be major sites of effector transfer, there is also the possibility that the RXLR-containing proteins might also be secreted from growing hyphae. Thus, a hypersensitive reaction can often be observed in epidermal cells in which a penetration hypha has begun to form underneath an appressorium but without production of a lateral haustorium into the cell (Koch and Slusarenko, 1990). Furthermore, in some incompatible genotype combinations (e.g., *H. arabidopsidis* isolate WELA with Col-0), the pathogen makes some growth into the mesophyll, and penetration is eventually stopped by an extended hypersensitive response (HR), although not all HR-responding cells seem to be penetrated by haustoria (Koch and Slusarenko, 1990). Similarly, the "trailing necrosis" phenomenon, which is observed in host tissues induced to systemic acquired resistance, is associated with a swathe of dead host cells that are not all penetrated by haustoria. Another indicator for release of effectors from intercellular hyphae might be the presence of race-specific elicitor active molecules in intercellular washing fluids from compatible interactions of *Arabidopsis* with *H. arabidopsidis* (Rethage et al., 2000; Soylu and Soylu, 2003). These observations suggest that effector molecules are also being released into the apoplast.

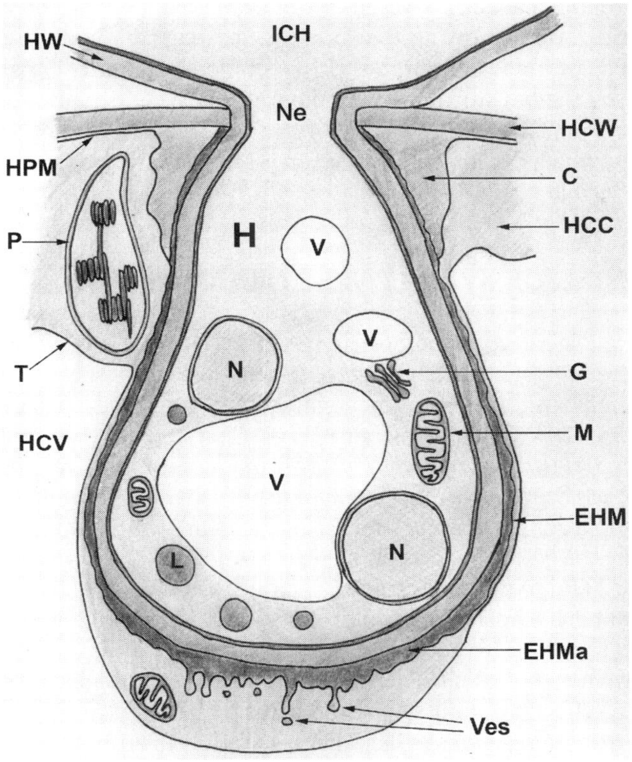

FIG. 13.1 Schematic representation of a haustorium of *Hyaloperonospora arabidopsidis* in an *Arabidopsis* host cell (based on electron micrographs from Mims et al. (2004)). C = collar of host material (including callose), CC = host cell cytoplasm, CV = host cell vacuole, CW = host cell wall, EHMa = electron dense extra haustorial matrix, EHM = extra haustorial membrane, G = Golgi body, H = haustorium, HM = hyphal membrane, HW = hyphal wall, Ne = neck region, PM = host plasma membrane, HW hyphal wall, ICH = intercellular hypha, L = electron dense lipid vesicle, M = mitochondrion, N = nucleus, Ne = constricted neck region, P = plastid, PM = invaginated host cell plasma membrane, T = tonoplast membrane of the host cell vacuole, V = vacuoles in the haustorium, Ves = vesicles either fusing with, or budding off from, the extra haustorial matrix.

The ultrastructure of the interface between *H. arabidopsidis* and *A. thaliana* was nicely detailed by Mims et al. (2004). Fig. 13.1 is a diagrammatic representation based on their findings. The authors observed vesicles (Ves) at the interface of the extrahaustorial matrix (EHMa) and the host cell cytoplasm (CC). It could be interpreted that these vesicles are budding off from, or coalescing with, the extrahaustorial membrane, or both. Thus, these vesicles may represent a major mechanism and site of information/nutrient exchange between host and pathogen. The collar of host-cell material could be decorated with immunogold particles labeled with a monoclonal antibody recognizing β-1,3-glucan epitopes showing that it is at least partially composed of callose. The host-cell plasma membrane (PM) could be clearly observed to be invaginated by the haustorium, becoming the EHM. Other interesting features of the haustorium are the presence of multiple vacuoles and several nuclei. Golgi bodies, which are absent in true fungi but present in oomycete cells, were also clearly visible.

13.3 CLONING OF AVIRULENCE EFFECTORS (ATRs)

Arabidopsis thaliana recognized proteins (ATRs) (Holub et al., 1994) are effectors made by *H. arabidopsidis* that trigger *R* gene-dependent defense responses in *Arabidopsis* cells and are thus functional homologs of the type three secretion system (TTSS)-delivered effector proteins of phytopathogenic bacteria [(e.g., *Pseudomonas syringae* pv. *tomato* (Pst)] (McCann and Guttman, 2008). Bacterial effectors like avrRpm1 or avrRpt2 are transported via a pilus-like structure into the plant cytoplasm (Buttner and Bonas, 2002) where they exert their effects aimed at diminishing the defense efforts of the plant (Mudgett, 2005; Nomura et al., 2005; Grant et al., 2006). According to models of coevolutionary warfare between the pathogen and the host plant (Chisholm et al., 2006; Jones and Dangl, 2006), plants acquired the capability to recognize pathogen manipulation of effector-targeted proteins in the host cell using *R* gene-encoded resistance proteins (see below). This recognition of effector action allows the plant to initiate active countermeasures that lead to defense reactions, which include a HR, effective in limiting pathogen spread from the site of ingress (Lamb, 1994; Bent, 1996; Hammond-Kosack and Jones, 1996). Currently, our knowledge of oomycete effectors is limited. However, RXLR-containing effector molecules secreted by the pathogen and targeted into host cells have been cloned from *Phytophthora infestans* and *H. arabidopsidis* (see below), and a detailed knowledge of these as well as of bacterial effectors and their targets will allow comparisons as to whether oomycete effectors have an overlapping set of targets in host cells.

13.3.1 ATR13

A major advance in the understanding of the *Arabidopsis–Hyaloperonospora arabidopsidis* coevolution at the molecular level came with the cloning of the first *H. arabidopsidis* effector protein ATR13, which acts as an avirulence factor in a certain set of *Arabidopsis* accessions (Allen et al., 2004). Effectors are secreted from the pathogen but recognized inside *Arabidopsis* cells. Recognition results in a form of programmed host cell death called the HR, which is recognizable at an early stage by the granulation of cytoplasm in the affected cells (Koch and Slusarenko, 1990).

Before cloning ATRs, some important but technically difficult groundwork needed to be carried out. This work was jointly performed by the Holub and

Beynon groups. Thus, the pathogen race/host–genotype interaction patterns needed to be established in detail for this pathosystem. At the very beginning, the inoculation of 11 *Arabidopsis* accessions with seven *H. arabidopsidis* (in those days still called *P. parasitica*) isolates identified an interaction pattern that suggested the presence of 12 different *ATRs* and their corresponding *RPP* genes in the host (Holub et al., 1994). For example, the *H. arabidopsidis* isolate MAKS9 has at least two avirulence determinants: *ATR13Nd*, which allows it to be recognized on Nd-1 (Niederzenz) plants that express *RPP13*, and *ATR1-WsB*, which allows recognition of MAKS9 by Ws-0 (Wassilewskija) plants that express *RPP1*. The next step was the identification of genes expressed by MAKS9 in susceptible Col-0 *Arabidopsis* plants, called *Ppat*, for *Peronospora parasitica* expressed in *Arabidopsis thaliana*, which was achieved by suppression subtractive hybridization (SSH). Total mRNA populations were isolated from mock-infected and MAKS9-infected cotyledons and from pure conidiospores, and cDNAs were generated from all samples. Mock-infected leaf and cotyledon cDNAs were subtracted from the MAKS9-infected leaf cDNA, thus enriching for genes expressed by *H. arabidopsidis*. From a total of more than 1,300 sequenced DNA clones, 25 *Ppat* genes could be identified, which underscored the difficulties of working with an obligate biotrophic pathogen (Bittner-Eddy et al., 2003). Finally, mapping these *Ppat* genes in the F_2 progeny of a cross between the *H. arabidopsidis* isolates MAKS9 (*ATR13*) and EMOY2 (*atr13*) revealed *Ppat17* to be cosegregating with the *ATR13*-conditioned avirulence phenotype (Allen et al., 2004). An assay was developed to test the hypothesis that *Ppat17* encodes *ATR13*. Thus, *in planta* expression of *Ppat17*, which is delivered by either biolistics or inducible expression in stably transformed plants, resulted in an HR in *RPP13Nd*–expressing plants but not in *rpp13* plants. Moreover, the nonfunctional EMOY2 allele of *Ppat17* did not elicit an HR in either *RPP13* or *rpp13* plants. This was taken as evidence that *Ppat17* indeed encodes *ATR13* (Allen et al., 2004).

ATR13 consists of five domains: a 19-amino-acid N-terminal secretion signal followed by the RXLR motif (see below), a heptad leucine/isoleucine repeat present five times, a stretch of 11 amino acids repeated three to four times, and a C-terminal domain. RPP13 is one of the most variable gene loci known in *Arabidopsis* (see below). A sequence comparison of the corresponding ATR13 proteins from 16 *H. arabidopsidis* isolates showed that there is a similar degree of polymorphism with 15 protein variants found. Furthermore, this analysis revealed high sequence variation in the 11-amino-acid repeat, which did not affect recognition specificity. The leucine/isoleucine residues were highly invariant, which suggests an important contribution to protein function. Interestingly, domain-swap experiments revealed that the C-terminal domain is responsible for recognition specificity in the host (Allen et al., 2008).

The activation of *RPP13-Nd* by *ATR13*^{EMCO} was also achieved by expressing *ATR13*^{EMCO} in Pst and using the bacteria to deliver ATR13^{EMCO} into plant cells by the TTSS. To this end, *ATR13*^{EMCO} was fused to a TTSS signal from the Pst effector molecules avrRpm1 and avrRPS4, so that the proteins made in Pst could be transported into *Arabidopsis* cells. In the plant cell, recognition by *RPP13-Nd* resulted in enhanced resistance against bacteria and viruses (Sohn et al., 2007; Rentel et al., 2008), whereas in *rpp13* plants, the susceptibility against Pst DC3000 was enhanced after ATR13 delivery (Sohn et al., 2007).

13.3.2 ATR1

The second *ATR* gene to be cloned was $ATR1^{NdWsB}$ (Rehmany et al., 2005). $ATR1^{NdWsB}$ derived from *H. arabidopsidis* isolate EMOY2 is detected by *RPP1-Nd* from the *Arabidopsis* accession Niederzenz (Nd-1) and by *RPP1-WsB* from accession Ws-0. A segregating F_2 population of a cross between EMOY2 ($ATR1^{Nd}$) and MAKS9 ($atr1^{Nd}$) allowed mapping of $ATR1^{Nd}$ (Rehmany et al., 2003, 2005). A sequence comparison between the *RPP1-Nd*-incompatible EMOY and the *RPP1-Nd*–compatible MAKS alleles at the ATR1 locus indicated five amino-acid polymorphisms between the pathogen isolates. Leaves of recombinant inbred (RI) lines from a Col-0×Nd cross, which either express or lack *RPP1-Nd*, were bombarded with constructs expressing *ATR1* alleles from eight *H. arabidopsidis* isolates. This allowed functional alleles to be grouped accordingly. Thus, the *ATR1* alleles from HIKS, WACO, and EMOY elicited an HR, whereas the $ATR1^{Nd}$ alleles from MAKS, EMCO, NOCO, CALA, and EMWA did not. This correlated well with the sequence analysis of the various alleles. Thus, HIKS, WACO, and EMOY had identical nucleotide sequences, whereas MAKS showed five amino-acid substitutions to the EMOY sequence, and EMCO, NOCO, CALA, and EMWA were highly divergent from the EMOY $ATR1^{Nd}$ protein. Thus, the single *RPP1* gene in accession Nd-0 recognized the $ATR1^{Nd}$ gene present in three *H. parasitica* isolates. However, in the alleles tested, the amino acid sequence variation was not tolerated (Rehmany et al., 2005).

Arabidopsis Ws-0 plants can recognize attempted infection by various *H. arabidopsidis* isolates by virtue of the complexity of the *RPP1* region, which contains at least three *RPP*-sequences with different recognition specificities (see below). Thus, resistance to the CALA, EMOY, HIKS, MAKS, and NOCO isolates is dependent on at least three different *RPP1*-related variants (*RPP1-WsA*, *-WsB*, and *-WsC*) in Ws-0 (Botella et al., 1998). To study the contribution of the *RPP1-WsB* gene in restricting growth of *H. arabidopsidis* in Ws-0, similar experiments were performed to those detailed for ATR13, in which leaves that express *RPP1-WsB* were bombarded with the various *ATR1* variant genes. These experiments revealed that *RPP1-WsB* could recognize the divergent *ATR1* alleles from EMOY, MAKS, EMCO, and NOCO but not the *ATR1* alleles from the CALA or EMWA isolates. Therefore, the authors renamed $ATR1^{Nd}$ to $ATR1^{NdWsB}$. The fact that $ATR1^{NdWsB}$ from the isolate EMCO can be recognized by Ws-0 was surprising because this accession is susceptible to the EMCO isolate (Rehmany et al., 2005). Likely, another factor produced by EMCO can suppress the HR in the *in vivo* situation, which explains the discrepancy between the "artificial" gene-for-gene interaction in the bombardment experiments and the "natural" infection (Sohn et al., 2007). As with ATR13, delivery of ATR1 into host cells by Pst TTSS enhances susceptibility to Pst DC3000 in susceptible backgrounds, which emphasizes again the general role in virulence of pathogen-derived effector molecules transferred into plant cells (Sohn et al., 2007).

The known oomycete avirulence determinants show a high degree of divergence. Thus, for ATR13 from *H. arabidopsidis* and Avr1b-1 from *Phytophthora sojae*, respectively, 10% and 13% polymorphic amino acids were found in various pathogen isolates. ATR1 in turn has more than 300 amino acids as compared with the 187 and 138 for ATR13 and avr1b-1, respectively. In addition, up to 30% of its amino acids showed polymorphisms between the different pathogen isolates so far tested (Rehmany et al., 2005). The more polymorphic an effector is, the more variable is the cognate resistance protein recognizing the effector likely to be. In fact, *RPP* genes are among the most polymorphic genes known in *Arabidopsis* (see below), which indicates a long, standing arms race in the coevolution between host and pathogen, in which a large collection of genes and alleles is maintained in the pathogen as well as the host, caused by reciprocal selection (Rose et al., 2004).

13.3.3 The "RXLR" and "EER" Motif and Identification of Potential ATRs

Once ATRs were cloned, sequence analysis identified the presence of a four-amino-acid motif "RXLR" (where "X" stands for any amino acid). This motif was also found in 13 other known oomycete effector proteins, like Avr3a or PiAvr4 from *P. infestans*, or Avr1b from *P. sojae* (Win et al., 2007). Moreover, analysis within the genome sequences available for these three oomycetes allowed the identification of a total of 43 possibly secreted proteins greatly varying in length that carry this motif. After cleavage of the potential secretion signal peptide, the "RXLR" motif would be located very close to the N-terminus of the mature protein. Interestingly, fusion of this motif, as well as a similar motif "RXLX(E/Q)," which is present in virulence proteins of malaria parasites of the genus *Plasmodium* to green fluorescent protein (GFP), facilitated uptake of this reporter into erythrocytes (Hiller et al., 2004; Marti et al., 2004; Bhattacharjee et al., 2006). Thus, it is likely that the "RXLR" motif, which is in the most conserved region of the various ATR1 proteins from different isolates, might have a similar function in uptake into plant cells. However, a group of haustorially expressed secreted proteins (HESPs) from *Melampsora lini* (flax rust), which included avirulence determinants, did not contain an RXLR motif (Catanzariti et al., 2006). Thus, the RXLR motif is not universal in effector proteins of hyphal plant pathogens but seems to be a characteristic of oomycetes.

About 20 amino acids downstream towards the C-terminus a second motif, which is the "EER" motif, was identified in 30 of the 43 RXLR-carrying proteins; however, a function for this motif has yet to be found (Win et al., 2007).

Thus, it is suggested that the effector proteins of oomycete plant pathogens have an N-terminal signal peptide followed by RXLR and EER motifs. Algorithms developed based on this positional information allowed identification of another 49 or 19 potential ATRs in the *H. arabidopsidis* genome, depending on which algorithm was used. Interestingly, little primary sequence similarity was found between the potential ATRs from *H. arabidopsidis*, *P. infestans*, and *P. sojae*, which points to independent or "birth and death" evolution of effectors in oomycetes (Win et al., 2007). In 16 potential ATRs, the RXLR motif is immediately followed by a "FLAK" motif, which is 100% conserved not only in the *H. arabidopsidis* proteins but also in 16 Crinkler family proteins. These proteins are a distinct class of cytoplasmic effector proteins from *P. infestans* (Win et al., 2007).

Obviously, it will be of greatest importance to clarify the mechanism of how the secreted ATRs end up in the cytoplasm of the host cell, where they can act as effectors in susceptible hosts or elicit strong defence reactions by being recognized by the corresponding *RPP* genes in resistant hosts. Also, the identification of more ATRs by bioinformatic approaches coupled to functional assays is needed to obtain an idea of the spectrum of effectors present in oomycetes. Having complete genome sequence data available should make this task feasible. Moreover, more functional/genetic assays need to be developed to identify ATRs in an unbiased way.

13.4 CLONING OF HOST RESISTANCE GENES (RPPS)

Recognition of pathogen-derived effectors is dependent on *R* gene-encoded resistance proteins most of which have a characteristic domain structure: At their N-terminus, they contain either a toll-interleukin-receptor (TIR)-like domain or a coiled-coil (CC) domain [also called leucine zipper (LZ)], followed by a nucleotide binding (NB) site and at their C-terminus a leucine rich repeat (LRR) domain (Ellis and Jones, 1998; van der Biezen and Jones, 1998; Ellis et al., 2000). Thus, they are often referred to as either TIR-NB-LRR or CC-NB-LRR (Takken et al., 2006). Identification of *R* genes recognizing *H. arabidopsidis* effectors was the target of intense research since the early 1990s; to date, seven *RPP* genes have been cloned from *Arabidopsis*, 4 TIR-NB-LRRs (RPP1, 2, 4, and 5), and 3 CC-NB-LRRs (RPP7, 8, and 13) (Parker et al., 1997; Botella et al., 1998; Bittner-Eddy et al., 2000; Cooley et al., 2000; van der Biezen et al., 2002; Sinapidou et al., 2004). Usually, mapping the *RPP* gene after a cross of a susceptible with a resistant line was used as a strategy to clone RPPs. In plants, *R* genes often occur in clusters of two or more genes located close to each other on the chromosomes. In some cases, these have been designated major recognition complexes (MRCs), where genes for recognition of diverse pathogens are often to be found (Holub and Beynon, 1997) (see also Fig. 13.2 in Slusarenko and Schlaich, 2003). It was suggested that this physical proximity increases the chances of recombination by unequal crossover and/or gene conversion, which provides a mechanism for generating new recognition specificities (Kuang et al., 2008). There are, however, also examples in which individual members of a gene family, that is, highly sequence-related genes (defined by Kuang et al. as > 80% identical), are located at separate sites within the genome (Initiative, 2000).

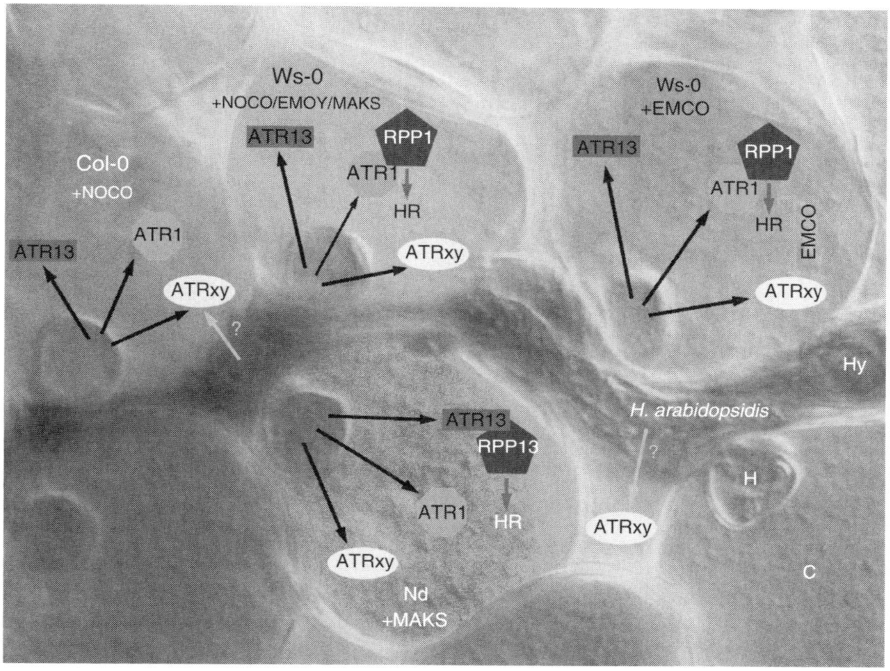

FIG. 13.2 Model illustrating the outcomes of interactions of *Arabidopsis thaliana*-recognized proteins (ATRs) with recognition of *Peronospora parasitica* RPPs. The background is formed by a light micrograph of a trypan blue-stained *Arabidopsis* leaf infected with *H. arabidopsidis*. Haustoria (H) emanating from the intercellular hypha (Hy) can be seen penetrating host cells (C) and invaginating the host cell membrane. *H. arabidopsidis* produces and secretes ATRs (in this example ATR1, ATR13, and the hypothetical ATRxy). Depending on the genetic makeup of the host cell, these ATRs are either not recognized by corresponding RPP proteins (e.g., in the "Col-0" cell with NOCO) or are recognized by corresponding RPPs [(e.g., Ws-0 (RPP1) + NOCO/EMOY/MAKS or Nd (RPP13) + MAKS)] which results in cytoplasmic granulation of the host cell undergoing a hypersensitive response (HR; digitally represented here for clarity). Although EMCO has a functional *ATR1* gene, an EMCO infection of Ws (RPP1) does not result in an HR because the theoretical ATRxy effector suppresses this response by an as yet unknown mechanism (Sohn et al., 2007). The possibility that the secretion of ATRs into the apoplast and into host cells occurs directly from the intercellular hypha is indicated with a "?".

The first *RPP* gene to be cloned was *RPP5* (At4g16950), which is required for recognition of the NOCO isolate in the *Arabidopsis* accession Landsberg *erecta* (L*er*). RPP5 belongs to the TIR-NB-LRR class of resistance proteins (Parker et al., 1997), which can also specify resistance against the EMOY and EMWA isolates of *H. parasitica* (van der Biezen et al., 2002). The MRC-H region that contains *RPP5* is complex with 10 *R* gene-like sequences (called

RPP5 *Ler*A-J) present in a 95-kb DNA region in *Arabidopsis* accession Landsberg *erecta* (*Ler*). In Col-0, only eight *R*-gene-like sequences (called Col-A to Col-H) were found in the equivalent 90-kb DNA region. In both accessions, the most centromeric member of these genes was found to encode the functional resistance protein recognizing *H. arabidopsidis* isolates NOCO and EMOY, respectively (Noel et al., 1999).

RPP4 (At4g16860) from Col-0 confers resistance to the *H. arabidopsidis* isolates EMOY and EMWA (Holub et al., 1994). *RPP4* is also a member of the *RPP5* multigene MRC-H region on chromosome 4 and shows more than 60% identity on the amino-acid level to RPP5 from Landsberg. Transformation of EMOY-susceptible *Arabidopsis* plants with *RPP4* resulted in resistance specifically to EMOY and EMWA, which indicates that *RPP4* is in fact the Col-0 gene responsible for resistance to these *H. arabidopsidis* isolates (van der Biezen et al., 2002). Thus, resistance to the same two pathogen isolates (EMOY and EMWA) in two different *Arabidopsis* ecotypes is conditioned by two related but not identical *RPP* genes (*RPP5* in Landsberg and *RPP4* in Col-0) clustered in the same region of chromosome 4.

The region that contains *RPP1* in Ws-0 proved to be a complex locus mediating resistance to five *H. parasitica* isolates: CALA, EMOY, HIKS, MAKS, and NOCO (as well as an oospore germling derived from NOCO, called NOKS). In this region (MRC-F on chromosome 3), at least three highly similar, functional *RPP1* genes were found: *RPP1-WsA*, *RPP1-WsB*, and *RPP1-WsC* (Botella et al., 1998). The transformation of plants susceptible to CALA, EMOY, MAKS, and NOCO with *RPP1-WsC* showed that this gene mediates full resistance in Ws-0, specifically to the NOCO isolate, whereas transformation with *RPP1-WsB* specified partial resistance to NOCO, EMOY, and MAKS. The Golgi apparatus — or endoplasmic reticulum-localized *RPP1-WsA* — mediated full resistance to CALA, EMOY, HIKS, MAKS, and NOCO (Botella et al., 1998). Thus, the *H. arabidopsidis* isolate NOCO must either carry three different effectors, each of which is specifically recognized by the three different RPP1 resistance proteins, or NOCO expresses only one effector that is uniquely able to trigger defense signaling by any of the three RPP1 proteins. At the other extreme, CALA, EMOY, HIKS, MAKS, and NOCO must express variants of ATR1 that are highly effective in triggering defence in RPP1-WsA-expressing plants. These ATR1 variants need not necessarily have identical amino acid sequences in the different isolates, as shown by the sequence-divergent ATR1 alleles, which are all recognized by RPP1-WsB (Rehmany et al., 2005).

Resistance genes must be in place for recognition to occur when a pathogen attacks and some, like the CC-NBS-LRR *R* genes RPM1 and RPS2 involved in the recognition of phytopathogenic bacteria, show a low level of expression, which is not increased by contact with pathogens. However, TIR-NB-LRR genes like *RPP1* were shown to be induced by salicylic acid (SA) (Shirano et al., 2002), which is a metabolite synthesized in response to various pathogens that is essential for local and systemic defense reactions against (hemi-)biotrophic pathogens (Loake and Grant, 2007).

In the *Arabidopsis* accession L*er*, the *RPP8* gene (At5g43470) specifies resistance to the EMCO isolate of *H. arabidopsidis*. The transformation of EMCO-susceptible Col-0 plants with *RPP8-Ler* was sufficient to confer resistance against this isolate. In contrast to RPP5, RPP4, and the RPP1 resistance proteins, *RPP8* encodes a CC-NB-LRR-type resistance protein (McDowell et al., 1998). A second highly similar CC-NB-LRR gene (*RPH8A*, 90% identity and 94% similarity at the amino acid level), which is located not even 4 kb from *RPP8-Ler*, did not confer the resistance phenotype. Interestingly, the gene encoded by At5g43470 in the accession Di-17 codes for a 105-kDa protein that is more than 91% identical at the amino acid level to RPP8, but it does not provide protection against *H. arabidopsidis*. Infact, this gene (*HRT*) is the only so far known *Arabidopsis R* gene recognizing the coat protein of turnip crinkle virus (Cooley et al., 2000). And in the accession C24, the gene at locus At5g43470 encodes an *R* gene (*RCY1*) that is again more than 91% identical at the amino acid level to RPP8; however, this protein recognizes the coat protein of the yellow strain of cucumber mosaic virus (Takahashi et al., 2002). Thus, different alleles of the same gene can recognize evolutionarily distinct pathogens.

Apart from the functionally characterized RPP8 proteins in L*er*, Di, and C24 (see above), there are five *RPP8*-like genes in the Col-0 genome that have more than 80% identity at the nucleotide level (Meyers et al., 2003). They are located as two and three single genes on chromosomes 1 and 5, respectively. The generation of the three loci on chromosome 5 involved repeated duplication of a genomic segment of less than 10 kb in length. The comparison of two of these three chromosome 5 loci between L*er* and Col-0 suggested a complex evolution of *RPP8*-like genes, which involved duplications, translocations, and crossovers (Meyers et al., 2003). Consensus primers from known *RPP8* family members were used to amplify *RPP8*-like genes from five *Arabidopsis* accessions (Cvi, Ei2, Hod, Sorb, and Tsu). A sequence analysis showed that *RPP8* gene families of three to five members were present. Using the same method, six and four *RPP8*-like genes could even be amplified from *Arabidopsis lyrata* and *Arabidopsis arenosa*, respectively (Kuang et al., 2008). Building a phylogenetic tree from all the *RPP8*-like sequences available allowed grouping of the sequences into five clades: two clades with a type I lineage and three clades with a type II lineage. A type I *R* gene lineage is characterized by the fact that the individual sequences were relatively diverse because of frequent sequence exchanges, whereas type II *R* gene lineages are characterized by highly similar sequences (Kuang et al., 2008). The type I *R* gene lineage contains the three functionally characterized genes *RPP8*, *HRT*, and *RCY*. It appears that the evolutionary forces involved in forming type I or type II *R* gene lineages are quite diverse. Thus, some *RPP8*-like sequences evolve rapidly through frequent sequence exchanges (type I *R* genes) whereas others have evolved independently and are rather conserved throughout the various genotypes (type II *R* genes) (Kuang et al., 2008).

In the *Arabidopsis* accession Niederzenz (Nd-1), *RPP13* specifies resistance against five isolates of *H. parasitica*: ASWA, EDCO, EMCO, GOCO, and MAKS. Mapping the *RPP13-Nd* gene in a cross of MAKS-susceptible

Col-0 × MAKS-resistant Nd plants allowed cloning of the resistance gene. Like *RPP8*, it encodes, a CC-NB-LRR-type resistance protein that confers resistance to the five isolates to Col-0 plants when transformed with the corresponding gene. The *RPP13-Nd* locus is a simple locus that contains only this *R* gene (Bittner-Eddy et al., 2000). In replicate-limiting dilution (RLD), *RPP13* specifies resistance to the isolate WELA. Using polymerase chain reaction (PCR), the *RPP13* gene from RLD was also cloned and found to be 89% identical at the amino acid level to RPP13-Nd. Identity as high as 96% was found between the two proteins when the more variable leucine rich repeats were excluded from the comparison (Bittner-Eddy et al., 2000). In fact, sequence analyses of the *RPP13* gene in two different studies comprising 24 and 60 *Arabidopsis* accessions, respectively, it was shown that *RPP13* is the most polymorphic protein in *Arabidopsis* known so far (Rose et al., 2004; Ding et al., 2007). Although the authors of the 24-accession study suggested that a diverse collection of alleles of this gene has been maintained in *Arabidopsis* by continuous reciprocal selection between the host plant and the pathogen (Rose et al., 2004), the authors who studied 60 alleles of *RPP13* came to the conclusion that directional as well as diversifying selection are working on this locus (Ding et al., 2007). A more general comparison of the leucine rich repeats of 27 *R* genes in 96 *Arabidopsis* accessions confirmed that this domain is highly polymorphic and suggested that transient or frequency-dependent selection maintains protein variants at a locus for variable periods of time (Bakker et al., 2006).

RPP2 specifies resistance in Col-0 plants to the *H. parasitica* isolate CALA (Holub et al., 1994). Like the other *RPP* genes, *RPP2* was cloned by crossing CALA-resistant Col-0 to CALA-susceptible Nd-1 plants. *RPP2* mapped to a locus in which four TIR-NB-LRR genes are located. Interestingly, two chromosomally adjacent TIR-NB-LRR genes *RPP2A* (At4g19500) and *RPP2B* (At4g19510) are required together for full CALA resistance. Although *RPP2A* shows an unusual TIR-NB-LRR structure in that it has two incomplete TIR-NB domains at the N-terminus followed by a short CC-domain and a short LRR domain at the C-terminus, *RPP2B* displays the classic TIR-NB-LRR structure (Sinapidou et al., 2004). Thus, *RPP2A* and *RPP2B* probably cooperate to specify resistance to the CALA isolate.

13.5 RPP-MEDIATED DEFENCE SIGNALLING

Many genetic screens have been aimed at finding components required for pathogen defense. These screens include inoculating mutant plant populations with either virulent or avirulent pathogens and screening them for altered phenotypes. This included the search for second site suppressors in already known defense mutants. Using virulent pathogens, genes that contribute to basal defense were identified, whereas avirulent pathogens helped to unravel genes in *R* gene-mediated defense pathways. These clear distinctions, however, were blurred because it was recognized that components of basal defense, like enhanced disease susceptibility1 (*EDS1*), are also contributing to *R* gene-mediated defenses and *vice versa*. Thus, today many mutants are known that impair or increase defense against virulent and avirulent pathogens, which include *H. arabidopsidis*. In this respect, it is interesting to note that all TIR-NB-LRR *R* genes analyzed to date require functional *EDS1*/*PAD4* genes to mediate resistance (Wiermer et al., 2005).

In Fig. 13.3 we provide recent information on signaling via the cloned *RPP* genes, and the reader is referred to excellent recent reviews covering specific aspects of pathogen defence (Slusarenko and Schlaich, 2003; Hardham, 2007; Loake and Grant, 2007; Robatzek, 2007; Wiermer et al., 2007).

This graphical representation of information doesn't intend to claim sequential action of the individual components, which might seem to be implied by the top-to-bottom appearance. It should be remembered that signaling is much more complex than a linear signaling relay. However, there is not yet enough genetic (let alone biochemical) information available to organize the individual components more precisely.

13.6 CELL BIOLOGY

The *Arabidopsis*–*H. arabidopsidis* pathosystem can also be used to study the cell biology of plant defense (Heath, 2000; Takemoto and Hardham, 2004). In this regard, the production of transgenic plants with green fluorescent protein (GFP)-tagged organelles, combined with high resolution fluorescence microscopy, has allowed recent rapid progress (Mano et al., 2002; Takemoto et al., 2003; Takemoto and Hardham, 2004; Koh et al., 2005; Hardham, 2007).

Organellar migration to the point of pathogen penetration of the cell was shown by Tomiyama for potato and *Phytophthora* (Tomiyama, 1956). *Hyaloperonospora* was used to study the defense reactions/movements of organelles in *Arabidopsis* after attack by an avirulent, a virulent, and of a nonhost isolate (Takemoto et al., 2003). The authors showed that in *Arabidopsis*, the endoplasmic reticulum and the Golgi bodies translocate toward the site of contact with the pathogen along reorganized actin cables. Only the microtubule cytoskeleton remained largely unchanged (Takemoto et al., 2003). These observations were extended to mitochondria and peroxisomes, and it was shown that cytoplasmic aggregation is a defense response that functions independently of important signaling components, such as *RAR1*, *SGT1b*, *PAD4*, and *NPR1*, which are required to express full resistance to *H. arabidopsidis* (Hermanns et al., 2008). Similar cellular responses were also observed in *Arabidopsis* with other pathogens (Koh et al., 2005; Shimada et al., 2006). Furthermore, pharmacological interference showed a functional relevance for cytoskeleton reorganization in pathogen defense (Kobayashi et al., 1997; Shimada et al., 2006). The fact that organelle movements toward the site of cellular contact with the pathogen were observed as a response of various plants to different pathogens (*R* gene-mediated defense against

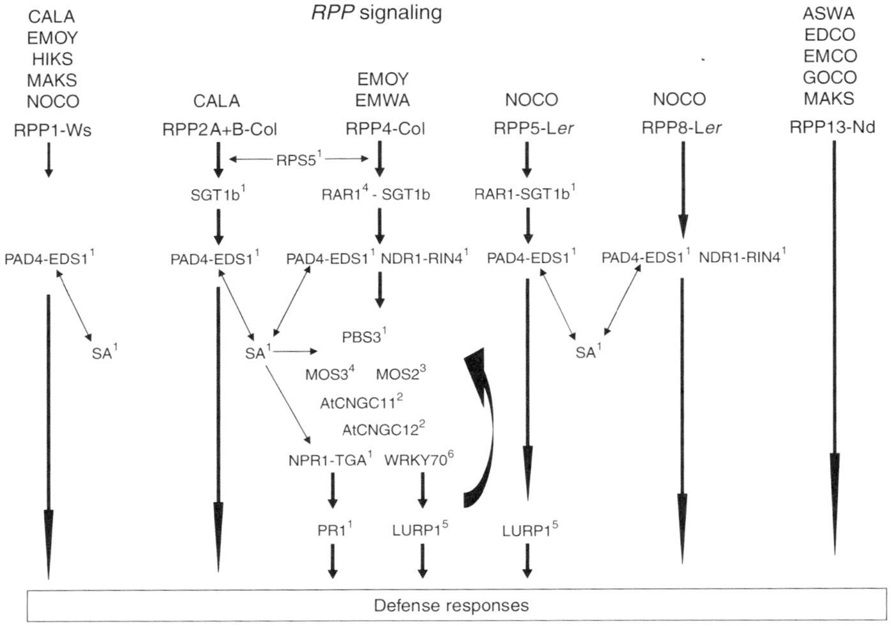

FIG. 13.3 Schematic representation of *RPP*-conditioned signal transduction. *H. arabidopsidis* isolates are shown above the corresponding RPPs that recognize them below the isolates (hyphenated with the *Arabidopsis* accession from which they were cloned). RPP1 from Ws (RPP1-Ws) is dependent on PAD4 and EDS1. RPP2A + B from Col-0 is quantitatively influenced by the resistance protein RPS5 that recognizes the bacterial pathogen *Pseudomonas* and requires SGT1b, PAD4, and EDS1. RPP4 from Col-0 is particularly sensitive to changes in resistance signaling. Thus, mutations in many defense-related genes lower the defense responses to EMOY and EMWA. RPP5 from L*er* requires RAR1 and SGT1b as well as PAD4 and EDS1. Furthermore, RPP5 was shown to be dependent on LURP1. RPP8 from L*er* is an exception, because it encodes a CC-NB-LRR-type resistance protein, yet is dependent not only on NDR1 and RIN4 (like other CC-NB-LRRs) but also on PAD4 and EDS1. No downstream signaling components of RPP13-Nd have so far been reported. Please note: Signaling by the TIR-NB-LRR-type *RPP* genes *RPP1*-Ws, *RPP2A* + *B*-Col-0, *RPP4*-Col-0, and *RPP5*-L*er* was dependent on PAD4 and EDS1, which have a positive interlocked feedback regulation with salicylic acid (SA). Footnote 1: Slusarenko and Schlaich (2003); Footnote 2: Yoshioka et al. (2006); Footnote 3: Zhang et al. (2005); Footnote 4: Zhang and Li (2005); Footnote 5: Knoth and Eulgem (2008); Footnote 6: Knoth et al. (2007).

incompatible isolates, basal defense against compatible isolates, nonhost interactions, as well as in important defense-signal transduction mutants) suggests that these responses are part of an evolutionarily ancient defense response (Hermanns et al., 2008), which is likely initiated by the mechanical forces exerted by the pathogen when displacing host tissue during penetration (Gus-Mayer et al., 1998; Hardham et al., 2008).

13.7 HOW IS ORGAN SPECIFICITY DETERMINED IN THIS PATHOSYSTEM?

Although we now understand in detail the regulation of host specificity in terms of *R* genes and ATR genes, little is known about the regulation of organ specificity within the host for a given pathosystem. Thus, *H. arabidopsidis* is regarded as a leaf pathogen, but it has been reported that leaves and cotyledons of a single *Arabidopsis* genotype can show different responses to the same isolate of *H. arabidopsidis*. For example, Aarts et al. (1998) showed that the impairment of *RPP2*- and *RPP4*-mediated resistance observed in Col-0 *ndr1-1* cotyledons (Century et al., 1995) was not observed in true leaves. Furthermore, there is evidence that the gene-for-gene specificity, which determines the outcome of interacting host and pathogen genotypes in leaves and cotyledons, is inoperative in root tissues (Hermanns et al., 2003). The authors showed that *Arabidopsis* roots appeared equally susceptible to infection by ostensibly avirulent as well as virulent isolates of *H. arabidopsidis*. Reverse transcription PCR (RT-PCR) showed that transcripts of *R*-genes and signaling components were present, and that the phenomenon could not be explained in terms of developmental/organ specific differences in gene expression. One possibility that would explain these observations would be if transfer of effectors from the pathogen into root cells were somehow compromised. However, transient biolistic transformation experiments with *ATR* genes and GFP constructs showed that the expected *R* gene-determined specificities were observed in leaves, whereas in roots the ATR constructs were inefficient at inducing HR (Müller, Deisen, Rehmany, et al., unpublished). This would suggest that the susceptibility of roots to ostensibly genetically incompatible isolates of the pathogen may not be explained by an inefficient/absent transfer of the *ATR* gene into the root cells by the pathogen, but it must have some other as yet unknown cause.

13.8 FUTURE RESEARCH DIRECTIONS

Six years after the publication of our last review on this pathosystem (Slusarenko and Schlaich, 2003), it is satisfying to report on the enormous progress, particularly with the pathogen. Thus, in the interim, the milestones of the *H. arabidopsidis* genome and *ATR* cloning reflect the pace at which this field is moving. An immediate need to better understand the functions of ATRs is clear, and for that to occur more ATRs have to be cloned. Having both host and pathogen genome sequences is a great help, and understanding the cellular and biochemical facets of the interaction between these two organisms remains a challenge. Thus, for example, it will be interesting to learn more details about the transfer (location and mechanism) of ATRs from *H. arabidopsidis* into host cells. How phenomena like organ specificity and age-related resistance are regulated at the molecular level are still unclear. Thus, we think it is still safe to say that in the coming years we shall learn an enormous amount from this pathosystem that will have wider implications in the field of plant pathology as a whole.

ACKNOWLEDGMENTS

Ovidiu Constantinescu (Uni Uppsala), Walter Gams (formerly CBS), Marcus Göker (Uni Tübingen), Jürgen Hammerstaedt (Uni Köln), and Paul Kirk (CABI) for contributions to the nomenclature debate of *H. arabidopsidis*; Eric Holub (HRI) for providing information prior to publication; Monika Hermanns for the micrograph used to make Fig. 13.2; Benjamin Slusarenko for technical advice on Fig.13.1; and Ulrike Noll for computer adaptation and proofreading.

REFERENCES

Aarts N, Metz M, Holub E, Staskawicz BJ, Daniels MJ, Parker JE 1998. Different requirements for EDS1 and NDR1 by disease resistance genes define at least two R gene-mediated signaling pathways in Arabidopsis. Proc Nat Acad Sci USA 95:10306–10311.

Allen RL, Bittner-Eddy PD, Grenville-Briggs LJ, Meitz JC, Rehmany, AP, Rose LE, Beynon JL 2004. Host-parasite coevolutionary conflict between Arabidopsis and downy mildew. Science 306:1957–1960.

Allen RL, Meitz JC, Baumber RE, Hall SA, Lee SC, Rose LE, Beynon JL 2008. Natural variation reveals key amino acids in a downy mildew effector that alters recognition specificity by an *Arabidopsis* resistance gene. Molec Plant Pathol 9:511–523.

Bakker EG, Toomajian C, Kreitman M, Bergelson J 2006. A genome-wide survey of *R* gene polymorphisms in *Arabidopsis*. Plant Cell 18:1803–1818.

Bent AF 1996. Plant disease resistance genes: function meets structure. Plant Cell 8:1757–1771.

Bhattacharjee S, Hiller NL, Liolios K, Win J, Kanneganti TD, Young C, Kamoun S, Haldar K 2006. The malarial host-targeting signal is conserved in the Irish potato famine pathogen. PLoS Pathogens 2:e50.

Birch PRJ, Boevink PC, Gilroy EM, Hein I, Pritchard L, Whisson SC 2008. Oomycete RXLR effectors: delivery, functional redundancy and durable disease resistance. Curr Opin Plant Biol 11:373–379.

Birch PRJ, Rehmany AP, Pritchard L, Kamoun, S, Beynon, JL 2006. Trafficking arms: oomycete effectors enter host plant cells. Trends Microbiol 14:8–11.

Bittner-Eddy PD, Allen RL, Rehmany AP, Birch P, Beynon JL 2003. Use of suppression subtractive hybridization to identify downy mildew genes expressed during infection of *Arabidopsis thaliana*. Molec Plant Pathol 4:501–507.

Bittner-Eddy PD, Crute IR, Holub EB, Beynon JL 2000. *RPP13* is a simple locus in *Arabidopsis thaliana* for alleles that specify downy mildew resistance to different avirulence determinants in *Peronospora parasitica*. Plant J 21:177–188.

Botella MA, Parker JE, Frost LN, Bittner-Eddy PD, Beynon JL, Daniels MJ, Holub EB, Jones JD 1998. Three genes of the *Arabidopsis RPP1* complex resistance locus recognize distinct *Peronospora parasitica* avirulence determinants. Plant Cell 10:1847–1860.

Buttner D, Bonas U 2002. Port of entry — the type III secretion translocon. Trends Microbiol 10:186–192.

Catanzariti AM, Dodds PN, Lawrence GJ, Ayliffe MA, Ellis JG 2006. Haustorially expressed secreted proteins from flax rust are highly enriched for avirulence elicitors. Plant Cell 18:243–256.

Century K, Holub E, Staskawicz B 1995. NDR1, a locus of *Arabidopsis thaliana* that is required for disease resistance to both a bacterial and a fungal pathogen. Proc Nat Acad Sci USA 92:6597–6601.

Channon AG 1981. Downy mildew of Brassicas. In: Spencer DM, editor. The downy mildews. Academic Press London. pp. 321–339.

Chisholm ST, Coaker G, Day B, Staskawicz BJ 2006. Host-microbe interactions: shaping the evolution of the plant immune response. Cell 124:803–814.

Clark JSC, Spencer-Phillips PTN 1993. Accumulation of photoassimilate by *Peronospora viciae* (Berk.) Casp. and leaves of *Pisum sativum* L.: evidence for nutrient uptake via intercellular hyphae. New Phytol 124:107–119.

Clark JSC, Spencer-Phillips PTN 2004. The compatible interaction in downy mildew infections. In: Spencer-Phillips PTN, Jeger M, editors. Advances in downy mildew research. Vol. 2. Kluwer Academic Publishers. Dordrecht, Germany. pp. 1–34.

Constantinescu O, Fatehi J 2002. Peronospora-like fungi (Chromista, Peronosporales) parasitic on Brassicaceae and related hosts. Nova Hedwigia 74:291–338.

Cooley MB, Pathirana S, Wu HJ, Kachroo P, Klessig DF 2000. Members of the *Arabidopsis HRT/RPP8* family of resistance genes confer resistance to both viral and oomycete pathogens. Plant Cell 12:663–676.

Dangl JL, Holub EB, Debener T, Lehnackers H, Ritter C, Crute IR 1992. Genetic definition of loci involved in *Arabidopsis*-pathogen interactions. In: Koncz C, Nam-Hai Chua, Schell J, editors. Methods in *Arabidopsis* research. World Scientific Press. Singapore. pp. 393–418.

Ding J, Cheng H, Jin X, Araki H, Yang Y, Tian D 2007. Contrasting patterns of evolution between allelic groups at a single locus in *Arabidopsis*. Genetics 129: 235–242.

El Gariani NK, Spencer-Phillips PTN 2004. Isolation of viable *Peronospora viciae* hyphae from infected *Pisum sativum* leaves and accumulation of nutrients *in vitro*. In: Spencer-Phillips PTN, Jeger M, editors. Advances in downy mildew research. Vol 2. Kluwer Academic Publishers. Dordrecht, Germany. pp. 249–264.

Ellis J, Dodds P, Pryor T 2000. Structure, function and evolution of plant disease resistance genes. Curr Opin Plant Biol 3:278–284.

Ellis J, Jones D 1998. Structure and function of proteins controlling strain-specific pathogen resistance in plants. Curr Opin Plant Biol 1:288–293.

Gäumann E 1918. Ueber die Formen der *Peronospora parasitica* (Pers.) Fries. Beihefte zum Botanischen Zentralblatt 35:395–533.

Göker M, Riethmüller A, Voglmayr H, Weiss M, Oberwinkler F 2004. Phylogeny of *Hyaloperonospora* based on nuclear ribosomal internal transcribed spacer sequences. Mycol Prog 3:83–94.

Grant SR, Fisher EJ, Chang JH, Mole BM, Dangl JL 2006. Subterfuge and manipulation: type III effector proteins of phytopathogenic bacteria. Annu Rev Microbiol 60:425–449.

Gus-Mayer S, Naton B, Hahlbrock K, Schmelzer E 1998. Local mechanical stimulation induces components of the pathogen defense response in parsley. Proc Nat Acad Sci USA 95:8398–8403.

Hammond-Kosack KE, Jones JD 1996. Resistance gene-dependent plant defense responses. Plant Cell 8:1773–1791.

Hardham AR 2007. Cell biology of plant-oomycete interactions. Cell Microbiol 9:31–39.

Hardham AR, Takemoto D, White RG 2008. Rapid and dynamic subcellular reorganization following mechanical stimulation of *Arabidopsis* epidermal cells mimics responses to fungal and oomycete attack. BMC Plant Biol.

Heath MC 2000. Advances in imaging the cell biology of plant-microbe interactions. Annu Rev Phytopathol 38:443–459.

Hermanns M, Slusarenko A, Schlaich N 2002. Lack of active defence responses revealed in a soil-free *Arabidopsis*/*Peronospora* sterile co-cultivation system. Plant Protect Sci 38:136–138.

Hermanns M, Slusarenko AJ, Schlaich NL 2003. Organ-specificity in a plant disease is determined independently of R gene signaling. Mol Plant-Microbe Interact 16:752–759.

Hermanns M, Slusarenko AJ, Schlaich NL 2008. The early organelle migration response of *Arabidopsis* to *Hyaloperonospora arabidopsidis* is independent of *RAR1*, *SGT1b*, *PAD4* and *NPR1*. Physiol Mol Plant Pathol 72:96–101.

Hiller NL, Bhattacharjee S, vsn Ooij C, Liolios K, Harrison T, Lopez-Estrano C, Haldar K 2004. A host-targeting signal in virulence proteins reveals a secretome in malarial infection. Science 306:1934–1937.

Holub EB 2008. Natural history of *Arabidopsis thaliana* and oomycete symbioses. Eur J Plant Pathol. In press.

Holub E, Beynon JL 1997. Symbiology of mouse ear cress (Arabidopsis) and oomycetes. Adv Bot Res 24:227–273.

Holub EB, Beynon JL, Crute IR 1994. Phenotypic and genotypic characterization of interactions between isolates of *Peronospora parasitica* and accessions of *Arabidopsis thaliana*. Mol Plant-Microbe Interact 7:223–229.

Initiative AG 2000. Analysis of the genome sequence of the flowering plant *Arabidopsis thaliana*. Nature 408:796–815.

Jones JDG, Dangl JL 2006. The plant immune system. Nature 444:323–329.

Knoth C, Eulgem T 2008. The oomycete response gene *LURP1* is required for defense against *Hyaloperonospora parasitica* in *Arabidopsis thaliana*. Plant J. In press.

Knoth C, Ringler J, Dangl JL, Eulgem T 2007. Arabidopsis WRKY70 is required for full RPP4-mediated disease resistance and basal defense against *Hyaloperonospora parasitica*. Mol Plant-Microbe Interact 20:120–128.

Kobayashi, I, Kobayashi, Y, Hardham AR 1997. Inhibition of rust-induced hypersensitive response in flax cells by the microtubule inhibitor oryzalin. Aust J Plant Physiol 24:733–740.

Koch E, Slusarenko AJ 1990. Arabidopsis is susceptible to infection by a downy mildew fungus. Plant Cell 5:437–445.

Koh S, Andre A, Edwards H, Ehrhardt D, Somerville S 2005. *Arabidopsis thaliana* subcellular responses to compatible *Erysiphe cichoracearum* infections. Plant J. 44:516–529.

Kuang H, Caldwell KS, Meyers BC, Michelmore RW 2008. Frequent sequence exchanges between homologs of *RPP8* in *Arabidopsis* are not necessarily associated with genomic proximity. Plant J 54:69–80.

Lamb CJ 1994. Plant disease resistance genes in signal perception and transduction. Cell 76:419–422.

Lindau G 1901. Hilfsbuch für das Sammeln Parasitischer Pilze. Bornträger Verlag. Berlin, Germany.

Loake GJ, Grant M 2007. Salicylic acid in plant defence—the players and protagonists. Curr Opin Plant Biol 10:466–472.

Mano S, Nakamori C, Hayashi M, Kato A, Kondo M, Nishimura M 2002. Distribution and characterization of peroxisomes in *Arabidopsis* by visualization with GFP: dynamic morphology and actin-dependent movement. Plant Cell Physiol 43:331–341.

Marti M, Good RT, Rug M, Knuepfer E, Cowman AF 2004. Targeting malaria virulence and remodeling proteins to the host erythrocyte. Science 306:1930–1933.

Mauch-Mani B, Slusarenko AJ 1993. Arabidopsis as a model host for studying plant-pathogen interactions. Trends Microbiol 1:265–270.

McCann HC, Guttman DS 2008. Evolution of the type III secretion system and its effectors in plant-microbe interactions. New Phytol 177:33–47.

McDowell JM, Dhandaydham M, Long, TA, Aarts MGM, Goff S, Holub EB, Dangl JL 1998. Intragenic recombination and diversifying selection contribute to the evolution of downy mildew resistance at the *RPP8* locus of *Arabidopsis*. Plant Cell 10:1861–1874.

Meyers BC, Kozik A, Griego A, Kuang H, Michelmore RW 2003. Genome-wide analysis of NBS-LRR-encoding genes in *Arabidopsis*. Plant Cell 15:809–834.

Mims CW, Richardson EA, Holt III BF, Dangl JL 2004. Ultrastructure of the host-pathogen interface in Arabidopsis thaliana leaves infected by the downy mildew *Hyaloperonospora parasitica*. Can J Bot 82:1001–1008.

Mudgett MB 2005. New insights to the function of phytopathogenic bacterial typeIII effectors in plants. Annu Rev Plant Biol 56:509–531.

Noel L, Moores TL, van der Biezen EA, Parniske M, Daniels MJ, Parker JE, Jones JDG 1999. Pronounced intraspecific haplotype divergence at the *RPP5* complex disease resistance locus of *Arabidopsis*. Plant Cell 11:2099–2112.

Nomura K, Melotto M, He SY 2005. Suppression of host defense in compatible plant-*Pseudomonas syringae* interactions. Curr Opin Plant Biol 8:361–368.

Parker JE, Coleman MJ, Szabo V, Frost LN, Schmidt R, van der Biezen EA, Moores T, Dean C, Daniels MJ, Jones JDG 1997. The *Arabidopsis* downy mildew resistance gene *RPP5* shares similarity to the toll and interleukin-1-receptors with *N* and *L6*. Plant Cell 9:879–894.

Rehmany AP, Gordon A, Rose LE, Allen RL, Armstrong MR, Whisson SC, Kamoun S, Tyler BM, Birch PRJ, Beynon JL 2005. Differential recognition of highly divergent downy mildew avirulence gene alleles by RPP1 resistance genes from two arabidopsis lines. Plant Cell 17:1839–1850.

Rehmany AP, Grenville LJ, Gunn ND, Allen RL, Paniwnyk Z, Byrne J, Whisson SC, Birch PRJ, Beynon JL 2003. A genetic interval and physical contig spanning the Peronospora parasitica (At) avirulence gene locus ATR1Nd. Fungal Genet Biol 38:33–42.

Rentel MC, Leonelli L, Dahlbeck D, Zhao B, Staskawicz BJ 2008. Recognition of the *Hyaloperonospora parasitica* effector ATR13 triggers resistance against oomycete, bacterial, and viral pathogens. Proc Nat Acad Sci USA 103:1091–1096.

Rethage J, Ward PI, Slusarenko AJ. 2000. Race-specific elicitors from the Peronospora parasitica/Arabidopsis thaliana pathosystem. Physiol Mol Plant Pathol 56: 179–184.

Robatzek S 2007. Vesicle trafficking in plant immune responses. Cell Microbiol 9:1–8.

Rose LE, Bittner-Eddy PD, Langley CH, Holub EB, Michelmore RW, Beynon JL 2004. The maintenance of extreme amino acid diversity at the disease resistance gene, *RPP13*, in *Arabidopsis thaliana*. Genetics 166:1517–1527.

Shimada C, Lipka V, O'Connell R, Okuno T, Schulze-Lefert P, Takano Y 2006. Nonhost resistance in *Arabidopsis-Colletotrichum* interactions acts at the cell periphery and requires actin filament function. Mol Plant-Microbe Interact 19:270–279.

Shirano Y, Kachroo P, Shah J, Klessig DF 2002. A gain-of-function mutation in an *Arabidopsis* toll interleukin1 receptor-nucleotide binding site-leucine-rich repeat type *R* gene triggers defense responses and results in enhanced disease resistance. Plant Cell 14:3149–3162.

Sinapidou E, Williams K, Nott L, Bahkt S, Tor M, Crute I, Bittner-Eddy P, Beynon J 2004. Two TIR:NB:LRR genes are required to specify resistance to *Peronospora parasitica* isolate Cala2 in *Arabidopsis*. Plant J 38:898–909.

Slusarenko AJ, Schlaich NL 2003. Downy mildew of *Arabidopsis thaliana* caused by *Hyaloperonospora parasitica* (formerly *Peronospora parasitica*). Mol Plant Pathol 4:159–170.

Sohn KH, Lei R, Nemri A, Jones JDG 2007. The downy mildew effector proteins ATR1 and ATR13 promote disease susceptibility in *Arabidopsis thaliana*. Plant Cell 19:4077–4090.

Soylu S, Soylu EM 2003. Preliminary characterization of race-specific elicitors from *Peronospora parasitica* and their ability to elicit phenolic accumulation in Arabidopsis. Phytoparasitica 31:381–392.

Spencer-Phillips PTN 1997. Function of fungal haustoria in epiphytic and endophytic infections. Adv Bot Res 24:309–333.

Takahashi H, Miller J, Nozaki Y, Takeda M, Shah J, Hase S, Ikegami M, Ehara Y, Dinesh-Kumar SP 2002. RCY1, an *Arabidopsis thaliana RPP8/HRT* family resistance gene, conferring resistance to cucumber mosaic virus requires salicylic acid, ethylene and a novel signal transduction mechanism. Plant J 32:655–667.

Takemoto D, Hardham AR 2004. The cytoskeleton as a regulator and target of biotic interactions in plants. Plant Physiol 136:3864–3876.

Takemoto D, Jones DA, Hardham AR 2003. GFP-tagging of cell components reveals the dynamics of subcellular re-organization in response to infection of *Arabidopsis* by oomycete pathogens. Plant J 33:775–792.

Takken FLW, Albrecht M, Tameling WIL 2006. Resistance proteins: molecular switches of plant defence. Curr Opin Plant Biol 9:383–390.

Tomiyama K 1956. Cell physiological studies on the resistance of potato plant to *Phytophthora infestans*. Ann Phytopathol Soc Jpn 21:54–62.

Uknes S, Mauch-Mani B, Moyer M, Potter S, Williams S, Dincher S, Chandler D, Slusarenko A, Ward E, Ryals J 1992. Acquired resistance in Arabidopsis. Plant Cell 4:645–656.

van der Biezen EA, Freddie CT, Kahn K, Parker JE, Jones JD 2002. *Arabidopsis RPP4* is a member of the *RPP5* multigene family of TIR-NB-LRR genes and confers downy mildew resistance through multiple signalling components. Plant J 29:439–451.

van der Biezen EA, Jones JD 1998. The NB-ARC domain: a novel signalling motif shared by plant resistance gene products and regulators of cell death in animals. Curr Biol 8:226–227.

Voegele RT, Mendgen K 2003. Rust haustoria: nutrient uptake and beyond. New Phytol 159:93–100.

Wiermer M, Feys BJ, Parker JE 2005. Plant immunity: the EDS1 regulatory node. Curr Opin Plant Biol 8:383–389.

Wiermer M, Palma K, Zhang Y, Li X 2007. Should I stay or should I go? Nucleocytoplasmic trafficking in plant innate immunity. Cell Microbiol 9:1880–1890.

Win J, Morgan W, Bos J, Krasileva KV, Cano LM, Chaparro-Garcia A, Ammar R, Staskawicz BJ, Kamoun S 2007. Adaptive evolution has targeted the C-terminal domain of the RXLR effectors of plant pathogenic oomycetes. Plant Cell 19:2349–2369.

Yerkes WD, Shaw CG 1959. Taxonomy of the Peronospora species on cruciferae and chenopodiaceae. Phytopathology 49:499–507.

Yoshioka K, Moeder W, Kang HG, Kachroo P, Masmoudi K, Berkowitz G, Klessig DF 2006. The chimeric *Arabidopsis* CYCLIC NUCLEOTIDE-GATED ION CHANNEL11/12 activates multiple pathogen resistance responses. Plant Cell 18:747–763.

Zhang Y, Cheng YT, Bi D, Palma K, Li X 2005. MOS2, a protein containing G-patch and KOW motifs, is essential for innate immunity in *Arabidopsis thaliana*. Curr Biol 15:1936–1942.

Zhang Y, Li X 2005. A putative nucleoporin 96 is required for both basal defense and constitutive resistance responses mediated by *suppressor of npr1-1 constitutive 1*. Plant Cell 17:1306–1316.

14

INTERACTIONS BETWEEN *PHYTOPHTHORA INFESTANS* AND *SOLANUM*

MIREILLE VAN DAMME, SEBASTIAN SCHORNACK, LILIANA M. CANO, EDGAR HUITEMA, AND SOPHIEN KAMOUN

The Sainsbury Laboratory, John Innes Centre, Norwich, United Kingdom

14.1 INTRODUCTION

Phytophthora infestans, the notorious Irish potato famine organism, causes the late blight disease of potato and tomato. Late blight is a global constraint for potato and tomato production with recurrent epidemics occurring worldwide, which often follow the migration or emergence of particularly aggressive strains (see Chapter 7 by Fry et al.). Worldwide losses in potato production and increased fungicide costs caused by late blight are estimated in the billions of dollars annually (Duncan, 1999). In the United States and other developed countries, chronic use of chemicals to manage late blight reduces the profit margins of farmers and is not always successful. In developing countries, late blight affects subsistence potato production. With the growing concerns about food prices and security, potato has emerged as an alternative to the major cereal crops for feeding the world's poor (Mackenzie, 2008). For these reasons, managing late blight is facing a renewed urgency not seen since the Irish potato famine.

P. infestans is an aggressive and destructive pathogen. Late blight epidemics spread rapidly when left unchecked and result in significant losses in product yield (Garelik, 2002; Schiermeier, 2001). Two examples illustrate the devastating nature of the disease. In the summer of 2007, the late blight season in Great Britain was the worst in 50 years mainly because of the emergence and rapid

Oomycete Genetics and Genomics: Diversity, Interactions, and Research Tools
Edited by Kurt Lamour and Sophien Kamoun
Copyright © 2009 John Wiley & Sons, Inc.

spread of an aggressive clone termed genotype_13 (see the section by Cooke in Chapter 7 by Fry et al.). British news outlets reported that chemical sprays had to be applied to organic potatoes to save the crop. In 2003, potato production was nearly eliminated in Papua New Guinea, which is one of the few countries in the world that was previously free of the disease. As vividly described by John Reader in his recent book on the world history of potato (Reader, 2008), the disease spread throughout the entire country within months of first incidence causing food shortages and social havoc reminiscent of the European epidemics of the mid-nineteenth century.

Similar to other oomycetes, *P. infestans* has long been considered an intractable organism by fungal genetic research standards (Kamoun, 2003). However, in the last 10–15 years, technical developments, such as DNA transformation, genetic manipulation using gene silencing, high-throughput *in planta* expression systems, and genomics resources, have greatly accelerated the pace of research and facilitated the discovery and functional analyses of a myriad of interesting genes (Birch et al., 2006; Judelson and Blanco, 2005; Kamoun, 2006). With the completion of the genome sequence of several oomycete species, including *P. infestans*, and the discovery of complex repertoires of pathogenicity effector genes, *P. infestans* and other oomycetes have moved to center stage of research on plant–microbe interactions. In this chapter, we provide the basic background information and review the current state of understanding in molecular studies on *P. infestans* pathogenicity.

14.2 PATHOGENICITY OF *P. INFESTANS*

14.2.1 Host Specificity

P. infestans is a relatively specialized pathogen that causes disease on various tissues of potato and tomato crops (Erwin and Ribeiro, 1996). However, *P. infestans* was also reported on a variety of species in the genus *Solanum* (Adler et al., 2002, 2004; Flier et al., 2003; Ordonez et al., 2000), and reports of infection on non-solanaceous plants occur periodically in the literature (Erwin and Ribeiro, 1996). The closely related species *Phytophthora mirabilis* and *Phytophthora ipomoeae* infect plants as diverse as four-o'clock (*Mirabilis jalapa*) and morning glory (*Ipomoea longipedunculata*), respectively (Flier et al., 2002; Goodwin and Fry, 1994). These findings suggest significant adaptive flexibility within the *P. infestans* lineage that led to virulence on diverse hosts.

14.2.2 Infection Cycle of *P. infestans*

The infection stages of *P. infestans* are illustrated in Fig. 14.1. *P. infestans* adopts a two-step infection style, which is typical of hemibiotrophs. An early phase of infection, in which the pathogen requires living host cells, is followed

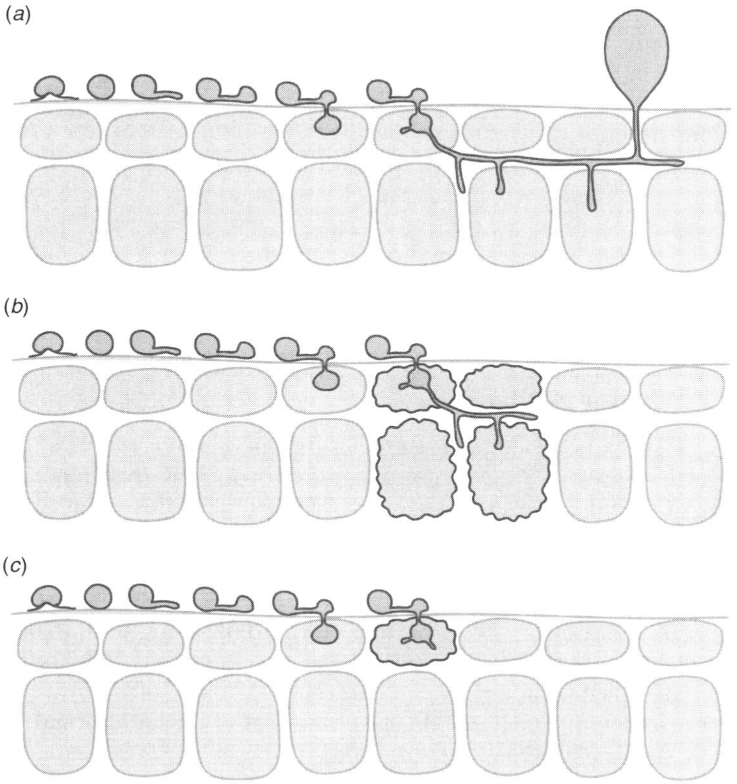

FIG. 14.1 Schematic view of the infection cycle of *Phytophthora infestans* on (a) susceptible and (b, c) resistant plants. (a) The *P. infestans* developmental stages displayed are from left to right: zoospore, cyst, germinating cyst, germinated cyst with appressorium, germinated cyst with appressorium and infection vesicle, haustorium, and sporangium. (b, c) Similar to panel, (a) except that plant cells undergoing hypersensitive cell death are shown in red. The hypersensitive response can include groups of plant cells (b) or 1–2 cells (c) depending on the genotypes of the plant and pathogen.

by extensive necrosis of host tissue that results in colonization and sporulation. The infection events and host responses that accompany *P. infestans* attack are well understood at the cellular level (Coffey and Wilson, 1983; Vleeshouwers et al., 2000). Infection generally starts when motile zoospores encyst and germinate on the leaf surface. Occasionally, sporangia can also initiate infections. Germ tubes form an appressorium followed by penetration peg, which pierces the cuticle and penetrates an epidermal cell where it establishes an infection vesicle. Branching hyphae with narrow digit-like haustoria expand from the site of penetration to neighboring cells through the intercellular space. Subsequently, infected tissue necrotizes, and the mycelium develops sporangiophores that emerge through the stomata to produce numerous asexual spores called sporangia. Pathogen dispersal usually occurs through the sporangia, which release zoospores under cool and humid conditions (Judelson and Blanco, 2005). Most plant species are not colonized during *P. infestans* infection, and they are known as nonhost plants. Investigations of *P. infestans* interactions on both resistant host and nonhost plant species revealed that the hypersensitive response (HR), a cell-death defense response in plants, is frequently associated with resistance (Kamoun et al., 1999b; Vleeshouwers et al., 2000).

14.3 PLANT RESISTANCE TO *P. INFESTANS*

14.3.1 Late Blight Resistance Genes

Nonhost plant species and resistant *Solanum* genotypes mount effective resistance responses to *P. infestans* generally through R (resistance) proteins, which are the innate immunity receptors within a plant (Kamoun et al., 1999b; Vleeshouwers et al., 2000). Numerous late blight resistance genes (*R* genes) were introgressed from the wild species *Solanum demissum*, *Solanum bulbocastanum*, and *Solanum berthaultii* into crop potatoes using classic breeding strategies (Ballvora et al., 2002; Huang et al., 2005; Song et al., 2003). Some of these *R* genes, notably *R1*, *R3a*, *Rpi-blb1* (also known as *RB*), and *Rpi-blb2*, have been identified (Ballvora et al., 2002; Huang et al., 2005; Song et al., 2003; van der Vossen et al., 2003, 2005). The cloning of several additional late blight *R* genes is imminent (Kuang et al., 2005; Park et al., 2005a and b; Smilde et al., 2005). So far, all cloned late blight *R* genes encode polypeptides that belong to the coiled coil motif, nucleotide binding site, and leucine-rich repeat domain (CC-NBS-LRR) class of plant R proteins (van Ooijen et al., 2007). As with several other *R* genes, late blight *R* genes often belong to complex loci that carry several *R* gene analogs (RGAs) of the CC-NBS-LRR class (Meyers et al., 2003; Michelmore and Meyers, 1998). Some of these complex loci, such as those on chromosomes 4 and 11, carry several genes effective against *P. infestans* and are known as major late blight resistance loci (Huang et al., 2005; Park et al., 2005a).

14.3.2 "Broad-Spectrum" Resistance Genes

The *Rpi-blb1* and *Rpi-blb2* genes have been described as *P. infestans* resistance genes that confer a broad-spectrum of late blight disease resistance (Song et al., 2003; van der Vossen et al., 2003, 2005) in contrast to *R1* and *R3a*, for instance, which are only effective against particular races of *P. infestans*. *Rpi-blb1* and *Rpi-blb2* are effective against all (so far) tested races of the *P. infestans* pathogen and have a great potential to achieve long-term sustainable late blight management.

TABLE 14.1 Cloned *Phytophthora infestans Avr* genes.

Avr gene	Matching *R* gene, species	Number of homologs in T30-4 genome
Avr1	*R1, Solanum demissum*	1
Avr2	*R2, Solanum* spp.	18
Avr3a	*R3a, Solanum demissum*	3
Avr4	*R4, Solanum demissum*	1
Avr-blb1	*Rpi-blb1, Solanum bulbocastanum*	1
Avr-blb2	*Rpi-blb2, Solanum bulbocastanum*	8

In fact, somatic hybrids between potatoes and one *Rpi-blb1* (= *RB*) *S. bulbocastanum* clone PT29, as well as a number of backcrossed progenies, exhibited broad spectrum and persistent resistance in a variety of field trials over several years (Helgeson et al., 1998; Song et al., 2003). The discovery that *Rpi-blb1* and *Rpi-blb2* are effective in the field in several geographical areas and over several growing seasons has resulted in commercial interest and prompted the deployment of resistant potato cultivars. Most recently, a transgenic potato variety that carries *Rpi-blb1* and *Rpi-blb2* has entered the commercialization pipeline in Europe.

14.3.3 Can *R* Genes Mediate Durable Resistance?

The deployment of "broad spectrum" *Rpi-blb1* and *Rpi-blb2* as well as other emerging *R* genes challenges the traditional view that *R* genes are ineffective in the field over long periods of time. There are plausible hypotheses to support the view that *R* genes could mediate durable resistance (Kamoun et al., 1999b; Kamoun and Smart, 2005), but it remains difficult to predict the outcome of the large-scale deployment of such resistant varieties. Cloning of the *Avr* genes recognized by *Rpi-blb1* and *Rpi-blb2* (Vleeshouwers et al., 2008) (Table 14.1) will help to evaluate to which extent the *Rpi-blb* genes have wide-spectrum activity against *P. infestans* (see below for background on *Avr* and effector genes). This will also allow researchers to study the mechanisms by which *P. infestans* evades recognition in the event of emergence of races that overcome these *R* genes. In summary, further investigation of how *P. infestans* genes evolve in order to circumvent or overcome *R* gene-mediated resistance will be required to fully appreciate resistance and to estimate durability.

14.4 *P. INFESTANS* PROTEINS REQUIRED FOR SUCCESSFUL INFECTION

14.4.1 Effectors Are Key Pathogenicity Determinants

For *P. infestans* to infect and colonize its hosts successfully, a series of pathogenic processes are necessary (Huitema et al., 2004; Kamoun, 2003).

These include adhesion to the host surface, penetration, suppression of host immune responses, uptake of nutrients, and colonization of host tissue. Pathogenesis involves the secretion of proteins and other molecules by *P. infestans* (Huitema et al., 2004; Kamoun, 2006). Some of these participate in helping the pathogen attach to plant surfaces, whereas others help in breaking down physical barriers to infection (e.g., plant membranes or cell walls), and some proteins could manipulate the plants recognition system. Secreted proteins, which are known as effectors, manipulate biochemical and physiological processes of host plants presumably by interacting with host proteins known as effector targets (ETs) (Kamoun, 2006, 2007)). In susceptible plants, these effectors promote infection by suppressing defense responses, enhancing susceptibility, or inducing disease symptoms. On the other hand in resistant plants, effectors are typically recognized by resistance proteins from the plant resulting in effective defenses, including hypersensitive cell death. Such effectors are usually termed avirulence (AVR) proteins, although it is assumed that these AVR proteins have virulence functions in susceptible plants.

14.4.2 Identification of *P. infestans* Effectors

Traditionally, effectors have been identified by biochemical purification and genetic analyses. With the advent of genomics, novel strategies have emerged (Fig. 14.2). Application of biochemical, genetic, and bioinformatic strategies are nowadays mostly combined to identify effector genes from *P. infestans* (Kamoun, 2006). The implementation of functional genomics pipelines to identify effector genes from sequence datasets has proven to be particularly successful. A common pipeline consists of two major steps: (1) use of data mining tools to identify candidate genes that fulfill a list of specific criteria and (2) followed by analysis and validation of these candidate genes by functional assays, such as expression *in planta* and evaluation for effector-like activities. In *P. infestans*, a first functional genomics pipeline was used for the discovery of the *Crinkler* gene family (Torto et al., 2003). Later, variations on the original pipeline led to the identification of several other gene-encoding effectors like the protease inhibitors EPI1, EPI10, and EPIC2B and the avirulence protein AVR3a and AVR-blb1 (Armstrong et al., 2005; Tian et al., 2004, 2007; Vleeshouwers et al., 2008).

14.4.3 *P. infestans* Effectors Carry Signal Peptides

A critical milestone in the identification of oomycete effector genes was the validation of the concept that effector proteins must be secreted to reach their cellular targets at the intercellular interface between the plant and pathogen or inside the host cell (Torto et al., 2003). Identification of putative secreted oomycete proteins was facilitated by the fact that, as in other eukaryotes, most secreted proteins are exported through the general secretory pathway, via short N-terminal amino acid sequences, which are known as signal peptides

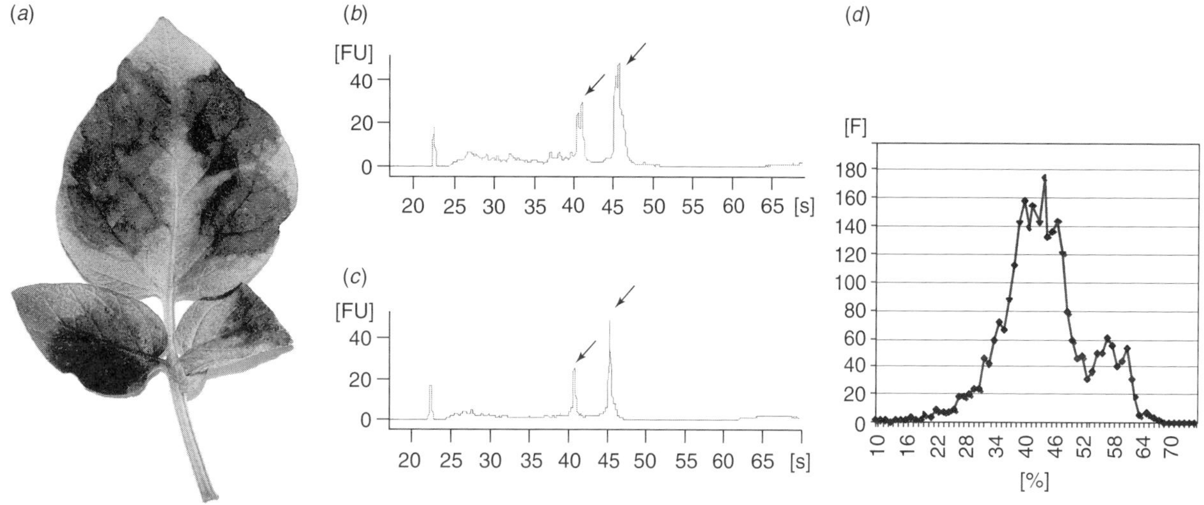

FIG. 14.2 Molecular and computational characterization of *P. infestans*–host interactions. (a) Lesions produced at 4 days postinoculation in potato (*Solanum tuberosum*) leaves infected with *P. infestans*. (b,c) Microcapillary electrophoresis chromatograms of RNA isolated from diseased leaves (b) and isolated from *P. infestans* zoospores (c). In (b), each arrow points to two double-high fluorescent intensity (FU) peaks (b) that reflects the presence of both plant and pathogen rRNAs, whereas in (c) the arrows point two single peaks (c) that correspond to the pathogen rRNAs only. (d) Guanine cytosine (GC) counting analyses of expressed sequence tag (EST) data, which is generated by sequencing of interaction cDNA libraries, reveals two distinct classes of mRNAs. EST populations differ in GC content (X-axis) as can be observed by the frequency (Y-axis) of content in a *P. infestans*–host derived EST dataset (2,805 sequences). Oomycete spp. are known to exhibit high GC content in gene coding sequences, which allows preliminary assignment of EST origin by simple GC counting methods.

(Torto et al., 2003). Signal peptides are highly degenerate and cannot be identified using DNA hybridization or polymerase chain reaction (PCR)-based techniques. Nonetheless, computational tools, particularly the SignalP program that was developed using machine learning methods (Nielsen et al., 1999), can assign signal peptide prediction scores and cleavage sites to unknown amino acid sequences with a high degree of accuracy (Menne et al., 2000; Schneider and Fechner, 2004). Therefore, with the accumulation of oomycete cDNA and genome sequences, lists of candidate secreted proteins could be readily generated using bioinformatics tools. In an early study, Torto et al. (2003) developed PexFinder (with Pex standing for *Phytophthora* extracellular protein), an algorithm based on SignalP v2.0 (Nielsen et al., 1999) to identify proteins that contain putative signal peptides from expressed sequence tags (ESTs). PexFinder was subsequently applied to ESTs from *P. infestans*. Proteomic identification of secreted proteins collected from culture filtrates of *P. infestans* matched PexFinder predictions, which convincingly validated the algorithm performance (Torto et al., 2003).

14.4.4 Several Apoplastic Effectors Are Inhibitors of Plant Enzymes

Effector proteins with inhibitory activities for protection against host apoplastic proteases and glucanases have been reported in *P. infestans* (Kamoun, 2006, 2007). EPI1 and EPI10 are multidomain secreted serine protease inhibitors of the Kazal family that bind and inhibit the pathogenesis-related (PR) protein P69B, a subtilisin-like serine protease of tomato that functions in defense (Tian et al., 2004, 2005). *P. infestans* also secretes the cystatin-like cysteine protease inhibitors EPIC1 and EPIC2B into the apoplast that target PIP1 and other apoplastic cysteine proteases of tomato (Tian et al., 2007). Another set of apoplastic effectors produced by *P. infestans* are inhibitors of tomato glucanases, hydrolytic enzymes that can degrade oomycete cell walls and release glucan oligosachharide elicitors (Damasceno et al., 2008).

14.4.5 AVR Proteins Belong to the RXLR Class of Cytoplasmic Effectors

Oomycete effectors of the RXLR class are defined by an N-terminal motif RXLR (arginine, any amino acid, leucine, and arginine) and are delivered inside plant cells where they alter host defenses (Birch et al., 2006; Kamoun, 2006; Rehmany et al., 2005; Whisson et al., 2007). Genome-wide catalogs of RXLR effectors, which are generated using computational approaches, unraveled a remarkably complex and divergent set of hundreds of candidate genes. The *P. infestans* effectors AVR3a, AVR4, and AVR-blb1 (also known as IPIO) are the most intensely studied effectors and belong to the RXLR family (Armstrong et al., 2005; van Poppel et al., 2008; Vleeshouwers et al., 2008). Expression of these AVR proteins inside the cytoplasm of plants that contain the matching R protein results in hypersensitive cell death, indicating that recognition occurs inside the host cells where the effectors are translocated by *P. infestans* (Bos et al., 2006; Rehmany et al., 2005; Whisson et al., 2007).

14.4.6 RXLR Effectors Are Modular Proteins

RXLR effectors are modular proteins with two main functional domains (Bos et al., 2006; Kamoun, 2006). While the N-terminal domain that encompasses the signal peptide and conserved RXLR region functions in secretion and targeting, the remaining C-terminal domain carries the effector activity and operates inside plant cells. The RXLR motif defines a domain that enables translocation of the effector proteins into the host cell (Whisson et al., 2007). Interestingly, the RXLR motif is similar in sequence and position to the plasmodial host translocation (HT)/Pexel motif that functions in delivery of parasite proteins into the red blood cells of mammalian hosts (Hiller et al., 2004; Marti et al., 2004). The RXLR domains of *P. infestans* AVR3a and another RXLR protein PH001D5 can mediate the export of the green fluorescent protein (GFP) from the *Plasmodium falciparum* parasite to red blood cells (Bhattacharjee et al., 2006), indicating that plant and animal eukaryotic pathogens share similar secretion and translocation signals for effector delivery into host cells (Haldar et al., 2006) and that the RXLR domain is by itself sufficient for translocation within the host cell.

14.4.7 Virulence Functions of RXLR Effectors

The avirulence activity of AVR3a is most likely not the main function of this effector. Bos et al. (2006) reported that the *P. infestans* RXLR effector AVR3a suppresses the hypersensitive cell death induced by another *P. infestans* protein INF1 elicitin, which points to a possible virulence function. The finding that some effectors have such activity was expected because the occurrence of cell-death-suppressing effectors has been hypothesized for biotrophic fungal and oomycete pathogens (Panstruga, 2003) based on cytological observations of susceptible interactions and the prevalence of cell-death suppressors among bacterial effectors (Jamir et al., 2004; Janjusevic et al., 2006). At this stage, the mechanisms through which AVR3a and other RXLR effectors interfere with plant immunity remain to be elucidated.

14.4.8 Crinklers: A Second Class of Cytoplasmic Effectors

P. infestans CRN ("Crinkler") proteins form a distinct class of secreted proteins that alter host responses and are thought to play important roles in disease progression (Kamoun, 2007; Torto et al., 2003; Win et al., 2007). Two *Crinkler* genes were identified following an *in planta* functional expression screen of candidate-secreted proteins of *P. infestans* based on a vector derived from *Potato virus X* (Torto et al., 2003). Ectopic expression of both genes in *Nicotiana* spp. and in the host plant tomato results in a leaf-crinkling and cell-death phenotype accompanied by an induction of defense-related genes. Torto et al. (2003) proposed that the CRNs function as effectors that perturb host cellular processes based on an analogy to bacterial effectors, which typically cause macroscopic phenotypes such as cell death, chlorosis, and tissue browning when expressed in host cells. *In planta* expression of a collection of deletion mutants of *crn2*, including ones that miss the signal peptide sequence, indicates that the CRN2 protein activates defense responses inside plant cells. This result suggests that the CRNs are cytoplasmic effectors (unpublished data from the Kamoun laboratory).

14.4.9 The Crinklers Are Also Modular Proteins

Computational analyses revealed that the CRNs form a complex family of relatively large proteins (about 400–850 amino acids) in *Phytophthora*. Interestingly, the CRNs are defined by a distinct conserved N-terminal motif characterized by the consensus LXLFLAK (Kamoun, 2007; Win et al., 2007). In *H. parasitica*, LXLFLAK overlaps with the RXLR motif and results in RXLRLFLAK. This finding, along with the observation that the N-terminal region of the CRNs is dispensable for cell death induction when these proteins are directly expressed *in planta*, suggests that the N-terminus of the CRNs contributes to host targeting similar to RXLR effectors, perhaps through the LXLFLAK motif. Thus, similar to RXLR effectors, the CRNs are modular proteins that consist of distinct N-terminal targeting and C-terminal effector domain. However, unlike the RXLR effectors, the CRNs show high rates of gene conversion and/or recombination with the most prominent recombination site occuring after the conserved N-terminal motifs (unpublished, Kamoun Lab). Therefore, a number of the *P. infestans* CRNs are chimeras that show unique associations between a conserved N-terminal domain and a variety of divergent C-terminal regions. The discovery of the chimeras by genome mining is valuable to gain insight into the CRN effector functions and points to the plasticity of the *crn* genes in the *P. infestans* genome.

14.5 GENOMICS RESOURCES

The *P. infestans* genome is sequenced and the already initiated data mining confirms the presence of several effector classes, and it will likely reveal more classes of genes important in pathogenicity. The genome size of *P. infestans* is estimated to be 237 Mb (Tooley and Therrien, 1987). Microscopic analyses indicate that *P. infestans* has 8–10 chromosomes (Sansome and Brasier, 1973). The genome of *P. infestans* strain T30-4 was sequenced to 8 × coverage and assembled into 4,921 supercontigs (scaffolds) by Chad Nusbaum and collaborators at the Broad Institute (see Chapter 27 by Zody and Nusbaum). The genome assembly was released to GenBank (accession AATU01000000). Release 2.1 of the gene annotation comprises 18,111 annotated genes.

On September 18–19, 2007, ~30 scientists attended the *P. infestans* genome analysis meeting in Boston. Publications that describe the genome sequence are planned for the near future. A full description of the project is available at http://www.broad.mit.edu/annotation/genome/phytophthora_infestans/Home.html

Additional genomics resources that are generated for *P. infestans* consist of >80,000 expressed sequence tags (ESTs) from a variety of developmental and infection stages (Kamoun et al., 1999a; Randall et al., 2005). In addition an Affymetrix GeneChip was designed, based on ~15,000 unique sequences obtained from ESTs (Judelson et al., 2008).

14.6 FUTURE PROSPECTS

Remarkable progress has been made in understanding the pathology of *P. infestans* with the completion of the genome sequence and the subsequent discovery of a large repertoire of effector proteins. Yet, many questions remain unanswered. The focus now is on dissecting the molecular mechanisms that enable *P. infestans* to successfully infect plants and the plant processes that are perturbed and manipulated by this pathogen further. In turn, the mechanisms that underlie *P. infestans–Solanum* interactions will need to be interpreted in the appropriate evolutionary context to translate the findings into significant conceptual advances.

Below, we list some key questions that we predict will drive research on *P. infestans* effectors in the coming years.

- What are the biochemical activities of effectors and their host targets? How do the effectors perturb their targets? Do effectors and their targets interact directly or indirectly to establish a pathogen-friendly environment inside the host plant?
- What is the temporal and spatial dimension of effector activity in both the pathogen and the host? Are there waves of effector secretion? Are effectors secreted at particular sites at the interface between microbe and plant? Do they have distinct functions depending on the stage of the infection process?
- Do effectors act collectively or independently in perturbing host cell processes? Also, are effectors redundant in their functions?
- How are cytoplasmic effectors translocated inside host cells? To which organelles do they go and where do they actually function?
- What are the levels of natural variation of effectors and their targets? How does natural variation affect function? Will similar effectors have a different function in different pathogen–host interactions?
- Are there undiscovered classes of effectors, and how can they be identified?
- How did *P. infestans* and its sister species adapt to their respective hosts?

- Can *P. infestans* evolve to overcome "broad spectrum" *R* genes? If so, how does it achieve this? Is it possible to engineer durable resistance to *P. infestans*?

ACKNOWLEDGMENTS

The Kamoun Laboratory is supported by The Gatsby Charitable Foundation. We thank past and present members of the laboratory for their contributions.

REFERENCES

Adler NE, Chacon G, Flier WG, Forbes GA 2002. The andean fruit crop, pear melon (*Solanum muricatum*) is a common host for A1 and A2 strains of *Phytophthora infestans* in Ecuador. Plant Pathol 51:802.

Adler NE, Erselius LJ, Chacon MG, Flier WG, Ordofiez ME, Kroon LPNM, Forbes GA 2004. Genetic diversity of *Phytophthora infestans* sensu lato in Ecuador provides new insight into the origin of this important plant pathogen. Phytopathol 94: 154–162.

Armstrong MR, Whisson SC, Pritchard L, Bos JI, Venter E, Avrova AO, Rehmany AP, Bohme U, Brooks K, Cherevach I, et al. 2005. An ancestral oomycete locus contains late blight avirulence gene *Avr3a*, encoding a protein that is recognized in the host cytoplasm. Proc Nat Acad Sci USA 102:7766–7771.

Ballvora A, Ercolano MR, Weiss J, Meksem K, Bormann CA, Oberhagemann P, Salamini F, Gebhardt C 2002. The *R1* gene for potato resistance to late blight (*Phytophthora infestans*) belongs to the leucine zipper/NBS/LRR class of plant resistance genes. Plant J 30:361–371.

Bhattacharjee S, Hiller NL, Liolios K, Win J, Kanneganti TD, Young C, Kamoun S, Haldar K 2006. The malarial host-targeting signal is conserved in the Irish potato famine pathogen. PLoS Path 2:e50.

Birch PR, Rehmany AP, Pritchard L, Kamoun S, Beynon JL 2006. Trafficking arms: oomycete effectors enter host plant cells. Trends Microbiol 14:8–11.

Bos JI, Kanneganti TD, Young C, Cakir C, Huitema E, Win J, Armstrong MR, Birch PR, Kamoun S 2006. The C-terminal half of *Phytophthora infestans* RXLR effector AVR3a is sufficient to trigger R3a-mediated hypersensitivity and suppress INF1-induced cell death in *Nicotiana benthamiana*. Plant J 48:165–176.

Coffey MD, Wilson UE 1983. Histology and cytology of infection and disease caused by *Phytophthora*. In: Erwin DC, Bartnicki-Garcia S, Tsao PH, editors. *Phytophthora*. American Phytopathological Society. St. Paul, MN. pp. 289–301.

Damasceno CM, Bishop JG, Ripoll DR, Win J, Kamoun S, Rose JK 2008. Structure of the glucanase inhibitor protein (GIP) family from *Phytophthora* species suggests coevolution with plant endo-beta-1,3-glucanases. Mol Plant Microbe Interact 21:820–30.

Duncan JM 1999. *Phytophthora*-an abiding threat to our crops. Microbiol Today 26:114–116.

Erwin DC, Ribeiro OK 1996. *Phytophthora* diseases worldwide. APS Press. St. Paul, MN.

Flier WG, Grunwald NJ, Kroon LPNM, van den Bosch TBM, Garay-Serrano E, Lozoya-Saldana H, Bonants PJM, Turkensteen LJ 2002. *Phytophthora ipomoeae* sp. nov., a new homothallic species causing leaf blight on *Ipomoea longipedunculata* in the Toluca Valley of central Mexico. Mycol Res 106:848–856.

Flier WG, van den Bosch GBM, Turkensteen LJ 2003. Epidemiological importance of *Solanum sisymbriifolium*, *S. nigrum* and *S. dulcamara* as alternative hosts for *Phytophthora infestans*. Plant Pathol 52:595–603.

Garelik G 2002. Agriculture. Taking the bite out of potato blight. Science 298:1702–1704.

Goodwin SB, Fry WE 1994. Genetic analyses of interspecific hybrids between *Phytophthora infestans* and *Phytophthora mirabilis*. Exp Mycol 18:20–32.

Haldar K, Kamoun S, Hiller NL, Bhattacharje S, van Ooij C 2006. Common infection strategies of pathogenic eukaryotes. Nat Rev Microbiol 4:922–931.

Helgeson JP, Pohlman JD, Austin S, Haberlach GT, Wielgus SM, Ronis D, Zambolim L, Tooley P, Mcgrath JM, James RV, et al. 1998. Somatic hybrids between *Solanum bulbocastanum* and potato; a new source of resistance to late blight. Theoret App Genet 96:738–742.

Hiller NL, Bhattacharjee S, van Ooij C, Liolios K, Harrison T, Lopez-Estrano C, Haldar K 2004. A host-targeting signal in virulence proteins reveals a secretome in malarial infection. Science 306:1934–1937.

Huang S, van der Vossen E, Kuang H, Vleeshouwers V, Zhang N, Borm T, van Eck H, Baker B, Jacobsen E. Visser R 2005. Comparative genomics enabled the isolation of the *R3a* late blight resistance gene in potato. Plant J 42:251–261.

Huitema E, Bos JIB, Tian M, Win J, Waugh ME, Kamoun S 2004. Linking sequence to phenotype in Phytophthora-plant interactions. Trends Microbiol 12:193–200.

Jamir Y, Guo M, Oh HS, Petnicki-Ocwieja T, Chen S, Tang X, Dickman MB, Collmer A, Alfano JR 2004. Identification of Pseudomonas syringae type III effectors that can suppress programmed cell death in plants and yeast. Plant J 37:554–565.

Janjusevic R, Abramovitch RB, Martin GB, Stebbins CE 2006. A bacterial inhibitor of host programmed cell death defenses is an E3 ubiquitin ligase. Science 311:222–226.

Judelson HS, Ah-Fong AM, Aux G, Avrova AO, Bruce C, Cakir C, da Cunha L, Grenville-Briggs L, Latijnhouwers M, Ligterink W, et al. 2008. Gene expression profiling during asexual development of the late blight pathogen Phytophthora infestans reveals a highly dynamic transcriptome. Mol Plant Microbe Interact 21:433–447.

Judelson HS, Blanco FA 2005. The spores of *Phytophthora*: weapons of the plant destroyer. Nature Revi Microbiol 3:47–58.

Kamoun S 2003. Molecular genetics of pathogenic oomycetes. Eukaryot Cell 2:191–199.

Kamoun S 2006. A catalogue of the effector secretome of plant pathogenic oomycetes. Ann Rev Phytopathol 44:41–60.

Kamoun S 2007. Groovy times: filamentous pathogen effectors revealed. Curr Opin Plant Biol 10:358–365.

Kamoun S, Hraber P, Sobral B, Nuss D, Govers F 1999a. Initial assessement of gene diversity for the oomycete pathogen *Phytophthora infestans* based on expressed sequences. Fungal Genet Biol 28:94–106.

Kamoun S, Huitema E, Vleeshouwers VGAA 1999b. Resistance to oomycetes: a general role for the hypersensitive response? Trends Plant Sci 4:196–200.

Kamoun S, Smart CD 2005. Late blight of potato and tomato in the genomics era. Plant Dis 89:692–699.

Kuang H, Wei F, Marano MR, Wirtz U, Wang X, Liu J, Shum WP, Zaborsky J, Tallon LJ, Rensink W, et al. 2005. The R1 resistance gene cluster contains three groups of independently evolving, type I R1 homologues and shows substantial structural variation among haplotypes of *Solanum demissum*. Plant J 44:37–51.

Mackenzie D 2008. Let them eat spuds. New Scientist 2667:30–33.

Marti M, Good RT, Rug M, Knuepfer E, Cowman AF 2004. Targeting malaria virulence and remodeling proteins to the host erythrocyte. Science 306:1930–1933.

Menne KM, Hermjakob H, Apweiler R 2000. A comparison of signal sequence prediction methods using a test set of signal peptides. Bioinformatics 16:741–742.

Meyers BC, Kozik A, Griego A, Kuang H, Michelmore RW 2003. Genome-wide analysis of NBS-LRR-encoding genes in Arabidopsis. Plant Cell 15:809–834.

Michelmore RW, Meyers BC 1998. Clusters of resistance genes in plants evolve by divergent selection and a birth-and-death process. Genome Res 8:1113–1130.

Nielsen H, Brunak S, von Heijne G 1999. Machine learning approaches for the prediction of signal peptides and other protein sorting signals. Protein Eng 12:3–9.

Ordonez ME, Hohl HR, Velasco JA, Ramon MP, Oyarzun PJ, Smart CD, Fry WE, Forbes GA, Erselius LJ 2000. A novel population of *Phytophthora*, similar to *P. infestans*, attacks wild *Solanum* species in Ecuador. Phytopathol 90:197–202.

Panstruga R 2003. Establishing compatibility between plants and obligate biotrophic pathogens. Curr Opin Plant Biol 6:320–326.

Park TH, Gros J, Sikkema A, Vleeshouwers VG, Muskens M, Allefs S, Jacobsen E, Visser RG, van der Vossen EA 2005a. The late blight resistance locus Rpi-bib3 from *Solanum bulbocastanum* belongs to a major late blight R gene cluster on chromosome 4 of potato. Mol Plant Microbe Interact 18:722–729.

Park TH, Vleeshouwers VG, Huigen DJ, van der Vossen EA, van Eck HJ, Visser RG 2005b. Characterization and high-resolution mapping of a late blight resistance locus similar to R2 in potato. Theor Appl Genet 111:591–597.

Randall TA, Dwyer RA, Huitema E, Beyer K, Cvitanich C, Kelkar H, Ah Fong AMV, Gates K, Roberts S, Yatzkan E, et al. 2005. Large-scale gene discovery in the oomycete *Phytophthora infestans* reveals likely components of phytopathogenicity shared with true fungi. Mol Plant-Microbe Interact 18:229–243.

Reader J 2008. Propitious esculent: the potato in world history. William Heinemann. pp. 320.

Rehmany AP, Gordon A, Rose LE, Allen RL, Armstrong MR, Whisson SC, Kamoun S, Tyler BM, Birch PR, Beynon JL 2005. Differential recognition of highly divergent downy mildew avirulence gene alleles by *RPP1* resistance genes from two Arabidopsis lines. Plant Cell 17:1839–1850.

Sansome E, Brasier CM 1973. Diploidy and chromosomal structural hybridity in *Phytophthora infestans*. Nature 241:344–345.

Schiermeier Q 2001. Russia needs help to fend off potato famine, researchers warn. Nature 410:1011.

Schneider G, Fechner U 2004. Advances in the prediction of protein targeting signals. Proteomics 4:1571–1580.

Smilde WD, Brigneti G, Jagger L, Perkins S, Jones JD 2005. Solanum mochiquense chromosome IX carries a novel late blight resistance gene Rpi-moc1. Theor Appl Genet 110:252–258.

Song J, Bradeen JM, Naess SK, Raasch JA, Wielgus SM, Haberlach GT, Liu J, Kuang H, Austin-Phillips S, Buell CR, et al. 2003. Gene RB cloned from *Solanum bulbocastanum* confers broad spectrum resistance to potato late blight. Proc Nat Acad Sci USA 100:9128–9233.

Tian M, Benedetti B, Kamoun S 2005. A Second Kazal-like protease inhibitor from *Phytophthora infestans* inhibits and interacts with the apoplastic pathogenesis-related protease P69B of tomato. Plant Physiol 138:1785–1793.

Tian M, Huitema E, da Cunha L, Torto-Alalibo T, Kamoun S 2004. A Kazal-like extracellular serine protease inhibitor from *Phytophthora infestans* targets the tomato pathogenesis-related protease P69B. J Biol Chem 279:26370–26377.

Tian M, Win J, Song J, van der Hoorn R, van der Knaap E, Kamoun S 2007. A *Phytophthora infestans* cystatin-like protein targets a novel tomato papain-like apoplastic protease. Plant Physiol 143:364–377.

Tooley PW, Therrien CD 1987. Cytophotometric determination of the nuclear DNA content of 23 Mexican and 18 non-Mexican isolates of *Phytophthora infestans*. Exp Mycol 11:19–26.

Torto T, Li S, Styer A, Huitema E, Testa A, Gow NAR, van West P, Kamoun S 2003. EST mining and functional expression assays identify extracellular effector proteins from *Phytophthora*. Genome Res 13:1675–1685.

van der Vossen E, Sikkema A, Hekkert BL, Gros J, Stevens P, Muskens M, Wouters D, Pereira A, Stiekema W, Allefs S 2003. An ancient *R* gene from the wild potato species *Solanum bulbocastanum* confers broad-spectrum resistance to *Phytophthora infestans* in cultivated potato and tomato. Plant J 36:867–882.

van der Vossen EA, Gros J, Sikkema A, Muskens M, Wouters D, Wolters P, Pereira A, Allefs S 2005. The *Rpi-blb2* gene from *Solanum bulbocastanum* is an *Mi-1* gene homolog conferring broad-spectrum late blight resistance in potato. Plant J 44: 208–222.

van Ooijen G, van den Burg HA, Cornelissen BJ, Takken FL 2007. Structure and function of resistance proteins in solanaceous plants. Annu Rev Phytopathol 45:43–72.

van Poppel PMJA, Guo J, van de Vondervoort PJI, Jung MWM, Birch PRJ, Whisson SC, Govers F 2008. The *Phytophthora infestans* avirulence gene Avr4 encodes an RXLR-dEER effector. Molec Plant Microbe Interact 21:1460–1470.

Vleeshouwers VG, Rietman H, Krenek P, Champouret N, Young C, Oh SK, Wang M, Bouwmeester K, Vosman B, Visser RG, et al. 2008. Effector genomics accelerates discovery and functional profiling of potato disease resistance and *Phytophthora infestans* avirulence genes. PLoS ONE 3:e2875.

Vleeshouwers VG, van Dooijeweert W, Govers F, Kamoun S, Colon LT 2000. The hypersensitive response is associated with host and nonhost resistance to *Phytophthora infestans*. Planta 210:853–864.

Whisson SC, Boevink PC, Moleleki L, Avrova AO, Morales JG, Gilroy EM, Armstrong MR, Grouffaud S, van West P, Chapman S, et al. 2007. A translocation signal for delivery of oomycete effector proteins into host plant cells. Nature 450:115–118.

Win J, Morgan W, Bos J, Krasileva KV, Cano LM, Chaparro-Garcia A, Ammar R, Staskawicz BJ, Kamoun S. 2007. Adaptive evolution has targeted the C-terminal domain of the RXLR effectors of plant pathogenic oomycetes. Plant Cell 19: 2349–2369.

15

PHYTOPHTHORA SOJAE AND SOYBEAN

MARK GIJZEN AND DINAH QUTOB
Agriculture and Agri-Food Canada, London, Ontario, Canada

15.1 INTRODUCTION

It was in the early 1950s that soybean growers in the United States and Canada were alarmed by a new and devastating disease that was spreading and destroying their crop. Outbreaks of root rot in Illinois, Indiana, Ohio, Missouri, North Carolina, and Ontario, Canada, were attributed to a type of *Phytophthora*, but pathologists could not initially agree whether this should be classified as a new species (Hildebrand, 1959; Kaufmann and Gerdemann, 1958). Nearly 40 years passed before there was good evidence and a general consensus that *Phytophthora sojae* was the most appropriate name for this distinct species that is highly host specific and aggressive on soybeans (Erwin and Ribeiro, 1996). Root rot caused by *P. sojae* remains an endemic disease in soybean production regions around the globe. In terms of prevalence and yield loss, it is among the most damaging problems that confront producers (Wrather and Koenning, 2006). This chapter will review knowledge and emphasize recent advances in the genetics and molecular biology of the *P. sojae*–soybean interaction.

15.2 THE SOURCE AND NATURAL HISTORY OF *P. SOJAE* IS UNCERTAIN

The center of origin of *P. sojae* is unresolved. One hypothesis is that the species resided in North America on indigenous plants and then jumped hosts to soybean. It is known that *P. sojae* can infect many lupine species that are native

Oomycete Genetics and Genomics: Diversity, Interactions, and Research Tools
Edited by Kurt Lamour and Sophien Kamoun
Copyright © 2009 John Wiley & Sons, Inc.

to North America, so it is plausible that the pathogen spread to soybean after the introduction and widespread cultivation of the crop on the continent (Jones, 1969; Jones and Johnson, 1969). Certainly, a center of diversity of *P. sojae* seems to be in the American midwest. Alternatively, it is possible that *P. sojae* coevolved with wild (*Glycine soja*) and domesticated (*Glycine max*) soybean in East Asia. Taxonomically, *P. sojae* is closely related to the cowpea pathogen *Phytophthora vignae* (Cooke et al., 2000). In fact, a report of *P. sojae* × *P. vignae* hybrids suggests that the two species may share a common origin in East Asia (May et al., 2003).

Soybeans are an ancient crop that have been cultivated for over 3,000 years in Asia. With such a long history, it is surprising there is not more evidence or historical records that could help to distinguish these alternative hypotheses whether *P. sojae* is native to East Asia or North America. For example, in places where soybean has been grown for centuries, has this disease been a long-standing or more recent problem? Judging from the scientific literature, *P. sojae* seems to be a new disease in Asia. But the science of plant pathology is young and there is uncertainty. Farmers have always had an appreciation of diseases that affect their crops, as evidenced by Biblical references to blights and blasts. Perhaps there are historical or anecdotal reports in China, Korea, or Japan that we are unaware of in the West, which can help inform us about the natural history of *P. sojae*. Regardless, an extensive worldwide population genetic analysis that employs numerous *P. sojae* isolates, especially from North America and throughout Asia, is required and would likely resolve the question.

15.3 *P. SOJAE* IS A MAJOR DISEASE PROBLEM ON ONE OF THE WORLD'S LARGEST CROPS

The narrow host range of *P. sojae* is somewhat unusual among *Phytophthora* species, because multiple-host lifestyles are more common in members of this genus. Defining host range determinants remains elusive despite that whole genome sequences are now available for several species of *Phytophthora*, which includes those with narrow and broad host ranges. In any case, the limited host range of *P. sojae* does not diminish the economic damage potential of this organism, as soybeans are one of the largest production crops in the world. Annual losses that result from *P. sojae* infestations in North America alone are in the range of 10^9 kg of soybeans (Wrather and Koenning, 2006). At current prices over \$10 per bushel, this is equivalent to more than \$300 million in lost yearly production. Extrapolating these estimates to worldwide production figures suggests that the disease is responsible for \$1–2 billion in lost revenue per year. Despite these huge losses, there is sometimes complacency about *P. sojae* because it is a familiar disease that is endemic and widespread, and it does not capture attention like new and invasive problems. This is understandable but perilous because *P. sojae* is a relentless and persistent disease that pathologists need to monitor and guard against continually.

P. sojae is a soilborne pathogen that causes damping off in addition to stem and root rot (Erwin and Ribeiro, 1996; Tyler, 2007; Ward, 1990). It is considered a hemibiotrophic plant pathogen that infects living plant cells but also can live as a saprophyte off dead tissue. Infections are debilitating and may kill the plant outright, especially seedlings. The disease normally begins below the soil line and spreads up the stem, but aerial parts of the plant can be directly infected by rain splashes. *P. sojae* can attack plants at any stage, so it is a constant threat throughout the growing season. Wet conditions, poor drainage, soil compaction, and no-till practices tend to foster the disease in the field. Asexually produced zoospores are considered the primary infectious agent. These biflagellate, water-motile spores are attracted to isoflavones and other compounds secreted by soybean roots. The zoospores encyst and germinate on the root surface. The resulting germ tube penetrates the plant either directly or after growing along the surface. Sometimes, it forms an appressorium-like structure at the site of penetration. Epidermal cells are breached periclinally, and *P. sojae* hyphae spread intercellularly to form small finger-like and bulbous haustoria into host cells during this initial biotrophic phase of growth (Enkerli et al., 1997b; Ranathunge et al., 2008). Invading hyphae ramify throughout the cortex and penetrate into the stele. The transition from biotrophy to necrotrophy occurs as the pathogen builds up biomass and overwhelms the host. Gene expression patterns change rapidly in host and pathogen cells during colonization. Massive proliferation of hyphae within the stele is a feature of rapidly spreading infections that characterize the necrotrophic stage. Infected plant tissues lose their turgor and develop water-soaked lesions that eventually rot below ground or that dry out and shrivel up in aerial parts. *P. sojae* is dipoid throughout most of its life cycle; meiosis occurs in oogonia and antheridia that fuse to produce an oospore. These large sexual spores are abundantly produced in infected tissues and can survive for long periods in the soil. There are no known mating types of *P. sojae*, so it is considered to be a homothallic and essentially clonally propagating organism. Nonetheless, outcrosses can occur between different strains that will provide important sources of variation (Förster et al., 1994). The movement of soil by animals or machinery can spread *P. sojae* inoculums and introduce the disease to new areas. The disease may also be spread by contaminated seed.

15.4 CONTROL OF *P. SOJAE* RELIES ON DEVELOPMENT OF RESISTANT CULTIVARS

The management of *P. sojae* has relied on the introduction of soybean cultivars with enhanced resistance or tolerance. Chemical control, such as by seed treatments with antioomycete "fungicides" such as metalaxyl, is an effective alternative in heavily infested soils, but the cost of this approach precludes its widespread use. Soil treatments with $CaCl_2$ and $Ca(NO_3)_2$ have also been shown to reduce the severity of disease, but this method is not widely employed

(Sugimoto et al., 2005). The integration of *P. sojae* resistance screening into soybean breeding programs is standard practice and by far the most universal approach to dealing with the disease. There are two main types of genetically controlled resistance of soybean to *P. sojae* that are commonly recognized. Partial resistance is mediated by quantitative trait loci (QTL) that lessen the severity of disease symptoms and attenuate the growth of the pathogen in host tissues. Race- and cultivar-specific resistance follows the canonical gene-for-gene interaction model, in which the presence of host resistance genes and pathogen avirulence genes determine the disease outcome.

15.4.1 Partial Resistance

This may also be called rate-limiting or rate-reducing resistance, field tolerance, QTL-mediated resistance, horizontal resistance, and non-race-specific resistance. Although these are grouped together for simplicity, there are likely multiple underlying genetic and mechanistic differences that operate to produce partially resistant or tolerant phenotypes. Partial resistance has been studied most extensively in the cultivar Conrad. This cultivar has no known *Rps* (resistance to *P. sojae*) genes but displays high levels of partial resistance whether assayed in the greenhouse or the field. The mapping of QTL from Conrad controlling partial resistance has led to identification of loci on soybean multiple linkage groups (MLG) F, MLG Db1 + W, and on MLG J (Burnham et al., 2003; Han et al., 2008; Mideros et al., 2007; Weng et al., 2007).

It is not known how these QTL mediate resistance, but recent work has implicated root suberin content with partial resistance (Ranathunge et al., 2008; Thomas et al., 2007). Suberin and the closely related polymer cutin provide protection and control water movement in plant tissues. Cutin is made primarily of aliphatic esters, whereas suberin is a biopolyester that contains aliphatic and aromatic domains. Suberin is deposited to varying degrees in the epidermis, cortical cells, and endodermis of soybean roots, so it is well positioned to provide defense against *P. sojae*. In a comparison of nine different cultivars and lines with varying levels of partial resistance, a positive association between root aliphatic suberin content and partial resistance was noted (Thomas et al., 2007). Additional analysis of recombinant inbred lines from a cross between Conrad and a highly susceptible line seemed to substantiate the correlation between partial resistance and root aliphatic suberin content. Additional work is necessary to map QTL that control aliphatic suberin content in the root and to determine whether genes that control suberin content cosegregate with any of the known partial resistance QTL.

Besides root suberin, another possible mechanism for control of partial resistance relies on differential recognition of *P. sojae*. Secreted factors from *P. sojae*, or specific pathogen-associated molecular patterns (PAMPs), could activate resistance mechanisms without providing absolute protection. In this model, soybean cultivars with effective partial resistance have better surveillance systems and enhanced effector- or PAMP-triggered innate immunity to

P. sojae. Evidence so far for this model is thin, but results suggesting that PAMP-triggered immune responses are crucial to resistance lend weight to this hypothesis (Graham et al., 2007).

15.4.2 Race- and Cultivar-Specific Resistance

Single dominant soybean *Rps* genes condition resistance to *P. sojae* strains that carry the cognate avirulence (*Avr*) gene. Soybean *Rps* genes are believed to activate effector-triggered immune responses like canonical resistance (*R*) genes that have been well characterized in other pathosystems (Jones and Dangl, 2006). There are at least 14 known *Rps* genes localized to four different MLGs as follows: MLG N, *Rps1-a, -b, -c, -d, -k*, and *Rps7*; MLG J, *Rps2*; MLG F, *Rps3-a, -b, -c*, and *Rps8*; MLG G, *Rps4, Rps5*, and *Rps6* (Demirbas et al., 2001; Sandhu et al., 2005; Weng et al., 2001). Additional *Rps* genes undoubtedly exist but have not been genetically characterized (Gordon et al., 2007). Soybean *Rps* genes that condition resistance to particular *P. sojae* strains usually provide absolute protection against infection. One exception is the *Rps2* gene that provides incomplete resistance and is root specific (Mideros et al., 2007). *R* genes to diseases other than *P. sojae* sometimes cluster together with *Rps* genes, such as in MLG J and MLG F (Graham et al., 2002; Hayes et al., 2004). Despite the usefulness of *Rps* genes, their durability in the field is limited to about 8–15 years, because new *P. sojae* strains develop that evade *Rps*-mediated immunity (Dorrance et al., 2003). The effectiveness and lifespan of discrete *Rps* genes is variable for reasons that are not understood. It may depend on their mode of action and how they are deployed.

The *Rps1* locus is the best characterized and shown to harbor a cluster of coiled-coil (CC), nucleotide binding site (NBS), leucine-rich repeat (LRR) type *R* genes (Bhattacharyya et al., 2005; Gao and Bhattacharyya, 2008; Gao et al., 2005). Four closely related copies of CC-NBS-LRR genes isolated from *Rps1-k* plants indicate that at least two of these copies, which are *Rps1-k-1* and *Rps1-k-2,* function in conditioning race-specific resistance. Similarly, the *Rps4-Rps6* locus contains multiple copies of an NBS-LRR gene. The loss of *Rps4*-mediated resistance is associated with gene deletions from this cluster (Sandhu et al., 2004). Together, work on the *Rps1* and *Rps4-Rps6* loci has suggested that it may be useful to consider particular *Rps* specificities to be conditioned by haplotypes or copy number variants rather than by single genes.

15.5 KNOWLEDGE OF *P. SOJAE* GENETICS AND MOLECULAR BIOLOGY IS ADVANCING RAPIDLY

15.5.1 The Advent of DNA Markers Enables Genetic Analyses of *P. sojae*

The genetic analysis of *P. sojae* presented many challenges to early researchers and still does today. Obtaining the basic genetic facts about *P. sojae* has been surprisingly difficult. For example, the chromosome number of *P. sojae* is still not known for certain, but most studies suggest a haploid number of 12–15 chromosomes (Hansen et al., 1986; Howlett, 1989; Tooley and Carras, 1992). It is astonishing that scientists could not even agree until the 1970s that oomycetes are diploid organisms. Likewise, it was not until the 1990s that oomycetes were appropriately placed taxonomically, together with heterokont algae and separate from fungi, which reflects their natural phylogeny. Many other obstacles have hampered the genetic studies of oomycetes over the years. Pigmentation phenotypes and other mutants that are amenable to laboratory investigations are generally not available for oomycete researchers. Another impediment for homothallic oomycetes like *P. sojae* is that it is difficult to distinguish self-fertilized from hybrid progeny without appropriate strains or markers. Nonetheless, hybrid progeny of *P. sojae* were reported in the 1980s (Layton and Kuhn, 1988), and tracking segregation in F_2 progeny was accomplished by several laboratories in the 1990s (Bhat and Schmitthenner, 1993; Tyler et al., 1995; Whisson et al., 1994). The development of DNA markers as an efficient means of identifying hybrids was an important advance that has greatly aided genetic analyses. The isolation of oospores requires skill and practice, and it remains a laborious process to create large segregating populations. Oospores of *P. sojae* germinate asynchronously, so it is necessary to examine plates repeatedly to select those that are alive and germinating. The steps involved in oospore isolation are outlined in Fig. 15.1.

Crosses between different *P. sojae* strains were used to construct genetic maps of linkage groups and to follow segregation of *Avr* loci in F_2 progeny (Gijzen et al., 1996; May et al., 2002; Tyler et al., 1995; Whisson et al., 1995, 2004). Markers and traits may follow Mendelian segregation ratios, but many exceptions have been noted. In practice, segregation distortion is often observed (MacGregor et al., 2002). The occurrence of parasexual genetic mechanisms in *P. sojae* and other oomycetes has been suspected for many years because of the observations of variation in virulence of single-zoospore derived cultures (Rutherford et al., 1985). An in-depth study of DNA markers in F_1 and F_2 progeny and in zoospores of *P. sojae* suggested that high-frequency gene conversion can be a major factor in skewing inheritance patterns (Chamnanpunt et al., 2001). The conversion of heterozygous loci to homozygosity in vegetative hyphae can distort segregation ratios. Gene conversion can also result in heterokayrons that can produce zoospores that are not genetically uniform, despite that these asexual propagules are considered "clonal." Mechanistically, it remains unknown how gene conversion operates in *P. sojae*, but models have been proposed (Chamnanpunt et al., 2001).

Forward genetic screens have generally not been employed thus far on *P. sojae*, but it is possible to perform these screens given sufficient resources. Certainly, the feasibility of generating *P. sojae* culture libraries with targeted induced local lesions in their genomes (TILLING) has been demonstrated (Lamour et al., 2006). This method provides a powerful tool for identifying mutants for particular genes and therefore will be valuable for experimentation into gene function and its relation to phenotype.

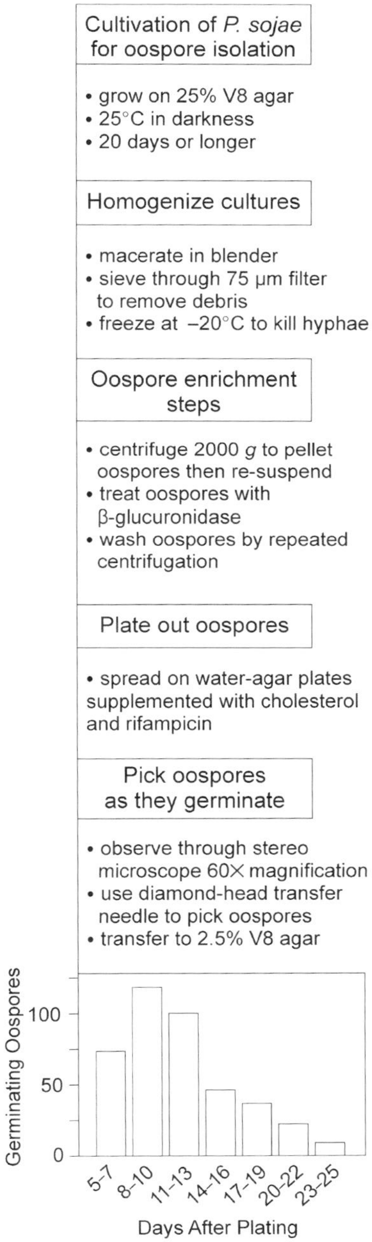

FIG. 15.1 Protocol for isolation of oospores from *P. sojae* for genetic studies. The primary steps for oospore isolation are shown in this flow chart. The graph at the base of the figure shows the distribution of germination oospores over time; most oospores germinate between 8 to 10 days after plating out on water agar.

15.5.2 *P. sojae* and Soybean Enter the Post-Genomic Age

The sampling of expressed sequence tags (ESTs) for *P. sojae* and soybean heralded the beginning of large-scale gene sequence characterization for these species (Qutob et al., 2000; Shoemaker et al., 2002). Currently, more than 28,000 *P. sojae* and 392,000 soybean ESTs have been deposited to GenBank, which represents approximately 7,800 and 45,000 distinct transcripts, respectively. Initial comparative sampling of ESTs from *P. sojae* growing axenically or on its host revealed insight into gene expression patterns during different life cycle stages (Qutob et al., 2000). This work showed that *P. sojae* and soybean transcripts differed in their nucleotide compositions, because soybean has a lower average guanine cytosine (GC) content (46%) compared with *P. sojae* (58%). This has been followed up by wider and deeper sampling of *P. sojae* transcripts (Torto-Alalibo et al., 2007) and by studies that employ differential screening techniques to identify infection-specific transcripts (Chen et al., 2007; Wang et al., 2006). The first microarray-based experiments of *P. sojae* gene expression were performed using custom-made cDNA arrays, but now commercially available platforms offer tens of thousands of soybean and *P. sojae* targets on a single slide. The analysis of rRNA subunit profiles is an effective method to monitor the growth of the pathogen in the host (Moy et al., 2004). Pathogen growth may also be visualized by simply plotting raw hybridization intensities for all *P. sojae* and soybean targets for arrays that contain a mix of genes from each organism, as illustrated in Fig. 15.2. A limitation of infection-site sampling is that pathogen biomass at the early stages of infection is too low to detect gene expression except for the most highly expressed genes. Laser capture microdissection, or massively parallel sequencing of cDNA transcripts, present methods that may overcome this limitation and characterize the pathogen's transcriptome at early infection stages.

Databases of ESTs, gene transcripts, and associated expression patterns are important and insightful, but the ultimate genetic resource for an organism is its genome sequence. For *P. sojae*, sequencing of the 95-Mb genome to an average depth of 9× was recently accomplished (Govers and Gijzen, 2006; Tyler et al., 2006). The 1.2-Gb soybean genome has also been sequenced to a depth of 7×, and a draft assembly was released in early 2008.

Integration of *P. sojae* sequence contigs from the whole genome shotgun together with bacterial artificial chromosome (BAC) contigs assembled by restriction fragment length polymorphisms resulted in genome-wide physical map that contained 79 superscaffolds (Zhang et al., 2006). There are now plans to finish and improve on this initial draft assembly to produce a high-quality reference sequence. This ongoing work includes gap closure, high-density BAC end sequencing, targeted BAC sequencing, and construction of a single nucleotide polymorphism (SNP)-based genetic map. To help correct gene models, an additional 400,000 *P. sojae* ESTs will be generated by pyrosequencing (B. Tyler, personal communication).

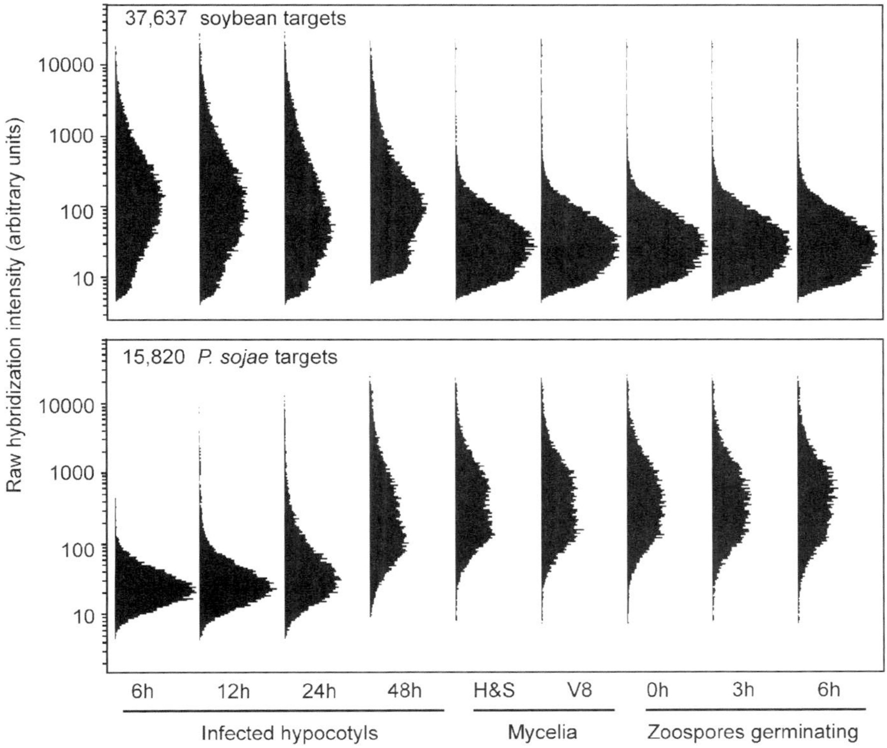

FIG. 15.2 Comprehensive microarray gene expression data for *P. sojae* and soybean. The distribution of raw hybridization intensities for 37,637 soybean and 15,820 *P. sojae* targets are shown. The nine treatments are as follows: four different time points of infection corresponding to 6, 12, 24, and 48 h after inoculation of soybean hypocotyls with mycelia plugs of *P. sojae*; H&S, *P. sojae* mycelia grown on synthetic media; V8, *P. sojae* mycelia grown on rich media; and *P. sojae* germinating zoospores at 0, 3, and 6 h after encystment. No normalization steps were performed on this data to illustrate the range of signal intensities depending on the occurrence or ratio of *P. sojae* and soybean messenger RNA (mRNA) in the various treatments. Values represent the mean of three independent experiments (platform: Affymetrix GeneChip® Soybean Genome Array).

It is hard to overstate the importance of genome sequence information. The data are of enormous value and durability, because they will be relied on by countless researchers for years to come. Genome sequence data not only furnishes strings of nucleotides and predicted gene lists, but also it provides insight into the evolution of an organism and may be used to assess emerging concepts or to formulate new hypotheses. In fact, scientists have predicted that in the future, the acquisition of genome sequence data will be considered a necessary starting point for most biological investigations (Lander and Weinberg, 2000). For studies on *P. sojae* and soybeans, the completion of whole genome sequences has launched a new era of research that is now beginning.

15.5.3 Proteomics of *P. sojae*

Proteomic studies of *P. sojae* have been facilitated by completion of the whole genome sequence because peptide masses can now be matched to predicted open reading frames (ORFs). For example, in a proteomic study of *P. sojae*, a total of 457 proteins were identified at a 99.9% confidence level, which includes 433 that matched predicted proteins from the whole genome annotation (Padliya et al., 2007). Proteomic approaches can also contribute to better annotation of the genome sequence just as EST information aids gene prediction. This has been demonstrated in a survey of mycelia and germinating cysts of *P. sojae*, in which a total of 3,967 peptide tags could be matched to predicted ORFs, representing 20.8% of the total predicted proteome of 19,027 proteins (Savidor et al., 2006). An additional 1,256 peptide tags matched ORFs that were not included in the proteome annotation, which indicates room for improvement in oomycete gene calling programs. An in-depth comparative analysis of the proteins present in *P. sojae* and *Phytophthora ramorum* led to identification of species-specific patterns that likely develop from host–range and lifestyle differences (Savidor et al., 2008). Protein profiles from this study also suggested a major shift in metabolism in different life cycle stages. Germinating cysts seem to rely primarily on β-oxidation of fatty acids for energy generation, whereas the glycolytic pathway predominates in mycelia (Savidor et al., 2008).

15.6 SOYBEAN RESPONSES TO *P. SOJAE* INFECTION

The defense responses of soybean to infection by *P. sojae* have been studied for many years, especially the production and accumulation of the isoflavonoid phytoalexins called the glyceollins. The hypothesis that glyceollins play a role in limiting the pathogen is compelling but has been a point of controversy because these phytoalexins are produced in both resistant and susceptible interactions, and their presence has not always been well correlated with resistance (Abe et al., 2002; Hahn et al., 1985). The isolation of *isoflavone synthase* (*IFS*) and other genes in the pathway to the glyceollins has allowed researchers to more thoroughly probe the upregulation of these transcripts and the role of phytoalexin biosynthesis in the disease process (Dhaubhadel et al., 2003; Lindermayr et al., 2002; Moy et al., 2004; Schopfer and Ebel, 1998; Schopfer et al., 1998; Steele et al., 1999). It has been shown that silencing of *IFS* severely impairs resistance to *P. sojae* and other pathogens (Graham et al., 2007; Lozovaya et al., 2007; Subramanian et al., 2005). These results substantiate the view that isoflavonoid phytoalexins are crucial to resistance. More surprisingly, the silencing of *IFS* also compromises the ability of soybean cells to mount an effective hypersensitive response (HR), which suggests a general role for 5′deoxy-isoflavonoids or other products that result from IFS activity in the development of HR in soybean (Graham et al., 2007). This is a departure from the conventional view that phytoalexin molecules simply inhibit pathogen growth through their direct toxicity. The authors point out that another product of the phenylpropanoid pathway, which is salicylic acid (SA), is considered to play a similar but reciprocal role.

Although SA is generally regarded as a signaling molecule, it may have phytoalexin-like properties as well. In the interaction between *P. sojae* and soybean, both SA and glyceollins are produced in abundance in resistant and susceptible interactions. Nonetheless, metabolite profiling at 48 h after infection indicated that SA and related compounds reach higher concentrations in susceptible interactions, whereas the level of glyceollins is greater in resistant interactions (Gijzen and McGarvey, unpublished). Apparently, *P. sojae* strongly triggers SA responses in soybeans whereas the octadecanoid and jasmonic acid (JA)-signaling pathways are dormant or repressed. This hypothesis is supported by transcriptome analysis and by studies that demonstrate *P. sojae* β-glucans do not activate JA pathways (Fliegmann et al., 2003; Moy et al., 2004).

Over the years, numerous studies have compared host responses in resistant versus susceptible interactions. These studies are often performed using genetic isolines for particular *Rps* genes to eliminate genetic background effects. Isolines have also facilitated the mapping of *Rps* genes in soybean (Kasuga et al., 1997). Incompatible interactions between *P. sojae* and soybean are characterized by rapid activation of defense responses, synthesis of pathogenesis related proteins, localized cell death, callose accumulation, actin cytoskeleton rearrangements, and other features consistent with *R*-gene mediated HR (Cahill et al., 2002; Enkerli et al., 1997a and b; Liu et al., 2001; Mithofer et al., 2002; Mohr and Cahill, 2001; Valer et al., 2006; Ward et al., 1989; Yi and Hwang, 1997). In aggregate, these responses are supposed to contain the pathogen and halt its ingress. Compatible interactions display many of the same responses but often with a delay or lower intensity than in the incompatible type. Host cell death is not feature of compatible interactions until the infection has progressed into the necrotrophic stage, in which it serves to accelerate rather than contain pathogen growth.

15.7 *P. SOJAE* ELICITORS, TOXINS, ENZYMES, AND EFFECTORS

The search for *P. sojae* molecules that elicit defense responses in plants has a rich history that helped to formulate our present concept of PAMP-triggered immunity in plants. Among the first well-characterized elicitors from plant pathogenic organisms were the β-glucan and transglutaminase elicitors from *P. sojae*.

15.7.1 The β-Glucan Elicitor Triggers PAMP-Mediated Immune Responses

The *P. sojae* β-glucan elicitor was initially purified based on its ability to induce isoflavonoid phytoalexin (glyceollin) biosynthesis in soybean cotelydons and cell cultures (Ayers et al., 1976; Sharp et al., 1984). Although the investigators at the time were hoping to find Avr factors, we now understand that the β-glucan elicitor intercepts PAMP-triggered rather than effector-triggered immunity in plants. The β-glucan elicitor is a branched heptaglucoside that is liberated from the cell walls of *P. sojae* by endoglucanase enzymes secreted by soybean cells. As a counterdefense, *P. sojae* produces glucanase inhibitor proteins (GIPs) to slow cell wall digestion and β-glucan release (Rose et al., 2002). Work has shown that β-glucan molecules cause cation fluxes, membrane depolarization, oxidative burst, and mitogen-activated protein kinase (MAPK) activity in soybean cells (Daxberger et al., 2007; Mithofer et al., 2001, 2005). The soybean β-glucan binding protein is proposed to act as part of an initial catalytic and perception system localized on the plasma membrane (Fliegmann et al., 2004, 2005). The inhibition of β-glucan release by *P. sojae* GIPs can be considered a pathogen counterdefense that, in turn, imposed diversifying selection on host endoglucanases (Bishop et al., 2005). Sequence comparisons indicate that *P. sojae* GIPs developed from the trypsin class of serine proteinases. Thus, GIPs are likely ancient proteinases that lost their peptide hydrolysis activity but developed new protein–protein interaction specificities that now define their functionality. The perception of β-glucans by soybean cells seems to be a crucial component in immunity to *P. sojae*, because RNA interference (RNAi) silencing of a β-glucan-releasing endoglucanase results in heightened susceptibility to infection and may even compromise *Rps* gene function (Graham et al., 2007). These results highlight the importance of PAMP-triggered immune responses in the resistance of soybean to *P. sojae*.

15.7.2 A *P. sojae* Transglutaminase is a PAMP in Nonhost Species

Another well-known PAMP initially isolated from *P. sojae* is the transgutaminase elicitor. The *P. sojae* transglutaminase elicitor was purified based on its capacity to induce phenylcoumarin phytoalexin biosynthesis in parsley cells (Nurnberger et al., 1994). The elicitor activates MAPK pathways in parsley, and the active portion of the protein can be reduced to a 13-amino acid peptide (Pep-13) that corresponds to a region crucial for the enzymatic activity of the transglutaminase (Brunner et al., 2002; Ligterink et al., 1997). Thus, parsley and other plants have developed their immune systems to recognize this motif as a PAMP. However, defining a role for Pep-13 in the interaction between *P. sojae* and soybean has been elusive because soybean does not seem to recognize this PAMP.

15.7.3 Necrosis and Ethylene-Inducing-Like (NLP) Proteins

The *P. sojae* necrosis inducing protein (PsojNIP) constitutes a pathogen molecule that is a powerful elicitor of defense responses and cell death in soybeans (Qutob et al., 2002). PsojNIP belongs to a group of related proteins that have been described in a wide variety of organisms that include fungi, oomycetes, and bacteria (Gijzen and Nurnberger, 2006). The proteins are similar to the *Fusarium oxysporum* necrosis- and ethylene-inducing peptide 1 (Nep1) and so they have been named Nep1-like proteins (NLPs). The function of NLPs is not known, but the proteins are common in plant pathogens and they share many characteristics with toxins and PAMPs (Qutob et al., 2006b). An analysis of the *P. sojae* genome sequence revealed at least 55 separate NLP sequences; however, about half of these ORFs are fragmentary or represent pseudogenes. Regardless, the NLP family in *P. sojae* is large and diverse and has evolved substantially in the time since *P. sojae* and *P. ramorum* diverged. These observations suggest that NLPs are under strong selective pressure and that they play an important role in pathogenesis.

15.7.4 Elicitins

The elicitins comprise yet another large family of proteins that were first characterized based on their ability to induce plant defense responses in tobacco (Ricci et al., 1989). Elicitins are small secreted proteins that are restricted to oomycetes and that likely have a role in the transport, perception, or metabolism of sterols and lipids (Osman et al., 2001; Ponchet et al., 1999). Plant species vary in their responsiveness to elicitins, and there is some evidence that these proteins may act as PAMPs in the interaction between *P. sojae* and its host (Becker et al., 2000). Furthermore, *P. sojae* is a sterol auxotroph, like other species of *Phytophthora*, and so it may rely on elicitins for the acquisition of host sterols (Marshall et al., 2001).The sequence diversity and evolution of the *P. sojae* elicitin gene family has been extensively characterized (Jiang et al., 2006b; Qutob et al., 2003). These studies have shown that *P. sojae* elicitins may be abundant transcripts that display stage-specific expression patterns.

15.7.5 Secreted Hydrolytic Enzymes, Uptake, and Export Mechanisms

As an invasive pathogen and an osmotrophic organism that assimilates nutrients from its immediate environment, *P. sojae* must rely on an arsenal of enzymes and transport proteins. These factors are required for perception, ingress, nutrition, and for handling toxins and wastes. This field is ripe for future study because there have been comparatively few investigations that deeply probe the mechanisms of nutrition and metabolic homeostasis of *Phytophthora* species.

Sequencing and proteomic studies of *P. sojae* have provided some insight into this area, at least for certain families of genes that could be annotated based on sequence similarities. For example, large gene families that encode glycosyl hydrolases, pectinases, lipases, and proteinases can be identified (Costanzo et al., 2006; Qutob et al., 2000; Torto-Alalibo et al., 2007; Tyler et al., 2006). Identification at the protein level of particular enzymes, transporters, and adhesion proteins has provided another layer of data that has been interpreted in the context of metabolic and nutritional strategies of the organism (Savidor et al., 2008).

The genome sequence also revealed several esterases, which include genes with high similarity to the cutin degrading enzymes, or cutinases. At least 14 genes predicted to encode cutinases are present in *P. sojae*, whereas only four such genes are in *P. ramorum* (Belbhri et al., 2008; Jiang et al., 2006a). Because cutin and suberin are closely related polymers, it is likely that cutinases may have activity toward suberin. Esterases that degrade suberin have not been biochemically characterized, so there is no way to annotate the genes, but one can predict that they would share features or overlap with cutinases. A cluster of 10 cutinase genes that occur in a contiguous region in *P. sojae* has apparently resulted from recent duplication events, because three pairs of genes in this cluster are nearly identical in nucleotide sequence. Transcripts for at least three different *P. sojae* cutinase genes are represented by EST matches and may be highly expressed during zoospore germination and infection. Compared with other gene families, the difference in size of the cutinase gene family between *P. sojae* and *P. ramorum* is unusual, and it seems to be the result of duplications that have happened since the two species diverged (Jiang et al., 2006a). Perhaps the expansion and diversification of this gene family in *P. sojae* is caused by selection pressure, which developed from variation in root suberin content and associated partial resistance in soybean cultivars.

In addition to hydrolytic enzymes, *P. sojae* must employ transporters to partition nutrients, toxins, and wastes to their proper location, inside or outside the cell. An adenosine triphosphate binding cassette (ABC)-type protein with homology to the pleiotropic drug resistance family of transporters has been described (Connolly et al., 2005). This zoospore-specific transport protein may have a role in toxin extrusion. Other zoospore transport and perception systems have also been studied. The sensitivity and responsiveness of zoospores to K^+ and Ca^{++} ions, polyamines, isoflavonoids, and other compounds demonstrates the importance of environmental signals for the proper development and taxis of these motile propagules (Chibucos and Morris, 2006; Connolly et al., 1999; Morris et al., 1998; Sugimoto et al., 2007; Tyler et al., 1996).

15.7.6 Avirulence Genes and RxLR Effector Proteins

For many years, the desire to isolate Avr determinants was a major driver in the hunt for pathogen factors that cause defense responses and HR in plant cells. The Avr factors have been long sought because these molecules control race–cultivar compatibility and effector-triggered immunity through their interaction with host *R* genes. Biochemical approaches based on purification of elicitor-active molecules from pathogen cultures were successful in identifying toxins and PAMPs, but they failed to yield any *Avr* genes in the case of oomycetes. The identification of the first Avr effectors from oomycetes resulted from genetic and positional cloning strategies, and *P. sojae* played a primary role in these advances as well.

In fact, the first oomycete Avr effector to be described was Avr1b-1 from *P. sojae* (Shan et al., 2004; Tyler, 2002). This finding was contentious and not widely accepted until additional Avr effectors from *Hyaloperonospora parasitica* and *Phytophthora infestans* were identified shortly afterward and found to share similarities with Avr1b-1 from *P. sojae* (Allen et al., 2004; Armstrong et al., 2005; Rehmany et al., 2005). A common denominator among these small, secreted oomycete Avr effector proteins is the presence of a host-targeting sequence downstream from the signal peptide, which is known as the RxLR motif (Whisson et al., 2007; Win et al., 2007). Currently, the study of oomycete RxLR effectors is drawing a lot of attention as researchers try to define these proteins and determine how they function and intersect with plant immune systems (Dou et al., 2008; Jiang et al., 2008). The discovery of RxLR effectors, together with the release of whole genome sequence information, has transformed investigative methods. Oomycete *Avr* genes are now being discovered at a rapid pace. A summary of the current state of knowledge of *P. sojae Avr* genes is shown in Table 15.1.

The *P. sojae Avr1b* gene was isolated by a map-based approach, relying on several hundred F_2 progeny from two separate crosses (Shan et al., 2004). This resulted in the localization of the gene to a 60-kb interval. Screening for transcriptional polymorphisms using probes derived from this region led to the identification of a predicted gene *Avr1b*-1 encoding a small secreted protein of 138 amino acids. The heterologous expression of this protein and infiltration into soybean leaves triggered an HR that was dependent on the presence of *Rps1b*. Characterization of various *P. sojae* races showed that virulent alleles of *Avr1b*-1 were either transcriptionally silent or had accumulated numerous mutations that altered the ORF and evaded *Rps1b* surveillance. Another locus that occurred in the vicinity of *Avr1b*-1 was proposed as a transcriptional regulator and was named *Avr1b*-2.

Two pieces of evidence indicated that transcriptional silencing of *Avr1b*-1 could be released by outcrossing to strains that are transcriptionally active. First, in F_2 progeny that were heterozygous for *Avr1b*-1, transcripts from both parents were detected despite that one of the parents and its matching homozygous F_2 progeny did not express the gene whatsoever. Thus, the transcriptional silencing of *Avr1b*-1 in strain P6497 was released when it was in a heterozygous state with the corresponding allele from P7064. Second, a *P. sojae* strain (P7076) that carries a transcriptionally active but virulent allele of the *Avr1b*-1 ORF was crossed with a strain (P6497) that carries a transcriptionally silent allele but with a predicted ORF identical to the original

TABLE 15.1 Predicted and identified *Avr* genes from *Phytophthora sojae* controlling race-cultivar compatibility on soybean.

Avr gene	Status	GenBank	Characteristics[a]	Polymorphisms and variant alleles in virulent strains, or inheritance data, or *Avh* number[b]	References
Avr1a	Identified	EF463064	SP + RxLR, 121 aa	Copy number variants, transcriptional polymorphisms	Qutob et al., 2009
Avr1b-1	Identified	AY426744	SP + RxLR, 138 aa	Transcriptional polymorphisms, aa changes	Shan et al., 2004
Avr1b-2	Mapped			Linked to, and predicted to control transcription of, *Avr1b*-1	Shan et al., 2004
Avr1c	Inheritance studied			Inherited as dominant or recessive gene, depending on cross	Tyler et al., 1995; May et al., 2002
Avr1d	Mapped			Inherited as single dominant gene in two separate crosses	Tyler et al., 1995; May et al., 2002
Avr1k	Identified	EU282487	SP + RxLR, 280 aa	*Avh331*	Tyler, unpublished; Dou et al., 2008
Avr2	Uncharacterized				
Avr3a	Identified	EF587759	SP + RxLR, 111 aa	Copy number variants, transcriptional polymorphisms, aa changes	Qutob et al., 2009
Avr3b	Mapped			Inherited as dominant or single dominant gene, three separate crosses	May et al., 2002
Avr3c	Identified	FJ705360	SP + RxLR, 220 aa	aa changes	Dong et al., 2009
Avr4/6	Identified	ABS50087	SP + RxLR, 123 aa	*Avh171*	Dou et al., 2008
Avr5	Mapped			Inherited as single dominant gene closely linked or allelic to *Avr3a*	May et al., 2002; Qutob et al., 2009
Avr7	Uncharacterized				
Avr8	Uncharacterized				

[a] Signal peptide, SP; host targeting motif, RxLR; total length of preprotein amino acids, aa.
[b] *Avirulence homolog*, *Avh* (Jiang et al., 2008).

Avr1b-1 protein from avirulent isolates. Each of the parental strains displayed a virulent phenotype on *Rps1b* plants, but the resulting F_1 hybrids were avirulent. These surprising results led the authors to propose complementary genes *Avr1b*-1 and *Avr1b*-2, which encode an effector and its transcriptional regulator, respectively (Shan et al., 2004). Mapping data suggests that the *Avr1b*-2 locus is genetically linked and in close proximity to *Avr1b*-1 itself.

Unfortunately, the complexity of this model inhibited its easy digestion and contributed to the initial uncertainty about *Avr1b*-1. However, as more *Avr* genes are discovered in *P. sojae*, it seems that silencing of effector gene transcription may be a widespread and common mechanism of evading *R* gene-mediated immunity. Conceptually, a switch-regulator that qualitatively controls transcription of effector genes would provide a powerful and useful element for the pathogen in its cat-and-mouse game with host plants.

Besides *Avr1b*, several other *Avr* genes from *P. sojae* are in various stages of discovery and elucidation (Table 15.1). The *Avr1a* locus, like *Avr1b*, was positionally cloned (Qutob et al., 2009). The mapping of *Avr1a* to a 30 kb interval required approximately 2,000 F_2 individuals from two separate crosses, which illustrates the relatively low recombination frequency of the region. The structure of the *Avr1a* locus and the copy number of the gene itself varies among *P. sojae* strains. The *Avr1a* transcript is expressed during infection but only in *P. sojae* strains that trigger *Rps1a* resistance; strains that are virulent on *Rps1a* do not express the gene. These transcript differences are caused by *Avr1a* gene deletions or gene silencing, depending on the *P. sojae* strain.

The *Avr3a* locus also occurs as a gene cluster with copy number variants depending on the *P. sojae* strain (Qutob et al., 2009). This gene was identified by transcriptional screens using microarrays and applying bulked segregant analysis. Allelic diversity within the coding sequence of *Avr3a*, and evidence of diversifying selection, is apparent among *P. sojae* strains. Evasion of *Rps3a*-mediated immunity is accomplished by gene deletions or by transcriptional silencing, which demonstrates commonalities with *Avr1b*-1 and *Avr1a*.

We are still in the early stages of characterizing *P. sojae Avr* genes, but it is worthwhile to review any emerging trends. Thus far, all *P. sojae Avr* genes identified encode RxLR effector proteins. Nonetheless, the proposed *Avr1b*-2 locus suggests that regulators of gene transcription may also emerge as bona fide *Avr* genes. The transcriptional polymorphisms observed for *Avr1a* and *Avr3a* in certain *P. sojae* strains display a similar pattern to *Avr1b*-1. This suggests that variation in effector transcription, rather than variation in the effector itself, may be a common mechanism of *R*-gene evasion. Falling in with this model is the *P. infestans Avr3b-Avr10-Avr11* locus, because it has been proposed to operate as a transcriptional regulator (Jiang et al., 2006c; Qutob et al., 2006a). Copy number polymorphisms that occur at *Avr* loci is another feature that is shared among *P. sojae Avr1a* and *Avr3a*, as well as the *P. infestans Avr3b-Avr10-Avr11* locus. Overall, it is clear that *Phytophthora* species adapt to the pressures of a parasitic life through a variety of genetic changes to meet the challenges imposed by plant immune systems. The diploid nature of oomycetes offers routes of genetic selection that are not available to fungi and bacteria, so differences in the mechanisms of *Avr* gene evolution among these disparate organisms should be expected.

15.8 CONCLUSIONS AND FUTURE PROSPECTS

P. sojae is among the best studied of oomycete species, and it is sometimes promoted as a model organism. The *P. sojae*–soybean interaction has even been studied on space shuttle missions, where it was discovered that growth in microgravity increases susceptibility to disease (Ryba-White et al., 2001). In this review, we have examined how research on *P. sojae* is entering a new phase, facilitated by advances in technology. But research on oomycetes is swiftly progressing on several species in parallel, and it is unlikely that a single species will become a "super model" and dominate the field like *Arabidopsis* dominates plant biology. Oomycetes are unusual and somewhat obscure organisms that command our attention as plant pathogens that threaten crops, landscapes, and natural ecosystems. Interest in *P. sojae* is driven by its destructive potential as a pathogen on one of the world's largest and most versatile crops. We are motivated to understand how *P. sojae* causes disease on soybean plants because this will enable us to control it.

Contemporary descriptions of plant defense and immunity generally delineate three major layers of protection that pathogens need to breach to colonize a host: preformed physical and chemical barriers, PAMP-triggered immune responses, and *R* gene surveillance and effector-triggered immune responses. In studies of soybean resistance to *P. sojae*, effector-triggered immune responses have probably received the most attention. Recent work that implicates preformed suberin in partial resistance and that demonstrates the importance of β-glucan signaling in PAMP-triggered immunity provides a broader view of the interaction and illustrates how initial lines of protection are important components to consider. It is also reassuring to fit our observations of this pathosystem into a larger theoretical framework of plant defense and immunity.

There is an emerging consensus that oomycetes such as *P. sojae* depend heavily on secreted protein effectors as enabling factors for pathogenicity and colonization of host tissues, more so than do fungal plant pathogens. The absence of polyketide synthase genes in oomycetes, coupled with a relatively small cytochrome P450 family, fits with the hypothesis that oomycetes are limited or lack the ability to biosynthesize nonprotein toxins. There remains much to learn about the basic metabolic capabilities of oomycetes, but transcriptome, proteome, and metabolome analyses are beginning to provide insight in this area. Currently, bioinformatic methods are limited by the relatively poor annotation of oomycete genes because these organisms are so distantly removed from the best-studied eukaryotes, which include the animals, plants, and fungi. Improved annotation will evolve over time, as the functional properties of oomycete gene products are defined and cataloged.

What are the priorities for the future? It is imperative to define the molecular and genetic determinants that govern compatibility between *P. sojae* and soybean. The complete repertoire of known *Avr* and *Rps* genes needs to be identified and characterized from *P. sojae* and soybean, respectively. After more than 50 years of research, progress in this area is now brisk. It is foreseeable that most *Avr* and *Rps* genes in this pathosystem will be delimited in short order. Likewise, research on partial resistance is now moving forward, and fresh models are emerging to account for this. Genome databases, microarrays, high-throughput sequencing methods, and other developments have altered the research landscape dramatically. These advances have provided such powerful tools for genetic analyses that it seems many longstanding puzzles are soluble. Now that we can contemplate possessing a full inventory of *Avr* and *Rps* genes, in addition to genes that control partial resistance, how will this impact control of *P. sojae* in the field? Clearly, a more rational approach to breeding, diagnostics, and cultivar deployment will develop from such solid genetic information. It will also lead to a better mechanistic understanding of disease and how it may be managed.

It is essential to define fully the existing and naturally occurring resistance mechanisms, but we also need to look beyond these. Innovative methods to create *P. sojae*-resistant soybean plants are required. Recent discoveries bring new questions to the forefront and offer challenges and opportunities for researchers. Can host plants be engineered to disable, neutralize, or more efficiently recognize oomycete RxLR effectors, or to produce products curtailing the growth of the pathogen? How are RxLR effectors delivered into host cells? What are the most important effector and metabolic entities in the pathogen that are crucial for its growth in host tissues? What host targets are most vulnerable to perturbation or control by oomycete effectors? The use of *Arabidopsis* as a model to study oomycete effectors is actively being pursued and will hasten progress in many of these areas. Perhaps it will be feasible to develop a technology that provides broad protection against oomycetes and that can be deployed in a range of crop plants based on our intimate molecular knowledge of disease mechanisms.

Studies of oomycetes have a long history that spans more than 150 years, but it seems that we are only now beginning to understand how these organisms have excelled as pathogens. In light of the recent pace of discovery, there is reason to be optimistic that effective new methods for control of *P. sojae* and other oomycetes will soon be developed.

ACKNOWLEDGMENTS

The authors thank their colleagues in the *Phytophthora* research community for sharing information for this review, the current laboratory comembers working on *Phytophthora sojae*, who include Suomeng Dong, Aldona Gaidauskas-Scott, Kuflom Kuflu, and Jennifer Tedman-Jones, for their ongoing assistance and enthusiasm, and Agriculture and Agri-Food Canada for providing support to enable the research and the writing of this review.

REFERENCES

Abe K, Fujikawa E, Takeuchi Y, Takada Y, Yamaoka N 2002. The possibility of factors other than phytoalexin accumulation preventing fungal growth in the incompatible interactions of soybean with *Phytophthora megasperma f.sp.* glycinea. Plant Pathol 51:237–241.

Allen RL, Bittner-Eddy PD, Grenville-Briggs LJ, Meitz JC, Rehmany AP, Rose LE, Beynon JL. 2004. Host-parasite coevolutionary conflict between *Arabidopsis* and downy mildew. Science 306:1957–1960.

Armstrong MR, Whisson SC, Pritchard L, Bos JIB, Venter E, Avrova AO, Rehmany AP, Bohme U, Brooks K, Cherevach I, et al. 2005. An ancestral oomycete locus contains late blight avirulence gene *Avr3a*, encoding a protein that is recognized in the host cytoplasm. Proc Nat Acad Sci USA 102:7766–7771.

Ayers AR, Ebel J, Finelli F, Berger N, Albersheim P 1976. Host-pathogen interactions: IX. Quantitative assays of elicitor activity and characterization of the elicitor present in the extracellular medium of cultures of *Phytophthora megasperma* var. sojae. Plant Physiol 57:751–759.

Becker J, Nagel S, Tenhaken R 2000. Cloning, expression and characterization of protein elicitors from the soyabean pathogenic fungus *Phytophthora sojae*. J Phytopathol 148:161–167.

Belbhri L, Calmin G, Mauch F, Andersson JO 2008. Evolution of the cutinase gene family: evidence for lateral gene transfer of a candidate *Phytophthora* virulence factor. Gene 408:1–8.

Bhat RG, Schmitthenner AF 1993. Genetic crosses between physiologic races of *Phytophthora sojae*. Exp Mycol 17:122–129.

Bhattacharyya MK, Narayanan NN, Gao H, Santra DK, Salimath SS, Kasuga T, Liu Y, Espinosa B, Ellison L, Marek L, et al. 2005. Identification of a large cluster of coiled coil-nucleotide binding site-leucine rich repeat-type genes from the *Rps1* region containing *Phytophthora* resistance genes in soybean. Theor Appl Genet 111:75–86.

Bishop JG, Ripoll DR, Bashir S, Damasceno CMB, Seeds JD, Rose JKC 2005. Selection on glycine beta-1,3-endoglucanase genes differentially inhibited by a *Phytophthora* glucanase inhibitor protein. Genetics 169:1009–1019.

Brunner F, Rosahl S, Lee J, Rudd JJ, Geiler C, Kauppinen S, Rasmussen G, Scheel D, Nurnberger T 2002. Pep-13, a plant defense-inducing pathogen-associated pattern from *Phytophthora* transglutaminases. EMBO J 21:6681–6688.

Burnham KD, Dorrance AE, Van Toai TT, St. Martin SK 2003. Quantitative trait loci for partial resistance to *Phytophthora sojae* in soybean. Crop Sci 43:1610–1617.

Cahill D, Rookes J, Michalczyk A, McDonald K, Drake A 2002. Microtubule dynamics in compatible and incompatible interactions of soybean hypocotyl cells with *Phytophthora sojae*. Plant Pathol 51:629–640.

Chamnanpunt J, Shan WX, Tyler BM 2001. High frequency mitotic gene conversion in genetic hybrids of the oomycete *Phytophthora sojae*. Proc Nat Acad Sci USA 98:14530–14535.

Chen X, Shen G, Wang Y, Zheng X, Wang Y 2007. Identification of *Phytophthora sojae* genes upregulated during the early stage of soybean infection. FEMS Microbiol Lett 269:280–288.

Chibucos MC, Morris PF 2006. Levels of polyamines and kinetic characterization of their uptake in the soybean pathogen *Phytophthora sojae*. Appl Environ Microbiol 72:3350–3356.

Connolly MS, Sakihama Y, Phuntumart V, Jiang YJ, Warren F, Mourant L, Morris PF 2005. Heterologous expression of a pleiotropic drug resistance transporter from *Phytophthora sojae* in yeast transporter mutants. Curr Genet 48:356–365.

Connolly MS, Williams N, Heckman CA, Morris PF 1999. Soybean isoflavones trigger a calcium influx in *Phytophthora sojae*. Fungal Genet Biol 28:6–11.

Cooke DEL, Drenth A, Duncan JM, Wagels G, Brasier CM 2000. A molecular phylogeny of *Phytophthora* and related oomycetes. Fungal Genet Biol 30:17–32.

Costanzo S, Ospina-Giraldo MD, Deahl KL, Baker CJ, Jones RW 2006. Gene duplication event in family 12 glycosyl hydrolase from *Phytophthora* spp. Fungal Genet and Biol 43:707–714.

Daxberger A, Nemak A, Mithofer A, Fliegmann J, Ligterink W, Hirt H, Ebel J 2007. Activation of members of a MAPK module in beta-glucan elicitor-mediated non-host resistance of soybean. Planta 225:1559–1571.

Demirbas A, Rector BG, Lohnes DG, Fioritto RJ, Graef GL, Cregan PB, Shoemaker RC, Specht JE 2001. Simple sequence repeat markers linked to the soybean *Rps* genes for *Phytophthora* resistance. Crop Sci 41:1220–1227.

Dhaubhadel S, McGarvey BD, Williams R, Gijzen M 2003. Isoflavonoid biosynthesis and accumulation in developing soybean seeds. Plant Mol Biol 53:733–743.

Dong S, Qutob D, Tedman-Jones J, Kuflu K, Wang Y, Tyler BM, Gijzen M 2009. The *Phytophthora sojae* avirulence locus *Avr3c* encodes a multi-copy RXLR effector with sequence polymorphisms among pathogen strains. PLoS ONE (in press).

Dorrance AE, McClure SA, Silva Ad 2003. Pathogenic diversity of *Phytophthora sojae* in Ohio soybean fields. Plant Dis 87:139–146.

Dou D, Kale SD, Wang X, Chen Y, Wang Q, Jiang RH, Arredondo FD, Anderson RG, Thakur PB, McDowell JM, et al. 2008. Conserved C-terminal motifs required for avirulence and suppression of cell death by *Phytophthora sojae* effector Avr1b. Plant Cell 20: 1118–1133.

Enkerli K, Hahn MG, Mims CW 1997a. Immunogold localization of callose and other plant cell wall components in soybean roots infected with the oomycete *Phytophthora sojae*. Can J Bot 75:1509–1517.

Enkerli K, Hahn MG, Mims CW 1997b. Ultrastructure of compatible and incompatible interactions of soybean roots infected with the plant pathogenic oomycete *Phytophthora sojae*. Can J Bot 75:1493–1508.

Erwin DC, Ribeiro OK 1996. *Phytophthora* diseases worldwide. The American Phytopathological Society. St. Paul, MN.

Fliegmann J, Mithofer A, Wanner G, Ebel J 2004. An ancient enzyme domain hidden in the putative beta-glucan elicitor receptor of soybean may play an active part in the perception of pathogen-associated molecular patterns during broad host resistance. J Biol Chem 279:1132–1140.

Fliegmann J, Montel E, Djulic A, Cottaz S, Driguez H, Ebel J 2005. Catalytic properties of the bifunctional soybean beta-glucan-binding protein, a member of family 81 glycoside hydrolases. FEBS Letters 579:6647–6652.

Fliegmann J, Schuler G, Boland W, Ebel J Mithofer A 2003. The role of octadecanoids and functional mimics in soybean defense responses. Biol Chem 384:437–446.

Förster H, Tyler BM, Coffey MD 1994. *Phytophthora sojae* races have arisen by clonal evolution and by rare outcrosses. Mol Plant-Microbe Interact 7:780–791.

Gao H, Bhattacharyya MK 2008. The soybean-*Phytophthora* resistance locus *Rps1-k* encompasses coiled coil-nucleotide binding-leucine rich repeat-like genes and repetitive sequences. BMC Plant Biol 8:29.

Gao H, Narayanan NN, Ellison L, Bhattacharyya MK 2005. Two classes of highly similar coiled coil-nucleotide binding-leucine rich repeat genes isolated from the *Rps1-k* locus encode *Phytophthora* resistance in soybean. Mol Plant-Microbe Interact 18:1035–1045.

Gijzen M, Förster H, Coffey MD, Tyler B 1996. Cosegregation of *Avr4* and *Avr6* in *Phytophthora sojae*. Can J Bot 74:800–802.

Gijzen M, Nurnberger T 2006. Nep1-like proteins from plant pathogens: Recruitment and diversification of the NPP1 domain across taxa. Phytochemistry 67:1800–1807.

Gordon SG, Berry SA, Martin SKS, Dorrance AE 2007. Genetic analysis of soybean plant introductions with resistance to *Phytophthora sojae*. Phytopathol 97:106–112.

Govers F, Gijzen M 2006. *Phytophthora* genomics: The plant destroyers' genome decoded. Mol Plant-Microbe Interact 19:1295–1301.

Graham TL, Graham MY, Subramanian S, Yu O 2007. RNAi silencing of genes for elicitation or biosynthesis of 5-deoxyisoflavonoids suppresses race-specific resistance and hypersensitive cell death in *Phytophthora sojae* infected tissues. Plant Physiol 144:728–740.

Graham MA, Marek LF, Shoemaker RC 2002. Organization, expression and evolution of a disease resistance gene cluster in soybean. Genetics 162:1961–1977.

Hahn MG, Bonhoff A, Grisebach H 1985. Quantitative localization of the phytoalexin glyceollin I in relation to fungal hyphae in soybean roots infected with *Phytophthora megasperma f. sp. glycinea*. Plant Physiol 77:591–601.

Han Y, Teng W, Yu K, Poysa V, Anderson T, Qiu L, Lightfoot DA, Li W 2008. Mapping QTL tolerance to *Phytophthora* root rot in soybean using microsatellite and RAPD/SCAR derived markers. Euphytica 162:231–239.

Hansen EM, Brasier CM, Shaw DS, Hamm PB 1986. The taxonomic structure of *Phytophthora megasperma*: evidence for emerging biological species groups. Trans Br Mycol Soc 87:557–573.

Hayes AJ, Jeong SC, Gore MA, Yu YG, Buss GR, Tolin SA, Maroof MA 2004. Recombination within a nucleotide-binding-site/leucine-rich-repeat gene cluster produces new variants conditioning resistance to soybean mosaic virus in soybeans. Genetics 166:493–503.

Hildebrand AA 1959. A root and stalk rot of soybeans caused by *Phytophthora megasperma* Drechsler var. *sojae* var. nov. Can J Bot 37:927–957.

Howlett BJ 1989. An electrophoretic karyotype for *Phytophthora megasperma*. Exp Mycol 13:199–202.

Jiang RH, Tyler BM, Govers F 2006a. Comparative analysis of *Phytophthora* genes encoding secreted proteins reveals conserved synteny and lineage-specific gene duplications and deletions. Mol Plant Microbe Interact 19:1311–1321.

Jiang RH, Tyler BM, Whisson SC, Hardham AR, Govers F 2006b. Ancient origin of elicitin gene clusters in *Phytophthora* genomes. Mol Biol Evol 23:338–351.

Jiang RHY, Tripathy S, Govers F, Tyler BM 2008. RXLR effector reservoir in two *Phytophthora* species is dominated by a single rapidly evolving superfamily with more than 700 members. Proc Nat Acad Sci USA 105:4874–4879.

Jiang RHY, Weide R, de Vondervoort P, Govers F 2006c. Amplification generates modular diversity at an avirulence locus in the pathogen *Phytophthora*. Gen Res 16:827–840.

Jones JD, Dangl JL 2006. The plant immune system. Nature 444:323–329.

Jones JP 1969. Reaction of *Lupinus* species to *Phytophthora megasperma* var. *sojae*. Plant Dis Rep 53:907–909.

Jones JP, Johnson HW 1969. Lupine, a new host for *Phytophthora megasperma* var. *sojae*. Phytopathol 59:504–507.

Kasuga T, Salimath SS, Shi J, Gijzen M, Buzzell RI, Bhattacharyya MK 1997. High resolution genetic and physical mapping of molecular markers linked to the *Phytophthora* resistance gene *Rps1-k* in soybean. Mol Plant-Microbe Interact 10:1035–1044.

Kaufmann MJ, Gerdemann JW 1958. Root and stem rot of soybeans caused by *Phytophthora sojae* n. sp. Phytopathol 48:201–208.

Lamour KH, Finley L, Hurtado-Gonzales O, Gobena D, Tierney M, Meijer HJ 2006. Targeted gene mutation in *Phytophthora* spp. Mol Plant-Microbe Interact 19:1359–1367.

Lander ES, Weinberg RA 2000. Genomics: journey to the center of biology. Science 287:1777–1782.

Layton AC, Kuhn DN 1988. The virulence of interracial heterokaryons of *Phytophthora megasperma* f.sp. *glycinea*. Phytopathol 78:961–966.

Ligterink W, Kroj T, zur Nieden U, Hirt H, Scheel D 1997. Receptor-mediated activation of a MAP kinase in pathogen defense of plants. Science 276:2054–2057.

Lindermayr C, Mollers B, Fliegmann J, Uhlmann A, Lottspeich F, Meimberg H, Ebel J 2002. Divergent members of a soybean (*Glycine max* L.) 4-Coumarate:coenzyme A ligase gene family: primary structures, catalytic properties, and differential expression. Eur J Biochem 269:1304–1315.

Liu Y, Dammann C, Bhattacharyya MK 2001. The matrix metalloproteinase gene GmMMP2 is activated in response to pathogenic infections in soybean. Plant Physiol 127:1788–1797.

Lozovaya VV, Lygin AV, Zernova OV, Ulanov AV, Li SX, Hartman GL, Widholm JM 2007. Modification of phenolic metabolism in soybean hairy roots through down regulation of chalcone synthase or isoflavone synthase. Planta 225:665–679.

MacGregor T, Bhattacharyya M, Tyler B, Bhat R, Schmitthenner AF, Gijzen M 2002. Genetic and physical mapping of *Avr1a* in *Phytophthora sojae*. Genetics 160:949–959.

Marshall JA, Dennis AL, Kumazawa T, Haynes AM, Nes WD 2001. Soybean sterol composition and utilization by *Phytophthora sojae*. Phytochemistry 58:423–428.

May KJ, Drenth A, Irwin JAG 2003. Interspecific hybrids between the homothallic *Phytophthora sojae* and *Phytophthora vignae*. Australas Plant Pathol 32:353–359.

May KJ, Whisson SC, Zwart RS, Searle IR, Irwin JAG, Maclean DJ, Carroll BJ, Drenth A 2002. Inheritance and mapping of 11 avirulence genes in *Phytophthora sojae*. Fungal Genet Biol 37:1–12.

Mideros S, Nita M, Dorrance AE 2007. Characterization of components of partial resistance, Rps2, and root resistance to *Phytophthora sojae* in soybean. Phytopathol 97:655–662.

Mithofer A, Ebel J, Felle HH 2005. Cation fluxes cause plasma membrane depolarization involved in beta-glucan elicitor-signaling in soybean roots. Mol Plant-Microbe Interact 18:983–990.

Mithofer A, Fliegmann J, Daxberger A, Ebel C, Neuhaus-Url G, Bhagwat AA,Keister DL, Ebel J 2001. Induction of H(2)O(2) synthesis by beta-glucan elicitors in soybean is independent of cytosolic calcium transients. FEBS letters 508:191–195.

Mithofer A, Muller B, Wanner G, Eichacker LA 2002. Identification of defence-related cell wall proteins in *Phytophthora sojae*-infected soybean roots by ESI-MS/MS. Mol Plant Pathol 3:163–166.

Mohr PG, Cahill DM 2001. Relative roles of glyceollin, lignin and the hypersensitive response and the influence of ABA in compatible and incompatible interactions of soybeans with *Phytophthora sojae*. Physiol Mol Plant Pathol 58:31–41.

Morris PF, Bone E, Tyler BM 1998. Chemotropic and contact responses of phytophthora sojae hyphae to soybean isoflavonoids and artificial substrates. Plant Physiol 117:1171–1178.

Moy P, Qutob D, Chapman BP, Atkinson I, Gijzen M 2004. Patterns of gene expression upon infection of soybean plants by *Phytophthora sojae*. Mol Plant-Microbe Interact 17:1051–1062.

Nurnberger T, Nennstiel D, Jabs T, Sacks WR, Hahlbrock K, Scheel D 1994. High affinity binding of a fungal oligopeptide elicitor to parsley plasma membranes triggers multiple defense responses. Cell 78:449–460.

Osman H, Mikes V, Milat ML, Ponchet M, Marion D, Prange T, Maume BF, Vauthrin S, Blein JP 2001. Fatty acids bind to the fungal elicitor cryptogein and compete with sterols. FEBS Lett 489:55–58.

Padliya ND, Garrett WM, Campbell KB, Tabb DL, Cooper B 2007. Tandem mass spectrometry for the detection of plant pathogenic fungi and the effects of database composition on protein inferences. Proteomics 7:3932–3942.

Ponchet M, Panabieres F, Milat ML, Mikes V, Montillet JL, Suty L, Triantaphylides C, Tirilly Y, Blein JP 1999. Are elicitins cryptograms in plant-Oomycete communications? Cell Mol Life Sci 56:1020–1047.

Qutob D, Hraber PT, Sobral BW, Gijzen M 2000. Comparative analysis of expressed sequences in *Phytophthora sojae*. Plant Physiol 123:243–254.

Qutob D, Huitema E, Gijzen M, Kamoun S 2003. Variation in structure and activity among elicitins from *Phytophthora sojae*. Mol Plant Pathol 4:119–124.

Qutob D, Kamoun S, Gijzen M 2002. Expression of a *Phytophthora sojae* necrosis-inducing protein occurs during transition from biotrophy to necrotrophy. Plant J 32:361–373.

Qutob D, Kemmerling B, Brunner F, Kufner I, Engelhardt S, Gust AA, Luberacki B, Seitz HU, Stahl D, Rauhut T, et al. 2006b. Phytotoxicity and innate immune responses induced by Nep1-like proteins. Plant Cell 18:3721–3744.

Qutob D, Tedman-Jones J, Gijzen M 2006a. Effector-triggered immunity by the plant pathogen *Phytophthora*. Trends Microbiol 14:470–473.

Qutob D, Tedman-Jones J, Dong S, Kuflu K, Pham H, Wang Y, Dou D, Kale SD, Arredondo FD, Tyler BM, Gijzen M 2009. Copy number variation and transcriptional polymorphisms of *Phytophthora sojae* RXLR effector genes *Avr1a* and *Avr3a*. PLoS ONE 4:e5066.

Ranathunge K, Thomas RH, Fang X, Peterson CA, Gijzen M, Bernards MA 2008. Soybean root suberin and partial resistance to root rot caused by *Phytophthora sojae*. Phytopathology 98:1179–1189.

Rehmany AP, Gordon A, Rose LE, Allen RL, Armstrong MR, Whisson SC, Kamoun S, Tyler BM, Birch PRJ, Beynon JL 2005. Differential recognition of highly divergent downy mildew avirulence gene alleles by RPP1 resistance genes from two *Arabidopsis* lines. Plant Cell 17:1839–1850.

Ricci P, Bonnet P, Huet JC, Sallantin M, Beauvais-Cante F, Bruneteau M, Billard V, Michel G, Pernollet JC 1989. Structure and activity of proteins from pathogenic fungi *Phytophthora* eliciting necrosis and acquired resistance in tobacco. FEBS J 183:555–563.

Rose JKC, Ham KS, Darvill AG, Albersheim P 2002. Molecular cloning and characterization of glucanase inhibitor proteins: coevolution of a counterdefense mechanism by plant pathogens. Plant Cell 14:1329–1345.

Rutherford FS, Ward EWB, Buzzell RI 1985. Variation in virulence in successive single-zoospore propagations of *Phytophthora megasperma* f.sp. *glycinea*. Phytopathol 75:371–374.

Ryba-White M, Nedukha O, Hilaire E, Guikema JA, Kordyum E, Leach JE 2001. Growth in microgravity increases susceptibility of soybean to a fungal pathogen. Plant Cell Physiol 42:657–664.

Sandhu D, Gao HY, Cianzio S, Bhattacharyya MK 2004. Deletion of a disease resistance nucleotide-binding-site leucine-rich-repeat-like sequence is associated with the loss of the *Phytophthora* resistance gene *Rps4* in soybean. Genetics 168:2157–2167.

Sandhu D, Schallock KG, Rivera-Velez N, Lundeen P, Cianzio S, Bhattacharyya MK 2005. Soybean *Phytophthora* resistance gene *Rps8* maps closely to the *Rps3* region. J Hered 96:536–541.

Savidor A, Donahoo RS, Hurtado-Gonzales O, Land ML, Shah MB, Lamour KH, McDonald WH 2008. Cross-species global proteomics reveals conserved and unique processes in *Phytophthora sojae* and *Phytophthora ramorum*. Mol Cell Proteom 7:1501–1516.

Savidor A, Donahoo RS, Hurtado-Gonzales O, VerBerkmoes NC, Shah MB, Lamour KH, McDonald WH 2006. Expressed peptide tags: an additional layer of data for genome annotation. J Proteome Res 5:3048–3058.

Schopfer CR, Ebel J 1998. Identification of elicitor-induced cytochrome P450 s of soybean (*Glycine max* L.) using differential display of mRNA. Mol Gen Genet 258:315–322.

Schopfer CR, Kochs G, Lottspeich F, Ebel J 1998. Molecular characterization and functional expression of dihydroxypterocarpan 6a-hydroxylase, an enzyme specific for pterocarpanoid phytoalexin biosynthesis in soybean (*Glycine max* L.). FEBS Lett 432:182–186.

Shan WX, Cao M, Dan LU, Tyler BM 2004. The *Avr1b* locus of *Phytophthora sojae* encodes an elicitor and a regulator required for avirulence on soybean plants carrying resistance gene *Rps1b*. Mol Plant-Microbe Interact 17:394–403.

Sharp JK, Valent B, Albersheim P 1984. Purification and partial characterization of a beta glucan fragment that elicits phytoalexin accumulation in soybean. J Biol Chem 259:11312–11320.

Shoemaker R, Keim P, Vodkin L, Retzel E, Clifton SW, Waterston R, Smoller D, Coryell V, Khanna A, Erpelding J, et al. 2002. A compilation of soybean ESTs: generation and analysis. Genome 45:329–338.

Steele CL, Gijzen M, Qutob D, Dixon RA 1999. Molecular characterization of the enzyme catalyzing the aryl migration reaction of isoflavonoid biosynthesis in soybean. Arch Biochem Biophys 367:146–150.

Subramanian S, Graham MY, Yu O, Graham TL 2005. RNA interference of soybean isoflavone synthase genes leads to silencing in tissues distal to the transformation site and to enhanced susceptibility to *Phytophthora sojae*. Plant Physiol 137:1345–1353.

Sugimoto T, Aino M, Sugimoto M, Watanabe K 2005. Reduction of *Phytophthora* stem rot disease on soybeans by the application of $CaCl_2$ and $Ca(NO_3)(2)$. J Phytopathol 153:536–543.

Sugimoto T, Watanabe K, Yoshida S, Aino M, Matsuyama M, Maekawa K, Irie K 2007. The effects of inorganic elements on the reduction of *Phytophthora* stem rot disease of soybean, the growth rate and zoospore release of *Phytophthora sojae*. J Phytopathol 155:97–107.

Thomas R, Fang X, Ranathunge K, Anderson TR, Peterson CA, Bernards MA 2007. Soybean root suberin: anatomical distribution, chemical composition, and relationship to partial resistance to *Phytophthora sojae*. Plant Physiol 144:299–311.

Tooley PW, Carras MM 1992. Separation of chromosomes of *Phytophthora* species using CHEF gel electrophoresis. Exp Mycol 16:188–196.

Torto-Alalibo TA, Tripathy S, Smith BM, Arredondo FD, Zhou L, Li H, Chibucos MC, Qutob D, Gijzen M, Mao C, et al. 2007. Expressed sequence tags from *Phytophthora sojae* reveal genes specific to development and infection. Mol Plant Microbe Interact 20:781–793.

Tyler BM 2002. Molecular basis of recognition between *Phytophthora* pathogens and their hosts. Ann Rev Phytopathol 40:137–167.

Tyler BM 2007. *Phytophthora sojae*: root rot pathogen of soybean and model oomycete. Mol Plant Pathol 8:1–8.

Tyler BM, Förster H, Coffey MD 1995. Inheritance of avirulence factors and restriction fragment length polymorphism markers in outcrosses of the oomycete *Phytophthora sojae*. Mol Plant Microbe Interact 8:515–523.

Tyler BM, Tripathy S, Zhang XM, Dehal P, Jiang RHY, Aerts A, Arredondo FD, Baxter L, Bensasson D, Beynon JL, et al. 2006. *Phytophthora* genome sequences uncover evolutionary origins and mechanisms of pathogenesis. Science 313:1261–1266.

Tyler BM, Wu M, Wang J, Cheung W, Morris PF 1996. Chemotactic preferences and strain variation in the response of *Phytophthora sojae* zoospores to host isoflavones. Appl Environ Microbiol 62:2811–2817.

Valer K, Fliegmann J, Frohlich A, Tyler BM, Ebel J 2006. Spatial and temporal expression patterns of *Avr1b-1* and defense-related genes in soybean plants upon infection with *Phytophthora sojae*. FEMS Micro Lett 265:60–68.

Wang Z, Wang Y, Chen X, Shen G, Zhang Z, Zheng X 2006. Differential screening reveals genes differentially expressed in low- and high-virulence near-isogenic *Phytophthora sojae* lines. Fungal Genet Biol 43:826–839.

Ward EWB 1990. The interaction of soya beans with *Phytophthora megasperma* f.sp. *glycinea*: pathogenicity. In Hornby D, editor. Biological control of soil-borne plant pathogens. CAB International. Wallingford, UK. pp. 311–327.

Ward EWB, Cahill DM, Bhattacharyya MK 1989. Early cytological differences between compatible and incompatible interactions of soybeans with *Phytophthora megasperma-f-sp-Glycinea*. Physiol Mol Plant Pathol 34:267–283.

Weng C, Yu K, Anderson TR, Poysa V 2001. Mapping genes conferring resistance to *Phytophthora* root rot of soybean, *Rps1a* and *Rps7*. J Hered 92:442–446.

Weng C, Yu K, Andersen TR, Poysa V 2007. A quantitative trait locus influencing tolerance to *Phytophthora* root rot in the soybean cultivar 'Conrad'. Euphytica 158:81–86.

Whisson SC, Basnayake S, Maclean DJ, Irwin JAG, Drenth A 2004. *Phytophthora sojae* avirulence genes *Avr4* and *Avr6* are located in a 24 kb, recombination-rich region of genomic DNA. Fungal Genet Biol 41:62–74.

Whisson SC, Boevink PC, Moleleki L, Avrova AO, Morales JG, Gilroy EM, Armstrong MR, Grouffaud S, van West P, Chapman S, et al. 2007. A translocation signal for delivery of oomycete effector proteins into host plant cells. Nature 450:115–118.

Whisson SC, Drenth A, Maclean DJ, Irwin JA 1994. Evidence for outcrossing in *Phytophthora sojae* and linkage of a DNA marker to two avirulence genes. Curr Genet 27:77–82.

Whisson SC, Drenth A, Maclean DJ, Irwin JA 1995. *Phytophthora sojae* avirulence genes, RAPD, and RFLP markers used to construct a detailed genetic linkage map. Mol Plant Microbe Interact 8:988–995.

Win J, Morgan W, Bos J, Krasileva KV, Cano LM, Chaparro-Garcia A, Ammar R, Staskawicz BJ, Kamoun S 2007. Adaptive evolution has targeted the C-terminal domain of the RXLR effectors of plant pathogenic oomycetes. Plant Cell 19:2349–2369.

Wrather JA, Koenning SR 2006. Estimates of disease effects on soybean yields in the United States 2003 to 2005. J Nematol 38:173–180.

Yi SY, Hwang BK 1997. Purification and antifungal activity of a basic 34 kDa beta-1,3-glucanase from soybean hypocotyls inoculated with *Phytophthora sojae f. sp. glycines*. Mol Cells 7:408–413.

Zhang X, Scheuring C, Tripathy S, Xu Z, Wu C, Ko A, Tian SK, Arredondo F, Lee MK, Santos FA, et al. 2006. An integrated BAC and genome sequence physical map of *Phytophthora sojae*. Mol Plant Microbe Interact 19:1302–1310.

16

PHYTOPHTHORA BRASSICAE AS A PATHOGEN OF ARABIDOPSIS

FELIX MAUCH, SAMUEL TORCHE, KLAUS SCHLÄPPI, LORELISE BRANCIARD, KHAOULA BELHAJ, VINCENT PARISY, AND AZEDDINE SI-AMMOUR

Department of Biology, University of Fribourg, Fribourg, Switzerland

16.1 INTRODUCTION

The analysis of Arabidopsis–pathogen interactions has in recent years significantly contributed to a better understanding of the molecular basis of plant immunity. Most breakthroughs in this field have been achieved with model pathosystems of Arabidopsis with either the bacterial pathogen *Pseudomonas syringae* or the oomycete pathogen *Hyaloperonospora arabidopsis*. The molecular analysis of disease resistance to Phytophthora is experimentally challenging because interactions of, for example, potato with *Phytophthora infestans* and soybean with *Phytophthora sojae*, are not ideal for genetic and molecular analysis. An Arabidopsis–Phytophthora pathosystem has the potential to ameliorate this situation. However, no Phytophthora species was known to infect Arabidopsis until some isolates of *Phytophthora porri* were shown to be pathogenic on Arabidopsis (Roetschi et al., 2001). *P. porri* is mainly known as a pathogen of plants belonging to the family of the Amarillidaceae (Foister, 1931). Interestingly, some isolates were reported to infect carrots (Ho, 1983; Semb, 1971; Stelfox and Henry, 1978) and cabbage (Geeson, 1976; Semb, 1971). Because isolates capable of infecting members of the Brassicaceae were not infectious on members of the Amarillidaceae (and *vice versa*) it was proposed that the former should be classified separately (De Cock et al., 1992). This proposal was later supported by additional molecular data, and the Brassicacae-specific *P. porri* isolates were grouped into the new species,

Oomycete Genetics and Genomics: Diversity, Interactions, and Research Tools
Edited by Kurt Lamour and Sophien Kamoun
Copyright © 2009 John Wiley & Sons, Inc.

Phytophthora brassicae (Man in't Veld et al., 2002). Besides *P. brassicae, Phytophthora cinnamomi* (Robinson and Cahill, 2003) and *Phytophthora palmivora* (Daniel and Guest, 2006) have been shown to infect Arabidopsis. In addition, defense responses of Arabidopsis to the non-host pathogen *P. infestans* have been described (Huitema et al., 2003; Kamoun et al., 1999).

16.2 ARABIDOPSIS AS A HOST OF *P. BRASSICAE*

The early events of infection are similar in both compatible and incompatible interactions. The zoospores rapidly encyst, germinate, and form an appressorium within 2–4 hpi (h post inoculation). Penetration into the leaf occurs preferentially between anticlinal walls of epidermal cells (Fig. 16.1a). Penetration directly into epidermal cells or through stomatal openings is rare. Differences between compatible and incompatible interactions become apparent immediately after penetration. In an incompatible interaction, the most conspicuous reaction of the host is the hypersensitive cell death at the site of attempted infection, which is known as hypersensitive response (HR). Around 6 hpi, the results of the HR process are readily visible when the plant cells are stained with lactophenol trypan blue. The HR normally occurs in mesophyll cells directly underlying the penetration site and encompasses a few host cells (Fig. 16.1b). Epidermal HR is normally restricted to sites of directly penetrated epidermal cells. Another cytological defense response is the formation of cell wall appositions known as papillae at the site of attempted cellular penetration (Fig. 16.1a). However, papillae formation is less frequently observed than HR. In incompatible interactions, the growth of *P. brassicae* is stopped within 24 hpi, whereas in compatible interactions, HR and papillae formation rarely occur and the hyphae can spread into the leaves (Fig. 16.1c). *P. brassicae* is a hemibiotrophic pathogen. The hyphae initially grow exclusively in the intercellular space and form, although rarely, finger-like haustoria into neighboring host cells. This biotrophic phase lasts until about 3 dpi (days post inoculation) at which time point host cells start to collapse and water soaked lesions become macroscopically visible. Finally, around 4–5 dpi, *P. brassicae* starts to produce zoosporangia and oospores (Fig. 16.1d and e), and the infected leaves disintegrate completely (Fig. 16.1f).

The inheritance of disease resistance is not well analyzed. The F2-progeny of a cross between the resistant accession Col-0 and the susceptible accession Ler showed intermediate disease phenotypes that differed from parental phenotypes, which indicates the involvement of multiple genes. The analysis of disease resistance to *P. brassicae* has been limited mainly to the accessions Col-0, Ws, and Ler. It will be necessary in the future to include more accessions and to analyze the progeny of more crosses between resistant and susceptible accessions. The model pathosystem would greatly benefit from the identification of combinations that show Mendelian segregation of parental disease phenotypes in the F2-progeny.

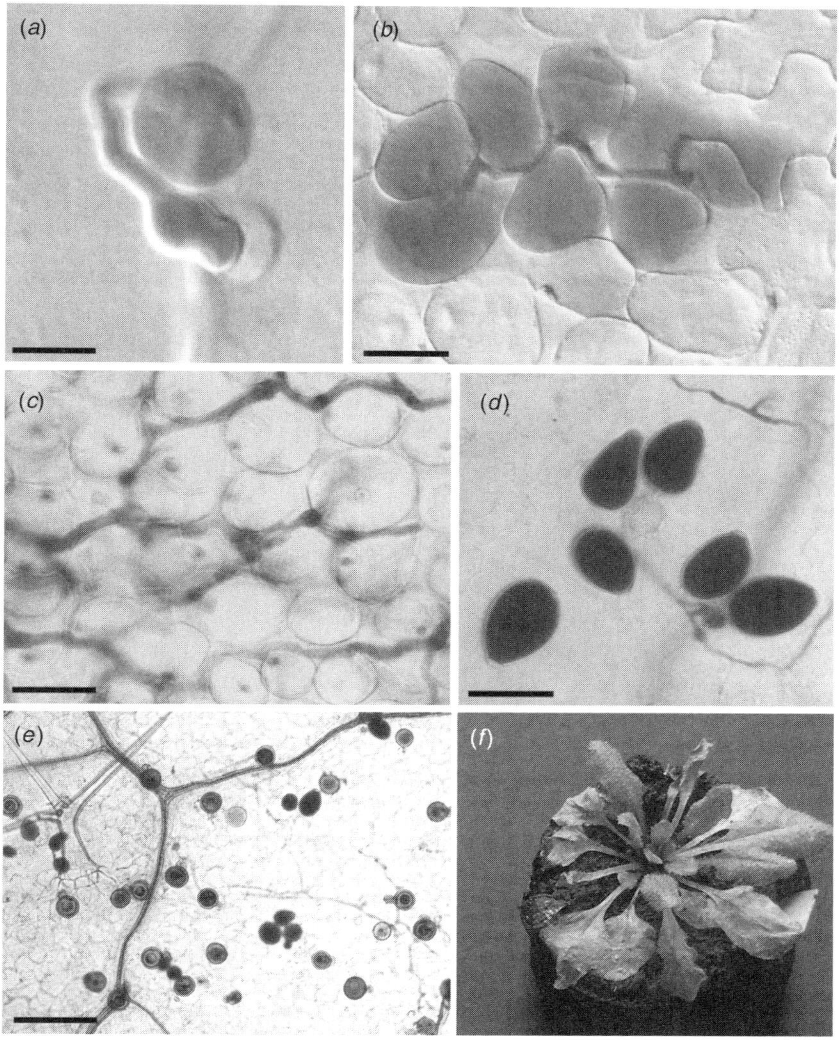

FIG. 16.1 (See color insert) Compatible and incompatible interactions of Arabidopsis with *P. brassicae*. Arabidopsis plants were spray-inoculated with a zoospore suspension of *P. brassicae* isolate D. (a, b) Incompatible interaction. (c–f) Compatible interaction. (a–e) Micrographs of lactophenol-trypan blue-stained leaves. (a) Germinated cyst which has formed an appresorium above the junction of two epidermal cells to penetrate in between these cells into the leave of the resistant accession Ws. The cell on the right has initiated the formation of a papillae. Bar = 5 μm. (b) Hypersensitive reaction of Arabidopsis accession Ws (12 hpi). The cells in contact with the hyphae have undergone an HR. Bar = 45 μm. (c) Hyphae spreading in the intercellular space of the susceptible mutant *pad2-1* during the biotrophic phase of the interaction. Bar = 55 μm. (d) Zoosporangia on the surface of a leaf of a susceptible Arabidopsis plant 4 dpi. Bar = 60 μm. (e) Infected leaf of *pad 2-1* with oospores and a few zoosporangia 7 dpi. Bar = 250 μm. (f) The leaves of the mutant *pad2-1* are disintegrated 7 dpi.

16.3 RESISTANCE OF ARABIDOPSIS ACCESSIONS TO DIFFERENT ISOLATES OF *P. BRASSICAE*

Screening of 40 Arabidopsis accessions with the plug-based inoculation method (see below) revealed that the majority are resistant to most of the tested isolates of *P. brassicae*. This contrasts with the pathosystem based on *P. cinnamomi* where most accessions were reported to be susceptible and only moderate resistance phenotypes were found (Robinson and Cahill, 2003). Table 16.1 presents a partial summary of tested combinations of Arabidopsis accessions with isolates of *P. brassicae*. It includes, besides Col-0 and Ws, the resistance response to six *P. brassicae* isolates of the five most resistant and the five most susceptible Arabidopsis accessions. The most susceptible accessions are Ler and Lu-1, which both were susceptible to varying degrees to all six isolates included in Table 16.1. The most virulent isolate D caused disease symptoms in 50% of the tested accessions. Extensive accession screening with the more natural zoospore inoculation method has not yet been performed.

16.4 ANALYSIS OF HOST GENE EXPRESSION

Arabidopsis responds to *P. brassicae* with dramatic changes in gene expression patterns. Expression analysis based on the 8 K Affymetrix Chip identified more

TABLE 16.1 Disease resistance of selected Arabidopsis accessions to 6 isolates of *P. brassicae*. Six-week-old Arabidopsis plants were plug-inoculated and resistance was scored at 7 dpi. A score of 4 corresponds to full resistance and a score of 0 to full susceptibility as described in the text.

Arabidopsis Accession	Isolate HH[a]	Isolate II[a]	Isolate 35[b]	Isolate Belot[b]	Isolate A[c]	Isolate D[c]
Col-0	4	4	4	4	3	4
Ws	4	3	4	3	3	3
Bl-1	4	4	4	4	4	4
Per	4	4	4	4	4	4
Kas-1	4	4	nd	3	4	4
St-0	4	4	nd	4	4	4
Est-0	4	4	nd	4	4	4
Ler	2	2	1	1	0	0
Lu-1	2	1	nd	1	2	0
Bla-1	4	1	1	1	1	0
Wa-1	3	3	3	3	2	1
Wei	4	4	2	3	2	1

[a] Isolates provided by Francine Govers, University of Wageningen, The Netherlands.

[b] Isolates provided by Emmanuel Pajot, Bretagne Biotechnologie Végétale (BBV), Saint Pol-de-Léon, France.

[c] Isolate A: CBS212.82; isolate D: CBS179.87; nd: not determined.

than 300 genes that are at least 2-fold upregulated and more than 200 genes that were at least 2-fold downregulated in an incompatible interaction with Col-0. Qualitatively similar, although delayed, changes were observed in the compatible interaction with the susceptible accession Ler. Figure 16.2 summarizes expression data from the incompatible Col-*P. brassicae* interaction for a selected group of genes involved in stress hormone production and signaling or the production of defense-related secondary metabolites. Arabidopsis reacts to *P. brassicae* with the increased production of the stress hormones salicylic acid, jasmonic acid, and ethylene. The enhanced stress hormone production is reflected by the enhanced abundance of transcripts of marker genes of these three stress hormones. Genes involved in either hormone production or marker genes of stress hormone action are strongly upregulated in response to *P. brassicae*. Coordinated changes of expression were observed in genes coding for a set of enzymes that catalyze the conversion of chorismate to tryptophan thus suggesting enhanced tryptophan production in challenged leaves. A strong induction was also observed for *CYP79B2* and to a lesser extent *CYP79B3*, which encode enzymes that catalyze the conversion of tryptophan to indole-3-acetaldoxime, the common precursor of the phytoalexin camalexin, indole-glucosinolates, and auxin. The expression profiling did not suggest an increased production of auxin in response to *P. brassicae* (data not shown). In contrast, camalexin was shown to accumulate in Arabidopsis in response to *P. brassicae* (Roetschi et al., 2001). Both camalexin and glucosinolates are sulfur-containing compounds. Interestingly, genes involved in sulfur assimilation were found to be activated by *P. brassicae*. Hence, the expression profiling predicts a role of camalexin and possibly indole glucosinolates in defense against *P. brassicae*.

16.5 ANALYSIS OF DISEASE RESISTANCE OF ARABIDOPSIS DEFENSE MUTANTS

An Arabidopsis-based pathosystem offers the opportunity to identify genetically the components of plant disease resistance by screening mutagenized populations of Arabidopsis for either gain or loss of resistance mutants and testing well characterized mutants with compromised resistance to other pathogens for resistance to *P. brassicae*. The second approach was used to test the importance of stress hormones for resistance by analyzing Arabidopsis mutants affected in stress hormone production and signaling such as the salicylic acid-deficient mutant *sid2* (Nawrath and Métraux, 1999), the salicylic acid pathway signaling mutant *npr1* (Cao et al., 1994), the jasmonate-resistant mutant *jar1* (Staswick et al., 1992), and the ethylene-insensitive mutant *ein2* (Guzmann and Ecker, 1990). Surprisingly, all these mutants (all in the genetic background of the resistant accession Col) remained resistant to *P. brassicae* (Roetschi et al., 2001). Although stress hormone signaling is enhanced in response to *P. brassicae* it does not seem to be of similar importance for resistance to *P. brassicae* as in many other plant diseases (Glazebrook, 2001). Similarly, the phytoalexin biosynthesis

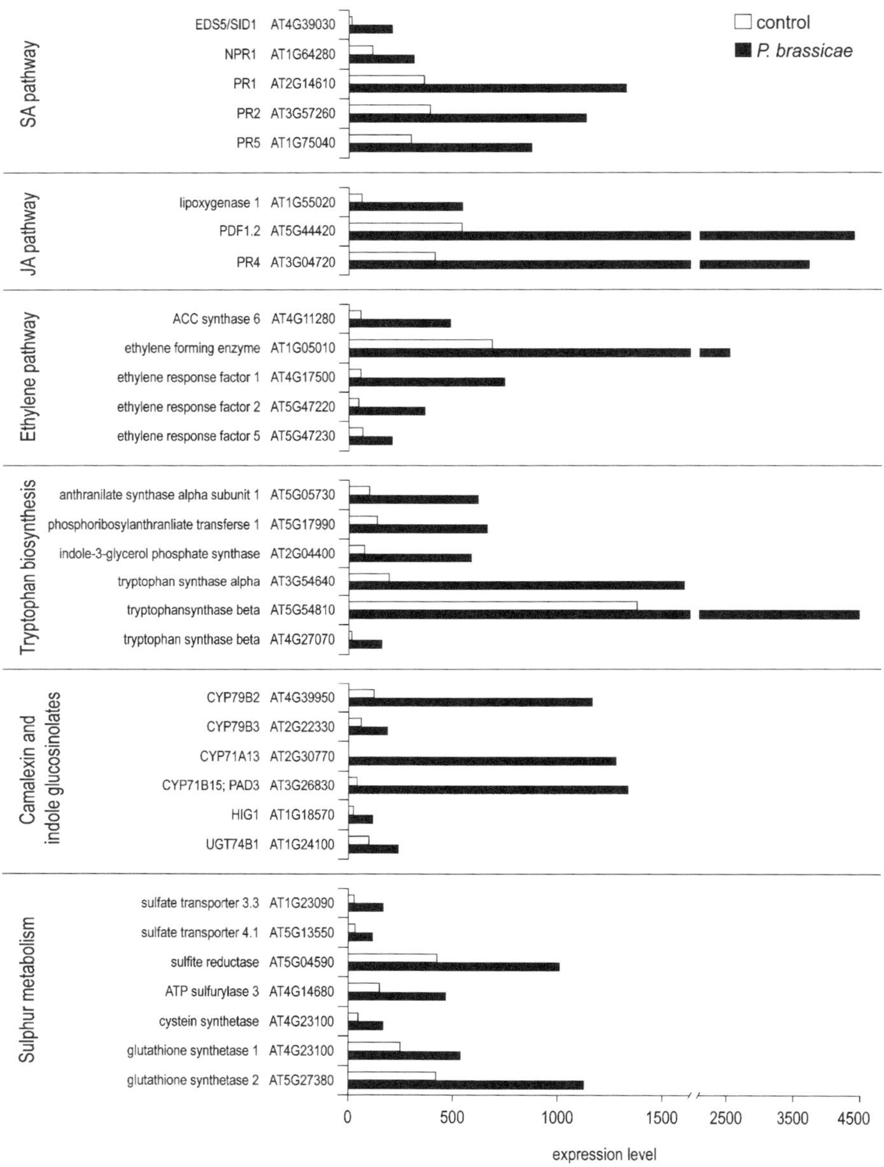

FIG. 16.2 Expression analysis of an incompatible interaction of Arabidopsis with *P. brassicae*. Col-0 plants were inoculated with empty plugs (control) or plugs that contain mycelia of *P. brassicae* isolate HH (*P. brassicae*). Global expression analysis was performed 24 hpi based on the 8 K Affymetrix Chip. RNA was extracted from three independent experiments and pooled prior to analysis. Quantitative polymerase chain reaction (PCR) and RNA blot analysis yielded qualitatively similar results. SA: salicylic acid; JA: jasmonic acid.

mutant *pad3* (Glazebrook and Ausubel, 1994) showed only slightly enhanced disease symptoms but remained resistant to *P. brassicae*, which indicates that loss of camalexin accumulation alone does not much compromise resistance. It is important to note that these results do not exclude a function of stress hormones or camalexin in disease resistance to *P. brassicae*. They indicate that the loss of only one of these defense components is not sufficient to let *P. brassicae* overcome the first layer of defense. Stress hormone signaling and camalexin might still be important by their combined effect or contribute to later resistance responses that become important in case the first layer of defense is breached.

Many other Arabidopsis defense mutants were tested, but only four showed enhanced susceptibility. The most susceptible mutant is *pad2-1*, which was originally identified in a screen of phytoalexin deficiency (Glazebrook et al., 1997). The *pad2-1* mutant is more susceptible than the susceptible natural accessions, which indicates that the latter still contain some residual resistance to *P. brassicae*. *PAD2* is a single-copy gene-encoding glutathione synthethase 1 that catalyzes the first step in the synthesis of the tripeptide glutathione (Parisy et al., 2007). The reduction to 20% of the wild-type glutathione level in *pad2-1* completely breaks disease resistance to *P. brassicae*. Glutathione plays multiple roles in cellular redox control and detoxification processes. In addition to the reduced accumulation of camalexin, the *pad2-1* mutant was shown to have a defect in pathogen-induced salicylic acid signaling and accumulation of the pathogenesis-related protein PR1 (Roetschi et al., 2001). However, it is not yet known whether these defects or other not yet identified deficiencies are ultimately responsible for the susceptibility of *pad2-1*. It will be interesting to observe whether glutathione plays a similar role in the interaction of crop plants with Phytophthora and whether enhanced glutathione levels correlate with enhanced disease resistance.

Intermediate disease-resistance phenotypes were observed in the Arabidopsis mutants *pad4* and *eds1*. PAD4 and EDS1 control, individually or in combination, various functions in disease resistance signaling (Wiermer et al., 2005). The effect was more pronounced with *eds1-1* and *pad4-5* in the accession Wassilewskija (Ws) than with *pad4-1* in the accession Col. In addition, the *eds1-1* mutant was more susceptible than *pad4-5*, which indicates an important function in disease resistance of EDS1 beyond its combined activity within the EDS1/PAD4 complex (Wiermer et al., 2005). Finally, the ABA-synthesis mutant *npq2-1* (Niyogi et al., 1998; Col background) showed an intermediate resistance phenotype suggesting that ABA-signaling might to some degree contribute to disease resistance against *P. brassicae*.

16.6 TOOLS AND RESOURCES

16.6.1 *In vitro* Culture and Storage of *P. Brassicae*

P. brassicae is routinely cultivated on 20% (v/v) V8 juice (Campbell Soups) agar (Erwin and Ribeiro, 1996) in the dark at 16–18°C. For short-term storage up to several months, the isolates can be cultivated at 4°C on potato carrot agar (Johnston and Booth, 1968). For long-term storage, agar plugs with mycelium are immersed in 10% glycerol followed by storage in liquid nitrogen (Smith, 1982). For liquid culture, *P. brassicae* is grown in either 10% (v/v) clarified V8 medium or the synthetic P1 medium of Hohl (Erwin and Ribeiro, 1996).

16.6.2 Zoospore Production

Production of zoospores from *P. brassicae* is more complicated than from other Phytophthora species such as *P. infestans* and includes several steps that need to be optimized. Twenty-five agar plugs (∅ 0.5 cm) from the edge of an expanding culture of *P. brassicae* grown on 20% (v/v) V8-agar plates are placed into a 250-mL Erlenmayer flask that contains 10 mL of clarified 10% (v/v) V8 medium. After 3–4 days of incubation at 16°C in the dark, the V8-medium is replaced with Schmitthenner mineral solution (Erwin and Ribeiro, 1996) and incubated for additional 4 days to induce the formation of zoosporangia. At the day of zoospore harvest, the Schmitthenner mineral solution is replaced with 10 mL of 4°C cold sterile water at pH 7. After 2–4 h at 4°C, the zoospores are released into the medium and can be harvested. Depending on the *P. brassicae* isolate, this protocol routinely yields concentrations of $1–3 \times 10^5$ zoospores per milliliter. *P. brassicae* isolates HH and D proved to be the best producers of zoospores. However, even with these isolates it is until now not possible to produce large amounts of zoospores during winter time.

16.6.3 Methods of Plant Inoculation

Two methods of plant inoculation are routinely used. Both the plug-inoculation and the zoospore-inoculation method give qualitatively similar results. However, the plug-inoculation method tends to cause higher susceptibility than zoospore inoculation when tested in the same accession-isolate combination. For plug inoculation, plugs (∅ 0.5 cm) of young mycelium growing on V8 agar are cut out using a cork borer and placed upside down on wet leaves of 5- to 7-week-old plants. For zoospore inoculations, 25-μL droplets of a zoospore suspension ($> 10^5$ spores per mL) are placed on the leaves of 3–4-week-old plants. Alternatively, plants can be inoculated by spraying at low pressure a zoospore suspension until run off. Inoculated plants are incubated at 18°C with a relative humidity of 100%. The disease is scored 3 to 7 days after inoculation. Inoculations are routinely started at the beginning of the dark period.

16.6.4 Disease Rating

Various methods can be used to quantify plant susceptibility to *P. brassicae*. The simplest method is to score inoculated plants based on symptom development. A score of zero corresponds to a completely susceptible leaf and a score of 4 to a fully resistant leaf, whereas scores of 1, 2, or 3 represent leaves with lesions of about 75%, 50%, or 25% of their surface, respectively. The second and more accurate method is based on polymerase chain reaction (PCR) quantification of a Phytophthora-specific gene as a measure of pathogen biomass. RNA-free DNA extracted from infected leaves is used for quantitative PCR. The *P. brassicae*-specific target gene brassicein 1 (AY244549) is amplified with the primers F-bra1: 5′-GTTCAACAAGTGCGTGGATG, and R-bra1: 5′-TCGTGGGTACAGTCAGATCG, and the results are normalized relative to the reference gene SAND (Czechowski et al., 2005) from Arabidopsis (F-Sand: 5′-AACTCTATGCAGCATTTGATCCACT and R-SAND: TGATTGCATATCTTTATCGCCATC). A third method of disease quantification is based on the use of transgenic *P. brassicae* that contains the p34GFN plasmid, which leads to the constitutive expression of green fluorescent protein (GFP) whose fluorescence can be determined in infected leaves as a measure of pathogen abundance (Si-Ammour et al., 2003).

16.6.5 Transformation of *P. brassicae*

A detailed procedure for the stable transformation of *P. brassicae* has been described (Si-Ammour et al., 2003). It is based on liposome-mediated transfer of DNA into protoplasts derived from young mycelia. This transformation protocol was much more efficient for the transformation of *P. brassicae* than for *P. infestans*. The transgenic progeny was stable for several years as exemplified with *P. brassicae* that contains p34GFN, which leads to the constitutive expression of GFP. However, transformants often showed a variable reduction in *in vitro* growth, which was not correlated to the expression level of transgenes. It seems that the transformation procedure can lead to a reduction of fitness of the transformants. Therefore, it is important for disease resistance experiments to screen the transgenic progeny for lines with wild-type fitness.

Transformation efficiency was found to be about 3-fold higher with double casette constructs containing the selectable marker and the gene of interest than with the frequently used cotransformation method (Si-Ammour et al., 2003). We have produced new transformation vectors by rearranging the genetic elements from *Bremia lactucae* described by Judelson (Judelson et al., 1991, 1992). The new transformation vectors have a pBluescript II SK backbone and are double cassette constructs that contain NPTII as a selectable marker gene. The construction of p34GFN that confers constitutive expression of GFP was described before (Si-Ammour et al., 2003). The maps of two additional vectors are shown in Fig. 16.3a. pTOR is a vector for constitutive sense or antisense expression of transgenes. It contains an expression cassette that consists of the *ham34* promoter and the termination sequence of *hsp70* of *B. lactucae* separated by a multiple cloning site with seven unique restriction sites. pTOR was successfully used by the *Phytophthora* community for overexpression of diverse genes (Avrova et al., 2007; Blanco and Judelson, 2005; Horner and Van West, 2007; Judelson and Tani, 2007; Whisson et al., 2007). pTOR has later been modified by replacement of the *ham34* promoter with an inducible promoter

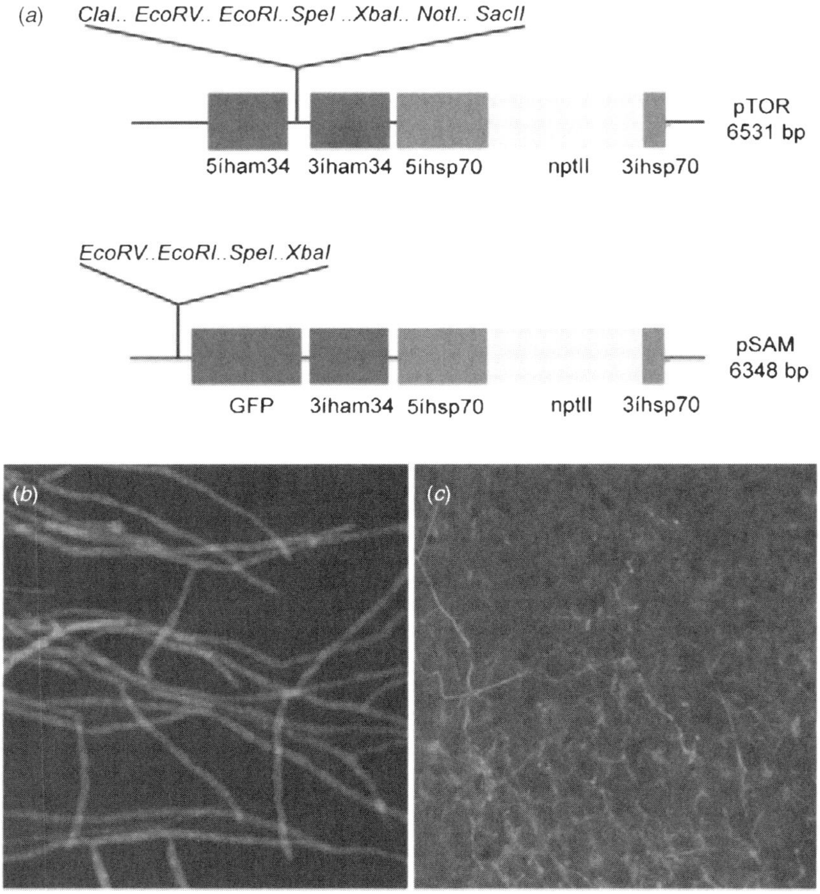

FIG. 16.3 Transformation vectors and examples of expression analysis. (a) Map of transformation vector pTOR suitable for over- and antisense expression constructs and pSAM suitable for expression analysis with *promoter::GFP* constructs, (b) in vitro grown hyphae of transgenic *P. brassicae* expressing a pSAM-derived *bra7::GFP* construct for the analysis of the expression of the *P. brassicae* elicitin brassicein 7 (Genebank AY244549), (c) *in planta* growth of *P. brassicae* expressing green fluorescent protein (GFP) under the control of the brassicein 1 promoter (AY244544). The red fluorescence is derived from chlorophyll. The absence of red fluorescence indicates necrotic areas that are full of GFP-expressing hyphae.

(Judelson et al., 2007). pSAM was designed for transcriptional fusions with GFP for expression analysis. It contains a codon-optimized GFP (Pang et al., 1996) behind a multiple cloning site that contains four unique restriction sites. Figures 16.3b and 16.3c show examples of transgenic lines of *P. brassicae* transformed with pSAM-derived constructs.

16.6.6 Sequence Information

Five cDNA libraries were produced of *P. brassicae* isolate HH at different developmental stages (mycelial growth and germinating cysts), from mycelia grown under varying nutritional conditions (carbon- or nitrogen-starvation) and from mycelia exposed to Arabidopsis. The cDNA libraries were used to generate sequence information of more than 12,000 expressed sequence tags (ESTs). The sequences are available at NCBI (NCBI ES277745-ES290687). The ESTs represent partial sequence information for about 3,200 *P. brassicae* genes.

REFERENCES

Avrova AO, Whisson SC, Pritchard L, Venter E, De Luca S, Hein I, Birch PRJ 2007. A novel non-protein-coding infection-specific gene family is clustered throughout the genome of *Phytophthora infestans*. Microbiol 153:747–759.

Belbahri L, Calmin G, Mauch F, Andersson JO 2008. Evolution of the cutinase gene family: evidence for lateral gene transfer of a candidate *Phytophthora* virulence factor. Gene 408:1–8.

Blanco FA, Judelson HS 2005. A bZIP transcription factor from *Phytophthora* interacts with a protein kinase and is required for zoospore motility and plant infection. Molec Microbiol 56:638–648.

Cao H, Bowling SA, Gordon AS, Dong XN 1994. Characterization of an Arabidopsis mutant that is nonresponsive to inducers of systemic acquired resistance. Plant Cell 6:1583–1592.

Czechowski T, Stitt M, Altmann T, Udvardi MK, Scheible W-R 2005. Genome-wide identification and testing of superior reference genes for transcript normalization in Arabidopsis. Plant Physiol 139:5–17.

Daniel R, Guest D 2006. Defence responses induced by potassium phosphonate in *Phytophthora palmivora*-challenged *Arabidopsis thaliana*. Physiol Molec Plant Pathol 67:194–201.

De Cock AWAM, Neuvel A, Bahnweg G, De Cock JCJM, Prell HH 1992. A comparison of morphology pathogenicity and restriction fragment patterns of mitochondrial DNA among isolates of *Phytophthora porri* Foister. Netherlands J. Plant Pathol 98:277–289.

Erwin D, Ribeiro OK 1996. Phytophthora diseases worldwide. APS Press. St Paul, MN.

Foister CE 1931. The white tip disease of leeks and its causal fungus, *Phytophthora porri* n.sp. Trans Proc Bot Soc Edinb 30:257–281.

Geeson J 1976. Storage rot of white cabbage caused by *Phytophthora porri*. Plant Pathol 25:115–116.

Glazebrook J 2001. Genes controlling expression of defense responses in Arabidopsis-2001 status. Curr Opin Plant Biol 4:301–308.

Glazebrook J, Ausubel FM 1994. Isolation of photoalexin-deficient mutants of *Arabidopsis thaliana* and characterization of their interaction with bacterial pathogens. Proc Natl Acad Sci USA 91:8955–8959.

Glazebrook J, Zook M, Merrit F, Kagan I, Rogers EE, Crute IR, Holub EB, Hammerschmidt R, Ausubel FM 1997. Phytoalexin-deficient mutants of Arabidopsis reveal that *PAD4* encodes a regulatory factor and that four PAD genes contribute to downy mildew resistance. Genetics 146:381–392.

Guzman P, Ecker JR. 1990. Exploiting the triple response of Arabidopsis to identify ethylene-related mutants. Plant Cell 2:513–523.

Heimann MF 1994. First report of Phytophthora rot of cabbage caused by *Phytophthora porri* Foister in Wisconsin. Plant Dis 78:1123.

Ho H 1983. *Phytophthora porri* from stored carrots in Alberta. Mycologia 75:747–751.

Horner N, van West P 2007. Successful transformation of the mycoparasitic oomycete *Pythium oligandrum*. Oomycete Molecular Genetics Network Workshop, Asilomar, CA.

Huitema E, Vleeshouwers VGAA, Francis DM, Kamoun S 2003. Active defence responses associated with non-host resistance of *Arabidopsis thaliana* to the oomycete pathogen *Phytophthora infestans*. Molec Plant Pathol 4:487–500.

Johnston A, Booth C 1968. Plant pathologist's pocketbook. Commonwealth Mycological Institute. Slough, UK.

Judelson HS, Tyler BM, Michelmore RW 1992. Regulatory sequences for expressing genes in oomycete fungi. Mol Gen Genet 234:138–146.

Judelson HS, Narayan R, Fong A, Tani S, Kim KS 2007. Performance of a tetracycline-responsive transactivator system for regulating transgenes in the oomycete *Phytophthora infestans*. Curr Genet 51:297–307.

Judelson HS, Tani S 2007. Transgene-induced silencing of the zoosporogenesis-specific *NIFC* gene cluster of *Phytophthora infestans* involves chromatin alterations. Eukaryo Cell 6:1200–1209.

Judelson HS, Tyler BM, Michelmore RW 1991. Transformation of the Oomycete pathogen *Phytophthora infestans*. Molec Plant-Microbe Interact 4:602–607.

Kamoun S, Huitema E, Vleeshouwers VGAA 1999. Resistance to oomycetes: a general role for the hypersensitive response? Trends Plant Sci 4:196–200.

Man in't Veld WA, de Cock AWAM, Ilieva E, Lévesque CA 2002. Gene flow analysis of *Phytophthora porri* reveals a new species: *Phytophthora brassicae* sp. nov. Eur J Plant Pathol 108:51–62.

Nawrath C, Métraux J-P 1999. Salicylic acid induction-deficient mutants of Arabidopsis express PR-2 and PR-5 and accumulate high levels of camalexin after pathogen inoculation. Plant Cell 11:1393–1404.

Niyogi KK, Grossman AR, Bjorkman O 1998. Arabidopsis mutants define a central role for the xanthophyll cycle in the regulation of photosynthetic energy conversion. Plant Cell 10:1121–1134.

Pang S, DeBoer DL, Wan Y, Ye G, Layton JG, Neher MK, Armstrong CL, Fry JE, Hinchee M, Fromm ME 1996. An improved green fluorescent protein gene as a vital marker in plants. Plant Physiol 112:893–900.

Parisy V, Poinssot B, Owsianowski L, Buchala A, Glazebrook J, Mauch F 2007. Identification of PAD2 as γ-glutamylcysteine synthetase highlights the importance of glutathione in disease resistance of Arabidopsis. Plant J 49:159–172.

Robinson L, Cahill DM 2003. Ecotypic variation in the response of *Arabidopsis thaliana* to *Phytophthora cinnamomi*. Aust. Plant Pathol 32:53–64.

Roetschi A, Si-Ammour A., Belbahri L, Mauch F, Mauch-Mani B 2001. Characterization of an Arabidopsis-Phytophthora pathosystem: resistance requires a functional *PAD2* gene and is independent of salicylic acid-, ethylene- and jasmonic acid-signaling. Plant J 28:293–305.

Semb L 1971. A rot of stored cabbage caused by Phytophthora sp. Acta Hort 20:32–35.

Si-Ammour A, Mauch-Mani B, Mauch F 2003. Quantification of induced resistance against Phytophthora species expressing GFP as a vital marker: β-aminobutyric acid but not BTH protects potato and Arabidopsis from infection. Molec Plant Pathol 4:237–248.

Smith D 1982. Liquid nitrogen storage of fungi. Trans Br Mycol Soc 79:415–421.

Staswick PE, Yuen GY, Lejman CC 1998. Jasmonate signaling mutants of Arabidopsis are susceptible to the soil fungus *Pythium irregulare*. Plant J 15:747–754.

Stelfox D, Henry AW 1978. Occurrence of the rubbery brown rot of stored carrots in Alberta Can Plant Dis Surv 58:87–91.

Whisson SC, Boevink PC, Moleleki L, Avrova AO, Morales JG, Gilroy EM, Armstrong MR, Grouffaud S, van West P, Chapman S 2007. A translocation signal for delivery of oomycete effector proteins into host plant cells. Nature 450:115–118.

Wiermer M, Feys BJ, Parker JE 2005. Plant immunity: the EDS1 regulatory node. Curr Opini Plant Biol 8:383–389.

17

APHANOMYCES EUTEICHES AND LEGUMES

Elodie Gaulin, Arnaud Bottin, and Christophe Jacquet
Université de Toulouse, UPS, Surfaces Cellulaires et Signalisation chez les Végétaux, 24 chemin de Borde-Rouge, Castanet-Tolosan, France; CNRS, Surfaces Cellulaires et Signalisation chez les Végétaux, Castanet-Tolosan, France

Bernard Dumas
CNRS, Surfaces Cellulaires et Signalisation chez les Végétaux, Castanet-Tolosan, France

17.1 GENERAL DESCRIPTION

17.1.1 Taxonomy and Morphology

Aphanomyces is a water mold that was first described by de Bary in 1860, and was the subject of a detailed monograph published by Scott a century later (de Bary, 1860; Scott, 1961). Phenetic analysis recently placed *Aphanomyces* within the *Leptolegniaceae* family erected from the *Saprolegniacea sensu lato* family (Dick et al., 1999). The term "water mold" is customarily used to designate the Saprolegniales because they most often colonize aquatic habitats by participating in organic matter decay and carbon recycling, and eventually attacking aquatic animals (Chapters 20 and 21). Accordingly, the genus *Aphanomyces* contains marine and freshwater saprophytic and zoopathogenic species (Chapter 21). However, among the 45 species or *formae speciales* that are registered at the Index Fungorum (http://www.indexfungorum.org/Names/Names.asp), 10 are plant pathogens that occur in terrestrial ecosystems, preferentially in moist soils. Among them, *Aphanomyces euteiches* and *Aphanomyces cochlioides*, which cause root rot diseases of sugar beet and legumes respectively are the most devastating (Martin, 2003; Gaulin et al., 2007).

Oomycete Genetics and Genomics: Diversity, Interactions, and Research Tools
Edited by Kurt Lamour and Sophien Kamoun
Copyright © 2009 John Wiley & Sons, Inc.

A. euteiches is represented by four entries at the Index Fungorum, because Pfender and Hagedorn (1982) described 3 *formae speciales* in addition to the initial species description of Jones and Drechsler (1925). As several recent molecular studies did not clearly associate host range with genetic markers, we will consider *A. euteiches* as a single species as originally described.

In contrast to many Saprolegniales, *Aphanomyces* sp. has relatively thin hyphae, between 3 and 12 μm in diameter and oogonia usually contain only one oospore or more rarely two (Beghdadi et al., 1992). Whereas sexual reproduction has never been observed in the zoopathogenic species *Aphanomyces astaci* and *Aphanomyces invadans*, *A. euteiches* and *A. cochlioides* are homothallic and form oospores of diameter 16 to 25 μm. The morphology and biology of these two species were well described in a review from Papavizas and Ayers (1974). One of the most striking morphological traits of *Aphanomyces* sp. is related to the process of asexual sporulation, which resembles the one occurring in *Achlya* and is therefore referred to as an achlyoid type of sporangium dehiscence (Leclerc et al., 2000). Sporangia occur in the form of hyphae that are morphologically indistinguishable from nonsporulating hyphae. The first step of sporulation corresponds to a cytoplasmic rearrangement and cleavage that leads to the formation of uninucleated protoplasts within the hyphal sporangium. These protoplasts migrate toward the tip of the sporangium, are discharged, and encyst immediately (Fig. 17.1a). Each spore cyst then germinates to develop into a motile zoospore, usually called "secondary zoospore" if one considers the original sporangium protoplast, although it is not flagellated (Hoch and Mitchell, 1972), as a degenerated "primary zoospore" equivalent to the primary zoospore that emerges from the sporangium in *Achlya* (Daugherty et al., 1998; Leclerc et al., 2000; Petersen and Rosendahl, 2000). The secondary zoospore encysts after a period of swimming and, in some species, can eventually germinate to produce a new zoospore (diplanetism) in the absence of nutrients or of potential hosts. Repeated zoospore emergence (polyplanetism) is thought to represent an adaptation to a parasitic life mode (Cerenius and Söderhäll, 1985). Procedures for efficient production of zoospores and their study are available in *A. euteiches* (Mitchell and Yang, 1966; Deacon and Saxena, 1998), but most data on zoospore behavior has been acquired on *A. cochlioides* (Islam and Tahara, 2001; Islam, 2008).

17.1.2 Phylogeny

Recent phylogenetic studies have suggested that, contrary to what was previously assumed (Daugherty et al., 1998), *Aphanomyces*, and not *Saprolegnia*, contains the most ancestral organisms of the Saprolegniales (Leclerc et al., 2000; Petersen and Rosendahl, 2000). Interestingly, phytopathogenic species of other genera, such as *Pachymetra chaunorhiza* (Verrucalvaceae) or *Plectospira myriandra* (Saprolegniaceae), clustered together with phytopathogenic or saprophytic *Aphanomyces* species, whereas the zoopathogenic *Aphanomyces laevis*

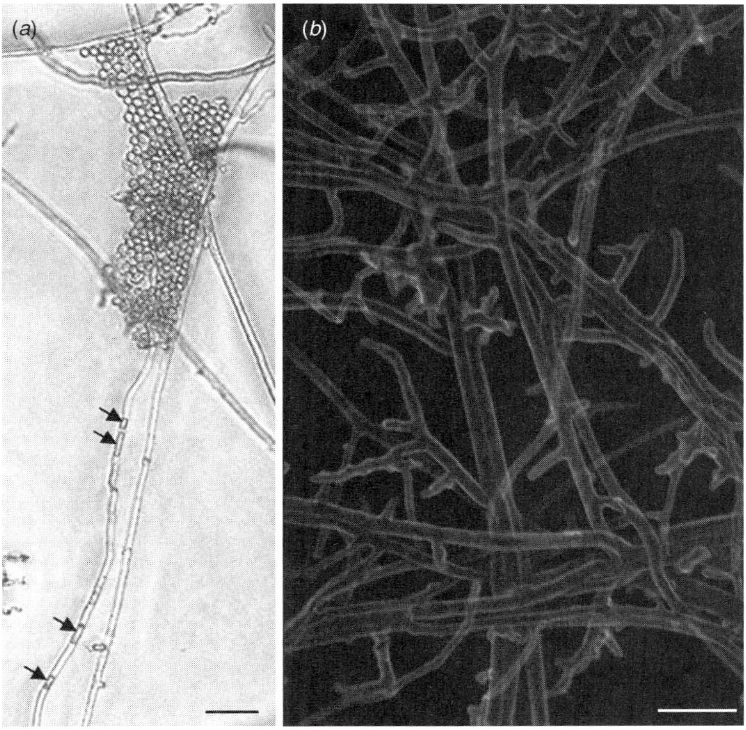

FIG. 17.1 *Aphanomyces euteiches* morphology. (a) Transmission microscopy observation of a sporulating culture on agar medium. Arrows show protoplasts moving toward the tip of a sporangium where many protoplasts have already been discharged and have encysted. (b) Epifluorescence microscopy observation of mycelium incubated in presence of a fluorescein isothiocyanate — wheat germ agglutinin conjugate. Bright labeling of the cell surface by the conjugate is observed. Bars, 50 μm.

and *A. astaci* appeared on a different branch (Riethmuller et al., 1999; Leclerc et al., 2000). This suggests that the *Aphanomyces* genus radiated early, and that the ability of *Aphanomyces* to infect plants might be ancient and roots to the origin of the Peronosporomycetes. Unfortunately, these studies included only three or four *Aphanomyces* species (Lilley et al., 2003; Levenfors and Fatehi, 2004). A detailed phylogenetic study of the *Aphanomyces* and related genera, considering phenotypic characters such as presence or absence of sexual reproduction and host range of pathogenic species, would therefore be of great interest to understand the evolution of pathogenicity in this group of organisms.

17.1.3 Metabolism and Cell Wall Biology

A. euteiches is a facultative pathogen that can be grown in various artificial media, such as Corn Meal Agar (CMA). Like other Saprolegniales, it shows no auxotrophy and, in contrast to *Phytophthora*, can be grown in defined synthetic medium lacking thiamine or sterols. Oospores can also be formed without the requirement for specific vitamins. As other Saprolegniales also, *A. euteiches* cannot use oxidized sulfur or nitrogen sources (such as SO_4^{2-} or NO_3^-), but it can use inorganic sulfur or nitrogen compounds such as elemental sulfur or ammonium. Only a limited collection of carbon sources allow *in vitro* growth, which include glucose, galactose, maltose, or glycerol, and no growth is obtained when for example mannose, sucrose, sorbitol, xylan, or pectin is supplied as sole carbon source (reviewed in Papavizas and Ayers, 1974; Gaulin et al., 2008). In contrast, sucrose or pectin alone supports growth of *Phytophthora*. Whether these differences result from different contents in genes involved in nutrient usage or from different regulation of common orthologous genes should be investigated.

The cell wall is an essential structure that is involved in development and interactions with the environment. It contains constituents that are widely distributed among microbes, such as structural β-glucans, as well as specific constituents involved in cellular interactions and host–parasite interactions. A feature often mentioned when distinguishing oomycetes from true fungi, which is mainly based on data obtained in *Phytophthora* and related species, is the presence of cellulose and the absence of chitin in their cell walls (Bartnicki-Garcia, 1968). However, some Leptomitales contain more chitin than cellulose (Lin et al., 1976), and the presence of low amounts of chitin was demonstrated in some Saprolegniales, which includes *Achlya* and *Saprolegnia*. A study of the *A. euteiches* cell wall was initiated at our laboratory, which showed the presence of an unexpectedly high amount of N-acetylglucosamine, averaging 10% of the cell wall dry weight. Biochemical and biophysical analyses showed, however, that *A. euteiches* does not contain chitin in crystalline form but rather noncrystalline chitooligosaccharides. Growth inhibition by the chitin synthase inhibitor Nikkomycine Z was observed, which suggests that these compounds are involved in cell wall function by being covalently linked to other structural polysaccharides, such as crystalline cellulose. This data highlights for the first time the potential of chitin synthesis inhibition in controlling a phytopathogenic oomycete (Badreddine et al., 2008). Another salient feature of *A. euteiches* is that the chitooligosaccharides are exposed at the cell surface, as shown by intense labeling of the hypha surface by wheat germ agglutinin (Fig. 17.1b). This finding has an important practical application of allowing fast and easy detection of the parasite in plant tissues (Fig. 17.2b). Whether the data obtained on cell walls of *A. euteiches* illustrate a general property of the *Aphanomyces* genus or of ancestral oomycetes requires more studies. Interestingly, the high specificity and absence of crossreactivity of antibodies raised against *A. euteiches* oospores led Petersen et al. (1996) to suggest that *Aphanomyces* has a different wall chemistry compared with other groups of oomycetes.

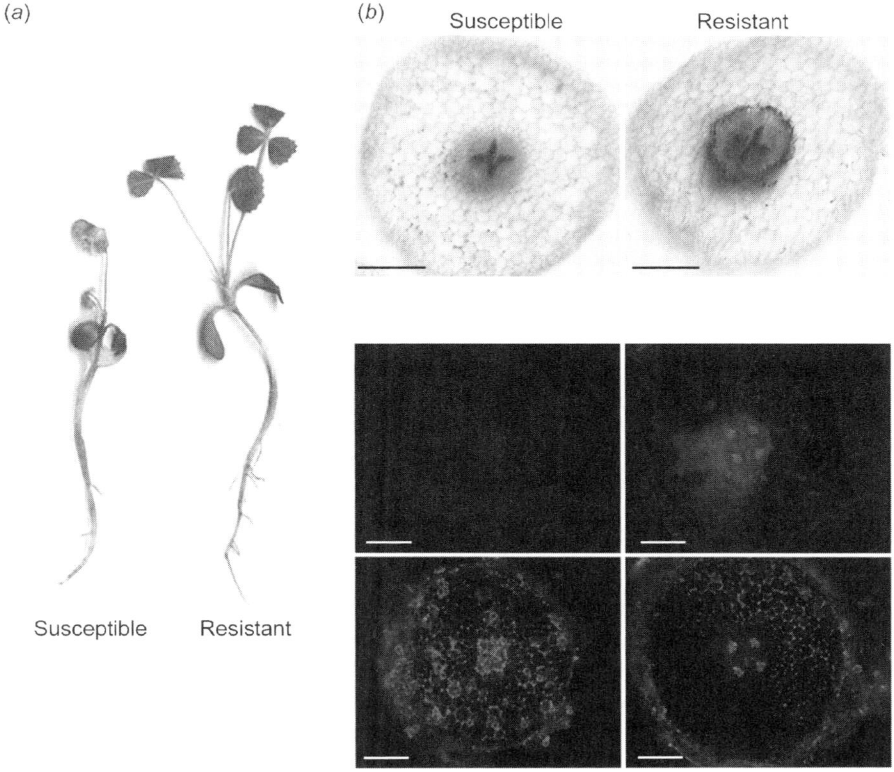

FIG. 17.2 (See color insert) Interaction between *A. euteiches* and the model legume *Medicago truncatula*. (a) Differential symptoms between the A17 resistant and F83005.5 susceptible lines, observed 15 days after *A. euteiches* inoculation. *In vitro* infection assays were performed on 2-week-old *M. truncatula* plants by drop inoculation of the roots with a zoospore suspension. Root browning detected in the resistant line was linked with an accumulation of defense compounds. Plant growth was generally not altered by pathogen infection. By contrast, the susceptible line showed typical root rot symptoms (soft brown macerated tissues). Fast pathogen development led to the onset of secondary symptoms on the aerial parts (such as leaf wilting) and finally to plant death. (b) Transversal root sections of inoculated susceptible and resistant *M. truncatula* lines. Top: staining with phloroglucinol to visualize lignin accumulation (6 days postinoculation). A ring of lignin was observed on the outer cell walls of pericycle cells only in A17 roots. In F83005.5 only xylem cells were stained. Middle and bottom: Root sections (15 dpi) were observed either in autofluorescence under ultraviolet light (middle), or in epifluorescence to visualize *A. euteiches* mycelium, stained by a fluorescein isothiocyanate — wheat germ agglutinin (bottom). Defense-associated fluorescent soluble phenolic compounds were detected only in the cortex cells of the A17 resistant line. (Note the autofluorescence of vascular cells.) Bars, 400 μm.

17.2 ROOT ROT DISEASES ON LEGUMES

17.2.1 Distribution and Symptoms

The pathogen was first reported in 1925 in the United States (Jones and Drechsler, 1925), where peas are often grown in monoculture. Within a few years, the disease was reported in many areas around the world and considered almost a century later as the most important and destructive disease of peas, notably in Europe (Allmaras et al., 2003; Levenfors et al., 2003). *A. euteiches* can cause also severe symptoms on different legumes (e.g., alfalfa, faba bean, vetch, lentil, and red clover), and distinct pathotypes with distinct legumes preferences have been described (Malvick and Grau, 2001; Wicker et al., 2001).

Root rot occurs to some extent wherever peas are grown, and the host plant is susceptible at any stage of its development. The first symptoms may be discerned on roots that are softened and water soaked. Then, the pathogen spreads rapidly in all directions through the cortical tissue and the fine branches of feeding rootlets are destroyed. In severe cases, the plants collapse and die before forming any pod. Late-infected plants seem almost normal aboveground and often produce seeds (Papavizas and Ayers, 1974). The intensity of symptoms of pea root rot varies with the environmental conditions. Disease in alfalfa is characterized by stunted seedlings with chlorotic cotyledons, damping off, and poor stand establishment. The pathogen may also cause a sublethal disease that results in chronic root infection of mature plants.

17.2.2 Life Cycle and Infection Process

A. euteiches is considered to be a strict soilborne pathogen, because its entire life cycle occurs only in soil. Long-distance dispersal is mediated by the transport of contaminated soil or materials, or of infected plants that contain oospores. Oospores are typically resistant to desiccation and capable of long-term survival. The infection is initiated by oospore germination in close vicinity to a plant host. The oospores form a germ tube and a long terminal zoosporangium, which can release hundreds of zoospores (Scott, 1961; Fig. 17.1a). Motiles zoospores have been shown to sense prunetin, which is a flavone compound from root exudates (Yokosawa et al., 1986). This chemotactic attraction probably enhances docking of the zoospores to infection sites on the host surface (Sekizaki et al., 1993). It is not known whether penetration of the host plant requires specialized infection structures, although appressorium-like structures have been occasionally observed (Cunningham and Hagedorn, 1962). After penetration of cortical cells, *A. euteiches* forms coenocytic hyphae that develop mainly extracellularly and ramify laterally into the root cortex. Within a few days of infection, the mycelium forms oogonia. Terminal oogonia are penetrated by fertilization tubes from antheridia, which deliver male nuclei to the oogonia, resulting in the formation of diploid oospores (Scott, 1961). The mycelium also can release new zoospores after forming sporangia at the surface of infected roots.

17.2.3 Pathogen Variability

Although *A. euteiches* is homothallic, occasional outcrossing might account for the considerable phenotypic and genetic variation observed between isolates (Malvick and Percich, 1998; Levenfors and Fatehi, 2004). A recent survey suggested that the *A. euteiches* population structure in the northern United States is patterned by regular selfing and asexual reproduction, with occasional migration of novel genotypes or outcrossing (Grünwald and Hoheisel, 2006). Using a differential set of six pea genotypes, Wicker and colleagues identified two virulence groups, including a main group to which most of the French isolates belong and another group specific to some U.S. isolates (Wicker et al., 2001, 2003). Genetic variation was also observed using random amplification of polymorphic DNA (RAPD), amplified fragment length polymorphism (AFLP), and restriction fragment length polymorphism (RFLP) markers but has not been related to the phenotypic variation (Malvick and Percich, 1998; Levenfors and Fatehi, 2004; Grünwald and Hoheisel, 2006). Evaluation of diversity and population genetics studies should improve with the use of simple sequence repeats (SSR), single nucleotide polymorphism (SNP), and sequence characterized amplified region (SCAR) markers that were recently developed for *A. euteiches* (Akamatsu et al., 2007).

17.2.4 Control and Disease Management

Resistance to *A. euteiches* has been studied in peas for several years; nevertheless no fully resistant pea cultivar is available (Pilet-Nayel et al., 2005). Efficient fungicides are also lacking. Avoiding highly infested fields and evaluation of potential inoculum in soil before sowing are control methods currently available (Vandemark et al., 2000). Crop rotation and cultivation of other legumes (e.g., faba bean) are often used in infested fields (Levenfors et al., 2003). The efficiency of seed dressing and of green manure methods still needs to be confirmed (Muechlchen et al., 1990; Smolinska et al., 1997; Stones et al., 2003). Bioprotection of pea roots by arbuscular mycorrhizal fungi such as *Glomus fasciculatum* was first reported by Rosendahl (1985), and further studies showed that suppression of *A. euteiches* infection on pea depends on a fully established symbiosis (Slezack et al., 2000). In this context, the development of resistant varieties and comprehension of *A. euteiches* pathogenicity are major objectives to control the disease.

17.3 LEGUME RESISTANCE TO *APHANOMYCES*: VARIABILITY AND DEFENSE COMPONENTS

17.3.1 Sources of Resistance

Several surveys have been performed within different legume germplasm collections to assess the biological variability of plant responses to *A. euteiches* infection. Major efforts were dedicated to screen collections of peas to search for sources of resistance that might be used in a conventional breeding program (Lockwood, 1960; Engqvist, 1992; Wicker et al., 2003). Despite intensive search, no accession was found fully resistant to *A. euteiches,* although a partial level of resistance was detected.

Screening for resistance to *A. euteiches* was also performed in several other legume species, including alfalfa (*Medicago sativa*), barrel medics (*Medicago truncatula*), lentil (*Lens culinaris*), faba bean (*Vicia faba*), chickpea (*Cicer arienatum*), common vetch (*Vicia sativa*), French bean (*Phaseolus vulgaris*), clover (*Trifolium repens* and *Trifolium pratense*), and lupin (*Lupinus alba*) (Moussart et al., 2007). In surveys performed with several strains, results indicated that some strains showed a broad host spectrum, whereas others were specific for a legume species (Levenfors et al., 2003). Inoculation of various legume species with a broad host spectrum strain allowed the distinction of different legume categories. For some legume species, all the cultivars tested were susceptible or highly susceptible (i.e., lentil), whereas for other species all the cultivars were found to be fully resistant and thus are probably non-host (i.e., chickpea) (Moussart et al., 2008). These studies suggested that several resistance mechanisms to *A. euteiches* could exist, and brought useful data to select legume species which could be included in crop rotations.

17.3.2 Breeding and Quantitative Trait Loci (QTL) Mapping

As neither cultural nor chemical methods, other than avoiding infested fields, can prevent spread of the disease, the genetic improvement of legume resistance remains the most promising approach to control *A. euteiches*. Alfalfa and peas are the main cultivated legumes for which significant breeding programs have been conducted to increase resistance to *A. euteiches*.

Aphanomyces alfalfa isolates are distributed in two races (race 1 and race 2). An alfalfa germplasm (WAPH-1) resistant to the most common race 1 was first released in 1989 (Grau, 1992) and was intensively used to incorporate resistance into commercial varieties. A few years later, resistance to race 1 was reported to be overcome by new isolates (therefore belonging to race 2) (Grau et al., 1991). New alfalfa lines WAPH-4 (resistant to race 2) and WAPH-5 (resistant to both race 1 and race 2) were identified (Malvick, 2002) and used in breeding programs to develop new lines resistant to both *Aphanomyces* races.

Resistance to *A. euteiches* in peas is expressed as a quantitative trait. Only low levels of resistance were found, and a few accessions that show a low level of partial resistance have been released (Davis, 1995). Breeding programs to improve resistance to root rot are hindered by several constraints: polygenic inheritance and weak levels of resistance, high level of pathogenic variability in populations of *A. euteiches,* interactions with pathogens of the root complex (including notably *Fusarium* sp.), and strong impact of environmental field conditions (Wicker et al., 2003). To overcome these limitations and support pea breeding programs, molecular markers associated with quantitative partial resistance have been identified by Pilet-Nayel and coworkers (Pilet-Nayel et al., 2002, 2005). Three stable QTLs associated with resistance to two *A. euteiches* strains in France and the United States have been localized and can be used for marker-assisted breeding.

Recently, the barrel medic *M. truncatula* has emerged as a model host to better understand *A. euteiches*–legumes interactions. This self-fertile, diploid plant has a relatively small genome (about 500 Mbp) and shows a high level of variability in response to *A. euteiches* (Vandemark and Grunwald, 2004; Moussart et al., 2007). Its high level of synteny with other crop legumes (notably peas), makes this plant agronomically relevant and useful for identifying new sources of resistance and to transfer the gained knowledge to legume crops (Tivoli et al., 2006). The genetic dissection of resistance to *A. euteiches* was achieved through the phenotyping of two recombinant inbred line (RIL) populations, which were generated from crosses between the susceptible parent (F83005.5) and either a partially resistant line (A17, whose sequencing is planned to be finished at the end of 2009), or a highly resistant line (DZA45.5). The former cross led to the "LR5" RIL population and the latter the "LR3" RIL population. Genetic analyses identified one major recessive QTL in LR5, named *Ae1* (Jacquet et al., unpublished data) and a single dominant gene "AER1" in LR3 (Pilet-Nayel et al., 2009).

17.3.3 Plant Defense Mechanisms

The wealth of genomic resources developed for *M. truncatula* made it possible to identify molecular components involved in legume defense and resistance to pathogens, notably to *A. euteiches* (Ameline-Torregrosa et al., 2006). A macroarray analysis comprising a collection of 560 expressed sequence tags (ESTs) enriched in *M. truncatula* defense-related sequences indicated the key role of pathogenesis-related (PR) proteins and ABA-regulated genes (Nyamsuren et al., 2003). These results were confirmed through a proteomic approach (Colditz et al., 2004). However, another study surprisingly demonstrated that silencing PR-10–like proteins in *M. truncatula* increased *A. euteiches* tolerance. This effect might be explained by induction of a new set of other PR-proteins genes that are normally repressed when PR-10 genes are expressed (Colditz et al., 2007). Other defense components were identified by comparative proteomic and by cytological and biochemical studies using *M. truncatula* lines that display differential responses during *A. euteiches* infection (Colditz et al., 2005). The phenylpropanoid pathway is clearly involved in plant resistance. Transverse thin sections of inoculated roots indicated that resistance was linked to the accumulation of soluble phenolic compounds, and a lignin deposit was also observed around the central cylinder (Fig. 17.2b). Microscopy observations also showed the formation of an additional pericycle cell layer. All these responses allowed protection of the root central cylinder against pathogen invasion, in contrast to the susceptible line that was entirely colonized (Jacquet et al., unpublished data). A transcriptomic approach confirmed the stimulation of the phenylpropanoid pathway during infection, and several hundred of genes showed an altered expression after inoculation (Krajinski and Jacquet, unpublished data). Functional analysis of selected induced genes is in progress.

17.4 PATHOGENICITY DETERMINANTS OF *A. EUTEICHES*

Recently, a large-scale sequencing project of *A. euteiches* cDNAs was undertaken, which resulted in the release of the first genomic database dedicated to the *Aphanomyces* genus (AphanoDB) (http://www.polebio.scsv.ups-tlse.fr/aphano/). The current version of AphanoDB includes a unigene set of 7,977 sequences obtained from 18,864 high-quality ESTs that correspond to two cDNA libraries [i.e., saprophytic mycelium and mycelium in contact to *M. truncatula* roots (Madoui et al., 2007)]. A comparative analysis of the *A. euteiches* unigene set to proteomes derived from fully sequenced organisms (oomycetes, plant, fungi, and diatoms) revealed that most sequences showed a high similarity to *Phytophthora*-predicted protein sequences (Gaulin et al., 2008). About 20% of unigenes presented an ortholog in *Arabidopsis thaliana*. Comparative analyses highlighted significant differences between *A. euteiches* and *Phytophthora* species, which included genes involved in pathogenicity and metabolic pathways.

New putative effectors of pathogenicity comprised a large family of genes that contain cellulose-binding domains (CBDs). These molecular motifs are known to act as pathogen-associated molecular patterns (PAMPs) on various plant species (Gaulin et al., 2006). Whereas CBDs are often linked to N/apple PAN domains in *Phytophthora* species, these domains in *A. euteiches* proteins are associated with acidic domains found in the N-terminal part of *Phytophthora infestans* cyst germination proteins (Gornhardt et al., 2000). This highly expressed family of proteins can represent a novel class of oomycete adhesins. Many genes that encode proteases were found, suggesting that protein degradation is a key function for *A. euteiches* pathogenicity. Protease diversity can be related to the fact that the *Aphanomyces* genus comprises plant and animal pathogens. Comparative genomic analyses between these different species should help clarify this point. Reciprocally, some enzyme sequences that are well represented in plant pathogens were not found in the *A. euteiches* transcriptome. This is the case for pectinolytic enzymes that constitute a large family in *Phytophthora* sp. No transcript encoding enzymes related to pectin metabolism was identified in the *A. euteiches* ESTs collection. However, the search for polymorphic DNA markers in *A. euteiches* led incidentally to the identification of a small genomic DNA sequence that showed homology to a *Phytophthora cinnamomi* polygalacturonase gene (Akamatsu et al., 2007). Thus, more investigations are needed to conclude definitely whether there are functional pectinase genes in the *A. euteiches* genome.

Saprolegniales possess biosynthetic pathways not found in the genus *Phytophthora*. These include the ability to synthesize thiamine. Accordingly, two sequences that correspond to thiamine biosynthetic genes were detected in *A. euteiches* as well as in the fish pathogen *Saprolegnia parasitica* (Torto-Alalibo et al., 2005). Several genes that encodes enzymes involved in sterol biosynthesis were also identified. This result confirmed earlier studies suggesting that unlike *Phytophthora* species, Saprolegniales synthesize their own sterols from simple carbon sources. Interestingly, very few sequences that correspond to putative sterol carriers, (elicitins) were obtained, whereas elicitin genes are abundantly expressed in *Phytophthora* species and form a large gene family of more than 50 elicitin and elicitin-like genes in the *Phytophthora sojae* genome (Tyler et al., 2006). Altogether, these data suggest that an ancient sterol biosynthetic pathway was conserved in *A. euteiches* and probably in other Saprolegniales. In the *Phytophthora* lineage, this pathway has been lost, and this loss was accompanied by expansion and diversification of the elicitin gene family (Gaulin et al., 2008).

17.5 CONCLUSIONS AND FUTURE PROSPECTS

Legumes play a critical role in natural ecosystems and agriculture, because their ability to fix nitrogen in symbiosis allows a consistent increase in crop yield. Unfortunately, a major factor of yield instability of peas and other legume crops has been the development of the common root rot disease caused by *A. euteiches*. Because of its destructiveness on legumes and the absence of efficient treatment and resistant cultivar, *A. euteiches* has to be considered as a major pathogen. Despite this, little attention has been given to this parasite in comparison with other oomycetes. Much work remains to be done to understand the population dynamics of *A. euteiches*. The recent development of molecular markers will strengthen population studies. The recent discovery of resistance QTLs in peas and in the model legume *M. truncatula* should significantly help legume resistance improvement by classic breeding. The development of *M. truncatula* genomic tools [i.e., mutant collections and DNAchip array (for review, see Ané et al., 2008)], will facilitate the understanding of plant defense reactions by deciphering signaling pathways. The recent large-scale analysis of an *A. euteiches* EST collection illustrates the diversity and the evolutionary distance that exists among oomycete species. It also provides the first sequences of genes involved in sterol metabolism and chitin synthesis in this organism, which are attractive targets for the development of new chemicals against this oomycete. The complete sequencing of the *A. euteiches* genome and other *Aphanomyces* species, particularly those that infect animal hosts, is essential to understand fully the pathogenicity and biology of this devastating parasite. Nevertheless, fruitful exploitation of these sequences will require the development of methods for functional analysis. Standard *Phytophthora* transformation protocols (e.g., protoplast permeabilization, electroporation, and microprojectile bombardment) will have to be adapted to *A. euteiches*. Regeneration of *A. euteiches* protoplasts in absence of transforming DNA was reported, but no genetic transformation procedure was developed (Weiland, 2001). Preliminary results using *Agrobacterium tumefaciens*-mediated transformation are encouraging (Gaulin et al., unpublished data). Reverse genetics approaches combined with the development of genomic resources and transcriptome profiling will be the basis for the rationale design of new strategies to combat this oomycete.

REFERENCES

Akamatsu HO, Grünwald NJ, Chilvers MI, Porter LD, Peever LT 2007. Development of codominant simple sequence repeat, single nucleotide polymorphism and sequence characterized amplified region markers for the pea root rot pathogen, *Aphanomyces euteiches*. J Microbiol Met 71:82–86.

Allmaras RR, Fritz VA, Pfleger FL, Copeland SM 2003. Impaired internal drainage and *Aphanomyces euteiches* root rot of pea caused by soil compaction in a fine-textured soil. Soil Till Res 70:41–52.

Ameline-Torregrosa C, Dumas B, Krajinski F, Esquerré-Tugayé MT Jacquet C 2006. Transcriptomic approaches to unravel plant-pathogen interactions in legume. Euphytica 147:25–36.

Ané JM, Zhu H, Frugoli J 2008. Recent advances in *Medicago truncatula* genomics. 2008. Intern J Plant Gen 60:2–11.

Badreddine I, Lafitte C, Heux L, Skandalis N, Spanou Z, Martinez Y, Esquerré-Tugayé MT, Bulone V, Dumas B, Bottin A 2008. Cell wall chitosaccharides are essential components and exposed patterns of the phytopathogenic oomycete *Aphanomyces euteiches*. Eukaryot Cell. 7:1980–1993.

Bartnicki-Garcia S 1968. Cell wall chemistry, morphogenesis, and taxonomy of fungi. Annu Rev Microbiol 22:87–108.

Beghdadi A, Richard C, Dostaler D 1992. L'*Aphanomyces euteiches* des luzernières du Québec: isolement, morphologie et variabilité de la croissance et du pouvoir pathogène. Can J Bot 70:1903–1911.

Cerenius L, Söderhäll K 1985. Repeated zoospore emergence as a possible adaptation to parasitism in *Aphanomyces*. Exp Mycol 9:259–263.

Colditz F, Braun HP, Jacquet C, Niehaus K, Krajinski F 2005. Proteomic profiling unravels insights into the molecular background underlying increased *Aphanomyces euteiches*-tolerance of *Medicago truncatula*. Plant Mol Biol 59:387–406.

Colditz F, Niehaus K, Krajinski F 2007. Silencing of PR-10-like proteins in *Medicago truncatula* results in an antagonistic induction of other PR proteins and in an increased tolerance upon infection with the oomycete *Aphanomyces euteiches*. Planta 226:57–71.

Colditz F, Nyamsuren O, Niehaus K, Eubel H, Braun HP, Krajinski F 2004. Proteomic approach: identification of *Medicago truncatula* proteins induced in roots after infection with the pathogenic oomycete *Aphanomyces euteiches*. Plant Mol Biol 55:109–120.

Cunningham JL, Hagedorn DJ 1962. Penetration and infection of pea roots by zoospore of *Aphanomyces euteiches*. Phytopathol 52:827–834.

Daugherty J, Evans TM, Skillom T, Watson LE, Money NP 1998. Evolution of spore release mechanisms in the saprolegniaceae (Oomycetes): evidence from a phylogenetic analysis of internal transcribed spacer sequences. Fungal Genet Biol 24: 354–363.

Davis DW, Fritz VA, Pfleger FL, Percich JA, Malvick DK 1995. MN144, MN313, and MN314: Garden pea lines resistant to root rot caused by *Aphanomyces euteiches* Drechs. Hort Sci 30:639–640.

Deacon J, Saxena, G 1998. Germination triggers of zoospore cysts of *Aphanomyces euteiches* and *Phytophthora parasitica*. Mycol Res 102:33–41.

de Bary H 1860. Einige neue Saprolegnieen. Jahrbücher Wissenschaftliche Bot 2:169–179.

Dick MW, Vick MC, Gibbings JG, Hedderson TA, Lopez-Lastra CC 1999. 18S rDNA for species of Leptolegnia and other Peronosporomycetes: justification for the subclass taxa Saprolegniomycetidae and Peronosporomycetidae and division of the Saprolegniaceae sensu lato into the Leptolegniaceae and Saprolegniaceae. Mycol Res 103:1119–1125.

Engqvist LG 1992. Studies on common root rot (*Aphanomyces euteiches*) of peas (*Pisum sativum*) in Sweden. Norw J Agri Sci 7:111–118.

Gaulin E, Drame, N, Lafitte C, Torto-Alalibo T, Martinez Y, Ameline-Torregrosa C, Khatib M, Mazarguil H, Villalba-Mateos F, Kamoun S, et al. 2006. Cellulose-binding domains of a *Phytophthora* cell wall protein are novel pathogen-associated molecular patterns. Plant Cell 18:1766–1777.

Gaulin E, Jacquet C, Bottin A, Dumas B 2007. Root rot disease of legumes caused by *Aphanomyces euteiches*. Mol Plant Pathol 8:539–548.

Gaulin E, Madoui MA, Bottin A, Jacquet C, Mathé C, Couloux A, Wincker P, Dumas B 2008. Transcriptome of *Aphanomyces euteiches*: new oomycete putative pathogenicity factors and metabolic pathways. PLoS One 3:e1723.

Gornhardt B, Rouhara I, Schmelzer E 2000. Cyst germination proteins of the potato pathogen *Phytophthora infestans* share homology with human mucins. Mol Plant Microbe Interact 13:32–42.

Grau CR 1992. Registration of WAPH-1 alfafa germplasm with resistance to *Aphanomyces* root rot. Crop Sci 32:287–288.

Grau CR, Muehlchen A, Tofte J 1991. Variability in virulence of *Aphanomyces euteiches*. Plant Dis 75:1153–1156.

Grünwald NJ, Hoheisel GA 2006. Hierarchical analysis of diversity, selfing and genetic differentiation in populations of the oomycete *Aphanomyces euteiches*. Phytopathol 96:1134–1141.

Hoch HC, Mitchell JE 1972. The ultrastructure of *Aphanomyces euteiches* during asexual spore formation. Phytopathol 62:149–160.

Islam MT 2008. Dynamic rearrangement of F-actin organization triggered by host-specific plant signal is linked to morphogenesis of *Aphanomyces cochlioides* zoospores. Cell Motil Cytoskeleton 65:553–562.

Islam MT, Ito T, Tahara S 2003. Host-specific plant signal and G-protein activator, mastoparan, trigger differentiation of zoospores of the phytopathogenic oomycete *Aphanomyces cochlioides*. Plant Soil 255:131–142.

Islam MT, Tahara S 2001. Chemotaxis of fungal zoospores, with special reference to *Aphanomyces cochlioides*. Biosci Biotechnol Biochem 65:1933–1948.

Jones FR, Drechsler C 1925. Root rot of peas in the United States caused by *Aphanomyces euteiches*. J Agr Res 30:293–325.

Leclerc MC, Guillot J, Deville M 2000. Taxonomic and phylogenetic analysis of Saprolegniaceae (Oomycetes) inferred from LSU rDNA and ITS sequence comparisons. Anton Leeuw Int J G 77:369–377.

Levenfors JP, Fatehi J 2004. Molecular characterization of *Aphanomyces species* associated with legumes. Mycol Res 108:682–699.

Levenfors JP, Wikström M, Persson L, Gerhardson B 2003. Pathogenicity of *Aphanomyces* spp. from different leguminous crops in Sweden. Eur J Plant Pathol 109: 535–543.

Lilley JH, Hart D, Panyawachira V, Kanchanakhan S, Chinabut S, Soderhall K, Cerenius L 2003. Molecular characterization of the fish-pathogenic fungus *Aphanomyces invadans*. J Fish Dis 26:263–275.

Lin CC, Sicher RC Jr, Aronson JM 1976. Hyphal wall chemistry in *Apodachlya*. Arch Microbiol 108:85–91.

Lockwood JL 1960. Pea introductions with partial resistance to *Aphanomyces* root rot. Phytopathol 50:621–624.

Madoui MA, Gaulin E, Mathé C, San Clemente H, Couloux A, Wincker P, Dumas B 2007. AphanoDB: a genomic resource for *Aphanomyces* pathogens. BMC Genomics 8:471.

Malvick DK, Grau CR 2001. Characteristics and frequency of *Aphanomyces euteiches* races 1 and 2 associated with alfafa in the Midwestern United States. Plant Dis 74:716–718.

Malvick D 2002. *Aphanomyces euteiches* Race 2 in central Illinois alfalfa fields. Plant Dis 86:560–570.

Malvick DK, Percich JA 1998. Genotypic and pathogenic diversity among pea-infecting strains of *Aphanomyces euteiches* from the central and western United States. Phytopathol 88:915–921.

Martin HL 2003. Management of soilborne diseases of beetroot in *Australia*: a review. Aust J Exp Agric 43:1281–1292.

Mitchell JE, Yang CY 1966. Factors affecting growth and development of *Aphanomyces euteiches*. Phytopathol 56:917–922.

Moussart A, Even MN, Tivoli B 2008. Reaction of genotypes from several species of grain and forage legumes to infection with a French pea isolate of the oomycete *Aphanomyces euteiches*. Eur J Plant Pathol, 122:321–333.

Moussart A, Onfroy C, Lesné A, Esquibet M, Grenier E, Tivoli B 2007. Host status and reaction of *Medicago truncatula* accessions to infection by three major pathogens of pea (*Pisum sativum*) and alfalfa (*Medicago sativa*). Eur J Plant Pathol 117:57–69.

Muechlchen A, Rand R, Parke J 1990. Evaluation of crucifer green manures for controlling *Aphanomyces* root rot of peas. Plant Dis 74:651–654.

Nyamsuren O, Colditz F, Rosendahl S, Tamasloukht MB, Bekel T, Meyer F, Kuester H, Franken P, Krajinski F 2003. Transcriptional profiling of *Medicago truncatula* roots after infection with *Aphanomyces euteiches* (oomycota) identifies novel genes upregulated during this pathogenic interaction. Physiol Mol Plant Pathol 63:17–26.

Papavizas G, Ayers W 1974. *Aphanomyces* species and their root diseases on pea and sugarbeet. A review. US Dept Agricul Res Tech Bull 1485:1–158.

Petersen AB, Olson LW, Rosendahl S 1996. Use of polyclonal antibodies to detect oospores of *Aphanomyces*. Mycol Res 100:495–499.

Petersen AB, Rosendahl S 2000. Phylogeny of the Peronosporomycetes (Oomycota) based on partial sequences of the large ribosomal subunit (LSU rDNA). Mycol Res 104:1295–1303.

Pfender WF, Hagedorn DJ 1982. *Aphanomyces euteiches f. sp. phaseoli*, a causal agent of bean root rot and hypocotyl rot. Phytopathol 72:306–310.

Pilet-Nayel ML, Prospéri JM, Hamon C, Lesné A, Lecointe R, Le Goff I, Hervé M, Deniot G, Delalande M, Huguet T, Jacquet C, Baranger A 2009. AER1, a major gene conferring resistance to *Aphanomyces euteiches* in *Medicago truncatula*. Phytopathol 99:203–208.

Pilet-Nayel L, Muehlbauer FJ, McGee RJ, Kraft JM, Baranger A, Coyne CJ 2002. Quantitative trait loci for partial resistance to *Aphanomyces* root rot in pea. Theor Appl Genet 106:28–39.

Pilet-Nayel ML, Muehlbauer FJ, McGee RJ, Kraft JM, Baranger A, Coyne CJ 2005. Consistent quantitative trait loci in pea for partial resistance to *Aphanomyces euteiches* isolates from the United States and France. Phytopathol 95:1287–1293.

Riethmuller A, Weiss M, Oberwinkler F 1999. Phylogenetic studies of Saprolegniomycetidae and related groups based on nuclear large subunit ribosomal DNA sequences. Can J Bot 77:1790–1800.

Rosendahl S 1985. Interactions between the vesicular-arbuscular mycorrhizal fungus *Glomus fasciculatum* and *Aphanomyces euteiches* root rot of peas. Phytopathol 114:31–40.

Scott W 1961. A monograph of the genus *Aphanomyces*. Virginia Agricul Exp Sta Tech Bull 151.

Sekizaki H, Yokosawa R, Chinen C, Adachi H, Yamane Y 1993. Studies on zoospore attracting activity: synthesis of isoflavones and their attracting activity to *Aphanomyces euteiches* zoospore. Biol Pharm Bull 16:698–701.

Slezack S, Dumas-Gaudot E, Paynot M, Gianinazzi, S 2000. Is a fully established arbuscular mycorrhizal symbiosis required for a bioprotection of *Pisum sativum* roots against *Aphanomyces euteiches*? Mol Plant Microbe Interact 13:238–241.

Smolinska U, Knudsen G, Morra M, Borek V 1997. Inhibition of *Aphanomyces euteiches* f. sp. *pisi* by volatiles produced by hydrolysis of *Brassica napus* seed meal. Phytopathol 81:288–292.

Stones A, Valllad G, Cooperland L, Rotenberg D, James R, Stevenson W, Goodman R 2003. Effect of organic amendments on soilborne and foliar diseases in field-grown snap bean and cucumber. Plant Dis 87:1037–1042.

Tivoli B, Baranger A, Sivasithamparam K, Barbetti MJ 2006 Annual Medicago: from a model crop challenged by a spectrum of necrotrophic pathogens to a model plant to explore the nature of disease resistance. Ann Bot 98:1117–1128.

Torto-Alalibo T, Tian M, Gajendran K, Waugh M, van West P, Kamoun S 2005. Expressed sequence tags from the oomycete fish pathogen *Saprolegnia parasitica* reveal putative virulence factors. BMC Microbiol 2:5–46.

Tyler BM, Tripathy S, Zhang X, Dehal P, Jiang RH, Aerts A, Arredondo F D, Baxter L, Bensasson D, Beynon JL, et al. 2006. *Phytophthora* genome sequences uncover evolutionary origins and mechanisms of pathogenesis. Science 313:1261–1266.

Vandemark G, Grünwald NJ 2004. Reaction of *Medicago truncatula* to *Aphanomyces euteiches* race 2. Arch Phytopathol Plant Protect 37:59–67.

Vandemark G, Kraft JM, Larsen R, Gritsenko M, Boge W 2000. A PCR based assay by sequence-chracterized DNA markers for the identification and detection of *Aphanomyces euteiches*. Phytopathol 90:1137–1144.

Weiland JJ 2001. Regeneration of pathogenic *Aphanomyces cochlioides* and *A. euteiches* from protoplasts. J Sugarbeet Res 38:139–151.

Wicker E, Hulle M, Rouxel F 2001. Pathogenic characteristics of isolates of *Aphanomyces euteiches* from pea in France. Plant Pathol 50:433–442.

Wicker E, Moussart A, Duparque M, Rouxel F 2003. Further contributions to the development of a differential set of pea cultivars (*Pisum sativum*) to investigate the virulence of isolates of *Aphanomyces euteiches*. Eur J Plant Pathol 109:47–60.

Yokosawa R, Kuninaga S, Sekizaki H 1986. *Aphanomyces euteiches* zoospore attractant isolated from pea root: prunetin. Ann Phytopathol Soc Jap 52:809–816.

Note added in proof: During the processing of the manuscript, a study describing thoroughly the phylogenetic relationship within the *Aphanomyces* genus was published, that showed the occurence of clades with distinct pathogenic or saprophytic properties (Uriebondo et al., 2009. Fungal Genet Biol, in press)

18

EFFECTORS

Brett M. Tyler
Virginia Bioinformatics Institute, Virginia Polytechnic Institute and State University, Blacksburg, Virginia 24061, USA

18.1 INTRODUCTION

Like all pathogens, the success of oomycetes as pathogens depends on their ability to overcome the extensive defenses of their hosts, as well as to gain nutrition and proliferate. This ability in turn depends on specific physiological functions in the pathogens that in many cases have been finely tuned to specific hosts or groups of hosts. A major force in shaping the physiological repertoire of pathogens is a coevolutionary battle with the host (e.g., Allen et al., 2004; de Wit et al., 2002). An evolving pathogen places increasing selection pressure on its host as its virulence increases, which results in improvements in the host's resistance mechanisms. This in turn places selection pressure on the pathogen population to adapt or lose a host.

Both plant and animal hosts of oomycete pathogens have complex multifaceted defense systems. Thus, the pathogens must be able to suppress or avoid a sufficient number of these mechanisms to be successful. In plants, defense mechanisms include structural defenses, constitutive and induced chemical defenses, and programmed cell death (e.g., Heath and Boller, 2002; Jones and Dangl, 2006). Because plant cells are sessile, every plant cell is programmed with the ability to resistant pathogen attack. In animals, defense against infection is principally mediated by specialized motile cells, which include lymphocytes, phagocytic cells, and killer cells. In plants and lower animals, the defense mechanisms are principally innate [i.e., the ability to defend against a particular (strain of a) pathogen is inherited]. In fact, there are significant mechanistic similarities in the mechanism of innate immunity in plants and

Oomycete Genetics and Genomics: Diversity, Interactions, and Research Tools
Edited by Kurt Lamour and Sophien Kamoun
Copyright © 2009 John Wiley & Sons, Inc.

animals (Nurnberger et al., 2004). In higher animals (fish and higher vertebrates), adaptive immunity is present in which the ability to detect and respond to particular pathogens can be acquired as a result of genetic rearrangements within lineages of lymphocytes.

Effector molecules are a key weapon used by pathogens to combat their hosts' defense systems. For the purposes of this review, effectors are defined as any molecules released by the pathogen that have a function to modify the physiology of the host to promote the success of the pathogen, which includes to suppress or neutralize some component of the host defense system.

In addition to the suppression or neutralization of host defenses, a pathogen must gain nutrition from the host and must reproduce. Thus, effectors may be produced by the pathogen that promote its nutrition through the production and/or release of specific nutrients or else through the formation of specialized feeding structures, such as haustoria. Reproduction may be fostered by inducing the host to participate in formation of reproductive structures of the pathogen or by inducing a physiological state that supports reproduction (e.g., suppression of leaf senescence by some rust fungi).

Within the phylum oomycota, most plant pathogens occur within the class Peronosporomycetidae, whereas most animal pathogens occur within the class Saprolegniomycetidae. This suggests that plant pathogenicity may have developed initially in the ancestor of the Peronosporomycetidae, whereas animal pathogenicity may have developed initially in the ancestor of the Saprolegniomycetidae. Therefore, by comparing pathogenicity mechanisms (including repertoires of effectors) within each class, it is expected that mechanisms common to all pathogens in the class could be identified, as well as specialized mechanisms could be acquired more recently. There are a few animal pathogens in the Peronosporomycetidae, for example, *Pythium insidiosum* (other *Pythium* species are plant pathogens). Likewise, there are a few plant pathogens in the Saprolegniomycetidae, for example, *Aphanomyces euteiches* and *Aphanomyces cochlioides* (other *Aphanomyces* species are animal pathogens).

The sequencing of the genomes of several oomycete plant pathogens has led to a rapid advance in our understanding of the repertoire of effectors produced by oomycetes and the mechanisms the effectors use to promote pathogen success (Kamoun and Goodwin, 2007; Tyler et al., 2006). The recent description of a large collection of expressed sequence tag (EST) sequences from the Saprolegniomycete plant pathogen *A. euteiches* has enabled fascinating comparisons of the effector repertoire of plant pathogens from two major branches of the oomycetes (Gaulin et al., 2008). Some of this information has been reviewed in detail elsewhere (Birch et al., 2006, 2008; Kamoun, 2006; Morgan and Kamoun, 2007; Tyler, 2002). In this chapter, I will summarize this knowledge and provide a review of the most recent developments concerning oomycete effectors. This review will not cover molecules that trigger defense responses in plants but do not seem to contribute directly to pathogenicity, for example pathogen-associated molecular patterns (PAMPs) such as elicitins, transglutaminase, and cellulose-binding proteins. These have been reviewed elsewhere (Kamoun, 2006). Almost all of our knowledge of oomycete effectors comes from studies of plant pathogens, and so most information covered will be for this group of pathogens. Hopefully soon, information about oomycete animal pathogens will also become available.

18.2 AVIRULENCE GENES AND RXLR-dEER-FAMILY EFFECTORS IN OOMYCETE PLANT PATHOGENS

18.2.1 Avirulence Genes

Most major resistance genes in plants encode receptors capable of detecting the presence of effectors produced by pathogens from diverse taxa (e.g., Jones and Dangl, 2006). Many of these receptors consist of proteins with nucleotide binding sites (NBS) and leucine-rich repeats (LRR) (NBS-LRR proteins). These proteins have intracellular locations and are specialized for detection on intracellular effectors. Viral plant pathogens produce intracellular effectors as a natural consequence of their life style. Many bacterial pathogens inject effector proteins into plant cells via type III or type IV secretion systems, whereas nematode and insect pests inject effectors into cells via their feeding stylets.

Major resistance genes against oomycete pathogens have been described in many host plants (Tyler, 2002), particularly in potatoes and tomatoes against *Phytophthora infestans*, in soybeans against *Phytophthora sojae*, in *Arabidopsis thaliana* against *Hyaloperonospora arabidopsidis* (formerly *Hyaloperonospora parasitica*), and in lettuce against *Bremia lactucae*. Resistance genes that have been characterized and demonstrated to encode NBS-LRR proteins include the *Arabidopsis Rpp*1, *Rpp*2, *Rpp*4, *Rpp*5, *Rpp*7, *Rpp*8, and *Rpp*13 genes against *H. arabidopsidis* (Slusarenko and Schlaich, 2003); the lettuce *Dm*3, *Dm*14, and *Dm*16 resistance genes against *B. lactucae* (Shen et al., 2002; Wroblewski et al., 2007); the potato *R*1 (Ballvora et al., 2002), *Rb*, or *Rpi-Blb*1 (Song et al., 2003; van der Vossen et al., 2003), *Rpi-Blb*2 (van der Vossen et al., 2005); and *R*3a (Huang et al., 2005) genes against *P. infestans*; and the soybean *Rps*2 (Graham et al., 2002), *Rps*1k (Gao et al., 2005), *Rps*4, and *Rps*6 (Sandhu et al., 2004) genes against *P. sojae*.

The intracellular location of these resistance gene products implies that oomycetes must also introduce effectors into the cytoplasm of their hosts. One of the most exciting recent advances in understanding oomycete pathogenicity has been the identification of a large superfamily of effector proteins that includes many avirulence gene products (Birch et al., 2006; Jiang et al., 2008; Rehmany et al., 2005; Tyler et al., 2006). The members of this superfamily share the protein motif RXLR-dEER (arginine, any residue, leucine, arginine-aspartate (lower frequency), glutamate, glutamate, arginine) that is necessary and sufficient for the proteins to enter host cells (Dou et al., 2008b; Whisson et al., 2007).

The first four avirulence genes to be cloned from oomycete plant pathogens were the *Avr1b*-1 gene from *P. sojae* (Shan et al., 2004), the *Avr*3a gene from *P. infestans* (Armstrong et al., 2005), and the *ATR*13 (Allen et al., 2004) and *ATR*1 (Rehmany et al., 2005) genes from *H. arabidopsidis*. Many paralogs with some similarity to the four proteins were found in the *P. sojae* and *P. infestans* EST collections and in the *P. sojae* and *Phytophthora ramorum* genome sequences. A comparison of the four avirulence proteins to each other and to the paralogs led to the initial identification of the RXLR-dEER motifs (Birch et al., 2006; Rehmany et al., 2005; Tyler et al., 2006). The recognition of the RXLR-dEER motif as a signature for candidate avirulence genes has led to the rapid identification of many more avirulence genes from the genome sequences of *P. sojae* and *P. infestans* based on genetic mapping information and polymorphism analysis and functional screening. Table 18.1 shows a current list of avirulence genes that have been confirmed or for which very strong candidates have been identified.

TABLE 18.1 Cloned oomycete avirulence genes.

Species	Gene	Nature, length	Reference
P. sojae	*Avr1a*	RXLR-dEER 121 aa	Qutob et al., 2009
	Avr1b-1	RXLR-dEER 139 aa Also recognized by *Rps*1k	Shan et al., 2004
	Avr1k	RXLR-dEER 279 aa	D. Dou, S. Kale, B. Tyler, unpublished
	Avr3a	RXLR-dEER 111 aa	Qutob et al., 2009
	Avr3c	RXLR-dEER 220 aa	Dong et al., 2009
	Avr4/6	RXLR-dEER 123 aa Recognized by both *Rps*4 and *Rps*6	D. Dou, S. Kale, B. Tyler, unpublished
P. infestans	*Avr1*	RXLR-dEER 208 aa	F. Govers, pers. comm.
	Avr2	RXLR-dEER 116 aa	S. Whisson, pers. comm.
	Avr3a	RXLR-dEER 147 aa	Armstrong et al., 2005
	Avr3b/10/11	Amplified gene cluster	Jiang et al., 2006
	Avr4	RXLR-dEER 287 aa	Van Poppel et al., 2008
	AvrBlb1/ AvrSto1/ ipiO	RXLR-dEER 152 aa recognized by Blb1 and Sto1	Vleeshouwers et al., 2008; Pieterse et al., 1994
	AvrBlb2	RXLR (dEER?)[a] 100 aa	S. Kamoun, pers. comm.
H. arabidopsidis	*Atr1*	RXLR-dEER 311 aa	Rehmany et al., 2005
	Atr13	RXLR (dEER?)[a] 187 aa	Allen et al., 2004

[a] AvrBlb2 and Atr13 have barely recognizable dEER motifs; it is not yet known whether these are functional or replaced by a different motif.

18.2.2 Function of the RXLR and dEER Motifs

The first evidence to support the hypothesis that the RXLR-dEER motif was involved in effector entry into host cells came from the observation that effector proteins from the malaria parasite *Plasmodium* use a similar host targeting sequence (RxLx$^{E}/_{Q}$) to cross the membrane of the parasitiphorous vacuole (Hiller et al., 2004; Marti et al., 2004). The hypothesis received more support when it was shown that the RXLR-dEER region of *P. infestans* Avr3a could replace the native HTS of *Plasmodium* proteins (Bhattacharjee et al., 2006). Whisson et al. (2007) then showed that the RXLR and dEER sequences of Avr3a were required to trigger an avirulence reaction when *P. infestans* transformants that expressed *Avr3a* transgenes were inoculated onto potato plants containing *R*3a but not when the protein was expressed in the plant cells. Furthermore, the fusion of Avr3a to beta-glucuronidase resulted in transfer of beta-glucuronidase into potato cells during infection with *P. infestans* transformants that express the fusion protein. Dou et al. (2008b) similarly showed that the RXLR and dEER sequences of Avr1b were required to trigger an avirulence reaction when *P. sojae* transformants that express *Avr1b*-1 transgenes were inoculated onto soybean plants containing *Rps*1b but not when the protein was expressed in the plant cells following particle bombardment with the *Avr1b*-1 gene. Dou et al. (2008b) also showed that fusion of an N-terminal fragment of Avr1b that contained the RXLR-dEER motif to green fluorescent protein (GFP) was sufficient to enable the purified GFP fusion protein to enter soybean root cells. Entry of the fusion proteins into the root cells required both the RXLR and dEER motifs. Dou et al. (2008b) also showed that the RXLR and dEER motifs themselves were not sufficient for host cell entry, and that sequences surrounding the two motifs that could be described with a hidden Markov model (HMM) also were required. A similar observation was made for *Plasmodium* effectors (Bhattacharjee et al., 2006); in fact Dou et al. (2008b) showed that N-terminal fragments of three *Plasmodium* effectors could functionally replace the Avr1b N-terminus enabling entry into soybean cells. Both the *Plasmodium* results and the results of mutagenesis revealed that the arginine in the fourth position of the RXLR motif was not essential for function (Dou et al., 2008b). Additional mutagenesis showed that the first arginine could be replaced with a lysine albeit with reduced function, but there was a strict requirement for the leucine in the third position; replacing the leucine with valine or moving it to position four (swapping it with the arginine) abolished the function of the RXLR motif (Dou et al., 2008b). The precise sequence requirements from the dEER motif have not yet been defined but seem more variable than RXLR. In some effectors, such as Atr13 and AvrBlb2, the dEER motif can barely be discerned, which suggests that a different sequence is functioning in its place. Together, these results establish that the RXLR and dEER motifs, along with surrounding sequences, are necessary and sufficient for oomycete effectors to enter plant cells. Another important finding by Dou et al. (2008b) was that no additional pathogen machinery is required for protein entry. Avr1b proteins could re-enter soybean and onion cells following secretion in a particle bombardment assay. Even more strikingly, fusing the RXLR-dEER region of Avr1b to GFP enabled the purified GFP to enter soybean root cells in the absence of the pathogen (Dou et al., 2008b).

Little is known about the mechanism by which the RXLR-dEER domain enables effector proteins to cross the host plasma membrane. One possibility, which is modeled on existing proteins such as the HIV TAT protein and the Drosophila antennapedia protein that can cross membranes, is that entry is enabled by a physicochemical interaction between the basic and hydrophobic residues of the translocation domain with the negatively charged phospholipid membrane (Dou et al., 2008b). However, given that the RXLR motif is intolerant of many mutations that preserve hydrophobicity and charge (e.g., RFLR -> RFVR or RFLR -> FRLR in Avr1b, and RLLR -> KMIK in Avr3a) (Dou et al., 2008b; Whisson et al., 2007), it seems more likely that RXLR-dEER is binding a specific molecule on the plant cell surface. If that is the case, then the plant molecule must be so essential that it has not been lost as a result of selection pressure exerted by the pathogen's use of it. In malaria, the RxLx$^{E}/_{Q}$ motif seems to be required to concentrate the effectors at the cell surface within specialized endocytotic vesicles called Maurer's clefts (Bhattacharjee et al., 2008), which suggests that the effector may be entering via receptor-mediated endocytosis. Birch et al. (2008) noted that many plant proteins associated with the endocytic pathway contain sequences that resemble RXLR-dEER motifs. Although the authors do not describe whether sequences flanking the RXLR-dEER motifs were considered in the search, the association with the endocytic pathway is intriguing because many bacterial toxins (as well as whole bacteria, viruses, and parasites) enter animal cells by receptor-mediated endocytosis, followed by retrograde passage through the endocytic pathway (Gruenberg and van der Goot, 2006; Lafont et al., 2004; Lord et al., 1999; Medina-Kauwe, 2007). Furthermore, efficient functioning of the secretory pathway is essential for host defense (Lipka and Panstruga, 2005), so any interference with the pathway by pathogen effectors could promote infection.

18.2.3 Bioinformatic Identification of Genomic Repertoires of RXLR-dEER Effectors

The existence of the RXLR-dEER motif has enabled a systematic search of the genome sequences of oomycete pathogens for candidate effector proteins. Initial attempts to do this (Bhattacharjee et al., 2006; Win et al., 2007) used string searches focused on the RXLR and dEER motifs combined with bioinformatic prediction of secreted proteins. These searches identified very large numbers of potential effectors (Bhattacharjee et al., 2006; Win et al., 2007). However, Jiang et al. (2008) used sequence permutation to demonstrate that the simple string searches also had a high rate of false positives (around 50%). With the demonstration that flanking sequences are also an essential component of the translocation domain (Dou et al., 2008b), a refinement of this search strategy became possible. Jiang et al. (2008) used a Hidden Markov Model search combined with sequence permutation to minimize the number of false positives. In parallel, these authors used a recursive sequence similarity search (Blast) to identify candidates based on C-terminal as well as N-terminal sequence similarity. Combined, the two approaches identified 374 high-quality candidate effectors encoded in the *P. ramorum* genome and 396 in the *P. sojae* genome (Jiang et al., 2008). EST collections from the Saprolegniomycetes *Saprolegnia parasitica* (Torto-Alalibo et al., 2005) and *A. euteiches* (Gaulin et al., 2008) contain a very small number of sequences encoding for secreted proteins with RXLR-dEER-like motifs, but it is unknown yet whether these motifs are functional or whether the proteins are involved in pathogenicity.

Jiang et al. (2008) used the 770 candidate effectors to search for other conserved motifs in the effector superfamily and identified three C-terminal motifs named W, Y, and L after the most conserved amino acids in each motif. Approximately 60% of the effectors contained at least one W motif, and 30% also contained Y and L motifs. In effectors that contained W, Y, and L motifs, the three motifs were usually arranged close together in the order W-Y-L, comprising a module. In some cases such as Avr1b and Avr3a, only part of a module, W-Y was present (Fig. 18.1a). Some effectors had up to eight W-Y-L modules. Interestingly, a much higher percentage of *P. ramorum* effectors (33%) have two or more W-Y-L modules than in *P. sojae* (17%) (Fig. 18.1b). Win et al. (2007) described that there was a much higher percentage of effectors with closely related paralogs in *P. ramorum* than in *P. sojae*, and it is possible that the two phenomena are related.

Secondary structure predictions suggest that in many effectors, the residues spanning the K, W, Y, and L motifs would form a series of alpha helices, one each for the K and Y motifs and two each for the W and L motifs. Dou et al. (2008a) showed that the predicted alpha helices spanning the K and W motifs of Avr1b would be amphipathic, with the hydrophobic residues concentrated on one side of each helix and the hydrophilic residues concentrated on the other side. In the case of the two alpha helices spanning the W motif, the hydrophilic sides of each helix contained all the polymorphic residues, whereas the hydrophobic sides carried the conserved residues; this suggests that the hydrophilic sides form an exposed surface that might interact with the product of the *Rps*1b gene.

18.2.4 Evolution of RXLR-dEER Effector Genes

It is expected that the regions of the effector proteins that (presumably) interact with plant gene products will show adaptive evolution (positive selection) as a result of host–pathogen coevolutionary conflict. Evidence for strong positive selection was observed from comparisons of different alleles of Avr1b (Shan et al., 2004), ATR1 (Rehmany et al., 2005), and ATR13 (Allen et al., 2004),

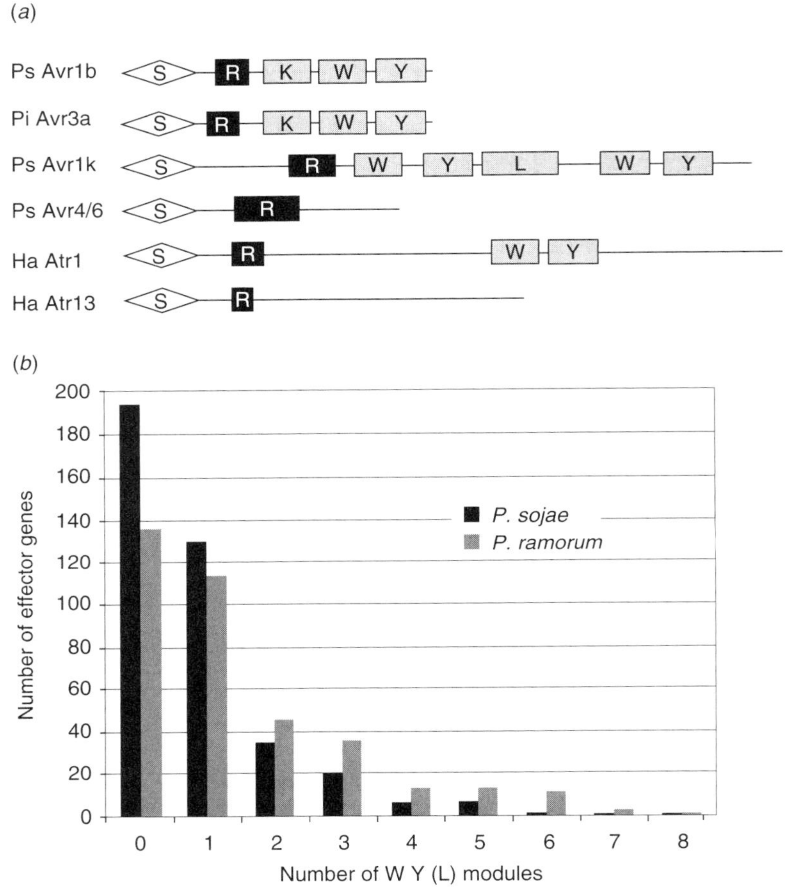

FIG. 18.1 Modular arrangement of conserved C-terminal motifs in RXLR-dEER effectors. (a) S = secretory leader, R = RXLR-dEER regions (Ha Atr13 has only an RXLR motif), K, W, Y and L = conserved C-terminal motifs (Dou et al., 2008b). (b) Distribution of numbers of W-Y or W-Y-L modules in *Phytophthora sojae* and *Phytophthora ramorum*, from data of Jiang et al. (2008).

which differed in their interactions with their cognate plant-resistance gene products. These three avirulence genes showed very extensive sequence polymorphism in their C-terminal regions, presumably because of selection pressure from the plant R genes. In *P. sojae* and *P. ramorum*, many families of closely similar paralogs of RXLR-dEER-effector genes show evidence for positive selection (Win et al., 2007). Presumably the force for the selection is caused by unrecognized *resistance* (*R*) genes or alterations in the targets of the effectors as a result of coevolutionary conflict between pathogen and plant gene products. The high rate of sequence polymorphisms in effector genes has been exploited to screen predicted secreted proteins directly for sequence polymorphisms (Bos et al., 2003).

Despite the high rate of evolution, Jiang et al. (2008) demonstrated that most RXLR-dEER effectors encoded in the *P. sojae* and *P. ramorum* genomes belonged to a single superfamily. Although the sequences were too diverged for conventional phylogenetic analysis, the similarities could be demonstrated by recursive BLAST similarity searches (an approach that involves several rounds of recurrent BLAST similarity searches). Most sequence similarities were located in the C-terminal regions of the proteins, which indicates that the family-wise similarities were not simply caused by the RXLR-dEER domain. The observation that the RXLR-dEER effectors in *P. sojae* and *P. ramorum* belong to single superfamily raises several important questions. For example, because *P. sojae* and *P. ramorum* are not closely related within the genus *Phytophthora*, it implies that the RXLR-dEER superfamily developed from a single ancestral gene very early in the evolution of the genus. In fact, because the superfamily is also represented in the *H. arabidopsidis* genome (Win et al., 2007), the implication is that the family originated very early in the evolution of plant pathogenic oomycetes. Another important question is whether all family members share the same function or whether functional diversification has occurred, with subfamilies acquiring new functions and potentially losing the original function. This question is especially important in the context of different species and genera of oomycetes, as well as their host specificity. For example, are there fundamental differences in the functions of effectors in broad host range species such as *P. ramorum* compared with *P. sojae*, or has the ability to infect host species in a particular taxonomic family developed because an effector subfamily has acquired a new function?

Related to the question of similarity of function is the question of how the effector repertoire is maintained by selection. If several hundred or even several tens of effector genes have the same function, then conventional evolutionary theory would predict that most genes will eventually be lost to mutational damage. It was hypothesized (Jiang et al., 2008; Tyler et al., 2008) that selection for increased virulence had resulted in expansion of the superfamily, and that pressure from the host had resulted in rapid diversification of the genes. The continuing process of expansion and diversification was proposed as a birth and death process in which damaged genes are continually replaced by an expanding pool of functional genes. Tyler et al. (2008) explored this idea by constructing a mathematical model that represented the process. The results of the modeling showed that the combined effects of gene duplication, mutation, and positive selection led to a rapid expansion of the effector gene pool until an equilibrium was reached in which loss to mutation and gain by gene duplication were balanced. The model showed that selection at the level of the pathogen individual would be sufficient to preserve those repertoires of effectors that contained the most functional genes and to eliminate those that contained the most damaged genes. Another prediction of the model is that oomycete genomes should contain large numbers of recent effector gene duplications as well as large numbers of pseudogenes. Both predictions are consistent with what is observed (Jiang et al., 2008; Tyler et al., 2006; Win et al., 2007; Qutob et al., 2009; Dong et al., 2009).

18.2.5 Functions of RXLR-dEER Effectors During Infection

The RXLR-dEER effectors are presumed to contribute to successful infection by the pathogen. Testing this hypothesis experimentally by means of gene knockouts or gene silencing is challenging because of the large size of the family and the presumptive redundancy of function. In fact, *P. sojae*, *P. infestans*, and *H. arabidopsidis* strains that escape *R* gene detection because they have lost the function of particular effector genes encoding avirulence proteins are still fully virulent. As an alternative approach, Dou et al. (2008a) showed that overexpression of *Avr1b*-1 in *P. sojae* transformants resulted in a small but significant increase in virulence (20% increase in lesion size). In obligate pathogens such as *H. arabidopsidis,* an additional difficulty is the lack of tools for genetic manipulation of the pathogen. Sohn et al. (2007) showed that expression of the *H. arabidopsidis Atr*1 and *Atr*13 genes in the bacterial pathogen *Pseudomonas syringae* and delivery of the encoded proteins into *Arabidopsis* cells via the type III secretion machinery enhanced the virulence of the bacteria, which increased the bacterial population about 30-fold by 4 days after infection.

The mechanisms by which RXLR-dEER effectors might promote virulence are beginning to emerge for several of the most well-characterized effectors. Bos et al. (2006a) showed that overexpression of Avr3a could suppress programmed cell death triggered by a PAMP, namely the elicitin INF1. Programmed cell death (PCD) is a very effective defense strategy against biotrophic pathogens such as *H. arabidopsidis,* and against hemibiotrophic pathogens such as *P. sojae* and *P. infestans*, which initiate infection biotrophically but later switch to a necrotrophic strategy once tissue colonization has occurred. PCD triggered as part of a successful defense response is generally referred to as the hypersensitive response (HR), although growing evidence indicates that several distinct signal transduction pathways can lead to cell death during infection (Beers and McDowell, 2001; Dangl et al., 1996; Hoeberichts and Woltering, 2002; Kanneganti et al., 2006; Qutob et al., 2006a; Williams and Dickman, 2008). Dou et al. (2008a) showed that *Avr1b*-1 expression in soybean and *Nicotiana benthamiana* cells also could suppress programmed cell death, in this case triggered by the proapoptotic mouse protein BAX. Many bacterial pathogenicity effectors share the ability to suppress BAX-triggered PCD (Jamir et al., 2004), which suggests that both bacterial and oomycete effectors target components of the signal transduction pathway that controls PCD at site(s) located downstream on the pathway from where BAX acts. Avr1b (Dou et al., 2008a) and also bacterial effectors (Jamir et al., 2004) can suppress PCD in the yeast *Saccharomyces cerevisiae*, which suggests that the targeted components are well conserved between fungi and plants. Interestingly, neither *P. infestans* Avr3a (Bos et al., 2006a) nor Avr1b (Kale et al., unpublished) could suppress

PCD triggered by the *Phytophthora* toxins NPP1 and CRN2 (see below), which suggests that the actions of the effectors have been refined over evolution to enable the pathogens to suppress PCD during biotrophic growth and to avoid suppressing PCD triggered by its own toxins, presumably during necrotrophic growth (Qutob et al., 2002).

Other PAMP-triggered defense responses include callose deposition and production of reactive oxygen species (ROS) (Jones and Dangl, 2006; Nurnberger et al., 2004). Sohn et al. (2007) demonstrated that expression of Atr13 in *P. syringae* reduced callose deposition by 100-fold during infection of *Arabidopsis* leaves. Callose deposition triggered by the bacterial PAMP flg2 was also reduced when Atr13 was expressed in transgenic *Arabidopsis* cells. The expression of Atr13 in *Arabidopsis* cells also significantly reduced production of ROS.

Some RXLR-dEER effectors may have acquired additional functions outside the cell. For example, the *P. infestans* effector ipiO (Pieterse et al., 1994) contains the cell adhesion motif RGD (Senchou et al., 2004). The role of RGD motifs is well characterized in mammalian systems, in which the plasma cell membrane is connected to the extracellular matrix via RGD motifs in integrin proteins in the matrix as well as RGD receptors in the membrane. In plants, synthetic RGD peptides can disrupt adhesion between the cell wall and the plasma cell membrane, and purified ipiO protein can cause similar disruptions (Senchou et al., 2004). Adhesion is important in a wide variety of physiological functions in plants including defense; overexpression of the proteins results in detachment of the plant plasma cell membrane from the cell wall. An RGD binding receptor (a lectin receptor kinase) that binds RGD peptides including ipiO has been identified in Arabidopsis (Gouget et al., 2006). The most similar paralog of ipiO in *P. sojae* contains the sequence RGE instead of RGD (Jiang et al., 2008), which is predicted not to have the same function. However in several other *P. sojae* effectors, RGD does overlap the RXLR motif (Jiang et al., 2008). Thus, expressing a certain number of RXLR effectors with an RGD motif may be important to *Phytophthora* infection. The importance of RGD effectors is underlined by the fact that ipiO is recognized as an avirulence protein by the broad spectrum *P. infestans* resistance gene *Rpi-Blb*1 as well as its functional homologs, such as *Rpi-Sto*1 (Song et al., 2003; van der Vossen et al., 2003; Vleeshouwers et al., 2008).

18.2.6 Structure Function Analysis of RXLR-dEER Effectors

Dou et al. (2008a) carried out a detailed mutational analysis of the C-terminal domain of Avr1b, focusing on the conserved W and Y motifs identified by Jiang et al. (2008) as well as an additional motif (the K motif) conserved in the effector subfamily that includes both Avr1b and Avr3a (Dou et al., 2008a). The results showed that both the W and Y motifs were required for the functional interaction with soybean resistance gene *Rps*1b that triggers HR, but the K motif was not (Dou et al., 2008a). The most extensive analysis was carried out on the W motif, which contains many residues that are polymorphic between different Avr1b alleles. Substitutions that altered either the conserved hydrophobic residues of the W motif or the polymorphic hydrophilic residues eliminated the functional interaction, which prevented Avr1b from triggering HR in plants that express *Rps*1b. Both the W and Y motif but not the K motif were also required for the ability of Avr1b to suppress PCD triggered by BAX. However, substitutions that altered the polymorphic hydrophilic residues did not affect suppression of BAX-triggered PCD, which identified a structural distinction between the two activities. More detailed mutational analysis identified an isoleucine residue at position 109 that was required for suppression of BAX-triggered PCD; replacement of this residue with a glycine or alanine abolished suppression activity but replacement with valine did not. The glycine substitution also abolished the functional interaction with *Rps*1b, but the alanine substitution did not, which identified a second structural distinction between the two activities of Avr1b (Dou et al., 2008a).

A structural analysis of Avr3a was carried out by Bos et al. (2006b). In that study, 5′ and 3′ deletions were used to show that the C-terminal 60 amino acid residues of Avr3a were sufficient for both the functional interaction with the *R*3a resistance gene product and the ability to suppress INF1-triggered PCD. The C-terminal 75 amino acid residues of Avr3a were fully sufficient for the functional interaction with the *R*3a resistance gene product, but the ability to suppress INF1-triggered PCD was partially reduced. The alleles of Avr3a from avirulent and virulent races of *P. infestans* were different at only two residues, positions 80 (lysine in avirulent and glutamate in virulent) and 103 (isoleucine in avirulent and methionine in virulent) (Bos et al., 2006b). In addition to the interaction with *R*3a, the KI allele could suppress INF1-triggered PCD. However, the EM allele could interact only weakly with *R*3a and could suppress INF1-triggered PCD only very weakly. Recombinants that contain the KM or EI combinations of substitutions showed intermediate interactions with *R*3a and ability to suppress INF1-triggered PCD, which suggests that the sites on the protein defined by these substitutions contribute equally to the two activities (Bos et al., 2006b).

18.3 POTENTIAL TOXIN GENES

Toxins may be considered to be proteins or other molecules released by the pathogen that trigger cell death in a way that benefits the pathogen, although we acknowledge that many molecules released by bacterial and fungal pathogens are called toxins even though they do not trigger cell death (e.g., HC toxin produced by *Cochliobolus carbonum*) (Wolpert et al., 2002). There is no evidence so far that oomycete plant pathogens produce small-molecular-weight toxins like the ones produced by fungal plant pathogens via the actions of polyketide synthases and nonribosomal peptide synthases. In fact, no detectable polyketide synthase genes exist in the *Phytophthora* genome or

EST sequences at all and only four genes exist that encode nonribosomal peptide synthase genes (Randall et al., 2005; Tyler et al., 2006). However, as summarized below, many proteins have been described that trigger cell death in a manner that could potentially benefit necrotrophic growth and hence are potential toxins. So far, a direct positive contribution to virulence has not been demonstrated for any of these potential toxin proteins.

18.3.1 NLP Proteins

The NEP1-like protein (NLP) superfamily shares substantial sequence similarity to the first discovered member of the family, which is the Necrosis and Ethylene-inducing protein of *Fusarium oxysporum* f.sp. *erythroloxyli* (Pemberton and Salmond, 2004; Qutob et al., 2006a). This protein is found not only in oomycetes but also in many bacterial species and in many fungi (Pemberton and Salmond, 2004; Qutob et al., 2006a). However, whereas bacterial and fungal species contain only 1–4 NLP genes in their genomes, the oomycete genomes contain very large numbers of rapidly diversifying NLP genes (Tyler et al., 2006). The *P. sojae* genome contains around 29 NLP genes, and *P. ramorum* genome contains around 40 NLP genes (Qutob et al., 2006b; Tyler et al., 2006). Furthermore, phylogenetic evidence indicated that only 7 genes were orthologous, and there were several large clades that were specific to each species (Qutob et al., 2006b; Tyler et al., 2006). As discussed in the previous section on RXLR-dEER effectors, the rapid amplification and divergence of the NLP family suggests that it has a direct role in the interaction with the plant host. The presence of many NLP pseudogenes in the genomes suggests that, like the RXLR-dEER family, the genes are evolving via a birth-and-death process (Tyler et al., 2006).

Functional evidence does not yet exist to suggest that the NLP proteins play a role in *Phytophthora* virulence. However, evidence from several bacterial and fungal systems supports this hypothesis (reviewed in Qutob et al., 2006a). Disruption of the gene in *Erwinia carotovora* subsp. *carotovora* and in *Erwinia carotovora* subsp. *atroseptica* reduced symptoms during infection of potato tubers and stems, respectively (Mattinen et al., 2004; Pemberton et al., 2005). The timing of NLP gene expression during *P. sojae* infection, around the switch from biotrophy to necrotrophy, is also consistent with the hypothesis (Qutob et al., 2002). Another interesting correlation is that the cell killing was triggered only in dicotyledonous plants (Qutob et al., 2006b), which is also the range of plants to which most *Phytophthora* pathogens are confined.

There are several unusual aspects to the phylogenetic distribution of NLP proteins. First, as mentioned, they are found in fungi and bacteria as well as oomycetes. Second, only a very small number of bacteria have the genes, and those genes have disparate phylogenetic affinities (Qutob et al., 2006b), which suggests they may have acquired them from eukaryotes by horizontal gene transfer. Third, although NLPs do not cause cell killing in grasses, the rice blast pathogen *Magnaporthe grisea* and the wheat scab pathogen *Fusarium graminearum* both have NLP genes. More curiously, the genes are found not only in necrotrophic and hemi-biotrophic pathogens but also in saprophytic non-pathogens such as *Streptomyces coelicolor*, *Bacillus halodurans*, *Bacillus licheniformis*, *Neurospora crassa*, and *Aspergillus nidulans* and also in the obligate biotroph *H. arabidopsidis*. This distribution suggests that NLPs may have a physiological function within the pathogen that is different than their role in plant infection. In contrast, NLP sequences were not found among a large collection of ESTs from the Saprolegniomycete *A. euteiches* (Gaulin et al., 2008), which suggests that NLPs are not essential for life or for pathogenicity by oomycetes.

The response of *Arabidopsis* plants to NLPs includes many components of innate immunity, which include phytoalexin and ethylene production, and many transcriptional responses in common with responses to PAMPS such as flagellin (Qutob et al., 2006a). However, cell death did not depend on any of the signaling pathways that normally regulate defense-related cell death, namely salicylate, ethylene, and jasmonate (Qutob et al., 2006a). It also did not involve caspases or Sgt1b (Qutob et al., 2006a), although Kanneganti et al. (2006) showed that in *N. benthamiana* the cell death induced by the NLPs is dependent on both SGT1 and HSP90. One possible explanation is that plants are responding to the tissue damage caused by NLPs rather than to NLPs themselves, and that the tissue damage response overlaps with the defense response. An alternative explanation is that, as discussed above, NLPs have or had a noninfection role in the pathogens and that plants evolved to recognize and respond to NLPs as PAMPs; *Phytophthora* species and other necrotrophic pathogens such as *Erwinia* may then have counterevolved to amplify and distort the plants' responses to NLPs to the detriment of the plant and the advantage of the pathogen.

18.3.2 PcF/Scr Toxin Family

The toxin PcF was originally identified as a component of culture filtrates from *Phytophthora cactorum* that caused cell death on strawberries (Orsomando et al., 2001). PcF is a small secreted hydroxyproline-containing protein with three disulfide bridges (Orsomando et al., 2001). Genome sequencing of *P. sojae* and *P. ramorum* (Tyler et al., 2006) together with EST sequencing of *P. infestans* (Randall et al., 2005) revealed a very heterogeneous evolution of gene families that encode PcF-like proteins. There were two genes for PcF-like proteins in *P. sojae*, four in *P. ramorum*, and five distinct unigenes (called Scr91) (Bos et al., 2003) in *P. infestans*. Furthermore, a related family of proteins with four disulfide bridges, which was termed the Scr74 family, was identified (Liu et al., 2005). This family was represented by completely nonoverlapping clades in *P. sojae* and *P. infestans* of 14 and 11 members, respectively, and it was completely absent from *P. ramorum*, which indicates distinct evolutionary histories in the three species (Liu et al., 2005). In addition, within *P. infestans*, the Scr74 family was shown to be continuing to evolve under strong diversifying selection, subsequent to speciation, which is consistent with a major ongoing role on virulence (Liu et al., 2005).

18.3.3 Crinkling and Necrosis-Inducing Proteins

Crinkling and necrosis-inducing proteins (crinklers; CRN proteins) were initially identified from a functional genomics screen of predicted secreted proteins encoded in the ESTs of *P. infestans* (Torto et al., 2003). The expression of cDNA clones that encode two related sequences (CRN1 and CRN2) in an *Agrobacterium*-mediated transient assay triggered necrosis at the site of inoculation as well as necrosis and crinkling of systemic leaves. The symptoms occurred both on the host plants, *N. benthamiana* and tomato, and also on non-host plants (tobacco) (Torto et al., 2003). Subsequently, very large complex superfamilies of sequences related to CRN1 and CRN2 were identified in the genome sequences and EST collections of *P. sojae*, *P. ramorum*, and *P. infestans* (Torto, 2003; Torto et al., 2003; Tyler et al., 2006; Win et al., 2007). CRN proteins were also found encoded in EST sequences of *A. euteiches* (Gaulin et al., 2008) and even in the genome sequence of the obligate biotroph *H. arabidopsidis* (Win et al., 2007). Expansion of the CRN superfamily was particularly striking in *P. sojae*, which had 40 *crn* genes compared with 8 in *P. ramorum*; this finding suggests that *crn* genes had acquired a special significance in soybean infection (Tyler et al., 2006). In *P. infestans*, 16 different CRN proteins could be identified from cDNA sequences (Win et al., 2007), but several additional families occur (see chapter by Kamoun et al.).

An intriguing characteristic of CRN proteins is that synthesis of the proteins inside the cytoplasm of *N. benthamiana* cells (Kamoun, 2006) and soybean cells (Gu, et al., unpublished) is sufficient to trigger a response. This implies that, like RXLR-dEER effectors, CRN proteins interact with their plant targets in the host cell cytoplasm and have a mechanism for crossing the host plasma membrane. CRN proteins from *P. sojae*, *P. ramorum*, and *P. infestans* do not have RXLR-dEER motifs but instead display the conserved motif LxLFLAK (Win et al., 2007). In contrast, many CRN proteins in *H. arabidopsidis* do have RXLR motifs, and these overlap with the LxLFLAK motif (e.g., RKLRLFLAK) to suggest a functional connection between the two motifs, related to host cell entry (Win et al., 2007). In *A. euteiches*, CRN sequences display another variant of the LxLFLAK motif, namely $^{F}/_{L}$xLYLALK (Gaulin et al., 2008). Thus it remains to be determined which portions of these motifs are actually responsible for cellular entry and whether any portions of them have other functions once the protein enters the cell.

The actual functions of CRN proteins during infection are unknown. CRN proteins seems to affect a wide range of plants, including so far *Nicotiana*, tomatoes (Torto et al., 2003), and soybean (Gu et al., unpublished). The presence of CRN proteins in an obligate biotroph *H. arabidopsidis* suggests that the proteins have a role other than causing cell death. Furthermore, all assays of CRN proteins that result in cell death and crinkling require strong over-expression, either in an *Agrobacterium*-mediated transient assay or in a bombardment assay. This finding raises the question of whether cell death is the normal function of these effectors.

18.4 HYDROLASES AND HYDROLASE INHIBITORS

Like many other plant pathogens, oomycetes secrete a broad array of hydrolytic enzymes, such as glucanases, pectinases, cellulases, and proteinases, as a strategy for acquiring nutrition and as a strategy for suppressing the plants' defense responses. The genome sequences of *P. sojae* and *P. ramorum* encode large numbers of extracellular proteases, glycosyl hydrolases, pectinases, cutinases, and lipases (Tyler et al., 2006). In all, 20–40% of the genes in each class lack identifiable orthologs in comparisons between the two genomes, which suggests that these genes also are evolving rapidly (Tyler et al., 2006). Concomitantly, the plant secretes its own glucanases and cellulases aimed at damaging the cell wall of the pathogen and secretes proteinases aimed at neutralizing the pathogens arsenal of proteinaceous effectors. As a result, the genomes of oomycete species also encode large numbers of hydrolase inhibitors (Tyler et al., 2006).

18.4.1 Proteinase Inhibitors

Numerous Kazal-class protease inhibitors were found in EST and genome survey sequences from *P. infestans, P. brassicae*, and *Plasmopara halstedii* (Tian et al., 2004). Furthermore, the genomes of *P. sojae* and *P. ramorum* contain 15 and 12 copies of genes, respectively, to encode serine protease inhibitors in Kazal family (Tyler et al., 2006). Of the genes found in *P. sojae* and *P. ramorum*, only eight were orthologous, which suggests that these gene families have been diversifying rapidly since speciation, for example because of pressure from interaction with plant proteases (Tyler et al., 2006). Consistent with this inference, Tian et al. (2004) found that many Kazal inhibitors in *Phytophthora* species were encoded by transcripts expressed during infection. In *P. infestans*, two particular inhibitors EPI1 and EPI10 bound to and inhibited the tomato protease P69B (Tian et al., 2004, 2005; Tian and Kamoun, 2005). P69B is strongly induced during defense, and EPI1 and EPI10 are expressed during infection, so it is likely that the interaction between the inhibitors and the protease is of direct physiological relevance (Tian et al., 2004, 2005; Tian and Kamoun, 2005). This hypothesis is supported by the observation that predicted protease contact residues in the EPI proteins show evidence of diversifying selection (Liu, 2003). It is not yet known whether specific *Phytophthora* virulence proteins are targeted by P69B and protected by EPI1 and EPI10. Kazal family protease inhibitors are also used by apicomplexan parasites, such as *Toxoplasma gondii*, which illustrates convergent evolution in these two eukaryotic pathogens (Kamoun, 2006).

The *P. sojae* and *P. ramorum* genomes each contain four genes, which include three orthologous pairs, that encode cystatin-like cysteine protease inhibitors (Tyler et al., 2006). *P. infestans* contains genes orthologous to two of these, but it also contains genes for a rapidly evolving family *EPIC1/2* that is not found in *P. sojae* or *P. ramorum* (Tian et al., 2007). *S. parasitica* and *A. euteiches* ESTs also include sequences with similarity to the *Phytophthora* cysteine protease inhibitors (Torto-Alalibo et al., 2005; Gaulin et al., 2008). In *P. infestans*, *EPIC1* and *EPIC2* genes are strongly transcribed during infection, and abundant levels of EPIC1 protein could be detected in the tomato apoplast during infection (Tian et al., 2007). A tomato cysteine protease PIP1 was identified as binding to EPIC2 and being inhibited by it (Tian et al., 2007). PIP1 is induced during infection and by salicylic acid, thus it is a defense protein. PIP1 is closely related to the tomato cysteine proteins Rcr3, which is involved in detection of a proteinase inhibitor Avr2 from the fungal pathogen *Cladosporium fulvum* by the tomato resistance gene product Cf2. Interestingly, Avr2 inhibits both PIP1 and Rcr3, which indicates that PIP1 is targeted by both oomycete and fungal pathogens (Shabab et al., 2008). Presumably as a result of targeting by these (and probably other) pathogens, PIP1 and Rcr3 are evolving rapidly under strong divergent selection (Shabab et al., 2008).

18.4.2 Glucanase Inhibitor Proteins

Plant defense responses include the production of secreted beta-1,3 glucanases that can damage fungal and oomycete cell walls directly and/or the release oligosaccharide cell wall fragments that act as elicitors of additional defense responses. Glucanase inhibitor proteins (GIPs) produced by *P. sojae* can inhibit the release of elicitor-active oligosaccharides from *P. sojae* cell walls *in vitro* (Rose et al., 2002, and references therein). Three soybean GIP sequences were cloned (*GIP*1, *GIP*2, and *GIP*3), and the protein encoded by *GIP*1 was shown to bind stably to and inhibit the soybean endoglucanase EGaseA (Rose et al., 2002). GIP1 closely resembles a serine protease, but it lacks the catalytic triad necessary for protease activity (Rose et al., 2002). There are at least three *GIP* genes in *P. sojae*, one each encoding *GIP*1, *GIP*2, and *GIP*3. In several cases, there seem to be multiple identical copies of the genes in the draft sequence assembly (Damasceno et al., 2008). There are also at least three other genes [designated serine protease homolog; (*SPH*)], also possibly present in multiple copies, that encode closely similar proteins that lack the catalytic triad and so might be GIPs (Damasceno et al., 2008). Three additional diverse *SPH*s were detected in the genome sequence (Damasceno et al., 2008). Homologs of the *P. sojae* sequences were found in the *P. ramorum* and *P. infestans* genome sequences and ESTs (Damasceno et al., 2008). Four *P. infestans GIP*s (*PiGIP*1 through *PiGIP*4), were confirmed to be expressed during infection; *PiGIP*1 and *PiGIP*3 were expressed only during infection (Damasceno et al., 2008). A comparison of GIP and SPH sequences from the three species identified amino acid residues that were under positive selection (Damasceno et al., 2008). Based on three-dimensional structure predictions for PsGIP1 and soybean 1,3 endoglucanase, the residues under positive selection were predicted to correspond to residues on PsGIP1 that contact endoglucanase (Damasceno et al., 2008). A similar comparison of glucanase sequences suggested that the residues predicted to contact PsGIP were also under positive selection. The combined results from the *P. sojae*–soybean and *P. infestans*–tomato pathosystems suggest that the interaction of pathogen GIPs with host glucanases is an important determinant of pathogen success (Damasceno et al., 2008).

18.5 SUMMARY AND FUTURE

The genomics resources that have been developed for oomycetes over the last 5–10 years, especially the genome sequences, have resulted in a dramatic acceleration in our understanding of the evolution and mechanisms of pathogenicity in these organisms, especially the repertoires of effectors. However, important gaps remain. For example, a relatively narrow sample of oomycete pathogens have been sequenced. Very little is known of the genomic repertoire of important broad host range oomycete plant pathogens, such as *Phytophthora cinnamomi* and *Phytophthora parasitica*; the diverse downy mildews and white rusts such as *B. lactucae, Plasmopara viticola*, and *Albugo candida*; and oomycete pathogens of animals such as *Pythium insidiosum, S. parasitica*, and *Aphanomyces invadans*. Nothing at all is known of other plant pathogens in the kingdom Stramenopila, including *Hyphochytriomycetes* and *Labyrinthulids*.

Our understanding of exactly how the immense effector repertoires of oomycetes contribute to infection is just the beginning (Fig. 18.2). For example, the very large size of the RXLR-dEER superfamily, the very high rate of divergence, and the fact that it contains almost all avirulence genes discovered so far point to this family as being the key to the pathogenicity of oomycete phytopathogens. But how the role of this family is integrated with other large families such as the crinklers and NLPs is still not understood, particularly with regard to biotrophic versus necrotrophic lifestyles. Key questions are as follows: What are the precise mechanisms of entry into host cells by RXLR-dEER and crinkler effectors? Is effector translocation restricted to haustoria?

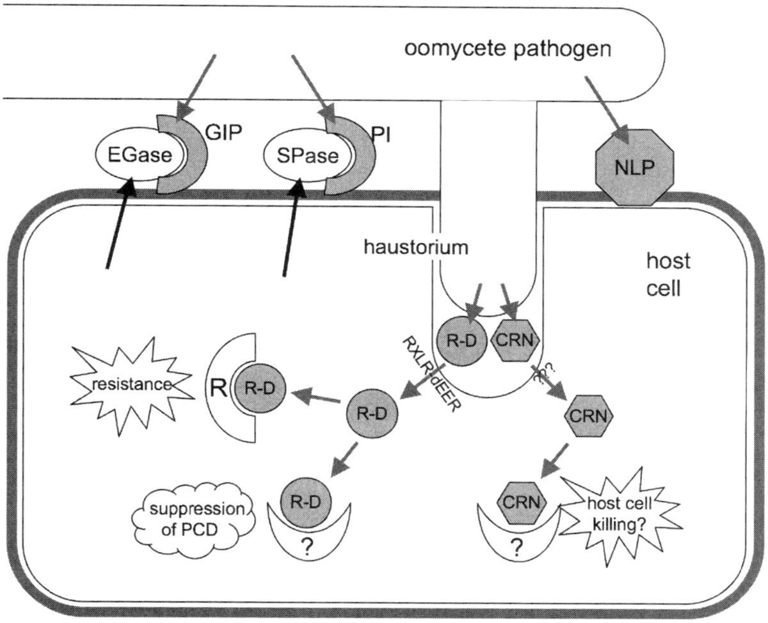

FIG. 18.2 Some extracellular and intracellular oomycete effectors. EGase = extracellular glucanase, GIP = glucanase inhibitor protein, SPase = secreted protease, PI = proteinase inhibitor, NLP = NEP-like protein, R-D = RXLR-dEER effector, CRN = crinkler protein, R = plant resistance protein. The plant targets of RXLR-dEER effectors and crinkler proteins are currently unknown, as is the mechanism of crinkler protein entry.

What is the range of functions affected by RXLR-dEER proteins — is it restricted to inhibition of defense signaling, or are transcriptional and developmental programs affected also? Do individual proteins have more than one function? What is the role of the effectors in host specificity? Do necrotrophs such as *Pythium* species and oomycete pathogens of animals use RXLR-dEER effectors, or do toxins play a more important role? A general issue with regard to determining the contributions of effector proteins to virulence is the question of the physiological concentration of effector proteins inside and outside host cells.

ACKNOWLEDGMENTS

We thank Teresa Jewell and Lisa Gunderman for manuscript assistance and F. Govers, S. Kamoun, S. Whisson, and M. Gijzen for communicating unpublished data. This work was supported by grants from the National Research Initiative of the USDA Cooperative State Research, Education and Extension

Service, grants, 2002-35600-12747, 2007-35600-18530, and 2007-35319-18100, and from the U.S. National Science Foundation, grants MCB-0242131 and MCB-0731969.

REFERENCES

Allen RL, Bittner-Eddy PD, Grenville-Briggs LJ, Meitz JC, Rehmany AP, Rose LE, Beynon JL 2004. Host-parasite coevolutionary conflict between *Arabidopsis* and downy mildew. Science 306:1957–1960.

Armstrong MR, Whisson SC, Pritchard L, Bos JI, Venter E, Avrova AO, Rehmany AP, Bohme U, Brooks K, Cherevach I, et al. 2005. An ancestral oomycete locus contains late blight avirulence gene *Avr3a*, encoding a protein that is recognized in the host cytoplasm. Proc Natl Acad Sci USA 102:7766–7771.

Ballvora A, Ercolano MR, Weiss J, Meksem K, Bormann CA, Oberhagemann P, Salamini F, Gebhardt C 2002. The R1 gene for potato resistance to late blight (*Phytophthora infestans*) belongs to the leucine zipper/NBS/LRR class of plant resistance genes. Plant J 30:361–371.

Beers EP, McDowell JM 2001. Regulation and execution of programmed cell death in response to pathogens, stress and developmental cues. Curr Opin Plant Biol:561–567.

Bhattacharjee S, Hiller NL, Liolios K, Win J, Kanneganti TD, Young C, Kamoun S, Haldar K 2006. The malarial host-targeting signal is conserved in the Irish potato famine pathogen. PLoS Pathog 2:e50.

Bhattacharjee S, van Ooij C, Balu B, Adams JH, Haldar K 2008. Maurer's clefts of *Plasmodium falciparum* are secretory organelles that concentrate virulence protein reporters for delivery to the host erythrocyte. Blood 111:2418–2426.

Birch PR, Boevink PC, Gilroy EM, Hein I, Pritchard L, Whisson SC 2008. Oomycete RXLR effectors: delivery, functional redundancy and durable disease resistance. Curr Opin Plant Biol 11:373–379.

Birch PR, Rehmany AP, Pritchard L, Kamoun S, Beynon JL 2006. Trafficking arms: oomycete effectors enter host plant cells. Trends Microbiol 14:8–11.

Bos JI, Kanneganti TD, Young C, Cakir C, Huitema E, Win J, Armstrong MR, Birch PR, Kamoun S 2006. The C-terminal half of *Phytophthora infestans* RXLR effector AVR3a is sufficient to trigger *R3a*-mediated hypersensitivity and suppress INF1-induced cell death in *Nicotiana benthamiana*. Plant J 48:165–176.

Bos JIB, Armstrong M, Whisson SC, Torto T, Ochwo M, Birch PRJ, Kamoun S 2003. Intraspecific comparative genomics to identify avirulence genes from *Phytophthora*. New Phytologist 159:63–72.

Damasceno CMB, Bishop JG, Ripoll RD, Win J, Kamoun S, Rose JKC 2008. Structure of the glucanase inhibitor protein (GIP) family from *Phytophthora* species suggests coevolution with plant endo-beta-1,3-glucanases. Molec Plant-Microbe Interact 21:820–830.

Dangl JL, Dietrich RA Richberg Ml 1996. Death don't have no mercy: cell death programs in plant-microbe interactions. Plant Cell 8:1793–1807.

de Wit PJ, Brandwagt BF, van den Burg HA, Cai X, van der Hoorn RA, de Jong CF, van Klooster J, de Kock MJ, Kruijt M, Lindhout WH, et al. 2002. The molecular basis of co-evolution between *Cladosporium fulvum* and tomato. Antonie Van Leeuwenhoek 81:409–412.

Dong S, Qutob D, Tedman-Jones J, Kuflu K, Wang Y, Tyler BM, Gijzen M 2009. The *Phytophthora sojae* avirulence locus Avr3c encodes a multi-copy RXLR effector with sequence polymorphisms among different pathogen strains. PLOS One: in press.

Dou D, Kale SD, Wang X, Chen Y, Wang Q, Wang X, Jiang RHY, Arredondo FD, Anderson R, Thakur P, et al. 2008a. Carboxy-terminal motifs common to many oomycete RXLR effectors are required for avirulence and suppression of BAX-mediated programmed cell death by *Phytophthora sojae* effector Avr1b. Plant Cell 20:1118–1133.

Dou D, Kale SD, Wang X, Jiang RHY, Bruce NA, Arredondo FD, Zhang X, Tyler BM 2008b. RXLR-mediated entry of *Phytophthora sojae* effector Avr1b into soybean cells does not require pathogen-encoded machinery. Plant Cell 20: 1930–1947.

Gao H, Narayanan NN, Ellison L, Bhattacharyya MK 2005. Two classes of highly similar coiled coil-nucleotide binding-leucine rich repeat genes isolated from the *Rps*1-k locus encode *Phytophthora* resistance in soybean. Mol Plant Microbe Interact 18:1035–1045.

Gaulin E, Madoui MA, Bottin A, Jacquet C, Mathe C, Couloux A, Wincker P, Dumas B 2008. Transcriptome of *Aphanomyces euteiches*: new oomycete putative pathogenicity factors and metabolic pathways. PLoS ONE 3:e1723.

Gouget A, Senchou V, Govers F, Sanson A, Barre A, Rouge P, Pont-Lezica R, Canut H 2006. Lectin receptor kinases participate in protein–protein interactions to mediate plasma membrane-cell wall adhesions in *Arabidopsis*. Plant Physiol 140:81–90.

Graham MA, Marek LF, Shoemaker RC 2002. Organization, expression and evolution of a disease resistance gene cluster in soybean. genetics 162:1961–1977.

Gruenberg J, van der Goot FG 2006. Mechanisms of pathogen entry through the endosomal compartments. Nat Rev Mol Cell Biol 7:495–504.

Heath MC, Boller T 2002. Editorial overview: levels of complexity in plant interactions with herbivores, pathogens and mutualists. Curr Opin in Plant Biol 5:277–278.

Hiller NL, Bhattacharjee S, van Ooij C, Liolios K, Harrison T, Lopez-Estrano C, Haldar K 2004. A host-targeting signal in virulence proteins reveals a secretome in malarial infection. Science 306:1934–7.

Hoeberichts FA, Woltering EJ 2002. Multiple mediators of plant programmed cell death: interplay of conserved cell death mechanisms and plant-specific regulators. BioEssays 25:47–57.

Huang S, van der Vossen EA, Kuang H, Vleeshouwers VG, Zhang N, Borm TJ, van Eck HJ, Baker B, Jacobsen E, Visser RG 2005. Comparative genomics enabled the isolation of the *R3a* late blight resistance gene in potato. Plant J 42:251–261.

Jamir Y, Guo M, Oh HS, Petnicki-Ocwieja T, Chen S, Tang X, Dickman MB, Collmer A, Alfano JR 2004. Identification of *Pseudomonas syringae* type III effectors that can suppress programmed cell death in plants and yeast. Plant J 37:554–565.

Jiang RHY, Tripathy S, Govers F, Tyler BM 2008. RXLR effector reservoir in two *Phytophthora* species is dominated by a single rapidly evolving super-family with more than 700 members. Proc Nat Acad Sci USA 105:4874–4879.

Jiang RHY, Weide R, van de Vondervoort PJI, Govers F 2006. Amplification generates modular diversity at an avirulence locus in the pathogen *Phytophthora*. Genome Research 16:827–840.

Jones JD, Dangl JL 2006. The plant immune system. Nature 444:323–329.

Kamoun S 2006. A catalogue of the effector secretome of plant pathogenic oomycetes. Annu Rev Phytopathol 44:41–60.

Kamoun S, Goodwin SB 2007. Fungal and oomycete genes galore. New Phytol 174: 713–717.

Kanneganti TD, Huitema E, Cakir C, Kamoun S 2006. Synergistic interactions of the plant cell death pathways induced by *Phytophthora infestans* Nepl-like protein PiNPP1.1 and INF1 elicitin. Mol Plant Microbe Interact 19:854–863.

Lafont F, Abrami L, van der Goot FG 2004. Bacterial subversion of lipid rafts. Curr Opin Microbiol 7:4–10.

Lipka V, Panstruga R 2005. Dynamic cellular responses in plant-microbe interactions. Curr Opin Plant Biol 8:625–631.

Liu Z, Bos JI, Armstrong M, Whisson SC, da Cunha L, Torto-Alalibo T, Win J, Avrova AO, Wright F, Birch PR, et al. 2005. Patterns of diversifying selection in the phytotoxin-like scr74 gene family of *Phytophthora infestans*. Mol Biol Evol 22: 659–672.

Lord JM, Smith DC, Roberts LM 1999. Toxin entry: how bacterial proteins get into mammalian cells. Cell Microbiol 1:85–91.

Marti M, Good RT, Rug M, Knuepfer E, Cowman AF 2004. Targeting malaria virulence and remodeling proteins to the host erythrocyte. Science 306:1930–1933.

Mattinen L, Tshuikina M, Mae A, Pirhonen M 2004. Identification and characterization of Nip, necrosis-inducing virulence protein of *Erwinia carotovora* subsp. *carotovora*. Mol Plant Microbe Interact 17:1366–1375.

Medina-Kauwe LK 2007. "Alternative" endocytic mechanisms exploited by pathogens: new avenues for therapeutic delivery? Adv Drug Deliv Rev 59:798–809.

Morgan W, Kamoun S 2007. RXLR effectors of plant pathogenic oomycetes. Curr Opin Microbiol 10:332–338.

Nurnberger T, Brunner F, Kemmerling B, Piater L 2004. Innate immunity in plants and animals: striking similarities and obvious differences. Immunol Rev 198:249–266.

Orsomando G, Lorenzi M, Raffaelli N, Dalla Rizza M, Mezzetti B, Ruggieri S 2001. Phytotoxic protein PcF, purification, characterization, and cDNA sequencing of a novel hydroxyproline-containing factor secreted by the strawberry pathogen *Phytophthora cactorum*. J Biol Chem 276:21578–21584.

Pemberton CL, Salmond GPC 2004. The Nep1-like proteins—a growing family of microbial elicitors of plant necrosis. Mol Plant Pathol 5:353–359.

Pemberton CL, Whitehead NA, Sebaihia M, Bell KS, Hyman LJ, Harris SJ, Matlin AJ, Robson ND, Birch PR, Carr JP, et al. 2005. Novel quorum-sensing-controlled genes in *Erwinia carotovora* subsp. *carotovora*: identification of a fungal elicitor homologue in a soft-rotting bacterium. Mol Plant Microbe Interact 18:343–353.

Pieterse CMJ, Van West P, Verbakel HM, Brasse PWHM, Van Den Berg-Velthuis GCM, Govers F 1994. Structure and genomic organization of the ipiB and ipiO gene clusters of *Phytophthora infestans*. Gene (Amsterdam) 138(1–2):67–77.

Qutob D, Kamoun S, Gijzen M 2002. Expression of a *Phytophthora sojae* necrosis-inducing protein occurs during transition from biotrophy to necrotrophy. Plant J 32:361–373.

Qutob D, Kemmerling B, Brunner F, Kufner I, Engelhardt S, Gust AA, Luberacki B, Seitz HU, Stahl D, Rauhut T, et al. 2006a. Phytotoxicity and innate immune responses induced by Nep1-like proteins. Plant Cell 18:3721–3744.

Qutob D, Tedman-Jones J, Gijzen M 2006b. Effector-triggered immunity by the plant pathogen *Phytophthora*. Trends Microbiol 14:470–473.

Qutob D, Tedman-Jones J, Dong S, Kuflu K, Wang Y, Dou D, Kale SD, Arredondo FD, Tyler BM, Gijzen M 2009. Copy number variation and transcriptional polymorphisms of *Phytophthora sojae* RXLR effector genes. PLoS ONE 4(4):e5066.

Randall TA, Dwyer RA, Huitema E, Beyer K, Cvitanich C, Kelkar H, Fong AM, Gates K, Roberts S, Yatzkan E, et al. 2005. Large-scale gene discovery in the oomycete *Phytophthora infestans* reveals likely components of phytopathogenicity shared with true fungi. Mol Plant Microbe Interact 18:229–243.

Rehmany AP, Gordon A, Rose LE, Allen RL, Armstrong MR, Whisson SC, Kamoun S, Tyler BM, Birch PR, Beynon JL 2005. Differential recognition of highly divergent downy mildew avirulence gene alleles by RPP1 resistance genes from two *Arabidopsis* lines. Plant Cell 17:1839–1850.

Rose JK, Ham KS, Darvill AG, Albersheim P 2002. Molecular cloning and characterization of glucanase inhibitor proteins: coevolution of a counterdefense mechanism by plant pathogens. Plant Cell 14:1329–1345.

Sandhu D, Gao H, Cianzio S, Bhattacharyya MK 2004. Deletion of a disease resistance nucleotide-binding-site leucine-rich- repeat-like sequence is associated with the loss of the *Phytophthora* resistance gene *Rps*4 in soybean. Genetics 168:2157–2167.

Senchou V, Weide R, Carrasco A, Bouyssou H, Pont-Lezica R, Govers F, Canut H 2004. High affinity recognition of a *Phytophthora* protein by *Arabidopsis* via an RGD motif. Cell Molec Life Sci 61:502–509.

Shabab M, Shindo T, Gu C, Kaschani F, Pansuriya T, Chintha R, Harzen A, Colby T, Kamoun S, van der Hoorn RA 2008. Fungal effector protein AVR2 targets diversifying defense-related Cys proteases of tomato. Plant Cell. In press.

Shan W, Cao M, Leung D, Tyler BM 2004. The *Avr1b* locus of *Phytophthora sojae* encodes an elicitor and a regulator required for avirulence on soybean plants carrying resistance gene *Rps1b*. Mol Plant Microbe Interact 17:394–403.

Shen KA, Chin DB, Arroyo-Garcia R, Ochoa OE, Lavelle DO, Wroblewski T, Meyers BC, Michelmore RW 2002. *Dm3* is one member of a large constitutively expressed family of nucleotide binding site-leucine-rich repeat encoding genes. Mol Plant Microbe Interact 15:251–261.

Slusarenko AJ, Schlaich NL 2003. Pathogen profile. Downy mildew of *Arabidopsis thaliana* caused by *Hyaloperonospora parasitica* (formerly *Peronospora parasitica*). Molec Plant Pathol 4:159–170.

Sohn KH, Lei R, Nemri A, Jones JD 2007. The downy mildew effector proteins ATR1 and ATR13 promote disease susceptibility in *Arabidopsis thaliana*. Plant Cell 19:4077–4090.

Song J, Bradeen JM, Naess SK, Raasch JA, Wielgus SM, Haberlach GT, Liu J, Kuang H, Austin-Phillips S, Buell CR, et al. 2003. Gene *RB* cloned from *Solanum bulbocastanum* confers broad spectrum resistance to potato late blight. Proc Nat Acad Sci USA 100:9128–9133.

Tian M, Benedetti B, Kamoun S 2005. A Second Kazal-like protease inhibitor from *Phytophthora infestans* inhibits and interacts with the apoplastic pathogenesis-related protease P69B of tomato. Plant Physiol 138:1785–1793.

Tian M, Huitema E, Da Cunha L, Torto-Alalibo T, Kamoun S 2004. A Kazal-like extracellular serine protease inhibitor from *Phytophthora infestans* targets the tomato pathogenesis-related protease P69B. J Biol Chem 279:26370–26377.

Tian M, Kamoun S 2005. A two disulfide bridge Kazal domain from *Phytophthora* exhibits stable inhibitory activity against serine proteases of the subtilisin family. BMC Biochem 6:15.

Tian M, Win J, Song J, van der Hoorn R, van der Knaap E, Kamoun S 2007. A *Phytophthora infestans* cystatin-like protein targets a novel tomato papain-like apoplastic protease. Plant Physiol 143:364–377.

Torto T 2003. Functional genomics of extracellular proteins of *Phytophthora infestans*. The Ohio State University. Columbus, OH.

Torto TA, Li S, Styer A, Huitema E, Testa A, Gow NA, van West P, Kamoun S. 2003. EST mining and functional expression assays identify extracellular effector proteins from the plant pathogen *Phytophthora*. Genome Res 13:1675–1685.

Torto-Alalibo T, Tian M, Gajendran K, Waugh ME, van West P, Kamoun S 2005. Expressed sequence tags from the oomycete fish pathogen *Saprolegnia parasitica* reveal putative virulence factors. BMC Microbiol 5:46.

Tyler BM 2002. Molecular basis of recognition between *Phytophthora* species and their hosts. Annual Rev Phytopath 40:137–167.

Tyler BM, Dou D, Kale SD, Jiang RHY, Chen Y, Wang Q, Wang X, Arredondo FD, Wang Y 2008. The effector repertoire of *Phytophthora sojae*: structure, function and evolution. In: Lorito M, Woo S, Scala F, editors. Biology of molecular plant-microbe interactions. International Society for Molecular Plant-Microbe Interactions. St. Paul, MN.

Tyler BM, Tripathy S, Zhang X, Dehal P, Jiang RH, Aerts A, Arredondo FD, Baxter L, Bensasson D, Beynon JL, et al. 2006. *Phytophthora* genome sequences uncover evolutionary origins and mechanisms of pathogenesis. Science 313:1261–1266.

van der Vossen E, Sikkema A, Hekkert BL, Gros J, Stevens P, Muskens M, Wouters D, Pereira A, Stiekema W, Allefs S 2003. An ancient R gene from the wild potato species *Solanum bulbocastanum* confers broad-spectrum resistance to *Phytophthora infestans* in cultivated potato and tomato. Plant J 36:867–882.

van der Vossen EAG, Gros J, Sikkema A, Muskens M, Wouters D, Wolters P, Pereira A, Allefs S 2005. The *Rpi-blb*2 gene from *Solanum bulbocastanum* is an *Mi*-1 gene homolog conferring broad-spectrum late blight resistance in potato. Plant J 44: 208–222.

van Poppel PMJA, Guo J, van de Vondervoort PJI, Jung MWM, Birch PRJ, Whisson SC, Govers F 2008. The *Phytophthora infestans* avirulence gene Avr4 encodes an RXLR-dEER effector. Molec Plant Microbe Interact 21:1460–1470.

Vleeshouwers VGAA, Rietman H, Krenek P, Champouret N, Young C, Oh S-K, Wang M, Bouwmeester K, Vosman B, Visser RGF, et al. 2008. Effector genomics accelerates discovery and functional profiling of potato disease resistance and *Phytophthora infestans* avirulence genes. PLOS One. In press.

Whisson SC, Boevink PC, Moleleki L, Avrova AO, Morales JG, Gilroy EM, Armstrong MR, Grouffaud S, West Pv, Chapman S, et al. 2007. A translocation signal for delivery of oomycete effector proteins into host plant cells. Nature 450:115–119.

Williams B, Dickman M 2008. Plant programmed cell death: can't live with it; can't live without it. Molec Plant Pathol 9:531–544.

Win J, Morgan W, Bos J, Krasileva KV, Cano LM, Chaparro-Garcia A, Ammar R, Staskawicz BJ, Kamoun S 2007. Adaptive evolution has targeted the C-terminal domain of the RXLR effectors of plant pathogenic oomycetes. Plant Cell 19:2349–2369.

Wolpert TJ, Dunkle LD, Ciuffetti LM 2002. Host-selective toxins and avirulence determinants: what's in a name?. Annu Rev Phytopathol 40:251–285.

Wroblewski T, Piskurewicz U, Tomczak A, Ochoa O, Michelmore RW 2007. Silencing of the major family of NBS-LRR-encoding genes in lettuce results in the loss of multiple resistance specificities. Plant J 51: 803–818.

19

PYTHIUM INSIDIOSUM AND MAMMALIAN HOSTS

LEONEL MENDOZA
Biomedical Laboratory Diagnostics Program, Department of Microbiology and Molecular Genetics, Michigan State University, East Lansing, Michigan

19.1 INTRODUCTION

Pythium insidiosum is a mammalian pathogen with morphological and physiological attributes in common with other *Pythium* species and with the other closely related Oomycetes (Peronosporomycetes). Although some members of this class also have been reported to cause disease in animals such as fish, insects, amphibian, crustaceous, and others (Dick, 2001; Neish, 1977; Sogin and Silberman 1998; Sosa et al., 2007; Tyler, 2001; Willoughby, 1969, 1985), *P. insidiosum* is the only member that infects mammals and birds (Dick, 2001; Pesavento et al., 2008). One exception seems to be *Blastocystis hominis* that infects humans (Sogin, 1998; Tyler, 2001). However, it is not clear whether this anaerobic stramenopilan is responsible for the array of clinical manifestations observed in some of the infected humans, even when the pathogen is present in the stool samples of the affected hosts. Because *B. hominis* does not invade the intestinal tract, the expression of virulence factors to cause disease in humans is controversial (Stenzel and Boreham, 1996). In contrast, *P. insidosum* penetrates mammalian and bird tissues efficiently causing life threatening infections in apparently healthy bovines, birds, canines, equines, felines, humans, sheep, and captive zoo animal species (Mendoza, 2005; Pesavento et al., 2008; Santurio et al., 2008). This suggests that *P. insidiosum* has evolved a novel approach to infect mammals (Fig. 19.1). The ability to invade mammalian hosts is unique among stramenopilan microbes and warrants additional genomic and phylogenetic studies.

Oomycete Genetics and Genomics: Diversity, Interactions, and Research Tools
Edited by Kurt Lamour and Sophien Kamoun
Copyright © 2009 John Wiley & Sons, Inc.

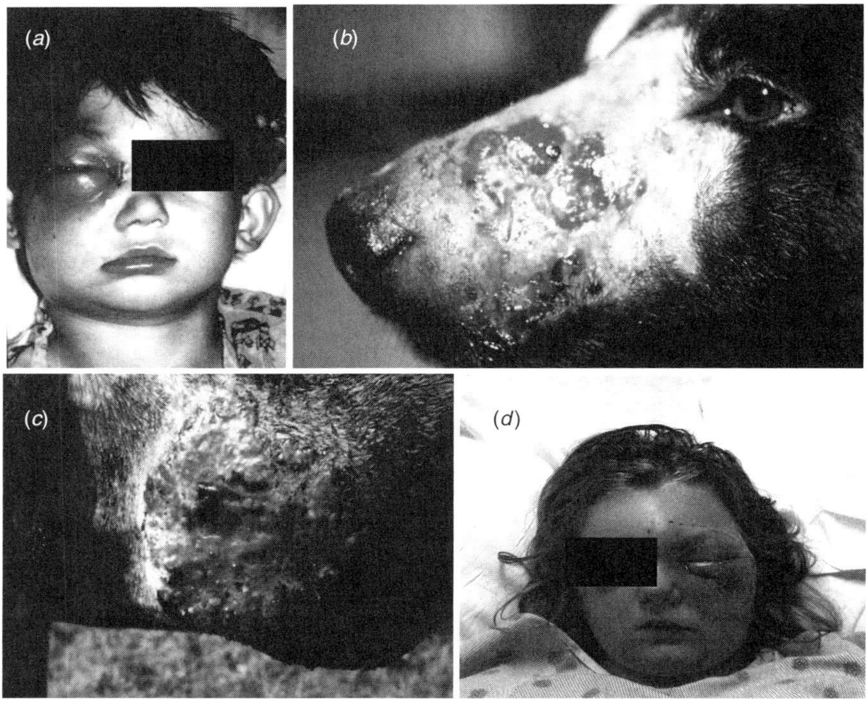

FIG. 19.1 (See color insert) The figure shows the most prominent clinical features observed in cases of pythiosis in mammals. Panels (a) and (d) show two U.S. children from Texas and Tennessee, respectively, with orbital pythiosis. Both were treated and cured by radical surgery (Courtesy of Drs. M.G. Rinaldi; S. Seidemseld; and S.L.R. Arnold). Panel (b) A skin infection affecting the subcutaneous tissues in a dog (courtesy of Dr. R.C. Thomas). (c) an infected mare with pythiosis. Two circular lesions with numerous "kunkers" are readily observed. The dog and the mare both finally succumbed to the disease.

19.2 *P. INSIDIOSUM* HISTORY, BIOLOGY, ECOLOGY, AND TAXONOMY

Pythium insidiosum, which is the etiologic agent of pythiosis, was first reported to cause skin infections in horses in the nineteenth century. The disease was known with regional names such as bursattee, bursatti (from the Indian term for rain bursat) in India (Smith 1884), espundia in Latin America (Gonzalez and Ruiz, 1975), granular dermatitis in Japan (Amemiya, 1982), leeches (Florida horse leeches) in the United States (Bitting 1894), swamp cancer in Australia (Austwick and Coplan, 1974), and equine phycomycosis by others (Connole, 1973). These classic studies reported the presence of aseptate hyphae in masses called "kunkers" present in the skin of the infected equines.

Initially, it was believed that the etiologic agent of equine cutaneous granulomas was a fungus (Smith, 1884; Bitting, 1894). Despite new reports on equine granulomas caused by this pathogen (De Haan and Hoogkamer 1901), early in the twentieth century this theory was abandoned. It took more than 60 years to determine the true etiologic agent of the equine granulomas (Bridges and Emmons, 1961). In the early 1960s, several investigators argued that the etiologic agent of pythiosis was a species of the genus *Mortierella*, and at one point it was known as *Hyphomyces destruens* (Bridges and Emmons, 1961). Its taxonomic relationship with the genus *Pythium* was found by serendipity on a New Zealand isolate from a horse with the disease (Austwick and Coplan, 1974). These investigators noted that their isolate developed biflagellate zoospores after placing the culture in water that contained rotten maize silage. Later, deCock et al. (1987) induced oogonia production in several Costa Rica strains, which caused equine pythiosis and successfully introduced the binomial *P. insidiosum*. Since then, several similar cases of pythiosis in humans and lower animals were diagnosed in different geographical areas (Mendoza, 2005).

Because *Pythium* species are well known as plant pathogens and as saprotrophic species that complete their life cycles in nature in decaying plants and soil, the finding that *P. insidiosum* has the ability to cause pathology in mammalian hosts came as a surprise (Dick, 2001; Mendoza, 2005). The *in vitro* development of zoospores by this pathogen has been well documented (Mendoza, 2005). The zoospores are morphologically identical to those of *Pythium* (see below), *Lagenidium*, and other closely related species (Fig. 19.2). The presence of *P. insidiosum* in aquatic environments was previously suspected. Recent studies in Thailand showed that *P. insidiosum*'s zoospores are frequently found in agricultural irrigation water and reservoirs (Supabandhu et al., 2007). The *in vitro* development of sexual spores called oogonia (the survival units in nature) is difficult and has been achieved only in a few instances (deCock et al., 1987; Ichitani and Amemiya, 1980; Shipton, 1987; Supabandhu et al., 2007). It is thought that *P. insidiosum* in nature might use a plant to complete its life cycle by the formation of zoospores and the production of oogonia to survive during the dry season (Ichitani and Amemiya, 1980; Mendoza et al., 1993; Shipton, 1985, 1987), but this hypothesis has yet to be supported by data. However, the recent finding of *P. insidiosum* zoospores recovered from Thai irrigation water tends to support this view (Supabandhu et al., 2007). The cell wall of *P. insidiosum* contains cellulose, minor quantities of chitin, and sugars, such as mannose, galactose, and rhamnose (Shipton et al., 1982). Recent ultrastructural studies of *P. insidiosum* hyphae revealed that this pathogen of mammals has several morphological attributes in common with *Pythium* species, which include a tubular mitochondria and the presence of large vesicles and electron-dense bodies (Garcia et al., 2007).

P. insidiosum is more frequently found in tropical and subtropical areas of the world, but cases in temperate areas such as Japan (Amemiya, 1982; Ichitani and Amemiya, 1980), Korea (Sohn et al., 1996), and the United States (Illinois, New York, Virginia, Wisconsin) (Mendoza, 2005) suggest that this microbial

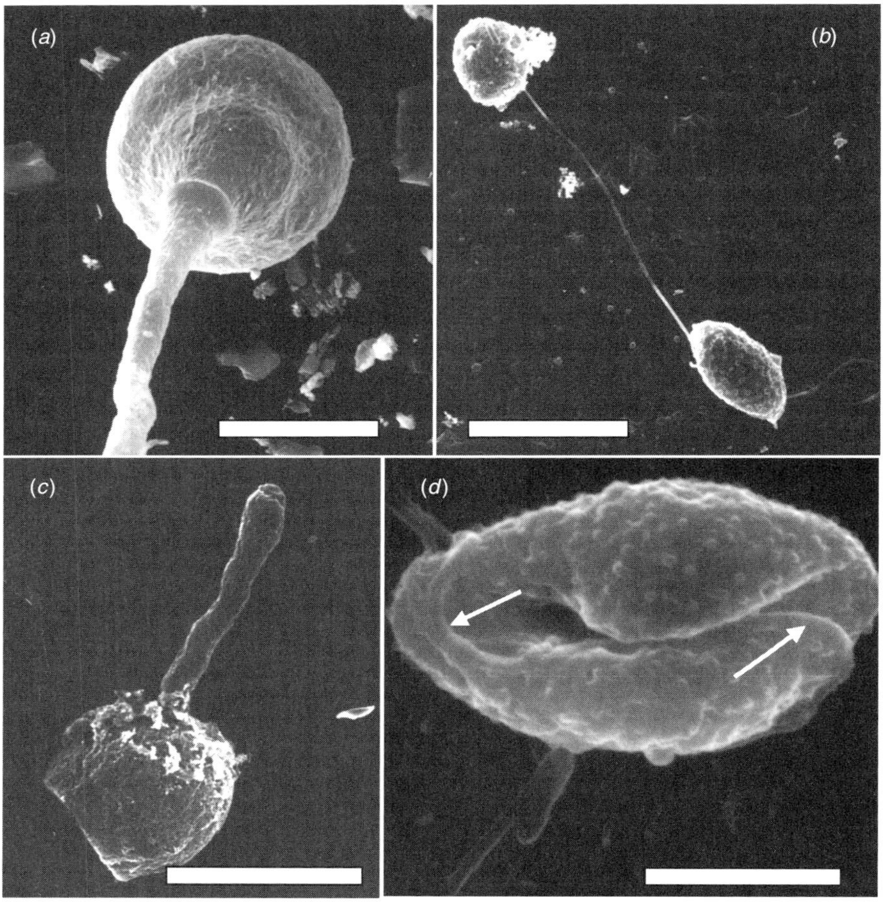

FIG. 19.2 Scanning electron micrographs of *Pythium insidiosum* zoospores and appressoria. (a) *P. insidiosum* appressorium attached to horse tissue (Bar = 10.0 μm). (b) depicts a *P. insidiosum* zoospore with two extended flagella (right lower section) and encysted zoospores expressing an adhesive substance (left upper section) (Bar = 10.0 μm). (c) an encysted *P. insidiosum* zoospore developing a germ tube (Bar = 5.0 μm). Close up of a kidney bean shape *P. insidiosum* zoospore (d). Note the two flagella (arrows) that originate from a ventral central groove a typical feature of the *Pythium* zoospores (Bar = 3.0 μm).

pathogen is possibly in the process of adapting to cold environments or may be expanding its ecological niche because of global warming. *P. insidiosum* seems to develop better in stagnant waters accumulated after warm rainy seasons. The fact that most equine and canine cases of pythiosis occur in such periods suggests that the resting spores of *P. insidiosum* present in soil from prior rainy seasons are stimulated by near-neutral pH and ions such as Ca^{2+}, Mg^{2+}, and K^{+} present in the soil of the damp areas that lead to the formation of sporangia and zoospores (Shipton, 1985). The development of sporangia and the release of thousand of zoopores ensure the pathogen colonization of new neighboring environments in which mammals seem to be only accidental hosts (see below). The finding that *P. insidiosum* zoospores have a high tropism for plant leaves (grass and lily) and mammalian tissues (Mendoza and Prendas, 1988; Miller, 1983) prompted some to hypothesize that *P. insidiosum* zoospore capabilities have adapted to attach and invade animal tissue (see below) (Mendoza et al., 1993; Miller, 1983).

The production of an adhesive amorphous material around the encysted zoospores has been reported previously in many plant pathogenic *Pythium* species (Henrix and Campbell, 1973). The development of this material aids attachment to natural substrates, such as decaying leaves and plants, which allows the development of specialized structures called appressoria to invade the affected substrate (Henrix and Papa, 1974). Additionally, *P. insidiosum* expresses these structures *in vitro* (deCock et al., 1987). Using scanning electron microscopy Mendoza et al. (1993) showed the presence of appressoria over grass leaves and horse tissues, which indicates a similar strategy to invade plants and animals (Fig. 19.2). Some authors theorized that *P. insidiosum* has used mammalian hosts to expand its ecological distribution (Mendoza et al., 1993; Miller, 1983). This hypothesis is supported by the observation that during equine infection, *P. insidiosum*'s hyphae are sequestered into small hard masses called "kunkers," which are expelled periodically from the infected tissues to new ecological niches distant from the ponds where the equines were infected (Mendoza, 2005). However, *P. insidiosum* development of "kunkers" in the infected hosts may be coincidental because other affected mammalian species do not develop these structures during infection (Mendoza, 2005; Thianprasit, 1999; Thomas and Lewis, 1998).

Prior to the advent of molecular phylogenetic analysis, the taxonomic position of the stramenopilans in the tree of life was solely based on morphological, physiological, and other related attributes. The production of tubular bodies similar to those in the fungi was used by numerous investigators to classify the Oomycetes and the other stramenopilans within the kingdom Fungi (Mendoza, 2005). Thus, *P. insidiosum* was considered an "aquatic fungus" and studied by medical mycologists. Phylogenetic analysis of the Oomycetes, however, showed that they are related more closely to the brown algae and the diatoms (see several other chapters in the book). Nonetheless, some investigators still treat this group of microbes as fungi (Dick, 2001), which has caused a great deal of confusion to those not familiar with *P. insidiosum* in the clinical environment. Fortunately, physicians are learning that the resistance of *P. insidiosum* to antifungal drugs is simply becuase this pathogen is not a fungus (Mendoza, 2005).

19.3 GENOMICS AND PHYLOGENETICS OF *P. INSIDIOSUM*

Data on the genomics of *P. insidiosum* are scarce. Few laboratories are dealing with this unique pathogen of mammals, thus studies to understand gene organization and structure are still to be undertaken. The genome of the oomycetes varies from 18 to 250 Mb using pulse field electrophoretic data (Kamoun, 2003). Within the oomycetes *Pythium* species seem to possess the smallest genomes (Martin, 1995). Preliminary data in our laboratory using contour-clamp homogeneous electric field (CHEF) estimates the *P. insidiosum* genome size at 35 to 38 Mb (unpublished data), which is within the range of most *Pythium* species (18.8 to 41.5 Mb) (Kamoun, 2003; Mort-Bontemps and Fevre, 1995). However, these results were not replicable using several isolates of *P. insidiosum* from different geographical locations. The estimated genome sizes in some of the *P. insidiosum* isolates may have been different because of deficiencies of enzymatic cell wall disruption, which results in the formation of unresolved bands. Despite this problem, it appears that *P. insidiosum* has a genome size similar to other *Pythium* species.

BLAST sequence similarity searches with *chitin synthase 2* (*CHS2*) genes from at least 18 *P. insidiosum* isolates showed that this pathogen clustered with the orthologous sequences available in GenBank of *Saprolegnia monoica* and *Phytophthora capsici* (unpublished data). A more comprehensive phylogenetic study was carried out by Schurko et al., (2003a and b), in which internal transcribed spacers (ITSs) were used to compare 23 *P. insidiosum* isolates from different geographical locations in Asia, Australia and the Americas. Although morphologically all studied strains were identical, at least three unforeseen species (cryptic species) within the investigated isolates were revealed (Fig. 19.3). Intriguingly, the cryptic isolates clustered according to their geographic preferences. For instance, all isolates from the Americas formed a well supported taxon (cluster I), whereas cluster II involved Asia, Australia, and New Zealand strains, and Cluster III comprised isolates from both the Americas and Asia. Oddly, cluster II harbored an isolate recovered in the United States from a case of human keratitis (Schurko et al., 2003b). A closer look into this anomaly revealed that this particular strain was isolated from a Middle-Eastern man who had received paraphernalia from that particular geographic area prior to the disease onset. Thus, it is likely that the patient acquired *P. insidiosum* from that geographic region explaining the placement of this strain within cluster II.

The arrangement of *P. insidiosum* into three well-defined clusters might be related to the geographical area where the isolates were recovered. The DNA genomic changes observed in the ITS sequences of the investigated *P. insidiosum* isolates suggest important genetic variations after their separation from a common ancestor. Clusters I and II shared a common ancestor and are placed in two well-defined phylogenetic groups (Schurko et al., 2003a), whereas cluster III diverged earlier from the two main clusters, and it comprises a mixture of strains from the Americas and Asia. The presence of isolates from

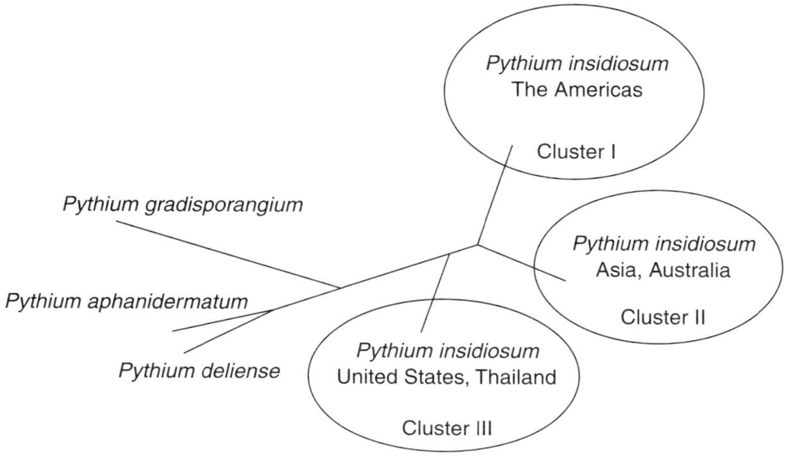

FIG. 19.3 Phylogeny of *P. insidiosum* clinical isolates recovered from different continents using ITS DNA sequences. The figure illsutrates the three clusters that define *P. insidiosum*: Cluster I and II represent the Americas and the Asian strains respectively, whereas Cluster III contains isolates from the United States and Thailand. Cluster III is closely related to *Pythium grandisporangium*, which is a marine saprotroph, and to the other *Pythium* species (modified from Schurko et al., 2003a).

Australia, India, Japan, New Guinea, and Thailand in cluster II suggests that the Asian strains may have the same ancestor of that in cluster I, and both have been separately evolving. Cluster III showed rapid genetic variation in both the Asian and the American strains. In contrast to clusters I and II, in which geography could have played an important role in shaping the population structure, the members of cluster III seem to be evolving independently of their geographic location into a new complex of cryptic species. Rapid genetic variation has been reported in other oomycetes such as *Phytophthora* species (Chamnanpunt et al., 2001). It is possible that *P. insidiosum* cryptic species variation is the result of meiotic recombination that is generating unique genetic types, because this species seems to be constantly expanding its host range capabilities (Schurko et al., 2003a and b).

Recent molecular phylogeny of the genus *Pythium* showed *P. insidiosum* as a unique member of the genus in a solitary phylogenetic group (Martin, 2000; Lévesque and deCock, 2004). Initially, Martin (2000) divided all *Pythium* isolates into three taxons and subdivided the third taxon into subgroups named with the letters A to D. Lavesque and deCock (2004), however, reported two taxons; taxon 1 comprised subgroups A, B, C, and D of Martin (Martin, 2000), whereas taxon 2 was formed by subgroups F, G, H, I, and J. The third taxon has only one subgroup, K. *P. insidiosum* and *Pythium grandisporangium* were placed alone in group C. The placement of *P. insidiosum* in group "C" with a marine saprotrophic microbe is intriguing because *P. insidiosum* is "fresh water" and a soil microbe. We are faced with many questions – What are the implications of *P. insidiosum* phylogenetic relationship to a marine saprotroph microbe? Did *P. insidiosum* evolve from a saprotroph marine ancestor to a mammalian pathogen? The fact that this pathogen of mammals is also phylogenetically linked to plant pathogens such as *Pythium deliense* and *Pythium aphanidermatum* suggests that *P. insidiosum* could have evolved with both saprotrophic and pathogenic capabilities to attack plants (Schurko et al., 2003a). The observation of appressoria development in the presence of plant tissue reported by some (deCock et al,, 1987; Mendoza et al., 1993) strongly supports this notion (Fig. 19.2). However, the finding that *P. insidiosum* might have evolved from a saprotrophic ancestor (probably extinct) is in agreement with the fact that this microbe still behaves as a saprothrophic microbe. It is thought that *P. insidiosum* is completing its life cycle in decaying substrates, as so far, no one has been able to show that *P. insidiosum* can cause disease in plants. If this hypothesis is correct, then it would be interesting to know when *P. insidiosum* acquired pathogenic attributes to attack mammals. The answer to that question may be found through comparative genomics with closely related ancestors such as those shared by the plant pathogens *P. deliense* and *P. aphanidermatum*.

Of interest from an evolutionary perspective is the finding that one of *P. insidiosum* isolates in a study by Schurcko et al. (2003a) was recovered from the infected larva of *Culex quinquefasciatus* in India (Martin, 1995; Schurcko et al., 2003a and b). The isolate was originally misidentified as a *Lagenidium* spp. However, based on morphological and other characteristics, deCock (personal communication), originally classified it as a *Pythium* spp., and based on Schurcko et al. (2003b) data it was classified later as *P. insidiosum*. This finding is intriguing and suggests that *P. insidiosum* might have the ability to invade aquatic insects similar to entomopathogenic oomycetes such as *Lagenidium* spp. (Dick, 2001; Willoughby, 1969). The report of *P. insidiosum* affecting mosquito larvae is also of epidemiological concern. The likelihood of *P. insidiosum* transmission through infected mosquitoes needs more investigation, especially in the tropical endemic areas of the disease where mosquitoes are prevalent.

So far, a phylogentic analysis has shown that *P. insidiosum* is the only etiologic agent of pythiosis in mammals (Schurko et al., 2003a and b; Martin, 2000; Levesque and deCock, 2004). However, first Grooters et al. (2003) and then Grooters (2003) reported an unusual infection caused by a putative species of *Lagenidium* in 40 dogs with clinical features identical to that observed in cases of dog pythiosis. They based their observations on phenotypic and molecular studies of isolates recovered from the infected dogs and introduced the disease name lagenidiosis (Grooters, 2003). Interestingly, no additional cases of *Lagenidium* infection in dogs have been published since these first two reports in 2003. In addition, no DNA sequences were deposited in any database from those studies. It is possible that these investigators misidentified an

American cluster III cryptic strain of *P. insidiosum* (Fig. 19.3) as an unusual strain they misclassified as *Lagenidium* spp. The fact that the last publication of these authors coincided with the appearance of Schurcko et al. (2003a and b) phylogenetic studies suggests that they had corrected their error. However, they have yet to inform the scientific community to clarify this issue. This type of mistake is understandable because *P. insidiosum* could embrace a phylogenetic complex of closely related species or a complex of different strains that belong to a single species (Cooke et al., 2000; Mendoza, 2005: Schurko et al., 2003b).

19.4 *P. INSIDIOSUM* LIFE CYCLE IN NATURE AND HOST PATHOGENESIS

P. insidiosum epidemics are highly dynamic and are characterized by sexual recombination and explosive clonal reproduction (McMeekin and Mendoza, 2000). It is thought that sporangia, which are the deciduous asexual spores that form on surfaces of infected plants, are spread by water. They may either germinate directly on a host to cause infection (Fig. 19.2) or germinate when immersed in water and release zoospores that swim chemotactically toward hosts and encyst on surfaces to produce infection structures. This cycle of infection, sporulation, and spread is favored by warm, wet conditions and can occur many times during a typical rainy season (Shipton, 1985; Mendoza and Prendas, 1988; Mendoza et al., 1993). Mammalian hosts that enter endemic areas of the disease may come in contact with the propagules (zoospores, hyphae, or oogonia) of *P. insidiosum* through skin lesions. Zoospores have been identified as the infecting units (Mendoza et al., 1993; Shipton, 1985), but pythiosis cases in humans and animals, including birds (Pesavento et al., 2008), acquired in dry areas suggest that propagules other than zoospores could also act as infecting units. After contact with the mammalian hosts, *P. insidiosum* propagules secrete an adhesive material to keep tight contact with the infected host (Mendoza et al., 1993). The reports that the apex of *P. insidiosum* hypha at 37°C exerted forces of only 2.3 μN to a maximum force of 6.9 μN over *in vitro* plant and animal tissues suggests that this pathogen needs a traumatic lesion to penetrate mammalian tissue successfully (MacDonald et al., 2002; Ravishankar et al., 2001). Although the stramenopilans and the fungi diverged early during eukaryotes differentiation, fungi and oomycetes both developed invasive hyphae that exert pressure against the target host tissues. This approach has been interpreted by some as evidence of convergent evolution in these phylogenetically distant microbes (Money, 1999; Money et al., 2004).

When *P. insidiosum* attaches to the host's tissue, it develops a germ tube that actively penetrates the invaded tissues (Fig. 19.1). The *in vitro* expression of several proteases by some *Pythium* species, which include *P. insidiosum*, at 37°C may correspond to one aspect of pathogenesis of this oomycete once in the tissues of a susceptible host (Davis, et al., 2006). Thus, *P. insidiosum* could express similar proteases *in vivo* to facilitate penetration in the infected tissue. At this point, *P. insidiosum* uses host proteins to nourish and then proliferate releasing several metabolites (exoantigens) that are then presented to host immune system by regional dendritic cells (Davis et al., 2006; Mendoza and Newton, 2005). Several lines of evidence suggest that *P. insidiosum* uses some of these metabolites to manipulate the host immune response by activating a T helper 2 (Th2) subset in the infected tissues. This in turn could trigger the release of numerous eosinophils, basophils, mast cells, and related inflammatory components to the infected areas (Mendoza, 2005; Mendoza and Newton, 2005). This strategy seems to favor the establishment of *P. insidiosum* in the affected hosts. At least two other groups of eukaryotic pathogens (the parasites and some fungi) have acquired a similar mechanism during evolution to a pathogenic microbe. The following genera of parasites: *Angiostrongylus, Schistosoma*, and *Strongyloides* (Roberts et al., 2004) to mention some, and at least two fungi: *Basidiobolus* and *Conidiobolus* (Rippon, 1988; Thomas et al., 2006) also trigger a Th2 response with eosinophils and mast cells in the infected tissues. These phylogenetically divergent groups of microbial pathogens have in common the expression of some key antigenic components that appear to manipulate the immune system of the infected hosts to a typical Th2 response within an eosinophilic granuloma. All above microbes trigger large quantities of eosinophils in the infected areas that usually degranulate over the pathogens with the formation of the Splendore-Heoppli-like phenomenon (Mendoza and Newton, 2005; Rippon, 1988; Roberts et al., 2004; Thomas et al., 2006). Intriguingly, in most cases the Th2 inflammatory response induced by these pathogens causes more damage to the host tissues than to the pathogens (Mendoza, 2005; Mendoza and Newton, 2005). The ability to overcome the Th2 host response by these pathogens is more evident in cases of pythiosis (Mendoza, 2005). The disease caused by *P. insidiosum* in humans and animals is so insidious that if the patients are not treated early during infection, the disease usually becomes life threatening (Mendoza, 2005, Thianprasit, 1990). It has been proposed that *P. insidiosum* could use the "kunkers" (horses), which are masses made of degranulate eosinophils and hyphae, and the Splendore-Hoeppli phenomenon to hide key antigenic components of the pathogen from the immune systems and ensure its survival.

19.5 *P. INSIDOSUM* DISEASE IN MAMMALS AND BIRDS, AN EVOLUTIONARY CHOICE?

The destructive nature of the infections caused by *P. insidiosum* suggests that this pathogen has acquired the necessary weaponry to invade mammalian tissue only recently. This finding is supported by recent phylogenetic analyses placing this pathogenic species away from important plant pathogens and sharing a unique phylogenetic group with the marine saprothroph *P. grandisporangium* (Martin, 2000; Lévesque and deCock, 2004). The first weapon in *P. insidosum* arsenal is its well-known zoospore chemotaxis toward damaged plant leaves and injured mammalian skin tissues (Mendoza et al., 1993; Miller, 1983). This feature was probably developed early during evolution to complete its life cycle using plant parts. Most likely, all *Pythium* species have been in contact for millions of years with animals accidentally entering their ecosystems. However, the adaptation of *P. insidiosum* to parasitism probably took place long after animals entered their ecological niches.

Pythiosis is not a communicable disease. Thus, the real incidence of the infection is unknown. According to published data in the United States, equines and dogs are the most affected species (Bridges and Emmons, 1961; Thomas and Lewis, 1998; Grooters, 2003; Mendoza and Netwon, 2005). Based on personal communications from veterinary physicians who work in endemic areas, we believe that in the United States alone several hundred of cases annually occurred in both species. In contrast, human cases of pythiosis in the United States are rare with only 12 known reported cases, some in children (Mendoza, 2005). The recent finding of a bird with cutaneous lesions caused by *P. insidiosum* suggests that in addition to mammals and insects, this pathogen can also affect birds and possibly aquatic animals (Pesavento et al., 2008). Pythiosis in humans and animals have been also reported in Latin America and other continents. Intriguingly, Thailand is the country with more reported cases of human pythiosis especially in thalassemic patients (Thianprasit, 1990). Although the disease occurred all year around, the infection is more frequently reported during the hot summer months in the endemic areas, where large bodies of water are formed during the rainy seasons (Mendoza, 2005). Equine and canine pythiosis is characterized by the involvement of subcutaneous, bone, and intestinal tissues (Mendoza, 2005). In humans, however, *P. insidiosum* induce a wide array of clinical manifestations from superficial (keratitis) (Imwidthaya, 1995) and subcutaneous (Thianprasit, 1990), to systemic disease that involves major arteries (Imwidthaya, 1994; Thianprasit, 1990). The disease has been reported with less frequency in other mammalian hosts, such as cats, cattle, sheep, and zoo animals that contain subcutaneous and intestinal disease (Mendoza, 2005). In these hosts, the main feature of the disease is the appearance of extensive lesions (sometime as large as 1 m in diameter) with severe tissue damage (Fig. 19.1).

Curiously, thus far experimental infection using naturally susceptible animals has been unsuccessful (Patino-Meza, 1988). Several investigators used hosts such as dogs, horses, and cattle inoculated with *P. insidosum*'s zoospores and other propagules, but the injected animals did not develop pythiosis (Mendoza, 2005; Patino-Meza, 1988). They even submerged the limbs of several horses with small skin lesions into water with thousands of *P. insidiosum* zoospores without success. In addition, the experimental infection using mice and rats has been equally unsuccessful (Miller, 1983).

Thus, pythiosis is a disease that does not comply with Koch's postulates. These findings suggest that *P. insidiosum* may require the presence of unknown factors to establish the initial infection. This may explain the sporadic occurrence of pythiosis in humans and animals in the endemic areas of the disease.

Oddly, the experimental infection using *P. insidiosum* propagules was achieved in rabbits, which is not a natural host of the disease (Amemiya, 1982; Mendoza, 2005; Patino-Meza, 1988; Santurio et al., 2003). Rabbits are very sensitive to experimental inoculation with any of the propagules of *P. insidiosum* (Amemiya, 1982; Patino-Meza, 1988). After inoculation of *P. insidiosum* propagules in rabbits, the disease becomes systemic and mimics some clinical features observed in humans and lower animals. This is a bizarre feature of *P. insidiosum* because naturally infected rabbits have not been reported. So, what are the unknown factors necessary for *P. insidiosum* to establish a successful infection in a susceptible host? We do not know yet the answer to this question. However, it is possible that most mammalian hosts (humans, horses, dogs, etc.) are resistant to *P. insidiosum* infection, and only few are susceptible to pythiosis (Mendoza and Newton, 2005). In summary, our working hypothesis is that (1) the tropism of *P. insidiosum* zoospores for leaves led to an accidental encounter with a mammalian host, (2) it adapted to this host by evolving novel pathogenic attributes to cause disease only in individuals that cannot eliminate the pathogen, and (3) the susceptibility to pythiosis may be related to undetectable defects in the host's immune response to *P. insidiosum* propagules, which could explain why the disease is sporadic and not all animals are infected in endemic areas.

19.6 *P. INSIDIOSUM* INNATE ANTIFUNGAL DRUG RESISTANCE AND IMMUNOTHERAPY

Localized skin infections caused by *P. insidiosum* in humans and lower animals are traditionally treated by surgery (Mendoza, 2005; Thianprasit, 1990). This procedure has been used for more than two centuries with limited results, because mammals have critical anatomical structures difficult to reach. Thus, this procedure has only a 45% success rate. Iodides and other compounds for the management of pythiosis were introduced early in the history of the disease. Although iodides are still in use and this procedure is partially successful, it causes severe toxic side effects in some treated mammals (Bridges and Emmons, 1961). Because reports in the 1960s indicated that *P. insidiosum* was probably a fungus, antifungal drugs were used to treat pythiosis but were met with little success (McMullan et al., 1977, Thianprasit, 1990). The reclassification of this pathogen to the Oomycetes in the 1970's helped explain the poor response of *P. insidiosum* to some antifungal drugs that targeted ergosterol. *P. insidiosum*, similar to other *Pythium* and *Phytophthora* spp., lacks ergosterol in its cytoplasmic membrane, which explains why antifungal drugs that target this compound, like amphothericin B and the imidazoles, may be ineffective against this stramenopilan microbe (Dick, 2001; Mendoza, 2005). Nonetheless, the reports of some cured human, equine, and dog pythiosis cases using amphotericin B and other similar antifungal drugs suggests that these drugs could target protein pathways other than ergosterol in the *P. insidiosum* cell wall (Mendoza, 2005).

Antifungal drugs such as caspofungin (Brayer-Pereira et al., 2007), Voriconazole, itraconazole, and terbinafine (Argenta et al., 2008; Shenep et al., 1998) showed limited *in vitro* and *in vivo* activity against *P. insidiosum*. Although some could argue that the inhibitory *in vitro* concentrations can be extrapolated to achieve inhibition of *P. insidiosum in vivo*, so far the poor response of these drugs in the clinical setting suggests that these extrapolations could have serious consequences for the infected host (unpublished data). The poor response to the available antifungal drugs showed the necessity for new approaches to treat pythiosis. Although immunotherapy, which is an unconventional immunological intervention to treat pythiosis, was introduced 20 years ago, it is still under investigation (Miller, 1981). This approach is unusual among most fungal and parasitic infections. Data collected from successfully treated humans and lower animals showed that several antigenic proteins of *P. insidosum*, which are yet to be characterized, injected into patients with active pythiosis appear to down regulate the Th2 response triggered in the infected hosts to a curative T helper 1 (Th1) subset (Hensel et al., 2003; Mendoza et al., 1992; Thitithanyanont et al., 1998). The induced Th1 response is thought to be responsible for the curative properties of this unusual approach (Mendoza et al., 2003; Mendoza and Newton, 2005; Wanachiwanawin et al., 2004). The curative rate of immunotherapy varies according to the host, but it is around 60% and sometimes higher ($\sim$87%), especially in bovines and equines with the disease (Mendoza and Newton, 2005). The high rate of cure is remarkable and does not have any parallel with other curative approaches in infectious diseases.

It is believed that *P. insidosum* expresses these antigens *in vitro*, but during infection the genes that encode these antigenic proteins are probably downregulated (Mendoza and Netwon, 2005). Interestingly, when these *in vitro* expressed proteins were injected back into an infected host, it triggered an immune response that eliminated the pathogen from the infected areas (Mendoza, 2005). The phylogenetic relationship of *P. insidiosum* with the oomycetes could perhaps explain the unusual curative properties of this approach. For instance, well-known pathogens of mammals eliminated genes with antigenic epitopes that, if expressed, could trigger a deleterious immune response against the pathogens (Sogin and Silberman, 1998). *P. insidiosum* may be in the process of becoming a more adapted pathogen, because *in vitro* it still expresses some antigenic components that seem to trigger a lethal immune response to the pathogen when injected in humans and animals with pythiosis (immunotherapy) (Hensel et al., 2003; Mendoza et al., 1992, 2003; Wanachiwanawin et al., 2004).

19.7 EVOLUTIONARY IMPLICATIONS OF A MAMMALIAN PATHOGEN AMONG OOMYCETES

Although some oomycetes can infect a great variety of multicellular organisms (Dick, 2001; Neish, 1977; Sogin and Silberman, 1998; Sosa et al., 2007; Stenzel and Boreham 1996; Tyler, 2001; Willoughby, 1969, 1985), *Pythium* species have expanded their host-range capabilities to include plants (Henrix and Campbell, 1973; Miller, 1983; Plaats-Niterink, 1981), fungi (Deacon, 1976), insects (Shurcko et al., 2003a and b), freshwater fish (Khulbe, 1982; Sati, 1991; Scott and O'Bien, 1962), crustaceae (Plaats-Niterink, 1972), and mammals (Amemiya, 1982; Austwick and Copland 1974; Bitting, 1894; Bridges and Emmons, 1961; Connole, 1973; Gonzalez and Ruiz, 1975; Ichitani and Amemiya, 1980; Smith, 1884; Thianprasit, 1990). The recent reports of a *Lagenidium* sp affecting dogs has yet to be validated with taxonomic and molecular data (see above) (Grooters, 2003; Grooters et al., 2003). Thus, *P. insidiosum* is the only Oomycete with the necessary pathogenic capabilities to cause life-threatening infections in mammals (Mendoza, 2005).

The implications of this evolutionary leap to the Oomycetes, in particular to the genus *Pythium*, are enormous. *P. insidiosum* has adapted particular morphological and physiological attributes to complete its life cycle in nature on mammals. For instance, the zoospore tropism for a particular substance in nature may overlap with a tropism for a currently unidentified substance in mammalian injured tissue. *P. insidiosum* zoospores then would have dual chemotaxis, one for a substrate in nature and the other in the mammalian host (Mendoza and Prendas, 1988; Mendoza et al., 1993; Miller, 1983). This may allow *P. insidiosum* frequent encounters with mammalian tissue. The sticky material expressed outside the zoospores after encysment to attach natural substrates could has been modified to attach the infecting units to injured mammalian skin (Mendoza et al., 1993). The *in vitro* formation of appressoria (Fig. 19.2) may also be an adaptation from natural substrates to the mammalian tissues, as supported by *in vitro* experiments (Fig. 19.2) (deCock et al., 1987). This new feature would allow the pathogen to penetrate the open mammalian tissues actively and cause disease. The fact that so far no other stramenopilan microbe has developed pathogenic attributes to cause life-threatening infection in mammals suggests that the genome of *P. insidiosum* could hold important clues about the evolutionary changes necessary to switch from a putative saprotrophic microbe (Chamnanpunt et al., 2001; Lavesque and deCock, 2004; Martin, 2000) to a mammalian pathogen (Amemiya, 1982; Austwick and Copland 1974; Bitting, 1894; Bridges and Emmons, 1961; Connole, 1973; Gonzalez and Ruiz, 1975; Ichitani and Amemiya, 1980; Smith, 1884; Thianprasit, 1990).

Currently, genome sequencing in *Phytophthora* spp. (Kamoun and Goodwin, 2007; Tayler et al., 2006), *Saprolegnia parasitica* (Torto-Alalibo et al., 2005), *Pythium ultimum* (underway), *Aphanomyces* spp. (Madoui et al., 2007), and other Oomycetes (Bouzide et al., 2007) have showed the importance of comparative genomics to understand genetic changes and evolution among pathogenic species (see various chapters in this book). Because *P. insidiosum* may hold important evolutionary clues to understanding gene adaptation in the Oomycetes involved in the expansion of host range, I believe that a comprehensive understanding of oomycetes genomics will be incomplete without the genome sequence of this mammalian pathogen. The genome sequence of *P. insidiosum* could provide information regarding the evolutionary path of this oomycete and its phylogenetic relationship with related microbes. Clearly, the most attractive aspect of this pathogen is its unique ability to invade mammalian hosts as this ability appears to be absent in the entire kingdom of straminopilan microbes.

REFERENCES

Amemiya J 1982. Granular dermatitis in the horse, caused by *Pythium gracile*. Bull Fac of Agriculure Kagoshima Univ 32:141–147.

Argenta JS, Santurio JM, Alvez SH, Pereira DIB, Cavalheiro AS, Spanemberg A, Ferreirole 2008. *In vitro* activities of voriconazole, itraconazol, and terbinafine alone or in combination against *Pythium insidiosum* isolates from Brazil. Antimicrob Agents Chemothe 52:767–769.

Austwick PKC, Copland JW 1974. Swamp cancer. Nature 250:84–85.

Bitting AW 1894. Leeches or leeching. Florida Univ AES Bull 25:37.

Bouzidi MF, Parlange F, Nicolas P, Mouzeyar S 2007. Expressed sequence Tags from the oomycete *Plasmopara halstedii* an obligate parasite of the sunflower. BMC Microbiol 7:110–120.

Brayer-Pereira DI, Santurio JM, Alves SH, Argenta JJ, Pötter L, Spanamberg A, Ferreiro L 2007. Caspofungin *in vitro* and *in vivo* activity against Brazilian *Pythium insidiosum* strains isolated from animals. J of Antimicrob Chemother 60:1168–1171.

Bridges CH, Emmons CW 1961. A phycomycosis of horses caused by *Hyphomyces destruens*. J Am Veter Med Asso 138:579–589.

Chamnanpunt J, Shan WX, Tyler BM 2001. High frequency mitotic gene conversion in genetic hybrid of the oomycete *Phytophthora sojae*. Proc Nat Acad Sci USA 98:14530–14535.

Connole PKC 1973. Equine phycomycosis. Austra Vet 49:214–215.

Cooke DEL, Drenth A, Duncan JM, Wayek G, Brasier CM 2000. A molecular phylogeny of *Phytophthora* and related Oomycetes. Fungal Genet Biol 30:17–32.

Davis DJ, Lanter K, Makselan S, Bonati C, Asbrock P, Ravishankar JP, Money NP 2006. Relationship between temperature optima and secreted protease activities of three *Pythium* species and pathogenicity toward plant and animal hosts. Mycol Res 110:96–103.

Deacon JW 1976. Studies of *Pythium oligandrum*, an aggressive parasite of other fungi. Trans Br Mycol Soc 66:383–391.

deCock AWAW, Mendoza L, Padhye AA, Ajello L, Kaufman L 1987. *Pythium insidiosum* sp. Nov. the etiological agent of pythiosis. J Clin Microbiol 25: 344–349.

de Haan, J Hoogkamer LJ 1901. Hyphomycosis destruens. *Veeartsenijk Bl v Ned Indie* 13:350–374.

Dick MW 2001. *Straminipilous fungi*: systematics of the *Peronosporomycetes* including accounts of the marine straminipilous protist, the plasmodiophorids and similar organisms. Kluwer Academic Publishers. London, UK. pp. 1–670.

Garcia RB, Pastor A, Mendoza L 2007. Mapping of *Pythium insidiosum* hyphal antigens and ultrastructural features using TEM. Mycol Res 111:1352–1360.

Gonzalez HE, Ruiz A 1975. Espundia equina. Etiología y patogénesis de una ficomicosis. Revista ICA 10:175–185.

Grooters AM 2003. Pythiosis, lagenidiosis, and zygomycosis in small animals. Vet Clin North America, Small Animal Practice 33:695–720.

Grooters AM, Hodgin EC, Bauer RW, Detrisac CJ, Znajda NR, Thomas RC 2003. Clinicopathologic findings associated with *Lagenidium* sp. infection in 6 dogs: initial description of an emerging oomycosis. J Vet Int Med 17:637–46.

Hendrix FF, Campbell WA. 1973. *Pythium* and plant pathogens. Ann Rev Phytopathol 11:77–98.

Hendrix FF, Papa KE 1974. Taxonomy and genetics of *Pythium*. Proc Am Phytopathol Soc 1:200–207.

Hensel P, Greene CE, Medleau L, Latimer KS, Mendoza L 2003. Immunotherapy for treatment of multicentric cutaneous pythiosis in a dog. J Am Vet Med Assoc 223:215–218.

Ichitani T, Amemiya J 1980. *Pythium gracile* isolated from the foci of granular dermatitis in the horse (*Equus caballus*). Trans Mycol Soc Jpn 21:263–265.

Imwidthaya P 1994. Systemic fungal infections in Thailand. J Med Vet Mycol 32: 395–399.

Imwidthaya P 1995. Mycotic keratitis in Thailand. J Med Vet Mycol 33:81–82.

Kamoun S 2003. Molecular genetics of pathogenic oomycetes. Eukaryot Cell 2: 191–199.

Kamoun S, Goodwin SB 2007. Fungal and oomycete genes galore. New Phytopathol 174:713–717.

Khulbe RD 1982. Pathogenicity of some species of *Pythium* Pringsheim on certain fresh water fishes. Mykosen 26:273–275.

Lévesque CA, deCock AW 2004. Molecular phylogeny and taxonomy of the genus *Pythium*. Mycol Res 108:1363–1383.

MacDonald E, Millward L, Ravishankar JP, Money NP 2002. Biomechanical interaction between hyphae of two *Pythium* species (Oomycota) and host tissues. Fungal Genet Biol 37:245–249.

Madoui M-A, Gaulin E, Mathe C, Clemente HS, Coo loox A, Wincker P, Dumar B 2007. AphanoDB: a genomic resource for *Aphanomyces* pathogens. BMC Genom 8:471–477.

Martin FN 1995. Electrophoretic karyotype in the genus *Pythium*. Mycologia 87: 333–353.

Martin FN 2000. Phylogenetic relationships among some *Pythium* species inferred from sequence analysis of the mitochondrially encoded cytochrome oxidase II gene. Mycologia 92:711–727.

McMeekin D, Mendoza L 2000. *In vitro* effect of streptomycin on clinical isolates of *Pythium insidiosum*. Mycologia 92:371–373.

McMullan WC, Joyce JR, Hansellca DV, Heitmann JM 1977. Amphotericin B for the treatment of localized subcutaneous phycomycosis in the horse. J Am Vet Med Assoc 170:1293–1298.

Mendoza, L 2005. *Pythium insidiosum*. In: Merz WG and Hay RJ, editors. Topley and Wilson's microbiology and microbial infections. 10th ed., ASM Press, London, UK. pp. 412–429.

Mendoza L, Hernandez F, Ajello L 1993. Life cycle of the human and animal oomycete pathogen *Pythium insidiosum*. J Clin Microbiol 31:2967–2973.

Mendoza L, Mandy W, Glass R 2003. An improved *Pythium insidiosum*-vaccine formulation with enhanced immunotherapeutic properties in horses and dogs with pythiosis. Vaccine 21:2797–804.

Mendoza L, Newton JC 2005. Immunology and immunotherapy of the infections caused by *Pythium insidiosum*. Med Mycol 43:477–86.

Mendoza L, Prendas J 1988. A method to obtain rapid zoosporogenesis of *Pythium insidiosum*. Mycopathologia 104:59–62.

Mendoza L, J Villalobos, Calija CE, Solis A 1992. Evaluation of two vaccines for the treatment of pythiosis in horses. Mycopathologia 119:89–95.

Miller R 1981. Treatment of equine phycomycosis by immunotherapy and surgery. Austra Vet J 57:377–382.

Miller R 1983. Investigation into the biology of the three phycomycotic agents pathogenic for horses in Australia. Mycopathologia 81:23–28.

Miller RI, Campbell RS 1983. Experimental pythiosis in rabbits. Sabouraudia 21:331–341.

Money NP 1999. Fungus punches its way in. Nature 401:332–333.

Money NP, Davis CM, Ravishankar JP 2004. Biomechanical evidence for convergent evolution of the invasive growth process among fungi and oomycete water molds. Fungal Genet Biol 41:872–876.

Mort-Bontemps M, Fevre M 1995. Electrophoretic karyotype of *Saprolegnia monoica*. FEMS Microbiol 131:325–328.

Neish GA 1977. Observations on saprolegniasis of adult sockeye salmon, *Oncorhynchus nerka* (Walbaum). J Fish Biol 10:513–522.

Patino-Meza F 1988. Role of the zoospores of *Pythium insidiosum* in the experimental reproduction of pythiosis in susceptible species. DVM thesis. National University. Heredia, Costa Rica. pp. 1–31.

Pesavento PA, Barr B, Riggs SM, Eigenheer AL, Pamma R, Walker RL 2008. Cutaneous pythiosis in a nestling white-faced ibis. Vet Pathol 45:538–541.

Plaats-Niterink AJ Vander 1972. The occurrence of *Pythium* in the Netherlands 3. *Pythium flevoence* sp. n. Acta Botanica Neerl 21:633–639.

Plaats-Niterink A.J Vander 1981. Monograph of the genus *Pythium*. Centralaalbureau Voor Schimmelcultures Baarn. Studies in Mycology No 21. Institute of the Royal Netherlands. pp. 1–242.

Ravishankar JP, Davis CM, Davis DJ, MacDoneld E, Makselvan SD, Millward L, Money NP 2001. Mechanics of solid tissue invasion by the mammalian pathogen *Pythium insidiosum*. Fungal Genet Biology 34:167–175.

Rippon JW 1988. Medical mycology. The athogenic fungi and the pathogenic actinomycetes. 3rd ed., WB Saunders Company. Philadelphia, PA. pp. 708–710.

Roberts LS, Janovy GD Jr 2004. Schmidt and Larry S. Roberts' foundations of parasitology. 7th ed. McGraw Hill. New York. pp. 1–670.

Sati SC 1991. Aquatic fungi parasitic on temperate fishes of Kumaun Himalaya, India. Mycoses 34:437–441.

Santurio JM, Argenta JS, Schwendler SE, Cavalheiro AS, Pereira DI, Zanette RA, Alves SH, Dutra V, Silva MC, Arruda LP, et al. 2008. Granulomatous rhinitis associated with *Pythium insidiosum* infection in sheep. Veter Res 163:276–277.

Santurio JM, Leal AT, Leal ABM, Festugatto R, Cobeck L, Sallis ES, Copetti MV, Alves SH, Ferreiro 2003. Three types of immunotherapeutics against pythiosis insidiosi developed and evaluated. Vaccine 21:2535–2540.

Schurko AM, Mendoza L, deCock AWAM, Klassen GR 2003a. Molecular genetic differences between strains of *Pythium insidiosum* from Asia, Australia, and the Americas: Evidence for geographic clusters. Mycologia 95:200–208.

Schurko AM, Mendoza L, Lévesque CA. Desaulniers NL, deCock AWAM, Klassen GR 2003b. A molecular phylogeny of *Pythium insidiosum*. Mycol Res 107:537–544.

Scott, WW, O'Bien, AH 1962. Aquatic fungi association with diseased fish and fish eggs. Progve Fish Cult 24:3–15.

Shenep JL, English BK, Kaufman L, Pearson TA, Thompson JW, Kaufman RA, Frisch G, Rinaldi MG 1998. Successful medical therapy for deeply invasive facial infection due to *Pythium insidiosum* in a child. Clin Infect Dis 27:1388–1393.

Shipton WA 1985. Zoospore induction and release in a *Pythium* causing equine phycomycosis. Trans Br Mycol Soc 84:147–155.

Shipton WA 1987. *Pythium destruens* sp. nov., an agent of equine pythiosis. J Med Vet Mycol 25:137–151.

Shipton WA, Miller RI, Lea IR 1982. Cell wall, zoospore and morphological characteristics of Australian isolates of a *Pythium* causing equine phycomycosis. Trans Br Mycol Soc 79:15–23.

Smith F 1884. The pathology of bursattee. Vet J 19:16–17.

Sogin ML, Silberman JD 1998. Evolution of the protists and protistan parasites from the perspective of molecular systematics. Int J Parasitol 28:11–20.

Sohn Y, Kim D, Oh K, Seol IB 1996. Enteric pythiosis in a Jindo dog. Korean J Vet Res 36:447–451.

Sosa ER, Landsberg JH, Stephenson CM, Forstchen AB, Vandersea MW, Litaker RW. 2007. *Aphanomyces invadans* and ulcerative mycosis in estuarine and freshwater fish in Florida. J Aqua Animal Health 19:14–26.

Stenzel DJ, Boreham PFL 1996. Blastocystis hominis revisited. Clin Microbiol Rev 9:563–584.

Supabandhu J, Fisher MC, Mendoza L, Vanittanakom N 2007. Isolation and identification of the human pathogen *Pythium insidiosum* from environmental samples collected in Thai agricultural areas. Med Mycol 18:1–12.

Tayler BM, Tripathy S, Zhang X, Dehal P, Jiang RHY, Aerts A, Arredondo FD, Baxter L, Bensasson D, Beynon JL 2006. *Phytophthora* genome sequences uncover evolutionary origins and mechanisms of pathogenesis. Science 313:1261–1266.

Thianprasit M 1990. Human pythiosis. Trop Dermatol 4:1–4.

Thitithanyanont A, Mendoza L, Chuansumrit A, Pracharktam R, Laothamatas J, Sathapatayavongs B, Lolekha S, Ajello L 1998. The use of an immunotherapeutic vaccine to treat a life threatening human arteritis infection caused by *Pythium insidiosum*. Clin Infectious Dis 27:1394–1400.

Thomas MM, Bai SM, Jayaprakash C, Jose P, Ebenezer R. 2006. Rhinoentomophthoromycosis. Indian J Dermatol Venereol and Leprol 72:296–9.

Thomas RC, Lewis DT 1998. Pythiosis in dogs and cats Compendium 20:63–69.

Torto-Alalibo T, Tian M, Gajendran K, Waugh ME, van West P, Kamoun S 2005. Expressed sequence tags from the oomycete fish pathogen *Saprolegnia parasitica* reveal putative virulence factors. BMC Microbiol 5:46–58.

Tyler BM 2001. Genetics and genomics of the oomycete-host interface. Trends Gene 17:611–614.

Wanachiwanawin W, Mendoza L, Visuthisakchai S, Motsikapan P, Sathapatayavongs B, Chaiprasert A, Suwangool P, Manuskiatt W, Ruangsetakit C, Ajello L 2004. Efficacy of immunotherapy using antigens of *Pythium insidiosum* in the treatment of vascular pythiosis in humans. Vaccine 22:3613–3621.

Willoughby, LG 1969. Pure culture studies of the aquatic phycomycete *Lagenidium giganteum*. Tran Br Mycol Soc 52:393–410.

Willoughby, LG 1985. Rapid preliminary screening of *Saprolegnia* on fish. J Fish Dis 8:473–476.

20

SAPROLEGNIA—FISH INTERACTIONS

EMMA J. ROBERTSON

Aberdeen Oomycete Group, University of Aberdeen, College of Life Sciences and Medicine, Institute of Medical Sciences, Foresterhill, Aberdeen, Scotland, United Kingdom; Department of Microbiology and Immunology, Albert Einstein College of Medicine, Bronx, New York

VICTORIA L. ANDERSON AND ANDREW J. PHILLIPS

Aberdeen Oomycete Group, University of Aberdeen, College of Life Sciences and Medicine, Institute of Medical Sciences, Foresterhill, Aberdeen, Scotland, United Kingdom; School of Biological Sciences, University of Aberdeen, Aberdeen, Scotland, United Kingdom

CHRIS J. SECOMBES

School of Biological Sciences, University of Aberdeen, Aberdeen, Scotland, United Kingdom

JAVIER DIÉGUEZ-URIBEONDO

Departamento de Micología, Real Jardín Botánico CSIC, Madrid, Spain

PIETER VAN WEST

Aberdeen Oomycete Group, University of Aberdeen, College of Life Sciences and Medicine, Institute of Medical Sciences, Foresterhill, Aberdeen, Scotland, United Kingdom

Emma Robertson and Vicky Anderson contributed equally to this work.

Oomycete Genetics and Genomics: Diversity, Interactions, and Research Tools
Edited by Kurt Lamour and Sophien Kamoun
Copyright © 2009 John Wiley & Sons, Inc.

20.1 SAPROLEGNIALES: FISH PATHOGENS

Many oomycetes species from the order Saprolegniales have the capacity to infect and cause disease in fish or shellfish. Examples include some *Achlya*, *Aphanomyces*, *Leptolegnia*, and *Saprolegnia* species (Atkins, 1954; Phillips et al., 2008). *Aphanomyces astaci* has been impacting the crayfish species throughout Europe since the end of the nineteenth century (Chapter 21 and Edgerton et al., 2004), and *Saprolegnia ferax* and *Saprolegnia diclina* are causing declines in amphibian populations (Kiesecker et al., 2001; Fernández-Benéitez et al., 2008). *Saprolegnia parasitica* is known to be a devastating pathogen on many freshwater species of fish, and *S. diclina* is a potent pathogen of fish eggs. Both are currently causing significant economic damage in the global fish farming industry (van West 2006; Phillips et al., 2008). Previously, *S. parasitica* and *S. diclina* infections were controlled with the biocide malachite green. However, there was a worldwide ban on the use of this chemical because of its potential toxicological and carcinogenic effects on both fish and its consumers (reviewed in Alderman, 1994; Marking et al., 1994a). As a result of this ban, there has been a significant increase in *Saprolegnia* infections (Marine Harvest and Landcatch Ltd, personal communications).

20.2 ECONOMIC IMPACT OF *S. PARASITICA*

Because of overfishing within the seas in recent years, fish production has become dependent on fish farming for an adequate supply, and as a result, aquaculture has become the world's fastest growing food sector. This fact is reflected by the astonishing increase in fish production through fish farming. In 2004, a staggering 45.5 million tons of fish were farmed worldwide, which are worth US$ 63.4 billion. This represents a 12.6% and 15.3% increase, respectively, over reported figures in 2002 (FAO Fishery Information: www.fao.org). Aquaculture has both social and economic importance in many countries, which include China, Canada, Chile, Japan, Norway, the United States, and the United Kingdom. The greatest cause of economic loss in aquaculture is a result of diseased fish, with bacterial diseases being the most common, closely followed by fungal and oomycete infections (Meyer, 1991). *S. parasitica*, which is a ubiquitous freshwater pathogen (Diéguez-Uribeondo et al., 2007), can infect not only a wide range of wild and farmed fish (Fig. 20.1), but also may cause infection in hobby fish tanks. Amazingly, 1 in 10 hatched salmon raised in fish farms die as a result of *S. parasitica* infection (Marine Harvest, personal communication). In Japanese freshwater ponds, *Saprolegnia* species are among the highest cause of infection in salmonid fish (Hussein and Hatai, 2002). *S. parasitica* has been reported to cause mass mortality of coho salmon in Japanese salmon farms (Hatai and Hoshiai, 1992). *S. parasitica* is also the cause of the disease termed "winter kill," which affects catfish in the United States and causes significant financial losses in the farming of this fish (Bly et al., 1992). In Chile, one of the main fish-producing countries, *S. parasitica* has

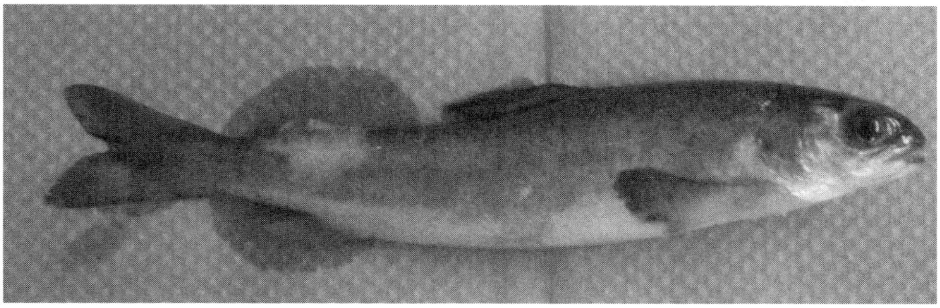

FIG. 20.1 *Saprolegnia parasitica* mycelial growth on Atlantic salmon (*Salmo salar*). *S. salar* parr infected with *S. parasitica*. Areas of mycelial growth are highlighted with arrows. Scale bar represents 1 cm.

become a major threat for farms of Atlantic salmon, coho salmon, and rainbow trout (Zaror et al., 2004). In addition to aquaculture, *S. parasitica* has been reported to have a major impact in wild fish populations globally. *Saprolegnia* species have been isolated from headburn lesions of wild adult Chinook and Steelhead salmon. Also, the disease is having a significant impact on the populations of these species in the Northwest United States, where up to 22% of returning salmon die as a result of headburn lesions that have become infected with *S. parasitica* (Neitzel et al., 2004).

20.3 THE LIFECYCLE OF *SAPROLEGNIA* SPECIES

The Saprolegniales (Oomycetes) are diploid organisms, and the life cycle of *S. parasitica* consists of both sexual and asexual stages. Sexual reproduction involves the fertilization of oospheres, which are located in the oogonium, with a male nucleus that is released from the antheridium (Hughes, 1994). In the asexual cycle, mycelia, which are visible to the eye, grow in and on the surface of the infected host (fish/eggs or colonized substrate). When there is a decrease in surrounding nutrients (Tiffney, 1939), the hyphal tips begin to swell and form sporangia, in which primary zoospores are generated (Fig. 20.2a–c). When the tips of the sporangia burst, apically biflagellate zoospores are released into the water and remain in this state for a short period of time. These spores, which are named primary zoospores, are weak swimmers, and their principal role is to disperse from the parent colony (Beakes et al., 1994). These primary zoospores encyst and form primary cysts, which subsequently release secondary zoospores (Fig. 20.2d). The secondary zoospores are laterally biflagellate, highly motile, and considered the infective stage of the *S. parasitica* life cycle (Diéguez-Uribeondo et al., 1994a, 2006; Andersson and Cerenius, 2002). Secondary zoospores can remain in this stage for a short period of time, which may be several hours or days, until they detect a potential host. The zoospore encysts to form a secondary cyst (Beakes et al., 1994; Diéguez-Uribeondo et al., 2006). The surface of secondary cysts are ornamented with

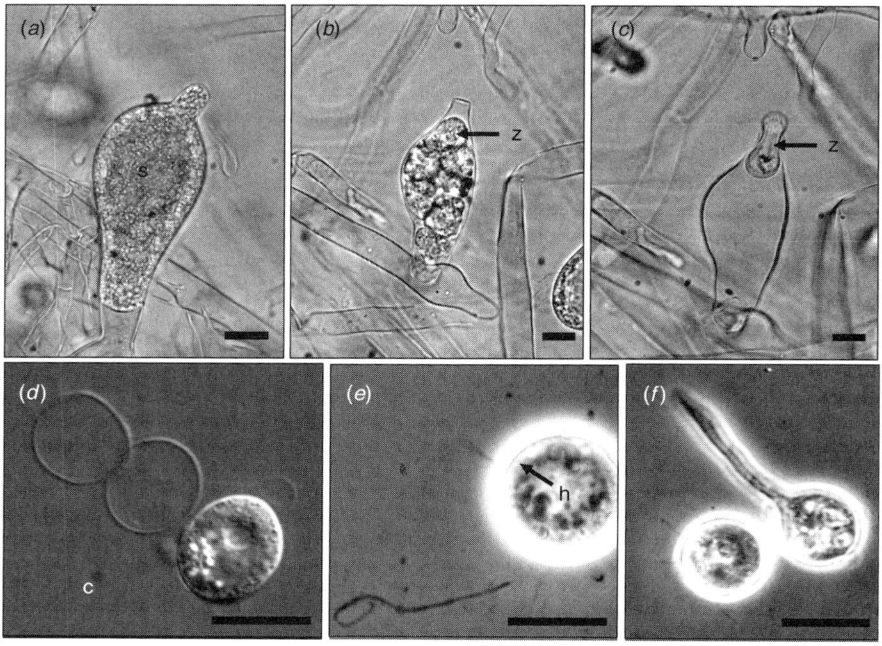

FIG. 20.2 Asexual life cycle stages of *Saprolegnia parasitica*. (a) A mycelial tip swelling to produce a sporangium (s) (scale bar represent 15 μm). (b, c) The sporangium cleaves to release motile zoospores (z) (scale bars represent 10 μm). (d) The zoospores can encyst. (c) A phenomenon known as repeated zoospore emergence allows the cycle between zoospore and cyst to occur many times for a suitable host to be found (Diéguez-Uribeondo et al., 1994a). (e) The secondary cyst has hairs (h) on its surface. (f) On attachment to a suitable host, the cyst germinates and penetrates into the fish epidermal tissues. (Panels d, e, and f scale bars represent 5 μm). From Robertson et al. (2008).

spines of varying lengths (Fig. 20.2e). In *S. parasitica*, these hairs are long; they measure over 2 μm in length and on the end are "boat hooks," which are thought to aid the attachment of cysts to the host (Beakes, 1983; Beakes et al., 1994; Fregeneda-Grandes et al., 2000; Diéguez-Uribeondo et al., 2007). After attachment, the cysts begin to germinate (Fig. 20.2f) and colonize the host (fish/eggs) or substrate. Some *Saprolegnia* spp. have a physiological pathway in their asexual life cycle that is unique to a range of oomycetes, which include *Aphanomyces* (Cerenius and Söderhäll, 1984, 1985; Diéguez-Uribeondo et al., 1994a, 1996, 2006, 2007; Bangyeekhun et al., 2003). Once a secondary cyst has developed, it may follow one of two different pathways: germination or formation of a new laterally biflagellate zoospore-infective unit. The formation of a new generation seems to occur when nonspecific stimuli, (mechanical, physical, etc.) have triggered encystment instead of specific stimuli from the host, such as an amino acid, proteins, and so on. This pathway is termed repeated zoospore emergence (RZE), or polyplanetism, and it is an efficient method for increasing the opportunity for individual zoospores to search, adhere, and infect a suitable host (Cerenius and Söderhäll, 1984, 1985; Beakes et al., 1994; Diéguez-Uribeondo et al., 1994a). Thus, in the *Aphanomyces* genus, it has been proposed as a mechanism of adaptation to a parasitic mode of life (Cerenius and Söderhäll, 1985; Diéguez-Uribeondo et al., 2006).

20.4 SAPROLEGNIOSIS

On fish, saprolegniosis is characterized by white or gray patches of cotton wool-like filamentous mycelia (Hatai and Hoshiai, 1992). In general, infection initially appears on epidermal tissues of the head, tail, and fins of the fish (Tiffney 1939; Hatai and Hoshiai, 1992; Fregeneda-Grandes et al., 2001; Hussein and Hatai, 2002), and it subsequently spreads to the rest of the body (Fig. 20. 1). Lesion areas may be soft, necrotic, and ulcerated, and the surrounding areas may show edema and necrosis (Gieseker et al., 2006). Saprolegniosis has been documented in a range of fish species, which includes Atlantic salmon (*Salmo salar* L.), brown trout, (*Salmo trutta* L.), coho salmon (*Oncorhynchus kisutch*), perch (*Perca fluviatilis*), masu salmon (*Oncorhynchus masou*), rainbow trout (*Oncorhynchus mykiss*), Japanese char (*Salvelinus leucomenis*), sockeye salmon (*Oncorhynchus nerka*), channel catfish (*Ictalurus punctatus*), and many more (Tiffney, 1939; Hatai and Hoshiai, 1992; Bly et al., 1993; Hussein and Hatai, 2002; Stueland et al., 2005). Saprolegniosis has also been reported in other aquatic animals, such as freshwater crayfish (Diéguez-Uribeondo et al., 1994b) and amphibians (Kieseker et al., 2001; Fernández-Benéitez et al., 2008).

Although saprolegniosis generally affects the epidermis of fish, infection has been identified in internal organs too (Fregenda-Grandes et al., 2001; Zaror et al., 2004; Gieseker et al., 2006). Gieseker et al. (2006), for example, identified hyphae of *S. parasitica* that had proceeded to invade the caudal blood vessels, spinal chord, and posterior kidney. It is possible that the pathogen may gain access to the bloodstream via penetration through the gills or invasion of blood vessels in the superficial lesions. A range of other histological changes can occur in the body of fish infected with *S. parasitica*. Hussein and Hatai (2002), for example, reported degenerative changes in musculature, edema between muscle bundles, and a reduction of the epidermal layer. It is believed that *S. parasitica* mycelium penetrates the epidermis of the fish, which disrupts the integrity of the skin. As a result of this destruction, death is most likely caused by a breakdown in osmoregulation and an isotonic imbalance, which leads to hemodilution (Richards and Pickering, 1979; Tiffney, 1939).

20.5 HOST–PATHOGEN INTERACTIONS

It is well established that stressed fish become predisposed to a range of bacterial, fungal, and oomycete infections, which are likely attributable to, among other things, an impaired immune response. A range of conditions may lead to stress, such as temperature changes, poor water quality, insufficient dissolved oxygen, sexual maturation, overcrowding of stocks, excessive handling, inadequate diets, physical abuse, or toxic substances (reviewed in Meyer, 1991; Pickering, 1994). Temperature in particular seems to play a role in causing fish susceptibility to saprolegniosis; catfish become immunocompromised during the winter months when the water temperatures drop significantly, which renders the fish more susceptible to infections (Bly et al., 1993). Quiniou et al. (1998) also demonstrated that an acute decrease in water temperature leads to a reduction in mucosal cells in the catfish epidermis.

Although *S. parasitica* was initially considered a secondary or opportunistic pathogen, some reports have described strains that have the capacity to act as a primary pathogen on brown trout and Atlantic salmon (Richards and Pickering, 1979; Stueland et al., 2005; van West, 2006). Indeed, Stueland et al. (2005) described two hypervirulent *Saprolegnia* strains that cause 89% and 31% cumulative mortality in challenged Atlantic salmon, which had been stressed by a combination of cold water temperature, starvation, and amimomi treatment (agitating the fish in nets). In addition, Puckeridge et al. (1989) also suggested that bony bream *Nematalosa erebi* (Günther) may suffer from primary saprolegniosis. Thus, if it is not an opportunistic pathogen, it is likely that, similar to other oomycetes (see Chapter 18), *Saprolegnia* species have evolved pathogenicity effectors that contribute toward its infection and colonization of fish.

20.6 CYST COAT ONTOLOGY: LONG, HOOKED HAIRS ON SOME *SAPROLEGNIA* ISOLATES

It has been suggested (Diéguez-Uribeondo et al., 2007) that specific *Saprolegnia* species have adapted to parasitism at the spore level, which may be reflected by their secondary cysts possessing bundles of long, hooked hairs; their ability to undergoing RZE; and their retracted germination pattern. These infection strategies may provide some strains with increased opportunities to infect and colonize susceptible fish.

Transmission electron microscopy (TEM) allows the visualization of long, hooked hairs on the surface of some *Saprolegnia* species. The structure of the hairs is highly variable; some hairs can have three or more hooks at their tips, and some may also have hooks on their shafts. Some long hairs begin to bend and curve and may even form hoops, whereas others are more rigid (Beakes et al., 1995). A study by Hatai et al. (1990) examined various *Saprolegnia* species that had been isolated from Japanese fish hatcheries and also demonstrated that different species of *Saprolegnia* can have very diverse cyst coat ontology. A linear relationship exists between mean bundle length and number of hairs per bundle. It should be noted that some *Saprolegnia* strains do not possess these hairs on their secondary cysts (Stueland et al., 2005; Fregenda-Grandes, 2007).

A review by Beakes et al. (1994) proposed potential roles for these hooked hairs: They may play a role in the buoyancy of the cysts in water, which slows down their sedimentation rate to maintain their circulation in the water and subsequently increases their chances of finding a host. The contact area of the pathogen with a potential host is increased by possessing long hairs, and the hairs themselves may be coated in some kind of adhesive material. Again, this property increases the likelihood of their attachment to a moving fish in the water.

A collection of studies by Fregeneda-Grandes et al. (2000, 2001, 2007) described the identification of hairs on *Saprolegnia* isolates collected from infected wild brown trout using TEM, their capacity to cause infection on rainbow trout, and attempts to detect them using monoclonal antibodies. In the first of these three studies, they collected isolates from infected trout, healthy trout, and water, and when these isolates were induced to produce secondary cysts, hairs that grouped into bundles were observed in all samples. They also identified a direct correlation among the number of bundles present on each cyst, the number of hairs per bundle, and the length of the hairs. Based on these details, isolates were sorted into two groups: Group I had a higher number of bundles per cyst, and these bundles consisted of a greater number of hairs, which were longer in length. Conversely, group II consisted of cysts with fewer bundles of shorter hairs. There was also a trend between the cysts morphological group and the origin of where the sample was collected. Despite the notion that pathogenicity is likely related to hair length and number, the samples that were taken from lesions on trout with saprolegniosis belonged to group II (less bundles per cysts with shorter hairs), whereas those that originated from water or healthy trout belonged to morphotype I. The subsequent study by these authors (2001) also demonstrated that isolates from group II are indeed more pathogenic and cause greater levels of mortality when used in infection assays with rainbow trout. It should be noted, however, that this research is correlative and does not provide causative evidence that secondary cyst coat ontology is responsible for the invasiveness of *Saprolegnia*. Finally, in 2007, the authors attempted to use monoclonal antibodies to distinguish between *Saprolegnia* species that possess/lack the hair bundles; however, none of the five antibodies tested bound specifically to either group I or group II. Interestingly, based on phylogenetic analysis of internal spacer (ITS) ribosomal sequences of *Saprolegnia* spp., Diéguez-Uribeondo et al. (2007) concluded that all *S. parasitica* isolates studied have bundles of long hairs, and that despite the cyst coat ontology, both morphotypes described by Fregeneda-Grandes et al. (2000, 2001) are members of the same phylogenetic species and do not group in separate clades.

Another study by Stueland et al. (2005) also demonstrated that when performing pathogenicity assays on Atlantic salmon, the most virulent strains possessed bundles of long, hooked hairs on their secondary zoospores.

However, they also report that two avirulent *S. parasitica* strains possess these hairs, and they, like Fregeneda-Grandes et al. (2001), conclude that the bundles of hairs seem to be a requisite for, but do not determine, pathogenicity of *Saprolegnia* strains.

20.7 REPEATED ZOOSPORE EMERGENCE AND RETRACTED GERMINATION

Diéguez-Uribeondo et al. (1994a, 1996) described that in the laboratory it is possible to mimic conditions that cause either cyst germination or RZE. They recorded that if the triggers required for germination were not given within 45 min post encystment, then a new zoospore was released from the cyst spontaneously. The strains tested could perform up to six consecutive rounds of RZE, but after this time they were observed to lyse. Bangyeekhun et al. (2003) also noted that if each zoospore generation was left to swim for 150 min, then four generations could be produced. This process of RZE may provide the pathogen with an increased opportunity to locate and infect hosts. RZE does seem to be strain dependant, however, and it is not conserved in all *Saprolegnia* species (Bangyeekhum, 2003; Diéguez-Uribeondo et al., 1996) or in all *S. parasitica* strains (Diéguez-Uribeondo et al., 2007).

Under low-nutrient conditions, *Saprolegnia* cysts seem to germinate in a very distinctive manner: Initially, narrow germ tubes that are frequently septated form, and after approximately 20–500 μm of this type of growth, normal growth resumes (reviewed in Beakes et al., 1994). This retracted germination process, which is also known as indirect germination, permits the pathogen to grow at very high elongation rates. Stueland et al. (2005) noted that two pathogenic strains of *S. parasitica* also showed high germination rates in sterilized tap water. The two mentioned strains generated 59% and 65% germlings, whereas two other strains that had low pathogenicity frequencies only exhibited 1–19% germination rates. Diéguez-Uribeondo et al. (2007) proposed that the retracted germination and high germination rates of some *Saprolegnia* strains may aid the colonization of the host by permitting faster growth on the host surface where nutrient levels may be low until growth into the deeper, more nutrient-rich tissues is established.

20.8 *S. PARASITICA* VIRULENCE FACTORS

Torto-Alalibo et al. (2005) analyzed sequences from a *S. parasitica* mycelial cDNA library and identified many expressed sequence tags (ESTs) that may play a role in pathogenicity. Four sequences appeared similar to a fungal-type I cellulose binding domain (CBD), which they propose may be involved in the binding of the pathogen to organic substances present on the host surface. Seven ESTs showed similarities to the glycoprotein cellulose binding, elicitor and lectin-like protein (CBEL). CBEL plays many roles in *Phytophthora parasitica*, which include binding to cellulose, lectin activities, and also necrosis of the host. It is possible that in *S. parasitica*, these CBELs may play a role in the initial attachment to the host. The authors also identified several cDNAs that exhibited similarity to the aspartyl, serine, and cysteine proteases. The proteases generated and secreted by *S. parasitica* may be used to break down the epidermal barrier of the fish host, for example. Peduzzi and Bizzozero (1977) also identified a chymotrypsin-like protease from three *Saprolegnia* species and hypothesized that this proteolytic activity may indeed be involved in pathogenicity. In addition to proteases, Torto-Alalibo et al. (2005) identified two types of protease inhibitors: a Kazal-like serine protease inhibitor and a cystatin class of cysteine protease inhibitor. It is possible that *S. parasitica* secretes these proteins to block the proteolytic activity of any proteases released by the threatened fish. As a result of preventing the actions of the fish proteases, this may permit the pathogen to infect and colonize the fish.

In recent years, much excitement has been generated in the oomycete research field following the discovery of a host translocation system for pathogen proteins with similarities to a system found in the malaria parasite *Plasmodium*. In *Plasmodium*, the proteins translocate into host cells, manipulate the red blood cells to suppress immune responses, and facilitate infection (Romisch, 2005). In plant pathogenic oomycetes, the translocation system is characterized by the presence of an N-terminal RxLR motif (Birch et al., 2006). Many putative RxLR-containing proteins have been identified in *S. parasitica* cDNA libraries, and it will prove interesting to examine whether this motif is essential for the protein to act as a disease effector (Phillips et al., 2008). It is clear that despite the fact that few virulence factors have been analyzed at the molecular level, *S. parasitica* seems to use several factors that may cumulatively result in its pathogenicity.

20.9 FISH RESPONSES

The cellular response of fish to *Saprolegnia* infection has not been well studied, and currently little is known about what happens when a host comes into contact with the pathogen. Fish have three layers of defense, of which the first are primarily external barriers that include the mucosal layer, which prevents the invasion or colonization of pathogens. In case the pathogen is able to pass the first barrier, both the innate and adaptive immune responses usually take over. In fish, neutrophils account for only a small proportion of circulating leukocytes, although Wood et al. (1986) showed they play a potentially important role during a *Saprolegnia* infection of brown trout. The role of macrophages during a *Saprolegnia* infection has also been implicated when Kales et al. (2007) observed that when challenging the RTS11 monocyte/macrophage cell line (derived from rainbow trout) with *S. parasitica*, within 48 h post challenge, most of the hyphal surface was coated with macrophages.

This suggests that macrophages may be chemotactically drawn toward the invading pathogen. The macrophages, however, could not ingest the *S. parasitica* cysts, most likely because the cysts are too large to be phagocytosed. *Saprolegnia* cysts are around 10 μM in size, whereas trout macrophages are only 7–15 μM in diameter (Kales et al., 2007).

There is a debate as to whether *S. parasitica* suppresses or modifies the immune system of the fish host. Many reports indicate that there is a lack of a leukocyte inflammatory response after the onset of saprolegniosis (Puckeridge et al., 1989 and references therein; Bly et al., 1992; Alvarez et al., 1995). However, it should be highlighted that in some cases, in which the onset of disease is related to a decrease in water temperature, the lack of an inflammatory response may be a direct result of the cold water shock on the fish, and perhaps it is not attributable to *Saprolegnia* infection. Indeed, Pickering (1986) described that during the winter months, leukocyte numbers in mature brown trout are significantly reduced, which correlates with their susceptibility to saprolegniosis. Other reports indicate that *Saprolegnia* infections stimulate an acute phase response in infected fish (Roberge et al., 2007; Bly et al., 1994). Using microarray analysis, Roberge et al. (2007) identified Atlantic salmon genes that were upregulated after *Saprolegnia* infection: Acute phase proteins were identified, including proteins from the complement pathways, proteases that exhibit collagenolytic activity that may aid leukocyte transmigration, a G-protein-coupled receptor involved in skin inflammation, and TAP2, which is an ATP-binding cassette transporter that plays important roles in MHC class I antigen presentation. Other acute phase proteins, such as cytokine receptors, leukocyte chemotaxins, and agglutination and aggregation factors, were also upregulated. Despite contradictory reports, it is clear that acute phase proteins are indeed induced by *Saprolegnia* infection, and perhaps the lack of an immune response observed in other studies is indeed a result of the cold-water effect on the fish immune system.

20.10 TOOLS FOR *SAPROLEGNIA* RESEARCH

20.10.1 Infection Models

Infection models are a vital tool when it comes to studying a whole range of aspects of disease. In the case of *Saprolegnia*, for example, they may be used to study host–pathogen interactions, to perform pathogenicity tests, to compare the effects of different species/strains of *Saprolegnia* on a range of fish species, or to perform drug screens and evaluate the efficacy of drugs against saprolegniosis.

In vitro experiments can be performed at the cellular level, as demonstrated by Kales et al. (2007). The authors used a monocyte/macrophage cell line RTS11, which originated from rainbow trout spleen, to determine the response of the cells to *S. parasitica*. At the Aberdeen Oomycete Group, we also perform assays using the RTG-2 cell line, which is derived from rainbow trout gonads. We have performed a range of experiments using these cell lines, which have subsequently been challenged with *S. parasitica*, including the generation of a cDNA library from infected cell lines, scanning electron microscopy and transmission electron microscopy to determine the mode of *Saprolegnia* infection (intercellular or intracellular growth), and transcript profiling to determine which host genes are upregulated or downregulated following the onset of infection (Phillips et al., in preparation).

Although *in vitro* fish cell line assays permit an insight into fish–pathogen interactions, what occurs at the cell line level is not necessarily reflective of the response of the fish as a whole. A range of methods has been described to induce stress in the test fish, which predisposes them to infection, including cutaneous scarification, exposure to toxic compounds, change of water temperature, a combination of scarification and a decrease in water temperature, administration of cortisol, and the implantation of an androgen to change hormone levels (Fregeneda-Grandes et al., 2001 and references therein). Ami-momi treatment has also been shown to be effective in predisposing fish to saprolegniosis (Fregenda-Grandes et al., 2001; Hussein and Hatai, 2002; Stueland et al., 2005). The technique involves shaking the fish in a net, which leads to damage and loss of the epidermis and a decrease in the number of mucous cells, which leads to scarification and stress, and is based on replicating the stress that may be a result of excessive handling of fish in aquaculture (Hussein and Hatai, 2002).

20.10.2 *Saprolegnia* EST Projects

A very limited amount of information is known about the genomes of any *Saprolegnia* species. Mort-Bontemps and Fevre (1995) used contour-clamped electric field (CHEF) gel electrophoresis to estimate the size of the *Saprolegnia monoica* genome as 51 Mb. The authors identified eight chromosomal bands, correlating to 16 chromosomes, which is significantly less than the 21 chromosomes of its relative *S. ferax* (Mort-Bontemps and Fevre, 1995). The size of the *S. parasitica* genome is unknown. Currently, 1,510 ESTs from *S. parasitica* mycelia have been sequenced and annotated, and they are contained within a publicly available database (www.oomycete.org) (Torto-Alalibo et al., 2005). These ESTs generated from mycelia provide the first insight into genes expressed from *S. parasitica*. The information gained from the *S. parasitica* library described above is extremely valuable, although it solely provides an insight into the vegetative stage of the pathogen. Currently in the Aberdeen Oomycete Group we are annotating a zoospore/cyst/germinating cyst cDNA library and two interactive cDNA libraries: One cDNA library consists of RTG-2 trout gonad cell lines that have been harvested 8 h post infection with *S. parasitica*. The second cDNA library originates from a colony of *S. parasitica* that was isolated from the skin of an infected Atlantic salmon.

20.10.3 Gene Silencing in *Saprolegnia* Species

It has proven to be challenging to transform all oomycete species, and *Saprolegnia* is no exception. The transformation of *Saprolegnia* species has only been reported in *S. monoica* (Mort-Bontemps and Fevre, 1997). Stable transformants that exhibit hygromycin B resistance were obtained, with a transformation frequency of 0.1–0.2 transformants per μg of DNA. This rate is comparable with that observed at the time in *Phytophthora infestans* and *Phytophthora megasperma* (Judelson et al., 1991, 1993). Southern blot analysis suggested that a single vector copy was incorporated into the *S. monoica* genome.

As an alternative to generating stable transformants in oomycetes, many studies have been performed to generate transiently silenced strains using the method of RNA interference (RNAi). RNAi is a posttranslational gene-silencing technique that can be induced in many eukaryotes (Clemens et al., 2000; Misquitta and Paterson, 1999; Ngo et al., 1998; Tabara et al., 2002). Although gene silencing via RNAi is not stable, normal gene expression returns after 15 days in *P. infestans* (Whisson et al., 2005), silencing is maintained long enough to perform functional characterisation studies. We are currently attempting to develop RNAi in *S. parasitica*. Because the transformation of *S. parasitica* is proving difficult, RNAi technology for *S. parasitica* would provide an alternate means to transiently silence genes from this oomycete.

20.11 CONTROL METHODS FOR SAPROLEGNIOSIS

When it comes to dealing with *S. parasitica* infection, ultimately the easiest and most cost-effective way to control the disease is good husbandry techniques. Meyer (1991) described the main areas that fish health management should focus on to prevent the onset of disease in fish farms. This includes minimizing stressful environmental and physiological conditions, maintaining water supply standards and nutritionally complete diets, and isolating infected fish from healthy stocks. In reality, however, preventing saprolegniosis is undoubtedly a near-impossible task, and the use of drugs to both prevent and control the disease has been inevitable.

20.11.1 Alternatives to Malachite Green

As a result of the ban of malachite green, many alternative control measures for saprolegniosis have been tested, such as glutaraldehyde; hydrogen peroxide; iodine; the salts potassium permanganate and copper sulphate; and the herbicides diquat, simazine, hydrothol 191, and aquathol K.

Three commercial products that contain the chemical formalin, or formaldehyde, have been granted approval from the United States Food and Drug

Administration to treat fish eggs that are infected with fungus or *Saprolegnia*. Formalin has been shown to control *Saprolegnia* infection effectively in both fish and their eggs (Marking et al., 1994b; Gieseker et al., 2006). Bly et al. (1996) showed that formalin can prevent *Saprolegnia* infection on catfish; however, the concentration of formalin used should be monitored carefully, as it is toxic to the catfish at high concentrations. Formalin is currently the method used in aquaculture to control saprolegniosis; however, its use is also likely to be banned in the future.

The effect of high concentrations of sodium chloride (NaCl) on the inhibition of *Saprolegnia* growth has been reported several times (Marking et al., 1994b; Ali, 2005). Increasing concentrations of NaCl (0.07–0.14 M) causes *Saprolegnia* hyphae to become branched and crippled initially, and subsequently to plasmolyse (Ali, 2005). Additionally, at NaCl levels of 0.03 M, the formation of sporangia and any subsequent release of zoospores was rarely observed. Despite its apparent efficacy in *Saprolegnia* control, increasing salt concentration is not routinely used in fish farms because of the amount of salt that would be required to treat the fish continuously (Finfish Ltd., personal communication).

Although *S. parasitica* is not a true fungus, two antifungal drugs (amphotericin B and chitosan) have proven effective at controlling *Saprolegnia* infections. Bly et al. (1996) described that Amphotericin B can be used to control infections on channel catfish. Amphotericin B binds to sterols, which are components of the cell membrane, and forms a pore that leads to the loss of potassium from the cell. This action disrupts a range of metabolic processes. In addition, Muzzarelli et al. (2001) described the use of a range of chitosan derivatives as fungistatic agents against *S. parasitica*. When exposed to these drugs, abnormal hyphal growth was observed, and radial growth was significantly inhibited. The authors state that chitosan itself has been deemed as safe for oral administration, and it can be used as a dietary supplement for fish. The modified chitosans used in this study are also regarded as nontoxic. Like Amphotericin B, chitosan may interact with a cell surface component(s) of *S. parasitica*, which leads to the instability of the cell wall and perhaps a change in osmotic balance that causes the cells to lyse, inhibiting the growth of *S. parasiti*ca.

Bronopol (2-bromo-2-nitropropane-1,3-diol), which is an antimicrobial compound, is an inhibitor of thiol-containing dehydrogenase enzymes that seems to be effective in the control of *Saprolegnia*. Like the antifungal drugs, it is believed that bronopol targets the cell membranes of the pathogen, which causes them to lose their integrity and result in *Saprolegnia* death. Pottinger and Day (1999) and Branson (2002) have reported its efficacy in controlling *Saprolegnia* infection in rainbow trout. Indeed, the study from Pottinger and Day (1999) showed the drug can protect susceptible fish from infection, and when used at a higher concentration it could also prevent infection in fertilized rainbow trout ova. As bronopol seems to cause no serious toxicological effects in either humans or fish, perhaps more research should be concentrated on this compound for its use in aquaculture as an alternative to formalin.

20.11.2 Vaccine Potential?

Clearly the currently available methods for controlling saprolegniosis are not entirely effective. A future direction could be the development of a vaccine against *S. parasitic*a. Currently, fish are routinely vaccinated against a range of bacterial, and sometimes viral, diseases while being reared on fish farms. For example, vaccines against the bacteria *Aeromonas salmonicida, Yersinia ruckeri, Vibrio anguillarum, Vibrio ordalii*, and *Vibrio salmonicida* have been used in aquaculture for many years (Reviewed in Gudding et al., 1999). Vaccination can be delivered in three ways: injection (generally intraperitoneally), immersion of the fish in a vaccine solution, or oral administration, although only the former two are used routinely in the commercial sector (Gudding et al., 1999). Vaccines, such as those used to prevent the three *Vibrio* strains named above, are generally inactivated bacterins, although testing has been performed on live, attenuated vaccines against, for example, *A. salmonicida* that causes the disease furunculosis (Marsden et al., 1998). DNA vaccines, in which the gene encoding the antigen, as opposed to the antigen itself, is injected into fish muscle, have also been tested (Reviewed in Lorenzen and LaPatra, 2005). The most efficient DNA vaccines seem to be against the salmonid pathogens infectious hematopoietic necrosis virus (IHNV) (Anderson et al., 1996) and viral hemorrhagic septicemia virus (VHSV) (Lorenzen et al., 1999). These DNA vaccines have been deemed safe for use in fish, and a vaccine against IHNV has been approved and granted a Canadian product license (Hensley, 2005).

20.12 FUTURE DIRECTIONS

The morphological characteristics of *S. parasitica* are now clearly understood, although the molecular mechanism of infection has yet to be fully elucidated. A complete genome sequencing project or the sequencing of a significantly large number of cDNAs would advance our potential for understanding the genes that are important in establishing an infection. The optimization of stable transformation and transient gene silencing with RNAi will, no doubt, prove to be crucial advances for the *Saprolegnia* research field, and it will us to silence genes of interest to both fundamental biology and pathogenicity.

To ease the pressure of *Saprolegnia* infections in aquaculture, the development of a new control strategy is essential. Currently, we are researching the potential for developing a vaccine that could be used as a prophylactic against saprolegniosis in the fish farming industry. The impact of an effective vaccine against *S. parasitica* would be immense and would once again keep this devastating pathogen under control in the fish farming industry.

REFERENCES

Alderman DJ 1994. Control of oomycete pathogens in aquaculture. In: Mueller GJ, editor. *Salmon saprolegniasis*, U.S. Department of Energy, Bonneville Power Administration. Portland, OR. pp. 111–129.

Ali EH 2005. Morphological and biochemical alterations of oomycete fish pathogen *Saprolegnia parasitica* as affected by salinity, ascorbic acid and their synergistic action. Mycopathologia. 159:231–243.

Alvarez F, Villena A, Zapata A, Razquin B 1995. Histopathology of the thymus in *Saprolegnia*-infected wild brown trout, *Salmo trutta* L. Veter Immunol. Immunopathol. 47:163–172.

Anderson ED, Mourich DV, Fahrenkrug SC, LaPatra S, Shepherd J, Leong JA 1996. Genetic immunization of rainbow trout *Oncorhynchus mykiss* against infectious hematopoietic necrosis virus. Mol Mar Biol Biotechnol 5:114–122.

Andersson MG, Cerenius L 2002. Pumilio homologue from *Saprolegnia parasitica* specifically expressed in undifferentiated spore cysts. Eukaryot Cell 1:105–111.

Atkins D 1954. Further notes on a marine member of the Saprolegniaceae, *Leptolegnia marina* n.sp., infecting certain invertebrates. J Mar Biol Assoc UK 33:613–625.

Bangyeekhun E, Pylkkö P, Vennerström P, Kuronen H, Cerenius L 2003. Prevalence of a single fish-pathogenic *Saprolegnia* sp. clone in Finland and Sweden. Dis Aquat Org 53:47–53.

Beakes G 1983. A comparative account of cyst coat ontogeny in saprophytic and fish-lesion pathogenic isolates of the *Saprolegnia diclina-parasitica* complex. Can J Bot 61:603–625.

Beakes GW, Burr AW, Wood SE, Hardham AR 1995. The application of spore surface features in defining taxonomic versus ecological groupings in oomycete fungi. Can J Bot 73 1:S701–S711.

Beakes GW, Wood SE and Burr AW 1994. Features which characterize *Saprolegnia* isolates from salmonid fish lesions — a review. In Mueller GJ, ed., *Salmon Saprolegniasis*, U.S. Department of Energy, Bonneville Power Administration. Portland, OR. pp. 33–66.

Birch PR, Rehmany AP, Pritchard L, Kamoun S, Beynon JL 2006. Trafficking arms: oomycete effectors enter host plant cells. Trends Microbiol 14:8–11.

Bly JE, Lawson LA, Abdel-Aziz ES, Clem LW 1994. Channel catfish, *Ictalurus punctatus*, immunity to *saprolegnia* sp. J Appl Aquaculture 3:35–50.

Bly JE, Lawson LA, Szalai AJ, Clem LW 1993. Environmental factors affecting outbreaks of winter saprolegniosis in channel catfish, *Ictalurus punctatus* Rafinesque. J Fish Dis 16:541–549.

Bly JE, Lawson LA, Dale DJ, Szalai AJ, Durborow RM, Clem LW 1992. Winter saprolegniosis in channel catfish, Dis Aquat Org 15:155–164.

Bly JE, Quiniou SMA, Lawson LA, Clem LW 1996. Therapeutic and prophylactic measures for winter saprolegniosis in channel catfish. Dis Aquat Org 24:25–33.

Branson E 2002. Efficacy of bronopol against infection of rainbow trout *Oncorhynchus mykiss* with the fungus *Saprolegnia* species. Vet Rec 151:539–541.

Cerenius L, Söderhäll K 1984. Repeated zoospore emergence from isolated spore cysts of *Aphanomyces astaci*. Exp Mycol 8:370–377.

Cerenius L, Söderhäll K 1985. Repeated zoospore emergence as a possible adaptation to parasitism in *Aphanomyces*. Exp Mycol 9:9–63.

Clemens JC, Worby CA, Simonson-Leff N, Muda M, Maehama T, Hemmings BA, Dixon JE 2000. Use of double-stranded RNA interference in *Drosophila* cell lines to dissect signal transduction pathways. PNAS 97:6499–6503.

Diéguez-Uribeondo J, Cerenius L, Dyková I, Gelder SR, Henttonen P, Jiravanichpaisal P, Lom J, 2006. Söderhäll. Pathogens, parasites and ectocommensals. In Souty-Grosset C, Holdich DM, Noël PY, Reynolds JD, Haffner P, editor. Atlas of Crayfish in Europe. Muséum National d'Histoire Naturelle. Paris, France. pp. 133–155.

Diéguez-Uribeondo J, Cerenius L, Söderhäll K 1994a. Repeated zoospore emergence in *Saprolegnia parasitica*. Mycol Res 98:810–815.

Diéguez-Uribeondo J, Cerenius L, Söderhäll K 1994b. *Saprolegnia parasitica* and its virulence on three different species of crayfish. Aquaculture 120:219–228.

Diéguez-Uribeondo J, Cerenius L, Söderhäll K 1996. Physiological characterization of *Saprolegnia parasitica* isolates from brown trout. Aquaculture 140:247–257.

Diéguez-Uribeondo J, Fregeneda-Grandes JM, Cerenius L, Pérez-Iniesta E, Aller-Gancedo JM, Tellería MT, Söderhäll K, Martín MP 2007. Re-evaluation of the enigmatic species complex *Saprolegnia diclina–Saprolegnia parasitica* based on morphological, physiological and molecular data. Fungal Gene Biol. 44:585–601.

Edgerton BF, Henttonen P, Jussila J, Mannonen A, Paasonen P, Tauglø I, Edsman L, Souty-Grosset C 2004. Understanding the causes of disease in European freshwater crayfish. Conserv Biol 6:1466–1474.

Fernández-Benéitez MJ, Ortiz-Santaliestra ME, Lizana M, Diéguez-Uribeondo J 2008. *Saprolegnia diclina*: another species responsible for the emergent disease '*Saprolegnia* infections' in amphibians. FEMS Microbiol Lett 279:23–29.

Fregeneda-Grandes JM, Díez F, Gancedo A 2000. Ultrastructure analysis of *Saprolegnia* secondary zoospore cyst ornamentation from infected wild brown trout, *Salmo trutta* L., and river water indicates two distinct morphotypes amongst long-spined isolates. J. Fish Dis. 23:147–160.

Fregeneda-Grandes JM, Díez F, Gancedo A 2001. Experimental pathogenicity in rainbow trout, *Oncorhynchus mykiss* Walbaum, of two distinct morphotypes of long-spined *Saprolegnia* isolates obtained from wild brown trout, *Salmo trutta* L., and river water. J Fish Dis 24:351–359.

Fregeneda-Grandes JM, Rodriguez-Cadenas F, Carbajal-Gonzalez MT, Aller-Gancedo JM 2007. Detection of 'long-haired' *Saprolegnia S. parasitica* isolates using monoclonal antibodies. Mycol. Res. 111:726–733.

Gieseker CM, Serfling SG, Reimschuessel R 2006. Formalin treatment to reduce mortality associated with Saprolegnia parasitica in rainbow trout, *Oncorhynchus mykiss*. Aquaculture 253:120–129.

Gudding R, Lillehaug A, Evensen E, 1999. Recent developments in fish vaccinology. Veter Immunol Immunopathol 72:203–212.

Hatai K, Hoshiai G 1992. Mass mortality in cultured coho salmon *Oncorhynchus kisutch* due to *Saprolegnia parasitica* Coker. J Wildl Dis 28:532–536.

Hatai K, Willoughby LG, Beakes GW 1990. Some characteristics of *Saproelngia* obtained from fish hatcheries in Japan. Mycol Res 2:182–190.

Hensley S 2005. Vaccines that keep salmon safe to eat may help humans. Wall Street J.

Hughes GC 1994. Saprolegniasis, then and now: a retrospective. In Mueller GJ, editor. *Salmon Saprolegniasis*, U.S. Department of Energy, Bonneville Power Administration. Portland, OR. pp. 3–32.

Hussein MMA, Hatai K 2002. Pathogenicity of *Saprolegnia* species associated with outbreaks of salmonid saprolegniosis in Japan. Fish Sci 68:1067–1072.

Judelson HS, Coffey MD, Arredondo FR, Tyler BM 1993. Transformation of the oomycete pathogen *Phytophthora megasperma* f. sp. Glycinea occurs by DNA integration into single or multiple chromosomes. Curr Genet 23:211–218.

Judelson HS, Tyler BM, and Michelmore RW 1991. Transformation of the oomycete pathogen, *Phytophthora infestans*. Mol Plant Microbe Interact 4:602–607.

Kales SC, DeWitte-Orr SJ, Bols NC, Dixon B 2007. Response of the rainbow trout monocyte/macrophage cell line RTS11 to the water molds *Achlya* and *Saprolegnia*. Mol Immunol 44:2303–2314.

Kiesecker JM, Blaustein AR, Belden LK 2001. Complex causes of amphibian population declines. Nature 410:681–684.

Lorenzen N, LaPatra SE 2005. DNA vaccines for aquacultured fish. Rev Sci Tech Off Int Epiz 24:201–213.

Lorenzen N, Lorenzen E, Einer-Jensen K, Heppell J, Davis HL 1999. Genetic vaccination of rainbow trout against viral haemorrhagic septicaemia virus: small amounts of plasmid DNA protect against a heterologous serotype. Virus Res 63:19–25.

Marking LL, Rach JJ, Schreier TM 1994a. Search for antifungal agents in fish culture. In: Mueller GJ, editor. *Salmon Saprolegniasis*, U.S. Department of Energy, Bonneville Power Administration. Portland, OR. pp. 131–148.

Marking LL, Rach JJ, Schreier TM 1994b. Evaluation of antifungal agents for fish culture. Prog Fish-Cult 56:225–231.

Marsden MJ, Vaughan LM, Fitzpatrick RM, Fosters TJ, Secombes CJ 1998. Potency testing of a live, genetically attenuated vaccine for salmonids. Vaccine 16:1087–1094.

Meyer FP 1991. Aquaculture disease and health management. J Anim Sci 69:4201–4208.

Misquitta L, Paterson BM 1999. Targeted disruption of gene function in *Drosophila* by RNA interference RNAi: a role for *nautilus* in embryonic somatic muscle formation. Proc Natl Acad Sci 96:1451–1456.

Mort-Bontemps M, Fevre M 1995. Electrophoretic karyotype of *Saprolegnia monoi*ca. FEMS Microbiol Lett 131:325–328.

Mort-Bontemps M, Fevre M 1997. Transformation of the oomycete *Saprolegnia monoica* to hygromycin-B resistance. Curr Genet 31:272–275.

Muzzarelli RA, Muzzarelli C, Tarsi R, Miliani M, Gabbanelli F, Cartolari M 2001. Fungistatic activity of modified chitosans against *Saprolegnia parasitica*. *Biomacromolecules* 2:165–169.

Neitzel DA, Elston RA, Abernethy CS 2004. DOE report. Contract: DE-AC06-76RL01830. Prevention of prespawning mortality: cause of salmon headburns and cranial lesions. pp. 1–B25.

Ngo H, Tschudi C, Gull K, Ullu E 1998. Double-stranded RNA induces mRNA degradarion in *Trypanosoma brucei*. Proc Natl Acad Sci 95:14687–14692.

Peduzzi R, Bizzozero S 1977. Immunochemical investigation of four *Saprolegnia* species with parasitic activity in fish: serological and kinetic characterization of a chymotrypsin-like activity. Microb Ecol 3:107–118.

Phillips AJ, Anderson VL, Robertson EJ, Secombes CJ, van West P 2008. New insights into animal-pathogenic oomycetes. Trends Microbiol 16:13–19.

Pickering AD 1986. Changes in blood cell composition of the brown trout, *Salmo trutta* L., during the spawning season. J Fish Biol 29:335–347.

Pickering AD 1994. Factors which predispose salmonid fish to saprolegniasis. In Mueller GJ, editor. *Salmon Saprolegniasis*, U.S. Department of Energy, Bonneville Power Administration. Portland, OR. pp. 67–84.

Pottinger TG, Day JG 1999. A *Saprolegnia parasitica* challenge system for rainbow trout: assessment of Pyceze as an anti-fungal agent for both fish and ova, Dis Aquat Org 36:129–141.

Puckeridge JT, Langdon JS, Daley C, Beakes GW 1989. Mycotic dermatitis in a freshwater gizzard shad, the bony bream, *Nematalosa erebi* Gunther, in the River Murray, South Australia. J Fish Dis 12:205–221.

Quiniou SMA, Bigler S, Clem W, Bly JE 1998. Effects of water temperature on mucous cell distribution in channel catfish epidermis: a factor in winter saprolegniasis. Fish and Shellfish Immunol 8:1–11.

Richards RH, Pickering AD 1979. Changes in serum parameters of *Saprolegnia*-infected brown trout, *Salmo trutta* L. J Fish Dis 2:197–206.

Roberge C, Paez DJ, Rossignol O, Guderley H, Dodson J, Bernatchez L 2007. Genome-wide survey of the gene expression response to saprolegniasis in Atlantic salmon. Mol Immunol 44:1374–1383.

Romisch K 2005. Protein targeting from malaria parasites to host erythrocytes. Traffic 6:706–709.

Stueland S, Hatai K, Skaar I 2005. Morphological and physiological characteristics of *Saprolegnia* spp. Strains pathogenic to Atlantic salmon, *Salmo salar* L. J Fish Dis 28:445–453.

Tabara H, Yigit E, Siomi H, Mello CC 2002. The dsRNA binding protein RDE-4 interacts with RDE-1, DCR-1, and a DExH-Box helicase to direct RNAi in *C. elegans*. Cell 109:861–871.

Tiffney WN 1939. The host range of *Saprolegnia parasitica*. Mycologia 31:310–321.

Torto-Alalibo T, Tian M, Gajendran K, Waugh M, van West P, Kamoun S 2005. Expressed sequence tags from the oomycete fish pathogen *Saprolegnia parasitica* reveal putative virulence factors. BMC Microbiol 5:46–58.

van West P 2006. *Saprolegnia parasitica*, an oomycete with a fishy appetite: new challenges for an old problem. Mycologist 20:99–104.

Wood SE, Willoughby LG, Beakes GW 1986. Preliminary evidence for inhibition of *Saprolegnia* fungus in the mucus of brown trout, *Salmo trutta* L., following experimental challenge. J Fish Dis 9:557–560.

Whisson SC, Avrova AO, van West P, Jones JT 2005. A method for double-stranded RNA-mediated transient gene silencing in *Phytophthora infestans*. Mol Plant Pathol 6:153–163.

Zaror L, Collado L, Bohle, H, Landskron E, Montaña J, Avendaño F 2004. *Saprolegnia parasitica* in salmon and trout from southern Chile. Arch Med Vet 36:71–78.

21

APHANOMYCES ASTACI AND CRUSTACEANS

Lage Cerenius, M. Gunnar Andersson, and Kenneth Söderhäll

Department of Comparative Physiology, Uppsala University, Uppsala, Sweden

21.1 INTRODUCTION

A relatively large number of *Aphanomyces* species is attacking a variety of aquatic animals, from the copepod *Bloekella dilatata* (Burns, 1985) water fleas to the freshwater dolphin *Inia geoffrensis* (Fowles, 1976), although most of them are only known anecdotally. Two species, which include the fish parasite *Aphanomyces invadans* (also known as *Aphanomyces piscida* or *Aphanomyces invaderis*) and, in particular, the crayfish pathogen *Aphanomyces astaci*, have been studied in more detail. These specialized parasites among the *Aphanomyces* species have caused extensive epizootic events that have involved huge losses of natural populations and farmed stocks of fish and shellfish. Two prominent examples of such devastating epizootics are crayfish plague caused by *A. astaci* and epizootic ulcerative syndrome (EUS) among many freshwater and estuarine fish species caused by *A. invadans* (Phillips et al., 2008). These two diseases are characterized by rapid spread over large distances fuelled by human trade and transport of live fish and shellfish. The effects on the European crayfish fauna by the plague have been disastrous, and EUS has impeded fish farming in large areas of South and East Asia. In this review, the focus will be on the molecular aspects of developmental biology and host–parasite interactions of *A. astaci* and its host, freshwater crayfish. For more descriptions of *A. astaci* and especially on older data on this mold, the reader is referred to previous reviews (Cerenius and Söderhäll, 1992; Söderhäll and Cerenius, 1992); a more detailed account on crustacean defense reactions

Oomycete Genetics and Genomics: Diversity, Interactions, and Research Tools
Edited by Kurt Lamour and Sophien Kamoun
Copyright © 2009 John Wiley & Sons, Inc.

toward oomycetes and other pathogens can be found in Cerenius and Söderhäll (2004) and in Cerenius et al. (2008). In addition to these highly specialized parasites species, other *Aphanomyces* species such as *Aphanomyces laevis*, *Aphanomyces stellatus*, and *Aphanomyces frigidophilus* (Ballesteros et al., 2006) are frequently found on fish and crustaceans eggs or on animals with an impaired immune system. Perhaps the recently described *Aphanomyces repetans* (Royo et al., 2004) belongs to this category; it was isolated from diseased crayfish but was not found to be highly virulent at least in aquarium set ups.

21.2 *A. ASTACI* — CLONAL PROPAGATION AND HOST RANGE

The crayfish plague is an introduced disease that was brought into Europe in the 1850s from North America. The parasite *A. astaci* is highly virulent on indigenous European freshwater crayfish. In contrast, it maintains a relatively stable interaction with North American crayfish species and survives encased within the cuticle. It causes little harm unless the host becomes in some way immunocompromised, in which case the parasite may cause a mortal infection. European freshwater crayfish exhibit low resistance to the disease. Therefore, the crayfish plague is the most serious threat to the survival of most remaining populations of native European crayfish. Mortalities among these animals often reach 100%, and *A. astaci* can effectively wipe out the whole crayfish population in a lake within a few weeks when it is on introduced. Unfortunately, large-scale introductions and successful establishment of several American crayfish species throughout most European countries have established reservoirs for the pathogen and have made it virtually impossible to eradicate the disease.

A. astaci, as with most other animal pathogenic *Aphanomyces* species, does not produce sexual spores and is thus clonally spread. Three prominent clones (genotypes) have gained widespread recognition in Europe. These have identical internal spacer (ITS) sequences but can be identified by other molecular markers, such as random amplified polymorphic DNA (RAPD) patterns (Huang et al., 1994). One genotype may represent the original introduction (or at least an early introduction) of the parasite, because it is found in the oldest isolates investigated so far and has been recorded throughout Europe and Turkey (Huang et al., 1994). A second genotype was introduced to Europe by wide-scale stockings in Northern and Central Europe with North American crayfish (*Pacifastacus leniusculus*) from Lake Tahoe. In recent years, this second genotype has gradually replaced the first one, at least in Scandinavia. A third genotype can be found in southern Europe, which shows some interesting ecophysiological adaptations to higher water temperatures that are not observed in any other *A. astaci* isolates (Dieguez-Uribeondo et al., 1995). The third genotype seems to originate from Louisiana or nearby areas in the southern United States from which the *Procambarus clarkii* crayfish has been introduced to southern Europe for aquaculture. Also, *A. invadans* has been introduced by human activities throughout eastern and southern Asia most likely by trade of ornamental fish (Lilley et al., 1997).

A diagnosis of crayfish plague (and most other animal pathogenic species of this genus) is dependent on molecular tools because of the lack of sexual structures. There are published protocols on the diagnosis of *A. astaci* (Oidtmann et al., 2004), although great care must be applied when using them because of extensive sequence similarities between different *Aphanomyces* species (Ballesteros et al., 2007).

21.3 SPORE DEVELOPMENT

A. astaci zoospores are the sole infective units and must find the host during their a relatively short-lived life that normally ends with encystment. The motile phase lasts only a few hours up to maximally a few days at low temperatures. During encystment, the spore drops or retracts its flagella and becomes encased in a cell wall. The cell wall is covered with sticky substances to allow spore adherence to the host. Shortly thereafter, adhered spores start to germinate, and the emerging hypha will penetrate the skin or cuticle. It is important that the spores receive correct signals to trigger germination. Germination can be triggered *in vitro* by exudates from the host. The putative host-specific molecular cues for inducing germination in *A. invadans* and *A. astaci* have not been identified. Zoospores are easily induced to encyst by mechanical stimuli, increased ionic strength, pH changes, amino acids, and so on. During such circumstances, spore encystment is not followed by germination, but if conditions are favorable a new zoospore will be released through a rupture of the old cyst wall a few hours later. Such behavior is named repeated zoospore emergence (RZE) or polyplanetism and is likely to increase the chances of finding the host. Zoospores of saprophytic *Aphanomyces* species are much less prone to undergo repeated zoospore emergence but tend to, at most occasions, germinate during encystment (Cerenius and Söderhäll, 1985; Royo et al., 2004). Although RZE is not strictly restricted to parasitic species, there is a strong tendency that specialized parasites have the capacity to produce several successive zoospore generations. This capacity does not come without a cost; the spore has a very limited ability for renewal of macromolecules. Depending on the species, maximally 2–5 consecutive generations of zoospores can be produced from a single "first-generation" zoospore (Dieguez-Uribeondo et al., 1994). The length of each period of swimming is strongly dependent on temperature; the lower the temperature the longer is the maximal swimming time. The time spent as a cyst does not seem to have much influence on the maximal time that can be spent swimming. Together, these observations strongly suggest that there is a limited supply that is not renewable within the spore, and when it is finished, active swimming can no longer be sustained.

21.4 ROLE OF PUMILIO-LIKE PROTEINS IN SPORE DEVELOPMENT

One potential key regulator of differentiation of oomycete cysts to zoospores has recently been identified as a homolog of the eukaryotic protein Pumilio. Pumilio was first identified in *Drosophila*, where it is involved in establishing polarity in early embryogenesis. Pumilio-like factors have since been found in many eukaryotes where they constitute a relatively large group of translational inhibitors. Several pumilio-like proteins have for some time been known to regulate specific steps of the cellular differentiation in animals, fungi, and other organisms. A pumilio-like transcript is also present in some stages during both asexual and sexual spore formation in *Phytophthora infestans* (Cvitanich and Judelson, 2003). In *Saprolegnia* (Andersson and Cerenius, 2002b) and in at least three different *Aphanomyces spp; A. astaci, A. laevis*, and *A. stellatus* (Andersson and Cerenius, unpublished), we have observed that a pumilio homolog *puf-1* is expressed during certain stages of asexual and sexual reproduction.

In *A. astaci* and *Saprolegnia parasitica*, where RZE is common, we have analyzed the kinetics of *puf-1* expression in more detail during the different stages of asexual reproduction (Andersson and Cerenius, 2002b). Here, *puf-1* is not expressed in swimming zoospores, but its expression can be detected by reverse transcription polymerase chain reaction within 15 min after inducing encystment. This transcript is quickly lost when inducing germination, whereas in cysts treated in such way that repeated zoospore emergence will commence, the *puf-1* transcript gradually disappears during about 60 min and is gone well before the release of a new zoospore. The transcript then reappears in each cyst generation when the spore undergoes repeated cycles of cyst-zoospore transformation. Interestingly, the newly formed cyst, which actively expresses *puf-1*, can be triggered to germinate by external cues, whereas some time after encystment both germination "induction competence" and *puf-1* expression is lost and neither can be made to re-appear until a new cycle of RZE has taken place. Germinating cysts and young mycelia not yet capable of sporulation do not express *puf-1*. *Puf-1* transcription seems to be intimately involved in one of the two alternative paths of development that a spore cyst may follow: repeated zoospore emergence in contrast to germination. There seems to be a tight active regulation of both these alternatives, and thus repeated zoospore emergence should not merely be considered a default pathway for cysts that are not triggered to germinate. *Puf-1* expression is a marker of cysts that have not been induced into the germination pathway but that, at least, potentially can once again release a zoospore. The oomycete puf-1 contains the highly conserved RNA-binding region present in other pumilio-like proteins, which suggests it can bind RNA. In contrast, the amino terminal of the pumilio-like proteins are very variable in size and sequence, and it is likely that the amino-terminal of the putative *Aphanomyces* puf-1 protein will interfere specifically with spore cyst transcripts that influence the developmental fate of the cyst. One may envision that the ability to regulate when and where to once again release a zoospore could be important for a specialized parasite, and it could give it more than one opportunity to find its designated host.

21.5 EXTRACELLULAR ENZYMES

It has become apparent from recent genomic data from fungal pathogens that the particular ecological niche taken by different species in a genus is in general reflected by the spectrum of extracellular enzymes produced by each individual species. This seems to be the case also in oomycetes. In the genus *Aphanomyces*, there are apparent differences between *A. astaci* and the plant pathogens *Aphanomyces cochlioides* and *Aphanomyces euteiches* with respect to which classes of proteolytic proteinases are produced. It is noteworthy that crayfish cuticular and blood proteinase inhibitors are more suited to inhibit the *A. astaci* proteolytic cocktail of enzymes than the corresponding preparations from the plant pathogenic *A. euteiches* and *A. cochlioides* species (Dieguez-Uribeondo and Cerenius, 1998). Proteinases from plant-pathogenic *Aphanomyces* are, in a similar way, likely to be inhibited by host-specific inhibitors (Weiland, 2004).

Once it has become established on the host, the spore starts to release extracellular enzymes to aid tissue penetration and nutrient uptake. In *A. astaci*, a close correlation between enzyme release and phase of host penetration has been recorded. Lipases and proteinases are released very early before spore germination, whereas chitinase is produced after several hours of growth at a stage the young mycelium has reached the chitinous layer of the cuticle (Söderhäll et al., 1978). The initiation of chitinase production is apparently preprogrammed. The major chitinase gene *AaChtl* is transcribed from about 10 h and onward after spore germination (Andersson and Cerenius, 2002a). Gene expression starts irrespective of whether chitin is present in the medium and is continuous as long as the mycelium is actively growing. The production of *AaChtl* mRNA is not repressed by glucose. In contrast, saprophytic or less-specialized pathogens among the *Aphanomyces* species such as *A. repetans*, *A. laevis*, or *A. stellatus* need to encounter chitin in the environment to release the enzyme and trigger chitinase gene expression (Royo et al., 2004). On the other hand, *A. invadans* and the plant pathogenic *Aphanomyces* spp seem to produce little if any chitinase under all conditions. The published expressed sequence tag (EST) sequences from *A. euteiches* do not contain any obvious *AaChtl* homolog in contrast to *A. laevis*.

One intriguing finding was that in *A. astaci*, *AaChtl* transcription increases dramatically during the sporulation phase. Normally, extracellular chitinase levels as well as *AaChtl* transcription ceases when mycelial growth retards in an aging mycelium. However, exposing the mycelium to lake water triggers sporangium formation as well as chitinase production, as measured both by *AaChtl* transcript levels and by chitinase activity measurements. A corresponding chitinase production during spore production was not observed in other *Aphanomyces* species such as *A. laevis* or *A. stellatus* species, which indicates that chitinase production is not required for development but might have a more specific role during pathogenesis (Andersson and Cerenius, 2002a). Neither the chitinase enzyme nor its transcript are stored in the spore, and chitinase is synthesized *de novo* once again when the parasite becomes established in the host as described above. We have suggested that the role of chitinase during sporulation is to allow the hyphal tips to penetrate through the cuticle from inside to allow spore release from the tips to the external milieu (Andersson and Cerenius, 2002a).

Other extracellular enzymes are regulated differently. A major secreted trypsin-like enzyme AaSP2 is strongly induced in the presence crayfish hemolymph (blood), but *AaSP2* transcription in contrast to *AaCht1* transcription will decrease dramatically in sporulating hyphae (Bangyeekhun et al., 2001). Vertebrate sera, skim milk, albumin, or other general protein sources are not effective in inducing *AaSP2* gene expression; this fact may indicate the presence of a specific inducer in the crayfish hemolymph. Such an expression pattern indicates that AaSP2 is functioning during hyphal growth deep inside the host below the cuticle. The lack of catabolite repression by ammonium or other low molecular nitrogenous of *AaSP2* expression also fits into this pattern because the parasite is growing in an environment rich in nitrogen.

21.6 HOST IMMUNE REACTIONS TOWARD *A. ASTACI*

The crustacean immune system is thoroughly investigated and is a convenient model system, because relatively large amounts of material (e.g., fractionated blood cells or plasma) can be obtained. Furthermore, this system is amenable to RNA interference (RNAi) techniques as well as *in vitro* blood cell culturing (Söderhäll et al., 2005). For some recent reviews on the crustacean immune system, see Cerenius et al. (2008), Cerenius and Söderhäll (2004), and Jiravanichpaisal et al. (2006). An important component of the immune response is the melanization reaction. Pathogens become covered with melanin and their growth and spread are prevented. During the melanization reaction, several short-lived toxic compounds are produced. Because of the toxic nature of the products catalyzed by active phenoloxidase, this cascade is carefully checked spatially and temporally to ensure that melanization is carried when and where there is a need. Furthermore, connected with the melanization cascade are other key immune responses, such as production of opsonins and encapsulation-promoting factors. This complex of connected cellular and humoral defense reactions is often called the prophenoloxidase activating system (Cerenius and Söderhäll, 2004).

Cell wall components from *A. astaci* can, in extremely low concentrations, trigger an extensive immune response that includes the melanization cascade, production of cytotoxic compounds, phagocytosis, and so on. The main constituent of the oomycete cell wall is a (1,3)-β-glucan with (1,6)-linked branches, and purified preparations of the *A. astaci* cell wall are efficient immune stimulants in this system. The minimum requirement for triggering the activation is a linear pentasaccharide (Söderhäll and Unestam, 1979). Specific host (1,3)-β-glucan-binding proteins present in plasma (Cerenius et al., 1994; Lee et al., 2000) mediate different glucan-induced immune reactions after being engaged by the carbohydrate. One example is the proteolytic cascade that terminates by restricted proteolysis of the zymogenic prophenoloxidase into active phenoloxidase. The (1,3)-β-glucan also induce the release of immune components from these cells by regulated exocytosis after binding directly to hemocyte (blood cell) membrane receptors (Jiravanichpaisal et al., 2006).

In resistant crayfish species, this parasite is often observed melanized in the cuticle or occasionally at other places. In contrast, *A. astaci* growth in susceptible species is rarely accompanied by melanization. Furthermore, compounds such as quinones known to be produced by an active phenol oxidase retards mycelial growth and inhibit extracellular proteinases when tested on *A. astaci* in vitro (Söderhäll and Ajaxon, 1982). Therefore, we tested whether expression of the gene responsible for production of the melanizing enzyme phenoloxidase differed in resistant and susceptible crayfishes, respectively (Cerenius et al., 2003). Resistant crayfish continuously produced high levels of the prophenoloxidase transcript, and these levels were not increased with immunostimulatory treatments. In contrast, the susceptible species expressed this transcript at lower levels, although injecting immunostimulatory glucans into the animals could enhance transcription. The resistance toward *A. astaci* was augmented by this treatment, and the infection was significantly delayed although the crayfish finally succumbed to the infection. We interpret these results by a model in which the susceptible host species cannot produce enough melanin and other immune factors to prevent mycelial growth by *A. astaci* (Cerenius et al., 2003).

In some recent experiments we tested the importance of the melanization reaction by introducing double-stranded RNA (dsRNA) for prophenoloxidase into the animals to demonstrate that a reduced melanization capacity decreased host resistance to another serious pathogen on crayfish called *Aeromonas hydrophila* (Liu et al., 2007). The partial knockout of the prophenoloxidase resulted in increased survival of the pathogen and decreased phagocytosis levels. On the contrary, RNA interference of the pacifastin, which is an endogenous regulatory inhibitor of the crayfish prophenoloxidase-activating proteinase, resulted in higher phenoloxidase levels, increased phagocytosis levels, lower bacterial counts, and increased survival toward this pathogen. These experiments clearly demonstrate the importance of the host prophenoloxidase system for overcoming the parasitic assault.

21.7 FUTURE PERSPECTIVES

Interactions between oomycetes and animals are little studied compared with plant–oomycete interactions. Recent advances in invertebrate immunology are providing new opportunities to study host–parasite interactions, such as those between *A. astaci* and its crustacean host. For example, by using RNAi and isolated immune components from the animal, it will be possible to study to what extent *A. astaci* as compared with other *Aphanomyces* species have adapted to the activities of different immune proteins from the host. Taking into account the apparently ancient relationship between North American crayfish species and *A. astaci*, it seems likely that this pathogen has adapted to its host not only by producing a specific mixture of extracellular enzymes, but also by producing physiological adaptations to withstand animal immune activities.

REFERENCES

Andersson MG, Cerenius L 2002a. Analysis of chitinase expression in the crayfish plague fungus *Aphanomyces astaci*. Dis Aqua Organ 51:139–47.

Andersson MG, Cerenius L 2002b. Pumilio homologue from *Saprolegnia parasitica* specifically expressed in undifferentiated spore cysts. Eukaryot Cell 1:105–111.

Ballesteros I, Martin MP, Cerenius L, Söderhäll K, Telleria MT, Dieguez-Uribeondo J 2007. Lack of specificity of the molecular diagnosis method for identification of *Aphanomyces astaci*. Bull Francais Peche Pisciculture 385:17–23.

Ballesteros I, Martin MP, Dieguez-Uribeondo J 2006. First isolation of *Aphanomyces frigidophilus* (Saprolegniales) in Europe. Mycotaxon 95:335–340.

Bangyeekhun E, Cerenius L, Söderhäll K 2001. Molecular cloning and characterization of two serine proteinase genes from the crayfish plague fungus, *Aphanomyces astaci*. J Inverte Pathol 77:206–16.

Burns CW 1985. Fungal parasitism in a fresh-water copepod — components of the interaction between *Aphanomyces* and *Boeckella*. J Invert Pathol 46:5–10.

Cerenius L, Bangyeekhun E, Keyser P, Söderhäll I, Söderhäll K 2003. Host prophenoloxidase expression in freshwater crayfish is linked to increased resistance to the crayfish plague fungus, *Aphanomyces astaci*. Cell Microbiol 5:353–357.

Cerenius L, Lee BL, Söderhäll K 2008. The proPO-system: pros and cons for its role in invertebrate immunity. Trends Immunol 29:263–271.

Cerenius L, Liang ZC, Duvic B, Keyser P, Hellman U, Palva ET, Iwanaga S, Söderhäll K 1994. Structure and biological activity of a 1,3-beta-D-glucan-binding protein in crustacean blood. J Biol Chem 269:29462–29467.

Cerenius L, Söderhäll K 1985. Repeated zoospore emergence as a possible adaptation to parasitism in *Aphanomyces*. Exp Mycol 9:259–263.

Cerenius L, Söderhäll K 1992. Crayfish diseases and crayfish as vectors for important diseases. Finn Fish Res 14:125–133.

Cerenius L, Söderhäll K 2004. The prophenoloxidase-activating system in invertebrates. Immunol Rev 198:116–126.

Cvitanich C, Judelson HS 2003. A gene expressed during sexual and asexual sporulation in Phytophthora infestans is a member of the Puf family of translational regulators. Eukaryot Cell 2:465–473.

Dieguez-Uribeondo J, Cerenius L 1998. The inhibition of extracellular proteinases from *Aphanomyces* spp by three different proteinase inhibitors from crayfish blood. Mycol Res 102:820–824.

Dieguez-Uribeondo J, Cerenius L, Söderhäll K 1994. Repeated zoospore emergence in *Saprolegnia parasitica*. Mycol Res 98:810–815.

Dieguez-Uribeondo J, Huang TS, Cerenius L, Söderhäll K 1995. Physiological adaptation of an *Aphanomyces astaci* strain isolated from the fresh-water crayfish *Procambarus clarkii*. Mycol Res 99:574–578.

Fowles B 1976. Factors affecting growth and reproduction in selected species of *Aphanomyces*. Mycologia 68:1221–32.

Huang TS, Cerenius L, Söderhäll K 1994. Analysis of genetic diversity in the crayfish plague fungus, *Aphanomyces astaci*, by random amplification of polymorphic DNA. Aquaculture 126:1–9.

Jiravanichpaisal P, Lee BL, Söderhäll K 2006. Cell-mediated immunity in arthropods: Hematopoiesis, coagulation, melanization and opsonization. Immunobiol 211:213–236.

Lee SY, Wang RG, Söderhäll K 2000. A lipopolysaccharide- and beta-1,3-glucan-binding protein from hemocytes of the freshwater crayfish *Pacifastacus leniusculus* — Purification, characterization, and cDNA cloning. J Biol Chem 275:1337–1343.

Lilley JH, Hart D, Richards RH, Roberts RJ, Cerenius L, Söderhäll K 1997. Pan-Asian spread of single fungal clone results in large scale fish kills. Veter Rec 140:653–654.

Liu H, Jiravanichpaisal P, Cerenius L, Lee BL, Söderhäll I, Söderhäll K 2007. Phenoloxidase is an important component of the defense against *Aeromonas hydrophila* infection in a crustacean, *Pacifastacus leniusculus*. J Biol Chem 282: 33593–33598.

Oidtmann B, Schaefers N, Cerenius L, Söderhäll K, Hoffmann RW 2004. Detection of genomic DNA of the crayfish plague fungus *Aphanomyces astaci* (Oomycete) in clinical samples by PCR. Veter Microbiol 100:269–82.

Phillips AJ, Anderson VL, Robertson EJ, Secombes CJ, van West P 2008. New insights into animal pathogenic oomycetes. Trends Microbiol 16:13–19.

Royo F, Andersson MG, Bangyeekhun E, Muzquiz JL, Söderhäll K, Cerenius L 2004. Physiological and genetic characterisation of some new *Aphanomyces* strains isolated from freshwater crayfish. Veter Microbiol 104:103–12.

Söderhäll I, Kim YA, Jiravanichpaisal P, Lee SY, Söderhäll K 2005. An ancient role for a prokineticin domain in invertebrate hematopoiesis. J Immunol 174:6153–6160.

Söderhäll K, Ajaxon R 1982. Effect of quinones and melanin on mycelial growth of *Aphanomyces spp.* and extracellular protease of *Aphanomyces astaci*, a parasite on crayfish. J Invert Pathol 39:105–109.

Söderhäll K, Cerenius L 1992. Crustacean immunity. Ann Rev Fish Dis 2:3–23.

Söderhäll K, Svensson E, Unestam T 1978. Chitinase and protease activities in germinating zoospore cysts of a parasitic fungus, *Aphanomyces astaci*, Oomycetes. Mycopathologia 64:9–11.

Söderhäll K, Unestam T 1979. Activation of serum prophenoloxidase in arthropod immunity. The specificity of cell wall glucan activation and activation by purified fungal glycoproteins of crayfish phenoloxidase. Can J Microbiol 25:406–414.

Weiland JJ 2004. Production of protease isozymes by *Aphanomyces cochlioides* and *Aphanomyces euteiches*. Physiol Molec Plant Patho 65:225–233.

22

PROGRESS AND CHALLENGES IN OOMYCETE TRANSFORMATION

HOWARD S. JUDELSON AND AUDREY M.V. AH-FONG

Department of Plant Pathology and Microbiology, University of California, Riverside, California

22.1 INTRODUCTION

Techniques for DNA-mediated transformation provide a foundation for a wide range of biological investigations. For example, the function of genes can be assessed through inactivation or forced expression assays, and studies of cell biology and growth can be enhanced by expressing fluorescent markers or protein tags. Stable transformation has now been achieved in several oomycete genera, including *Achlya, Phytophthora, Pythium*, and *Saprolegnia*, and has been used to study many genes relevant to growth or pathogenicity. There is also the prospect of extending transformation to the more difficult systems, such as the obligate pathogens.

22.2 HISTORICAL PERSPECTIVES

Minimum requirements for a transformation system are vectors expressing markers that enable recipients of DNA to be identified, and a means to introduce that DNA. Such techniques were developed for numerous ascomycetes and basidiomycetes in the 1980s (Fincham, 1989). Despite the lack of taxonomic affinity between oomycetes and true fungi (Baldauf et al., 2000), there has been a traditional association between these organisms as well as scientists working with them, so it was natural that many tried to adapt procedures and vectors used for true fungi to oomycetes. Although

Oomycete Genetics and Genomics: Diversity, Interactions, and Research Tools
Edited by Kurt Lamour and Sophien Kamoun
Copyright © 2009 John Wiley & Sons, Inc.

certain early papers claimed transformation, many authors described systems that were very inefficient or not reproducible. For example, the expression of a foreign thymidine kinase in *Achlya ambisexualis* was dependent on its insertion near endogenous promoters (Manavathu et al., 1988). Most other attempts to use nonoomycete transcriptional regulators to drive genes were also unsuccessful (Kinghorn et al., 1991).

It might seem surprising that these early efforts used vectors in which genes were driven by nonoomycete promoters. This is because work in other organisms had already shown that promoters from one taxonomic group often function poorly in others, including between different true fungi (McKnight et al., 1985). It would therefore seem unwise to employ a fungal or other heterologous promoter in an oomycete transformation system. However, it should be realized that no oomycete gene had yet been cloned that might provide such a promoter.

The first oomycete gene isolated was interestingly not from one of the more tractable species but instead from an obligate pathogen, *Bremia lactucae* (lettuce downy mildew; Judelson and Michelmore, 1989). That gene, which encoded heat-shock protein Hsp70, was cloned purposely to yield a promoter, as *hsp70* sequences had proved useful for driving transgenes in other taxa. A second *B. lactucae* promoter (*ham34*) was also obtained by searching for genes with strong and constitutive expression (Judelson and Michelmore, 1990). These were observed to drive expression of the β-glucuronidase (GUS) transgene in *B. lactucae* hyphae when lettuce cotyledons were inoculated with conidia treated with the DNA by microprojectile bombardment (Fig. 22.1a). However, stable transformants were not obtained in experiments that used plasmids containing drug-resistance genes combined with selection on drug-imbibed plant material.

Although this work in *B. lactucae* was not wholly successful, it proved critical for allowing other oomycetes to be transformed. This is because the

FIG. 22.1 (See color insert) Expression of reporter genes. (a) Staining of GUS in conidiophore of *Bremia lactucae* emerging from lettuce cotyledon, driven by the constitutive *ham34* promoter (Judelson and Michelmore, unpublished). (b) GUS in stable transformant of *Phytophthora infestans* under control of *ham34* promoter. (c) Stage-specific expression of GUS using sporulation-specific *Cdc14* promoter in *P. infestans,* showing expression in sporangiophore and developing sporangia. (d) GUS under control of mating-specific *M25* promoter in *P. infestans,* with staining evident in an antheridium (a) through which an oogonial initial (o) is emerging (Niu and Judelson, unpublished). (e) GUS expressed from zoosporogenesis-specific *NIFC* promoter in *P. infestans*, which is stained for activity in freshly harvested sporangia (left) or chilled sporangia initiating zoospore formation (right). (f) Sporangium from transgenic *P. infestans* expressing green fluorescent protein (GFP) from the *ham34* promoter, which is shown under brightfield (left) and fluorescence (right) conditions; the unstained central vacuole is labeled (l). (g) *P. infestans* expressing GFP from constitutive promoter and mRFP fused to the Avr3A effector protein, showing delivery of the latter to haustoria (courtesy of Whisson and Boevink).

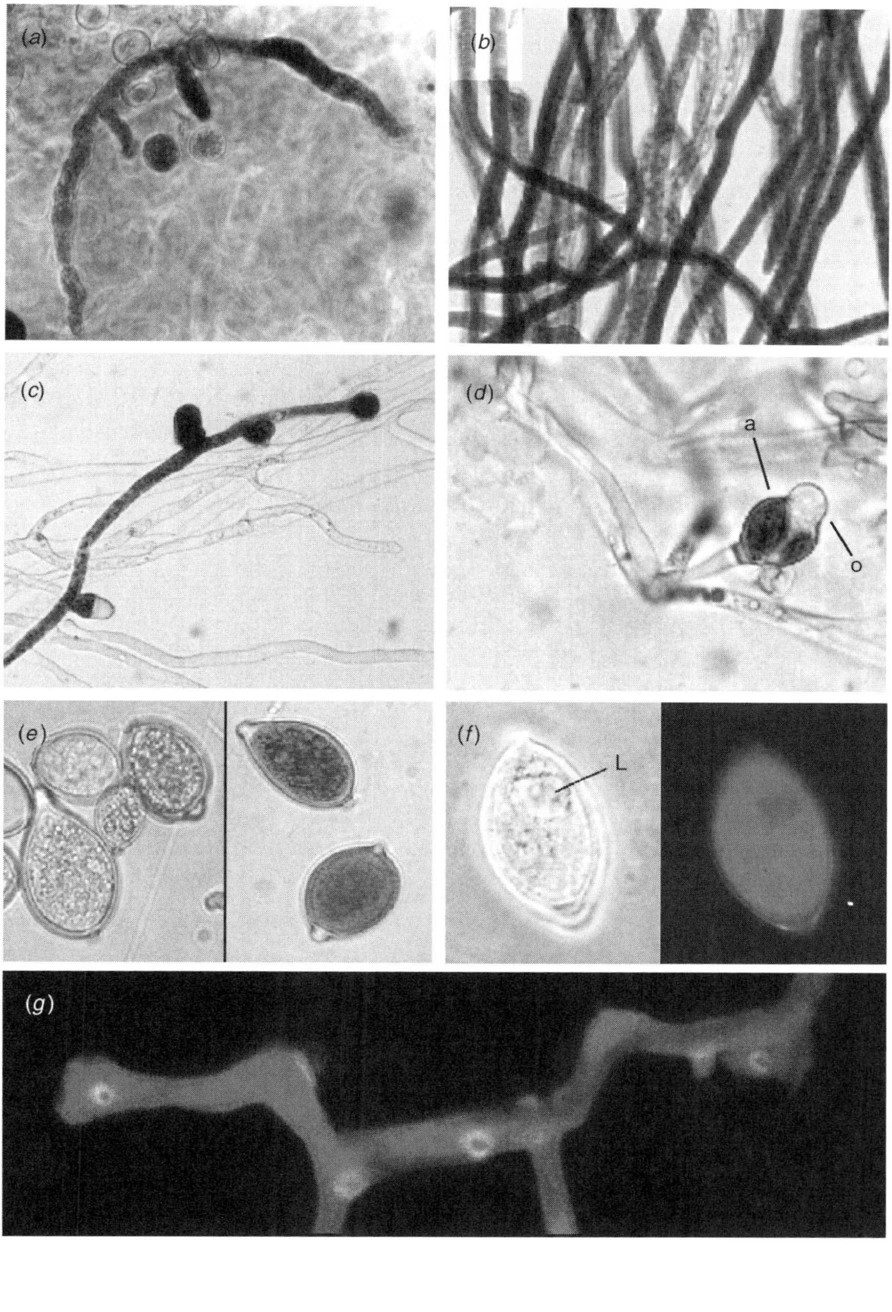
(a)
(b)
(c)
(d)
a
o
(e)
(f)
L
(g)

same vectors were later found to function in species more amenable to laboratory manipulation, starting with *Phytophthora infestans* and later others. Indeed, although other functional promoters are now identified, *ham34* and *hsp70* sequences are still the most widely used for the general expression of transgenes.

22.3 CURRENT GENE TRANSFER SYSTEMS

Protoplast-mediated DNA uptake, microprojectile bombardment, electroporation, and *Agrobacterium tumefaciens* methods have all been employed to introduce genes into oomycetes (Table 22.1). Multiple species have been stably transformed, although most published studies have targeted *P. infestans* and accomplished gene transfer using protoplasts. However, this chapter will avoid

TABLE 22.1 First reports of transformation using transgenes with oomycete promoters.

Species	Method	Reference
	Stable transformation	
Phytophthora brassicae	Protoplast	Si-Ammour et al., 2003
Phytophthora infestans	Protoplast	Judelson et al., 1991
"	Bombardment	Cvitanich and Judelson, 2003a
"	Electroporation	Latijnhouwers et al., 2004
"	Agrobacterium	Vijn and Govers, 2003
Phytophthora sojae	Protoplast	Judelson et al., 1993a
Phytophthora palmivora	Protoplast	van West et al., 1999b
"	Agrobacterium	Vijn and Govers, 2003
Phytophthora nicotianae	Protoplast	Bottin et al., 1999
Pythium aphanidermatum	Electroporation	Weiland, 2003
Pythium ultimum	Agrobacterium	Vijn and Govers, 2003
Saprolegnia monoica	Protoplast	Mort-Bontemps and Fevre, 1997
Pythium oligandrum	Protoplast	Masunaka and Takenaka, 2006
	Transient expression only	
Achlya ambisexualis	Protoplast	Judelson et al., 1992
Bremia lactucae	Bombardment	Judelson and Michelmore, 1989
Phytophthora citricola	Protoplast	McLeod et al., 2008
Plasmopara halstedii	Mechanoperforation	Hammer et al., 2007

concluding which is the superior transformation procedure or target species, because these are moving targets, and a method that works best with one species may not be superior for all others. Moreover, developing an efficient transformation system is not trivial, and many variables exist that require optimization. Transformation rates can be quoted from published papers using metrics such as the number of transformants per microgram, but these may not be representative and may have developed after publication. In the authors' laboratory, the most transformants of *P. infestans* are usually obtained using the protoplast method, which frequently yields up to 300 transformants per experiment or up to 10 per microgram of DNA. However, transformation requires meticulous skill, and often lower amounts are obtained.

22.3.1 Vector Systems

Illustrated in Fig. 22.2 are typical plasmids employed in contemporary experiments, which are derived from those used in the initial study of *P. infestans* transformation (Judelson et al., 1991). pHAMT35N/SK uses *ham34* promoter and transcriptional terminators to express *nptII*, which confers resistance to the aminoglycoside G418. The plasmid also contains the multiple cloning site from pBluescript SK(+) and a blue/white selection system for inserting other DNA. pTEP4 contains *nptII* driven by the *hsp70* promoter, plus a *ham34*-driven cassette for expressing other sequences. pNPGUS can be used to test promoters, as it contains a promoter-lacking GUS gene downstream of a multiple cloning site. pSTORA is designed for expressing hairpin RNAs, which efficiently trigger DNA-directed RNAi in *P. infestans* (Judelson and Tani, 2007). Besides *nptII*, pSTORA contains the intron from the *Pic20* gene of *P. infestans* flanked by sites for rare cutting-restriction enzymes into which the silencing target can be inserted. Many other useful vectors have also been developed, such as those that express green fluorescent protein (GFP) or other selectable markers, or cosmid or bacterial artificial chromosome (BAC) plasmids to enable transformable libraries to be constructed. Whether there is an upper size limit for DNA in transformation is not known, but interestingly 110-kb BAC plasmids resulted in much higher transformation rates than smaller plasmids (Randall and Judelson, 1999).

22.3.2 Selectable Markers

Several drugs have proved suitable for selecting transformants, based on their relative toxicity and the availability of a resistance gene. Appropriate drug concentrations for selection vary between species and strains, and it is necessary to titrate these carefully to avoid background growth while allowing authentic transformants to be recovered. The vectors described above contain the G418-inactivating *nptII* gene, because this is most-commonly used. However, other selections have used *hpt* (hygromycin phosphotransferase) or *spt* (streptomycin phosphotransferase). For example, *Saprolegnia monoica* was transformed

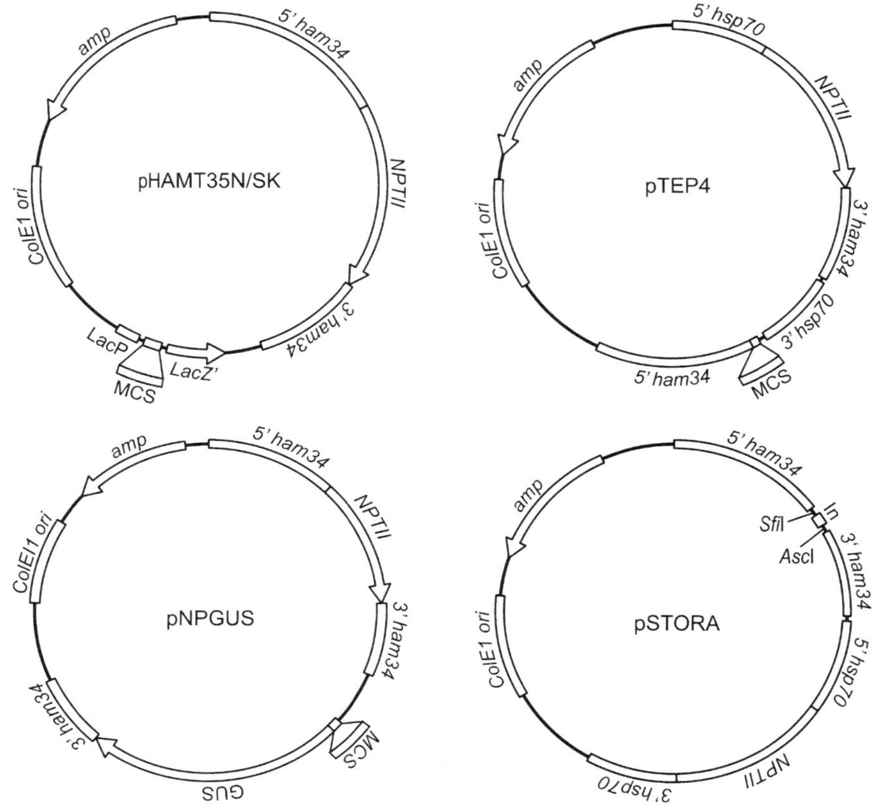

FIG. 22.2 Typical vectors used for transformation, containing oomycete promoters (*5' ham34, 5' hsp70*) and transcriptional terminators (*3' ham34, 3' hsp70*), and the *nptII* selectable marker. MCS, multiple cloning site; In, *Pic20* intron.

using hygromycin selection (Mort-Bontemps and Fevre, 1997). In *P. infestans, nptII* is favored as it yields slightly more transformants than *hpt* and because hygromycin is more toxic to laboratory workers. In *Pythium aphanidermatum,* experiments using G418 also proved to be more successful than hygromycin selections (Weiland, 2003). Streptomycin selection is problematic because resistance can develop spontaneously, and in tests with *P. infestans* only two thirds of resistant colonies were authentic transformants (unpublished data). Nevertheless, *spt* selection can be useful by enabling transgenes to be pyramided in the same strain.

Additional drugs exist that might be used for selection, such as bleomycin, and it may be helpful to test these as more diverse oomycetes are examined. Auxotrophic markers, which are frequently employed in true fungi (Fincham, 1989), would be challenging to use because the necessary mutants are not readily obtained. Also, most oomycetes grow poorly on the defined media needed for prototroph selection, and such markers might be ill-suited for studying pathogenic oomycetes because of fitness issues. A more useful marker might be a gene for resistance to a systemic fungicide such as metalaxyl, and these might also enable *in planta* selections with obligate pathogens, such as downy mildews and white rusts. However, such genes have not yet been cloned.

Another form of marker not established in oomycetes are genes that enable counterselection or two-way selection (positive and negative). These exist for bacteria, true fungi, mammalian cells, and plants, and they can be useful for selecting gene uptake or loss (Converse et al., 2004). Such markers might enable experiments that are not currently feasible, such as gene replacement or transposon excision studies.

22.3.3 Protoplast Transformation

This has been used to introduce DNA into *A. ambisexualis, Phytophthora citricola, P. infestans, Phytophthora nicotianae, Phytophthora palmivora, Phytophthora sojae, Pythium oligandrum*, and *S. monoica* (Bottin et al., 1999; Judelson et al., 1991, 1992, 1993a; Masunaka and Takenaka, 2006; McLeod et al., 2008; Mort-Bontemps and Fevre, 1997; van West et al., 1999b). The method typically starts by using cell-wall degrading enzymes to liberate protoplasts from hyphae using an osmoticum to maintain protoplast integrity. The protoplasts are then mixed with DNA in the presence of calcium chloride and then polyethylene glycol (PEG), which may promote membrane invagination and DNA endocytosis. Protoplasts are allowed to regenerate walls using osmotically stabilized media and transformants selected using the appropriate drug. Colonies appear after 4 to 9 days, depending on species.

Obtaining sufficient numbers of viable protoplasts is one of the challenges in this method. Superior results are obtained using young hyphae from germinated spores or hyphal mats, because older hyphae yield many anucleate protoplasts that bind DNA yet cannot regenerate. "Good" regeneration rates often range between 5% and 40%; this is generally lower than obtained with filamentous true fungi because protoplasts of the latter are generally multinucleate. However, a side benefit is that heterokaryotic transformants (containing transformed and nontransformed nuclei) are not common based on studies in *P. infestans* (Judelson et al., 1991). In contrast, primary transformants of true fungi are often heterokaryons (Fincham, 1989).

Another challenge to the protoplast method has been obtaining a reliable source of the cell-wall-degrading enzymes. Oomycete walls mostly contain various β-glucans and cellulose, unlike the chitinous walls of higher fungi, and enzymes must be chosen with this in mind. The best results in *Phytophthora* have been obtained using a cellulase with an enzyme mixture rich in β-glucanases. For the latter, the original transformation procedure used a concoction of hydrolytic enzymes from *Trichoderma* marketed as Novozyme 234, but when this stopped being sold, alternatives needed to be identified and protocols reoptimized. A variety of products have been satisfactory replacements, mostly β-glucanases made for clarifying wines such as Glucanex and Vinoflow. With *P. infestans* the β-glucanases can liberate protoplasts on their own, but adding cellulase accelerates wall digestion 10-fold. Whether this mixture is optimal for all oomycetes remains to be assessed.

Numerous modifications to the protoplast transformation procedure have been tested in *P. infestans*. Many were not helpful such as using nuclease inhibitors, chromosome-nicking agents, temperature shocks to increase DNA uptake, or cryogenically stored protoplasts, which makes transformation more convenient in some fungi (Fincham, 1989). However, cationic liposomes (Lipofectin) increased transformation rates three-fold (Judelson and Michelmore, 1991). This does not seem related to how liposomes cause DNA uptake in animal cells by fusing directly with the membrane, because PEG is still required for transformation. Unfortunately, batch-to-batch variation has been noted in Lipofectin, possibly because quality control is only performed for mammalian studies. Consequently, some workers have stopped using this reagent; most other formulations of cationic liposomes are toxic to *P. infestans*.

22.3.4 Agrobacterium-Mediated Transformation

A. tumefaciens, although used traditionally for plant transformation, can also mobilize DNA into a wide range of species. In true fungi, there has been a strong move toward using it for transformation (Mullins and Kang, 2001). Achieving transformation is relatively simple in theory, which entails mixing the target species with *A. tumefaciens* that have the transforming DNA on its Ti-plasmid, and then selecting transformants.

In oomycetes, the use of this method was reported for *P. infestans, P. palmivora*, and *Pythium ultimum* (Vijn and Govers, 2003). This involved cocultivating *A. tumefaciens* that carry the *nptII* and GUS genes with germinated zoospore cysts on filters. After 3 to 7 days, depending on the species, the material was transferred to media that contained cefotaxin and G418 to eliminate bacteria and select transformants, respectively. Many transformants expressed both *nptII* and GUS, and blot analysis indicated that only the T-DNA portion of the plasmid was integrated. Some isolates of *P. infestans* seemed more amenable to agrotransformation than others, but this might be related to their growth rates or antibiotic susceptibilities rather than differential recognition by the bacterium.

22.3.5 Transformation by Bombardment

This technique, which enjoys wide use in eukaryotes, entails coating DNA onto small gold or tungsten particles that are accelerated into tissue. In *P. infestans,* this process efficiently produced transformants using germinated sporangia or cysts as targets. Transformation was accomplished by spreading the tissues on polycarbonate membranes, which after bombardment were laid on

G418-amended media. Gene transfer also occurred using older hyphae but at a lower rate (Cvitanich and Judelson, 2003a).

One complication with bombardment is that transformants are generally heterokaryons; interestingly, only a fraction of transformed nuclei (5%) are required for vigorous growth on G418. For many applications, homokaryons would need to be isolated. This is not difficult, as it simply requires passage through the zoospore stage followed by G418 selection. A second challenge is that transformants are near each other on the bombardment membrane, and because colonies are rather diffuse, this complicates discerning one from another. This can also be an problem with the other transformation methods. For bombardment, this was resolved by cutting the membrane into small pieces, which were spread over many selection plates.

Despite these issues, the advantages of bombardment are that it bypasses the need to develop protoplasting and regeneration procedures, works with small amounts of tissue, and avoids potential problems with *Agrobacterium* host specificity. Bombardment may also be an option for oomycetes that cannot be cultured on artificial media. Indeed, the method succeeded for expressing GUS in conidia of *B. lactucae* (Fig. 22.1a). In theory, bombardment could also be used against other phases of obligate pathogen growth such as the *in planta* stage, but this was not successful with *Plasmopara halstedii* in sunflower (Hammer et al., 2007) as well as with *B. lactucae* in lettuce.

Other methods for physically disrupting the wall may also hold promise. For example, some mechanically perforated hyphae of *P. halstedii* treated with a GFP-containing plasmid were observed to express that marker transiently (Hammer et al., 2007). The approach is reminiscent of the use of glass beads or silica carbide whiskers for transformation in other taxa (Costanzo and Fox, 1988; Wang et al., 1995). It would be useful to learn whether such methods work with the more culturable oomycetes.

22.3.6 Electroporation

This method involves applying an electric pulse to tissue, which causes pores to form transiently across the plasma membrane through which DNA can enter. In theory, applying the method to a new species simply involves determining the voltage and ionic conditions that allow gene transfer while maintaining viability. In oomycetes, electroporation has been used with success against protoplasts and zoospores. With *P. aphanidermatum,* stable transformants were obtained using protoplasts (Weiland, 2003). A similar approach worked with *P. infestans*, but was not as efficient as the alternative protoplast-based methods (unpublished data). However, electroporation of *P. infestans* zoospores has proved to be convenient and effective, because zoospores lack cell walls. This involves liberating zoospores from sporangia, adding LiCl, concentrating zoospores on a density gradient, and then electroporating (Latijnhouwers and Govers, 2003). Adding LiCl is critical because this prevents the zoospores from encysting during the manipulations (Ersek et al., 1991).

There are no reports about the successful use of electroporation against intact oomycete tissue. However, electroporation has achieved gene transfer in the intact, walled cells of several plants and true fungi (Ruiz-Diez, 2002). Consequently, this technique may hold promise for oomycetes recalcitrant to the other methods.

22.4 FATES OF TRANSFORMING DNA

Based on studies in *Phytophthora, Pythium,* and *Saprolegnia*, transformation usually involves the chromosomal integration of plasmid DNA (Cvitanich and Judelson, 2003a; Judelson et al., 1991, 1993a; Mort-Bontemps and Fevre, 1997). Vectors that contain an autonomously replicating sequence (ARS), which have proved valuable in other systems because they allow high gene transfer rates, are not yet developed. An attempt to identify an ARS from *P. infestans* by testing 96 different 35-kb fragments for their ability to confer autonomous replication was unsuccessful (unpublished data). Transferring ARS sequences from *Aspergillus, Saccharomyces,* or *Ustilago* transformation plasmids into oomycete vectors did not enable the latter to replicate autonomously in *P. infestans.*

An exception to integrative transformation was discovered in *P. nicotianae,* in which plasmids similar to those used in other oomycetes yielded transformants in which transgene expression was unstable. Extensive analyses indicated that the plasmids had acquired the ability to persist extrachromosomally, with some seemingly altered in structure (Gaulin et al., 2007). The situation may resemble that noted in *Plasmodium*, in which DNA can be maintained as replicating concatamerized episomes (Kadekoppala et al., 2001). More investigations may lead to an autonomously replicating vector system for oomycetes.

The DNA that becomes chromosomally integrated represents only a fraction that entered the cell or protoplast. When *P. infestans* protoplasts were treated with a plasmid that expressed *nptII* and GUS, 24 h later about 20% of nucleated protoplasts were observed to express GUS (unpublished data). However, less than 0.001% of regenerated protoplasts were recoverable as G418-resistant transformants. This implies that only a subpopulation of nuclei is competent to integrate DNA or that most plasmids insert into transcriptionally silent heterochromatin.

When chromosomal integration does occur, it is common for plasmids to integrate as tandem arrays regardless of whether protoplast, microprojectile, or electroporation methods are used. For example, from 1 to more than 20 plasmid molecules were present within protoplast-derived transformants of *P. infestans, P. palmivora*, and *P. sojae*, most of which have multiple copies at single sites (Judelson, 1993; Judelson et al., 1991, 1993a; van West et al., 1999b). Electroporation and bombardment-derived *P. infestans* transformants also contained tandem arrays, as did strains obtained by electroporation in

P. aphanidermatum (Weiland, 2003). Some data suggest that copy numbers are lower in transformants obtained by electroporation compared with protoplasting (Latijnhouwers and Govers, 2003), but other studies indicated that these values are similar (unpublished data). Whether any differences simply reflect the concentrations of DNA employed or how DNA is handled by the tissues is unclear. Nevertheless, the common occurrence of single-site tandem arrays suggests that chromosomal entry is a limiting step in oomycete transformation. This differs from filamentous true fungi, in which multiple integration sites are typical (Ruiz-Diez, 2002). If this limiting step was better understood, it might be possible to increase transformation rates. Adding restriction enzymes to the transforming DNA (the REMI method), which has increased such rates in certain true fungi (Mullins and Kang, 2001), did not affect transformation in *P. infestans* (unpublished data).

The incorporation of multiple copies of DNA is not the rule in oomycetes. For example, most transformants of *Saprolegnia* only contained single copies, which suggests that different species process incoming DNA in varying ways (Mort-Bontemps and Fevre, 1997). Also, following a trend in true fungi, most *Agrobacterium*-derived transformants of *P. infestans* had only one or two copies (Mullins and Kang, 2001; Vijn and Govers, 2003). It remains to be determined whether such low-copy transformants tend toward lower levels of transgene expression or are less subject to repeat-mediated transgene silencing.

Studies in *P. infestans* indicated that transgene expression levels are strongly influenced by the integration site, as in most eukaryotes with complex genomes (Judelson and Whittaker, 1995). There is no evidence for homologous recombination between plasmid and chromosomal DNA, so integration sites are random and subject to position effects. The latter complicate quantitative analyses of gene or promoter function (Ah Fong et al., 2007). A related phenomenon is the occasional spontaneous silencing of transgenes, which possibly reflects their integration near heterochromatin (Judelson and Whittaker, 1995; Mort-Bontemps and Fevre, 1997; Vijn and Govers, 2003). Another consequence of the low rate of homologous integration is that gene-targeting methods, which include gene replacement, are not currently feasible. Attempts to stimulate homologous recombination using plasmids that contain pieces of *P. infestans* DNA from 0.5 to 100 kb, or vectors linearized or having long single-stranded termini within the homologous region, had no effect (unpublished data).

Although homologous recombination between plasmid and chromosomal DNA is infrequent, it is common between plasmids prior to integration. During an assay in *P. infestans* for recombination between two plasmids that have defects in the 5′ and 3′ regions of their *nptII* genes, respectively, intermolecular recombination restored function to about half of the plasmids (Judelson, 1993). This process likely explains the occurrence of tandem repeats in most transformants, although coinsertional replication might also be involved as in *Dictyostelium* (Barth et al., 1998).

Intermolecular recombination also influences cotransformation experiments, which involves the integration of selected and nonselected plasmids within the same strain. When circular plasmids that contain *nptII* or GUS markers were mixed and added to *P. infestans* protoplasts, about 10% of transformants expressed both markers with 20% of those bearing the plasmids at the same genomic location; it is likely that more had taken up both plasmids but the expression of one was hampered by position effects (Judelson, 1993). A strategy to achieve higher rates of cotransformation resulted from the finding that DNA with compatible ends was ligated efficiently within protoplasts. Using plasmids linearized with the same restriction enzyme, rates of cotransformation of 80-100% were achieved in *P. infestans, P. nicotianae*, and *P. sojae* (Bottin et al., 1999; Judelson, 1993; Judelson et al., 1993a).

22.5 APPLICATIONS OF TRANSFORMATION

Examples of the use of gene transfer techniques to study oomycete biology are presented below. The reader is referred to other chapters in this volume for additional information about many related topics, such as gene silencing, genes involved in asexual reproduction, and plant effector proteins.

22.5.1 Uses of Reporter Genes

Transformants expressing reporters such as GUS, or fluorescent proteins like GFP and monomeric red fluorescent protein (mRFP), have proved useful in examining growth and development by enabling more detailed analyses than traditional approaches (Fig. 22.1). For example, the use of a parent constitutively expressing GUS in *P. infestans* matings enabled the origin of gametangia to be traced, which allowed the sexual preference of isolates and outcrossed oospores to be identified (Judelson, 1997). Reporters also aided *in planta* studies of the proliferation of *Phytophthora brassica* in Arabidopsis, *P. infestans* in potatoes, *P. nicotianae* in tobacco and tomatoes, and *P. aphanidermatum* in sugarbeet. The measurements of reporter expression facilitated assays of natural resistance and chemical-induced systemic acquired resistance (Bottin et al., 1999; Kamoun et al., 1998; Le Berre et al., 2008; Si-Ammour et al., 2003; Weiland, 2003).

Tagging regulated promoters or proteins with reporters has also helped to define developmental structures in *P. infestans* (Fig. 22.1c–g). For example, expressing GUS from promoters activated early in asexual or sexual sporulation enabled sites of differentiation to be identified before they were morphologically distinguishable (Ah Fong and Judelson, 2003; Cvitanich and Judelson, 2003b). GUS fusions also helped to show that the IpiO effector is expressed during the biotrophic phase of potato colonization (van West et al., 1998), and mRFP fused to the Avr3a effector facilitated visualization of the extrahaustorial matrix into which it was secreted (Whisson et al., 2007).

22.5.2 Promoter Analysis

Tests of promoter function and structure are also enabled by transformation. Protoplast-based transient assays are often convenient for this because stable transformants are not required; tissue is simply treated with DNA, and transcription is measured using a reporter. For example, such assays demonstrated that the *ham34* and *hsp70* promoters from *B. lactucae* had strong activity in protoplasts of *Achlya, Phytophthora,* and *Pythium* and that they outperformed most *Phytophthora* promoters (Judelson et al., 1992, 1993b; Judelson and Michelmore, 1991; McLeod et al., 2008). The reliability of many of the latter results were subsequently verified in stable transformants (Judelson et al., 1993b). Transient assays also showed that promoters from plants, animals, or true fungi exhibited little activity in oomycetes (Judelson et al., 1992). Some nonoomycete promoters could still have some function, based on a report that the mammalian cytomegalovirus (CMV) promoter drove GUS in *P. aphanidermatum* (Weiland, 2003). However, whether normal CMV regulatory sequences or cryptic sites were used was not determined.

Both transient and stable transformation systems have been used to dissect several promoters from *P. infestans* using reporter plasmids such as pNPGUS (Fig. 22.2). Through transient assays against mutagenized promoters from the *Exo1* and *Endo1* genes fused to GUS, an Inr-like core promoter element recognized by general transcription factors was functionally defined (McLeod et al., 2004). Promoter mutagenesis also helped to identify transcription factor sites important in development, such as motifs that activate genes during asexual sporulation, zoosporogenesis, and mating (Ah Fong et al., 2007; Tani and Judelson, 2006). In such cases, stable transformants were studied because analyses in the relevant life stages were required. This made the work more time consuming and complex, with position effects necessitating the analysis of many transformants.

22.5.3 Forced Expression and Complementation Studies

The ability to express novel or modified proteins is a powerful application of transformation. Although such studies are common in prokaryotes and other eukaryotes, to date there are only a few examples in oomycetes. One defined the function of the RXLR-EER motif present in the N-terminus of many effectors expressed by pathogenic oomycetes. By altering residues in a mRFP-tagged effector and then following its movement by microscopy, the role of RXLR-EER in transport from the extrahaustorial space into plant cells was demonstrated (Whisson et al., 2007). Another example involved expressing an elicitin from *Phytophthora cryptogea* in *P. infestans,* which helped to confirm that the protein induces necrosis in tobacco (Panabieres et al., 1998).

A related approach that has not been used extensively is complementation, which involves introducing a gene into a strain that lacks a functional copy. This is a common strategy used in nonoomycete pathogens to test virulence and avirulence factors. The only illustration of this in oomycetes so far comes from *P. sojae*, in which inserting the Avr1b effector gene into an Avr1b-minus strain increased avirulence on *Rps1b* soybean (Dou et al., 2008). Such experiments would probably be more common if not for the availability of highly efficient viral and *Agrobacterium* systems for *in planta* tests of such proteins.

Cloning genes by shotgun transformation could be possible in an oomycete that has a high transformation efficiency. The rates in *P. infestans* are on the brink of enabling this, based on a proof-of-concept test in which DNA from a G418-resistant transformant was transformed into a wild-type strain and G418-resistant strains with *nptII* were recovered (unpublished data). One useful application of such an approach would be to isolate genes that determine fungicide resistance from a cosmid or BAC library.

A useful addition to expression studies would be a promoter that could be induced at will, for example by adding a small molecule. Toward this end, the Tet-On system that is used widely to control transgene expression on eukaryotic cells was adapted to *P. infestans* (Judelson et al., 2007). Although it functioned inefficiently, a better understanding of the oomycete transcriptional apparatus could lead to improvements. Other popular expression systems might also be adaptable to oomycetes.

22.5.4 Gene Silencing

This powerful tool for functional genomics will be addressed only briefly, because it is the focus of another chapter in this volume. Following approaches established first in plants and later in animals and filamentous fungi, several *Phytophthora* researchers showed that inserting sense, antisense, or inverted repeat versions of a native gene into stable transformants can trigger the silencing of both the native locus and transgene. Such methods effectively demonstrated the roles of several genes in sporulation, zoospore behavior, and plant infection (Blanco and Judelson, 2005; Gaulin et al., 2002; Judelson and Tani, 2007; Latijnhouwers and Govers, 2003), although silencing is still not routine or easily applied to all targets.

The molecular basis of silencing in *Phytophthora* is beginning to be revealed. Nuclear run-on assays indicated that silencing is generally transcriptional and associated with the compaction of chromatin (Judelson and Tani, 2007; Van West et al., 1999a, 2008). Silencing frequently eliminates detectable expression of the targeted gene, but partial knock-downs have also been reported (Ah-Fong et al., 2008; Gaulin et al., 2002). Silencing can spread between nuclei, based on analyses of heterokaryons established between silenced and nonsilenced nuclei, although some heterokaryons can persist for long periods without homogenization of the silenced state (Ah-Fong et al., 2008; Van West et al., 1999a). Double-stranded RNA (dsRNA) is a likely candidate for the internuclear signal based on its involvement in other taxa and the detection of 21-nt dsRNA complementary to a silenced gene in *P. infestans* (Ah-Fong et al., 2008). The importance of self-complementary RNA in triggering silencing is supported by recent optimization studies in *P. infestans* that indicated that the best results are obtained using inverted repeat constructs, which can be generated using plasmids like pSTORA (Fig. 22.2; Ah-Fong et al., 2008). Also, inverted repeat RNAs have been shown to induce transient silencing in *P. infestans* (Walker et al., 2008; Whisson et al., 2005).

A caveat when using transformation-based methods, such as silencing, to test gene function is that some strains may show abnormalities caused by chromosomal defects or epigenetic changes unrelated to the sequence being tested. For example, in one study of *P. infestans,* the infection frequency of general transformants was reduced by one third (Latijnhouwers et al., 2004), and in another study the pathogenicity was impaired in 20% of transformants (Judelson and Whittaker, 1995). In the latter study, differences were subtle and detectable only in competition assays with wild-type strains. Nevertheless, investigators should be careful to use multiple strains and controls to draw conclusions. Unintended effects of transformation were also problematic in one *P. nicotianae* study in which all transformants had defects in zoosporogenesis (Gaulin et al., 2002). Such issues are not unique to oomycetes, as transformation-induced variation is well described in many eukaryotes (Kaeppler et al., 2000).

22.6 OUTLOOK

The number of oomycetes that have been transformed is low but increasing, and of species for which transformation is reported, it has been exploited for significant biological investigations in only a few. As more species are studied, particularly from outside the Pythiaceae, appropriate systems will need to be chosen and optimized. Researchers should not assume that extant vectors, promoters, selection schemes, or DNA transfer procedures will be ideal for their system.

Even in the more-studied genera like *Phytophthora,* there is room for advancements. These include improved inducible promoters, proved technologies for expressing proteins with useful affinity tags, and easier methods for the high-throughput isolation of transformants. Some methods described above, such as gene silencing, are still not routine or trivial. New strategies for increasing transformation rates should also be considered, such as the use of transposable elements. Recently discovered viruses and other extrachromosomal elements in *Phytophthora* may lead to autonomously replicating vectors that have high transformation efficiencies (Hacker et al., 2005; Judelson and Fabritius, 2000).

Designing procedures for gene targeting is another important objective. These methods can enable gene knockouts or replacements and can improve expression studies by targeting plasmids to a reproducible location. A targeting system might have been considered unattainable a few years ago, but emerging technologies may make this possible. For example, custom zinc finger nucleases are making targeting within reach in several plant and animal systems; such proteins can be engineered to cause double-strand breaks within a gene of interest, which activates the endogenous homologous recombination machinery. A similar approach would enable targeted mutation, as such breaks are frequently misrepaired (Lloyd et al., 2005). Silencing the nonhomologous end-joining pathway may also stimulate homologous recombination, as demonstrated in true fungi (Ninomiya et al., 2004). A mycobacterial integrase system proved successful for both targeting insertions to a specific locus and increasing rates of stable transformation in apicomplexans, which have taxonomic affinity to oomycetes, where similar methods might be useful (Balu and Adams, 2007).

REFERENCES

Ah Fong A, Judelson HS 2003. Cell cycle regulator *Cdc14* is expressed during sporulation but not hyphal growth in the fungus–like oomycete *Phytophthora infestans*. Molec Microbiol 50:487–494.

Ah Fong A, Xiang Q, Judelson HS 2007. Architecture of the sporulation–specific Cdc14 promoter from the oomycete *Phytophthora infestans*. Eukaryot Cell 6:2222–2230.

Ah-Fong AM, Bormann-Chung CA, Judelson HS 2008. Optimization of transgene-mediated silencing in *Phytophthora infestans* and its association with small-interfering RNAs. Fungal Genet Biol 45:1197–1205.

Baldauf SL, Roger AJ, Wenk-Siefert I, Doolittle WF 2000. A kingdom-level phylogeny of eukaryotes based on combined protein data. Science 290:972–977.

Balu B, Adams JH 2007. Advancements in transfection technologies for *Plasmodium*. Int J Parasitol 37:1–10.

Barth C, Fraser DJ, Fisher PR 1998. Co-insertional replication is responsible for tandem multimer formation during plasmid integration into the *Dicytostelium* genome. Plasmid 39:141–153.

Blanco FA, Judelson HS 2005. A bZIP transcription factor from *Phytophthora* interacts with a protein kinase and is required for zoospore motility and plant infection. Molec Microbiol 56:638–648.

Bottin A, Larche L, Villalba F, Gaulin E, Esquerre-Tugaye M-T, Rickauer M 1999. Green fluorescent protein (GFP) as gene expression reporter and vital marker for studying development and microbe–plant interaction in the tobacco pathogen *Phytophthora parasitica* var. *nicotianae*. FEMS Microbiol Lett 176:51–56.

Converse AD, Belvi LR, Gori JL, Lio G, Amaya F, Agular-Cordova E, Hackett PB, Mcluor RS 2004. Counterselection and co-delivery of transposon and transposase functions for Sleeping Beauty-mediated transposition in cultured mammalian cells. Biosci Rep 24:577–594.

Costanzo MC, Fox TD 1988. Transformation of yeast by agitation with glass beads. Genetics 120:667–670.

Cvitanich C, Judelson H 2003a. Stable transformation of the oomycete, *Phytophthora infestans,* using microprojectile bombardment. Curr Genet 42:228–235.

Cvitanich C, Judelson HS 2003b. A gene expressed during sexual and asexual sporulation in *Phytophthora infestans* is a member of the Puf family of translational regulators. Eukaryot Cell 2:465–473.

Dou D, Kwes D, Wang X, Chen Y, Wang O, Wang X, Jiang RAY, Arredondo FD, Anderson RG, Thakur PB 2008. Conserved C-terminal motifs required for avirulence and suppression of cell death by *Phytophthora sojae* effector Avr1b. Plant Cell 20:1118–1133.

Ersek T, Hoelker U, Hoefer M 1991. Non-lethal immobilization of zoospores of *Phytophthora infestans* by lithium. Mycol Res 95:970–972.

Fincham JRS 1989. Transformation in fungi. Microbiol Rev 53:148–170.

Gaulin E, Haget N, Khatib M, Herbert C, Rickauer M, Bottin A 2007. Transgenic sequences are frequently lost in *Phytophthora parasitica* transformants without reversion of the transgene-induced silenced state. Can J Microbiol 53:152–157.

Gaulin E, Jauneau A, Villalba F, Rickauer M, Esquerre-Tugaye MT, Bottin A 2002. The CBEL glycoprotein of *Phytophthora parasitica* var. *nicotianae* is involved in cell wall deposition and adhesion to cellulosic substrates. J Cell Sci 115:4565–4575.

Hacker CV, Brasier CM, Buck KW 2005. A double-stranded RNA from a *Phytophthora* species is related to the plant endornaviruses and contains a putative UDP glycosyltransferase gene. J Gen Virol 86:1561–1570.

Hammer TR, Thines M, Spring O 2007. Transient expression of gfp in the obligate biotrophic oomycete *Plasmopara halstedii* using electroporation and a mechanoperforation method. Plant Pathol 56:177–182.

Judelson HS 1993. Intermolecular ligation mediates efficient cotransformation in *Phytophthora infestans*. Molec Gen Genet 239:241–250.

Judelson HS 1997. Expression and inheritance of sexual preference and selfing potential in *Phytophthora infestans*. Fungal Genet Biol 21:188–197.

Judelson HS, Coffey MD, Arredondo FR, Tyler BM 1993a. Transformation of the oomycete pathogen *Phytophthora megasperma* f. sp. *glycinea* occurs by DNA integration into single or multiple chromosomes. Curr Genet 23:211–218.

Judelson HS, Dudler R, Pieterse CMJ, Unkles SE, Michelmore RW 1993b. Expression and antisense inhibition of transgenes in *Phytophthora infestans* is modulated by choice of promoter and position effects. Gene 133:63–69.

Judelson HS, Fabritius AL 2000. A linear RNA replicon from the oomycete *Phytophthora infestans*. Molec Gen Genet 263:395–403.

Judelson HS, Michelmore RW 1989. Structure and expression of a gene encoding heat-shock protein Hsp70 from the Oomycete fungus *Bremia lactucae*. Gene 79:207–218.

Judelson HS, Michelmore RW 1990. Highly abundant and stage-specific messenger RNA in the obligate pathogen *Bremia lactucae*. Molec Plant-Microbe Interact 3:225–232.

Judelson HS, Michelmore RW 1991. Transient expression of genes in the oomycete *Phytophthora infestans* using *Bremia lactucae* regulatory sequences. Curr Genet 19:453–460.

Judelson HS, Narayan R, Fong AM, Tani S, Kim KS 2007. Performance of a tetracycline–responsive transactivator system for regulating transgenes in the oomycete *Phytophthora infestans*. Curr Genet 51:297–307.

Judelson HS, Tani S 2007. Transgene-induced silencing of the zoosporogenesis-specific PiNIFC gene cluster of *Phytophthora infestans* involves chromatin alterations. Eukaryot Cell 6:1200–1209.

Judelson HS, Tyler BM, Michelmore RW 1991. Transformation of the oomycete pathogen, *Phytophthora infestans*. Molec Plant-Microbe Interact 4:602–607.

Judelson HS, Tyler BM, Michelmore RW 1992. Regulatory sequences for expressing genes in oomycete fungi. Molec Gen Genet 234:138–146.

Judelson HS, Whittaker SL 1995. Inactivation of transgenes in *Phytophthora infestans* is not associated with their deletion, methylation, or mutation. Curr Genet 28:571–579.

Kadekoppala M, Cheresh P, Catron D, Ji DD, Deitsch K, Wellems TE, Seifert HS, Haldar K 2001. Rapid recombination among transfected plasmids, chimeric episome formation and trans gene expression in *Plasmodium falciparum*. Molec Biochem Parasit 112:211–218.

Kaeppler SM, Kaeppler HF, Rhee Y 2000. Epigenetic aspects of somaclonal variation in plants. Plant Mol Biol 43:179–188.

Kamoun S, van West P, Govers F 1998. Quantification of late blight resistance of potato using transgenic *Phytophthora infestans* expressing beta-glucuronidase. Eur J Plant Pathol 104:521–525.

Kinghorn JR, Moon RP, Unkles SE, Duncan JM 1991. Gene structure and expression in *Phytophthora infestans* and the development of gene-mediated transformation. In: Lucas JA, Shattock RC, Shaw DS, Cooke LR, editors. *Phytophthora*. Cambridge, University Press. New York. pp. 295–311.

Latijnhouwers M, Govers F 2003. A *Phytophthora infestans* G-protein ß subunit is involved in sporangium formation. Eukaryot Cell 2:971–977.

Latijnhouwers M, Ligterink W, Vleeshouwers VGAA, Van West P, Govers F 2004. A G-alpha subunit controls zoospore motility and virulence in the potato late blight pathogen *Phytophthora infestans*. Molec Microbiol 51:925–936.

Le Berre JY, Engler G, Panabieres F 2008. Exploration of the late stages of the tomato-*Phytophthora parasitica* interactions through histological analysis and generation of expressed sequence tags. New Phytol 177:480–492.

Lloyd A, Plaisier CL, Carroll D, Drews GN 2005. Targeted mutagenesis using zinc-finger nucleases in Arabidopsis. Proc Natl Acad Sci USA 102:2232–2237.

Manavathu EK, Suryanarayana K, Hasnain SE, Leung WC 1988. DNA-mediated transformation in the aquatic filamentous fungus *Achlya ambisexualis*. J Gen Microbiol 134:2019–2028.

Masunaka A, Takenaka S 2006. Transformation of the biocontrol agent *Pythium oligandrum* that induces plant defense reactions. Japan J Phytopath 72:267.

McKnight GL, Kato H, Upshall A, Parker MD, Saari G, O'Hara PJ 1985. Identification and molecular analysis of a third *Aspergillus nidulans* alcohol dehydrogenase gene. EMBO J 4:2093–2099.

McLeod A, Fry BA, Zuluaga P, Myers KL, Fry WE 2008. Toward improvements of oomycete transformation protocols. J Eukaryot Microbiol 55:103–109.

McLeod A, Smart CD, Fry WE 2004. Core promoter structure in the oomycete *Phytophthora infestans*. Eukaryot Cell 3:91–99.

Mort–Bontemps M, Fevre M 1997. Transformation of the oomycete *Saprolegnia monoica* to hygromycin B resistance. Curr Genet 31:272–275.

Mullins ED, Kang S 2001. Transformation: a tool for studying fungal pathogens of plants. Cell Mol Life Sci 58:2043–2052.

Ninomiya Y, Suzuki K, Ishii C, Inoue H 2004. Highly efficient gene replacements in *Neurospora* strains deficient for nonhomologous end-joining. Proc Natl Acad Sci USA 101:12248–12253.

Panabieres F, Birch PR, Unkles SE, Ponchet M, Laeourt I, Vernard P, Keller H, Allasia V, Ricci P, Duncan JM 1998. Heterologous expression of a basic elicitin from *Phytophthora cryptogea* in *Phytophthora infestans* increases its ability to cause leaf necrosis in tobacco. Microbiol 144:3343–3349.

Randall TA, Judelson HS 1999. Construction of a bacterial artificial chromosome library of *Phytophthora infestans* and transformation of clones into *P. infestans*. Fungal Genet Biol 28:160–170.

Ruiz–Diez B 2002. Strategies for the transformation of filamentous fungi. J Appl Microbiol 92:189–195.

Si-Ammour A, Mauch-Mani B, Mauch F 2003. Quantification of induced resistance against *Phytophthora* species expressing GFP as a vital marker: beta-aminobutyric acid but not BTH protects potato and *Arabidopsis* from infection. Molec Plant Pathol 4:237–248.

Tani S, Judelson HS 2006. Activation of zoosporogenesis-specific genes in *Phytophthora infestans* involves a 7-nucleotide promoter motif and cold-induced membrane rigidity. Eukaryot Cell 5:745–752.

van West P, De Jong AJ, Judelson HS, Emons AME, Govers F 1998. The *ipiO* gene of *Phytophthora infestans* is highly expressed in invading hyphae during infection. Fungal Genet Biol 23:126–138.

van West P, Kamoun S, Van 't Klooster JW, Govers F 1999a. Internuclear gene silencing in *Phytophthora infestans*. Molec Cell 3:339–348.

van West P, Reid B, Cambell TA, Sandrock RW, Fry WE, Kamoun S,Gown AD 1999b. Green fluorescent protein (GFP) as a reporter gene for the plant pathogenic oomycete *Phytophthora palmivora*. FEMS Microbiol Lett 178:71–80.

van West P, Shepherd SJ, Walker CA, Li S, Appiah AA, Grenville-Briggs LJ, Govers F, Gow NA 2008. Internuclear gene silencing in *Phytophthora infestans* is established via heterochromatin formation. Microbiol 54:1482–1490.

Vijn I, Govers F 2003. *Agrobacterium tumefaciens* mediated transformation of the oomycete plant pathogen *Phytophthora infestans*. Molec Plant Pathol 4:456–467.

Walker CA, Shepherd S, Walker CA, Li S, Appiah AA, Grenville-Brriggs LJ, Govers F, Gow War 2008. A putative DEAD-box RNA-helicase is required for normal zoospore development in the late blight pathogen *Phytophthora infestans*. Fungal Genet Biol 45:954–962.

Wang K, Drayton P, Frame B, Dunwell J, Thompson J 1995. Whisker-mediated plant transformation: An alternative technology. In Vitro Cell Develo Biol Plant 31:101–104.

Weiland JJ 2003. Transformation of *Pythium aphanidermatum* to geneticin resistance. Curr Genet 42:344–352.

Whisson SC, Avrova AO, van West P, Jones JT 2005. A method for double-stranded RNA-mediated transient gene silencing in *Phytophthora infestans*. Molec Plant Pathol 6:153–163.

Whisson SC, Boevink PC, Moleleki L, Aurova AO, Morales JG, Gilroy EM, Armstrong MR, Grouffard S, Van West P, Chapman S, et al. 2007. A translocation signal for delivery of oomycete effector proteins inside host plant cells. Nature 450:115–118.

23

IN PLANTA EXPRESSION SYSTEMS

VIVIANNE G.A.A. VLEESHOUWERS AND HENDRIK RIETMAN
Wageningen UR Plant Breeding, Wageningen, The Netherlands

23.1 INTRODUCTION

The pace of gene discovery in oomycetes is increasing tremendously (Kamoun et al., 1999b; Govers and Gijzen, 2006; Tyler et al., 2006), and a wealth of genes are waiting to be analyzed. For some genes, characteristics can be predicted based on sequence homology or presence of certain motifs, but in most cases, experimental systems are required to provide functional insights (Torto et al., 2003; Win et al., 2006; Whisson et al., 2007; Gaulin et al., 2008; Polesani et al., 2008). In the past, genes of interest were few and were often extensively studied by time-consuming processes such as stable transformation, but nowadays, more high-throughput (HTP) functional genomic techniques are required to exploit the huge amount of candidates (Win et al., 2006; Kamoun, 2007). In this chapter, we focus on recent developments for *in planta* testing of oomycete genes encoding effectors, which manipulate host cells during the plant–pathogen interaction.

The ability to perceive effector molecules of a pathogen is a vital defense strategy in plants (Staskawicz et al., 1995; Dangl and Jones, 2001). In interactions that follow the gene-for-gene model, there is a robust correlation between effector perception and disease resistance, with recognition only occurring in host genotypes that carry resistance (*R*) genes and the interacting effectors often being the direct products of avirulence (*Avr*) genes (Botella et al., 1998). After R-AVR recognition, a local cell death called the hypersensitive response (HR) is initiated, which is a powerful defense mechanism in oomycete–plant interactions such as *Phytophthora infestans*—potatoes (Kamoun et al., 1999a; Vleeshouwers et al., 2000). This phenomenon is extrapolated to functional

Oomycete Genetics and Genomics: Diversity, Interactions, and Research Tools
Edited by Kurt Lamour and Sophien Kamoun
Copyright © 2009 John Wiley & Sons, Inc.

genomic approaches for identifying effector candidates with AVR function, for example, by expressing *Avr* candidates *in planta* and functional profiling resistant host plants for occurrence of the HR (Champouret et al., 2007; Vleeshouwers et al., 2008), or by coexpressing both the *Avr* and *R* candidate genes in a heterologous model system (Oh et al., 2007).

Various methods for *in planta* expression of oomycete genes are available (Table 23.1). In contrast to stable transformation methods, transient assays are easier to perform, can be applied on differentiated plant tissue, and are not influenced by position effects of transgenes (Fischer et al., 1999). Expression systems based on potato virus X (PVX) and *Agrobacterium tumefaciens* are particularly simple and rapid, but plants may show defense or necrotic responses to these pathogens, and unnatural protein production rates can result in "false" responses. To cope with this, other vectors can be chosen or optimizations in the constructs can be made, for example, for altering the targeting of the protein. For handling extensive collections of candidate genes, HTP assays can reduce the number of candidates to a set that can be followed more closely with labor-intensive assays. Still, also in low-throughput assays, the engineered expression levels of the effector remain artificial, and indeed application of multiple techniques is recommended to get the best possible insight in gene function.

23.2 PVX AGROINFECTION

23.2.1 Introduction to PVX Agroinfection

PVX agroinfection is the most facile assay available for *in planta* testing of large numbers of effector candidates. The method is useful for various plant species that allow PVX replication and spread, such as *Nicotiana benthamiana,* tobacco (*Nicotiana tabacum*), tomatoes (*Lycopersicon esculentum*), and *Solanum* species (Takken et al., 2000; Qutob et al., 2002; Torto et al., 2003; Huitema et al., 2004; Vleeshouwers et al., 2006). It operates as a transient expression system based on delivery of PVX by the bacterium *A. tumefaciens* (see the next section of this chapter).

PVX is a member of the potex virus group of plant plus-strand RNA viruses (Batten et al., 2003). PVX contains a relatively small, single-genomic RNA that is functionally monocistronic and is available as an infectious clone. PVX requires four open reading frame (ORF) products for cell-to-cell and long-distance movement in infected plants (i.e., the overlapping ORF2–4, the so-called "triple gene block," and the coat protein ORF5). ORF1 is required for RNA replication. PVX replicates to high levels in plants, and it has been extremely useful as a model system in plant–pathogen research (Batten et al., 2003).

In addition to potatoes, PVX can infect most Solanaceous plant species, which include *Nicotiana*, tomatoes, and many *Solanum* species. The virus is

TABLE 23.1 Expression systems for testing oomycete genes for *in planta* effects. Listed are the possible systems for gene and protein expression, drawbacks, and efficiency by which the approach can be applied.

	Gene expression	Protein production/ exposure	Limitations of the system	Efficiency
PVX infection	PVX	*In planta*, mediated by PVX	Response to PVX, mechanical inoculation damage	Medium
PVX agroinfection	PVX (after *Agrobacterium tumefaciens*-mediated transient transformation)	*In planta*, mediated by PVX	Response to PVX	High
Agroinfiltration	*A. tumefaciens*-mediated transient transformation	*In planta*, infiltrated leaf panel with *A. tumefaciens*	Response to *A. tumefaciens*	Medium high
Agrosuppression	*A. tumefaciens*-mediated transient transformation	*In planta*, on petiole inoculated with *A. tumefaciens*	Optimized for *N. benthamiana*	High
Biolistic assay	Particle bombardment	*In planta*	Variation in transformation efficiency	Low
Stable transformation for protein production	Heterologous systems, e.g., *Pichia pastoris, Escherichia coli.*	*In planta*, infiltrated proteins	Protein stability	Low
Stable transformation plant	*In planta*	*In planta*	Long-lasting procedure	Very low
Stable transformation oomycete	Oomycete	During *in planta* infection	Laborious and difficult procedure	Extremely low

mechanically transmitted through contact, and symptoms on plants range from mild mottling of the leaf to a severe mosaic, with a dwarfing of the plant and reduced leaf size and crinkling. Plant resistance to PVX occurs as a HR or as extreme resistance (ER); plants with HR show either necrotic lesions or systemic necrosis, whereas plants with ER show no symptoms. However, a connection between ER and HR has been suggested and PVX-induced necrosis can occur in plants with ER genes and vice versa (Ross, 1958; Barker and Dale, 2006). Various resistance genes to PVX are genetically well studied or cloned, such as *Nb, Rx, Rx2, Rx3,* and *Rx4* (Ritter et al., 1991; de Jong et al., 1997; Bendahmane et al., 1999; Bendahmane et al., 2000; Butterbach et al., 2007). Additional resistance genes are reported to occur in various wild *Solanum* species (Ross, 1958, 1986; Hawkes, 1990; Valkonen et al., 1991; Rouppe van der Voort et al., 1999; Barker and Dale, 2006; Butterbach et al., 2007).

The PVX infection method is based on the fact that PVX allows the transient expression of heterologous genes in solanaceous plants by introducing the gene of interest after a duplicated copy of the viral promoter for the coat protein messenger RNA (mRNA). This way, genes of interest are systemically expressed in the plant, which can then be monitored for altered phenotypes (Chapman et al., 1992; Hammond-Kosack et al., 1995; Kooman-Gersmann et al., 1997; Kamoun et al., 1999c). Expected plant symptoms range from a systemic mosaic (no response to the inserted gene) to systemic necrosis (HR to an effector) or no symptoms (ER to an effector). The major disadvantage of this method is the low efficiency, because infectious transcripts of the recombinant PVX need to be generated for each effector gene of interest. In addition, inoculation wounds can mask the effector responses, because the PVX needs mechanic application on the leaves.

A breakthrough in developing HTP *in planta* expression systems was achieved by constructing binary PVX-expression vectors, in which the full PVX genome, flanked by the Cauliflower Mosaic Virus (CaMV) 35S promoter and the nopaline synthase terminator, was cloned in the T-DNA of an *A. tumefaciens* binary vector (Chapman et al., 1992; Jones et al., 1999; Lu et al., 2003a). In this PVX agroinfection system, numerous candidate genes or cDNAs of a library can efficiently be inserted into the PVX backbone, and the binary vector, such as the commonly used vector pGR106, facilitates transfer of the expression construct to the plant cells through local transformation by *A. tumefaciens* (Takken et al., 2000; Torto et al., 2003). The PVX genome with the inserted gene is transcribed from the 35S promoter, and virus particles can move from cell to cell and can spread systemically through the plant. This PVX-based expression of the genes of interest occurs at high levels *in planta*.

Current applications of PVX agroinfection involve large-scale initiatives to identify HR-inducing genes. The method can be used for overexpressing plant genes to identify positive regulators of cell death (Nasir et al., 2005; Takahashi et al., 2007; Coemans et al., 2008), or pathogen genes with HR-inducing activity (Tobias et al., 1999; Takken et al., 2000; Qutob et al., 2002; Torto et al., 2003;

Huitema et al., 2005). The unbiased approach of screening pathogen cDNA libraries has proven successful for identifying avirulence genes and other necrosis-inducing effectors (Takken et al., 2000; Torto et al., 2003; Kanneganti et al., 2006). Some of these nonspecific genes, such as the crinkling and necrosis-inducing gene *Crn2* (Torto et al., 2003), or the *Nep1*-like necrosis-inducing protein of *P. infestans* PiNPP1 (Kanneganti et al., 2006), also serve as reliable positive controls in PVX agroinfection experiments. Recent genomics initiatives have speeded up the process of *Avr* gene identification, and the efficiency of randomly screening cDNA libraries was 15 × improved by cloning a preselection of predicted extracellular proteins of *P. infestans* (Pex) (Kamoun et al., 1999b; Qutob et al., 2002; Torto et al., 2003). After the discovery of the RxLR motif as a translocation signal for delivery of oomycete effector proteins into host plant cells, the preselection of candidate effectors was enhanced (Kamoun, 2006, 2007; Whisson et al., 2007). Currently, large-scale screening of resistant germplasm with candidate Pex-RxLR that contains effector candidates has proven a successful functional strategy to identify *Avr* and *R* genes, by the discovery of *ipiO* as *Avr-blb1* interacting with *Rpi-blb1, Rpi-sto1*, and *Rpi-pta1* (Vleeshouwers et al., 2008; Oh et al., 2007; Champouret et al., 2007).

23.2.2 The PVX Agroinfection Method

The PVX agroinfection method for functional profiling of effector responses has been described in detail for *N. benthamiana* (Kanneganti et al., 2007a), and a protocol optimized for *Solanum* is displayed in Box 23.1. Some notes that relate to this protocol are listed below.

Note 1. Plant growth conditions. Plants should be healthy, not stressed, and not infected or challenged with pathogens before inoculation.

Note 2. Growing season. In spring and summer, the plant quality is better and the vegetative cycle is longer.

Note 3. Temperature. During the infection, a temperature within the range of 18–22°C is advised. Higher temperatures inhibit plant growth and PVX replication, and lower temperatures inhibit *A. tumefaciens* infection (Dillen et al., 1997; Kapilla et al., 1997; Lu et al., 2003a).

Note 4. Plant age. Preferably, young plants should be used. For *Solanum* plants generated from *in vitro* material, 2–3 weeks after transplanting to pots is appropriate (Fig 23.1a). Such plants show clear responses, and the development of symptoms can be followed during a long healthy phase of the leaf. Slightly older plants can provide 1–2 extra leaves for HTP screenings in the case of space limitation. Potato plants younger than 2 weeks should not be used because inoculated leaves will drop before the local symptoms appear.

Note 5. Leaf age. The youngest fully stretched leaves are preferred for inoculation. Younger leaves often show a tumor-like formation, possibly

BOX 23.1 PVX AGROINFECTION PROTOCOL IN *SOLANUM*

Plant material

Plants should be grown in controlled greenhouses or climate chambers within the temperature range of 18–22°C (Notes 1–3). For inoculation, use of young fully stretched leaves of 2–4-week-old young, medium-sized plants is advised (Notes 4 and 5).

Agrobacterium

Genes-of-interest are expressed in binary expression vectors e.g. pGR106, and transformed to *A. tumefaciens* strain GV3101 (Notes 6 and 7). Cultures of recombinant *A. tumefaciens* are grown for 2 days at 28°C on solid agar LB medium (10 g Bacto-tryptone, 5 g yeast extract, 10 g Nacl, in 1 liter water, pH 7.5) supplemented with antibiotics.

Inoculation

After dipping a toothpick in the bacteria culture, the excess of bacteria is inoculated by piercing the leaf at both sides of the midvein. To make a quantitative scoring possible, multiple inoculations sites should be made for each effector; for example, duplo inoculations on five different leaves preferably from different plants will yield 10 replicates. To exclude leaf position effects, plant effects, and so on, the replicates should be placed on each experimental unit.

Scoring

Symptoms are visually scored 1–2 times per week during 2–3 weeks, depending on the goal of the experiment. After finalizing the scoring, the responses for each inoculation spot are summarized, and the percentage of responding sites is calculated and compared with the controls.

because of tumor-forming activities of *A. tumefaciens*, but older leaves may senesce or drop down before scoring is finalized.

Note 6. Strains. We use *A. tumefaciens* strain GV3101 for PVX agroinfection in *Solanum* (Dinesh-Kumar et al., 2003; Kanneganti et al., 2007a).

Note 7. Positive and negative controls. For local scoring, good positive controls are general necrosis-inducing genes such as *Crn2* or *Npp1* in the expression vector (Torto et al., 2003; Kanneganti et al., 2006), and the empty vector can serve as a negative control (Fig. 23.1b). In the case of a systemic PVX agroinfection experiment, marker genes such as *Gfp* are preferred as a negative control, because PVX replication is reduced in the presence of alien inserts.

FIG. 23.1 PVX agroinfection in *Solanum*. (a) Young but well-developed *Solanum* plants with enough leaves were used for screening of local symptoms in *Solanum* species. The photograph shows a *Solanum okadae* plant with three labeled leaves, each inoculated with 6 *Agrobacterium tumefaciens* clones, on each side of the mid-veins. (b) Local cell death to the positive control pGR106-Crn2 *in duplo* but not to any of the other inoculated *A. tumefaciens* clones that express candidate effectors in a *Solanum stoloniferum* × *Solanum tuberosum* hybrid, 21 days after inoculation (dpi). (c) Specific local cell death for the four members of the RD39/40 family (top 5, *in duplo*) but not to RD36 (leaf basis), in *Solanum chacoense* at 20 dpi. (d) Local cell death to pGR106-INF in *Solanum microdontum* blocks the veins and causes yellowing, at 21 dpi. (e) Systemic mosaic symptoms at 14 dpi, inoculated with pGR106 in young *S. tuberosum* plants. (f) Systemic cell-death (scd) starting in the top of the plant, at 13 dpi with pGR106-RD39/40 in *Solanum pinnatisectum*.

23.2.3 Interpretation of Results

In *Solanum*, the symptoms become visible as a cell death zone around the inoculation wound at approximately 10 days after inoculation, and 1 to 2 weeks later most responses are very clear. Differences in timing occur between tested effector clones and plant genotypes. Also, phenotypes of the response may vary from an intense black necrosis surrounding the wound until faint necrotic trails near veins near the inoculation spot (Fig. 23.1b–d).

For interpreting the results, a few other features should be considered:

- Resistance to PVX is known to occur in *Solanum*, including *Solanum stoloniferum, Solanum brevidens, Solanum tuberosum* subsp *andigena,*

Solanum chacoense, Solanum acaule, and *Solanum sparsipilum* (Ross, 1958, 1986; Hawkes, 1990; Valkonen et al., 1991; Rouppe van der Voort et al., 1999; Barker and Dale, 2006; Butterbach et al., 2007). This will interfere with functional profiling by binary PVX. In a screening of 80 wild *Solanum* genotypes that belong to 31 different species, 50 clones were found to show cell death consistently to the positive control pGR106-Crn2 and no response to the empty pGR106 vector, which implies that 60–70% of the *Solanum* clones are expected to be applicable for PVX agroinfection (Vleeshouwers et al., 2006).

- The size of the necrotic lesion is no measure for R-AVR activity but is based on various other aspects, such as the virulence of PVX or transformation efficiency of *A. tumefaciens* in the genetic background of the plant.
- High expression levels of single effector candidates in plant cells may lead to "false positives" (e.g. by homologues of AVR proteins), perhaps by mimicking the homologous effector.
- In local scoring, ER can be misinterpreted as absence of response, and "false negatives" will occur. This can partly be solved by switching to a lower throughput systemic experiment that allows PVX quantification by enzyme-linked immunosorbent assay (ELISA) and scoring of systemic symptoms. In the absence of a cell death response, the virus will spread and cause mosaic symptoms (Fig. 23.1e), whereas the occurrence of cell death will either prevent viral spread (no systemic symptoms) or cause systemic cell death starting in the top of the plant (Fig. 23.1f).
- PVX vectors typically cannot accommodate inserts larger than 2 kb or intron-containing gene sequences. Because viruses are prone to mutate at high rates, large inserts can be thrown out, and a mixture of inoculated and mutated viruses may coexist and obscure the phenotypes, especially in systemic experiments.
- Virus-induced gene silencing, which is a common defense mechanism to plant viruses, may occur, and plants should only be inoculated once.
- During the genuine plant–pathogen infection, other effectors may interfere, and phenotypes of single applied effectors may differ from the natural situation.

In summary, PVX agroinfection is a highly sensitive HTP screening system and has provided comparable results with other assays, such as cobombardment studies (Armstrong et al., 2005), *A. tumefaciens* coinfiltrations (Armstrong et al., 2005; Bos et al., 2006), protein infiltrations (Vleeshouwers et al., 2006), and ELISA (Vleeshouwers et al, 2008). In the *P. infestans*–potato pathosystem, this method has proven successful in fast identification of *Avr* and *R* genes (Oh et al., 2007; Vleeshouwers et al., 2008), and it is currently adopted as a well-used system for routine HTP screenings. Still, because background responses and other implications of the two other potato pathogens are an issue, the identified R-AVR candidates should be tested or confirmed with other techniques.

23.3 AGROINFILTRATION

23.3.1 Introduction to Agroinfiltration

Agroinfiltration is an *A. tumefaciens*-based method for transient expression of genes of interest that can be applied in a rather HTP fashion *in planta*. Because of the broad host range of the bacterium, this assay works well on numerous dicot plant species, which include the model plant *N. benthamiana*, tomatoes, potatoes, *Arabidopsis*, and lettuce. Therefore, this assay is broadly applied in screenings, as well as detailed molecular plant–pathogen research (Bendahmane et al., 2000; van der Hoorn et al., 2000; Wroblewski et al., 2005; Lee and Yang, 2006; Kanneganti et al., 2007a).

The soilborne *A. tumefaciens* pathogen enters its host through natural wounds and manipulates the hormone balance, which causes a crown gall at the infection site. Acetosyringone, which is a natural compound released during plant wounding, induces *vir* gene expression and the virulence machinery is triggered. The bacterium integrates its transfer DNA (T-DNA) from a tumor-inducing plasmid (Ti plasmid) in the plant genome by horizontal gene transfer. This natural mechanism is exploited for development of transient (and stable) plant transformation. A range of strains has been disarmed for their tumor-inducing properties, and diverse binary Ti vectors have been developed for laboratory use (Hellens et al., 2000; Zupan et al., 2000).

Because *A. tumefaciens* is a plant pathogen, various defense responses from the host can be initiated. Indeed, numerous *A. tumefaciens* strains induce necrosis or cannot perform sufficient gene transfer during infiltration in certain hosts (Wroblewski et al., 2005). Phenotypic changes such as necrosis or chlorosis at infiltration sites do occur as well (Kjemtrup et al., 2000). Therefore, in case of examining other species than well-studied model plants, the choice of strains in relation to the host is essential (Wroblewski et al., 2005). In addition, the modification of expression vectors can improve applicability in broader series of host plants, which include wild *Solanum* species (Hein et al., 2007).

Agroinfiltration is currently the best developed and most reliable method for reconstructing the R-AVR interaction in plants (Armstrong et al., 2005; Bos et al., 2006). *A. tumefaciens* strains that express the *Avr* gene are coinfiltrated with *A. tumefaciens* strains that express the *R* gene, and consequently a HR occurs in the infiltrated leaf area (Fig. 23.2a and b). As also noted for PVX-based expression, background responses to the pathogen can occur or the transformation capability of *A. tumefaciens* can be too low, depending on the genetic background of the host plant (Fig. 23.2c and d). Currently, the agroinfiltration assay is routinely used in HTP screening for *R-Avr* genes. For example, *A. tumefaciens* strains that express *R* gene analog (RGAs) were infiltrated in *N. benthamiana* leaves. By subsequent PVX agroinfection of *Avr* candidates on the transiently transformed leaf panel, true R-AVR combinations were identified (Oh et al., 2007). Also, direct agroinfiltration of *A. tumefaciens* clones that express *Avr* genes on collections of resistant wild *Solanum*

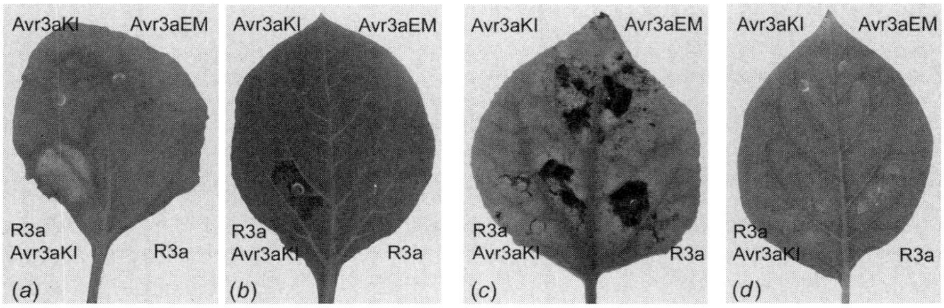

FIG. 23.2 Agroinfiltration in *Nicotiana benthamiana* and *Solanum*. Reconstruction of the R3a-AVR3a interaction in (a) *N. benthamiana* and (b) *Solanum avilesii*. Cell death occurs when pGRAB-Avr3a is co-infiltrated with pBIN19-R3a, but not when pBIN19-R3a, pGRAB-Avr3aKI, or pGRAB-Avr3aEM are singly infiltrated. (c) Background response to *A. tumefaciens* in *S. tuberosum MaR11*. (d) Low transformation efficiency and absence of R3a-Avr3a cell death in *S. tuberosum* MaR4. Photographs were courteously provided by Nicolas Champouret.

germplasm promises to detect a vast amount of *R* genes, as shown for the functional allele mining with *Avr3a* (Champouret et al., 2007; Hein et al., 2007). In addition to HTP screening, this method remains a valuable tool for detailed plant–microbe molecular research, and it was used to identify suppression of INF1-induced cell death mediated by *Avr3a* (Kanneganti et al., 2006). Virtually, agroinfiltration can be combined with many assays, which include disease tests (Rietman, in preparation), gene silencing (Lu et al., 2003a), and so on.

23.3.2 The Agroinfiltration Method

A protocol for agroinfiltration is described in Box 23.2. Transformation efficiency is largely dependent on various aspects, a few notes on this aspect are listed below, which relate to the protocol.

Note 8. Plant age. Typically 4–5-week-old *N. benthamiana* plants at the eight-leaf stage are well suited to agroinfiltration. For *Solanum*, 4–5-weeks-old transplants after transfer from *vitro* to pots, and for *Arabidopsis,* 4–5-week-old seedlings are well suited (Lee and Yang, 2006).

Note 9. Leaf age. To ensure good transgene expression, leaves must be young but well developed. Undeveloped young leaves do not allow efficient infiltration, whereas older leaves often have a "woody" shape, do not allow infiltration, and senesce quickly.

Note 10. Strain. The choice of the *A. tumefaciens* strain is often key to successful transformation. Some strains will initiate necrosis at site of infection, whereas others have a low infiltration efficiency (Wroblewski et al., 2005).

BOX 23.2 AGROINFILTRATION PROTOCOL

Plant material

Young plants are preferred, but the leaves need to be suitable for infiltrating a substantial amount of *A. tumefaciens* suspension (Notes 8 and 9).

Medium preparation

LB broth supplemented with the right antibiotics; Acetosyringone: 3′-5′ Dimethoxy-4′-hydroxy acetophenone, 200 mM stock, 39,3 mg/ml DSMO; MES buffer: 2-[N-Morpholino] ethane sulfonic acid (MES), 1 M stock, 195 g/L; YEB-medium: 5 g bacteriological peptone, 5 g beef extract, 5 g sucrose, 1 g yeast extract, and 2 mL 1 M $MgSO_4$/L (246 g/L); MMA medium: 5 g MS salts, 1.95 g MES, 20 g sucrose, pH adjusted to 5.6 with NaOH, and 200 uL acetosyringone/L.

Agrobacterium culture preparation

Inoculate 10 uL glycerol stock of the desired *A. tumefaciens* strain (Note 10) into 3 mL LB medium supplemented with the right antibiotics. Incubate for 24 h at 28°C at approximately 200 rpm. Inoculate part of this solution into the YEB medium that contains the appropriate antibiotics, 10 mM MES and 2 uM acetosyringone. You can make use of the logistic growth curve of your strain based on optical density (OD_{600}) to calculate the amount of bacteria to add to YEB to have enough bacteria the next day (make sure to have your solution within the range of OD_{600} 0.4). Harvest the cells by centrifugation (3600 rpm for 10 min), pour off the supernatant, and resuspend in MMA medium supplemented with 200 uM acetosyringone. Afterwards, dilute to your desired density (Note 11). Incubate bacteria at room temperature for 1–6 h. Also, the addition of transformation enhancers and silencing suppressors can improve expression (Notes 12 and 13).

Inoculation

Infiltrate your desired constructs into appropriate leaves using a 1 mL syringe and incubate plants at 22°C in a growing chamber (Notes 8, 9, 14, and 15).

Scoring

Responses can be scored after 2–3 days for Avr-R interactions, but timing can vary depending on the tested effectors and host plants (Note 16).

Note 11. Concentration of *A. tumefaciens* cell suspension. Normally, we use OD_{600} 0.4 for coinfiltration (OD_{600} 0.2 for each construct) in *N. benthamiana*. For other plant species, the dilution series can be made to select the best concentration (see Note 16).

Note 12. Additional compounds I. High levels of VirG promote T-DNA transfer by activating virulence gene transcription, which results in a higher transformation efficiency (McCullen and Binns, 2006).

Note 13. Additional compounds II. Transient *Agrobacterium*-mediated gene expression is often hampered by posttranscriptional gene silencing (PTGS), which can be overcome by coinfiltration with silencing inhibitors like P1/HC-Pro or P19 (Johansen and Carrington, 2001; Voinnet et al., 2003). In the case of *R* genes, care should be taken with using silencing suppressors, because overexpression can induce elicitor independent cell dead (Gabriels et al., 2007).

Note 14. Temperature. Agrotransformation is optimal at 22°C (Dillen et al., 1997).

Note 15. Light. Continuous light greatly enhances agrotransformation (Zambre et al., 2003).

Note 16. Host species. Since *N. benthamiana* is a host for *Phytophthora infestans* and has a good transformation efficiency compared with potatoes, it is an ideal model species for studying *P. infestans* genes. Also, lettuce has good transformation efficiency (Wroblewski et al., 2005), and is therefore well applicable for *Bremia lactucae*.

23.4 AGROSUPPRESSION

Agrosuppression is another *A. tumefaciens*-mediated assay, which can be applied for HTP functional screening of oomycete genes of interest in intact plants (Kamoun et al., 2003). This assay has been optimized in the model plant *N. benthamiana*, and it is based on simple mechanical wounding of petioles with a mixture of *A. tumefaciens* strains that carry a binary plasmid with one or more candidate HR-inducing genes and a tumor-inducing (oncogenic) T-DNA. In the absence of HR-inducing candidate genes, tumor formation is initiated and this results in a typical crown gall phenotype and a "crooked petiole phenotype." In contrast, during induction of the HR, tumor formation by the oncogenic T-DNA is suppressed, which does not result in any petiole deformation. Major advantages of this method are its adaptation to HTP screening and the possibility to accommodate larger inserts than PVX. Still, agrosuppression has not evolved into a widely used screening technique, perhaps because the assay is restricted to detecting HR-inducing factors in the particular *N. benthamiana* background and would require more adaptations when other genetic resources are to be studied.

23.5 BIOLISTIC ASSAY

Direct DNA transfer by biolistics does not depend on pathogens, and bypassing inherent effects is a great advantage. The biolistic delivery of genes of interest is a standard technique in many host plants for stable as well as transient transformation, and the latter is well applicable for examining the effect of *in-planta*-expressed-oomycete-effector candidates. Another advantage of this bombardment technique is that HR activity can be quantified by including a vector that expresses marker genes such as *gfp* or *uidA* next to the effector. In case the transiently coexpressed effector has AVR activity in a plant that contains the matching *R* gene, the HR will be induced, and consequently marker gene expression will be reduced compared with controls. The biolistic assay has provided elegant evidence for AVR-R interactions for *ATR13–RPP13* of *Peronospora parasitica* and *Arabidopsis thaliana* (Allen et al., 2004), as well as *Avr3a–R3a* of the *P. infestans* and potato interaction (Armstrong et al., 2005).

A limitation of the biolistic-mediated transient expression assay is that considerable variation in transformation efficiency occurs with respect to areas of target tissue and independent bombardments, and internal references or side-by-side treatments within in the same Petri dish need to be applied (Leister et al., 1996; Jia et al., 2000). Another disadvantage is that the technique requires extensive sample preparation. Despite absence of pathogens in biolistic DNA delivery, the spaciotemporal expression of introduced genes will most likely differ from the natural situation; therefore, care must be taken in drawing conclusions from observed effects.

23.6 VIRUS-INDUCED SILENCING IN PLANTS

Virus-induced gene silencing (VIGS) or RNA interference (RNAi) are functional techniques for testing unknown endogenous genes in plants or animals (Baulcombe, 1999; Dinesh-Kumar et al., 2003; Lu et al., 2003a; Brigneti et al., 2004). VIGS only requires a fragment of the candidate gene cloned into a suitable viral vector, such as PVX- or tobacco rattle virus (TRV), and infection of host plants with this recombinant vector will induce mRNA degradation of the expressed gene homologous to the cloned sequence. VIGS has in many cases proven a powerful system for studying genes required for disease resistance signaling pathways (Lu et al., 2003b; Brigneti et al., 2004), but also for unraveling the function of oomycete genes that act in concert with specific plant genes. For example, a VIGS-based assay in combination with agroinfiltration provided insight into the mechanisms that underlie the transport for effector proteins from cytoplasm to nucleus (Kanneganti et al., 2007b), and it illustrates the power of this technique.

23.7 STABLE EXPRESSION *IN PLANTA*

Years before transient gene expression systems were discovered, the stable transformation of pathogen genes has provided the first insights of gene function in plants. For example, the stable expression of elicitins in tobacco resulted in resistance to *Phytophthora parasitica* var. *nicotianae* (Tepfer et al., 1998), which suggests a role in HR-based resistance to oomycetes. This is in line with the finding that *Inf1* elicitin is downregulated during infection of susceptible plants, perhaps for evading defense responses (Kamoun et al., 1997; Vleeshouwers et al., 2006). Although the stable expression of candidate effectors in plants is a relatively laborious process, this approach is valuable for providing insight in the undisclosed functions of effectors, such as suppression of cell death (Bos et al., 2006), especially because challenge inoculations with various pathogens or effector genes in a vector are feasible in this case.

23.8 ENGINEERING EXPRESSION IN OOMYCETES

Altering the expression level of genes-of-interest using transient or stable transformations in oomycete pathogens provides another approach for phenotypic *in planta* evaluations during the host–pathogen interaction. Transient gene silencing triggered by double stranded RNA (dsRNA) has proven successful in *P. infestans* (Whisson et al., 2005). By delivering *in vitro* synthesized dsRNA into protoplasts, sequence-specific silencing was achieved, and detectable phenotypes in *P. infestans* strains were generated. Although the time range of transiently silencing the target genes may be rather limited, this approach is expected to provide a very valuable tool for functionally testing large amounts of genes, which include effector candidates that are effective during the first few days of the plant–oomycete interaction. Protocols for stable expression or silencing genes have also been described for oomycetes (Judelson et al., 1991; van West et al., 1999; Cvitanich and Judelson, 2003; McLeod et al., 2008). In one of the first studies, the stable transformation of *P. infestans* with sense, antisense, and promoterless constructs of *Inf1* elicitin resulted in transcriptional silencing of both the transgenes and the endogenous gene (van West et al., 1999), and it revealed a gain-of-virulence phenotype for the *Inf1*-silenced strains when inoculated on *N. benthamiana* (Kamoun et al., 1998). Recently, the use of *P. infestans* transformants engineered in the RXLR-EER motif of *Avr3a* provided very elegant functional evidence for this oomycete translocation signal for delivery of effector proteins inside plant cells (Whisson et al., 2007). Despite the unprecedented beauty of this approach, the transformation of most oomycete pathogens is recognized as an exceptionally elaborate process.

23.9 PROTEIN PRODUCTION IN HETEROLOGOUS SYSTEMS

Alternative to engineering expression of genes of interest during the oomycete *in planta* interaction, proteins can be produced in heterologous systems and subsequently infiltrated in plant tissue. This approach allows precisely controlled application and quantification of elicitor activity. By engineering an epitope tag sequence (e.g., FLAG), the produced proteins can routinely be collected by immuno-affinity purifications (Kamoun et al., 1998; Tian et al., 2004). Satisfactory amounts of correctly folded proteins that remain stable in the experimental environment were obtained for cysteine-rich extracellular proteins, such as elicitins or protease inhibitors. Heterologous protein production was achieved in the prokaryote *Escherichia coli* as well as the eukaryote *Pichia pastoris,* and the biological activities of recombinant purified elicitins were comparable with those obtained with agroinfiltration and PVX agroinfection (Kamoun et al., 1998; Vleeshouwers et al., 2006). In the case of unstable short-living proteins, this system is expected to be less suitable.

23.10 FUTURE PERSPECTIVES

Despite the considerable choice of *in planta* expression systems, this field is still relatively young, and optimizing the techniques will remain a challenge for the near future. The following areas are in most need of improvement.

1. Expression levels. The expression levels of oomycete effectors need to approach the natural situation, because engineered expression can result in phenotypic artifacts.
2. Efficiency. Expression methods and cloning strategies can be optimized for higher throughput applications (Maekawa et al., 2008).
3. Background artifacts. Necrotic responses, which are inherent to pathogen-based methods such as *A. tumefaciens* or viruses, should be minimized.
4. Transformation efficiency. Especially when working in diverse genetic backgrounds, transformation efficiency should be reliable and should reach sufficient levels. When applying *in planta* expression methods, preferably, a combination of different assays should be used (e.g. initial screening by a HTP assay such as PVX agroinfection and subsequently narrowing down using other methods).

In summary, functional assays provide a new dimension to the postgenomics biology and a first glimpse on *in planta* effects for the genes of interest.

ACKNOWLEDGMENTS

We thank Avebe, the Dutch Umbrella project, and WUR Plant Breeding for financing our work.

REFERENCES

Allen RL, Bittner Eddy PD, Grenville Briggs LJ, Meitz JC, Rehmany AP, Rose LE, Beynon JL 2004. Host-parasite coevolutionary conflict between *Arabidopsis* and downy mildew. Science 306:1957–1960.

Armstrong MR, Whisson SC, Pritchard L, Bos JIB, Venter E, Avrova AO, Rehmany AP, Bohme U, Brooks K, Cherevach I, et al. 2005. An ancestral oomycete locus contains late blight avirulence gene *Avr3a*, encoding a protein that is recognized in the host cytoplasm. Proc Natl Acad Sci USA 102:7766–7771.

Barker H, Dale MFB 2006. Resistance to viruses in potato. In: G. L, Carr JP, editors. Natural resistance mechanisms of plants to viruses. Springer. Dordrecht, The Netherlands.

Batten JS, Yoshinari S, Hemenway C 2003. Potato virus X: a model system for virus replication, movement and gene expression. Molec Plant Pathol 4:125–131.

Baulcombe DC 1999. Fast forward genetics based on virus-induced gene silencing. Curr Opin Plant Biol 2:109–113.

Bendahmane A, Kanyuka K, Baulcombe DC 1999. The *Rx* gene from potato controls separate virus resistance and cell death responses. Plant Cell 11:781–791.

Bendahmane A, Querci M, Kanyuka K, Baulcombe DC 2000. *Agrobacterium* transient expression system as a tool for the isolation of disease resistance genes: application to the *Rx2* locus in potato. Plant J 21:73–81.

Bos JIB, Kanneganti T-D, Young C, Cakir C, Huitema E, Win J, Armstrong M, Birch PRJ, Kamoun S 2006. The C-terminal half of *Phytophthora infestans* RXLR effector AVR3a is sufficient to trigger R3a-mediated hypersensitivity and suppress INF1-induced cell death in *Nicotiana benthamiana*. Plant J 48: 165–176.

Botella MA, Parker JE, Frost LN, Bittner EPD, Beynon JL, Daniels MJ, Holub EB, Jones JDG 1998. Three genes of the *Arabidopsis RPP1* complex resistance locus recognize distinct *Peronospora parasitica* avirulence determinants. Plant Cell 10:1847–1860.

Brigneti G, Martin-Hernandez AM, Jin H, Chen J, Baulcombe DC, Baker B, Jones JD 2004. Virus-induced gene silencing in *Solanum* species. Plant J 39:264–272.

Butterbach P, Slootweg E, Koropacka K, Spiridon L, Dees R, Roosien J, Bakker E, Arens M, Petrescu A, Smant G, et al. 2007. Functional constraints and evolutionary dynamics of the *Rx1/Gpa2* cluster in potato. 13th International Congress on Molecular Plant-Microbe Interactions. Sorrento, Italy.

Champouret N, Rietman H, Bos JIB, Kamoun S, van der Vossen EAG, Jacobsen E, Visser RGF, Vleeshouwers VGAA 2007. Functional allele mining: a new approach to identify *R* gene homologues in *Solanum*. 13th International Congress on Molecular Plant-Microbe Interactions. Sorrento, Italy.

Chapman S, Kavanagh T, Baulcombe D 1992. Potato virus X as a vector for gene expression in plants. Plant J 2:549–557.

Coemans B, Takahashi Y, Berberich T, Ito A, Kanzaki H, Matsumura H, Saitoh H, Tsuda S, Kamoun S, Sagi L, et al. 2008. High-throughput in planta expression screening identifies an ADP-ribosylation factor (ARF1) involved in non-host resistance and R gene-mediated resistance. Molec Plant Pathol 9:25–36.

Cvitanich C, Judelson HS 2003. Stable transformation of the oomycete, *Phytophthora infestans* using microprojectile bombardment. Curr Genet 42:228–235.

Dangl JL, Jones JD 2001. Plant pathogens and integrated defence responses to infection. Nature 411:826–833.

de Jong W, Forsyth A, Leister D, Gebhardt C, Baulcombe DC 1997. A potato hypersensitive resistance gene against potato virus X maps to a resistance gene cluster on chromosome 5. Theoret Appl Genet 95:1–2.

Dillen W, De Clercq J, Kapila J, Zambre M, Van Montagu M, Angenon G 1997. The effect of temperature on *Agrobacterium tumefaciens*-mediated gene transfer to plants. Plant J 12:1459–1463.

Dinesh-Kumar S, Anandalakshmi R, Marathe R, Schiff M, Liu Y 2003. Virus-induced gene silencing. In: Grotewold E, editor. Methods in molecular biology, Vol 236: plant functional genomics. Humana Press. Totowa, NJ.

Fischer R, Vaquero-Martin C, Sack M, Drossard J, Emans N, Commandeur U 1999. Towards molecular farming in the future: transient protein expression in plants. Biotechnol Appl Biochem 30:113–116.

Gabriels SH, Vossen JH, Ekengren SK, van Ooijen G, Abd-El-Haliem AM, van den Berg GC, Rainey DY, Martin GB, Takken FL, de Wit PJ, et al. 2007. An NB-LRR protein required for HR signalling mediated by both extra- and intracellular resistance proteins. Plant J 50:14–28.

Gaulin E, Madoui M-A, Bottin A, Jacquet C, Mathe C, Couloux A, Wincker P, Dumas B 2008. Transcriptome of *Aphanomyces euteiches*: new Oomycete putative pathogenicity factors and metabolic pathways. PLoS ONE 3:e1723,1–11.

Govers F, Gijzen M 2006. *Phytophthora* genomics: the plant destroyers' genome decoded. Molec Plant-Microbe Interact 19:1295–1301.

Hammond-Kosack KE, Staskawicz BJ, Jones JDG, Baulcombe DC 1995. Functional expression of a fungal avirulence gene from a modified potato virus X genome. Molec Plant-Microbe Interact 8:181–185.

Hawkes J 1990. The potato: evolution, biodiversity, and genetic resources. Belhaven Press. London, UK.

Hein I, Harrower B, Squires J, Birch PRJ, Bryan G 2007. Screening wild potato accessions for resistance to the virulent allele of the *Phytophthora infestans* effector *avr3aEM*. 13th International Congress on Molecular Plant-Microbe Interactions. Sorrento, Italy.

Hellens R, Mullineaux P, Klee H 2000. Technical Focus:A guide to *Agrobacterium* binary Ti vectors. Trends Plant Sci 5:446–451.

Huitema E, Bos JIB, Tian M, Win J, Waugh ME, Kamoun S 2004. Linking sequence to phenotype in *Phytophthora*-plant interactions. Trends Microbiol 12:193–200.

Huitema E, Vleeshouwers VGAA, Cakir C, Kamoun S, Govers F 2005. Differences in intensity and specificity of hypersensitive response induction in *Nicotiana* spp. by

INF1, INF2A, and INF2B of *Phytophthora infestans*. Molec Plant-Microbe Interact 18:183–193.

Jia Y, McAdams S, Bryan G, Hershey H, Valent B 2000. Direct interaction of resistance gene and avirulence gene products confers rice blast resistance. The EMBO J 19:4004–4014.

Johansen LK, Carrington JC 2001. Silencing on the spot. Induction and suppression of RNA silencing in the *Agrobacterium*-mediated transient expression system. Plant Physiol 126:930–938.

Jones L, Hamilton AJ, Voinnet O, Thomas CL, Maule AJ, Baulcombe DC 1999. RNA-DNA interactions and DNA methylation in post-transcriptional gene silencing. Plant Cell 11:2291–2301.

Judelson HS, Tyler BM, Michelmore RW 1991. Transformation of the oomycete pathogen, *Phytophthora infestans*. Mole Plant-Microbe Interac 4:602–607.

Kamoun S 2006. A catalogue of the effector secretome of plant pathogenic oomycetes. Annu Rev Phytopathol 44:1–20.

Kamoun S 2007. Groovy times: filamentous pathogen effectors revealed. Curr Opin Plant Biol 10:358–365.

Kamoun S, Hamada W, Huitema E 2003. Agrosuppression: a bioassay for the hypersensitive response suited to high-throughput screening. Molec Plant-Microbe Interact 16:7–13.

Kamoun S, Honée G, Weide R, Laugé R, Kooman-Gersmann M, de Groot K, Govers F, de Wit PJGM 1999a. The fungal gene *Avr9* and the oomycete gene *inf1* confer avirulence to potato virus X on tobacco. Molec Plant-Microbe Interact 12:459–462.

Kamoun S, Hraber P, Sobral B, Nuss B, Govers F 1999b. Initial assessment of gene diversity for the oomycete pathogen *Phytophthora infestans* based on expressed sequences. Fungal Genet Biol 28:94–106.

Kamoun S, Huitema E, Vleeshouwers VGAA 1999c. Resistance to oomycetes: a general role for the hypersensitive response? Trends Plant Sci 4:196–200.

Kamoun S, van West P, De Jong AJ, De Groot KE, Vleeshouwers VGAA, Govers F 1997. A gene encoding a protein elicitor of *Phytophthora infestans* is down-regulated during infection of potato. Mole Plant-Microbe Interact 10:13–20.

Kamoun S, van West P, Vleeshouwers VGAA, de Groot KE, Govers F 1998. Resistance of *Nicotiana benthamiana* to *Phytophthora infestans* is mediated by the recognition of the elicitor protein INF1. Plant Cell 10:1413–1425.

Kanneganti TD, Bai X, Tsai CW, Win J, Meulia T, Goodin M, Kamoun S, Hogenhout SA 2007a. A functional genetic assay for nuclear trafficking in plants. Plant J 50: 149–158.

Kanneganti TD, Huitema E, Cakir C, Kamoun S 2006. Synergistic interactions of the plant cell death pathways induced by Phytophthora infestans Nep1-like protein PiNPP1.1 and INF1 elicitin. Molec Plant-Microbe Interact 19:854–863.

Kanneganti TD, Huitema E, Kamoun S 2007b. In planta expression of oomycete and fungal genes. In: Ronald PC, editor. Methods mol biol. Humana Press Inc. Totowa, NJ. pp. 35–43.

Kapilla J, De Rijcke R, Van Montagu M, Angenon G 1997. An *Agrobacterium*-mediated transient gene expression system for intact leaves. Plant Sci 122:101–108.

Kjemtrup S, Nimchuk Z, Dangl JL 2000. Effector proteins of phytopathogenic bacteria: bifunctional signals in virulence and host recognition. Curr Opin Microbiol 3: 73–78.

Kooman-Gersmann M, Vogelsang R, Hoogendijk ECM, DeWit P 1997. Assignment of amino acid residues of the AVR9 peptide of *Cladosporium fulvum* that determine elicitor activity. Molec Plant-Microbe Interact 10:821–829.

Lee MW, Yang Y 2006. Transient expression assay by agroinfiltration of leaves. Methods Mol Biol 323:225–229.

Leister RT, Ausubel FM, Katagiri F 1996. Molecular recognition of pathogen attack occurs inside of plant cells in plant disease resistance specified by the *Arabidopsis* genes *RPS2* and *RPM1*. Proc Nat Acad Sci 93:15497–15502.

Lu R, Malcuit I, Moffett P, Ruiz MT, Peart J, Wu AJ, Rathjen JP, Bendahmane A, Day L, Baulcombe DC 2003a. High throughput virus-induced gene silencing implicates heat shock protein 90 in plant disease resistance. EMBO J 22:5690–5699.

Lu R, Martin Hernandez AM, Peart JR, Malcuit I, Baulcombe DC 2003b. Virus-induced gene silencing in plants. Methods 30:296–303.

Maekawa T, Kusakabe M, Shimoda Y, Sato S, Tabata S, Murooka Y, Hayashi M 2008. Polyubiquitin promoter-based binary vectors for overexpression and gene silencing in lotus japonicus. Molec Plant-Microbe Interact 21:375–382.

McCullen CA, Binns AN 2006. *Agrobacterium tumefaciens* and plant cell interactions and activities required for interkingdom macromolecular transfer. Ann Rev Cell Dev Biol 22:101–127.

McLeod A, Fry BA, Zuluaga AP, Myers KL, Fry WE 2008. Toward improvements of oomycete transformation protocols. J Eukaryot Microbiol 55:103–109.

Nasir KHB, Takahashi Y, Ito A, Saitoh H, Matsumura H, Kanzaki H, Shimizu T, Ito M, Fujisawa S, Sharma PC, et al. 2005. High-throughput in planta expression screening identifies a class II ethylene-responsive element binding factor-like protein that regulates plant cell death and non-host resistance. Plant J 43:491–505.

Oh S-K, Young C, Lee M, Win J, Bos JIB, Vleeshouwers VGAA, van der Vossen EAG, Kamoun S 2007. High-throughput in planta expression of *Phytophthora infestans* RXLR effectors reveals novel avirulence and virulence activities. 13th International Congress on Molecular Plant-Microbe Interactions. Sorrento, Italy.

Polesani M, Desario F, Ferrarini A, Zamboni A, Pezzotti M, Kortekamp A, Polverari A 2008. cDNA-AFLP analysis of plant and pathogen genes expressed in grapevine infected with *Plasmopara viticola*. BMC Genomics 9:142.

Qutob D, Kamoun S, Gijzen M 2002. Expression of a *Phytophthora sojae* necrosis-inducing protein occurs during transition from biotrophy to necrotrophy. Plant J 32:361–373.

Ritter E, Debener T, Barone A, Salamini F, Gebhardt C 1991. RFLP mapping on potato chromosomes of two genes controlling extreme resistance to potato virus X (PVX). Mol Gen Genet 227:81–85.

Ross H 1958. Inheritance of extreme resistance to virus Y in *Solanum stoloniferum* and its hybrids with *Solanum tuberosum* In: Quak F, Dijkstra J, Beemster ABR, van der Want JPH, editors. Lisse-Wageningen. Veenman en Zonen. pp. 204–211.

Ross H 1986. Potato breeding—Problems and perspectives. Horn W, Röbbelen G, editors. Verlag Paul Parey. Berlin, Germany. p. 132.

Rouppe van der Voort J, Kanyuka K, van der Vossen E, Bendahmane A, Mooijman P, Klein-Lankhorst R, Stiekema W, Baulcombe D, Bakker J 1999. Tight physical linkage of the nematode resistance gene *Gpa2* and the virus resistance gene *Rx* on a single segment introgressed from the wild species *Solanum tuberosum* subsp. *andigena* CPC 1673 into cultivated potato. Molec Plant-Microbe Interact 12:197–206.

Staskawicz BJ, Ausubel FM, Baker BJ, Ellis JG, Jones JDG 1995. Molecular genetics of plant disease resistance. Science 268:661–667.

Takahashi Y, Nasir KHB, Ito A, Kanzaki H, Matsumura H, Saitoh H, Fujisawa S, Kamoun S, Terauchi R 2007. A high-throughput screen of cell-death-inducing factors in *Nicotiana benthamiana* identifies a novel MAPKK that mediates INF1-induced cell death signaling and non-host resistance to *Pseudomonas cichorii*. Plant J 49:1030–1040.

Takken FLW, Luderer R, Gabriels SEJ, Westerink N, Lu R, de Wit PJGM, Joosten MHAJ 2000. A functional cloning strategy, based on a binary PVX-expression vector, to isolate HR-inducing cDNAs of plant pathogens. Plant J 24:275–283.

Tepfer D, Boutteaux C, Vigon C, Aymes S, Perez V, O. Donohue M, Huet JC, Pernollet JC 1998. *Phytophthora* resistance through production of a fungal protein elicitor (beta-cryptogein) in tobacco. Molec Plant-Microbe Interact 11:64–67.

Tian M, Huitema E, da Cunha L, Torto Alalibo T, Kamoun S 2004. A Kazal-like extracellular serine protease inhibitor from *Phytophthora infestans* targets the tomato pathogenesis-related protease P69B. J Biol Chem 279:26370–26377.

Tobias CM, Oldroyd GED, Chang JH, Staskawicz BJ 1999. Plants expressing the *Pto* disease resistance gene confer resistance to recombinant PVX containing the avirulence gene *AvrPto*. Plant J 17:41–50.

Torto TA, Li SA, Styer A, Huitema E, Testa A, Gow NAR, van West P, Kamoun S 2003. EST mining and functional expression assays identify extracellular effector proteins from the plant pathogen *Phytophthora*. Geno Res 13:1675–1685.

Tyler BM, Tripathy S, Zhang XM, Dehal P, Jiang RHY, Aerts A, Arredondo FD, Baxter L, Bensasson D, Beynon JL, et al. 2006. *Phytophthora* genome sequences uncover evolutionary origins and mechanisms of pathogenesis. Science 313:1261–1266.

Valkonen JP, Pehu E, Jones MG, Gibson RW 1991. Resistance in *Solanum brevidens* to both potato virus Y and potato virus X may be associated with slow cell-to-cell spread. J Gen Virol 72:231–236.

van der Hoorn RAL, Laurent F, Roth R, de Wit PJGM 2000. Agroinfiltration is a versatile tool that facilitates comparative analyses of *Avr9*/*Cf-9*-induced and *Avr4*/*Cf-4*-induced necrosis. Molec Plant-Microbe Interact 13:439–446.

van West P, Kamoun S, van 't Klooster JW, Govers F 1999. Internuclear gene silencing in *Phytophthora infestans*. Molec Cell 3:339–348.

Vleeshouwers VGAA, Driesprong JD, Kamphuis LG, Torto-Alalibo T, van 't Slot KAE, Govers F, Visser RGF, Jacobsen E, Kamoun S 2006. Agroinfection-based high throughput screening reveals specific recognition of INF elicitins in *Solanum*. Molec Plant Pathol 7:499–510.

Vleeshouwers VGAA, Rietman H, Krenek P, Champouret N, Young C, Oh S-K, Wang M, Bouwmeester K, Vosman B, Visser RGF, et al. 2008. Effector genomics accelerates discovery and functional profiling of potato disease resistance and phytophthora infestans avirulence genes. PLoS ONE 3:e2875.

Vleeshouwers VGAA, van Dooijeweert W, Govers F, Kamoun S, Colon LT 2000. The hypersensitive response is associated with host and nonhost resistance to *Phytophthora infestans*. Planta 210:853–864.

Voinnet O, Rivas S, Mestre P, Baulcombe D 2003. An enhanced transient expression system in plants based on suppression of gene silencing by the p19 protein of tomato bushy stunt virus. Plant J 33:949–956.

Whisson SC, Avrova AO, van West P, Jones JT 2005. A method for double-stranded RNA-mediated transient gene silencing in *Phytophthora infestans*. Molec Plant Pathol 6:153–163.

Whisson SC, Boevink PC, Moleleki L, Avrova AO, Morales JG, Gilroy EM, Armstrong MR, Grouffaud S, van West P, Chapman S, et al. 2007. A translocation signal for delivery of oomycete effector proteins into host plant cells. Nature 450:115–118.

Win J, Kanneganti TD, Torto Alalibo T, Kamoun S 2006. Computational and comparative analyses of 150 full-length cDNA sequences from the oomycete plant pathogen *Phytophthora infestans*. Fungal Gene Biol 43:20–33.

Wroblewski T, Tomczak A, Michelmore R 2005. Optimization of *Agrobacterium*-mediated transient assays of gene expression in lettuce, tomato and *Arabidopsis*. Plant Biotechnol J 3:259–273.

Zambre M, Terryn N, De Clercq J, De Buck S, Dillen W, Van Montagu M, Van Der Straeten D, Angenon G 2003. Light strongly promotes gene transfer from *Agrobacterium tumefaciens* to plant cells. Planta 216:580–586.

Zupan J, Muth TR, Draper O, Zambryski P 2000. The transfer of DNA from *Agrobacterium tumefaciens* into plants: a feast of fundamental insights. Plant J 23:11–28.

24

GENE EXPRESSION PROFILING

PAUL R.J. BIRCH
Division of Plant Sciences, College of Life Science, University of Dundee at SCRI, Invergowrie, Dundee, United Kingdom

ANNA O. AVROVA
Plant Pathology Programme, Scottish Crop Research Institute, Invergowrie, Dundee, United Kingdom

24.1 INTRODUCTION

Identification of genes that are expressed, or whose expression is modulated, in a tissue, life cycle stage, or physiological condition, has long been regarded as an important preliminary step in gaining information about the functions relevant to such processes as cell differentiation, morphological or metabolic change, responses to environmental cues and stresses, and in the case of disease development, pathogenicity and host specificity. Oomycetes go through many distinct developmental changes through their life cycle. This is never more apparent than during a successful infection cycle, when plant-pathogenic oomycetes undergo changes that include the formation of sporangia, release of motile zoospores, their encystment and germination to form hyphae and appressoria, production of primary and secondary infection hyphae, haustoria, and finally sporangiophores. As indicated in previous chapters, these various developmental processes facilitate dispersal, recognition of the host, adhesion, penetration and colonization, encompassing biotrophic and/or necrotrophic phases of infection, finally leading once again to dispersal. Our understanding of these processes and of the molecular events underlying sexual and asexual reproduction, particularly in the potato late blight pathogen *P. infestans*, has grown enormously in recent years largely because of initial observations of changes in gene expression.

Oomycete Genetics and Genomics: Diversity, Interactions, and Research Tools
Edited by Kurt Lamour and Sophien Kamoun
Copyright © 2009 John Wiley & Sons, Inc.

In this chapter, we review some approaches that have been used to profile gene expression in pathogenic oomycetes and how these observations have informed us about their biology. In recent years, we have observed a sharp increase in the generation of oomycete expressed sequence tags (ESTs). We have also witnessed a transition in the tools used to profile gene expression in oomycetes, from RNA (Northern) blotting, through a range of polymerase chain reaction (PCR)-based methods to microarrays. Although RNA blotting requires high quantities of RNA, for example 20 μg of total RNA to study just a few genes, microarrays require just 5 μg of total RNA per biological replicate to assess expression levels across the entire transcriptome. For accurate quantification of very rare messenger RNA (mRNA) transcripts, quantitative real-time (RT)-PCR (qRT-PCR) allows us to detect and measure expression levels in just a few cells. This has proven an advantage in assessing gene expression during the biotrophic stage of plant-oomycete interactions, when pathogen biomass may be relatively low compared with that of the host. While reviewing each of these developments, we also look to the future and the contributions that ever-cheaper next-generation sequencing technologies can make to our understanding of changes in gene expression in this important group of organisms.

24.2 EXPRESSED SEQUENCE TAGS—THE PLATFORM FOR INVESTIGATING OOMYCETE BIOLOGY

As indicated in Chapter 27, the genomes of many plant pathogenic oomycetes have been recently sequenced, which revealed the entire genetic blueprints for their biology and their pathogenic lifestyles. However, in the absence of such a wealth of information and to determine the expressed portions of those genome sequences that are available, a key starting point is the reverse transcription of RNA to cDNA and the deep sequencing of cDNA clones to generate ESTs. EST sequencing has provided initial assessments of gene diversity in a range of oomycetes (Table 24.1). Major EST sequencing efforts have been conducted for *P. infestans* (Judelson et al., 2005; Kamoun et al., 1999), the soybean pathogen *Phytophthora sojae* (Torto-Alalibo et al., 2007; Qutob et al., 2000), and the legume pathogen *Aphanomyces euteichus* (Gaulin et al., 2008; Madoui et al., 2007). In each case, the depth of EST sequencing has been significant enough to inform us considerably about the range of biochemical, metabolic, and physiological processes that occur in the cells of oomycete plant pathogens. Moreover, where a range of cDNA libraries were used to represent different developmental stages and growth conditions, it has been possible also to infer changes in gene expression that indicate candidate genes with roles in specific life cycle stages, in responding to conditions of stress or nutrient availability, or that may contribute to pathogenicity.

This is particularly the case for *P. infestans*, where 75,757 ESTs were generated from 20 cDNA libraries to represent 18 diverse growth conditions,

TABLE 24.1 Large-scale cDNA sequencing of expressed sequence tags (ESTs) from different oomycete species.

Species	ESTs in NCBI Genbank	Libraries	ESTs from this study	Unisequences	Reference
Phytophthora infestans	94,091	20 cDNA libraries representing a broad range of growth conditions, stress responses, and developmental stages.	75,757	18,256	Randall et al., 2005
Phytophthora sojae	28,381	6 cDNA libraries representing a range of growth conditions, developmental stages and infected soybean hypocotyl	26,943	7,863	Torto-Alalibo et al., 2007
Phytophthora parasitica	6,328	Tomato roots 4 dpi with *P. parasitica*	2,989	1,689	Le Berre et al., 2008
Aphanomyes euteiches	18,864	2 cDNA libraries from mycelium and infected material	18,684	7,977	Madoui et al., 2007; Gaulin et al., 2008
Saprolegnia parasitica	1,513	cDNA library from mycelium	1,510	1,279	Torto-Alalibo et al., 2005

stress responses, and developmental stages (Randall et al., 2005). The assembled EST and additional candidate genes from genomic sequences predicted a set of 18,256 unisequences, which potentially represents a significant proportion of the *P. infestans* transcriptome. The most common Pfam motifs represented in proteins encoded by *P. infestans* genes differed from the most abundant represented domains in other eukaryotic species. Many of the most prevalent ESTs showed no similarities to genes or motifs in public databases, and thus they can be regarded as oomycete specific, which probably reflects the evolutionary distance between *P. infestans* and the most intensively studied eukaryotes. In the case of each growth condition, stress response, and developmental stage that was examined, ESTs were observed that were specific to that cDNA library, which represent candidates for genes involved in a particular developmental stage or metabolic condition. In total, 5,119 library-specific unisequences were observed, which includes 1,682 from asexual life cycle stages that precede the establishment of plant infection. The proportions of the proteins encoded by these genes with matches in Pfam databases was lower than the overall unisequence assembly (20% versus 43%), which suggests that such proteins may participate in novel aspects of oomycete development.

In a further in-depth study, 26,943 ESTs from cDNA libraries that represent four developmental stages of *P. sojae* and three different mycelial growth conditions were analyzed, clustering into 7,863 unisequences (Torto-Alalibo et al., 2007). Again, ESTs were revealed that were specific to, or highly represented in, cDNA libraries from particular developmental stages or growth conditions. Many genes potentially involved in pathogenesis were identified, which includes 13 encoding RXLR effectors, 25 crinkling and necrosis-inducing (CRN) proteins, and a range of hydrolytic enzymes (Torto-Alalibo et al., 2007).

The third in-depth study generated 18,684 ESTs, which assembled into 7,977 unisequences, from *A. euteichus* (Madoui et al., 2007; Gaulin et al., 2008). Only two cDNA libraries were used for this study: one from mycelium grown on synthetic medium and the second from mycelium grown in contact with root tissues of the model legume *Medicago truncatula*. *A. euteichus* is distantly related to the *Phytophthora* spp, and the study revealed many differences in gene content, which includes biosynthetic pathways specific to *Aphanomyces* and many novel candidate pathogenicity determinants (Gaulin et al., 2008). The two cDNA libraries showed clear differences in EST content, which indicates that they reflected divergent expression profiles that, particularly, revealed several genes that were apparently upregulated following perception of host roots. Although the large numbers of proteases, protease inhibitors, and adenosine triphosphate binding cassette (ABC) transporters in the "host interaction" cDNA library was suggestive of these genes contributing to pathogenicity, in contrast to the phytophthoras, there were no pectinase genes. Moreover, the number of candidate secreted effectors that contained the RXLR motif for targeting to the host cytoplasm (Chapter 18) was negligible, whereas several genes encoding members of the CRN family were identified, which suggests that this is a more ancient class of effectors (Gaulin et al., 2008).

A study of the late stages of tomato infection by *P. parasitica* was aimed at identifying oomycete genes involved in pathogenicity (Le Berre et al., 2008). In this study, the authors identified members of the predicted classes of cytoplasmic (RXLR and CRN) and candidate apoplastic [small cysteine rich, elicitin, cellulose binding elicitor and leatin-like protein (CBEL)] effectors, plant cell-wall-degrading enzymes, proteases and protease inhibitors, and genes with a likely role in protection against reactive oxygen species that are generated during plant defense.

Only one study has investigated gene expression using ESTs in an oomycete animal pathogen: the economically important fish pathogen *Saprolegnia parasitica* (Torto-Alalibo et al., 2005). A total of 1,279 unisequences were generated from a mycelium cDNA library and revealed that *S. parasitica* genes tend to be relatively divergent from their *Phytophthora* counterparts. Nevertheless, many similar candidate pathogenicity determinants were observed, which include proteins that contain fungal-type cellulose-binding domains, glycosyl hydrolases, proteases, and protease inhibitors.

On the basis that many candidate pathogenicity determinants need to be secreted from the pathogen to interact with a host cell, ESTs can be mined for sequence encoding proteins with secretion signal peptides. Such a simple bioinformatic strategy was applied to 2,147 ESTs from *P. infestans*, identifying a set of 142 nonredundant *Pex* (*Phytophthora extracellular protein*) genes (Torto et al., 2003). A proteomic analysis of *P. infestans* secreted proteins was performed to validate this approach. To identify *Pex* genes that manipulate host processes, 63 of these candidates were expressed in plant cells using a potato virus X (PVX)-agrobacterium binary vector. This led to the discovery of the *crn* gene family, members of which induced strong necrosis when expressed in plants (Torto et al., 2003).

All of the above studies have generated an immense amount of data as well as a framework of expressed genes with which to investigate the processes occurring in oomycete cells. They have provided, in many cases, indications of differential gene expression suggestive of regulatory processes tailored to specific developmental stages or growth conditions. However, to investigate fully and verify such regulatory changes, a range of differential gene expression profiling methods has been used, which is the focus of the remainder of this chapter. These methods reveal a great deal about host–oomycete interactions and developmental stage-specific gene regulation.

24.3 EXPRESSION PROFILING TO IDENTIFY GENES WITH POSSIBLE ROLES IN INFECTION

Most genes with a key role either in defense in the host or development of disease in the pathogen and, certainly, those involved in forming cell structures, such as haustoria, which are specific to infection, will be upregulated during host–oomycete interactions. A range of approaches has been used to identify such differentially expressed genes, from traditional subtractive hybridization through to PCR-based profiling methods. A key challenge in identifying such genes is the need to differentiate pathogen genes from those of the host, which often entails additional steps following differential gene expression profiling.

Possibly the earliest studies of differential gene expression in oomycetes involved the isolation of *in planta induced* (*ipi*) genes from *P. infestans*. These genes were identified by differential hybridization screening of a *P. infestans* genomic DNA library with cDNA prepared from infected potatoes and cDNA prepared from *in vitro* grown *P. infestans* mycelium (Pieterse et al., 1991, 1993, 1994a and b). Genes that were upregulated during the interaction included the *ipiO* family that is clustered in the genome and upregulated in invading hyphae during infection (van West et al., 1998). Recently, it has been shown that *P. infestans* IPIO is the avirulence protein AvrBlb1 that is recognized by the *Solanum bulbocastanum* resistance protein Rpi-Blb1 (Vleeshouwers et al., 2008).

Some studies have used suppression subtractive hybridization (SSH); this method generates cDNA libraries that are enriched for differentially expressed genes. SSH is like traditional subtractive hybridization methods, in that the cDNA population from which upregulated transcripts are sought is termed the "tester," and the cDNA population for comparison is termed the "driver." Both cDNA populations are digested with a regularly cutting restriction enzyme. The tester population is subdivided into two equal portions, and a different adaptor is ligated to each. An excess of driver cDNA is hybridized to each tester subpopulation, which leads to subtraction of cDNA sequences common to both the tester and the driver. Each hybridization mixture is then combined and allowed to hybridize. After an extension reaction to fill in sticky ends, the primers that anneal to each adaptor are used to PCR amplify only those hybrid molecules that possess both adaptors (enriched for sequences specific to the tester); these are then directly cloned for study.

SSH has been used to isolate oomycete genes that are induced during the interaction with the host. This cDNA material contains a proportion of sequences that are derived from the host, in addition to the target pathogen sequences. To circumvent this problem, Beyer et al. (2002) induced *P. infestans* mycelium by contact with the host plant, which are potatoes, and then removed the host prior to SSH to avoid having to eliminate host sequences from the subtracted material. In a more recent publication, Avrova et al. (2007) used SSH to indentify *P. infestans* genes upregulated during the early, biotrophic phase of the potato–*P. infestans* interaction [15 h postinoculation (hpi)]. At this stage of infection, very limited amounts of pathogen biomass exist, and direct cloning and sequencing of subtracted cDNA revealed only sequences derived from the host (unpublished data). To enrich for pathogen sequences, the subtracted cDNA was hybridized to a *P. infestans* bacterial artificial chromosome (BAC) library (Whisson et al., 2001), and the hybridizing BACs were subcloned for sequencing. In this way, a nonprotein coding, infection-specific gene family was identified that is clustered throughout the *P. infestans* genome (Avrova et al., 2007).

SSH has also been used to identify genes upregulated in the obligate biotrophic *Arabidopsis* pathogen called *Hyaloperonospora arabidopsidis* (previously *Hyaloperonospora parasitica*) (Bittner-Eddy et al., 2003). In this case, the proportion of pathogen biomass was sufficient to identify pathogen-derived cDNAs directly, which obiviated the need for hybridization screens to genomic DNA libraries to identify transcripts from the pathogen. Many *H. parasitica* genes that are upregulated during infection were identified, which includes one that was subsequently shown to be the avirulence gene *ATR13* (Allen et al., 2004).

In a recent study, SSH was used to compare gene expression in two near-isogenic lines of *P. sojae* (Wang et al., 2006). In this study, the authors took an isolate (PS2) with low virulence on a resistant soybean cultivar and, after 14 successive inoculations, obtained a high-virulence line called PS2-vir. SSH, in combination with a cDNA macroarray, was used to identify genes upregulated in the PS2-vir line. Many genes that had increased expression in PS2-vir were involved in energy production, cell signaling, cell-wall biogenesis, and transcriptional regulation. Many upregulated genes were also found to be similar to cDNA sequences that were enriched in *P. sojae* ESTs from infected soybean Wang et al. (2006).

An alternative method, which is cDNA-amplified fragment length polymorphism (AFLP), has also been used to identify genes upregulated during oomycete infections. This method involves the selective PCR amplification of subsets of a cDNA population using primers that anneal to adaptors ligated to the ends of double-stranded cDNA molecules following restriction digestion. cDNA-AFLP has been used to profile differential gene expression in both the host and pathogen during the *Arabidopsis–H. arabidopsidis* compatible interaction (van der Biezen et al., 2000).

In addition to profiling gene expression during the infection cycle, cDNA-AFLP has been used to identify transcriptional differences in a segregating F1 population of *P. infestans* (Guo et al., 2006). cDNA from progeny with defined avirulence characteristics was pooled and used in a bulked segregant analysis to identify 99 avirulence-associated transcript-derived fragments (TDFs). cDNA-AFLP analysis on individual F1 progeny revealed 100% cosegregation of four TDFs with particular avirulence phenotypes, two of which match the same *P. infestans* EST and provide a candidate *Avr4* gene.

24.4 PROFILING DIFFERENTIAL GENE EXPRESSION IN ASEXUAL DEVELOPMENTAL STAGES THAT ARE REQUIRED FOR PATHOGEN DISPERSAL AND INITIATION OF INFECTION

The asexual stages prior to infection include the development of sporangia, which are the source of motile zoospores and are required for dissemination of the pathogen to new hosts. Differentiated zoospores, after release from sporangia, encyst and germinate to form, in many cases, appressoria-like structures for the direct penetration of host tissues. Taking steps back from the infected host, first to germinating cysts, then zoospores, then sporangia, several examples of the different studies of differential gene expression in these stages will be reviewed.

A range of approaches has been used to profile gene expression in the life cycle stages just prior to infection, particularly in zoospores and germinating cysts. In one such study, Gornhardt et al. (2000) used traditional cDNA subtractive hybridization to identify a family of mucin-like *car* genes from *P. infestans* that are upregulated in germinating cysts, just prior to host cell penetration, which may contribute to adhesion. In another study, SSH was used to reveal that amino acid biosynthesis genes from *P. infestans* are upregulated in germinating cysts with appressoria (Grenville-Briggs et al., 2005).

cDNA-AFLP has been used to identify transcripts upregulated in *P. infestans* germinating cysts (Avrova et al., 2003). In this case, qRT-PCR was used to quantify the expression of genes identified by cDNA-AFLP not only in germinating cysts but also during infection, at 15, 48, and 72 hpi, relative to expression levels of the constitutively expressed *ActA* and *ActB* genes. All the genes studied were upregulated in germinating cysts relative to mycelium, but many were also upregulated at various time points during infection, which indicates that the germinating cyst life cycle stage is a good source of candidate pathogenicity genes. During the biotrophic phase of interaction, as stated in the previous sections, often very limited pathogen biomass is available to study gene expression, and few pathogen-derived ESTs have been identified from cDNA libraries prepared from such material. Real-time qRT-PCR is the only method sensitive enough to measure changes in expression at such early infection stages and has been used recently to profile the expression of *P. infestans* RXLR effectors, many of which are most highly upregulated during biotrophy (Whisson et al., 2007). It was also used to demonstrate that a family of cellulose synthase genes, which are essential for pathogenicity, are upregulated in germinated cysts and germinated cysts with appressoria (Grenville-Briggs et al., 2008). Another important advantage of qRT-PCR is that it enables us to distinguish between closely related members of the same gene family.

Two studies have sought to identify genes expressed in zoospores and germinating cysts of *Phytophthora nicotianae*. In one, a cDNA library was constructed from zoospores; then it was arrayed and screened by hybridization using probes derived from mycelium or zoospore mRNA. More than 400 clones that hybridize more strongly to zoospore mRNAs were sequenced to generate ESTs representing 240 different genes (Skalamera et al., 2004). To characterize these genes even more, a colony array was created and screened with cDNA probes derived from various *P. nicotianae* developmental stages, which include zoospores, germinating cysts, vegetative mycelium, and sporulating hyphae, and from infected and uninfected tobacco. Genes that were preferentially expressed in zoospores were identified for functional analyses. In the second study by the same group, Shan et al. (2004) generated arrays of more than 12,000 random cDNA clones from a germinated cyst library and again hybridized these with probes derived from the four developmental stages, infected, and uninfected tobacco material, referred to above. The sequencing of more than 300 clones that represent upregulated genes in germinating cysts yielded 146 unisequences. An annotation of the sequences revealed genes encoding proteins that contribute to a range of biochemical processes, which includes novel genes involved in adhesion, cell-wall biogenesis, and transcriptional regulation (Shan et al., 2004).

Differential gene expression during sporangial cleavage to form zoospores and the identification of genes specific to sporangia has been studied in *P. infestans* by Judelson and coworkers (Tani et al., 2004; Kim and Judelson, 2003). Tani et al. (2004) made an array in which 4,174 clones, which were later shown to represent 2,617 unique sequences, were spotted and hybridized with cDNA probes from cleaving sporangia incubated at 10°C and cDNA from undifferentiated sporangia. A total of 69 genes were identified as upregulated more than five-fold during zoosporogenesis. Their specificity to zoosporogenesis and the timing of their induction during zoospore formation was characterized. Zoosporogenesis-specific genes included protein kinases, transcription factors, ion channels, and other regulators, which provided targets for the detailed functional characterization of this developmental stage. In a similar study, Kim and Judelson (2003) made an array of 5,184 clones, which represent 1,927 independent gene sequences, from a sporangia cDNA library and compared their induction in sporangia with expression in hyphae. Following repeated hybridization screens, 61 *pisp* (*P. infestans* sporangia) genes were identified. Again, many regulators (protein kinases and phosphatases, transcription factors, and G-protein subunits), transporters, and metabolic enzymes were identified.

Perhaps the definitive study to date of differential gene expression during asexual development involved a collaborative project between six different laboratories to provide independent biological replicate mRNA samples from *P. infestans* (Judelson et al., 2008). Overall, 15,650 *P. infestans* unisequences were included on an Affymetrix microarray and were hybridized with multiple cDNA probes from nonsporulating hyphae, asexual sporangia, cleaving sporangia, swimming zoospores, and germinating cysts that contain appressoria from isolate 88069. Altogether, expression was detected of 12,656 genes in at least one of the five developmental stages. The expression of 74 genes was tested using real-time RT-PCR and demonstrated the robustness of the Gene-Chip data. Expression levels of more than half of the genes changed significantly during developmental transitions, which reflects a striking degree of transcriptome remodeling compared with that observed in fungal plant pathogens (Fig. 24.1). Although this figure shows that many genes change between each life cycle stage, the true extent of mRNA

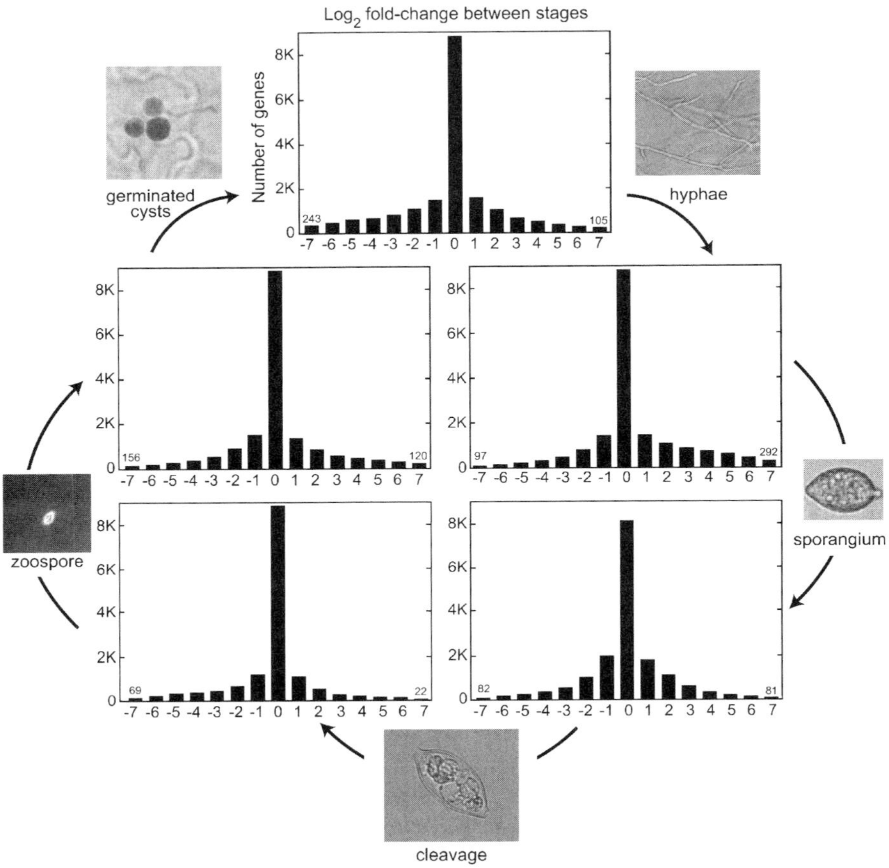

FIG. 24.1 Overview of expression changes during *Phytophthora infestans* asexual development. Illustrated are the stages examined (hyphae, sporangia, cleavage, zoospore, and the germinated cysts that contains appressoria) and the number of upregulated or downregulated genes during each transition, filtered for significance at $P < 0.05$. The number of genes changing are graphed at the indicated thresholds (1 to 7 in $\log_2$ format, which is equivalent to a >2-fold to >128 change). For changes of 128-fold or more, the actual numbers of genes are written above the bars to aid the reader. This figure is adapted from Judelson et al. (2008).

modulation is greater because numerous genes exhibited incremental changes. The expression patterns of genes were ordered on the basis of hierarchical clustering, and 11 selected developmentally regulated clusters were described. The fractions within these groups of differentially expressed genes that resembled non-oomycete proteins ranged from 14% to 66%. By comparison, 69% of genes whose expression was unaltered through the developmental stages encode proteins that are related to sequences in nonoomycete species.

Therefore, developmentally regulated genes are more likely to be oomycete specific.

Expression profiling methods, which include qRT-PCR and RNA blotting require constitutively expressed genes for normalization purposes. Although ribosomal RNA is by far the most popular endogenous control for RNA blotting, *actin A* is the most widely used endogenous control gene for qRT-PCR (Avrova et al., 2003, 2007; Grenville-Briggs et al., 2005, 2008; Walker et al., 2008; Whisson et al., 2007). However, according to the GeneChip data for *P. infestans*, many standard endogenous control genes used for qRT-PCR varied in their levels of expression in different developmental stages (Judelson et al., 2008). Thus, *actA* varied up to 3.5-fold between stages, EF-1 by 5-fold, actin B up to 12-fold, cyclophilin 5.8-fold, β-tubulin 2.1-fold, and ubiquitin-conjugating enzyme 2-fold (Judelson et al., 2008). More than 200 genes exhibited changes in mRNA levels of less than 50% between the five stages ($P < 0.05$). Most common were ribosomal proteins. Many ribosomal protein genes are upregulated during infection (Avrova, unpublished data). Eight genes exhibited less than 25% variation between asexual developmental stages, which includes a Mago nashi RNA-binding protein homolog (Pi000681), a ubiquitin-protein ligase (Pi002585), a ribonuclease (Pi011686), and protoporphyrinogen oxidase (Pi004155) (Judelson et al., 2008). Nevertheless, the expression of all of these genes requires careful evaluation during infection before they can be recommended as optimal endogenous controls for qRT-PCR.

24.5 PROFILING DIFFERENTIAL GENE EXPRESSION IN SEXUAL DEVELOPMENTAL STAGES

The Sexual reproduction by oomycetes generates thick-walled sexual spores called oospores that can survive in soil or plant debris for years, resisting harsh environmental treatments such as extreme cold, enzymatic degradation by other microbes, and chemical control agents. In many cases, the initial inoculum for epidemics is provided by germination of oospores. As such, they are important not only in generating recombinant genotypes that may be more pathogenic or resistant to control chemicals but also in providing a survival propagule in the face of harsh environmental conditions.

Differential gene expression during sexual development in oomycetes has almost exclusively been studied in *P. infestans* by the Judelson group. One of the earliest studies employed SSH to identify eight genes that are upregulated specifically in sexual development (Fabritius et al., 2002). Two genes were induced at the earliest stages of mating, whereas the remaining genes were induced later in sexual development. Five genes were single copy, whereas three were members of gene families (Fabritius et al., 2002). One of the latter was similar in sequence to the Pep-13 elicitor family in *P. sojae*. Although this *P. infestans* gene was predicted to have the transglutaminase activity of the Pep13 proteins, it lacked elicitor activity (Fabritius and Judelson, 2003).

Another gene encodes a member of the Puf family of translational regulators. This gene was subsequently shown to be induced in both sexual and asexual sporulation (Cvitanich and Judelson, 2003).

A large-scale screening using the *P. infestans* Affymetrix gene chip referred to above was performed to identify genes induced during sexual development (Prakob and Judelson, 2007). Of more than 15,650 unisequences represented on the chip, 87 were induced more than 10-fold during mating, which was confirmed in independent matings using RT-PCR and RNA blots. Several genes were induced during both mating and asexual sporogenesis, which suggests crosstalk between those pathways. In the homothallic close relative of *P. infestans*, which is *Phytophthora phaseoli*, 20% of the 87 genes were expressed at higher levels during conditions conducive to oosporogenesis than in other developmental stages, whereas expression levels of the remainder were unchanged. Many of the latter were already at significantly higher levels of expression in RNA from nonmating *P. phaseoli*, which suggests that components of the sexual pathway are active constitutively in homothallics when compared with the heterothallic *P. infestans*.

24.6 CONCLUSIONS AND THE NEXT STEPS FORWARD

A variety of approaches has been used to provide information about the genes expressed in oomycetes and of their patterns of differential expression in a range of developmental and physiological conditions. From these studies, multiple candidate targets have been provided for detailed functional analyses that can tell us about the roles of oomycete genes in sexual and asexual development and in pathogenicity. A key, understudied component of oomycete transcriptomes involves changes in levels of transcription during early, biotrophic stages of infection, in which biological material from the pathogen is limited compared with that from the host. Although much has been learned about the genes involved in these stages from studies of genes that are induced in germinating cysts, just prior to host cell penetration, many genes that are specific to biotrophy may remain unidentified until new, more sensitive methods of investigation have been developed.

One way forward in this regard involves the use of new generation sequencing (NGS) technologies, such as GS-Flex/454 pyrosequencing or Solexa/Illumina. Although these approaches currently generate only short sequence reads, at relatively low cost they also provide vast quantities of sequence coverage. The 454 technology has been applied for massively parallel sequencing from cDNA libraries of *Drosophila melanogaster*, which provided a scale of sequence redundancy that correlated well with microarray experiments in terms of quantifying transcript abundance (Torres et al., 2008). Such a scale of sequencing may be sufficient to detect most pathogen transcripts from early stages of infection. Moreover, the depth of sequence coverage from a cDNA library of *M. truncatula* was sufficient to aid the annotation significantly of predicted genes in genomic sequences of this plant (Cheung et al., 2006). This observation is particularly pertinent to the annotation of candidate RXLR- effector genes in the genomes of oomycete pathogens. In the case of *P. infestans*, for example, bioinformatic approaches predict more than 400 RXLR-EER effectors in the genome, but just 10% of these have been observed in existing EST libraries (Whisson et al., 2007). The fact that these genes are upregulated during the early stages of infection may explain the low numbers of corresponding ESTs generated from non-host-interaction cDNA libraries by conventional Sanger sequencing. To confirm that the predicted RXLR effector genes in the sequenced *P. infestans* isolate T30-4 are expressed, Solexa or 454 sequencing from the biotrophic stage of infection may be the way forward. This approach could be extended, at low cost, to additional pathogen isolates, which represent diverse genotypes and from different geographical locations, to profile the gene composition and allelic diversity within effector complements. Finally, to investigate more definitively the differential gene expression in specific cell types, from sexual or asexual developmental structures through to those, such as haustoria, which are specific to infection, NGS technologies could be combined with laser microdissection (LM). LM allows the isolation of certain cell types from heterogeneous tissues, which permits the global analyses of gene expression to be more finely tuned to the biological processes of interest. Such approaches have been used successfully to profile cell-type-specific gene expression in plants (Ohtsu et al., 2007).

ACKNOWLEDGMENTS

This work was supported by the Scottish Government Rural and Environment Research and Analysis Directorate (RERAD) (P.R.J.B.), as well as by the Biotechnology and Biological Sciences Research Council (BBSRC) (A.O.A.).

REFERENCES

Allen RL, Bittner-Eddy PD, Grenville-Briggs LJ, Meitz JC, Rehmany AP, Rose LE, Beynon JL 2004. Host-parasite coevolutionary conflict between Arabidopsis and Downy Mildew. Science 306:1957–1960.

Avrova AO, Venter E, Birch PRJ, Whisson SC 2003. Identification and quantification of *Phytophthora infestans* genes expressed prior to or during potato infection. Fungal Genet Biol 40:4–14.

Avrova AO, Whisson SC, Pritchard L, Venter E, de Luca S, Hein I, Birch PRJ 2007. A novel, non-coding infection-specific gene family is clustered throughout the genome of *Phytophthora infestans*. Microbiol 153:747–759.

Beyer K, Jimenez Jimenez S, Randall TA, Lam S, Binder A, Boller T, Collinge MA 2002. Characterisation of *Phytophthora infestans* genes regulated during the interaction with potato. Molec Plant Pathol 3:473–485.

Bittner-Eddy P, Allen R, Rehmany A, Birch PRJ, Beynon J 2003. Use of suppression subtractive hybridisation to identify downy mildew genes expressed during infection of *Arabidopsis thaliana*. Molec Plant Pathol 4:501–507.

Cheung F, Haas BJ, Goldberg SMD, May GD, Xiao Y, Town CD 2006. Sequencing *Medicago truncatula* expressed sequence tags using 454 life science technology. BMC Genomis 7:272–281.

Cvitanich C, Judelson HS 2003. A gene expressed during sexual and asexual sporulation in *Phytophthora infestans* is a member of the Puf family of translational regulators. Eukaryot Cell 2:465–473.

Fabritius AL, Cvitanich C, Judelson HS 2002. Stage-specific gene expression during sexual development in *Phytophthora infestans*. Mol Microbiol 45:1057–1066.

Fabritius AL, Judelson HS 2003. A mating-induced protein of *Phytophthora infestans* is a member of a family of elicitors with divergent structures and stage-specific patterns of expression. Molec Plant-Microbe Interact 16:926–935.

Gaulin E, Madui MA, Bottin A, Jacquet C, Mathé C, Couloux A, Wincker P, Dumas B 2008. Transcriptome of *Aphanomyces euteichus*: new oomycete putative pathogenicity factors and metabolic pathways. PLoS One 3:e1723.

Görnhaldt B, Rouhara I, Schmelzer E 2000. Cyst germination proteins of the potato pathogen *Phytophthora infestans* share homology with human mucins. Molec Plant-Microbe Interact 13:32–42.

Grenville-Briggs LJ, Anderson VL, Fugelstad J, Avrova AO, Bouzenzana J, Williams A, Wawra S, Whisson SC, Birch PRJ, Bulone V, van West P 2008. Cellulose synthesis in *Phytophthora infestans* is required for appressorium formation and successful infection on potato leaves. Plant Cell 20:720–738.

Grenville-Briggs LJ, Avrova AO, Bruce CR, Williams A, Whisson SC, Birch PRJ, van West P 2005. Elevated amino acid biosynthesis in *Phytophthora infestans* during appressorium formation and potato infection. Fungal Genet Biol 42: 244–256.

Guo J, Jiang RHY, Kamphuis LG, Govers F 2006. A cDNA-AFLP based strategy to identify transcripts associated with avirulence in *Phytophthora infestans*. Fungal Genet Biol 43:111–123.

Judelson HS, Ah-Fong AM, Aux G, Avrova AO, Bruce C, Cakir C, da Cunha L, Grenville-Briggs L, Latijnhouwers M, Ligterink W, et al. 2008. Profiling the asexual development and pre-infection transcriptome of the late blight pathogen *Phytophthora infestans*. Molec Plant-Microbe Interact 21:433–447.

Kamoun S, Hraber P, Sobral B, Nuss D, Govers F 1999. Initial assessment of gene diversity for the oomycete pathogen *Phytophthora infestans* based on expressed sequences. Fungal Genet Biol 28:94–106.

Kim KS, Judelson HS 2003. Sporangium-specific gene expression in the oomycete *Phytophthora infestans*. Eukaryot Cell 2:1376–1385.

Le Berre JY, Engler G, Panabières F 2007. Exploration of the late blight stages of tomato-*Phytophthora parasitica* interactions through histological analysis and generation of expressed sequence tags. New Phytol 177:480–492.

Madui MA, Gaulin E, Mathé C, San Clemente H, Couloux A, Wincker P, Dumas B 2007. AphanoDB: a genomic resource for *Aphanomyces* pathogens. BMC Genomi 8:471–477.

Ohtsu K, Takahashi H, Schnable PS, Nakazono M 2007. Cell type-specific gene expression profiling in plants using a combination of laser microdissection and high-throughput technologies. Plant Cell Physiol 48:3–7.

Pieterse CMJ, Derken A-MCE, Folders J, Govers F 1994b. Expression of the *Phytophthora infestans ipiB* and *ipiO* genes *in planta* and *in vitro*. Mol Gen Genet 244:269–277.

Pieterse CMJ, Risseeuw EP, Davidse LC 1991. An *in planta* induced gene of *phytophthora infestans* codes for ubiquitin. Plant Mol Biol 17:799–811.

Pieterse CMJ, Verbakel HM, Spaans JH, Davidse LC, Govers F 1993. Increased expression of the calmodulin gene of the late blight fungus *Phytophthora infestans* during pathogenesis on potato. Molec Plant-Microbe Interact 6:164–172.

Pieterse CMJ, van West P, Verbakel HM, Brasse PWHM, van den Berg-Velthuis GCM, Govers F 1994a. Structure and genomic organisation of the *ipi*B and *ipi*O gene clusters of *Phytophthora infestans*. Gene 138:67–77.

Prakob W, Judelson HS 2007. Gene expression during oosporogenesis in heterothallic and homothallic *Phytophthora*. Fugal Genet Biol 44:726–739.

Qutob D, Hraber PT, Sobral BW, Gijzen M 2000. Comparative analysis of expressed sequences in *Phytophthora sojae*. Plant Physiol 123:243–254.

Randall TA, Dwyer RA, Huitema E, Beyer K, Cvitanich C, Kelkar H, Ah Fong A, Gates K, Roberts S, Yatzkan E 2005. Large-scale gene discovery in the oomycete *Phytophthora infestans* reveals likely components of phytopathogenicity shared with true fungi. Molec Plant-Microbe Interact 18:229–243.

Shan W, Marshall JS, Hardham AR 2004. Gene expression in germinated cysts of *Phytophthora nicotianae*. Mol Plant Pathol 5:317–330.

Skalamera D, Wasson AP, Hardham AR 2004. Genes expressed in zoospores of *Phytophthora nicotianae*. Mol Gen Genom 270:549–557.

Tani S, Yatzkan E, Judelson HS 2004. Multiple pathways regulate the induction of genes during zoosporogenesis in *Phytophthora infestans*. Molec Plant-Microbe Interact 17:330–337.

Torres TT, Metta M, Ottenwälder B, Schlötterer C 2008. Gene expression profiling by massively parallel sequencing. Gen Res 18:172–177.

Torto TA, Li S, Styer A, Huitema E, Tesata A, Gow NAR, van West P, Kamoun S 2003. EST mining and fuinctional expression assays identify effector proteins from the plant pathogen *Phytophthora infestans*. Gen Res 13:1675–1685.

Torto-Alalibo T, Tian M, Gajendran K, Waugh ME, van West P, Kamoun S 2005. Expressed sequence tags from the oomycete fish pathogen *Saprolegnia parasitica* reveal putative virulence factors. BMC Microbiol 5:46–58.

Torto-Alalibo TA, Tripathy S, Smith BM, Arredondo FD, Zhou L, Li H, Chicubos C, Qutob D, Gijzen M, Mao C, Sobral BWS, Waugh ME, et al. 2007. Expressed sequence tags from *Phytophthora sojae* reveal genes specific to development and infection. Molec Plant-Microbe Interact 20:781–793.

van der Biezen EA, Juwana H, Parker JE, Jones JD 2000. cDNA-AFLP display for the isolation of *Peronospora parasitica* genes expressed during infection of *Arabidopsis thaliana*. Mol Plant-Microbe Interact 13:895–898.

van West P, de Jong AJ, Judelson HS, Emons AM, Govers F 1998. The *ipiO* gene of *Phytophthora infestans* is highly expressed in invading hyphae during infection. Fungal Genet Biol 23:126–138.

Vleeshouwers VG, Rietman H, Krenek P, Champouret N, Young C, Oh, SK, Wang M, Bouwmeester K, Vosman B, Visser RG 2008. Effector genomics accelerates discovery and functional profiling of potato disease resistance and *Phytophthora infestans* avirulence genes. PLoS One 3:e2875.

Walker CA, Köppe M, Grenville-Briggs LJ, Avrova AO, Horner NR, McKinnon AD, Whisson SC, Birch PRJ, van West P 2008. A putative DEAD-box RNA-helicase is required for normal zoospore development in the potato late blight pathogen *Phytophthora infestans*. Fungal Genet Biol 45:954–962.

Wang Z, Wang Y, Chen X, Shen G, Zhang Z, Zheng X 2006. Differential screening reveals genes differentially expressed in low- and high-virulence near isogenic *Phytophthora sojae* lines. Fungal Genet Biol 43:826–839.

Whisson SC, Boevink PC, Moleleki L, Avrova AO, Morales J, Gilroy EM, Armstrong MR, Grouffaud S, van West P, Chapman S 2007. A translocation signal for delivery of oomycete effector proteins inside host plant cells. Nature 450:115–118.

Whisson SC, van der Lee T, Bryan G, Waugh R, Govers F, Birch PRJ 2001. Physical mapping across an avirulence locus of *Phytophthora infestans* using a high representation, large insert bacterial artificial chromosome library. Mol Gen Genom 266:289–295.

25

MECHANISMS AND APPLICATION OF GENE SILENCING IN OOMYCETES

STEPHEN C. WHISSON AND ANNA O. AVROVA
Plant Pathology Programme, Scottish Crop Research Institute, Invergowrie, Dundee, United Kingdom

LAURA J. GRENVILLE BRIGGS AND PIETER VAN WEST
Aberdeen Oomycete Group, University of Aberdeen, Institute of Medical Sciences, Foresterhill, Aberdeen, United Kingdom

25.1 INTRODUCTION

Prior to the late 1990s, the phenomenon of gene silencing had been observed for many years without an understanding of the underlying biological processes. Only recently has the full scope and role of this cellular process been elucidated and exploited. Early observations of gene silencing were made in tobacco plants infected with tobacco ringspot virus that apparently overcame the virus infection and were immune to additional infection by that virus (Wingard, 1928, cited in Baulcombe, 2004). This is now known to be caused by virus-induced gene silencing. Other early instances of gene silencing include cosuppression in *Petunia hybrida* (Napoli et al., 1990) and quelling in the filamentous ascomycete fungus *Neurospora crassa* (Romano and Macino, 1992). These later observations of gene silencing became apparent only with the ability to introduce homologous DNA molecules into eukaryotic cells, leading to a suppression of the transgene and homologous native gene. Later experimental evidence from the model animal *Caenorhabditis elegans* demonstrated the pivotal role of double-stranded RNA (dsRNA) in the initiation of homology-dependent gene silencing (Fire et al., 1998). Clues as to the endogenous function of gene silencing have been revealed from

Oomycete Genetics and Genomics: Diversity, Interactions, and Research Tools
Edited by Kurt Lamour and Sophien Kamoun
Copyright © 2009 John Wiley & Sons, Inc.

mutants deficient in this capability. Silencing-deficient mutants typically have heightened levels of transposon activity and susceptibility to viruses (Tabara et al., 1999; Vastenhouw et al., 2003; Yang et al., 2004). Thus, gene silencing is a defense strategy employed at both the cellular and genome level to control damage caused by virus and transposon activity. Defense against viruses may be considered as cytoplasmic, whereas transposon control may be a nuclear process (Baulcombe, 2004). Gene silencing has since been found to operate in most eukaryotic organisms studied, although not all possible components are present in all organisms. This points toward gene silencing being an ancient process present in basal eukaryotes, elements of which have been lost or gained in evolutionary history. In oomycetes, as in other organisms, gene silencing of an endogenous gene was first observed when the elicitin *inf1* gene was transformed into *Phytophthora infestans*, which resulted in transcriptional inactivation of both the introduced and endogenous copies of *inf1* (Kamoun et al., 1998; van West et al., 1999). These experiments also led to the discovery of a phenomenon termed internuclear gene silencing (van West et al., 1999). A defining feature of internuclear gene silencing was the ability of a silencing signal to spread the silenced state from a silenced nucleus to a previously nonsilenced nucleus in a sequence-specific manner.

One of the most revealing ways to determine the role of a specific gene in the biology of an organism is to remove or reduce the levels of the encoded protein. The presence of an endogenous process that can suppress expression of both introduced and endogenous gene copies in a sequence-specific manner has led to the exploitation of this process to determine gene function in many organisms. In this respect, research that employs gene silencing in oomycetes has also followed this path (Kamoun et al., 1998; Gaulin et al., 2002; Latijnhouwers et al., 2004; Blanco and Judelson, 2005; Avrova et al., 2008; Grenville-Briggs et al., 2008; Walker et al., 2008). Oomycetes are typically diploid in their asexual stages, which signifies that mutagenesis strategies (mutagen or targeted knockout) for gene disruption must either target both copies of a gene or that homozygosity of a mutation must be achieved through sexual outcrossing or self-fertilization. However, some species such as *P. ramorum* do not readily produce oospores in culture, whereas others such as *P. infestans* produce oospores that are frequently recalcitrant to laboratory germination, further complicating attempts at mutagenesis (Knapova et al., 2002; Brasier and Kirk, 2004). In contrast to fungi, homologous recombination has not been reported in oomycetes, and thus, gene knockouts by this method have not yet been achieved. In oomycetes, most studies that involve gene silencing have been conducted in *P. infestans*, and much of this chapter will focus on that species.

25.2 GENE SILENCING — THE BASIC PARTS LIST

Two protein families, dicer and argonaute, are common to gene silencing pathways from most eukaryotes where gene silencing has been reported (Cerutti and Casas-Mollano, 2006). Additional proteins are frequently also associated with the silencing pathways of specific organisms, and these proteins determine specific properties depending on the organism. For example, in many eukaryotes, at least one RNA-directed RNA polymerase (RdRP) is involved, except for some groups of animals such as human and *Drosophila*, where this protein is absent (Verdel and Moazed, 2005). Other proteins involved in silencing are reviewed by Meister and Tuschl (2004) and include DEAD box RNA helicases, dsRNA binding proteins, chromodomain proteins, dsRNA-specific exonucleases, DNA methyltransfersases, histone deacetylases, and histone methyltransferases. Genes encoding proteins that exhibit similarity to RNA interference (RNAi) components have been identified in oomycete sequence databases (either expressed sequence tag or genome) (see below for more details).

25.3 MECHANISMS OF SILENCING

Gene silencing is triggered by dsRNA that, under normal cellular conditions, may develop from transposon transcripts, gross overexpression of transcripts (as for viruses), aberrant messenger RNAs, or from non–protein-coding-premicroRNA (miRNA) transcripts (Ruiz et al., 1998; Jensen et al., 1999; Gazzani et al., 2004; Luo and Chen, 2007; Ghildiyal et al., 2008). Alternatively, dsRNA molecules can be introduced into cells either as *in vitro* synthesized molecules (Timmons and Fire, 1998; Clemens et al., 2000) or from antisense or inverted repeat expression vectors (for example, Guo et al., 2003). The basic mechanism of gene silencing (reviewed in Baulcombe, 2004; Meister and Tuschl, 2004; Collins and Cheng, 2005) follows that dsRNA is bound and cut by the dicer enzyme into shorter dsRNA molecules, typically of 21–24 bp and termed short interfering RNAs (siRNAs). Dicer is a multidomain protein that comprises at least two RNase III domains and additional domains such as dsRNA binding, PAZ, and DEAD box helicase. The PAZ domain may act in concert with the RNaseIII domains as a molecular measure to define the number of nucleotides of RNA between the two RNaseIII cut sites on the dsRNA (Zhang et al., 2004). The presence of a helicase domain may assist in unwinding complex secondary structures, processing of long dsRNAs, or unwinding of siRNAs for the next stage of the silencing pathway. Following dicer digestion, the siRNAs strands are separated, and the antisense strand is incorporated into the RNA-induced silencing complex (RISC), the core of which is the multidomain argonaute protein. Argonaute proteins contain two major domains, PAZ and PIWI. The antisense strand of the siRNA duplex is bound into the RISC via the PAZ domain and acts as the guide strand, which targets the homologous sequence in the native messenger RNA (mRNA) and degrades it through the slicer activity of the PIWI domain (Collins and Cheng, 2005).

A related class of small RNA species, called miRNA, also develops through dicer or other RNaseIII enzyme (such as Drosha) digestion of non–protein-coding premiRNA transcripts containing a stem-loop secondary structure (reviewed in Baulcombe, 2004; Collins and Cheng, 2005). Similar to siRNAs, the mature miRNA may also associate with argonaute proteins to degrade endogenous mRNAs, thus achieving a level of posttranscriptional regulation. Alternatively, the mature miRNA may bind to the homologous messenger RNA (mRNA) to inhibit translation (Brodersen et al., 2008).

SiRNAs and fragments released from RISC degradation of mRNAs can act to prime synthesis of more dsRNA from the native mRNA through the action of the RdRP enzyme. RdRP can act both to spread the silencing target signal along the mRNA and to amplify the signal through production of more dsRNA (Alder et al., 2003). The synthesis of dsRNA by RdRP has been shown to occur in both a primer-dependent and primer-independent manner (Makeyev and Bamford, 2002). Secondary siRNAs in *C. elegans* have been shown to exhibit sequence direction bias suggestive of unidirectional priming (Sijen et al., 2001), whereas aberrant mRNAs can be substrates for primer-independent RdRP synthesis of dsRNA (Baulcombe, 2004; Luo and Chen, 2007).

The major cytoplasmic steps of gene silencing are frequently described as posttranscriptional gene silencing (PTGS) or RNAi. That is, the native gene is transcribed, but the mRNA is degraded before translation can occur. The components of PTGS also provide a link to silencing at the transcriptional level (Verdel et al., 2004; Bühler et al., 2007). In many organisms where PTGS operates, it can also lead to epigenetic changes at the DNA level, principally through *de novo* methylation of cytosines or through deacetylation and methylation of histone proteins. The outcome of these modifications is silencing at the transcriptional level. Plant proteins involved in these processes are RNA polymerase IV, DNA methyltransferase, histone methyltransferase, histone deacetylase, and chromodomain protein DRD1 (Huettal et al., 2007); RdRP, dicer and argonaute are also involved in initiating and maintaining TGS. Similar processes also operate in fungi and are best characterized in fission yeast *Schizosaccaromyces pombe*. In *S. pombe*, silencing operates at the transcriptional level through formation of heterochromatin. Similar to plants, this involves dicer, RdRP, and argonaute proteins. *S. pombe* argonaute forms part of the RNA induced transcriptional silencing (RITS) complex, together with proteins CHP1 and TAS3 (Verdel and Moazed, 2005). TAS3 has no discernable domains, but CHP1 is a chromodomain protein. The RITS complex, through CHP1, targets DNA sequences homologous to the siRNA bound in the RITS complex to modulate the deacetylation and methylation of histones, and thus the formation of heterochromatin (Verdel and Moazed, 2005). The methylation/acetylation status of histones is thought to then mediate access to the DNA for RNA polymerase II, with any mRNA formed then targeted by RDRP to form dsRNA, then siRNAs; this process maintains the silenced state (Grewel and Jia, 2007). Similar to *S. pombe*, TGS that develops from RNAi in the nematode *C. elegans* can persist through several generations, with its basis being the activity of histone deacetylase, histone acetyltransferase, a chromatin remodelling ATPase, and a chromodomain protein (Vastenhouw et al., 2006).

PTGS in *C. elegans* and other organisms such as *N. crassa* has been observed as a transient phenomenon that decreases in intensity with time after exposure to the initial silencing stimulus. A component of this "release" from PTGS has been identified from *C. elegans* as a dsRNA-specific RNase termed ERI1, which degrades siRNAs (Kennedy et al., 2004). ERI1 mutants exhibit enhanced and more persistent RNAi and TGS (Kennedy et al., 2004; Iida et al., 2006).

25.4 GENE SILENCING IN OOMYCETES: TALES FROM THE LABORATORY AND CLUES FROM GENOMES

As stated previously, the primary report of stable internuclear gene silencing in oomycetes indicated that silencing operated at the transcriptional level (van West et al., 1999, 2008). However, it has since been shown that exogenously applied *in vitro* synthesized dsRNA can initiate short-term silencing in *P. infestans* protoplasts that can persist for up to 15 days after exposure to dsRNA (Whisson et al., 2005). It is unlikely at this time after exposure to the dsRNA that any of the originally applied dsRNA is still present, and so the silencing signal must either be secondary and amplified, perhaps through the action of an RdRP, or maintained at the DNA level by chromatin modification or DNA methylation (although DNA methylation seems not to occur in *Phytophthora*). A recent study to optimize initiation of TGS in *P. infestans* (Ah Fong et al., 2008) used inverted repeat constructs of the *inf1* gene in stable transformants to initiate silencing. This process generated a mixture of fully and partially silenced transformants. Some partially silenced transformants became fully silenced after several serial subcultures, and siRNAs were detected in partially silenced transformants but not in fully silenced transformants. These pieces of evidence suggest that both PTGS and TGS are operational in oomycetes (Fig. 25.1).

TGS in oomycetes also can spread from silenced nuclei to nonsilenced nuclei, as has been shown in *P. infestans* and *P. parasitica* (van West et al., 1999; Gaulin et al., 2007). Although Ah Fong et al. (2008) could not identify siRNAs in fully silenced *P. infestans* transformants, some form of sequence-specific silencing signal must exist to propagate the silenced state from silenced nucleus to non-silenced nucleus. TGS in *S. pombe* also employs siRNAs to maintain the silenced state (Iida et al., 2006), so it remains possible that siRNAs contribute to TGS in *P. infestans* and other oomycetes, albeit at very low concentrations.

At the basis of TGS in *P. infestans* is the formation of heterochromatin at and surrounding the silenced DNA sequence. Judelson and Tani (2007) demonstrated that the formation of heterochromatin could extend several hundred base pairs outward from the targeted gene(s) in a cluster of NIF transcription factors. In their study, this likely resulted in the concurrent silencing of a neighboring gene family member. However, the more closely related (by DNA sequence) NIFS gene, which is more distant from the silenced NIFC1, 2, and 3 loci, remained unaffected, and it was speculated that this was

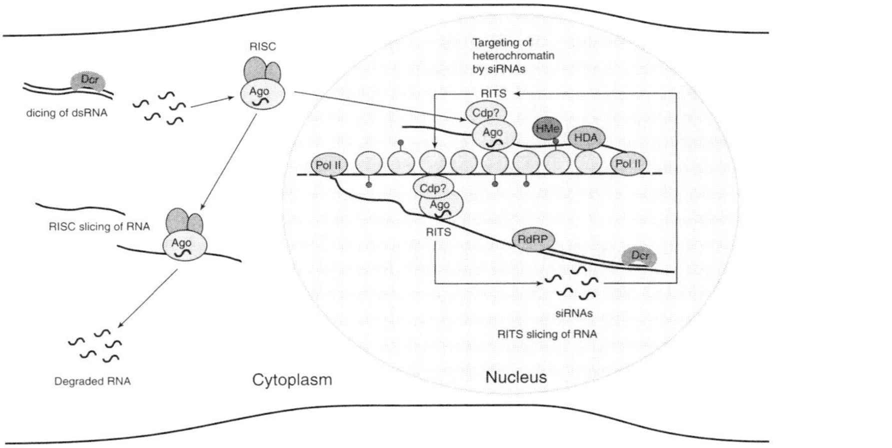

FIG. 25.1 Proposed model for post transcriptional gene silencing (PTGS) and transcriptional gene silencing (TGS) in *Phytophthora*, based on PTGS/TGS in *Schizosaccharomyces pombe* and adapted from Grewel and Jia (2007). PTGS is initiated in the cytoplasm when double-stranded RNA (dsRNA) is digested to short-interfering RNA (siRNAs) by dicer (Dcr). The RISC, containing argonaute (Ago) binds the siRNAs and can slice the homologous messenger RNA (mRNA) and degrade it. siRNAs may also be bound into an Ago-containing RNA-induced transcriptional silencing (RITS)-like complex within the nucleus, which leads to mRNA slicing and targeting of the homologous DNA sequences for heterochromatin formation. Heterochromatin is formed by deacetylation and methylation (dots) of histone proteins (circles) by histone deacetylases (HDA) and histone methyltransferases (HMe), respectively, to limit access by RNA polymerase II (PolII). The sequence recognition for HDA and HMe may be mediated through interaction with a chromodomain protein (Cdp), possibly in the RITS-like complex. Transcripts that are formed by PolII are either substrates for siRNA-primed dsRNA synthesis by RNA-directed RNA polymerase (RdRP) and subsequent dicing, or slicing by the RITS-like complex. The transcriptionally silenced state is maintained through successive cycles of PolII, RdRP, Dcr, and RITS activity.

caused either by insulator sequences or by inaccessibility of the gene to chromatin-modifying proteins. An alternative explanation for this observation may be that silencing of the NIFS gene is lethal for *P. infestans*, hence it does not permit the recovery of silenced transformants. The significance of these findings, however, is that when undertaking silencing experiments in *P. infestans* and other oomycetes, the expression of neighboring and sequence-related genes in silenced lines should also be determined to verify that any observed phenotype is caused by the silencing of only the target gene. The consideration of neighboring genes is of particular importance in oomycetes, as genes are often closely spaced (Whisson et al., 2004; Armstrong et al., 2005; Jiang et al., 2006; Judelson and Tani, 2007).

Unlike plants, *de novo* methylation of DNA sequences does not seem to occur or play a role in silencing of transgenes or their endogenous homologs in *P. infestans* TGS (Judelson and Tani, 2007; van West et al., 2008). The digestion of DNA from silenced transformants with methylation sensitive restriction endonucleases and the sequencing of bisulfite-treated DNA did not show any evidence of DNA methylation at silenced loci. Somewhat contradictory results were obtained from studies using the DNA methylation inhibitor 5-azacytidine (van West et al., 2008). Despite the lack of detectable cytosine methylation in transgene and homologous endogenous sequences in silenced transformants, 5-azacytosine could reverse silencing in *P. infestans*. It is possible that the 5-azocytosine also inhibits other methyltransferase enzymes, such as histone methyltransferase, which may in part explain the reversal of silencing after treatment with this chemical. Histone deacetylases have been implicated in formation of heterochromatin and TGS in other organisms (Kim et al., 2004; Vastenhouw et al., 2006; Huettal et al., 2007). The treatment of *inf1*-silenced transformants with the histone deacetylase inhibitor trichostatin A also led to a reversal of silencing, which indicates the key role that this enzyme plays in TGS in *P. infestans* (van West et al., 2008). Van West et al. (2008) also proposed that the histone acetylation/methylation status is maintained in a constant state by histone acetylation/deacetylation/methylation reactions and that perturbations to this status are very slowly returned to the "steady state" when removed from the silencing stimulus, being either exogenous dsRNA or transgene expression. Thus, it may be speculated from laboratory studies that TGS in *P. infestans* and likely in other oomycetes, is based on the formation of non–DNA-methylated but histone deacetylated and methylated heterochromatin, the formation of which is triggered by siRNAs from the action of PTGS (Fig. 25.1).

The recent sequencing of the genomes of oomycetes *P. sojae*, *P. ramorum*, *Phytophthora capsici*, *P. infestans*, and *Hyaloperonospora arabidopsidis* (previously *Hyaloperonospora parasitica*) (Tyler et al., 2006; http://www.broad.mit.edu/annotation/genome/phytophthora_infestans; http://phytophthora.vbi.vt.edu/) has and will continue to assist the discovery of numerous genes and pathways associated with oomycete biology. Identifying the components of oomycete gene silencing has also been assisted by the availability of genome sequences. Using the *P. infestans* genome sequence as an example, genes encoding the key components of PTGS are all present: dicer-like, argonaute, and RdRP (Ah Fong et al., 2008). Genes encoding other potential PTGS accessory proteins, for example that contain domains characteristic of proteins that are involved in PTGS in other organisms, such as dsRNA binding proteins and DEAD-box RNA helicases, are also present (Walker et al., 2008). Using both plant, *S. pombe*, and *C. elegans* as reference systems where TGS has been dissected, genes encoding the major components, such as chromodomain proteins and histone modifying proteins, are all present within oomycete genome sequences in multiple copies as in other organisms (Whisson and Grenville-Briggs, unpublished). It remains to be demonstrated which of these protein family members contribute to TGS in oomycetes. In support of the observations from laboratory-based studies, genes encoding proteins for *de novo* methylation of DNA were not identified (Judelson and Tani, 2007). Similarly, subunits of RNA polymerase IV, a plant-specific protein apparently involved in *de novo* DNA methylation, were not identified in the *P. infestans* genome (Whisson and Grenville-Briggs, unpublished).

Model gene-silencing systems *C. elegans, S. pombe*, and *Arabidopsis thaliana* all contain a protein called ERI1, which is an exoribonuclease that digests siRNAs, ultimately leading to a release of PTGS (Kennedy et al., 2004; Iida et al., 2006). The *P. infestans* genome does not seem to contain the gene encoding this protein. The lack of an ERI homolog may in part explain the persistence of gene silencing in the absence of any homologous transgene (van West et al., 1999, 2008; Whisson et al., 2005; Gaulin et al., 2007) and the observation that partial silencing in some *P. infestans* transformants can convert to complete silencing after some time (Ah Fong et al., 2008).

25.5 STRATEGIES FOR APPLICATION OF GENE SILENCING IN OOMYCETES

25.5.1 Stable Gene Silencing

Most studies that involve gene silencing in oomycetes have used the TGS phenomenon observed in a percentage of stable transformants. In many organisms in which gene silencing has been employed in determining gene function, transformation constructs for expression of sense, antisense, or inverted repeat with spacer have been used to induce PTGS or TGS. Silencing that results from introduction of sense constructs presumably develops either through gross overexpression of the transgene or expression of defective transgene mRNAs. Defective mRNA may develop if the terminator region of the transgene contains elements, for example, that may prevent correct polyadenylation (Luo and Chen, 2007). Such substrates may act as templates for RdRP to synthesize dsRNA, followed by processing through the PTGS and TGS pathways. Antisense constructs aim to initiate silencing through the hybridization of the antisense mRNA with the sense native mRNA to yield dsRNA that can be processed by dicer. Hairpin constructs use both sense and antisense copies of a gene or portions of a gene in a single expression cassette to anneal and form dsRNA. A spacer such as an intron is often included between the inverted repeats to maximize construct stability when plasmids are maintained in *Eschertichia coli*. The intron is spliced out of the hairpin pre-mRNA to yield mature dsRNA for dicer processing. All of these expression construct types have been used successfully in *P. infestans* to initiate silencing (Judelson et al., 1993b; Van West et al., 1999; Ah Fong et al., 2003, 2008; Blanco and Judelson, 2005; Judelson and Tani, 2007). Until recently, no specifically targeted study has determined which format of expression construct to be most efficient at initiating gene silencing in stably transformed *P. infestans*. Ah Fong et al. (2008) tested sense, antisense, and hairpin constructs of the *inf1* elicitin in stable transformants to demonstrate that hairpin constructs were most efficient in initiating silencing, followed by antisense and sense orientations of the *inf1* gene (Fig. 25.2). Partial silencing was also observed in this study and was most prevalent in transformants that expressed the hairpin construct. Furthermore, some partially silenced transformants became fully silenced after several serial transfers over a 7-month time period.

With one exception (Latijnhowers et al., 2004), all silencing experiments in oomycetes have used the *Ham34* promotor from *Bremia lactuca* (Judelson et al., 1992) to overexpress the transgene for silencing. From the oomycete promotors tested to date, the Ham34 promoter element directs the highest level of gene expression (McLeod et al., 2008). The only report where the *Ham34* promoter was not used for oomycete transgene overexpression to induce silencing used the native promoter of the G-protein being targeted for silencing (Latijnhowers et al., 2004).

The frequency of gene silencing in stable transformants has been highly variable, depending on which gene is being targeted for silencing and the expression construct used to initiate silencing. For example, up to 25% of *P. infestans* CDC14 transformants were silenced by expression of the gene in the sense orientation (Ah Fong and Judelson, 2003), compared with 72% for a bZIP transcription factor (Blanco and Judelson, 2005) and 11% for *inf1* (Ah Fong et al., 2008). This suggests that some genes may be more recalcitrant to silencing than others, according to an observation made by Judelson and Tani (2007).

The transformation method used to introduce expression constructs has also been considered as a factor involved in initiating silencing. Oomycetes have been transformed using a variety of methods: PEG-$CaCl_2$ transformation of protoplasts (Judelson et al., 1991, 1993a), electroporation of zoospores or protoplasts (Weiland, 2003; Ah Fong et al., 2008), *Agrobacterium tumefaciens* transformation (Vijn and Govers, 2003), and microprojectile bombardment (Cvitanich and Judelson, 2003). More detailed descriptions of transformation methods are given in Chapter 22 on oomycete transformation. The technique of protoplast transformation has been shown to yield the highest frequency of silenced transformants with genes expressed as sense, antisense, and hairpin configurations (Ah Fong et al., 2008; Fig. 25.2). In the same study, the electroporation of zoospores and microprojectile bombardment yielded useful numbers of silenced transformants only with hairpin constructs but at a reduced frequency compared with that of protoplast transformation. Each method of transformation results in specific types of genomic integration events. Protoplast transformation frequently leads to arrays of multiple integrations, with the transgene copy number shown to be correlated with gene silencing (Judelson et al., 1993a; Ah Fong et al., 2008).

Although gene silencing in stable transformants is the ideal situation to assess the function of the silenced gene, the disadvantages to this strategy are that it relies on stable transformation, which although routine in numerous oomycete laboratories, is highly labor intensive and restricts the numbers of genes that can be functionally assessed at any one time. Furthermore, this

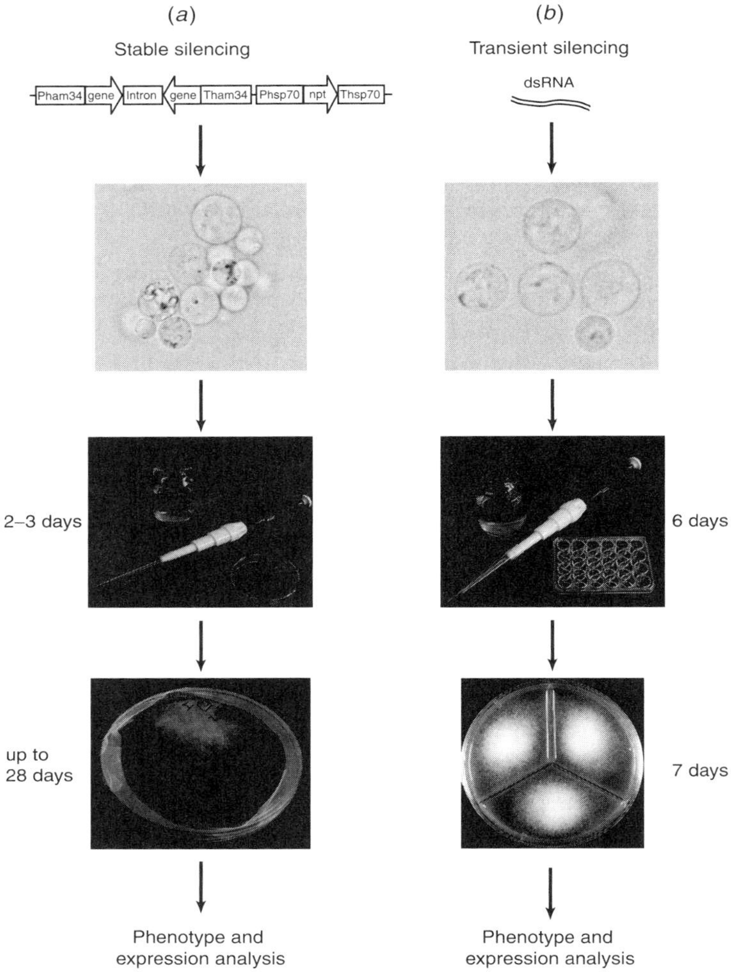

FIG. 25.2 Optimal strategies for gene silencing in *Phytophthora*. (a) Stable gene silencing. The most reliable strategy for generating silenced stable transformants uses an inverted repeat with intron spacer construct, with expression controlled by the constitutive Ham34 promotor. The optimal transformation procedure for generating silenced transformants uses PEG-$CaCl_2$ transformation of protoplasts (see chapter on transformation). Transformed protoplasts are regenerated in liquid medium for 2–3 days, followed by the selection of transformants on agar medium for up to 28 days. (b) Transient gene silencing. dsRNA is synthesized *in vitro* from polymerase chain reaction (PCR) products tagged with T7 RNA polymerase initiation sites. Protoplasts are treated for 24 h with exogenously applied dsRNA. Colonies are regenerated in liquid medium in 24-well plates for approximately 6 days to ensure individual colonies are recovered, then transferred to agar for further growth for approximately 7 days. Following either stable transformation or dsRNA treatment, subsequent phenotyping and gene expression analysis are carried out to confirm the association of gene silencing with any observed phenotype.

strategy relies on cloning each gene into an oomycete expression vector prior to transformation; this process is routine for sense and antisense expression constructs, but it requires at least two cycles of cloning for hairpin constructs. Added to this is also the erratic frequency with which silencing occurs, which is in part alleviated by the identification of a high frequency of partial silenced transformants obtained through using hairpin constructs. That is, partial silencing will allow the functional characterization of genes that are essential for cellular survival, the complete silencing of which would otherwise be lethal.

25.5.2 Transient Gene Silencing

Transient gene silencing is a relatively new addition to the functional genomics toolbox for oomycetes. Developed first in *P. infestans*, it has been used to silence nine, and demonstrate functions for six, genes (see Table 25.1; Whisson et al., 2005; Grenville-Briggs et al., 2008; Walker et al., 2008; Avrova et al., 2008). This strategy for gene silencing relies on the treatment of protoplasts with high concentrations of *in vitro* synthesized dsRNA, typically 1 μg/μl dsRNA for approximately 24 h (Fig. 25.2). Protoplasts are regenerated in liquid medium until colonies appear, which is approximately 4–6 days for *P. infestans*, and then transferred to solid agar medium for growth and subsequent phenotyping. In assessing gene expression knockdown using this silencing strategy, the effect of silencing is most readily demonstrated in the cell type that exhibits the highest level of gene expression for the gene under assay. That is, transient silencing of a gene most highly expressed in, for example zoospores, will be most clearly demonstrated in that life cycle stage. A remarkable phenomenon of transient gene silencing in *P. infestans* is its persistence for up to 15 days after treatment with the original dsRNA species. It is highly unlikely that the originally applied dsRNA would still be intact and present in the regenerated hyphae 15 days after dsRNA treatment, as dsRNA is removed by dilution from the regenerating protoplasts after 24 h. It may be that the application of exogenously applied dsRNA leads to partial transcriptional silencing of the target gene, which is also consistent with the persistence of silencing in studies of internuclear gene silencing (van West et al., 1999, 2008; Gaulin et al., 2007). Alternatively, the exogenously applied dsRNA may lead to amplification of the siRNA silencing signal through the RdRP-mediated synthesis of secondary dsRNA and subsequent siRNAs. These questions remain open until experiments that demonstrate either the involvement of heterochromatin or secondary siRNAs have been conducted on *P. infestans* transiently silenced for an endogenous gene. Oomycete hyphae are coenocytic and multinucleate. Thus, from exogenous application of dsRNA to protoplasts, it is highly likely that the resulting colonies will contain a population of nuclei, some of which will be silenced, and some of which will not be silenced. The result of this mixture of nuclei with differing silencing status is that silencing will seem to be partial, both in terms of gene expression and phenotype. Early experiments in the development of this technique

TABLE 25.1 Silenced genes in oomycetes and associated phenotypes.

Oomycete/gene	Function	Silencing phenotype	Silencing method	References
P. infestans inf1	Elicitin	Nonsecretion of INF1	Stable,	van West et al., 1999; Ah Fong et al., 2008
			Transient	Whisson et al., 2005
P. parasitica CBEL	Cellulose binding elicitor lectin	Reduced binding to cellulosic substrates	Stable	Gaulin et al., 2002
P. infestans PiGpb1	G protein beta subunit	Densely packed aerial mycelium, reduced sporulation, aberrant sporangia	Stable	Latijnhouwers and Govers, 2003
P. infestans Picdc14	protein phosphatase	No sporulation	Stable,	Ah Fong and Judelson, 2003
			Transient	Whisson et al., 2005
P. infestans PiGpa1	G protein alpha subunit	No zoospore autoaggregation, large aberrant zoospores, increased zoospore turning, aberrant appressorium formation, defective zoospore chemotaxis, loss of pathogenicity	Stable	Latijnhouwers et al., 2004

P. infestans gfp (transgenic)	Fluorescent protein	Reduced GFP fluorescence	Transient	Whisson et al., 2005
P. infestans Pibzp1	Transcription factor	Zoospore "spinning," increased cyst germination, appressorium formation defective, host penetration defective	Stable	Blanco and Judelson, 2005
P. infestans NIFC1, NIFC2, and *NIFC3*	Transcription factors	Reduced cyst germination	Stable	Judelson and Tani, 2007
P. infestans Pirnh1	RNA helicase	Aberrant sporangial cleavage, aberrant large osmotically sensitive zoospores	Transient	Walker et al., 2008
P. infestans PicesA1, PicesA2, PicesA3, and *PicesA4*	Cellulose synthases	Aberrant or burst appressoria, loss of pathogenicity	Transient	Grenville-Briggs et al., 2008
P. infestans Pihmp1	Novel transmembrane protein; possibly haustorium structure	Loss of pathogenicity	Transient	Avrova et al., 2008

used protoplasts from a transformant expressing the green fluorescent protein (GFP) treated with dsRNA homologous to the *gfp* gene (Whisson et al., 2005). The resulting colonies at 17 days after *gfp* dsRNA exposure showed reduced GFP fluorescence; however, both fluorescing and non-fluorescing hyphal tips were identified within a single colony. The experimental implications of partial silencing are that phenotypes from transient gene silencing are frequently scored as proportions of the total number of cells that exhibit a phenotype differing from the wild type (treated with a nonhomologous dsRNA) (Grenville-Briggs et al., 2008; Walker et al., 2008).

When treating protoplasts with dsRNA and regenerating colonies, it is important to dilute the dsRNA-treated protoplasts sufficiently to ensure that individual colonies are recovered. It follows that an excess of protoplasts should be avoided in dsRNA treatments. The outcome of excess numbers of protoplasts in dsRNA treatment is typically colonies that inextricably merge into one another, which mixes colonies that may exhibit significantly different levels of gene silencing. In turn, this can cause difficulties in associating gene expression knockdown from silencing with observed phenotypes; the colony with higher level expression masks the colony with silencing.

Transient gene silencing has the ability to accept multiple dsRNA species in a single experiment to silence multiple genes simultaneously. Grenville-Briggs et al. (2008) simultaneously silenced four DNA sequence-unrelated cellulose synthase genes in *P. infestans*. This allowed the potential problem of functional compensation by other proteins to be overcome and demonstrated the role of cellulose synthesis in *P. infestans* development and pathogenesis. Furthermore, complete silencing of cellulose synthase genes is likely to be lethal, because the cell wall of *P. infestans* is predominantly cellulose, which makes transient partial silencing the best-suited strategy to study this set of four genes.

25.6 VALIDATION: LINKING GENE SILENCING TO PHENOTYPE

Assays for gene expression are vital to correlate observed phenotypes with the level of reduction in gene expression caused by gene silencing. Where complete TGS is employed in stable transformants, a qualitative reverse transcription polymerase chain reaction (RT-PCR) or northern hybridization may yield sufficient supporting evidence; fully transcriptionally silenced transformants typically yield no DNA product from RT-PCR and no detectable homologous mRNA in northern hybridization (van West et al., 1999; Ah Fong et al., 2003; Blanco and Judelson, 2005). In *Entamoeba histolytica*, the quantitation of mRNA levels of the targeted gene in transcriptionally silenced stable transformants demonstrated that the mRNA was 5,000-fold less abundant compared with the wild type (Mirelman et al., 2006). Partial silencing that developed from transient gene silencing, or stable transformation using hairpin constructs, is best assayed using quantitative real-time RT-PCR or alternatively a semiquantitative RT-PCR, because transient gene silencing yields a variable range of gene expression knockdown in regenerated protoplast lines, varying from wild-type levels to as low as 1% of wild type levels (Whisson et al., 2005; Avrova et al., 2008; Grenville-Briggs et al., 2008; Walker et al., 2008). It is also crucial to assess sufficient control and silenced transformants/transient silenced lines in establishing a link between silencing of a target gene and the phenotype observed. Rare instances of spontaneous silencing of the *P. infestans inf1* gene in control transformants were observed by Ah Fong et al. (2008), which prompted caution in attempting to associate silencing of a specific gene with a phenotype when using low numbers of control and silenced lines. This may be of particular importance when associating partial silencing with a phenotype, because slight differences in the timing of development may lead to variation in expression of the gene(s) of interest in control and partially silenced lines.

Gene silencing relies entirely on homology of the transgene or exogenous dsRNA to the endogenous sequence. As such, it is possible that unintentional off-target silencing of other genes may occur if sufficient sequence homology exists (over 20 bp). Therefore, it is vital that target sequences for silencing be checked for sequence homology against genome sequence databases or extensive expressed sequence tag (EST) databases to ascertain the likelihood of off-target silencing. Any gene that exhibits DNA sequence similarity to the targeted gene should also be assayed for expression levels in parallel with the target gene, to ensure that any observed phenotype is caused by the silencing of the target gene alone.

25.7 WHICH STRATEGY? STABLE VERSUS TRANSIENT GENE SILENCING

Each of the two strategies for achieving gene silencing in oomycetes has inherent advantages and disadvantages. Stable, transcriptional silencing has the obvious advantages in maintaining undetectable levels of gene expression, such that effectively null genotypes and their associated phenotypes can be studied. This allows very subtle differences in phenotypes to be identified and studied in detail, such as the 50% reduction in cyst germination observed for silencing of NIF genes (Judelson and Tani, 2007). However, transcriptional gene silencing is effectively a gene knockout in terms of gene expression, which signifies that genes for which silencing leads to a lethal phenotype are difficult to study by this approach. Transient gene silencing, in which a range of gene expression knockdown levels are observed, may be better suited to the study of these "core biology" genes. The obvious disadvantage of transient silencing is that the genotype and phenotype apparently does not persist for more than 20 days, and thus experiments require repetition on each occasion that the gene under study is to be characterized phenotypically. However, the recent reporting of stable but partial gene silencing may alleviate this difficulty (Ah Fong et al., 2008). Stable gene silencing, whether partial or complete, relies on transformation of oomycetes. Although transformation is routine for model oomycete species

P. infestans and *P. sojae*, this is not the situation for many other oomycetes such as *Aphanomyces* sp., and *Saprolegnia* sp., which are genera that are becoming studied more intensively at the molecular genetic level. Notwithstanding, stable transformation of the two model species remains time consuming and of relatively low efficiency and throughput. Time taken for a single transformation experiment (although multiple genes may be transformed singly in parallel) can be up to 6 weeks before all transformants are recovered and sufficiently cultured for genotyping and phenotyping to commence. By comparison, it is possible to screen fully (gene expression and phenotype) up to 15 genes in parallel within this 6-week time period by transient gene silencing (Avrova, personal communication). Multiple genes may also be assayed simultaneously by transient silencing, which allows functional redundancy or compensation to be addressed. Although this may be possible by stable silencing, it would require either cotransformation or multiple cloning steps to insert the multiple hairpin constructs into a single-oomycete transformation vector. In summary, transient gene silencing is a useful screening tool for identifying "extreme or obvious" phenotypes, although more subtle phenotypes may not be identified. By comparison, stable and complete transcriptional silencing seems ideally suited to detailed studies of phenotypes associated with silencing of individual genes.

25.8 CONCLUSIONS AND SCOPE FOR FUTURE WORK

Despite the identification and adoption of gene silencing for elucidation of gene function in oomycetes, relatively few genes have been characterized by this strategy, and the mechanisms that lead to gene silencing are relatively unstudied. Significant technical challenges and opportunities remain in optimization of silencing procedures, especially in the utility of gene silencing to assay gene function in biotrophic oomycetes.

It is likely that PTGS and TGS are both functional in oomycetes, and that this leads to or is characterized by the formation of heterochromatin via histone modifications. However, the genes and their encoded proteins that lead to the establishment of gene silencing in oomycetes are relatively unstudied. For instance, *P. infestans* contains genes that encode a dicer-like protein, and multiple argonaute, chromodomain, putative RNA helicase, and histone deacetylase proteins. Are all of these involved in establishing gene silencing or are specific family members specialized to other genome tasks, such as suppression of transposon activity? The sequenced oomycete genomes, and the *P. infestans* genome particularly, contain large numbers of transposable elements. Oomycetes seem to use their gene-silencing machinery to control these repetitive elements, because siRNAs specific to some classes of transposons have been identified (Vetukuri et al., 2008). A broader view of the small RNA populations in oomycetes may be addressed through sequencing of small RNAs by new generation nucleic acid sequencing technologies (Ghildiyal et al., 2008).

The latter strategies would also reveal any microRNAs that may be encoded in the sequenced oomycete genomes.

Preparation of hairpin transformation vectors for establishment of stable TGS can be time consuming. A useful improvement would be the development of new vectors that exploit some innate silencing components and associated pathways. For example, in *P. infestans*, incorporation as a transcriptional fusion of a section of transposon for which siRNAs are known to exist in the cytosol or nucleus may silence the fused transgene and its endogenous copy through the activity of RdRP. Transposon sequences have been found to enhance silencing of closely neighboring genes in *A. thaliana* and *E. histolytica* (Mirelman et al., 2006; Fujimoto et al., 2008). The limitation of the application of this strategy to many oomycetes would be the restriction of the developed vector to the species and strain that contains the transposon-specific siRNAs.

An additional pathway that has been linked to gene silencing processes is that of nonsense mRNA decay (NMD) and aberrant mRNA decay. This process has been demonstrated to require the action of RNA helicases and RdRP, both of which oomycetes possess. In other organisms, NMD targets RNAs encoding premature stop codons for decay via decapping, and dsRNA formation by RdRP, dicer, and argonaute degradation (Gazzani et al., 2004; Arciga-Reyes et al., 2006; Isken and Maquat, 2007). This may be exploited in oomycetes by construction of vectors that express nonsense transcripts homologous to endogenous genes to initiate silencing. Alternatively, it has been shown that specific sequence motifs or structures in the 3′ UTR of mRNAs can inhibit translation and lead to mRNA readthrough (nonpolyadenylation) as well as subsequent action of RdRP to form dsRNA that is degraded by dicer and argonaute action (Furth and Baker, 1991; Furth et al., 1994; Fortes et al., 2003; Bühler et al., 2007; Luo and Chen, 2007). Again, such processes could be exploited by incorporation of these sequence motifs into the 3′ UTR of oomycete transformation vectors to target the introduced aberrant transcript and endogenous gene under study.

An advantage to the development of these types of transformation vectors to initiate silencing is that these would require only a single cloning step for construction and use in oomycetes, which alleviates the necessity for hairpin constructs to initiate silencing. Furthermore, this opens up the possibility that multiple genes may be silenced by concatenation of multiple, sequence-unrelated genes or fragments of genes in a single transformation vector.

The plant pathogenic oomycetes in which functional characterization of genes have been carried out have been those species that are culturable in the laboratory. That is, biotrophic species such as *H. arabidopsidis* and *Albugo candida* are obligate biotrophs, cannot be cultured in the laboratory, and no stable transformation system exists yet. The biotrophic oomycete species and some hemibiotrophic species such as *P. infestans* and *P. sojae*, form haustoria in infected plant cells. The intracellular haustoria are formed from the intercellular hyphae by invagination of the host cell membrane, and so are in intimate contact with host cells. It has been shown that pathogen effector proteins can cross this barrier into host cells (Whisson et al., 2007), but it is presently unknown whether any host molecules cross into the pathogen haustorium. If host molecules do cross into the haustorium, then it is possible that plant-derived small RNAs may also be transferred by oomycetes in this manner, which suggests the possibility of silencing oomycete genes from the host plant. This could be tested by expression of a silencing construct that contains an oomycete gene in a host plant, followed by infection with an oomycete. If successful, such an experiment would permit functional genomic studies of the biotrophic oomycete species in a relevant biological context.

The sequencing of oomycete genomes (Tyler et al., 2006), discovery of many novel genes, and their expression profiles via microarrays (Randall et al., 2005; Torto-Alalibo et al., 2007; Judelson et al., 2008) is necessitating the development and adoption of many different functional genomic assays for determining oomycete gene function. Most revealing in understanding oomycete gene and encoded protein function will be assays such as those afforded by gene silencing that are performed in the relevant biological context in the source organisms.

ACKNOWLEDGMENTS

The authors are grateful to those colleagues in the oomycete research community who willingly shared their unpublished data with us to assist in writing this chapter. This work was supported by the Scottish Government Rural and Environment Research and Analysis Directorate (A.O.A. and S.C.W.), the Biotechnology and Biological Sciences Research Council (L.J.G.B.), and The Royal Society (P.vW.).

REFERENCES

Ah-Fong AMV, Bormann-Chung CA, Judelson HS 2008. Optimization of transgene-mediated silencing in *Phytophthora infestans* and its association with small interfering RNAs. Fungal Genet and Biol 45:1197–1205.

Ah Fong AM, Judelson HS 2003. Cell cycle regulator Cdc14 is expressed during sporulation but not hyphal growth in the fungus-like oomycete *Phytophthora infestans*. Molec Microbiol 50:487–494.

Alder MN, Dames S, Gaudet J, Mango SE 2003. Gene silencing in *Caenorhabditis elegans* by transitive RNA interference. RNA 9:25–32.

Arciga-Reyes L, Wootton L, Kieffer M, Davies B 2006. UPF1 is required for nonsense-mediated mRNA decay (NMD) and RNAi in Arabidopsis. Plant J 47:480–489.

Armstrong MR, Whisson SC, Pritchard L, Bos JI, Venter E, Avrova AO, Rehmany AP, Bohme U, Brooks K, Cherevach I, et al. 2005. An ancestral oomycete locus contains late blight avirulence gene *Avr3a*, encoding a protein that is recognized in the host cytoplasm. Proc Nat Acad Sci, USA 102:7766–7771.

Avrova AO, Boevink PC, Young V, Grenville-Briggs LJ, van West P, Birch PR, Whisson SC 2008. A novel *Phytophthora infestans* haustorium-specific membrane protein is required for infection of potato. Cell Microbiol. 10:2271–2284.

Baulcombe D 2004. RNA silencing in plants. Nature 431:356–363.

Blanco FA, Judelson HS 2005. A bZIP transcription factor from *Phytophthora* interacts with a protein kinase and is required for zoospore motility and plant infection. Molec Microbiol 56:638–648.

Brasier C, Kirk S 2004. Production of gametangia by *Phytophthora ramorum in vitro*. Mycol Res 108:823–827.

Brodersen P, Sakvarelidze-Achard L, Bruun-Rasmussen M, Dunoyer P, Yamamoto YY, Sieburth L, Voinnet O 2008. Widespread translational inhibition by plant miRNAs and siRNAs. Science 320:1185–1190.

Bühler M, Haas W, Gygi SP, Moazed D 2007. RNAi-dependent and -independent RNA turnover mechanisms contribute to heterochromatic gene silencing. Cell 129:707–721.

Cerutti H, Casas-Mollano JA 2006. On the origin and functions of RNA-mediated silencing: from protists to man. Curr Genet 50:81–99.

Clemens JC, Worby CA, Simonson-Leff N, Muda M, Maehama T, Hemmings BA, Dixon JE 2000. Use of double-stranded RNA interference in *Drosophila* cell lines to dissect signal transduction pathways. Proc Nat Acad Sci USA 97:6499–6503.

Collins RE, Cheng X 2005. Structural domains in RNAi. FEBS Lett 579:5841–5849.

Cvitanich C, Judelson HS 2003. Stable transformation of the oomycete, *Phytophthora infestans*, using microprojectile bombardment. Curr Genet 42:228–235.

Fire A, Xu S, Montgomery MK, Kostas SA, Driver SE, Mello CC 1998. Potent and specific genetic interference by double-stranded RNA in *Caenorhabditis elegans*. Nature 391:806–811.

Fortes P, Cuevas Y, Guan F, Liu P, Pentlicky S, Jung SP, Martínez-Chantar ML, Prieto J, Rowe D, Gunderson SI 2003. Inhibiting expression of specific genes in mammalian cells with 5′ end-mutated U1 small nuclear RNAs targeted to terminal exons of pre-mRNA. Proc Nat Acad Sci USA 100:8264–8269.

Fujimoto R, Kinoshita Y, Kawabe A, Kinoshita T, Takashima K, Nordborg M, Nasrallah ME, Shimizu KK, Kudoh H, Kakutani T 2008. Evolution and control of imprinted FWA genes in the genus *Arabidopsis*. PLoS Genetics 4:e1000048.

Furth PA, Baker CC 1991. An element in the bovine papillomavirus late 3′ untranslated region reduces polyadenylated cytoplasmic RNA levels. J Viro 65:5806–5812.

Furth PA, Choe WT, Rex JH, Byrne JC, Baker CC 1994. Sequences homologous to 5′ splice sites are required for the inhibitory activity of papillomavirus late 3′ untranslated regions. Molec Cell Biol 14:5278–5289.

Gaulin E, Haget N, Khatib M, Herbert C, Rickauer M, Bottin A 2007. Transgenic sequences are frequently lost in *Phytophthora parasitica* transformants without reversion of the transgene-induced silenced state. Can J Microbiol 53:152–157.

Gaulin E, Jauneau A, Villalba F, Rickauer M, Esquerré-Tugayé MT, Bottin A 2002. The CBEL glycoprotein of *Phytophthora parasitica* var *nicotianae* is involved in cell wall deposition and adhesion to cellulosic substrates. J Cell Sci 115: 4565–4575.

Gazzani S, Lawrenson T, Woodward C, Headon D, Sablowski R 2004. A link between mRNA turnover and RNA interference in *Arabidopsis*. Science 306:1046–1048.

Ghildiyal M, Seitz H, Horwich MD, Li C, Du T, Lee S, Xu J, Kittler EL, Zapp ML, Weng Z, et al. 2008. Endogenous siRNAs derived from transposons and mRNAs in *Drosophila* somatic cells. Science 320:1077–1081.

Grenville-Briggs LJ, Anderson VL, Fugelstad J, Avrova AO, Bouzenzana J, Williams A, Wawra S, Whisson SC, Birch PR, Bulone V, van West P 2008. Cellulose synthesis in *Phytophthora infestans* is required for normal appressorium formation and successful infection of potato. Plant Cell 20:720–738.

Grewal SI, Jia S 2007. Heterochromatin revisited. Nature Revi Genet 8:35–46.

Guo HS, Fei JF, Xie Q, Chua NH 2003. A chemical-regulated inducible RNAi system in plants. Plant J 34:383–392.

Huettel B, Kanno T, Daxinger L, Bucher E, van der Winden J, Matzke AJ, Matzke M 2007. RNA-directed DNA methylation mediated by DRD1 and Pol IVb: a versatile pathway for transcriptional gene silencing in plants. Biochim Biophys Acta 1769:358–374.

Iida T, Kawaguchi R, Nakayama J 2006. Conserved ribonuclease, Eri1, negatively regulates heterochromatin assembly in fission yeast. Curr Biol 16:1459–1464.

Isken O, Maquat LE 2007. Quality control of eukaryotic mRNA: safeguarding cells from abnormal mRNA function. Genes Develop 21:1833–1856.

Jiang RH, Tyler BM, Govers F 2006. Comparative analysis of *Phytophthora* genes encoding secreted proteins reveals conserved synteny and lineage-specific gene duplications and deletions. Molec Plant-Microbe Interact 19:1311–1321.

Jensen S, Gassama MP, Heidmann T 1999. Taming of transposable elements by homology-dependent gene silencing. Nature Genet 21:209–212.

Judelson HS, Ah-Fong AM, Aux G, Avrova AO, Bruce C, Cakir C, da Cunha L, Grenville-Briggs L, Latijnhouwers M, Ligterink W, et al. 2008. Gene expression profiling during asexual development of the late blight pathogen *Phytophthora infestans* reveals a highly dynamic transcriptome. Molec Plant-Microbe Interact 21:433–447.

Judelson HS, Coffey MD, Arredondo FR, Tyler BM 1993a. Transformation of the oomycete pathogen *Phytophthora megasperma* f. sp. *glycinea* occurs by DNA integration into single or multiple chromosomes. Curr Genet 23:211–218.

Judelson HS, Dudler R, Pieterse CM, Unkles SE, Michelmore RW 1993b. Expression and antisense inhibition of transgenes in *Phytophthora infestans* is modulated by choice of promoter and position effects. Gene 133:63–69.

Judelson HS, Tani S 2007. Transgene-induced silencing of the zoosporogenesis-specific NIFC gene cluster of *Phytophthora infestans* involves chromatin alterations. Eukaryo Cell 6:1200–1209.

Judelson HS, Tyler BM, Michelmore RW 1991. Transformation of the oomycete pathogen, *Phytophthora infestans*. Molec Plant-Microbe Interac 4:602–607.

Judelson HS, Tyler BM, Michelmore RW 1992. Regulatory sequences for expressing genes in oomycete fungi. Molec Gen Genet 234:138–146.

Kamoun S, van West P, Vleeshouwers VGAA, de Groot KE, Govers F 1998. Resistance of *Nicotiana benthamiana* to *Phytophthora infestans* is mediated by the recognition of the elicitor protein INF1. Plant Cell 10:1413–1426.

Kennedy S, Wang D, Ruvkun G 2004. A conserved siRNA-degrading RNase negatively regulates RNA interference in *C. elegans*. Nature 427:645–649.

Kim HS, Choi ES, Shin JA, Jang YK, Park SD 2004. Regulation of Swi6/HP1-dependent heterochromatin assembly by cooperation of components of the mitogen-activated protein kinase pathway and a histone deacetylase Clr6. J Biol Chemi 279:42850–42859.

Knapova G, Schlenzig A, Gisi U 2002. Crosses between isolates of *Phytophthora infestans* from potato and tomato and characterization of F-1 and F-2 progeny for phenotypic and molecular markers. Plant Pathol 51:698–709.

Latijnhouwers M, Govers F 2003. A *Phytophthora infestans* G-protein beta subunit is involved in sporangium formation. Eukaryo Cell 2:971–977.

Latijnhouwers M, Ligterink W, Vleeshouwers VG, van West P, Govers F 2004. A G-alpha subunit controls zoospore motility and virulence in the potato late blight pathogen *Phytophthora infestans*. Molec Microbiol 51:925–936.

Luo Z, Chen Z 2007. Improperly terminated, unpolyadenylated mRNA of sense transgenes is targeted by RDR6-mediated RNA silencing in Arabidopsis. Plant Cell 19:943–958.

Makeyev EV, Bamford DH 2002. Cellular RNA-dependent RNA polymerase involved in posttranscriptional gene silencing has two distinct activity modes. Molec Cell 10:1417–1427.

Mcleod A, Fry BA, Zuluaga AP, Myers KL, Fry WE 2008. Toward improvements of oomycete transformation protocols. J Eukaryo Microbiol 55:103–109.

Meister G, Tuschl T 2004. Mechanisms of gene silencing by double-stranded RNA. Nature 431:343–349.

Mirelman D, Anbar M, Nuchamowitz Y, Bracha R 2006. Epigenetic silencing of gene expression in *Entamoeba histolytica*. Archi Med Res 37:226–233.

Napoli C, Lemieux C, Jorgensen R 1990. Introduction of a chimeric chalcone synthase gene into petunia results in reversible co-suppression of homologous genes in trans. Plant Cell 2:279–289.

Randall TA, Dwyer RA, Huitema E, Beyer K, Cvitanich C, Kelkar H, Fong AM, Gates K, Roberts S, Yatzkan E, et al. 2005. Large-scale gene discovery in the oomycete *Phytophthora infestans* reveals likely components of phytopathogenicity shared with true fungi. Molec Plant-Microbe Interact 18:229–243.

Romano N, Macino G 1992. Quelling: transient inactivation of gene expression in *Neurospora crassa* by transformation with homologous sequences. Molec Microbiol 6:3343–3353.

Ruiz MT, Voinnet O, Baulcombe DC 1998. Initiation and maintenance of virus-induced gene silencing. Plant Cell 10:937–946.

Sijen T, Fleenor J, Simmer F, Thijssen KL, Parrish S, Timmons L, Plasterk RH, Fire A 2001. On the role of RNA amplification in dsRNA-triggered gene silencing. Cell 107:465–476.

Tabara H, Sarkissian M, Kelly WG, Fleenor J, Grishok A, Timmons L, Fire A, Mello CC 1999. The *rde-1* gene, RNA interference, and transposon silencing in *C. elegans*. Cell 99:123–132.

Timmons L, Fire A 1998. Specific interference by ingested dsRNA. Nature 395:854.

Torto-Alalibo TA, Tripathy S, Smith BM, Arredondo FD, Zhou L, Li H, Chibucos MC, Qutob D, Gijzen M, Mao C, et al. 2007. Expressed sequence tags from *Phytophthora sojae* reveal genes specific to development and infection. Molec Plant-Microbe Interact 20:781–793.

Tyler BM, Tripathy S, Zhang X, Dehal P, Jiang, RH, Aerts A, Arredondo FD, Baxter L, Bensasson D, Beynon JL, et al. 2006. *Phytophthora* genome sequences uncover evolutionary origins and mechanisms of pathogenesis. Science 313:1261–1266.

Van West P, Kamoun S, Van 't Klooster JW, Govers F 1999. Internuclear gene silencing in *Phytophthora infestans*. Molec Cell 3:339–348.

Van West P, Shepherd SJ, Walker CA, Li S, Appiah AA, Grenville-Briggs LJ, Govers F, Gow NA 2008. Internuclear gene silencing in *Phytophthora infestans* is established through chromatin remodelling. Microbiol 154:1482–1490.

Vastenhouw NL, Brunschwig K, Okihara KL, Müller F, Tijsterman M, Plasterk RH 2006. Gene expression: long-term gene silencing by RNAi. Nature 442:882.

Vastenhouw NL, Fischer SE, Robert VJ, Thijssen KL, Fraser AG, Kamath RS, Ahringer J, Plasterk RH 2003. A genome-wide screen identifies 27 genes involved in transposon silencing in *C. elegans*. Curr Biol 13:1311–1316.

Verdel A, Jia S, Gerber S, Sugiyama T, Gygi S, Grewal SI, Moazed D 2004. RNAi-mediated targeting of heterochromatin by the RITS complex. Science 303:672–676.

Verdel A, Moazed D 2005. RNAi-directed assembly of heterochromatin in fission yeast. FEBS Lett 579:5872–5878.

Vetukuri R, Dixelius C, Savenkov E 2008. Evidence of siRNAs specific to three classes of retrotransposon elements in *Phytophthora infestans*. 9th European Conference of Fungal Genetics, University of Edinburgh, United Kingdom.

Vijn I, Govers F 2003. *Agrobacterium tumefaciens* mediated transformation of the oomycete plant pathogen *Phytophthora infestans*. Molec Plant Pathol 4:459–467.

Walker CA, Köppe M, Grenville-Briggs LJ, Avrova AO, Horner NR, McKinnon AD, Whisson SC, Birch PR, van West P 2008. A DEAD-box RNA helicase is required for normal zoospore development in the potato late blight pathogen *Phytophthora infestans*. Fungal Genet Biol 45:954–962.

Weiland JJ 2003. Transformation of *Pythium aphanidermatum* to geneticin resistance. Curr Genet 42:344–352.

Wingard SA 1928. Hosts and symptoms of ring spot, a virus disease of plants. J. Agricul Res 37:127–153.

Whisson SC, Avrova A, van West P, Jones JT 2005. A method for double-stranded RNA mediated transient gene silencing in *Phytophthora infestans*. Molec Plant Pathol 6:153–163.

Whisson SC, Basnayake S, Maclean DJ, Irwin JAG, Drenth A 2004. *Phytophthora sojae* avirulence genes *Avr4* and *Avr6* are located in a 24 KB, recombination-rich region of genomic DNA. Fungal Genet Biol 41:62–74.

Whisson SC, Boevink PC, Moleleki L, Avrova AO, Morales JG, Gilroy EM, Armstrong MR, Grouffaud S, van West P, Chapman S, et al. 2007. A translocation signal for delivery of oomycete effector proteins into host plant cells. Nature 450:115–118.

Yang SJ, Carter SA, Cole AB, Cheng NH, Nelson RS 2004. A natural variant of a host RNA-dependent RNA polymerase is associated with increased susceptibility to viruses by *Nicotiana benthamiana*. Proc Nat Acad Sci USA 101:6297–6302.

Zhang H, Kolb FA, Jaskiewicz L, Westhof E, Filipowicz W 2004. Single processing center models for human Dicer and bacterial RNase III. Cell 118:57–68.

26

GLOBAL PROTEOMICS AND *PHYTOPHTHORA*

Alon Savidor
Tel Aviv University, Tel Aviv, Israel

26.1 INTRODUCTION

Proteins complete most processes in living cells. A typical cell contains many different classes of proteins, each mediating different functions in the cell. For example, enzymes catalyze chemical reactions, structural proteins provide mechanical support to the cell, and chaperones assist with protein folding. Characterizing the existence and abundance of specific proteins in the cell provides useful information about the molecular processes that occur in the cell. Until recently, it was difficult to investigate protein dynamics on a large scale, and much of our knowledge concerning the presence/abundance of proteins was derived from experiments aimed at characterizing messenger RNA (mRNA) levels and not the actual proteins present in a biological sample. Since the 1990s, the field of transcriptomics has developed dramatically with high-throughput techniques, such as microarrays (Schena et al., 1995), SAGE (Velculescu et al., 1995), and most recently, the next-generation sequencing of cDNA (Margulies et al., 2005; Wold and Myers, 2008), becoming popular and widely used. These experiments measure the abundance of mRNA transcripts in the cell, which reflects the upregulation or downregulation of certain genes under different circumstances. Although these experiments provide information about transcription dynamics of genes, they provide only indirect evidence for the corresponding protein abundance. The abundance of protein in the cell is a function of how much mRNA transcript is present and how quickly it is translated, as well as how quickly the protein is degraded or postranslationally regulated. This explains the far-from-perfect correlation

Oomycete Genetics and Genomics: Diversity, Interactions, and Research Tools
Edited by Kurt Lamour and Sophien Kamoun
Copyright © 2009 John Wiley & Sons, Inc.

between transcript level and protein abundance (Conrads et al., 2005; Cox et al., 2005; Gygi et al., 1999; Mootha et al., 2003; Nie et al., 2007). To measure protein levels directly, it is best to explore the proteome itself.

The "proteome" refers to the entire set of proteins in a given sample such as a cell culture or tissue. The objective of the field of proteomics is to measure as completely as possible the proteome of the subject of study. As opposed to the reductionist approach that examines one or few genes/proteins at a time, global proteomics attempts to measure the entire set of proteins expressed in the cell at a given time. A single global proteomics experiment can yield data on thousands of proteins. The data obtained from such experiments can be both qualitative and quantitative. In other words, in addition to revealing the identity of proteins present in the sample, these experiments can measure the relative abundance of specific proteins between different treatments, life stages, and time points of infection. Thus, information regarding entire cellular processes, pathways, and even their interdependency can be obtained by proteomic approaches.

However, accurately measuring the proteome of an organism is no easy task. The complexity of even the simplest of bacteria is enormous; it contains thousands of different species of proteins that range widely in abundance, structure, and chemical properties. In the eukaryotic *Phytophthora*, this complexity is compounded by the large number of protein coding genes ($\sim$15–20,000 in the currently sequenced species) (Tyler et al., 2006), complex gene structure with introns and alternative splicing, posttranslational modification of proteins, and other factors. In recent years, technological developments and innovations in different proteomic tools provided the ability to deal with such complexity and enabled large-scale proteomic studies of eukaryotic organisms. In addition to the improvement in the volume, quality, and accuracy of the proteomic data, advances have also been made by automating large parts of the process. Thus, in experiments like multidimensional protein identification technology (MudPIT) (discussed below), data can be continuously acquired for 24 h on thousands of proteins with essentially no intervention of the operator, as the entire experiment is controlled computationally.

26.2 MASS SPECTROMETRY

Over the last two decades, mass spectrometry has become the main tool for proteomic investigation. In mass spectrometry, the analyte molecules (proteins and peptides in the case of proteome analysis) are ionized, enter the gas phase (get separated from their liquid solvents or their solid matrix), and analyzed by the mass analyzer in the mass spectrometer. The mass spectrometer measures the mass-to-charge ratio of the analyte ions [reviewed in Yates, (1998)]. Different ionization techniques are available for delivery of gas-phase ions into the mass spectrometer. For analysis of proteins, two of the most popular ionization techniques are electrospray ionizaion (ESI) (Fenn et al., 1989) and matrix-assisted laser desorption ionization (MALDI) (Hillenkamp et al., 1991; Karas et al., 1987). After ionization, the ions can be analyzed by different analyzers, including time-of-flight (TOF) mass analyzer, ion traps, or triple-quad mass analyzer (QqQ). Depending on the type of experiment performed and the analyzer used, mass spectrometry (MS) can be used to measure intact proteins or peptides, or to fragment those molecules and measure their fragmentation products in a process called "tandem mass spectrometry" (MS/MS). The resulting mass spectra are then typically searched against a known protein database to reveal the identity of the proteins or peptides which correspond to those spectra. Thus, large-scale proteomic studies are typically limited to organisms whose genome is sequenced and a comprehensive protein database is available.

Protein identification by mass spectrometry is typically preceded by protein or peptide separation and followed by analysis of the mass spectra. Significant improvement in all parts of the process has been achieved in recent years, which allows an in-depth proteomic examination of complex organisms including the Oomycetes.

26.3 TWO-DIMENSIONAL POLYACRYLAMIDE GEL ELECTROPHORESIS (2D-PAGE) AND MASS SPECTROMETRY

For a quality measurement of individual proteins, it is crucial to efficiently separate the proteins or their peptides prior to ionization and analysis by the mass spectrometer. Because of their complexity, an analysis of proteomes requires separation in more than one dimension. The first approach that allowed analysis of an entire proteome was 2D-PAGE in the mid-1970s (O'Farrell, 1975). In 2D-PAGE, proteins are separated based on their isoelectric point in the first dimension and then based on their molecular weight in the second dimension. Hundreds and even thousands of proteins can be separated and visualized on a single gel. Gel spots can then be excised, treated, and analyzed to reveal the identity of the protein in the spot.

2D-PAGE is particularly useful for comparative analysis of the proteomes of an organism under different conditions, which includes different treatments and different life stages. Studies that employ such a comparative strategy using 2D-PAGE followed by mass spectrometry analysis have been carried out on several Oomycetes. For *Phytophthora palmivora*, 2D-PAGE was applied to the proteomes of the mycelium, sporangium, zoospore, cyst, and germinating cyst stages of the asexual life cycle (Shepherd et al., 2003). The resolved gels suggested that 1% of the proteins were specific for each of these life stages.

Similar studies have been performed with *P. infestans* (Ebstrup et al., 2005; Grenville-Briggs et al., 2005; Kramer et al., 1997). Stage-specific spots on the gels were excised and digested, and the resulting peptides were measured using ESI-TOF or MALDI-TOF. Many life-stage-specific proteins that were identified in these studies were involved in essential processes such as protein synthesis, amino acid metabolism, and energy metabolism. Specifically, the life-stage-specific proteins suggested the importance of *de novo* protein synthesis (Ebstrup et al., 2005) and elevated amino acid synthesis (Grenville-Briggs et al., 2005) in the appresoria-forming germinating cyst of *P. infestans*. In addition, other stage-specific proteins were identified that may explain the unique properties of each life stage and contribute to the understanding of infection progression. For example, a protein specific to the cyst and germinating cyst life stages was identified as a crinkling and necrosis inducing protein (CRN2) (Ebstrup et al., 2005). CRNs are secreted proteins that are unique to *Phytophthora* and are thought to be important for interaction with the plant host (Torto et al., 2003). Their temporal expression in the cyst and germinating cyst suggests that they play a role in early infection.

2D-PAGE can also be employed without the subsequent identification of the spots using mass spectrometry for qualitative and comparative analysis of proteomes. Although the protein in the gel spot is not directly identified, the resolved 2D-PAGE gel provides information about the protein (molecular weight and pI) that can direct the investigator toward the identity of the protein. The suspected identity can be confirmed using other means such as immunoblotting. In addition, when used in a comparative manner, resolved gels can give a qualitative sense of the differences between proteomes. One study that employed such a strategy was done on *Phytophthora nicotianae*. When compared with each other, resolved 2D-PAGE gels of microsomal fractions of *P. nicotianae* zoospores and cysts revealed 14 zoospore-specific proteins and 39 cyst-specific proteins (Mitchell et al., 2002). The number of stage-specific proteins was remarkable considering the developmental stages were separated by only 5 min, which underscores the rapidity of zoospore encystment.

Another study that employed 2D-PAGE as a "standalone" comparative tool used parallel sporangial protein preparations of *P. palmivora* and *P. infestans* (Shepherd et al., 2003). The resolved gels showed similar complexity with 30% of the spots positioned in a similar or identical location, which suggests the existence of many orthologous proteins between the different species.

Although comparative proteomic studies provide information about treatment or life-stage-specific proteins, proteomic studies need not necessarily be comparative nor include whole proteomes. They can be tailored for specific applications and focus purely on identification of all proteins from a proteome or a fraction of it. For example, a total of nine proteins were identified using 2D-PAGE followed by MALDI-TOF in a screen for *P. infestans* extracellular proteins, which validated predictions of an algorithm predicting such proteins (Torto et al., 2003).

26.4 LIQUID CHROMATOGRAPHY AND MASS SPECTROMETRY

Although 2D-PAGE has been widely used, it has several drawbacks. Although many spots can be visualized on a gel, only a fraction of them can be identified using mass spectrometry or other techniques because of sample loss during treatment of the excised gel spot and comigration of different proteins. Comigration complicates both the interpretation of the gels and also interferes with protein quantification. Also, because of the dimensions of separations, 2D-PAGE is biased against proteins with extreme pH, molecular weight, or low abundance. As a result, no more than tens of proteins are typically identified in a single 2D-PAGE experiment. For comparative analysis, the reproducibility of 2D-PAGE is a challenge and posttranslational modification can shift the position of a protein on the gel. Finally, excision, treatment, and analysis of each individual gel spot are time consuming and resource demanding.

Another milestone improvement in the analysis of whole proteomes was the development of two dimensional liquid chromatography (LC/LC) that could be directly interfaced with a mass spectrometer (LC/LC-MS or LC/LC-MS/MS). Multidimensional protein identification technology (MudPIT), which was published in 2001 (Washburn, 2001), is one strategy that employs automated LC/LC-MS/MS and is becoming an increasingly popular tool for proteomic investigation. MudPIT falls under the category of "shotgun" or "bottom-up" proteomics. Instead of separating intact proteins as in 2D-PAGE, the shotgun proteomic approach relies on measurement of peptides generated by protein digestion with proteases or chemicals. Identification of proteins is done subsequently based on the identified peptides. A typical proteome contains thousands of different proteins. The digestion of a proteome results in a peptide mixture that is at least an order of magnitude more complex than the original protein mixture. Although it may seem counterintuitive to make a complex mixture even more complex, peptides are much easier to separate by LC/LC and analyze using mass spectrometry than intact proteins. A MudPIT experiment usually starts with digestion of a proteome with a protease, typically trypsin, followed by an offline loading of the resulting peptide mixture onto a biphasic fused-silica column that contains a reverse phase (RP) resin followed by a strong cation exchange (SCX) resin. After sample loading, the column is placed on the back end of a front column (a column placed directly in front of the mass spectrometer) packed with RP, which is online with the mass spectrometer. Separation of the peptides occurs on the SCX in the first dimension and the front column's RP in the second dimension. The additional RP on the back column is used for cleaning and desalting purposes only. As peptides elute off the front column, they enter directly into the mass spectrometer. Then they are subjected to tandem mass spectrometry where they are isolated and fragmented, and the resulting peptide fragments are measured. Finally, the accumulated spectra of fragmented peptides are searched against a defined protein database to identify their corresponding peptides, and therefore the proteins those peptides originated from. In the best case scenario, the protein database to be searched against contains a complete and accurate list of all the proteins (and peptides) coded for in a sequenced genome. Clearly, the accuracy of the genome assembly and annotation are critical to the success of a MudPIT-based proteomic investigation, and these types of investigations are typically limited to organisms with sequenced genomes.

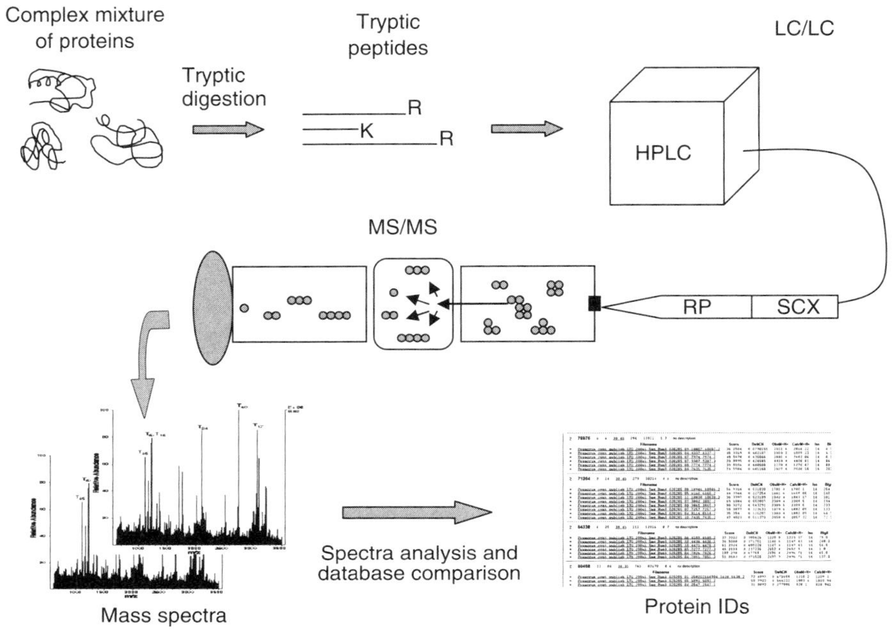

FIG. 26.1 Multidimensional protein identification technology (MudPIT). The objective of MudPIT is to sample and identify as many proteins as possible from the original sample. The starting protein mixture is digested with trypsin, which cleaves the proteins into peptides at the C-terminus of cysteine and arginine residues. The resulting peptide mixture is then separated in two dimensions by high-performance liquid chromatography (HPLC). The eluting peptides directly enter the mass spectrometer, where they are isolated and fragmented. The peptide fragments are then measured, resulting in spectra that are then computationally searched against a given protein database to reveal the identity of the peptides giving rise to them. Identified peptides lead to identification of the proteins they originated from.

A diagram of the entire MudPIT process is shown in Fig. 26.1. The use of MudPIT has several advantages over 2D-PAGE, which includes better sensitivity and dynamic range, less bias, higher throughput, and better protein-quantification capabilities.

26.5 MudPIT AND *PHYTOPHTHORA*

MudPIT experiments on different life stages of *Phytophthora sojae*, *Phytophthora ramorum*, *Phytophthora infestans*, and *Phytophthora capsici* have resulted in identification of thousands of proteins in each experiment (Savidor et al., 2008; Savidor, unpublished data). A comparative study of the germinating cyst and the mycelium of *P. sojae*, *P. ramorum*, and *P. capsici* revealed numerous proteins that were either life stage specific or organism specific, as well as many commonly expressed proteins. The germinating cyst is the asexual life stage that represents early infection, whereas the mycelium is the representative life stage of the latter part of the infection process. Thus, identification of life-stage-specific proteins by these proteomic experiments suggested different molecular strategies for infection initiation and infection establishment. For example, an increase in abundance of almost all enzymes in the β-oxidation pathway in the germinating cyst was observed in *P. sojae*, *P. ramorum*, and *P. capsici*. Because β-oxidation is the pathway of lipid catabolism for energy production, this increase in abundance supports the hypothesis that in early infection, the germinating cyst derives its energy from the catabolism of internal lipid reserves, as it is devoid of an external energy source when it is outside of the plant tissue at that point of infection. The energy that is produced via the β-oxidation pathway is used to derive extensive protein synthesis, which includes substantial synthesis of cytoskeletal proteins, as evident from the classes of proteins that are upregulated at that life stage. However, the increased abundance of multiple glycolytic enzymes and sugar transporters in the mycelia of these organisms suggested that when found in a nutrient-rich environment, the mycelium uses external energy sources and transports sugars (presumably host sugars during infection) that are then catabolized via glycolysis. The classes of proteins that are upregulated in the mycelium suggests that glycolysis-derived energy is used for synthesis of small molecules that include amino acids, cofactors, and secondary metabolites.

26.5.1 *In-Planta* Proteomic Studies of *Phytophthora*

Although several transcriptomic studies have been completed on Oomycete-infected plants (Avrova et al., 2003; Grenville-Briggs et al., 2005; Le Berre et al., 2008; Moy et al., 2004; Pieterse et al., 1993; Qutob et al., 2002; Torto-Alalibo et al., 2007), there is currently no published data on the proteomes of Oomycetes and their hosts during the infection process. Although an understanding of the cellular processes that occur in the pathogen *in planta* is most desirable, the proteomic experiments that involve both the host and the pathogen face several challenges.

The first challenge is the requirement of having sequenced and annotated genomes of both the pathogen and plant host. Until recently, complete genome sequences from *Phytophthora* species or their hosts were not available. In 2004, the genomes of *P. sojae* and *P. ramorum* became available (Tyler et al., 2006). However, the genomes of their hosts (soybean for *P. sojae*, and a variety of trees and shrubs for *P. ramorum*) were not available. In 2007 and 2008, sequencing of the genomes of *P. capsici* and *P. infestans* were completed, although the assembly and annotation of these genomes is still a work in progress. Both of these *Phytophthora* species are natural pathogens of tomatoes (*Lycopersicon* species). Currently, the tomato sequencing project is underway via an international initiative known as the "International Solanaceae Genome Project" (SOL) (http://www.sgn.cornell.edu/index.pl) (Shibata, 2005).

Although sequencing of the tomato genome is incomplete, a large portion of the gene-dense euchromatin regions has been sequenced. Based on a homology similarity to *Arabidopsis thaliana*, ~34,000 tomato genes are predicted in the available sequence. Recently, the Oomycete *Hyaloperonospora parasitica*, which is a natural pathogen of *Arabidopsis thaliana* (one of two completely sequenced plants), has been sequenced. This provides the unique opportunity for a large-scale proteomic investigation of this phyto-pathogenic system *in planta*.

A second challenge facing *in planta* investigations is simply that the overall mass of the microbial pathogen is only a small fraction of the total infected tissue. This is especially true in early infection, in which the pathogen has not grown significantly and is localized to small areas. Whereas some of the most important cellular processes and chemical warfare between the pathogen and the host are thought to occur at the earliest stage of infection, it is difficult to identify pathogen proteins because of their low abundance relative to the plant's proteins.

A third challenge involves protein homology. Typically, the identification of proteins in a proteomic experiment is based on identification of peptides through mass spectrometry and database searching. However, peptides can be redundant and are found in multiple proteins encoded in the genome of the same organism. That is especially true for eukaryotes with multiple paralogs and gene families. The situation in proteomic experiments on infected tissue is even more complex, where not only paralogs of the same species but also orthologs and other homologs between two eukaryotic species exist in the same sample. Thus, when peptides of conserved proteins or protein families are identified, it is sometimes impossible to assign the original protein or even organism source from which those peptides originated.

A large-scale *in planta* proteomic study of *Phytophthora* was recently initiated using the *P. capsici*–tomato system (Savidor et al., unpublished). In this study, the so-called "interactome" (the proteome of both host and pathogen during infection) was measured over a time course of 3 days of infection. Five-weeks old tomato plants were sprayed with a concentrated solution of *P. capsici* zoospores, and infected leaves were harvested 1, 2, and 3 days after inoculation. Leaf tissue from healthy plants was also collected and used as a control, and both the intercellular and soluble proteins were analyzed via MudPIT. Not surprisingly, most proteins identified at all time points were tomato proteins. At day 1, only 7% of identified proteins originated from *P. capsici* (195 *P. capsici* proteins versus 2,508 tomato proteins). This low number is expected considering the short amount of time the pathogen was allowed to grow in the plant tissue. The number of *P. capsici* proteins identified continuously increased as the pathogen spread through the tissue until the third day after inoculation, in which 25% of all identified proteins originated from the pathogen (594 *P. capsici* proteins versus 1,803 tomato proteins).

Different trends in protein expression were observed for both the plant and the pathogen. *P. capsici* proteins involved in protein synthesis, and cytoskeleton proteins were abundant early in infection and declined in abundance over the course of infection. This observation supported the hypothesis suggested previously that protein synthesis and cytoskeleton formation are important processes for cyst germination and early infection (Ebstrup et al., 2005; Savidor et al., 2008).

Later in infection, a different set of proteins was upregulated. For example, proteins involved in scavenging of reactive oxygen species (ROS), such as superoxide dismutases and catalases, increased in abundance over the course of infection and reached their peak expression at day 3. These may contribute to the ability of *P. capsici* to fight off defense responses such as the oxidative burst mounted by the plant.

Other *P. capsici* proteins upregulated in the late stages of infection may contribute to the necrotrophic phase of the pathogen. For example, an extracellular pectate lyase and a protein with similarity to phenolic acid decarboxylases were both upregulated in the late stages of infection. Both of these enzymes can degrade different constituents of the plant cell wall, and therefore they may contribute to necrosis of the plant tissue.

These experiments also allowed a view into the processes that occur in the host during infection. In response to *P. capsici* infection, the tomato seems to shift from normal house-keeping metabolism to defense mode, as evident by a decrease in abundance of house-keeping proteins such as Ribulose-1,5-bisphosphate carboxylase/oxygenase (Rubisco) proteins and an increase of proteins involved in defense (e.g., pathogenesis related proteins). Interestingly, a group of tomato lipoxygenases (LOXs) seemed to be down regulated as infection progressed. These enzymes catalyze the oxidation of polyunsaturated fatty acids and play a key role in the biosynthetic pathway of jasmonic acid (JA) (Brash, 1999). Typically, for defense against necrotrophs, plants induce the jasmonic acid pathway in which accumulation of jasmonic acid results in activation of proteins with potent antifungal activities (Kessler and Baldwin, 2002; Lorenzo et al., 2003; Penninckx et al., 1996; Spoel et al., 2007). Thus, if perceived by the tomato as a necrotroph, it was expected that *P. capsici* would trigger plant defense responses including an increase in LOXs and a subsequent production of JA. The decline in tomato LOXs suggests that *P. capsici* may evade recognition as a necrotroph by the tomato or that the pathogen is actively inhibiting the JA pathway in the tomato by an unknown mechanism.

26.5.2 Proteomics and Genome Annotation

Genome annotation refers to the process of identification of genes and their function in a sequenced genome. Once a genome is sequenced, gene prediction algorithms are typically applied to locate regions in the sequence that define genes. These algorithms use different strategies, which include searches for homology to known genes, nucleotide composition of the sequence (i.e., G + C content), and location of start codons, stop codons, and splice sites (reviewed in Mathe, (2002)). Thus, regardless of the strategy that they employ, the gene prediction algorithms rely on certain assumptions about the nature of the genes. As a result, genes that contain nontypical sequences might be missed or missannotated by the computational genome annotation, whereas noncoding sequences that comply with the above assumptions to a sufficient degree may be identified as genes.

Although computational genome annotation is essential considering the genome sizes of living organisms, it is not foolproof and remains a significant scientific challenge, especially when applied to higher organisms with large genomes and complex gene architectures (Claverie et al., 1997). One way to increase confidence in the computational gene prediction is to sequence expressed sequence tags (ESTs) and map them onto the genome. An accurate gene prediction would be in agreement with ESTs that are mapped to that region of the genome. Similarly, the confidence in the computational gene prediction could increase by using proteomic data.

The data from proteomic experiments is typically used to explore which proteins are present in the sample. As explained above, this is typically done by acquiring spectra of tryptic peptides by mass spectrometry and searching those spectra against a known protein database. Another possibility is to search the spectra against a six-frame translation of the genome instead (Arthur, 2004; Fermin et al., 2006; Giddings, 2003; Jaffe, 2004; Kalume DE, 2005; Küster, 2001; Savidor et al., 2006; Smith, 2005). Using this approach, no assumptions about the nature and location of protein-coding genes in the genome are made. Peptides that are identified and mapped back to the genome can then be compared to the computational gene predictions. These so-called "expressed peptide tags" (EPTs) (Savidor et al., 2006) can increase the confidence in the gene predictions and refine them, much in the same way that ESTs do.

A proteomic study that employs EPTs on the genomes of the first two sequenced *Phytophthora* species *P. sojae* and *P. ramorum* (Tyler et al., 2006) confirmed ~4,000 computational gene models for *P. sojae* and ~5,000 for *P. ramorum* when analyzing the mycelial and germinating cyst life stages (Savidor et al., 2006). In addition, hundreds of gene models that were missed or missannotated by the computational annotation were identified. Similar work has been done on *P. capsici* and *P. infestans* as the first drafts of their genomes became available and showed similar results. The large number of gene models that were missed or missannotated by the computational genome annotations may reflect the imperfect quality of those computational annotations. Several factors may contribute to the shortcomings in the annotation of *Phytophthora* genomes, which include the evolutionary uniqueness of the Oomycetes. *Phytophthora* genomes contain many unique genes with sequence and gene features that have not been previously sequenced and documented in other organisms. Homology searches to known genes by the computational annotation may not find significant matches to these unique *Phytophthora* genes, which results in their missannotation. The use of proteomics for annotation purposes is particularly attractive for organisms like the Oomycetes as they complement the shortcomings of the computational annotation.

26.6 FUTURE PERSPECTIVES

Proteomics is an important tool for the investigation of many organisms, including the *Phytophthora* species. Depending on the design of the experiment, proteomics can provide knowledge about different life stages of the pathogen, differences between species, the process of infection, and they can even assist in genome annotation. When performed on the pathogen *in planta*, proteomic experiments can reveal information not only about the pathogen but also about the host in response to the pathogen. Although the body of proteomic work carried out on *Phytophthora* to date is limited, this situation is bound to improve, as additional Oomycetes are sequenced and the quality of annotation improves. In addition, completion of genome sequencing of *Phytophthora* hosts such as the tomato and soybean will shed more light on the process of infection. As the components of the proteomic process (sequenced genomes, accurate annotations, mass spectrometry instrumentation, and database search software) improve, so will the knowledge gleaned from them.

REFERENCES

Arthur JW, Wilkins, MR 2004. Using proteomics to mine genome sequences. J Prote Res 3:393–402.

Avrova AO, Venter E, Birch PR, Whisson SC 2003. Profiling and quantifying differential gene transcription in Phytophthora infestans prior to and during the early stages of potato infection. Fungal Genet Biol 40:4–14.

Brash AR 1999. Lipoxygenases: occurrence, functions, catalysis, and acquisition of substrate. J Biol Chem 274:23679–23682.

Claverie JM, Poirot O, Lopez F 1997. The difficulty of identifying genes in anonymous vertebrate sequences. Comput Chem 21:203–214.

Conrads KA, Yi M, Simpson KA, Lucas DA, Camalier CE, Yu LR, Veenstra TD, Stephens RM, Conrads TP, Beck GR Jr. 2005. A combined proteome and microarray investigation of inorganic phosphate-induced pre-osteoblast cells. Mol Cell Proteomics 4:1284–1296.

Cox B, Kislinger T, Emili A 2005. Integrating gene and protein expression data: pattern analysis and profile mining. Methods 35:303–314.

Ebstrup T, Saalbach G, Egsgaard H 2005. A proteomics study of in vitro cyst germination and appressoria formation in Phytophthora infestans. Proteomics 5:2839–2848.

Fenn JB, Mann M, Meng CK, Wong SF, Whitehouse CM 1989. Electrospray ionization for mass spectrometry of large biomolecules. Science 246:64–71.

Fermin D, Allen BB, Blackwell TW, Menon R, Adamski M, Xu Y, Ulintz P, Omenn GS, States DJ 2006. Novel gene and gene model detection using a whole genome open reading frame analysis in proteomics. Genome Biol 7:R35.

Giddings MC, Shah, AA, Gesteland, R, Moore, B 2003. Genome-based peptide fingerprint scanning. Proc Nat Acad Sci USA 100:20–25.

Grenville-Briggs LJ, Avrova AO, Bruce CR, Williams A, Whisson SC, Birch PR, van West P 2005. Elevated amino acid biosynthesis in Phytophthora infestans during appressorium formation and potato infection. Fungal Genet Biol 42:244–256.

Gygi SP, Rochon Y, Franza BR, Aebersold R 1999. Correlation between protein and mRNA abundance in yeast. Mol Cell Biol 19:1720–1730.

Hillenkamp F, Karas M, Beavis RC, Chait BT 1991. Matrix-assisted laser desorption/ionization mass spectrometry of biopolymers. Anal Chem 63:1193A–1203A.

Jaffe JD, Berg, HC, Church, GM 2004. Proteogenomic mapping as a complementary method to perform genome annotation. Proteomics 4:59–77.

Kalume DE PS, Reddy R, Zhong J, Okulate M, Kumar N, Pandey A 2005. Genome annotation of Anopheles gambiae using mass spectrometry-derived data. BMC Genom 6:128.

Karas M, Bachmann D, Bahr U, Hillenkamp F 1987. Matrix-assisted ultraviolet laser desorption of non-volatile compounds. Int J Mass Spectrom Ion Phys 78:53–68.

Kessler A, Baldwin IT 2002. Plant responses to insect herbivory: the emerging molecular analysis. Ann Rev Plant Biol 53:299–328.

Kramer R, Freytag S, Schmelzer E 1997. *In vitro* formation of infection structures of *Phytophthora infestans* is associated with synthesis of stage specific polypeptides. Eur J Plant Pathol 103:43–53.

Küster B, Mortensen, P, Andersen, JS, Mann, M 2001. Mass spectrometry allows direct identification of proteins in large genomes. Proteomics 1:641–650.

Le Berre JY, Engler G, Panabieres F 2008. Exploration of the late stages of the tomato-Phytophthora parasitica interactions through histological analysis and generation of expressed sequence tags. New Phytol 177:480–492.

Lorenzo O, Piqueras R, Sanchez-Serrano JJ, Solano R 2003. ETHYLENE RESPONSE FACTOR1 integrates signals from ethylene and jasmonate pathways in plant defense. Plant Cell 15:165–178.

Margulies M, Egholm M, Altman WE, Attiya S, Bader JS, Bemben LA, Berka J, Braverman MS, Chen YJ, Chen Z, et al. 2005. Genome sequencing in microfabricated high-density picolitre reactors. Nature 437:376–80.

Mathe C, Sagot, M, Schiex, T, Rouze, P 2002. Current methods of gene prediction, their strengths and weaknesses. Nuc Acid Res 30:4103–4117.

Mitchell HJ, Kovac KA, Hardham AR 2002. Characterisation of Phytophthora nicotianae zoospore and cyst membrane proteins. Mycol Res 106:1211–1223.

Mootha VK, Bunkenborg J, Olsen JV, Hjerrild M, Wisniewski JR, Stahl E, Bolouri MS, Ray HN, Sihag S, Kamal M, et al. 2003. Integrated analysis of protein composition, tissue diversity, and gene regulation in mouse mitochondria. Cell 115:629–640.

Moy P, Qutob D, Chapman BP, Atkinson I, Gijzen M 2004. Patterns of gene expression upon infection of soybean plants by Phytophthora sojae. Mol Plant-Microbe Interact 17:1051–1062.

Nie L, Wu G, Culley DE, Scholten JC, Zhang W 2007. Integrative analysis of transcriptomic and proteomic data: challenges, solutions and applications. Crit Rev Biotechnol 27:63–75.

O'Farrell PH 1975. High resolution two-dimensional electrophoresis of proteins. J Biol Chem 250:4007–4021.

Penninckx IA, Eggermont K, Terras FR, Thomma BP, De Samblanx GW, Buchala A, Metraux JP, Manners JM, Broekaert WF 1996. Pathogen-induced systemic activation of a plant defensin gene in Arabidopsis follows a salicylic acid-independent pathway. Plant Cell 8:2309–2323.

Pieterse CMJ, Riach MBR, Bleker T, van den Berg-Velthuis GCM, Govers F 1993. Isolation of putative pathogenicity genes of the potato late blight fungus Phytophthora infestans by differential hybridization of a genomic library Physiol Mole Plant Pathol 43:69–79.

Qutob D, Kamoun S, Gijzen M 2002. Expression of a Phytophthora sojae necrosis-inducing protein occurs during transition from biotrophy to necrotrophy. Plant J 32:361–373.

Savidor A, Donahoo RS, Hurtado-Gonzales O, Land ML, Shah MB, Lamour KH, McDonald WH 2008. Cross-species global proteomics reveals conserved and unique processes in Phytophthora sojae and P. ramorum. Mol Cell Proteom.

Savidor A, Donahoo RS, Hurtado-Gonzales O, Verberkmoes NC, Shah MB, Lamour KH, McDonald WH 2006. Expressed peptide tags: an additional layer of data for genome annotation. J Proteome Res 5:3048–3058.

Schena M, Shalon D, Davis RW, Brown PO 1995. Quantitative monitoring of gene expression patterns with a complementary DNA microarray. Science 270:368–371.

Shepherd SJ, van West P, Gow NA 2003. Proteomic analysis of asexual development of Phytophthora palmivora. Mycol Res 107:395–400.

Shibata D 2005. Genome sequencing and functional genomics approaches in tomato. J Gen Plant Pathol 71:1–7.

Smith JC, Northey, JGB, Garg, J, Pearlman, RE, Michael Siu, KW 2005. Robust method for proteome analysis by MS/MS using an entire translated genome: Demonstration on the ciliome of Tetrahymena thermophila. J Proteome Res.

Spoel SH, Johnson JS, Dong X 2007. Regulation of tradeoffs between plant defenses against pathogens with different lifestyles. Proc Nat Acad Sci USA 104:18842–18847.

Torto-Alalibo TA, Tripathy S, Smith BM, Arredondo FD, Zhou L, Li H, Chibucos MC, Qutob D, Gijzen M, Mao C, et al. 2007. Expressed sequence tags from phytophthora sojae reveal genes specific to development and infection. Mol Plant Microbe Interact 20:781–793.

Torto TA, Li S, Styer A, Huitema E, Testa A, Gow NA, van West P, Kamoun S 2003. EST mining and functional expression assays identify extracellular effector proteins from the plant pathogen Phytophthora. Genome Res 13:1675–1685.

Tyler BM, Tripathy S, Zhang X, Dehal P, Jiang RH, Aerts A, Arredondo FD, Baxter L, Bensasson D, Beynon JL, et al. 2006. Phytophthora genome sequences uncover evolutionary origins and mechanisms of pathogenesis. Science 313:1261–1266.

Velculescu VE, Zhang L, Vogelstein B, Kinzler KW 1995. Serial analysis of gene expression. Science 270:484–487.

Washburn MP, Wolters, D, Yates III, JR 2001. Large-scale analysis of the yeast proteome by multidimensional protein identification technology. Nature Biotechnol 19:242–247.

Wold B, Myers RM 2008. Sequence census methods for functional genomics. Nat Methods 5:19–21.

Yates III JR 1998. Mass spectrometry and the age of the proteome. J Mass Spectrom 33:1–19.

27

STRATEGY AND TACTICS FOR GENOME SEQUENCING

Michael C. Zody and Chad Nusbaum
Broad Institute of MIT and Harvard University, Cambridge, Massachusetts

27.1 INTRODUCTION

The genome sequence has become a standard and essential tool for studying an organism. The genome provides a "parts list" of an organism and facilitates studies of function and genetics by providing a template from which to design reagents and a framework against which to analyze results. Reductions in cost and increases in throughput of DNA sequencing in recent years have made full genome sequences available to a rapidly increasing number of research communities. However, the process of sequencing, assembling, annotating, and analyzing a genome remains a sizable task (a schematic overview of this process is presented as Fig. 27.1). A genome sequence is an enduring resource, and annotation and experimental data will build on the reference sequence over time. Moreover, it requires significant cost and effort to generate. Thus, it is important to plan the sequence generation carefully. Here we hope to cover some of the choices and challenges faced in planning and executing the sequencing of a genome using as an example the genome sequencing of the oomycete potato pathogen *Phytophthora infestans*.

27.2 CHOICE OF INDIVIDUAL FOR SEQUENCING

Generating a high-quality draft sequence of a genome is a costly proposition. To maximize the value of the data, one must consider carefully the choice of strain or isolate of an organism to be sequenced. Ideally, the individual would

Oomycete Genetics and Genomics: Diversity, Interactions, and Research Tools
Edited by Kurt Lamour and Sophien Kamoun
Copyright © 2009 John Wiley & Sons, Inc.

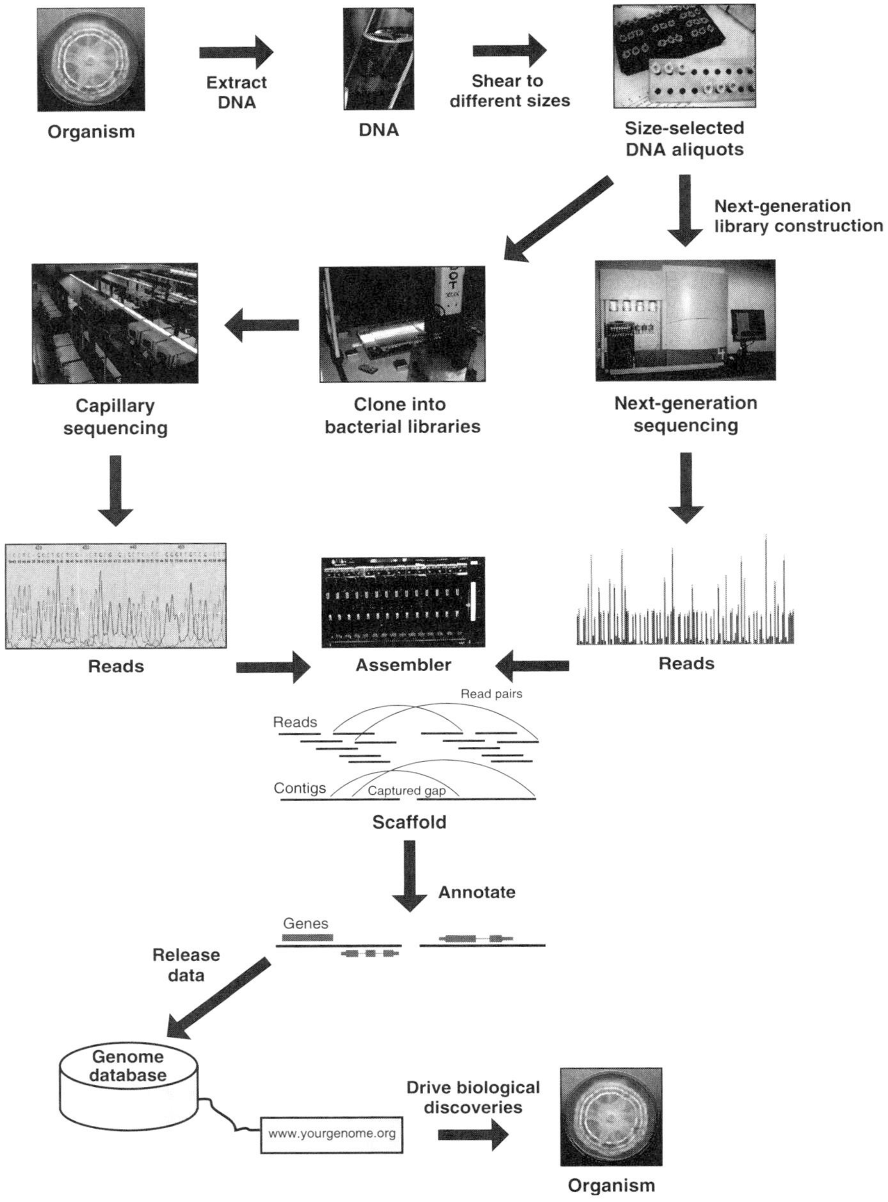

FIG. 27.1 Schematic overview of the process of genome sequencing, assembly, annotation, and release. Individual elements of the process are described in the main text or defined in the glossary.

be chosen to take advantage of existing genome resources and to best serve the scientific needs of the research community. For some organisms, particularly microbes, there can be significant differences between different isolates of the same species. It may not always be possible to target a single isolate that satisfies the optimal needs of all researchers, so it may be appropriate to plan follow-up projects to sequence related isolates.

Preexisting genomic resources are an important consideration. If there are large insert clone libraries that are commonly used or if genetic or physical mapping resources exist, there will be an advantage in matching the genome sequence to those existing resources. There may be strains that are particularly facile for genetic crosses, transfection, knockouts, RNA interference (RNAi), transgenics, and so on, or for studying key biological processes for which having the exact genome sequence of that strain would be especially valuable. For a diploid genome, extensive heterozygosity can create serious challenges to genome assembly; this should be considered in the choice of individual for sequencing (see below).

We selected the *P. infestans* strain T30-4 for sequencing on the basis of several issues described above. The existing genetic maps of the organism had been built from a cross of which T30-4 was one of the offspring (van der Lee et al., 1997, 2004), and our goal was to tie the sequence to the existing genetic map. A bacterial artificial chromosome (BAC) library constructed from this isolate exists and is available to the community as a research tool (Whisson et al., 2001). The T30-4 strain was selected for the BAC library because it contained the avirulence allele (see glossary) of all the known virulence/avirulence loci segregating in the cross (van der Lee et al., 2001). In this case, we felt our choice of organism was an easy one; however, there is not always an unambiguous best choice. In the case of *Saccharomyces cerevisiae*, many resources have been developed on the sequenced S228c strain, but much laboratory work is done on other strains, and there has been recent interest in developing a catalog of genetic differences (Gresham et al., 2006; Schacherer et al., 2007).

In some cases, it may be tempting to try to get more information out of the genome assembly by using multiple individuals, isolates, or even closely related species as part of a single genome assembly. Although there have been some examples of this in organisms with low levels of heterozygosity (the Celera assemblies of mouse and human, for example), it becomes much more difficult to assemble a genome accurately as more haplotypes (see glossary) are included. The initial efforts to sequence both tetraodon (pufferfish) (Jaillon et al., 2004) and zebrafish (www.sanger.ac.uk/Projects/D_rerio) used multiple individuals and encountered difficulty assembling highly divergent regions of the genome. In both cases, it was necessary to go back and generate the bulk of the sequence from a single individual at significant added cost. As a general rule, generating sequence from a single individual (if possible) yields the highest quality assembly.

In some cases, the question may not be which organism but rather which organisms. Although it is important to get the reference genome sequence, a genome alone is still very difficult to interpret. One of the most powerful tools for genome analysis is the comparison with other genomes. If the organism of interest is only very distantly related to other sequenced organisms, it can be of great analytic value to sequence related organisms. *S. cerevisiae* (Goffeau et al., 1996) and *Drosophila melanogaster* (Adams et al., 2000; Myers et al., 2000) are two examples of key model organisms whose sequences were generated early in the genomic era. However, the later sequencing of additional species of both fungi and flies at various distances from these models has yielded great insights in the gene content, regulatory machinery, and evolution of these organisms that would not have been possible from a single genome sequence (Cliften et al., 2003; Kellis et al., 2003; Drosophila 12 Genomes Consortium, 2007). Similarly, in the case of the filamentous fungal genus Aspergillus, three different species were sequenced: *Aspergillus nidulans*, the key model organism (Galagan et al., 2005); *Aspergillus fumigatus*, a human pathogen (Nierman et al., 2005); and *Aspergillus oryzae*, an important industrial organism (Machida et al., 2005). The availability of all three genomes again enabled a far deeper understanding of the gene and regulatory repertoire of all three species and allowed better understanding of the differences between the laboratory model and its nonlaboratory relatives (Galagan et al., 2005).

It is important to note that the reference genome represents only a single individual. The exact sequences of other individuals or isolates of that species will vary, possibly extensively. With this in mind, it may be desirable to generate additional sequence data from other individuals of the same species, using the reference as a framework for comparative studies. For example, even a small amount of whole genome shotgun (WGS) sequence from additional individuals can usually be readily aligned to an assembled reference to call single nucleotide polymorphisms (SNPs, see glossary) (Altshuler et al., 2000) or to identify variable simple sequence repeats (SSRs). Using assisted assembly methods, genomes from individuals of the same or a closely related species can be assembled at low ($\geqslant 2\times$) coverage using a reference genome as a guide (Chimpanzee Sequencing and Analysis Consortium, 2005; Pontius et al., 2007). This can be very useful in sampling diversity within a species or tight clade. For example, light sample sequencing was used to find SNPs between different strains of malaria and the reference strain (Volkman et al., 2007).

In the case of *P. infestans*, we were fortunate that two other Phytophthora species (*Phytophthora sojae* and *Phytophthora ramorum*) (Tyler et al., 2006) had already been sequenced, providing the opportunity for comparative analysis. Long-range synteny (see glossary) is well conserved among these three species (Tyler et al., 2006), although they are diverged and provide limited capability for fine-scale comparative analysis and no meaningful support for assisted assembly. New efforts are now underway to generate sequence for several species that are much more closely related to *P. infestans* as well as for additional wild isolates of *P. infestans*.

27.3 SEQUENCING STRATEGIES

Once the organism has been selected, there are several factors that may affect sequencing strategy, which include genome size, repeat content, ploidy, cloning bias, and availability and quality of genomic DNA.

27.3.1 Genome Size

The size of the genome is the strongest cost driver in any genome sequencing project, as it directly determines the amount of sequence required to assemble the organism at a given target coverage. Many methods exist to estimate genome sizes (e.g., www.genomesize.com; Gregory et al., 2007), but available estimates may be subject to significant errors. Inaccuracy in genome size estimation is most likely to occur if the organism lacks a close relative that has been sequenced, if the genome is small, or if the organism has a high content of heterochromatin (which generally will be neither cloned nor sequenced). There are plenty of stories of how genomes "grew dramatically" in the sequencing process. For example, despite being sized by restriction digest, the assembled size of the genome of the arachaeon *Methanosarcina acetivorans* turned out to be nearly twice as large as estimated (Galagan et al., 2002). Similarly, the genome of the mosquito *Culex pipiens* was estimated at ~800 Mb, and in fact was assembled at 1.2 Gb (Nusbaum, unpublished data). The genomes of *P. ramorum* and *P. sojae* also turned out to be 50% larger than expected (Tyler, personal communication). *P. infestans* itself seems to have been remarkably well estimated at 237 Mb (Tooley and Therrien, 1987). Our assembly contains 190 Mb of sequence in assembled contigs and an additional 38 Mb of gaps of known size and position ("captured gaps," see glossary, which consist of unassembled high-copy-number repeat sequences), for 228 Mb of total length. The repeat content of unassembled reads suggests that unassembled sequence in gaps of unknown size ("uncaptured gaps," see glossary) may raise the total euchromatic genome size to 240–250 Mb. This is in line with the genome size of 249 Mb seen in the optical map (see glossary; Zhou, personal communication). Underestimating the genome size can have dire consequences for assembly. Given fixed resources for a project, any growth in the genome size will result in a reduction of assembly coverage and, therefore, quality and contiguity.

In contrast to those genomes that were undersized prior to sequencing, reliable estimates of the genome size of Drosophila ranged from 156–176 Mb (Kurnick and Herskowitz, 1952; Rasch et al., 1971; Mulligan and Rasch, 1980), but the actual genome assembly contains only 120 Mb (Adams et al., 2000) because of the high content of heterochromatin. The full genome size was estimated at 180 Mb after assembly, 60 Mb of which seems to be unclonable by standard techniques. This, of course, presents a different kind of problem, as a significant fraction of the genome is not assembled. However, experience with Drosophila and other genomes suggest that, in most cases, DNA that cannot be cloned and sequenced tends to be repetitive and poor in gene content. Although the Drosophila assembly contained only two thirds of the total DNA, it contained almost all of the protein-coding genes.

In addition to the physical methods of estimating genome size, reasonably accurate sequence-based techniques are available. A relatively costly method is to generate a significant subset of the whole genome using shotgun sequencing and then assemble and evaluate. Starting with a conservative guess of genome size based on related species, we can generate sufficient sequence to assemble and then use this assembly to estimate the true size of the genome. This can be done iteratively to avoid overshooting the target coverage (thus incurring undue expense). With 3–4 × average sequence coverage, an assembly will span most of the genome in contigs at least a few kilobases in size. One can then consider the total length of genome assembled (including both sequence contigs and sized gaps) as being *most* of the genome, which provides a reasonable estimate. For example, three assemblies recently generated at the Broad Institute gave good estimates. A 4 × assembly of *Capsaspora owczarzaki* yielded an estimated genome size of ~30 Mb, whereas the final, full-coverage assembly was 29 Mb. Similarly, for *Spizellomyces punctatus* the estimated size was 28 Mb, and the final assembly was 24 Mb. In draft assemblies of repetitive genomes, some regions can be "overcollapsed." When this occurs, similar repeats are assembled on top of each other, and very highly repetitive regions may be excluded from the assembly altogether because the assembly algorithm cannot resolve them. For the highly repetitive *Allomyces macrogynus*, the estimated size from a 3 × assembly was 50 Mb, which underestimated the repeat content of the genome. The 8 × final assembly, which allowed us to account for unassembled high-copy sequences and identify overcollapsed regions, indicated that the genome size was 56 Mb.

27.3.2 Genome Repeat Content

With a sensible estimate of genome size in hand, the next key factor that affects assembly is repeat content. Genomes with a high content of mobile element sequence or that exhibit recent whole-genome or large-scale segmental duplication are much harder to assemble than similarly sized genomes that consist of a unique sequence. However, there are some adjustments to a sequencing strategy that can be made for repetitive genomes that will increase total costs somewhat but provide a much better quality product. As with estimation of genome size, we can estimate the amount of repetitive sequences in a genome by analyzing the ~0.1 × WGS data (see below). We align all reads to each other and identify the fraction of reads that align to more other reads than we would expect by chance. At this level of coverage, one would expect that only about 15% of reads should overlap in actual genome placement (the number is higher than 10% because reads are not point events and can overlap by less than 100%). If the alignment of this sample sequence to itself shows significantly more overlap, this can be used to estimate the repetitiveness. This will also give some idea of how similar the repeats may be. Copies of sequences more than 98% identical can be difficult to assemble by WGS regardless of technique, whereas copies of less than 90% similarity are essentially unique for purposes of sequencing by capillary (these will pose a greater challenge for short-read "next-generation" technologies).

An analysis of this type performed for the *P. infestans* pilot study used 32,000 capillary sequencing reads (after vector trimming and quality clipping), giving 0.1 × coverage of the estimated genome size. An alignment of the reads showed that 26,000 reads (> 80%) had hits to other reads. Many reads had tens or even hundreds of hits. Modeling of the expectations showed that this was inconsistent not only with a largely unique genome but also with a whole-genome duplication (which would be expected to simply look like a genome half the size). In fact, we concluded the best fit was a pattern in which a large fraction of the genome was moderate- to high-copy repeat (some with hundreds or thousands of copies) and very little (~25%) truly unique sequence. This indicated that assembling the *P. infestans* genome would be a challenge as it would be the most repetitive genome yet assembled. In fact, an analysis of the genome assembly showed this prediction to be accurate (see Fig. 27.2), with a minimum repeat content of 75%. The actual value may be even higher, as recently inserted elements are likely underrepresented in the assembly.

Our experience with other genomes with very large numbers of recent repeats suggested a way of mitigating this challenge by optimizing the mix of subclones that would be sequenced. In the case of the unusually repetitive genome of the zygomycete fungus *Rhizopus oryzae* (Ma, personal communication), we originally approached this project as if it were a typical fungal genome and generated our standard sequence coverage of 8 × data. In fact, the assembly had significantly lower contiguity than a typical 8 × assembly, with high counts and small sizes for contigs and scaffolds. An analysis of the assembly revealed that its lower quality was caused by a high level of repeated sequences. Further, although the size of the initial assembly of the *R. oryzae* genome was ~40 Mb; a comparison to an optical map (see glossary) indicated that the true size of the genome was 46 Mb. We addressed this at the level of both the data set and the assembly software. Our standard 8 × sequence data set includes ~30-fold physical coverage in long linking information from read pairs from 40 kilobase (kb) Fosmids (see glossary). These long-linking data are critical to building long-range contiguity of a genome assembly as they can reach across repeat sequences and can also reach into the repeated areas to add repeat sequences to the assembly that could not otherwise be placed. We reasoned that additional coverage of Fosmid pairs would improve the assembly and, accordingly, added 30 × physical coverage for a total of 85 × [including data from plasmids (see glossary)]. We also took advantage of a newly available version of the Arachne assembler that includes extensive optimizations in the handling of repeat sequences (Grabherr, personal communication). These steps resulted in greatly increased contiguity and completeness of the *R. oryzae* genome assembly to cover > 97% of the optical map.

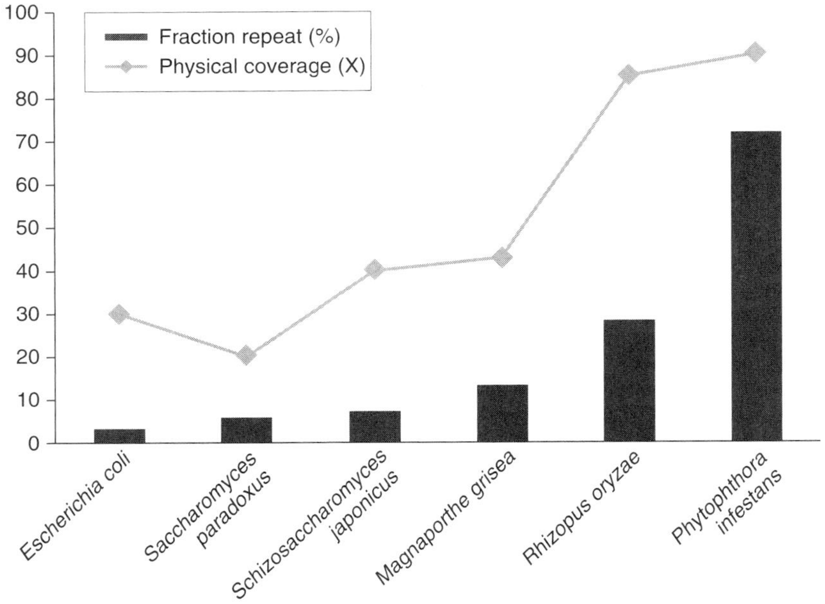

FIG. 27.2 For genomes with high repeat content, increased physical coverage enables delivery of high-quality assemblies. Data are shown for six genomes that range from 3.3–75% repeat content. Bars show repeat content for each genome calculated as follows: all overlapping 25-mers were counted within the draft genome sequence, any 25-mer occurring more than once (nonuniquely) was marked as repetitive, and repeat content was computed by summing the number of bases in the genome spanned by one or more nonunique k-mers and dividing by the total number of nongap bases. Solid diamonds show physical coverage that was used for each genome. Physical coverage is defined as the total number of bases contained in successfully sequenced paired end fragments [from plasmids, Fosmids, and bacterial artificial chromosome (BACs)] divided by the length of the genome as estimated from the genome assembly.

We designed our sequencing strategy for *P. infestans* to maximize long-linking information in the assembly based on our analyses of preliminary data and our experiences with *R. oryzae*. The large size of the genome made it expensive and impractical to generate excess sequence coverage as we did for *R. oryzae*, so we elected to generate a much larger fraction of the data in long links from 40 kb Fosmids (55 × physical coverage as opposed to 30 ×). We also added sequences from a second group of 10 kb plasmids (providing 15 × physical coverage) that deliver medium-range linking information. Along with 4 kb insert plasmids and a small number of BAC clones, the total physical coverage of the genome is 90 × . The additional Fosmids would provide good long-range contiguity across regions that could not be assembled, whereas the 10-kb plasmids would help with assembly of individual repeat elements, which our pilot BAC sequencing had shown ranged in size from about 5–15 kb.

27.3.3 Heterozygosity and Polymorphism

Another important consideration for organisms that are diploid (or have higher ploidy) and outbred is the degree of polymorphism. Low levels of SNPs usually have minimal impact on whole genome assembly. Organisms with levels of polymorphism up to 1 in 350 bases have been successfully assembled, but lower rates of heterozygosity will translate into better assemblies as the haplotypes are more similar than the expected error rates in individual read-read alignments and will assemble together as a single sequence. Structural variations (e.g., inversions, polymorphic repeat insertions, and copy number polymorphisms) pose a greater challenge and often cause an assembly break with neither haplotype adequately represented. Very high local divergence between haplotypes (1 in 100 or higher) may force the assembler to build both haplotypes individually, as if the region were duplicated. In addition to creating a misleading assembly of the genome (such regions will appear twice in the supposedly haploid assembly), this may require higher coverage to prevent the separated haplotypes from being sparsely assembled. Thus, for organisms with moderate-to-high rates of heterozygosity, it is preferable to use an inbred individual if inbred lines exist or can be generated. Where only outbred individuals exist, it is generally beneficial to sample several individuals and pick one with low heterozygosity. In our initial interactions with the Phytophthora community, based on our experience sequencing genomes, we argued strongly that the individual sequenced should be selected from an inbred line as that reduces heterozygosity and increases the quality of the assembly. However, we learned from our collaborators that crossing *P. infestans* is nontrivial, and, more importantly, that backcrossing quickly results first in a loss of virulence and then in a loss of viability so that an inbred individual would be an unsuitable representative of infectious strains even if one could be generated. This conversation exemplifies the need for close communication between the sequencing center and the community.

We can estimate heterozygosity with the same whole-genome sample sequencing and clone-based sequencing data that we used above to estimate the genome size and repeat content. Reads from light coverage WGS that align to each other in pairs (but not in large clusters, which are likely to be repeats) and have high-identity alignments are likely to derive from the same genomic locus. One can then use methods such as neighborhood quality standard (NQS) (Altshuler et al., 2000) to call SNPs and estimate rates. Reads can also be aligned to large insert clone assemblies to estimate heterozygosity rates for both SNPs and insertion/deletion polymorphism. For *P. infestans*, we also had the advantage of having identified several pairs of BACs that mapped to different haplotypes of the same locus. These confirmed our genomewide estimate of

SNP heterozygosity, which was very low (estimated between 1 in 1,500 and 1 in 2,500 by read-read and BAC-BAC, respectively) and showed significant large-scale insertion-deletion (indel) polymorphism between haplotypes. Subsequent assembly of the genome and SNP analysis confirms that the polymorphism rate in unique regions of the genome is very low.

27.3.4 Cloning Bias

Finally, there may be regions of the genome that are not easily cloned using standard techniques and vectors, either because of highly elevated content of G + C or A + T nucleotides or because the sequence is harmful to *Escherichia coli*. These artifacts may be more difficult to recognize during pilot sequencing but may, in more extreme cases, manifest as an apparent genome size much smaller than estimated by other methods. We describe here two examples of genomes in which significant fractions of the genome were recalcitrant to cloning and, therefore, to sequencing by traditional cloning in bacterial vectors and sequencing by Sanger chemistry. For both genomes, most missing sequences were recovered by sequencing with 454 technology (Margulies et al., 2005), which does not employ a bacterial cloning step. The Illumina (Bentley et al., 2008) and SOLiD (Schuster, 2008) methods also have this advantage, but assembly of these reads is at an earlier stage of development (see below). In *Neurospora crassa*, the genome was assembled to 38 Mb by traditional means, which included extensive manual finishing. Sequencing and assembly of the genome by 454 revealed an additional 4 Mb of sequence of extremely high A + T content (Young, personal communication). *Listeria monocytogenes* showed a different type of cloning bias that could nonetheless be remedied in a similar manner. Four separate isolates of *L. monocytogenes* were sequenced in our group by traditional means, and in all cases the assembled sequences covered only about 80% of the genome compared with a finished reference genome (Glaser et al., 2001). By comparison a typical bacterial species coverage of more than 97% is usually obtained. As with *N. crassa*, we overcame this cloning bias by employing 454 sequencing, which in combination with traditional sequencing delivered 98% of the genome sequence for *L. monocytogenes*. An analysis of the regions excluded from the traditional assemblies showed them to be greatly enriched in promoter sequences, which seem to be toxic to *E. coli* when propagated in plasmid vectors (Sorek et al., 2007).

27.3.5 Organelle and Plasmid DNA

Genomic DNA preparations can sometimes contain significant amounts of other DNA. DNA preps from eukaryotes can contain significant amounts of DNA from mitochondria or plastids. Some prokaryotic genomes contain native multicopy plasmids. Most DNA preparations from organelle-containing organisms carry enough organelle DNA so that, as a by-product of WGS assembly, a good representation of the organelle genome is obtained. In cases where the relative amount of organelle DNA present is high, artifacts can occur in the assembly process (such as false genomic insertions or large concatamerized contigs of mitochondrial sequence; Mikkelsen et al., 2007). The amount of organelle DNA present can vary widely depending on the organism's biology as well as the conditions of growth and the DNA preparation used. In the more than 70 eukaryotic genomes sequenced at the Broad Institute, we have observed mitochondrial sequence "contamination" ranging from 0.1% to more than 10% of shotgun sequencing reads.

27.3.6 Contaminating Sequences

Contaminating DNA from other organisms may also be present. Organisms that are obligate pathogens (such as *Pneumocystis* or *Plasmodium*) may need to be grown on host cells or sampled from native infections and may be difficult to purify fully from the host DNA. Similarly, microbes that are obligate biovores may be impossible to grow in the absence of feeder species. The sequencing target organism may itself be infected by other organisms (such as *Wolbachia* spp. common in insects and other edysozoa; Foster et al., 2005).

How the contamination presents itself in the data is partially dependent on the degree of contamination. At very low levels (and low coverage of the contaminating genome), it may appear only as a slightly higher than usual rate of unassembled reads. At higher levels, the contaminant reads will start to assemble together into small contigs and possibly scaffolds (connected units linked by read pairs; see Genome Assembly, below) but with low coverage and no linkage to the rest of the genome. At higher levels of coverage, the full genome of the contaminant will start to assemble, possibly even assembling better than the target if the relative coverage is higher (referred to by some as "bonus genomes"; Shapiro, personal communication).

At the Broad Institute, our standard sequencing quality-control process performs a BLAST (Altschul et al., 1997) analysis of a sampling of reads against the NCBI nonredundant database, and sequence matches are examined for unexpected (typically bacterial) hits. For pathogens of specific hosts, we BLAST against the host genomic sequence when available. We also BLAST assembly sequences when results are suspicious. If found, contaminating sequences need to be identified and removed from the assembly. Candidates for removal include contigs with BLAST hits to bacterial or host genomes and those with a base composition (percentage of G+C content) that differs significantly from that of most of the genome.

We faced a problem of contaminating sequences in a recent project to sequence 18 isolates of *Coccidiodes*, which is a fungal pathogen of human lung. In the first *Coccidioides* isolate sequenced, we found a subset of sequences that did not align to the reference strain and were apparently prokaryotic in origin, although not from a known species. After assembling the genome, we found a set of contigs that did not align to the genome sequence from the reference isolate, and we noted that these contigs had a significantly higher percentage of

G + C composition than the rest of the assembly. These sequences were present in seven of the isolates sequenced, and ranged from 2% of the reads to as much as 38% in one particularly bad case. In the end, we generated a high-quality assembly with more than 50-fold coverage of the "bonus" genome and identified it as being most closely related to *Nocardia*, which is a gram-positive bacterium from the order Actinomycetales.

It should be noted that these problems of biological contamination are distinct from laboratory mix-ups. In any sequencing operation, samples may on rare occasions get mixed, but such mistakes are usually readily apparent as assembled sequences that all share the same laboratory origin and thus can be easily filtered out. Because biological contamination of the source DNA will be randomly spread through the whole library, it is more difficult to recognize and remediate up front.

27.3.7 Choosing an Appropriate Sequencing Strategy

The final sequencing strategy should be based on the desired quality of the product, balanced against cost considerations along with the information gathered about the genome from pilot studies. Given the costs and the capabilities of today's sequencing instruments and protocols, it is almost certain that the strategy will be some form of WGS using paired-end sequencing. It has been demonstrated that this technique can assemble a wide range of large and complex genomes. Adjustments are likely to be only in the sequencing libraries used and the size and number of large-linking libraries. For example, knowing that the *P. infestans* genome contained a high fraction of repeats that were likely to be large in size (7–15 kb), we added large plasmids (~ 10 kb) and used a much higher fraction of Fosmids (~40 kb) to guarantee long-range connectivity and give ourselves the ability to assemble through repeats larger than the span of our workhorse plasmid libraries (~4 kb). There is, of course, a trade off. Shorter links provide more accurate placement information for ordering assemblies through regions that have sequencing gaps caused by stochastic variation in coverage or difficult to sequence regions [such as homopolymers, SSRs, and hairpin structures] and are more useful for properly assembling short-repeat copies. Long links are complementary; they provide long-range linking information and the ability to join the assembly accurately across longer repeats.

27.4 GENOME ASSEMBLY

Once data have been generated, many programs are available for assembling large genomes from Sanger chemistry data. We used Arachne (Batzoglou et al., 2002; Jaffe et al., 2003) for *P. infestans* and other genomes assembled by our group. Other programs that have been used successfully in published whole genome assemblies include PCAP (Huang et al., 2003), Atlas (Havlak et al., 2004),

Phusion (Mullikin and Ning, 2003), and the Celera assembler (Myers et al., 2000). Other programs, such as Phrap (Green, 1994), which perform very well on small projects, do not scale well to the number of reads and pairwise comparisons required to do whole genome assemblies of more than a few megabases (although both Atlas and Phusion use Phrap as an underlying assembly engine to generate layout and consensus once they have partitioned reads). Assembling genomes of more than a few megabases in size by any method requires significant resources in terms of compute power, core memory, and disk space. It is also worth noting that although all the programs mentioned work "out of the box," generating high-quality whole-genome assemblies for almost any sequence is not a push-button process but one that requires optimization of parameters for the specifics of the genome, which includes the degree of repetitiveness, amount of polymorphism, and uniformity of sequence representation. In absence of perfect data and completely unique genomes, all assembly algorithms eventually make a tradeoff between sensitivity (size and connectivity of assembled contigs, fraction of total sequence assembled) and specificity (how accurate the assembly is, accounting for regions that are misassembled, mislinked, overcollapsed, or undercollapsed because of misplacement of duplicated or polymorphic sequence).

Once an assembly is complete, it can be evaluated by several criteria. The most common ones are measures of length and connectivity of sequence contigs and scaffolds, which are typically cited as "N50," a weighted median value (see glossary). Other useful statistics include base error rates and estimated misassemblies. The latter makes use of markers mapped by other means to look for misjoins (regions where two otherwise correct stretches of sequence are erroneously fused in regions of poor sequence quality or genome duplication). Finally, one would like to estimate how much of the genome is present in the assembly, which can be done by comparing to finished clones (for short gaps), an estimated genome size (for large gaps), or an external set of known markers, such as mapped sequence tag site (STS) or gene lists from transcriptome sequencing (for estimate of total coverage of known features).

How good an assembly needs to be is dependent on the analyses planned, but a good assembly will generally contain more than 95% (often >99% for a nonrepetitive prokaryote) of the genome or at least 95% of the known genic sequences or mapping markers, and it should have scaffolds and contigs of sufficient length that most genes are contained within a single contig, and multiple consecutive genes lie within a scaffold. If this is not true, the annotation of genes (see below) is made much more challenging by the fragmentation of gene sequences in the assembly.

In the case of *P. infestans*, largely because of the high repeat content, we estimate that we have sequence contigs containing ~80% of the total genome length. Although this is low, comparison to extensive expressed sequence tags (ESTs) suggests we captured more than 95% of protein-coding genes. The fraction of genome assembled is equivalent to that for the *P. sojae* (78 of 95 Mb, 82%) and *P. ramorum* (54.4 of 65 Mb, 84%) genomes (Tyler et al., 2006), which are not as repetitive as *P. infestans* but contain an unusually large fraction of very recent repeat. Contiguity statistics are quite good, with an N50 scaffold of 1.57 Mb and an N50 contig size of 44 kb. Because the average *P. infestans* gene has a coding sequence ~ 1,200 bp long and spans less than 2,000 bp of genomic sequence [including untranslated regions (UTRs), introns, and promoters], we expect almost all genes will be completely assembled.

The next step is to link the assembled sequence to other map information if available or to the chromosome layout of the genome. Many methods exist that can be used for this, which include genetic or radiation hybrid (RH) linkage maps, fluorescent in-situ hybridization (FISH), or optical mapping. Correlating the maps requires some kind of markers (STS, physical clone with assembled end reads, or restriction fragment pattern) that can be positioned on both the map and the genome assembly. Because these non-sequence maps often span large distances, up to whole chromosomes, they can be used to order and orient sequence scaffolds into larger units and provide context to link mapping information developed with other resources onto the reference sequence. For *P. infestans*, we generated an optical map and added SNP markers to the existing amplification fragment length polymorphism (AFLP) map to provide additional long-range contiguity and a cross reference between reference scaffolds and existing genetic linkage groups (van der Lee et al., 2004; Govers, personal communication). It is important to note that the amount of sequence that can be anchored in this way is highly dependent on both the density of the map markers and the size and completeness of the assembly components.

27.5 FINISHING

Do you need to finish your genome? Many key model organisms are "finished," which means they contain zero or very few gaps of any type (other than heterochromatin, which is still refractory to cloning and sequencing techniques), have had all their low-quality sequence confirmed in a targeted manner, and are considered to have a base accuracy of less than one error in 10,000. Exhaustive sequence finishing is in fact a very expensive process, generally more costly than the entire draft sequencing of the organism, and often not required for the scientific goals of the project. At 7× or more shotgun sequence coverage, one will generate a high-quality draft, with the possible exception of organisms with very unusual genomes. Generally about 95–99% of the genome (or the euchromatin) will be assembled, and most bases present in the consensus sequence will already be of essentially "finished" accuracy. Connectivity will usually be good, and most genes and other important genomic features will be completely and accurately assembled. Some regions of recent duplication, high repeat content, or low sequence complexity are likely to be systematically missing from the assembly, and some regions of high sequence divergence in polymorphic samples are likely to be spuriously duplicated.

However, only a small amount of new biological information is likely to result from taking the sequence to a truly finished state.

An intermediate approach would be "genome improvement." This is a partial finishing method in which some fraction of the genome is selected for targeted finishing by high-throughput methods. For example, if large insert libraries were end sequenced as part of the WGS, one could use those end sequences to pick clones that span regions of poor assembly or that contain genes of particular interest to the biological functions one studies in the organism. These clones can then be individually shotgun sequenced, finished (possibly using the whole genome sequence as well), and reinserted into the genome assembly to produce a locally finished product. In absence of large insert clones, or if the desired regions of improvement are smaller, individual gaps can be targeted by polymerase chain reaction (PCR) either from whole genomic DNA or clones generated for the sequencing. Regions of the genome that are of no *a priori* biological interest or do not contain genes would not be targeted by such methods, and regions that prove recalcitrant to a single round of targeted finishing could be abandoned and left in their draft state or reevaluated as to importance. Genome improvement still represents a significant expense in terms of both sequencing and the support and informatics required to track improvement targets and reintegrate them into the genome. In this way, genome improvement can be an attractive compromise between a true finished genome and a pure draft assembly.

27.6 GENOME ANNOTATION

Once a genome assembly has been produced, it will need to be annotated. This process includes identifying protein-coding genes, noncoding RNAs, and other features, such as known repeats. Despite recent increased interest in the role of noncoding RNAs of various types, the bulk of standard genome annotation focuses on predicting protein-coding genes. Many methods exist for doing this. These methods can be broken down into roughly three classes, which include evidence-based, *de novo*, and comparative. Because all gene-calling methods are flawed, an appropriate strategy includes running multiple gene callers; combining their output by a sensible, systematic method; and manually reviewing a reasonable sampling of the gene calls.

The most reliable method is to use full-length cDNA sequences that align to the genome with proper splicing patterns and high identity. These should perfectly predict genes. However, one rarely has complete sets of these. ESTs, which are essentially incomplete cDNA sequences, can also be used for this, although additional tools are required (such as combining with a *de novo* predictor) to build complete genes. Alignment of cDNAs or ESTs is typically done with programs such as BLAT (Kent, 2002), which try to account for splicing rules in aligning transcript sequences to a genome. A secondary method is to rely on proteins predicted from other organisms and align these using programs such as blastx (Altschul et al., 1997) or genewise (Birney et al., 2004), which generate an alignment between the translated genomic sequence and the protein sequence and, at least in the case of genewise, take into account splicing to try to build valid gene models on the genome. All evidence-based methods are dependent not only on the presence of evidence but also the relevance of it. Often, much evidence (such as nucleotide or protein-level homology) comes from species different than the individual reference genome sequenced, and accuracy of gene prediction diminishes as the distance between the species increases. Genes that are highly altered or unique to the sequenced genome or lineage will be mispredicted or missed altogether.

De novo gene prediction methods mostly circumvent the problems of evidence-based predictors. These work by finding patterns of sequence that look like genes based only on the primary sequence content of the genome and patterns of gene stops, starts, and splicing. Most use hidden Markov models (HMMs) to assign a probability of a sequence coding for protein. Almost all require a known training set to distinguish gene-like sequence motifs from nongenic motifs, which means even *de novo* prediction will require some amount of evidence-based gene prediction to build training sets for the predictors. For simple prokaryotes, the accuracy of *de novo* prediction is often very good (not surprising because these genomes are mostly coding sequence), but as genes become larger, with more complicated splicing, and sparser on the genome, the accuracy of *de novo* predictors declines rapidly. Regardless, the accuracy of any predictor is limited by the quality and representation of the training set. For *P. infestans*, we used the program GeneID (Parra et al., 2000). Other common callers for eukaryotic organisms include Fgenesh and Fgenesh+ (Salamov and Solovyev, 2000) as well as GenScan (Burge and Karlin, 1997). The ability to train the gene caller specifically for the organism is a major consideration in selecting the best tool. Although these programs can each be run on any genome, calling genes using a model trained on a very distant organism will generally yield poorer performance than using a caller trained specifically for the target organism.

Comparative gene calling can be thought of as a combination of evidence-based and *de novo* methods. Most comparative callers use signals in the primary sequence recognized by *de novo* callers and then use information about the level of conservation of sequence in predicted exons in alignments to other species to increase or decrease the probability of a sequence being protein coding. The most commonly used comparative caller is Twinscan/Nscan (Korf et al., 2001; Gross and Brent, 2006). For *P. infestans*, we developed our own comparative caller Orthosearch (Handsaker, personal communication), which takes advantage of the relatively small gene and intron size to combine a near-exhaustive search of start, stop, and splice signals with patterns of conservation. This method was made possible not only by the relatively strictly defined gene structure of oomycetes but also by the high conservation information content of the related species *P. ramorum* and *P. sojae*, which had already been sequenced. A drawback of this method, however, is that, unlike Nscan and similar programs that combine an HMM with comparative data, Orthosearch cannot call a gene in absence of comparative information. Like *de novo* methods, all comparative callers require some amount of training on known gene structures to set the appropriate weighting to the comparative data and infer the conservation differences between coding and noncoding sequence. It is also worth noting that comparative methods work best at evolutionary distances at which a neutrally evolving sequence can barely be aligned. At closer distances, the power of sequence similarity to distinguish conserved from neutral evolution declines, whereas at further distances noncoding sequences will frequently fail to align, which leads to little positive predictive power for noncoding regions. We note that *P. ramorum* and *P. sojae* are both beyond the point at which neutral sequence aligns, but the short size (median ~70 bp) of introns means that between the anchoring exons and the presence of conserved splice signaling, it is usually possible to bridge introns with alignments. Because Orthosearch makes calls only in the presence of positive comparative support (unlike Nscan), it does not make spurious predictions in unconserved regions; however, it also fails to predict species-specific genes.

Because all gene callers have biases, it is preferable to use multiple callers and combine their results. In the case of *P. infestans*, we used both GeneID and Orthosearch and evaluated all of our calls against the available full-length cDNA and EST sequence data, which were not used in the prediction process. We note that even with a significant amount of such data for *P. infestans*, most genes are not completely covered. In cases where both callers agreed, we took that model as our call. In cases where the callers disagreed but some transcript evidence existed at the locus, we chose the call that matched the transcript evidence with the least discrepancy. Overall, Orthosearch performed better in this regard than GeneID, so in cases where there was disagreement and either no transcript evidence or both callers matched the transcript evidence equally, we took the Orthosearch call. In the end, 55% of automated gene calls came from Orthosearch, 36% came from GeneID, and 7% were identically called by both (the remaining 2% were known and predicted small secreted proteins, which were mostly missed by both of these methods).

An important consideration in gene annotation is the degree to which genes will be manually annotated or curated. Like sequence finishing, manual gene annotation can be a very expensive and time-consuming process; it requires specialized software and expertise, but it remains the best method to generate a complete and accurate gene set. For *P. infestans*, we chose to take an intermediate approach to manual curation. Because we had extensive transcript evidence, we started with examining cases in which our best-predicted gene had severe discrepancies with transcript evidence at the site. This allowed us to both fix those gene annotations most likely in need of adjustment and work on sites with the most evidence. We also undertook a second path in conjunction with the oomycete research community to target annotation of gene families of particular interest, which included several families involved in pathogenicity and host interaction, and to invite any researchers with special expertise or interest in certain gene families to submit revised annotations. It is also important during this process to track those genes that have been reviewed and found to be correct, so that these manually reviewed annotations will not be overwritten in any subsequent rounds of gene annotation. The curation process involved not only fixing genes that were already called but also identifying cases in which gene family members might have been missed. This process is typically carried out by blastx (Altschul et al., 1997) or genewise (Birney et al., 2004) searches of protein sequences or domains to identify loci that seem to have protein homology to the family but no annotated gene. An important part of this process was the assignment of meaningful gene names and functional annotation to genes as they were being reviewed.

At the Broad Institute as at other publicly funded sequencing centers, we are committed to prompt, public release of genome data. Sequence traces are submitted to the NCBI trace repository (http://www.ncbi.nlm.nih.gov/Traces/home/; also mirrored at EBI, http://trace.ensembl.org/). Genome assembly consensus sequences are submitted to GenBank at regular intervals, as are annotations. Sequences and annotations are also provided at our website (http://www.broad.mit.edu/tools/data/seq.html) along with basic whole genome analyses and comparative tools, when appropriate. In this way, we ensure that the genomic data we generate is of maximum utility to the scientific community.

27.7 FUTURE PROSPECTS

DNA sequencing technology has entered a period of rapid change. In the past 3 years, new technologies have been developed that deliver DNA sequence data at substantially lower cost per base than the standard Sanger chemistry. In particular, the machines available from Roche/454 (Margulies et al., 2005), Illumina/Solexa (Bentley et al., 2008) and Applied Biosystems (Schuster, 2008) are gaining increasingly wide use.

The new technologies offer both advantages and challenges, in particular for genome sequencing. They have a faster, more streamlined laboratory process and deliver data at a rate of 100 Mb to more than 2 Gb per day. Genomic DNA is prepared directly for sequencing rather than going through a lengthy process of subcloning into bacteria, so the process goes more quickly and avoids the representational biases associated with bacterial subcloning. However, currently, the sequence reads delivered by these new low-cost technologies offer some challenges. First, the new reads are shorter; they range from 25–400 bases in length compared with more than 700 bases for Sanger. Second, the base calls have a lower accuracy, with most individual bases having an error rate of roughly 1/1,000 compared with 1/10,000 or better with Sanger chemistry. Third, the new technologies do not yet offer the long-distance read pairing that provides long-range contiguity in larger assemblies. Finally, because bacterial subclone libraries are not made, no subclone templates are available for

TABLE 27.1 Comparison of *E. coli* assemblies with several technologies.

Chemistry	Sanger[c]	454	Illumina
Read length (bases)	700	250	36
Sequence coverage (x)	8	15[a]	147
Genome covered (%)	99	98	99
N50 contig (kb)[b]	150	98	187
N50 Scaffold (kb)[b]	1000	690	697
Error rate	1/10,000	1/100,000	1/600,000

[a] Paired-read coverage not included.

[b] See glossary for definition of N50.

[c] Sanger numbers are representative of a bacterial assembly of this size and complexity. Actual Sanger sequencing of *E. coli* was done long ago using techniques no longer commonly applied.

genome improvement or finishing. Table 27.1 shows an example comparison of three assemblies of *E. coli* done with traditional Sanger chemistry capillary sequencing and the new 454 and Illumina technologies. The new technologies yield assemblies of similar, and in some ways superior, quality at a fraction of the cost of the old.

As new sequencing technologies continue to develop (for example, in increasing read length) and the tools to use the data improve, using them to generate genome sequence assemblies is becoming a reality. The 454 technology has already been used to deliver genome sequence assemblies of many bacterial and fungal genomes (see references and www.broad.mit.edu/tools/data/seq.html for examples). The assembly of similar genomes has also recently proven possible with the advent of new assembly tools optimized to very short reads like those from Illumina (Butler et al., 2008; Chaisson and Pevzner, 2008; Zerbino and Birney, 2008). So far, the large size and especially the very high repeat content of the *P. infestans* genome would seem to put it outside the practical target range of new sequencing technologies, but given the rapid pace of development in both data generation and assembly algorithms and the ongoing reduction in costs, it may not be long before it will be practical to generate high-quality genome sequences of large, repetitive genomes like those of oomycetes.

ACKNOWLEDGMENTS

We are grateful to the following colleagues from the Broad Institute: Sarah Young, Sean Sykes, Evan Mauceli, LiJun Ma, Bob Handsaker, Manfred Grabherr, and Mark Borowsky (now at Massachusetts General Hospital) for sharing data prior to publication; Brian Haas for computational support, technical advice, and useful comments on the manuscript; and Leslie Gaffney for help with figures, formatting, and careful editing. We also thank Francine Govers, Wageningen University; Harris Shapiro of JGI; Brett Tyler of the

Virginia Bioinformatics Institute; and Shiguo Zhou of the University of Wisconsin, Madison for sharing stories.

REFERENCES

Adams MD, Celniker SE, Holt RA, Evans CA, Gocayne JD, Amanatides PG, Scherer SE, Li PW, Hoskins RA, Galle RF, et al. 2000. The genome sequence of *Drosophila melanogaster*. Science 287:2185–2195.

Altschul SF, Madden TL, Schäffer AA, Zhang J, Zhang Z, Miller W, Lipman DJ 1997. Gapped BLAST and PSI-BLAST: a new generation of protein database search programs. Nucl Acids Res 25:3389–3402.

Altshuler D, Pollara VJ, Cowles CR, Van Etten WJ, Baldwin J, Linton L, Lander ES 2000. An SNP map of the human genome generated by reduced representation shotgun sequencing. Nature 407:513–516.

Batzoglou S, Jaffe DB, Stanley K, Butler J, Gnerre S, Mauceli E, Berger B, Mesirov JP, Lander ES 2002. Arachne: a whole-genome shotgun assembler. Genome Res 12:177–189.

Bentley DR, Balasubramanian S, Swerdlow HP, Smith GP, Milton J, Brown CG, Hall KP, Evers DJ, Barnes CL, Bignell HR, et al. 2008. Accurate whole human genome sequencing using reversible terminator chemistry, Nature 456:53–59.

Birney E, Clamp M, Durbin R 2004. Genewise and genomewise. Genome Res 14:988–995.

Burge C, Karlin S 1997. Prediction of complete gene structures in human genomic DNA. J Mol Biol 268:78–94.

Butler J, MacCallum I, Kleber M, Shlyakhter IA, Belmonte MK, Lander ES, Nusbaum C, Jaffe DB 2008. ALLPATHS: *de novo* assembly of whole-genome shotgun micro-reads. Genome Res 18:810–820.

Chaisson MJ, Pevzner PA 2008. Short read fragment assembly of bacterial genomes. Genome Res 18:324–330.

Chimpanzee Sequencing and Analysis Consortium 2005. Initial sequence of the chimpanzee genome and comparison with the human genome. Nature 437:69–87.

Cliften P, Sudarsanam P, Desikan A, Fulton L, Fulton B, Majors J, Waterston R, Cohen BA, Johnston M 2003. Finding functional features in *Saccharomyces* genomes by phylogenetic footprinting. Science 301:71–76.

Drosophila 12 Genomes Consortium 2007. Evolution of genes and genomes on the Drosophila phylogeny. Nature 450:203–218.

Ewing B, Green P 1998. Base-calling of automated sequencer traces using phred. II. Error probabilities. Genome Res 8:186–194.

Ewing B, Hillier L, Wendl MC, Green P 1998. Base-calling of automated sequencer traces using phred. I. Accuracy assessment. Genome Res 8:175–185.

Foster J, Ganatra M, Kamal I, Ware J, Makarova K, Ivanova N, Bhattacharyya A, Kapatral V, Kumar S, Posfai J, et al. 2005. The Wolbachia genome of *Brugia malayi*: endosymbiont evolution within a human pathogenic nematode. PLoS Biol 3:e121.

Galagan JE, Calvo SE, Cuomo C, Ma LJ, Wortman JR, Batzoglou S, Lee SI, Baştürkmen M, Spevak CC, Clutterbuck J, et al. 2005. Sequencing of *Aspergillus nidulans* and comparative analysis with *A. fumigatus* and *A. oryzae*. Nature 438:1105–1115.

Galagan JE, Nusbaum C, Roy A, Endrizzi MG, Macdonald P, FitzHugh W, Calvo S, Engels R, Smirnov S, Atnoor D, et al. 2002. The genome of *M. acetivorans* reveals extensive metabolic and physiological diversity. Genome Res 12:532–543.

Glaser P, Frangeul L, Buchrieser C, Rusniok C, Amend A, Baquero F, Berche P, Bloecker H, Brandt P, Chakraborty T 2001. Comparative genomics of *Listeria* species. Science 294:849–852.

Goffeau A, Barrell BG, Bussey H, Davis RW, Dujon B, Feldmann H, Galibert F, Hoheisel JD, Jacq C, Johnston M, et al. 1996. Life with 6000 genes. Science 274: 546–567.

Green P. 1994. PHRAP documentation. http://www.phrap.org.

Gregory TR, Nicol JA, Tamm H, Kullman B, Kullman K, Leitch IJ, Murray BG, Kapraun DF, Greilhuber J, Bennett MD 2007. Eukaryotic genome size databases. Nucl Acids Res 35:D332–338.

Gresham D, Ruderfer DM, Pratt SC, Schacherer J, Dunham MJ, Botstein D, Kruglyak L 2006. Genome-wide detection of polymorphisms at nucleotide resolution with a single DNA microarray. Science 311:1932–1936.

Gross SS, Brent MR 2006. Using multiple alignments to improve gene prediction. J Comput Biol 13:379–393.

Havlak P, Chen R, Durbin KJ, Egan A, Ren Y, Song XZ, Weinstock GM, Gibbs RA 2004. The Atlas genome assembly system. Genome Res 14:721–732.

Huang X, Wang J, Aluru S, Yang SP, Hillier L 2003. PCAP: a whole-genome assembly program. Genome Res 13:2164–2170.

Jaffe DB, Butler J, Gnerre S, Mauceli E, Lindblad-Toh K, Mesirov JP, Zody MC, Lander ES 2003. Whole-genome sequence assembly for mammalian genomes: Arachne 2. Genome Res 13:91–96.

Jaillon O, Aury JM, Brunet F, Petit JL, Stange-Thomann N, Mauceli E, Bouneau L, Fischer C, Ozouf-Costaz C, Bernot A, et al. Schachter V, Quétier F, Saurin W, Scarpelli C, Wincker P, Lander 2004. Analysis of the *Tetraodon nigroviridis* genome reveals the protokaryotype of bony vertebrates and its duplication in teleost fish. Nature 431:946–957.

Kellis M, Patterson N, Endrizzi M, Birren B, Lander ES 2003. Sequencing and comparison of yeast species to identify genes and regulatory elements. Nature 423:241–254.

Kent WJ 2002. BLAT — the BLAST-like alignment tool. Genome Res 12:656–664.

Korf I, Flicek P, Duan D, Brent MR 2001. Integrating genomic homology into gene structure prediction. Bioinformatics 17:S140–S148.

Kurnick NB, Herskowitz IH 1952. The estimation of polyteny in Drosophila salivary gland nuclei based on determination of desoxyribonucleic acid content. J Cell Comp Phys 39:281–299.

Lander ES, Waterman MS 1988. Genomic mapping by fingerprinting random clones: a mathematical analysis. Genomics 2:231–239.

Machida M, Asai K, Sano M, Tanaka T, Kumagai T, Terai G, Kusumoto K, Arima T, Akita O, Kashiwagi Y, et al. 2005. Genome sequencing and analysis of *Aspergillus oryzae*. Nature 438:1157–1161.

Margulies M, Egholm M, Altman WE, Attiya S, Bader JS, Bemben LA, Berka J, Braverman MS, Chen YJ, Chen Z, et al. 2005. Genome sequencing in microfabricated high-density picolitre reactors. Nature 437:376–380.

Mikkelsen TS, Wakefield MJ, Aken B, Amemiya CT, Chang JL, Duke S, Garber M, Gentles AJ, Goodstadt L, Heger A, et al. 2007. Genome of the marsupial *Monodelphis domestica* reveals innovation in non-coding sequences. Nature 447:167–177.

Mitra RD, Church GM 1999. *In situ* localized amplification and contact replication of many individual DNA molecules. Nucl Acids Res 27:e34.

Mulligan PK, Rasch EM 1980. The determination of genome size in male and female germ cells of *Drosophila melanogaster* by DNA-Feulgen cytophotometry. Histochemistry 66:11–18.

Mullikin JC, Ning Z 2003. The phusion assembler. Genome Res 13:81–90.

Myers EW, Sutton GG, Delcher AL, Dew IM, Fasulo DP, Flanigan MJ, Kravitz SA, Mobarry CM, Reinert KH, Remington KA, et al. 2000. A whole-genome assembly of Drosophila. Science 287:2196–2204.

Nierman WC, Pain A, Anderson MJ, Wortman JR, Kim HS, Arroyo J, Berriman M, Abe K, Archer DB, Bermejo C, et al. 2005. Genomic sequence of the pathogenic and allergenic filamentous fungus *Aspergillus fumigatus*. Nature 438:1151–1156.

Parra G, Blanco E, Guigó R 2000. GeneID in Drosophila. Genome Res 10:511–515.

Pontius JU, Mullikin JC, Smith DR; Agencourt Sequencing Team, Lindblad-Toh K, Gnerre S, Clamp M, Chang J, Stephens R, Neelam B, et al. 2007. Initial sequence and comparative analysis of the cat genome. Genome Res 17:1675–1689.

Rasch EM, Barr HJ, Rasch RW 1971. The DNA content of sperm of *Drosophila melanogaster*. Chromosoma 33:1–18.

Salamov AA, Solovyev VV 2000. *Ab initio* gene finding in Drosophila genomic DNA. Genome Res 10:516–522.

Schacherer J, Ruderfer DM, Gresham D, Dolinski K, Botstein D, Kruglyak L 2007. Genome-wide analysis of nucleotide-level variation in commonly used *Saccharomyces cerevisiae* strains. PLoS ONE 2:e322.

Schuster SC 2008. Next-generation sequencing transforms today's biology. Nat Methods 5:16–18.

Schwartz DC, Li X, Hernandez LI, Ramnarain SP, Huff EJ, Wang YK 1993. Ordered restriction maps of *Saccharomyces cerevisiae* chromosomes constructed by optical mapping. Science 262:110–114.

Sorek R, Zhu Y, Creevey CJ, Francino MP, Bork P, Rubin EM 2007. Genome-wide experimental determination of barriers to horizontal gene transfer. Science 318: 1449–1452.

Tooley PW, Therrien CD 1987. Cytophotometric determination of the nuclear DNA content of 23 Mexican and 18 non-Mexican isolates of *Phytophthora infestans*. Exp Mycol 11:19–26.

Tyler BM, Tripathy S, Zhang X, Dehal P, Jiang RH, Aerts A, Arredondo FD, Baxter L, Bensasson D, Beynon JL 2006. *Phytophthora* genome sequences uncover evolutionary origins and mechanisms of pathogenesis. Science 313:1261–1266.

van der Lee T, De Witte I, Drenth A, Alfonso C, Govers F 1997. AFLP linkage map of the Oomycete *Phytiophthora infestans*. Fungal Genet Biol 21:278–291.

van der Lee T, Robold A, Testa A, van 't Klooster JW, Govers F 2001. Mapping of avirulence genes in *Phytophthora infestans* with amplified fragment length polymorphism markers selected by bulked segregant analysis. Genetics 157:949–956.

van der Lee T, Testa A, Robold A, van 't Klooster J, Govers F 2004. High-density genetic linkage maps of *Phytophthora infestans* reveal trisomic progeny and chromosomal rearrangements. Genetics 167:1643–1661.

Volkman SK, Sabeti PC, DeCaprio D, Neafsey DE, Schaffner SF, Milner DA Jr, Daily JP, Sarr O, Ndiaye D, Ndir O, et al. 2007. A genome-wide map of diversity in *Plasmodium falciparum*. Nat Genet 39:113–119.

Whisson SC, van der Lee T, Bryan GJ, Waugh R, Govers F, Birch PR 2001. Physical mapping across an avirulence locus of *Phytophthora infestans* using a highly representative, large-insert bacterial artificial chromosome library. Mol Genet Genomics 266:289–295.

Zerbino DR, Birney E 2008. Velvet: algorithms for *de novo* short read assembly using de Bruijn graphs. Genome Res 18:821–829.

Zebrafish sequencing consortium. www.sanger.ac.uk/Projects/D_rerio/.

GLOSSARY: Genome Sequencing Terminology

The field of genome sequencing is highly specialized and uses terminology that can be unfamiliar to the general biologist. Here, we attempt to explain some common terms you will find used both in this chapter and in interactions and collaborations with genomicists.

Read a single piece of sequence as determined by the sequencing instrument. Reads from current technologies range from ~25 bp (AB SOLiD) to almost 1,000 bp (AB 3730 capillary sequencers).

Read pair two reads generated from known loci (typically the ends) of a single DNA fragment and thus expected to have a known separation and relative orientation in the assembly. These are also sometimes referred to as "mate pairs."

Template a piece of DNA from which reads are generated. Often (but not always) cloned. Two reads from opposite ends of the same template form read pairs (see above), and their expected relative orientation and position is dependent on the template and the sequencing technology. See also vectors, below.

Contig a unit of assembled sequence that contains no sequence gaps (although it may contain individual ambiguous bases).

Scaffold a collection of contigs of known order and orientation, separated by captured gaps (see below) that are spanned by read pairs. Scaffolds are also often referred to as "supercontigs."

N50 a weighted median measure. The N50 size (of contigs or scaffolds) is determined by sorting the sequence units by size and assigning the size of the unit that contains the median base as the N50 size. This measure is considered preferable to an average or unweighted median because it is very robust to the degree of filtering applied to remove small or unplaced contigs or scaffolds from the assembly and also to changes in average contig size caused by incorrect joining of sequences.

Captured gap (or spanned gap) a gap spanned by one or more read pairs (and thus part of a scaffold). This allows sizing of the gap, which is often very accurate (the global error in gap sizing for a genome can be <1% of the estimated gap length, but individual gaps will have larger variance), in addition to providing ordering information.

Uncaptured gap (or unspanned gap) a gap not spanned by read pairs, which occurs at the end of scaffold. The order and orientation of sequences that flank an uncaptured gap can only be known if the scaffolds are ordered by information outside the assembly such as physical, genetic, RH, or optical maps, which can be used to order and orient scaffolds. Uncaptured gaps are also referred to as map gaps or clone gaps.

Sequence coverage The number of individual reads that cover each base in an assembly, which is usually given as a average "x" or "-fold" of coverage. There is no single standard for reporting. Total bases may be reported as raw targeted coverage (average read length times number of reads), delivered coverage (total number of bases in reads passing some quality filter), or assembled coverage (total number of bases in reads which were assembled into contigs). Coverage is often filtered by quality, for example, coverage in Q20 bases (see below). The length of the genome may be the estimated total length, the total assembled scaffold length, or the length of assembled contigs, which depends on the metric for total bases used. One should be careful in reporting or reading these numbers to understand which metrics are being used. Higher sequence coverage correlates with longer contigs and fewer sequence gaps (Lander and Waterman, 1988), although cloning and sequencing bias begins to dominate statistical effects by ~6–8×.

Physical coverage physical coverage (or "template coverage") is similar to sequence coverage but is based on the total physical length of DNA fragments in a library rather than the sequenced bases. For large insert libraries, one may report this independently of sequencing, but in the context of a genome assembly, one usually reports the total length of clones with logically pairing end reads. High physical coverage correlates with longer scaffolds and fewer map or uncaptured gaps, although at very high coverage cloning bias and repeat content begin to dominate statistical effects.

Optical map an optical map is an ordered restriction fragment map generated by randomly shearing the genome into large (100–500 kb) pieces, which are immobilized on a substrate and restriction digested while being imaged (Schwartz et al., 1993). This gives both the size and order of the restriction fragments in the piece. These can be shotgun assembled in the same sort of way sequence can, by finding identity in the restriction patterns of overlapping DNA units. Because the optical map fragments are much larger than any repeat element, most segmental or tandem duplications, and some heterochromatic regions, it is often possible to assemble whole chromosomes. Optical maps can only be linked to sequence assemblies by performing a "virtual restriction digest" of the assembly. Alignment is thus highly dependent on the contiguity and completeness of the sequence assembly.

Synteny literally, the property of being on the same chromosome. Markers with "conserved synteny" would therefore be on the same chromosome as each other in two different genomes. Common usage is that conserved synteny means being in the same order and orientation ("directed synteny") or the same general region with only local rearrangement ("undirected synteny") in two different genomes. Conserved synteny is also sometimes incorrectly referred to as simply "synteny," so that if gene B lies between gene A and gene C in species 1 and species 2, the orthologous copies of B may be referred to as "syntenic between 1 and 2." Regions of genes with conserved syntenic relationships are typically referred to as "blocks" or "segments" of conserved synteny (some authors use block versus segment to refer to a directed or undirected syntenic relationship, but the usage is inconsistent). The presence of extensive conserved synteny between two species can be very useful in studying gene gain and loss, identifying rapidly diverging orthologous genes, and detecting artifacts of gene annotation, such as retroposed pseudogenes.

Quality score also often "Phred score" or "Q," is a measure of the likelihood that a base is correct. Following phred (Ewing et al., 1998; Ewing and Green 1998), this is typically reported as $-10 \times \log_{10}(P(\text{error}))$, so Q10 would be 90% accurate, whereas Q40 would be 99.99% accurate. Some programs still use other methods, such as direct probabilities of all bases, but the Phred-like log-error quality is most common. Single-base qualities are usually trained for a given instrument type and software from sequencing of known templates. Base qualities for assembled reads are somewhat more heuristic. Base qualities often poorly represent the probability that an insertion or deletion error lies near the base.

Sanger sequencing a sequencing technique by which a mixture of single strands of DNA is extended from a known priming site and randomly terminated and labeled to identify the terminating base. These products are then size separated at single base resolution in a gel or other sieving medium (now typically done in a capillary tube using a flowing buffer rather than a fixed gel) by electrophoresis and imaged to determine the base at each position. Sanger reads are typically 700 bases or more in length, which is limited by the tradeoff between resolution at high run speeds and diffusion effects at longer run times.

Sequencing by synthesis a generic term for any of several different techniques that sequence by imaging the actual incorporation of nucleotides or by incorporating one or more nucleotides followed by a temporary termination and an imaging step. Sequencing by synthesis is done on an immobilized sample and as such is amenable to massively parallel operation. However, the individual samples are randomly spotted in each experiment, which makes it impossible to correlate a single read with a specific physical sample.

Most sequencing by synthesis techniques use clusters amplified from a single molecule (essentially variations on the polony methodology [Mitra and Church, 1999]) and are limited by various problems of asynchrony during the sequence synthesis steps (failure to incorporate, overincorporation, or failure to cleave terminators or dye labels). Read lengths vary from about 25 bp to more than 400 bp, depending on the exact method.

Vector a piece of DNA that will hold a foreign insert and be replicated by a host organism. Although other systems exist, virtually all vectors in use for sequencing today are circular, double-stranded DNA vectors propagated in *Escherichia coli*. These differ mainly in their origins of replication, copy number per cell, and the size of insert they will tolerate. Cloning bias may also differ between vector types, and is largely dependent on copy number and vector origin of replication.

Plasmid the most common sequencing vectors, plasmids are typically high copy per cell and will accommodate insertion of fragments from a few hundred bases to ~10 kb in length, although cloning is less efficient for large fragments. Their high copy makes them easier to sequence, but they are more prone than single copy vectors to bias against or deletion of specific sequences.

Fosmid a single copy vector that packages itself for insertion into the cell. Because they are single copy, they can clone some sequences that are not well represented in plasmids, whereas their packaging mechanism requires that the inserts be very tightly sized near 40 kb, which makes them very useful for accurate ordering of assembly fragments and sizing of gaps.

Bacterial artificial chromosome (BAC) a single copy vector capable of accepting very large DNA inserts (typically 100–200 kb in size, but with a wider size distribution than Fosmids) and replicating them with high fidelity. BAC end sequences are useful in assembly for providing very long links. Further, very large insert size makes them convenient as a distributable source of DNA, so end sequencing of these libraries in a genome project is often as much to link the clone resource to the assembly as for its contribution to the assembly itself.

Polymorphism a sequence that varies between different individuals of a single species. These may range from single nucleotide polymorphisms (SNPs), in which one nucleotide is substituted for another at a given position, to insertion-deletion (indel) polymorphisms, which can range from 1 base to over a megabase in length. Simple sequence length polymorphisms (SSLPs) and copy number variants (CNVs) are both specific forms of indel, although in theory any indel can be thought of as a CNV.

Allele one copy of two or more forms of a polymorphic sequence. In traditional genetics, an allele is defined by a heritable phenotype, and the mapping between alleles and distinct DNA polymorphisms may be many-to-many. In genomics, an allele usually refers to a single form of a distinct polymorphic locus, such as "C" versus "'T" at a SNP or an (AC)n repeat of length 10,12, or 14.

Haplotype the concatenation of multiple alleles on a single chromosome. The length of a haplotype is context specific. For example, in sequencing a diploid genome, we will refer to any combination of alleles from two adjacent polymorphisms up to the length of a chromosome as a "haplotype" if we can know clearly which alleles of each polymorphism belong to which chromosome in the sequenced individual or strain (this assignment of alleles to a haplotype is known as "phasing" the alleles). If the sample is diploid, there are only two haplotypes that span each chromosome, but it is generally not possible to phase haplotypes over this full length purely from the sequencing data.

INDEX

Oomycete Genetics and Genomics: Diversity, Interactions, and Research Tools
Edited by Kurt Lamour and Sophien Kamoun
Copyright © 2009 John Wiley & Sons, Inc.